Environmental Instruments and Institutions

Environmental Analysis and Economic Policy

Series Editors: Kenneth Button
                  Professor of Public Policy
                  The Institute of Public Policy, George Mason University, USA

                  Peter Nijkamp
                  Professor in Regional, Urban and Environmental Economics
                  Free University, Amsterdam, The Netherlands

Wherever possible, the articles in these volumes have been reproduced as originally published using facsimile reproduction, inclusive of footnotes and pagination to facilitate ease of reference.

For a list of all Edward Elgar published titles visit our site on the World Wide Web at
http://www.e-elgar.co.uk

Environmental Instruments and Institutions

Edited by

Thomas Tietenberg

Mitchell Family Professor of Economics, Colby College, USA

Kenneth Button

Professor of Public Policy, The Institute of Public Policy, George Mason University, USA

and

Peter Nijkamp

Professor in Regional, Urban and Environmental Economics, Free University, Amsterdam, The Netherlands

ENVIRONMENTAL ANALYSIS AND ECONOMIC POLICY

An Elgar Reference Collection
Cheltenham, UK • Northampton, MA, USA

333.7

E6128

Published by
Edward Elgar Publishing Ltd
Glensanda House
Montpellier Parade
Cheltenham
Glos GL50 1UA
UK

Edward Elgar Publishing, Inc.
136 West Street
Suite 202
Northampton
Massachusetts 01060
USA

A catalogue record for this book is available from the British Library.

Library of Congress Cataloguing in Publication Data

Environmental instruments and institutions / edited by Thomas Tietenberg,
   Kenneth Button and Peter Nijkamp.
     (Environmental analysis and economic policy: 6)
     Includes index.
     1. Environmental impact charges. 2. Pollution prevention.
   3. Pollution control equipment. 4. Environmental economics.
   I. Tietenberg, Thomas H. II. Button, Kenneth. III. Nijkamp, Peter.
   IV. Series: An Elgar reference collection.
   HJ5316.E578   1999
   333.7—dc21   99–19519
                                           CIP

ISBN 1 85898 732 6

Printed and bound in Great Britain by Biddles Ltd, Guildford and King's Lynn

Contents

B Tradable Permits

C Command-and-control

Acknowledgements

The editors and publishers wish to thank the authors and the following publishers who have kindly given permission for the use of copyright material.

Academic Press, Inc. for articles: W. David Montgomery (1972), 'Markets in Licenses and Efficient Pollution Control Programs', *Journal of Economic Theory*, **5** (3), 395–418; Zvi Adar and James M. Griffin (1976), 'Uncertainty and the Choice of Pollution Control Instruments', *Journal of Environmental Economics and Management*, **3** (3), October, 178–88; Jon D. Harford (1978), 'Firm Behavior Under Imperfectly Enforceable Pollution Standards and Taxes', *Journal of Environmental Economics and Management*, **5** (1), March, 26–43; Scott E. Atkinson and T.H. Tietenberg (1982), 'The Empirical Properties of Two Classes of Designs for Transferable Discharge Permit Markets', *Journal of Environmental Economics and Management*, **9** (2), June, 101–21; Eugene P. Seskin, Robert J. Anderson, Jr. and Robert O. Reid (1983), 'An Empirical Analysis of Economic Strategies for Controlling Air Pollution', *Journal of Environmental Economics and Management*, **10** (2), June, 112–24; William O'Neil, Martin David, Christina Moore and Erhard Joeres (1983), 'Transferable Discharge Permits and Economic Efficiency: The Fox River', *Journal of Environmental Economics and Management*, **10** (4), December, 346–55; Bruno S. Frey, Friedrich Schneider and Werner W. Pommerehne (1985), 'Economists' Opinions on Environmental Policy Instruments: Analysis of a Survey', *Journal of Environmental Economics and Management*, **12** (1), March, 62–71; Brian Beavis and Ian Dobbs (1987), 'Firm Behaviour under Regulatory Control of Stochastic Environmental Wastes by Probabilistic Constraints', *Journal of Environmental Economics and Management*, **14** (2), 112–27; Mary E. Deily and Wayne B. Gray (1991), 'Enforcement of Pollution Regulations in a Declining Industry', *Journal of Environmental Economics and Management*, **21** (3), 260–74; H. Landis Gabel and Bernard Sinclair-Desgagné (1993), 'Managerial Incentives and Environmental Compliance', *Journal of Environmental Economics and Management*, **24** (3), May, 229–40; Joseph E. Swierzbinski (1994), 'Guilty Until Proven Innocent – Regulation with Costly and Limited Enforcement', *Journal of Environmental Economics and Management*, **27** (2), September, 127–46; Robert N. Stavins (1995), 'Transaction Costs and Tradeable Permits', *Journal of Environmental Economics and Management*, **29** (2), September, 133–48; Jonathan D. Rubin (1996), 'A Model of Intertemporal Emission Trading, Banking, and Borrowing', *Journal of Environmental Economics and Management*, **31** (3), November, 269–86.

American Economic Association for articles: James M. Buchanan and Gordon Tullock (1975), 'Polluters' Profits and Political Response: Direct Controls Versus Taxes', *American Economic Review*, **65** (1), March, 139–47; Howard K. Gruenspecht (1982), 'Differentiated Regulation: The Case of Auto Emissions Standards', *American Economic Review*, **72** (2), May, 328–31; Albert L. Nichols (1982), 'The Importance of Exposure in Evaluating and Designing Environmental Regulations: A Case Study', *American Economic Review*, **72** (2),

May, 214–19; Wallace E. Oates, Paul R. Portney and Albert M. McGartland (1989), 'The *Net* Benefits of Incentive-Based Regulation: A Case Study of Environmental Standard Setting', *American Economic Review*, **79** (5), December, 1233–42; Gloria E. Helfand (1991), 'Standards versus Standards: The Effects of Different Pollution Restrictions', *American Economic Review*, **81** (3), June, 622–34.

Blackwell Publishers Ltd for articles: William J. Baumol and Wallace E. Oates (1971), 'The Use of Standards and Prices for Protection of the Environment', *Swedish Journal of Economics*, **73** (1), March, 42–54; W.J. Baumol and David F. Bradford (1972), 'Detrimental Externalities and Non-Convexity of the Production Set', *Economica*, **XXXIX** (154), May, 160–76.

Canadian Journal of Economics for articles: Susan Rose-Ackerman (1973), 'Effluent Charges: A Critique', *Canadian Journal of Economics*, **VI** (4), November, 512–28; John Pezzey (1992), 'The Symmetry Between Controlling Pollution by Price and Controlling It By Quantity', *Canadian Journal of Economics*, **XXV** (4), November, 983–91.

Elsevier Science BV for excerpt: Peter Bohm and Clifford S. Russell (1985), 'Comparative Analysis of Alternative Policy Instruments', in Allen V. Kneese and James L. Sweeney (eds), *Handbook of Natural Resource and Energy Economics*, Volume I, Chapter 10, 395–460.

Elsevier Science Ltd for articles: Marc J. Roberts and Michael Spence (1976), 'Effluent Charges and Licenses Under Uncertainty', *Journal of Public Economics*, **5** (3–4), 193–208; Wallace E. Oates and Diana L. Strassmann (1984), 'Effluent Fees and Market Structure', *Journal of Public Economics*, **24** (1), June, 29–46; David Besanko (1987), 'Performance versus Design Standards in the Regulation of Pollution', *Journal of Public Economics*, **34**, October, 19–44; Winston Harrington (1988), 'Enforcement Leverage When Penalties Are Restricted', *Journal of Public Economics*, **37**, 29–53; Kenneth E. Train, William B. Davis and Mark D. Levine (1997), 'Fees and Rebates on New Vehicles: Impacts on Fuel Efficiency, Carbon Dioxide Emissions, and Consumer Surplus', *Transportation Research E*, **33** (1), 1–13.

Journal of Law and Economics, University of Chicago Law School for article: Wesley A. Magat and W. Kip Viscusi (1990), 'Effectiveness of the EPA's Regulatory Enforcement: The Case of Industrial Effluent Standards', *Journal of Law and Economics*, **XXXIII** (2), October, 331–60.

Journal of Transport Economics and Policy for article: Erik Verhoef, Peter Nijkamp and Piet Rietveld (1995), 'Second-best Regulation of Road Transport Externalities', *Journal of Transport Economics and Policy*, **XXIX** (2), May, 147–67.

MIT Press Journals and the President and Fellows of Harvard College for articles: Thomas H. Tietenberg (1973), 'Specific Taxes and the Control of Pollution: A General Equilibrium

Analysis', *Quarterly Journal of Economics*, **LXXXVII** (4), November, 503–22; Robert W. Hahn (1984), 'Market Power and Transferable Property Rights', *Quarterly Journal of Economics*, **XCIX** (4), November, 753–65.

Natural Resources Journal for article: Gardner M. Brown, Jr. and Ralph W. Johnson (1984), 'Pollution Control by Effluent Charges: It Works in the Federal Republic of Germany, Why Not in the U.S.', *Natural Resources Journal*, **24** (4), October, 929–66.

Oxford University Press for article: T.H. Tietenberg (1990), 'Economic Instruments for Environmental Regulation', *Oxford Review of Economic Policy*, **6** (1), Spring, 17–33.

Policy Studies Organization for article: Hans Th. A. Bressers (1988), 'A Comparison of the Effectiveness of Incentives and Directives: The Case of Dutch Water Quality Policy', *Policy Studies Review*, **7** (3), Spring, 500–518.

Resources for the Future for excerpt: Clifford S. Russell (1990), 'Monitoring and Enforcement', in Paul R. Portney (ed.), *Public Policies for Environmental Protection*, Chapter Seven, 243–74.

Review of Economic Studies Ltd for article: Martin L. Weitzman (1974), 'Prices *vs.* Quantities', *Review of Economic Studies*, **XLI** (128), October, 477–91.

University of Wisconsin Press for article: Randolph M. Lyon (1982), 'Auctions and Alternative Procedures for Allocating Pollution Rights', *Land Economics*, **58** (1), February, 16–32.

Every effort has been made to trace all the copyright holders but if any have been inadvertently overlooked the publishers will be pleased to make the necessary arrangement at the first opportunity.

In addition the publishers wish to thank the Library of the London School of Economics and Political Science and the Marshall Library of Economics, Cambridge University for their assistance in obtaining these articles.

Series Preface

Kenneth Button and Peter Nijkamp

The environment has been a dominant theme for research and public policy during the latter part of the twentieth century and there is no sign that this will change as we move to the new millennium. The global interest in environmental matters in part stems from the increased pressures that a mounting population and increased production puts on the planet's natural resource base. Additionally, as personal incomes rise and leisure time becomes more freely available in the developed world, concern with more immediate human needs give way to an interest in preservation and conservation for future generations and for others. Science has also contributed by both highlighting emerging environmental problems and discovering the importance of many components of the natural environment for medical and other uses.

Historically, the interest in the environment can be traced through several phases, each with its own particular emphasis. The interest is also far from new and in antiquity there were often laws governing such things as the removal of night soil and the use of noisy wagons into towns. Traditionally, also the natural environment was seen as a source of the necessities of life and was controlled through farming techniques to ensure that food was adequate to maintain the population. Land use was controlled and water resources were often regulated at the local and regional levels.

With the Industrial Revolution came population growth, urbanization and the accelerated exploitation of natural resources. Malthus advanced ideas of over population while others, such as Jevons, became concerned about the finite nature of such resources as coal. The new technologies contaminated water courses and the widespread use of carbon fuels led to atmospheric pollution. In cities, dirt, noise and urban decay led to new ideas of town planning and ideal cities.

Much of the interest in the latter part of the last century centred around public health. As knowledge of how disease was spread – urban authorities in particular sought to improve the local environment by such measures as sewage control and clean water supply to reduce the spread of germs and infection. This trend, much later, spread to policies embodied in various pieces of clean air legislation to reduce local atmospheric pollution that causes smog and other harmful effects.

Wealthy societies, and the better off within poorer societies, with the time and resources to expand, became concerned with the built environment and in shaping nature in ways they found aesthetically pleasing. Over the centuries this has led to landscaping of the countryside and the provision of parks and gardens in urban areas.

The twentieth century has seen many of these trends continue although some of the problems, such as depletion of natural resources, initially appeared less severe as new sources were found in Africa, South America and Asia. Difficulties in feeding and otherwise sustaining the burgeoning populations of these regions, however, resurrected Malthusian concerns over whether the planet could sustain itself. Increasing reliance on fossil fuels for energy and the geographical concentration of the main reserves of oil in politically unstable areas resurrected concern over natural resource reserves in the post Second World War period.

More fundamental to modern debates has been the global orientation of recent concerns. This has been accompanied by a much more popular interest in the state of the natural environment. Kenneth Boulding's development of the idea of 'spaceship earth' attracted much attention, as did the rather doom ladened prophesies of the Club of Rome, but this was subsequently overshadowed by the publication of the Brundtland Report, *Our Common Future*, with its emphasis on the need to maintain resources for future generations. With the report came the much touted, but universally ambiguous, concept of sustainable development.

Just as the academic interest in the environment has evolved so has the institutional setting in which the topic is debated and researched. In a very general way as our understanding of the environment has progressed and as technology has developed so there has been a movement out from micro-policy making in the hands of city councils, lords of the manor and the like to macro-policy making, involving global agencies such as the United Nations and World Bank. Indeed, one of the pressing issues of contemporary environmental debate is the appropriate jurisdiction for different levels of government.

For those interested in researching environmental topics, either from the standpoint of the beginner or of that of an expert wishing to expand an established basis of knowledge, there are practical problems. Environmental analysis extends across virtually all academic disciplines although many such as economics and engineering have specific sub-areas of interest specifically devoted to environmental issues. The multidisciplinary and transdisciplinary nature of environmental research makes it difficult to keep track of the key literature. To those new to the field there is the added problem of tracing back to find classic papers or research results. There are now a large number of specialist journals that deal with environmental topics but this is often adding to the problem of collecting information rather than reducing it. While these journals and periodicals provide explicitly environmental research findings, other general journals in economics, sociology, geography, engineering, medicine, biology and so on still continue to publish important findings.

The aim of this series of edited books is to provide both a set of reference volumes for those already in the field and to facilitate newcomers with accessible core material. The aim is not to offer readings on the purely scientific issues but rather to limit the papers to those concerned with social science aspects of the environment. This does not mean that we have excluded papers with a natural science orientation but rather the focus is on the softer debates and research. There are seven volumes in the series. The papers have all been previously published but do not constitute a set of traditional classics. While many of the articles have an established pedigree, equally there are less often referenced pieces. The aim of each volume is to paint a picture of how a topic area has developed, provide an indication as to some of the key contributions to this development but also look at some of the more recent literature that covers policy considerations.

The books cover a wide range of topic areas. At one extreme there are the macro-oriented volumes edited with Bob Ayres looking at the *Global Aspects of the Environment* while, in conjunction with Kerry Turner, there is a collection concerned with *Ecosystems and Nature*. Also at the aggregate level, edited with Hans Opschoor, there is a volume which focuses on *Environmental Economics and Development*. At the other extreme there are volumes concerned with micro-issues such as that along with Ken Willis that deals with *Environmental*

Valuation and with David Banister on *The Environment, Land Use and Urban Policy*. Transport's impact on the environment is multifaceted and can be severe hence there is a volume edited with Yoshi Hayashi, *The Environment and Transport* dealing exclusively with that subject. Finally, edited with Tom Tietenberg there is a collection that looks more generally at *Environmental Instruments and Institutions*.

We accept that there are many omissions from the collections and that others may have made different selections or have structured the material differently. Space limits what can be included and there is much more that could have been justified in a larger exercise (although this would add further to deforestation). We have also had to omit work by many eminent and important scholars for the same reason. As for structuring the material, a number of possible cuts are possible, the ones selected here just seem logical to us and would seem to fit with the needs of both researchers and, possibly, teachers.

Introduction

Thomas Tietenberg, Kenneth Button and Peter Nijkamp

And it ought to be remembered that there is nothing more difficult to take in hand, more perilous to conduct, or more uncertain in its success, than to take the lead in the introduction of a new order of things. Because the innovator has for enemies all who have done well under the old conditions, and lukewarm defenders in those who may do well under the new. This coolness arises partly from fear of the opponents, who have the laws on their side, and partly from the incredulity of men, who do not really believe in new things until they have long experience with them.

Nicolò Machiavelli, *The Prince* (1513)

Environmental Economics

Background

In the beginning the power of environmental economics was perceived to be captured by a relatively few insights. These included:

- The efficient level of pollution minimized the sum of the costs of controlling pollution and the damage caused by the pollution.
- The efficient allocation of control responsibility would occur when the marginal cost of control was equal to the marginal damage caused by each pollution source.
- Normal market processes would not produce this outcome. Since the damage caused by pollution was an external cost, but the cost of control was a private cost, polluters would control too little pollution. (Indeed minimization of private costs would occur when controlling no pollution at all!)
- Governments could correct the problem by imposing a per unit emission charge such that the level of the charge was equal to the marginal social damage at the efficient level of pollution.

At the time it seemed to early practitioners that these insights were rather universal in their applicability and quite powerful in their potential to transform environmental policy. Whereas policy at the time was focused exclusively on mandating the installation of specific control equipment, the economic approach appeared to offer the opportunity for a much more flexible and efficient policy regime. So the early practitioners set out to introduce the new paradigm and to guide the transition to a more efficient, rational policy.

It was not to be. The logic which was so compelling to proponents turned out to be less so to those who were in a position to shape environmental policy (Kneese and Schultze 1975). During the early years traditional legal regulation proved immune to reform.

Some 30 years or so later the picture is rather different. Environmental economics is beginning to have a much larger impact on environmental policy. To touch on just a few examples:

- Environmental taxation has become rather common in Europe;
- Tradable permits have become common in the control of air pollution in the United States and a few other countries;
- Liability law is now routinely used to internalize the external costs of environmental accidents;
- Deposit/refund systems are used to reduce the external costs of littering;
- Volume pricing is used to reduce waste disposal costs;
  and on and on.

An evolution of institutions has paralleled the evolution of instrument adoption. At the beginning of the period domestic environmental institutions dominated the landscape and the typical environmental problems that confronted those institutions were sufficiently local that both control costs and damages fell within domestic political boundaries.

Environmental problems have now intensified in both scale and scope. Global pollution problems such as climate change or destruction of the ozone layer have become an important part of the policy agenda. In response, organizations such as the World Trade Organization and the various global environmental conventions have risen in importance as they attempt to resolve the transboundary problems. National institutions no longer suffice.

An Overview

This seems an appropriate time to step back and to examine this evolution. Not only is it possible to gain some sense of what worked and what did not, but the information gained from this experience can set the stage for the future. Not only can we profit from these insights as the policies are applied to new pollutants, but also as they applied to new settings such as the developing countries. To the extent the new applications can avoid the mistakes made in the past, they may be able to make the transition to a more effective policy regime more rapidly than was the case historically.

In this book our aim is to let the writings that occurred during this evolution and were influential in helping to shape it speak for themselves.[1] In this chapter we attempt to provide a broad overview to serve as a context for those readings.

Since other books in this series focus on the global and developing country perspectives, the focus of this volume is on the evolution of instruments and the domestic institutions which shaped and implemented those choices.

---

[1] Naturally the complexity of evolution cannot be completely captured by any reasonable-sized book. The most we can hope for is to hit a few of the highlights. To the large number of authors whose contributions helped to shape the field but, due to the constraints of space, did not make the cut, we apologize in advance.

Deepening Perspectives

Appreciating the Complexities

One of the earliest challenges to the orthodoxy was posed by the discovery that the very presence of externalities introduced the possibility of nonconvexities in the feasible set of outcomes (Baumol and Bradford 1972; Starrett 1972; Starrett and Zeckhauser 1974). Since convexity is the cornerstone of proving the uniqueness of equilibria, the existence of nonconvexities opened up the possibility of multiple equilibria. The importance of the point was underscored by the discovery that setting taxes equal to marginal social damage could even result in minimum, rather than maximum, net benefits. Policy, it began to appear, was not as simple as it had first appeared.

In the beginning control authorities were thought of as omnipotent, omniscient institutions that acted in the public interest. It soon became clear that this view was not very accurate and so the analysis began to focus on the ability of control authorities to act efficiently.

The initial inquiry focused on the amount and type of information required for administrators to implement the efficient set of charges. Upon closer inspection the burden turned out to be very large. Indeed it became clear that the amount and nature of information required to implement an efficient emissions charge was the same as the amount and nature of information required to establish an efficient set of emission standards. In this realm at least, the economists' prescription offered no advantage over the traditional approach.

As long as the marginal damage function was not constant, the regulator seeking an efficient allocation would have to know both the cost and damage functions for each pollution source (Rose-Ackerman 1973; Tietenberg 1973). This is a tall order, especially since the evidence which would flow from an unregulated allocation (on which empirical investigations would be based) involved much lower control levels than needed to establish the efficient level of control. Furthermore, acquiring control cost information, which seemed the easiest piece of the puzzle, was far from certain since pollution sources, who possessed the more precise knowledge of those costs, did not have an incentive to reveal them truthfully. (If a source revealed relatively low marginal costs, that revelation would invite regulators to assign more of the control responsibility to that source.) Finally, the existence of market power invalidated the normal prescription for how the tax rate should be set (Oates and Strassman 1984).

Recognizing the importance of uncertainty in instrument design added yet another layer of complexity (Weitzman 1974; Adar 1976; Roberts and Spence 1976; Adar and Griffin 1976). If imperfect information was the norm rather than the exception, then the possibility of mistakes (in the sense of producing an outcome which was not efficient) had to be considered. In particular it became necessary to consider the cost of being wrong. And the literature that examined that issue quickly found that when uncertainty entered the picture the preference for price-based (as opposed to quantity-based) instruments could not longer be sustained. The dominance of one over the other depended on the circumstances.

Institutional Objectives

As the analysis proceeded it began to question not only the omniscience of institutions, but their objectives as well. Was it true that institutions could always be counted upon to act 'in the social interest'?

One important strand of this literature suggested that regulatory agencies could be 'captured' by the very firms they were seeking to regulate (Stigler, 1971). In this world the agencies might end up serving narrowly defined special interests rather than broad social interests. And a considerable amount of resources could be wasted in 'rent seeking' activities (McKenzie, 1981).[2] In other words 'market failure' could be accompanied by 'government failure'.

The next strand of analysis examined how polluter preferences could shape the instrument choice. Specifically the profession began to examine a question which should have been more obvious from the beginning – namely, if pollution charges are so good, why do they have so few supporters in the community at large? One answer (Buchanan and Tullock, 1975) came from a close examination of the incentives of one particularly influential group – industrial polluters. Although the costs associated with installing control equipment or process changes to meet a similar environmental target were lower for an emissions charge than a command-and-control regime, an emission charge also involves payments to the government for uncontrolled emissions. These payments can be sufficiently large that they more than offset the lower cost of controlling emissions. The financial burden on the polluter could be higher for an emissions charge even when the burden for society as a whole was lower. The explanation, of course, was that while the financial transfers to the government were a cost to the pollution source, they were not to society as a whole (since the gain to the recipients completely offset the cost to payers).

As the complexities associated with implementing the traditional economists' prescription mounted, the profession could no longer be characterized as unanimously of the opinion that efficient charges were the preferred option. One prominent survey found that while more theoretically inclined economists in universities maintained their preference for efficient charges, those who were employed by the public sector in countries with more traditional support for government regulation preferred the regulatory approach (Frey, Schneider et al. 1985).

The Reform Movement

Redefining the Target

One turning point in the quest for political acceptability came with the refocusing of the objective of much of the empirical economic analysis from efficiency to cost/effectiveness. Whereas efficiency attempts to determine simultaneously the target (the efficient level of pollution) and the efficient means of reaching that target (the least cost means of allocating

---

[2]   Rent seeking involves investments in activities which are designed to produce a favorable political or administrative outcome for the investor.

control responsibility among the sources), cost/effectiveness attempts to find the lowest cost means of reaching a *predetermined* target. If, and only if, the predetermined target is itself efficient could the cost-effective allocation also be efficient.

The theory of how instrument design was related to the cost-effectiveness criterion developed rather rapidly. For the uniformly mixed case (where the location of the emissions did not matter) the design was both simple and straightforward (Baumol and Oates 1991). For the more complicated case the design was not simple, but still tractable, at least in principle (Montgomery 1972). And, in theory, cost/effective price-based (for example, emissions charges) and quantity-based (for example, tradable permits) instruments exhibited some remarkable symmetries (Pezzey 1992).

Political acceptability of a cost/effectiveness-based approach grew. Cost/effectiveness seemed more consistent with, and a less radical departure from, traditional environmental policy. Existing pollution targets (specified as ambient standards in most countries) could be retained. Accepting an efficiency framework required not only replacing those standards, but also replacing the process and the conceptual framework which guided their selection.

A pivotal point in the reform movement occurred when empirical cost/effectiveness studies were able to show that it was possible to reach the predetermined standards at a *much* lower cost than was the case with the traditional command-and-control regime (Atkinson and Lewis 1974; Atkinson and Tietenberg 1982; O'Neil, David et al. 1983; Seskin, Anderson et al. 1983). This rather consistent finding, produced for a number of different pollutants and geographic settings, offered the politically salable prospect of either achieving the existing environmental objectives at a much lower cost or of obtaining a much higher level of environmental quality for the same expenditure. While theory made it clear that command-and-control regulation was typically not cost-effective, empirical work made it clear that the degree of inefficiency was very large indeed. Since all reforms incur transition costs, this work suggested that the gains from reform would be large enough to justify the costs.

Implementing Economic Instruments

Both in Europe and the United States economic instruments began to make an appearance. In Europe emission charges began to appear in a variety of contexts (Bressers 1988; Brown and Johnson 1984), while in the United States the revealed preference was for tradable permits (Tietenberg 1985). Interestingly, neither approach bore much resemblance to the efficiency-based charges that consumed the early interest of the profession. Raising revenue rather than modifying polluting behaviour seemed to be the dominant purpose for instituting charges (Opschoor 1989). And in most cases the revenue was earmarked for environmental improvement rather than for the general treasury. Earmarking the revenue seemed to be the price reformers paid for being able to implement the system (Tietenberg 1990).

In the United States tradable permits began as a complement to the existing system rather than as a substitute (Hahn and Hester 1989; Tietenberg 1985). They were seen as a way to make the existing system more flexible and, hence, more resilient in the face of change. Since the baseline for trading in the mid 1970s was the command-and-control standards, in practice the permits were given to polluters on the basis of these preexisting standards. Whereas the permits system which would mimic an efficient system as closely as possible would involve auctioning off the permits (Lyon 1982), the 'grandfathering' approach, which has come to dominate existing programmes, transfers all of the rent associated with the right to pollute to the pollution sources rather than to the government. Political acceptability was gained, but at a rather high price.

Implementation Experience

The Many Dimensions of Pollution Control

As economic approaches began to enter the policy arena new design issues began to emerge. Cost/effectiveness turned out to be only one of many criteria in which policy makers were interested. In addition to the traditional concerns such as mitigating adverse distributional consequences, new concerns such as administrative ease, flexibility in the face of change, dynamic incentives for innovation, macro-economic consequences and enforceability became important (Bohm and Russell 1985).

One of the first concerns dealt with how emission charges or tradable permits could incorporate spatial considerations. While, in principle, the permit system which would resolve this problem had already been established in theory, in practice the complexity of the solution thwarted attempts to implement it. Furthermore various simple 'rules of thumb' used to approximate the optimal design (such as zonal systems) turned out (in the light of deeper analysis) to result typically in allocations which fell far short of the least cost allocation (Atkinson and Tietenberg 1982; Tietenberg 1995).

A deeper, but related, concern was the fact that least cost instrument design for the nonuniformly mixed case achieved some of its potentially large cost advantage over command-and-control by requiring less emission reduction. Tailoring the instruments to meet standards at particular receptors meant that sources that were somewhat distant from the binding receptors could be controlled less in a least cost allocation than required by a command-and-control regulation, while proximate receptors would be controlled more. On balance less emission reduction was needed in a cost/effective (spatially tailored) system than the typical command-and-control regime.

The increase in emissions that could be expected from moving from the existing command-and-control system was politically unpalatable (Krupnick, Oates et al. 1983; McGartland and Oates 1985; Oates and McGartland 1985). Furthermore, a subsequent investigation of the net benefits of the additional reductions achieved by command-and-control regulation suggested that politicians may have been right to oppose their elimination (Oates, Portney et al. 1989).

Observers also noted that treating ambient standards as given had the side effect of causing an inefficiently high level of pollution-related health problems because it failed to take human exposure into account (Nichols 1982). In the case where ambient standards are not sufficient to protect all humans from adverse health effects, efficiency analysis makes it clear that higher levels of control should be implemented where more humans are exposed. Since ambient standards regulate concentrations, not exposures, they fail to achieve the largest possible reduction in adverse health consequences per dollar expended on pollution control.

The Symmetry Between Instruments Breaks Down

One of the important notions that had come out of duality theory was the high degree of symmetry between quantity-based and price-based instruments. Under appropriate conditions they were simply the duals of a cost/effective allocation. Since either one could produce a cost/effective outcome policy makers should, at least in theory, be indifferent to the choice between them. This symmetry began to break down in practice as separate implementation issues began to arise for both charges and tradable permits.

For charges, for example, the optimal set of road charges would have to take into account the actual marginal costs of each trip, the length of the trip, the time of driving, the route followed and the type of vehicle used. Technically it was quite possible to implement systems of this sort due to the availability of electronic road pricing, but the entire notion of differentiated road pricing encountered considerable social and political resistance. As a result a number of 'second-best' strategies were derived (Verhoef, Nijkamp et al. 1995). These included establishing differentiated fees and rebates for new vehicles (Train et al. 1997). A similar turn to 'second best' strategies was necessitated by the regulatory infrastructures in some developing countries (Eskeland and Jimenez 1992).

Other issues arose specifically with tradable permits. These included the consequences of market power (Hahn 1984), transactions costs (Stavins 1995), and temporal issues such as when permits could be banked or borrowed (Cronshaw and Kruse 1996; Rubin 1996).

Command-and-Control Revisited

During this period the profession returned to a more serious consideration of the traditional command-and-control policies. Many of the previous studies had dealt with a caricature of emissions standards, such as assuming that all emission standards required a proportional reduction from all permitted sources. In fact a menu of different types of emissions standards existed and they had rather different effects (Gruenspecht, 1982; Besanko 1987; Beavis and Dobbs 1987; Helfand 1991).

The investigation of institutions was also expanded to consider a much wider array of tasks than merely designing instruments. Increasing familiarity with the details of both command-and-control and economic incentive approaches made it clear, for example, that the success or failure of both systems depended rather heavily on the monitoring and enforcement regimes within which they operated (Harford 1978; Russell 1990). The evidence suggested that this might be a weak link in all of environmental policy (Magat and Viscusi 1990). Particular issues addressed in this literature, among others, included the effectiveness of enforcement in declining industries (Deily and Gray 1991), incentives for compliance by individuals in complex organizations (Segerson and Tietenberg 1992; Gabel and Sinclair-Desgané 1993), and various innovative ways of getting around some of the practical difficulties associated with prevailing enforcement regimes (Harrington 1988; Swierzbinski 1994).

Assessing the State of the Art

Economic incentive approaches are currently enjoying a popularity that is clearly unprecedented. And in retrospect the high rate at which innovative ideas flowed from the environmental economics community over a relatively short period was rather remarkable.

In part necessity was the mother of innovation. The original ideas which motivated the field in the beginning were not ultimately the ideas that have transformed the field. As environmental economists became more aware of the complexities of pollution control and structured their theoretical and empirical work to account for them, the influence of the discipline increased. Environmental economics became more relevant, at least in the eyes of those who set the policy agenda. And with relevance came influence.

It is also important, however, to keep this rising role for economic incentive policies in perspective. Command-and-control regulation, not economic incentives, remains the predominant form of environmental regulation around the world. And in the developed world the traditional legal approach has certainly been effective, if not cost/effective. In the OECD, pollution levels for many conventional pollutants are unquestionably lower than they were despite considerable population and income growth during the intervening years.

We live in an age when the call for tighter environmental controls intensifies with each new discovery of yet another injury modern society is inflicting on the planet. But resistance to additional controls is also growing with the recognition that compliance with each new set of controls is more expensive than the last. While economic incentive approaches to environmental control offer no panacea, they frequently do offer a practical way to achieve environmental goals more flexibly and at lower cost than more traditional regulatory approaches. That is a compelling virtue.

References

Adar, Z. and J.M. Griffin (1976), 'Uncertainty and the Choice of Pollution Control Instruments', *Journal of Environmental Economics and Management*, **3** (3), 178–88.

Atkinson, S.E. and D.H. Lewis (1974), 'A Cost-Effectiveness Analysis of Alternative Air Quality Control Strategies', *Journal of Environmental Economics and Management*, **1** (3), 237–50.

Atkinson, S.E. and T.H. Tietenberg (1982), 'The Empirical Properties of Two Classes of Designs for Transferable Discharge Permit Markets', *Journal of Environmental Economics and Management*, **9** (2), 101–21.

Baumol, W.J. and D.F. Bradford (1972), 'Detrimental Externalities and Non-Convexity of the Production Set', *Economica*, **XXXIX**, 160–76.

Baumol, W.J. and W.E. Oates (1971), 'The Use of Standards and Prices for Protection of the Environment', *Swedish Journal of Economics*, **73** (1), 42–54.

Beavis, B. and I. Dobbs (1987), 'Firm Behaviour Under Regulatory Control of Stochastic Environmental Wastes by Probabilistic Constraints', *Journal of Environmental Economics and Management*, **14** (2), June, 112–27.

Besanko, D. (1987), 'Performance versus Design Standards in the Regulation of Pollution', *Journal of Public Economics*, **34**, 19–44.

Bohm, P. and C.S. Russell (1985), 'Comparative Analysis of Alternative Policy Instruments', in A.V. Kneese and J.L. Sweeney (eds), *Handbook of Natural Resources and Energy Economics*, Volume I, Chapter 10, Amsterdam: North-Holland, 395–460.

Bressers, H.Th.A. (1983), 'The Role of Effluent Charges in Dutch Water Quality Policy', in P. Downing and K. Hanf (eds), *International Comparisons in Implementing Pollution Laws*, Boston: Kluwer Nijhoff, 143–68.

Brown, G.M. Jr. and R.W. Johnson (1984), 'Pollution Control by Effluent Charges: It Works in the Federal Republic of Germany, Why Not in the United States?', *Natural Resources Journal*, 24 (4), 929–66.

Buchanan, J.M. and G. Tullock (1975), 'Polluters' Profits and Political Response: Direct Controls versus Taxes', *American Economic Review*, LXV (1), 139–47.

Cronshaw, M. and J.B. Kruse (1996), 'Regulated Firms in Pollution Permit Markets with Banking', *Journal of Regulatory Economics*, 9, 179–89.

Deily, M.E. and W.B. Gray (1991), 'Enforcement of Pollution Regulations in a Declining Industry', *Journal of Environmental Economics and Management*, 21 (3), 260–74.

Eskeland, G.S. and E. Jimenez (1992), 'Policy Instruments for Pollution Control in Developing Countries', *The World Bank Research Observer*, 7 (2), 145–69.

Frey, B.S., F. Schneider, et al. (1985), 'Economists' Opinions on Environmental Policy Instruments: Analysis of a Survey', *Journal of Environmental Economics and Management*, 12 (1), 62–71.

Gabel, H.L. and B. Sinclair-Desgagné (1993), 'Managerial Incentives and Environmental Compliance', *Journal of Environmental Economics and Management*, 24 (3), 229–40.

Hahn, R.W. (1984), 'Market Power and Transferable Property Rights', *Quarterly Journal of Economics*, XCIX (4), 753–65.

Hahn, R.W. and G.L. Hester (1989), 'Where Did All the Markets Go? An Analysis of EPA's Emission Trading Program', *Yale Journal of Regulation*, 6 (1), 109–53.

Harford, J.D. (1978), 'Firm Behavior Under Imperfectly Enforceable Pollution Standards and Taxes', *Journal of Environmental Economics and Management*, 5 (1), 26–43.

Harrington, W. (1988), 'Enforcement Leverage When Penalties Are Restricted', *Journal of Public Economics*, 37, 29–53.

Helfand, G.E. (1991), 'Standards versus Standards: The Effects of Different Pollution Restrictions', *American Economic Review*, 81 (3), 622–34.

Kneese, A.V. and C.L. Schultze (1975), *Pollution, Prices, and Public Policy*, Washington, DC: Brookings Institution.

Krupnick, A.J., W.E. Oates, et al. (1983), 'On Marketable Air Pollution Permits: The Case for a System of Pollution Offsets', *Journal of Environmental Economics and Management*, 10 (3), 233–47.

Lyon, R.M. (1982), 'Auctions and Alternative Procedures for Allocating Pollution Rights', *Land Economics*, 58 (1), 16–32.

McKenzie, R.B. and G. Tullock (1981), 'Rent Seeking', *The New World of Economics: Explorations into the Human Experience*, Homewood, IL: Richard D. Irwin: Chapter 15.

Magat, W.A. and W.K. Viscusi (1990), 'Effectiveness of the EPA's Regulatory Enforcement: The Case of Industrial Effluent Standards', *Journal of Law and Economics*, XXXIII (2), 331–60.

McGartland, A.M. and W.E. Oates (1985), 'Marketable Permits for the Prevention of Environmental Deterioration', *Journal of Environmental Economics and Management*, 12 (3), 207–28.

Montgomery, W.D. (1972), 'Markets in Licenses and Efficient Pollution Control Programs', *Journal of Economic Theory*, 5 (3), 395–418.

Nichols, A.L. (1982), 'The Importance of Exposure in Evaluating and Designing Environmental Regulations: A Case Study', *American Economic Review*, 72 (2), 214–19.

O'Neil, W., M. David, et al. (1983), 'Transferable Discharge Permits and Economic Efficiency: The Fox River', *Journal of Environmental Economics and Management*, 10 (4), 346–55.

Oates, W.E. and D.L. Strassmann (1984), 'Effluent Fees and Market Structure', *Journal of Public Economics*, 24 (1), 29–46.

Oates, W.E. and A.M. McGartland (1985), 'Marketable Pollution Permits and Acid Rain Externalities: A Comment and Some Further Evidence', *Canadian Journal of Economics*, 18 (3), 668–75.

Oates, W.E., P.R. Portney, et al. (1989), 'The Net Benefits of Incentive-Based Regulation: A Case Study of Environmental Standard Setting', *American Economic Review*, 79 (5), 1233–242.

Opschoor, J.B. and H.B. Vos (1989), *The Application of Economic Instruments for Environmental Protection in OECD Countries*, Paris: OECD.

Pezzey, J. (1992), 'The Symmetry Between Controlling Pollution by Price and Controlling It by Quantity', *Canadian Journal of Economics*, **XXV** (4), 983–91.

Roberts, M.J. and M. Spence (1976), 'Effluent Charges and Licenses Under Uncertainty', *Journal of Public Economics*, **5** (3–4), 193–208.

Rose-Ackerman, S. (1973), 'Effluent Charges: A Critique', *Canadian Journal of Economics*, **VI**, 512–28.

Rubin, J.D. (1996), 'A Model of Intertemporal Emission Trading, Banking, and Borrowing', *Journal of Environmental Economics and Management*, **31** (3), 269–86.

Russell, C.S. (1990), 'Monitoring and Enforcement', in P.R. Portney (ed.), *Public Policies for Environmental Protection*, Washington: Resources for the Future, Inc., 243–74.

Segerson, K. and T. Tietenberg (1992), 'The Structure of Penalties in Environmental Enforcement: An Economic Analysis', *Journal of Environmental Economics and Management*, **23** (2), 179–200.

Seskin, E.P., R.J. Anderson, Jr., et al. (1983), 'An Empirical Analysis of Economic Strategies for Controlling Air Pollution', *Journal of Environmental Economics and Management*, **10** (2), 112–24.

Starrett, D. (1972), 'Fundamental Nonconvexities in the Theory of Externalities', *Journal of Economic Theory*, **4** 180–99.

Starrett, D. and R. Zeckhauser (1974), 'Treating External Diseconomies-Markets or Taxes?', in J.W. Pratt (ed.), *Statistical and Mathematical Aspects of Pollution Problems*, New York: Marcel Dekker.

Stavins, R.N. (1995), 'Transaction Costs and Tradeable Permits', *Journal of Environmental Economics and Management*, **29** (2), 133–48.

Stigler, G. (1971), 'The Theory of Economic Regulation', *The Bell Journal of Economics and Management Science*, **2** (1), 3–21.

Swierzbinski, J.E. (1994), 'Guilty Until Proven Innocent – Regulation with Costly and Limited Enforcement', *Journal of Environmental Economics and Management*, **27** (2), 127–46.

Tietenberg, T.H. (1973), 'Specific Taxes and the Control of Pollution: A General Equilibrium Analysis', *Quarterly Journal of Economics*, **LXXXVII** (4), 503–22.

Tietenberg, T.H. (1985), *Emissions Trading: An Exercise in Reforming Pollution Policy*, Washington, DC: Resources for the Future.

Tietenberg, T.H. (1990), 'Economic Instruments for Environmental Regulation', *Oxford Review of Economic Policy*, **6** (1), 17–33.

Tietenberg, T.H. (1995), 'Tradable Permits for Pollution Control When Emission Location Matters: What Have We Learned?', *Environmental and Resource Economics*, **5** (2), 95–113.

Train, K.E. et. al. (1997), 'Fees and Rebates on New Vehicles', *Transportation Research E*, **33** (1), 1–14.

Verhoef, E., P. Nijkamp, et al. (1995), 'Second-best Regulation of Road Transport Externalities', *Journal of Transport Economics and Policy*, **XXIX** (2), 147–67.

Weitzman, M.L. (1974), 'Prices *vs.* Quantities', *Review of Economic Studies*, **XLI**, 477–91.

Part I
General Comparative Theory

[1]

JOURNAL OF ENVIRONMENTAL ECONOMICS AND MANAGEMENT **12**, 62–71 (1985)

Economists' Opinions on Environmental Policy Instruments: Analysis of a Survey

Bruno S. Frey, Friedrich Schneider, and Werner W. Pommerehne[1]

Institute for Empirical Economic Research, University of Zurich, Kleinstr. 15, 8008 Zurich, Switzerland

Received September 8, 1982; revised May 1984

Incentive instruments (effluent charges) are theoretically expected to be preferred to direct regulations (individual emission ceilings) by university-employed, theoretically inclined politically right-wing, professional economists living in a market-oriented country, *ceteris paribus*. Public sector employed, politically left-wing, professional economists living in a country with a long tradition of government involvement prefer the regulatory approach, *ceteris paribus*. These theoretical hypotheses tend to be supported by a bivariate probit analysis based on a survey of more than 1400 economists in five countries. © 1985 Academic Press, Inc.

I. INTRODUCTION

With few exceptions, the economic literature, as evidenced in journal articles and textbooks,[2] is strongly in favor of using incentive-based instruments for environmental policy, in particular effluent taxes, but also corresponding subsidies or tradeable pollution certificates. The "regulatory approach," on the other hand, the direct prescription of pollution ceilings to individual firms is, with few exceptions, rejected. However, there are but few cases in which incentive-based instruments have been applied in practical environmental policy; regulation is the dominating approach.[3]

In this context, it is important to inquire whether the view held by professional economists about the desirability of various environmental policy instruments is really so monolithic as it appears from the literature. Are there factors which in a theoretically predictable way induce economists to prefer incentive-based instruments over regulation, or the reverse? More specifically: do, for example, an economist's occupation, professional orientation, political ideology, and country of

---

[1] All authors from Institute for Empirical Economic Research, University of Zurich. The paper was written during the first author's stay as a Visiting Fellow at All Souls College, Oxford. We are grateful for helpful comments to our study from Peter Bernholz (University of Basel), Mark Blaug (University of London), René L. Frey (University of Basel), Guy Gilbert (University of Paris-X), Gebhard Kirchgaessner (Federal Polytechnic Institute in Zurich), Jean-Dominique Lafay (University of Poitiers), Mico Loretan (University of Zurich), Willi Nagel (University of Konstanz), Burton Weisbrod (University of Wisconsin), Hannelore Weck (University of Zurich), and Alain Wolfelsperger (University of Paris-I), as well as to two anonymous referees.

[2] Ethridge [8], Orr [16], Siebert [20], Kneese [12], and Baumol and Oates [2].

[3] Kneese *et al.* [13], Kneese and Schultze [14], Johnson and Brown [10], Baumol and Oates [3, Chap. 20], and Downing [6].

62

living systematically influence his or her preferences toward desirable environmental policy instruments?

This paper develops hypotheses on the likely influences of such factors (Section II) and tests them with the help of bivariate probit analyses on data collected by a survey among more than 1400 professional economists in five countries (Section III). The results suggest that the economists' opinions on the desirability of environmental policy instruments depend in fact on the factors mentioned: Working at a university, being theoretically inclined, politically right-wing, and living in a market-oriented country each, *ceteris paribus*, lead to a support of the use of incentive-based instruments (effluent taxes), while being employed in the public service, being politically left-wing, and living in a country with a long tradition of government intervention (France and Austria) each, *ceteris paribus*, lead to a support of the use of regulatory instruments.

II. HYPOTHESES

It is theoretically expected that academic economists' views on environmental policy instruments depend on (1) occupation, (2) professional orientation, (3) political ideology, and (4) country of residence. The hypotheses advanced in the following are based on the economists' self-interest model of behavior, as well as on influences of tradition and culture.

1. Occupation

The place of work is one of the important determinants of every person's welfare. The views expressed on the desirability of environmental policy instruments therefore depend (among other things, i.e., *ceteris paribus*) on the extent to which an instrument serves to augment the utility derived from one's own work. Mostly, these views are not chosen consciously, but are "internalized" (to use a sociological term) in the course of daily life. The adoption of views based on occupational advantages may also be psychologically explained by the desire of individuals to minimize the tension between what they proclaim in their occupational roles and what they privately think (reduction of "cognitive dissonance"). The position adopted by economists is thus taken to correspond closely to the interests of their particular occupation.

Two occupations are distinguished:

(*a*) *University scholars* (professors and other teaching or research staff). This occupational group has a vested general interest in maintaining accepted economic science as it is. The university scholars have attained their position by proving that they are competent in mastering a subject, they have (partly) contributed to its creation, and would therefore lose part of their accumulated "intellectual capital" if it would be radically put into doubt, changed, or even rejected. Moreover, most of them teach (or have to teach) what is regarded as accepted economics to their students, and it would put them under too much strain to privately hold another opinion than the one they teach and expect students to know in examinations. This basically conservative tendency of the university based economists[4] leads to the hypothesis that they tend to support the "textbook" view that incentive-based

---

[4] The view that economists are basically conservative has been stressed by Stigler [21].

instruments are better suited to fight environmental degradation than are regulations.

(*b*) *Public officials.* Following the economic theory of bureaucracy (see Tullock [22], Downs [7], Niskanen [15], Breton and Wintrobe [5]) it may be assumed that economists in public employment are quite interested in increasing the influence of the public sector and the number and intensity of government interventions. Increasing public activities give them the opportunity to be more influential, to enjoy higher prestige and to, at least indirectly, raise their income. The hypothesis is that they thus tend to support the regulatory rather than the incentive approach to environmental policy.

No specific hypotheses about the opinions of economists in other occupations are advanced.

2. Professional Orientation

For similar reasons as mentioned above, it may be expected that more *theoretically* inclined economists generally support incentive instruments, whereas economists who consider themselves to be *practically* oriented support direct government interventions by regulations.

3. Political Ideology

In the most simple terms, the following distinction of political ideologies along the right–left spectrum is possible: The right has a long record of distrust of government intervention (as it infringes on individual liberty) and prefers the use of anonymous market forces. On the other hand, the left is suspicious of the market (mainly because of its undesirable distributional consequences) and calls for government intervention to cure social ills such as environmental degradation. The hypothesis is that professional economists who consider themselves to be of the right will tend to support incentive instruments, while economists of the left will tend to support regulations.

4. Country

The country an economist lives and performs in must be expected to have some effect on his or her opinions. It is difficult or even impossible to disassociate oneself from the prevailing general outlooks and traditions of one's society. Two groups of countries may be distinguished for our purposes:

(*a*) *Market-oriented countries,* such as the Federal Republic of Germany and Switzerland, which since World War II have been strongly influenced by the United States. This influence also relates to the prevailing type of economic theory as advanced in American neoclassics-dominated textbooks (e.g. Samuelson [18]). It is theoretically expected that economists from countries of this group tend to prefer the incentive over the regulatory approach.

(*b*) *Countries with a long history of government intervention.* In Austria, for instance, the government's role in economic affairs goes back to the Hapsburg monarchy and has been intensified in the interwar years, i.e., long before the

ECONOMISTS ON ENVIRONMENTAL POLICY   65

"Keynesian Revolution." The Austrian government has also played a very important role in the postwar economy, exemplified by the fact that it owns more than half the shares of industrial corporations.[5] Another prominent case is France where the government's involvement in the economy has an even longer and more pronounced record, which dates (at least) back to the Mercantilists ("Colbertism") and is vigorous up to the present, as exemplified by the French "planification." It is hypothesized that economists living in such countries *ceteris paribus* tend to prefer the regulatory over the incentive approach.

It goes without saying that these hypotheses are always thought to hold when all other influences are kept constant. Support or preference for either incentive or regulatory instruments should not be taken in any absolute sense but relative to the opposite kind of instrument.

To summarize: It is expected that professional economists *ceteris paribus* are more likely to support the incentive approach to environmental policy (instead of the regulatory approach), the more they are

   -occupied at a university;

   -theoretically oriented;

   -ideologically on the right;

   -of a market-oriented country strongly influenced by the United States (such as West Germany, Switzerland or, of course, the United States itself).

On the other hand, it is theoretically expected that professional economists *ceteris paribus* more likely support the regulatory approach (over the incentive approach) when they are

   -occupied in the public sector;

   -practically oriented;

   -ideologically on the left;

   -of a country with a long tradition of government intervention (such as Austria and France).

III. EMPIRICAL TESTS

1. The Survey

In spring and summer 1981, a stratified random sample of 1472 economists was anonymously asked in writing to respond to two (among other)[6] propositions concerning the desirability of using the incentive and the regulatory approach to environmental policy. The first proposition stated was "Effluent taxes represent a better approach to pollution control than imposition of pollution ceilings." The second was "The maximum emission of pollutants should be prescribed to the

---

[5]See, for example, Abele *et al.* [1].

[6]The survey contains 44 propositions. The general results referring to the individual countries have been published in French and German in Bobe and Etchegoyen [4], Frey *et al.* [9], Schneider *et al.* [19], and Pommerehne *et al.* [17].

FREY, SCHNEIDER, AND POMMEREHNE

TABLE I

Responses to the Proposition "Effluent Taxes Represent a Better Approach
to Pollution Control than the Imposition of Pollution Ceilings"[a]

	Generally agree (%)	Agree with provisions (%)	Generally disagree (%)	No answer (%)	Total number (= 100%)
Occupations					
University staff	29.0	28.7	40.1	2.2	314
Public sector	25.1	29.8	40.9	4.2	191
Private sector	27.3	30.0	39.5	3.2	220
					$\chi^2 = 13.3^b$
Countries					
Austria	20.9	22.0	54.9	2.2	91
France	27.2	27.2	40.7	4.9	162
Germany (FR)	34.4	29.7	33.0	2.9	273
Switzerland	21.1	34.2	42.7	2.0	199
					$\chi^2 = 41.7^b$
All occupations and countries					
%	27.4	29.4	40.2	3.0	
Total number	199	213	291	22	725

[a] Results according to three occupations and four countries ($N = 725$).

[b] The likelihood-ratio test (χ^2) over the frequencies at the four response categories indicates that there is a significant difference between occupations as well as countries in the responses.

individual firms." This second proposition serves as a check on how well the first proposition was understood, and how seriously it is answered. Naturally, an answer contrary to the one given on the first proposition is expected. The respondents could either "generally agree," "agree with provisions," "generally disagree," or they could refuse to answer one or both of the statements; 725 complete replies were sent back and could be used for the study, implying a quite satisfactory return rate of 49.3% for such a type of survey, for which due to the anonymity no reminder was possible. The survey covers economists from the four countries Austria, France, Germany (Federal Republic) and Switzerland, chosen from a list of the members of professional associations of the respective countries.[7] The responses to the first statement on effluent taxes vs pollution controls are listed in Table I, divided according to three occupational groups and to four countries.

Table II shows the corresponding results for the second statement on the desirability of direct emission controls. The two tables are self-explanatory. It may be seen at once that the answers differ with respect to the occupation and nationality of the respondents.

[7] The professional associations are in Austria and Germany the Gesellschaft fuer Wirtschafts- und Sozialwissenschaft, Verein fuer Socialpolitik, in Austria furthermore the Nationaloekonomische Gesellschaft; in France the Association Nationale des Docteurs en Sciences Economiques, the Association Française de Science Economique and the Centre National de la Recherche Scientifique; in Switzerland the Schweizerische Gesellschaft fuer Statistik und Volkswirtschaft.

Environmental Instruments and Institutions

TABLE II

Responses to the Proposition "The Maximum Emission of Pollutants Should
Be Prescribed to the Individual Firms"[a]

	Generally agree (%)	Agree with provisions (%)	Generally disagree (%)	No answer (%)	Total number (= 100%)
Occupations					
University staff	59.9	23.9	14.9	1.3	314
Public sector	67.0	23.6	9.4	0.0	191
Private sector	58.2	29.5	11.8	0.5	220
					$\chi^2 = 10.6^b$
Countries					
Austria	69.2	25.3	4.4	1.1	91
France	77.8	14.8	6.8	0.6	162
Germany (FR)	51.6	29.4	18.3	0.7	273
Switzerland	57.3	29.1	13.1	0.5	199
					$\chi^2 = 24.0^b$
All occupations and countries					
%	61.2	25.5	12.6	0.7	
Total number	444	185	91	5	725

[a] Results according to three occupations and four countries ($N = 725$).
[b] See Table I, footnote b.

2. Probit Analysis

The theoretically derived hypotheses set forth are tested with the help of bivariate probit analyses. (Probit, rather than simple linear regression, is in order because the set of responses to a proposition is restricted to two categories[8] and sums to 1). For this purpose, the data contained in Tables I and II on occupations and nations are used. Moreover, the survey has collected data on the respondents' self-evaluation as to whether they consider themselves to be "practically" (coded $+1$) or "theoretically" (coded -1) oriented: 59.6% of the respondents indicated that they are "practically" and 40.4% that they are "theoretically" inclined. The respondents could also indicate their self-evaluated political position on a normalized right–left scale, running from $+1$ (right) to -1 (left).

The logit estimates for the two propositions are presented in Table III, Eqs. (1) and (2). (Due to missing information 42 and 27 respondents, respectively, have to be excluded, so that $N = 683$ and 698, respectively.)

According to the statistical test criteria, the two equations perform well. The χ^2 statistic indicates that the answers given ("agree"/"disagree") can be explained to a high degree by the variables used here. The values of ρ indicate the approximate likelihood of the influence of the independent variables on the dependent one. It may also be seen that the theoretically expected change of signs (at least of the statistically significant parameters) holds true in each case, i.e., the estimates of the two equations reinforce each other.

[8] For the sake of estimation, the categories "generally agree" and "agree with provisions" are put together and confronted with "generally disagree." The no answer category is omitted.

TABLE III

Probit Estimate of the Determinants of the Responses to the Two Propositions (in Parentheses: Approximate t-Values)[a]

Proposition	Constant	Occupation			Professional orientation Practic. (+1) Theoret. (−1)	Ideological position Right (+1) Left (−1)	Country					Interaction term[b]		Test Statistics	
		University staff	Public sector	Private sector			Austria	France	Germany	Switzerland	United States (only[1'])	Public service in France	University staff in Germany	χ^2	ρ
European countries															
(1) "Effluent taxes represent a better approach to pollution control than imposition of pollution ceilings" (N = 683)	0.27* (2.06)	0.31** (3.41)	−0.26* (2.34)	0.04 (0.62)	−0.41* (−1.98)	0.96** (2.91)	−0.21* (2.16)	−0.10 (−1.10)	0.18* (2.34)	0.11 (0.99)	—	−0.16* (−2.34)	0.14* (2.34)	264.3	0.75
(2) "The maximum emission of pollutants should be prescribed to the individual firms" (N = 698)	1.70** (12.9)	−0.08 (−0.81)	0.23** (2.89)	−0.10 (−0.36)	0.12 (1.06)	−0.23** (2.89)	0.12 (1.23)	0.09 (1.04)	−0.77** (−4.07)	−0.46* (−2.60)	—	0.12** (3.89)	—	275.2	0.89
European countries and United States															
(1') "Effluent taxes represent a better approach to pollution control than imposition of pollution ceilings" (N = 903)	0.17 (0.86)	0.08 (0.94)	−0.11 (−1.90)	−0.04 (−0.81)	—	—	−0.26 (−1.90)	−0.08 (−1.03)	0.20* (2.56)	0.04 (1.07)	0.23** (5.41)	—	—	20.7	0.15

[a] One asterisk indicates statistically significant parameters at the 95%, two asterisks at the 99% confidence level.
[b] In order to keep the table clear, only the statistically significant interaction terms are reproduced.

The first proposition on effluent taxes has in the aggregate been agreed to by 59%, and rejected by 41% of the respondents. An ex post forecast of the answers on the basis of the independent variables and the estimated parameters predicts 76% of the "agree" and "disagree" responses of the individual economists correctly, i.e., an ex post forecast on the basis of the estimated logit equation leads to a significantly better result than a forecast by chance (51% correctly predicted responses)[9] or by projecting an "agree" for everyone in the sample (59% correctly predicted responses). The second proposition on maximum emissions of pollutants has an *a priori* distribution of 87% "agree" and 13% "disagree." The estimation equation allows a correct prediction of 94% of the answers which is again significantly superior to "guesstimates" on the basis of a forecast by a chance or a projection of an "agree" response for everyone.

The estimates shown in Table III, Eqs. (1) and (2), support most of the theoretical hypotheses about the causal factors influencing the responses given:

(1) The university staff economists have a higher tendency to support effluent charges rather than pollution ceilings, while economists in the public service tend to reject effluent charges and to support the direct imposition of emission ceilings. The corresponding estimated parameters are statistically highly significant.

(2) Also, the theoretical expectations about the influence of practical and theoretical orientation are supported by the estimates.

(3) The theoretical hypotheses on the influence of an economist's political ideology are strongly supported by the estimates: Right-wing economists tend to support effluent taxes, left-wing economists favor the regulatory approach.

(4) The hypotheses about the influence of the country of residence are supported by the signs of the estimated coefficient. However, of the two countries expected to be in favor of regulatory activities, only Austria yields a statistically significant sign in the case of proposition (1). The hypotheses for the market-oriented countries Germany and Switzerland are, on the other hand, quite well supported. The interaction terms suggest that the predilection of the regulatory approach does not hold in general in France but is heavily concentrated among the public officials in that country. The strongest supporters of effluent taxes seem to be the university-based economists in Germany.

3. Including the United States

A similar survey on the opinions of professional economists has some time earlier been undertaken for the United States (Kearl *et al.* [11]), based on a stratified random sample from the members of the American Economic Association. However, the survey is restricted to the first proposition, and no data on the self-evaluated professional orientation and on political ideology have been collected. Therefore, the responses to the propositions are explained by occupation and country of residence only. Table IV presents the survey results including the answers of 211 American economists. It may be seen that the American economists' opinions differ from those of European economists.

[9]The sample probability of an "agree" is 0.59 and of a "disagree" 0.41. Hence the percentage that would be correctly predicted by chance using these aggregate probabilities is $(0.59)^2 + (0.41)^2 = 0.51$.

FREY, SCHNEIDER, AND POMMEREHNE

TABLE IV

Responses to the Proposition "Effluent Taxes Represent a Better Approach to Pollution Control
than the Imposition of Pollution Ceilings," for Five Countries Including the United States (N = 936)

	Generally agree (%)	Agree with provisions (%)	Generally disagree (%)	No answer (%)	Total number (= 100%)
Occupations (all countries)					
University staff	36.1	29.2	32.5	2.2	415
Public sector	27.4	29.0	39.0	4.6	241
Private sector	29.6	30.4	35.4	4.6	280
					$\chi^2 = 25.6^a$
United States	47.9	29.9	18.0	4.2	211
					$\chi^2 = 70.2^a$
All occupations and countries					
%	31.9	29.6	35.0	3.5	
Total number	299	276	328	33	936

^{a}See Table I footnote *b*.

The results of the probit estimate[10] are shown as Eq. (1') in Table III. The equation is able to forecast 69% correct responses (related to an *a priori* distribution of 64% "agree" and 36% "disagree"), and only two parameters are statistically significant. However, the signs of all independent variables are the same as in Eq. (1). This weaker performance may partially be due to the fact that the estimation equation is misspecified due to the exclusion of the professional orientation and ideological position. Nevertheless, the estimation equation tends to support the hypothesis that economists in the United States are particularly strong supporters of an effluent tax, the corresponding parameters being highly significant.

IV. CONCLUDING REMARKS

Our analysis indicates that the views of professional economists on the desirability of the incentive as opposed to the regulatory approach is significantly influenced by occupation, professional oriculation, political ideology, and country of residence. This influence has been predicted on the basis of the economic model of self-interest. It turns out that economists employed in a university, being theoretically inclined, with an ideology biased to the right, and living in a market-oriented country such as Germany, Switzerland, or the United States, *ceteris paribus* prefer the use of an effluent tax to individual prescriptions of emission levels. On the other hand, economists working in the public sector, with an ideology to the left, and living in a country with a long tradition of government intervention, such as Austria or France, *ceteris paribus* prefer a regulatory approach in environmental policy.

The results of this analysis may be useful for several different purposes. They may, for instance, suggest that the setting of individual emission standards widely

[10] In total, 33 questionnaires had to be excluded due to missing data, so that $N = 903$.

practiced in practical environmental policy is not due to lack of information, or to insufficient training and knowledge of economics by public officials, because as our analysis shows, even professional economists when employed in the public sector tend to support the regulatory approach. Rather, the preference for regulations should be interpreted to be the result of the self-interest (rational) behavior of individuals working in the public sector. This may imply further that incentive-based instruments will only have a chance of being more widely used in practical environmental policy when they can be made to conform to the self-interest of those designed to implement them.

REFERENCES

1. H. Abele, E. Nowotny, S. Schleicher, and G. v. Winckler, "Handbuch der oesterreichischen Wirtschaftspolitik," Manz, Vienna (1982).
2. W. J. Baumol and W. E. Oates, "The Theory of Environmental Policy: Externalities, Public Outlays, and the Quality of Life," Prentice–Hall, Englewood Cliffs, N.J. (1975).
3. W. J. Baumol and W. E. Oates, "Economics, Environmental Policy, and the Quality of Life," Prentice–Hall, Englewood Cliffs, N.J. (1979).
4. B. Bobe and A. Etchegoyen, "Economistes en désordre: consensus et dissension," Economica, Paris (1981).
5. A. Breton and R. Wintrobe, "The Logic of Bureaucratic Control," Cambridge Univ. Press, Cambridge (1982).
6. P. B. Downing, A political economy model of implementing pollution laws. *J. Environ. Econ. Manag.* **8**, 255–271 (1981).
7. A. Downs, "Inside Bureaucracy," Little, Brown, Boston (1967).
8. D. Ethridge, User charges as a means of pollution control: The case of sewer surcharges. *Bell J. Econ. Manag. Sci.* **3**, 346–354 (1972).
9. B. S. Frey, W. W. Pommerehne, F. Schneider, and H. Weck, Welche Ansichten vertreten Schweizer Oekonomen? *Schweiz. Z. Volkswirtsch. Statist.* **118**, 1–40 (1982).
10. R. W. Johnson and G. M. Brown, "Cleaning Up Europe's Waters," Praeger, New York (1976).
11. J. R. Kearl, C. L. Pope, G. C. Whiting, and L. T. Wimmer, A confusion of economists? *Amer. Econ. Rev., Papers Proc.* **69**, 28–37 (1979).
12. A. V. Kneese, "Economics and the Environment," Penguin, Harmondsworth (1977).
13. A. V. Kneese, S. E. Rolfe, and J. W. Harned, Eds., "Managing the Environment: International Cooperation for Pollution Control," Praeger, New York (1971).
14. A. V. Kneese and C. L. Schultze, "Pollution, Prices and Public Policy," Brookings Institution, Washington, D.C. (1975).
15. W. A. Niskanen, "Bureaucracy and Representative Government," Aldine, Chicago/New York (1971).
16. L. Orr, Incentive for innovation as the basis for effluent charge strategy. *Amer. Econ. Rev., Papers Proc.* **66**, 441–447 (1976).
17. W. W. Pommerehne, F. Schneider, and B. S. Frey, Quot homines, tot sententiae? A Survey among Austrian Economists, *Empirica* **13**, 93–127 (1983).
18. P. A. Samuelson, "Economics," 10th ed., McGraw-Hill, New York (1976).
19. F. Schneider, W. W. Pommerehne, and B. S. Frey, Relata referimus: Ergebnisse und Analyse einer Befragung deutscher Oekonomen. *Zeitschr. ges. Staatswiss.* **139**, 19–66 (1983).
20. H. Siebert, "Analyse der Instrumente der Umweltpolitik," Schwarz, Goettingen (1976).
21. G. Stigler, The politics of political economists, *Quart. J. Econ.* **73**, 522–532 (1959).
22. G. Tullock, "The Politics of Bureaucracy," Public Affairs Press, Washington, D.C. (1965).

[2]

Detrimental Externalities and Non-Convexity of the Production Set

By W. J. Baumol and David F. Bradford[1]

This paper undertakes to show that detrimental externalities tend to induce non-convexity of the social production possibility set. In particular we show that if externalities are sufficiently strong, convexity conditions must break down.

It is not our objective here to review in any detail the difficulties caused by non-convexity. Some of these consequences have long been recognized and are widely known.[2] However, until the recent appearance of papers by Starett [8], Portes [5], Kolm [3] and Baumol [2], it was apparently not recognized that externalities themselves are a source of non-convexity. These more recent writings suggest more than one connection between the two phenomena. However, one particularly straightforward relationship seems to have received little or no attention. With sufficiently strong interactive effects non-convexity follows from the simple fact that if *either* of two mutually interfering activities is operated at zero level the other suffers no hindrance. The goal of this paper is to explore this phenomenon and show very clearly how it is that sufficiently severe detrimental externalities of the form described and non-convexity necessarily go together.

In the first three sections we show both with the aid of illustrative examples and more general analysis that detrimental externalities of sufficient magnitude must always produce non-convexity in the production possibility set for two activities: one generating the externality and one affected by it. In Section IV we show that the problem is reduced but not generally eliminated by the possibility of spatial separation of offender and offended. Achievement of the "right" spatial separation turns out not always to be a simple matter, however. Section V contains some speculations about the way in which the number of local peaks in

---

[1] We would like to express our deep gratitude to Mrs. E. E. Bailey for her thorough review of an earlier version of this paper and her very helpful comments, and to the National Science Foundation for its generous support of our research.

[2] Pigou, for example, commented that ". . . if several arrangements are possible, all of which make the values of the marginal social net products equal, each of these arrangements does, indeed, imply what may be called a *relative maximum* for the [national] dividend but only one of these maxima is the unequivocal, or absolute, maximum. . . . It is not necessary that all positions of relative maximum should represent larger dividends than all positions which are not maxima. On the contrary, a scheme of distribution approximating that which yields the absolute maximum, but not itself fulfilling the condition of equal marginal yields, would probably imply a larger dividend than most of the schemes which do fulfil this condition and so constitute relative maxima of a minor character." Pigou [4], p. 140. References in square brackets are listed on p. 176, below.

the production function grows with the number of interacting activities. In Section VI we discuss the possibility of using Pigovian taxes to sustain desirable behaviour, and in a concluding section we review briefly the problems for social policy inherent in the sort of non-convexity we have been analysing.

An appendix contains a formal demonstration of the workability of Pigovian taxes in this context. It is shown that as long as individual production sets are convex, all socially efficient output vectors *can* be

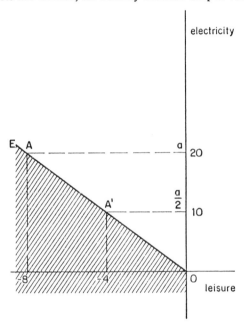

 = production set of electricity industry

FIGURE 1a

sustained as a sum of profit-maximizing output choices under taxes designed to equate marginal social and private costs.

I. A SIMPLE MODEL

Consider a two-output, one-input economy in which each output is produced by a single industry. To avoid compounding problems we shall assume that each industry has a convex technology in terms of its own inputs and outputs.[1] However, the presence of detrimental externalities

[1] Thus, if v_i is the quantity of input to industry i and x_i is its output, and if (v_i^*, x_i^*) and (v_i^{**}, x_i^{**}) are two feasible input–output combinations (holding

means that increases in the output of one of the industries raises the other's costs of production, which is to say the amount of input required to produce any given output. What we wish to show is that if this detrimental externality is strong enough, then the social production set must be non-convex.[1]

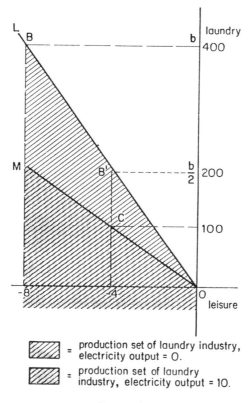

FIGURE 1b

For consistency with the general analysis in the Appendix let us carry through this example following the practice of measuring inputs as

---

constant inputs and outputs in other sectors), then $0 < \alpha < 1$ implies that $(\alpha v_i^* + (1-\alpha)v_i^{**}, \alpha x_i^* + (1-\alpha)x_i^{**})$ is also a feasible input–output combination. Convexity of a production set is sometimes referred to as "generalized non-increasing returns", by which is meant that a convex technology cannot exhibit increasing returns to scale and that it obeys the law of diminishing marginal rates of substitution among factors and among outputs, and diminishing marginal productivity of outputs by factors.

[1] In the notation of the preceding footnote, the social production set is the set of all vectors $(v_1 + v_2, x_1, x_2)$ such that (v_1, x_1) and (v_2, x_2) are *simultaneously* feasible for their respective industries.

negative outputs. Hence, our economy can be described as having three outputs: for concreteness—leisure, electricity and laundry. The shaded region of Figure 1a shows the production set (the set of attainable net output vectors) for the electricity industry, bounded by the ray OE. Figure 1b displays the production set for the laundry industry under two alternative assumptions about output in the electricity industry. The detrimental externality generated by electricity means that, for a given input of labour to laundry, less will be produced when electricity output is positive. Thus, OM, the ray serving as the laundry production frontier when some electricity is produced, must lie below OL, the laundry frontier when no electricity is produced. To make things easy for ourselves we have assumed constant returns to scale for each of the industries taken alone—hence the straight-line boundaries.

The non-convexity of the social production set for this economy is easily demonstrated. Consider two social production vectors on frontiers OE and OL: vector A on OE (-8 leisure, 20 electricity, 0 laundry), and vector B on OL (-8 leisure, 0 electricity, 400 laundry). Obviously both of these are technically feasible, as are (by constant returns to scale) the vectors A': (-4, 10, 0), and B': (-4, 0, 200), which are, respectively, half way to the origin from A and B. However, the vector $V=(-8, 10, 200)$, which is a convex combination of A and B since $V=A'+B'=\frac{1}{2}A+\frac{1}{2}B$, is *not* feasible technically. If we wish to give up 8 units of leisure altogether and insist on 10 units of electricity, requiring 4 of these units of leisure, the most we can obtain is 100 units of laundry (point C). More generally, if L is the amount of leisure devoted to the two outputs and a and b represent the respective outputs of electricity and laundry if L is devoted exclusively to the one or the other, then the assignment of $\frac{1}{2}L$ to each output must necessarily provide less than $b/2$ of laundry output if there is any detrimental externality present. Point B' is never attainable under these conditions, and non-convexity *must* follow.

II. AN ALTERNATIVE VERSION OF THE NON-CONVEXITY ARGUMENT

In Figure 2 let X_1 and X_2 represent, respectively, quantities of electricity and laundry. Dropping our earlier assumption of constant marginal rate of transformation between outputs, let $ORAR'$ represent the convex set of output combinations attainable from a fixed amount of labour in the absence of externalities. For expository convenience, introduce a parameter k measuring the strength of the externality. In terms of our example, k can be taken to measure the mean addition to the resources cost of cleaning a given batch of laundry which occurs when an added unit of electricity output causes smoke to increase. By definition, then, along RAR', which corresponds to the absence of external effects, the value of k (call it k_a) is zero.

Consider what happens to the production possibility locus as the value of k is increased. We will show that the position of the end-points

R and R' will be totally unaffected, while all other points on the locus will be shifted downward. Point R will be unaffected by a rise in the value of k since, whatever the social cost of smoke, at that point there will be no increase in damage because, by assumption, there is no smoke produced in the absence of any electricity output. Similarly, the location of R' is invariant with k since at that point no resources are devoted to laundry production, and hence there can be no increase in the resources cost of laundry output. There simply is no laundry to be damaged, so that electricity can smoke away without causing any harm to others. However, consider some intermediate level of electricity output, say x_1^*. Here an increase in k means that with a given amount of electricity and a given quantity of resources, a smaller quantity of clean

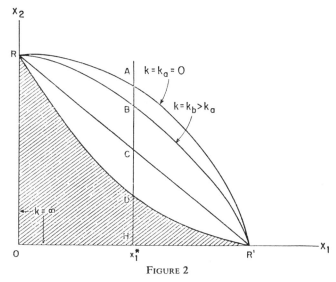

FIGURE 2

laundry can be produced than before. Consequently point A must shift downward to some lower point, B, and the entire possibility locus becomes something like RBR'. With further increases in the value of k, point A will be shifted lower still. If at some value of k it is pulled below line segment RCR', the possibility set becomes a non-convex region[1] such as shaded region $ORDR'$. But we know that if the external damage is sufficiently serious (i.e. for sufficiently high values of k), A must cut below C, for if the marginal smoke output is so great and so noxious that no quantity of resources can get laundry as clean as it would be

[1] Figure 2 can be connected directly to the inter-related individual production sets of Figures 1a and 1b. Points R and R' respectively represent the social output vectors $(-8, 0, 400)$, i.e. point B in Figure 1a, and $(-8, 20, 0)$, i.e. point A in Figure 1a. With constant returns to scale and a single input the production frontier in the absence of externalities must be the line segment RCR'. However, with electricity output at $x_1^* = 10$ in Figure 2, the most laundry we can obtain in the presence of the externality is $HD = 100$, not $HC = 200$.

in the absence of smoke, then A must fall all the way to the horizontal axis (point H). That is, in the limit, the possibility locus then must consist simply of the axis segments ROR'.

The simplicity of the preceding argument may belie its generality and rigour. The point is that with any pair of commodities one of which interferes with the production of the other there will be no such interference if one or the other is not produced. On the other hand if the interference is sufficiently great, the maximal output of the activity suffering the external damage will approach zero for *any* non-zero level of output of the other, and a non-convexity in the feasible set is unavoidable. Note finally that if there is a non-convexity in the production set for *any* pair of commodities, the full n-dimensional production set is also non-convex.

III. A FURTHER ILLUSTRATION

Some readers may prefer to deal with a concrete algebraic example explicitly relating a measure of the degree of detrimental externality with the "wrong" curvature of a production possibility frontier of the type displayed in Figure 2. We therefore offer a case in which the separate production sets of the two industries are strictly convex. Again let v_i stand for the amount of labour (negative leisure) used in industry i, x_1 for the output of electricity, and x_2 for the output of laundry services, and suppose

(1) $v_1 = x_1^2/2$

   $v_2 = x_2^2/2 + kx_1x_2.$

Each industry is separately subject to strictly diminishing returns to scale. The coefficient k now measures the strength of the effect of electricity output on laundry costs; the effect is detrimental if $k>0$. If a total of v units of labour is made available, we can write the implicit equation for the laundry–electricity possibility frontier as

(2) $v = x_1^2/2 + kx_1x_2 + x_2^2/2,$

   $x_1 \geqq 0, x_2 \geqq 0.$

We can deal with any such differentiable possibility locus in an obvious manner, calculating its second derivative and showing generally that when the externality parameter k becomes sufficiently large that derivative must take positive values. The present illustration, however, permits us to show this result more directly. If $k=0$ (no externality), (2) describes a quarter circle in an (x_1, x_2) co-ordinate system. This boundary obviously has the "right" curvature. For small positive k, the boundary continues to be concave to the origin. However, when $k=1$, (9) becomes the equation of a straight line $[(x_1 + x_2)^2 = 2v]$; and for larger values of k, non-convexity of the production set occurs.

In the preceding example non-convexity only happens to appear with a fixed large value of k, i.e. for $k=1$. However, generally the appearance of the non-convexity will depend both on the magnitude of the externality parameters and on the values of x_1 and x_2. For example, suppose in the preceding example we leave the electricity cost function unchanged but make the laundry resource requirement function

$$v_2 = x_2 + kx_1x_2.$$

Then the production possibility locus is given by

$$v = v_1 + v_2 = x_1^2/2 + kx_1x_2 + x_2.$$

A straightforward calculation of the second derivative shows that convexity will now be violated if and only if

$$2k^2x_2 + kx_1 > 1.$$

Clearly, for k or x_1 or x_2 sufficiently large, this requirement will not be satisfied. In this illustrative example, the maximum feasible values of x_2 occur in the vicinity of $x_1 = 0$. Here we have $x_2 \cong v$, and it is not difficult to imagine values of k and v that will violate the preceding convexity requirement. If v is very large, say of the order of thousands or millions of units, even a very small value of k will violate the second-order conditions. For example, if $v = 10{,}000$, then any $k > 0.01$ will have this effect.

IV. SPATIAL SEPARATION AS A PALLIATIVE

A lower bound to the degree of non-convexity in the social production set arising from detrimental externalities is provided by the possibility

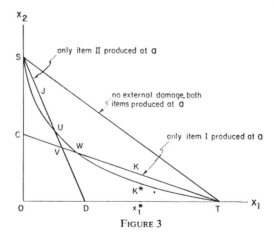

FIGURE 3

of separating the generators and their victims geographically, moving the laundries from the vicinity of the electricity producers, or *vice versa*. This is illustrated by the following example.

Assume once more that we have two outputs, 1 and 2 (electricity and laundry), and that these can be produced at either of two locations, *a* and *b*, with respective output levels, x_{1a}, x_{2a}, x_{1b} and x_{2b}. To begin with, we take all substitution relationships in the absence of externalities to be perfectly linear. Let us assume also that, were there no externalities, it would pay to produce both items at the same location, say *a*. In Figure 3 line segment *ST* represents the production possibility locus for our two items when external damage is zero and both are manufactured at the more economical location, *a*. *SD* represents the more restricted set of output levels[1] that remains possible if x_2 were still produced at *a* but the production of x_1 were moved to *b*. Since *b* is assumed to be a less suitable site, all of *SD* must lie below *ST*, with the exception of end-point *S* which corresponds to production of x_2 alone, and which, since all of that activity takes place at *a*, must provide the same output level as can be achieved when there is no restriction on the use of site *a*. Similarly, line segment *CT* represents the production possibilities when manufacture of x_2 is moved to *b* while that of x_1 takes place at *a*.

Now suppose that externalities generated by the production of x_1 at *a* grow serious, so that the locus corresponding to manufacture of both items at *a* shifts from the line segment *ST* to the convex locus *SUWT*, by the process described in the discussion of Figure 1. Then, if society wishes to produce, say, quantity x_1^* of item 1, it can only obtain $x_1^*K^*$ of x_2 if both goods continue to be produced at *a*. However, by separating the two production processes—shifting the manufacture of item 2 to site *b*—the community can increase its output of commodity 2 to x_1^*K.

Obviously, then, if we take into account the possibility of spatial separation of output processes, the production possibility locus becomes *SJUWKT*. In no event can externalities force this locus to retreat closer to the origin than *SVT*. However, even here, the feasible region *OSVT* cannot be convex, because the boundary point *V* must lie below the line *ST*. Figures 4a and 4b generalize the argument of Figure 3 to the case of non-linear substitution relationships in which it is no longer necessarily true that one location, *a*, is the best place for both outputs. Once again, *ST* is the possibility locus in the absence of externalities. However, some of one or both items may now be produced at *b* as well as at *a*. The two possibility curves corresponding to the two ways of separating the two outputs are *PR* and *CD*. These two curves need no longer have even a point in common with *ST*. Nor, as Figure 4b shows, need they intersect. They will limit the extent to which externalities can pull the possibility locus toward the origin, but they cannot prevent the appearance of a non-convexity in the feasible region, as Figures 4a and 4b indicate. For suppose externalities transform the locus *ST*, along which the activities are not separated, into the curve *SJT*. The true possibility locus will now be *SWVUT*, yielding a feasible region *OSWVUT* (shaded areas) that is non-convex.

_____

[1] This shrinking of the possibility set takes into account any resources which must be devoted to transport as a result of the separation of activities.

These diagrams illustrate the proposition that sufficiently severe externalities make locational specialization economical. An example of the application of this point is seen in the Ruhr region in Germany, where the Emscher River valley has been completely devoted to waste disposal, while two other river basins have been preserved free from pollution.

The diagrams also bring out the disconcerting possibility that *which locational specialization is optimal may well depend upon the desired output proportions*. Thus, in Figure 4a, with fairly strong externalities the production possibility function is $SWVUT$. For output combinations

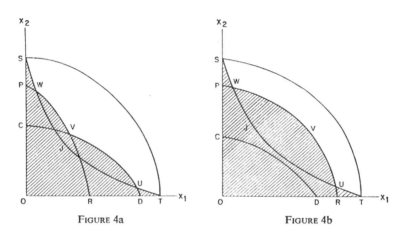

FIGURE 4a   FIGURE 4b

along segment WV all of x_2 is produced at a, all x_1 at b. Along segment VU the specialization is reversed. The danger of an incorrect choice by planners in this context appears clear. If it should turn out that, unpolluted, the Emscher River valley is uniquely well suited to growing marijuana it may turn out to have been a mistake to pick that one rather than one of the others for the area's sewer.

V. GENERALIZATION TO n ACTIVITIES

The arguments of the preceding sections have dealt with a world in which there are only two activities. Generalization of the argument to deal with more than two activities is immediate.[1] In a world of n outputs convexity can be guaranteed only if *each* of the partial possibility loci representing substitution between a pair of commodities is concave. Any single exception like that in Figure 2 means that at least two local

---

[1] Our discussion has also confined itself only to *detrimental* externalities. In principle, the presence of external benefits can also produce a multiplicity of local maxima, but here it is not so clear that the problem is likely to be serious. On this see Baumol [2], pp. 366–7.

maxima become possible. Thus the analysis holds whether the economy encompasses two outputs or n.

There is, however, one aspect of the matter that does require explicit analysis in terms of n activities. One may well ask how the number of local maxima is likely to grow with the number of activities. Here we can offer a few observations about polar cases which suggest that in at least some cases the number of local maxima may grow very rapidly with the number of activities involved. First, however, we deal with a case in which a proliferation of activities does not necessarily increase the number of local maxima.

Polar case (*a*): If one activity imposes external costs on m other activities, even if the detrimental effects are very great, no more than two local maxima need result.[1]

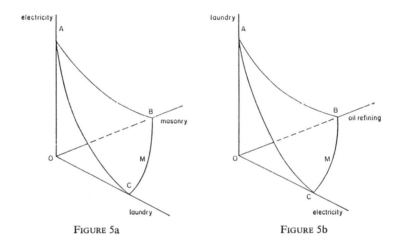

FIGURE 5a   FIGURE 5b

Polar case (*b*): A similar result holds when n activities impose external costs on one other activity.

In Figure 5a, the smoke from electricity production increases laundry cost and makes it more expensive to produce deterioration-resistant masonry. The production locus will then tend toward the form indicated (surface *ABC*). Since laundry and masonry activity impose no adverse external effects upon one another, their production possibility

---

[1] This does not preclude the possibility that there will be more than two maxima if the relevant functions violate the appropriate concavity–convexity conditions in the absence of externalities. Even where the maximum would otherwise be unique, externalities which are of intermediate strength may lead to three (or more) local maxima—characteristically two corner maxima produced by the externalities, and one interior maximum, a vestige of the unique maximum that would occur in the absence of externalities. Complications such as these and the possibility of irregularities in the relevant hypersurfaces probably limit the profit-ability of a more rigorous discussion of the subject of this section.

locus will have the normal shape (concave to the origin) illustrated by curve *CB*. However, for the reasons indicated in the discussion of Figure 2, if smoke damage grows sufficiently serious the other two partial loci will have shapes like those of *AC* and *AB*. We may then expect two local maxima, one at *A* and perhaps another at a point such as *M*. The interpretation of Figure 5b is exactly the same, and we merely pause to draw the reader's attention to the remarkable similarity between the diagram for the two-victim-one-polluter case and that for one victim and two polluters.

Next we come to cases involving more complex patterns of inter-dependence and show that here the number of local maxima may indeed increase with the number of activities involved. We have the following case.

Polar case (*c*): If each of *n* activities produces and suffers from very strong detrimental externalities and spatial separation is not possible, *n* local optima can be expected.

Here, in the limit as external damage becomes sufficiently great, it will be optimal (indeed it will only be possible) to carry on just one of the *n* activities. The choice of the activity to continue in operation clearly gives us our *n* local maxima (i.e. there are exactly *n* such choices available). If matters are not quite so serious, so that only a smaller number, *k*, of activities need be discontinued, it may be con-jectured that the number of local maxima will increase (to the number of combinations of *n* activities chosen *k* at a time).

Finally, we deal with the possibility of spatial separation, which, unlike its role in our earlier discussion, seems to aggravate the growth in number of maxima with the number of activities involved. We have the following case.

Polar case (*d*): If there are *n* activities each of which produces and suffers from externalities and there are just *n* discrete locations into which they can be separated, then, if the externalities are sufficiently severe, we can expect at least *n*! local maxima. For we have *n* candidates for the first location and, for each such choice, there remain $n-1$ candidates for the second location, then $n-2$ candidates for the third, etc., i.e. there are altogether *n*! different ways of achieving the desired isolation.

In practice, in some respects, this probably exaggerates the number of possibilities; in other ways it understates them. There really is no fixed finite number of discrete locations, and so one will normally have more than *n* geographic areas in which to locate *n* activities. If that is the right way of looking at the matter, it is clear that the number of local maxima (i.e. the number of ways of isolating each activity) will exceed *n*!. On the other hand, airborne pollution is known to travel over enormous distances. In that sense we may have no hiding-place from one another's emissions. We may then find ourselves back at the one-location case with its smaller number of local maxima but its more difficult problems of social damage.

VI. Convexity in Social and Individual Possibility Sets

In one respect the externality-induced non-convexity poses a less serious problem for social control than one might expect. For, as all of our examples indicate (see, notably, Figure 1), non-convexity in the social production possibility set is compatible with convexity in the sets over which individual producers make their choices. This has the consequence that it is possible through the use of prices and taxes alone to induce any individual firm to choose any designated point on its production possibility frontier, and hence to use these devices to sustain any designated point on the social possibility frontier, despite its "wrong" curvature. This may be contrasted, for example, with the case of non-convexity due to increasing returns to the scale of individual producers' production. Here, if every producer's average costs decline continually with scale over some substantial range, a producer confronted by a fixed price will either turn out zero output or some large quantity of output. Output combinations calling for intermediate levels of production of the good in question cannot be attained with the aid of the price mechanism alone.

The general principle may be illustrated with the example of Section III, involving two producers with input cost functions, (1). As it happens, if the input is inelastically supplied, in this case *any* pair of output choices by the two producers will be on the production possibility frontier. It need, then, only be demonstrated that any attainable (x_1, x_2) combination will be chosen in some price situation. Let the prices p_1 for electricity (x_1) and p_2 for laundry (x_2) be chosen, and let labour be given a price of unity. The profit functions of the two firms are given by

$$
\begin{aligned}
\pi_1 &= p_1 x_1 - x_1^2/2 \\
\pi_2 &= p_2 x_2 - x_2^2/2 - k x_1 x_2.
\end{aligned}
\tag{3}
$$

The individual production sets being strictly convex in "own" variables, the profit functions are strictly concave in own variables. Hence (as may be checked by calculating second partials) first-order conditions (4) are sufficient as well as necessary:

$$
\begin{aligned}
x_1 &= p_1 \\
k x_1 + x_2 &= p_2.
\end{aligned}
\tag{4}
$$

Eqs. (4) are invertible, which means that any desired pair (x_1, x_2) can be obtained as a solution to them for some combination of prices.

More generally, it is not true that all sets of profit-maximizing choices by producers will produce a socially efficient output combination, for we must usually worry about equating marginal rates of substitution and transformation by all producers, where any given pair of producers is likely to have many input–input, input–output and output–output pairs in common (e.g. all firms use some labour *and* some capital). The required adjustments to the prices facing an individual producer are, of course, Pigovian taxes, adding to the cost of inputs and subtracting from amounts received for outputs sums designed to bring into equality marginal private and marginal social cost. We reserve for an Appendix a formal demonstration of this fact.

VII. CONCLUDING COMMENTS: WHAT'S WRONG WITH NON-CONVEXITY?

There is only limited comfort to be derived from the knowledge that a sufficiently ingenious use of Pigovian taxes can keep a competitive economy at any desired point that is technologically efficient so long as detrimental externalities are the only source of non-convexity. Pigovian taxes cannot change the shapes of the technological relationships in the economy, and hence cannot remove the problems of evaluation of efficiency which non-convexity introduces. It seems appropriate therefore to conclude by reviewing this problem briefly.

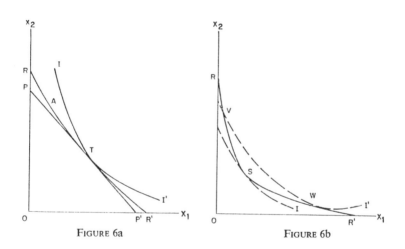

FIGURE 6a   FIGURE 6b

Let us, then, bring consumers into the picture. In Figure 6a let II' be a social indifference curve, so constructed that along it social welfare is constant and that its slope at any point equals the common slope of all consumers' indifference curves at the corresponding distribution of the two goods (see Samuelson [7] or Baumol [1] ch. 3, sec. IX, and Appendix). A social-welfare-maximizing state involving positive outputs of the two goods must be characterized by tangency of a social indifference curve with the production possibility frontier, as at point T in Figure 6a. As we have just suggested, so long as the only source of non-convexity is the presence of detrimental externalities, such a point can be sustained as a tax-adjusted competition equilibrium, in which producers are maximizing profits and individuals utility in the small and in the large.

However, note that, while convexity of preferences suffices to assure us that an equilibrium which maximizes the value of output at consumer prices over all feasible output vectors is Pareto-optimal (i.e. social-welfare-maximizing for some individualistic welfare function), the converse does not hold. It is no longer true that the availability of outputs that are more valuable (at current equilibrium consumer prices) implies that current output is not Pareto-optimal. In Figure 6a, for example, feasible outputs to the north-east of *PP'* are more valuable than output *T*, yet are socially inferior. A Pigovian-tax-compensated competitive equilibrium may thus be globally Pareto-optimal, as is *T* of Figure 6a; or a global minimum of social welfare among outputs which are technically efficient, as is *S* in Figure 6b; or a local, but not a global, social welfare maximum, as is *W* in Figure 6b.

In a world in which detrimental externalities are sufficiently severe to cause non-convexity of the social production possibility set, prices can no longer be depended upon to give us the right signals. *Even if we know the entire set of feasible output vectors, equilibrium prices usually tell us nothing about the Pareto-optimality of current output or even the direction in which to seek improvement.* While tax instruments will be effective in guiding the economy, the choice of the equilibrium point at which to settle must be made collectively by cost–benefit techniques.

Princeton University, Princeton, N.J.

APPENDIX

A Formal Model of External Effects and Corrective Taxes

In this Appendix we put forward a general definition of detrimental externalities of the sort discussed above. There is, of course, no single clear favourite among the possible generalizations of the effect whereby one single producer's expansion of his single output affects adversely the input requirements of other producers. However, the definition we propose does seem appropriate and does contain the simple definition as a special case Armed with this definition we show that social efficiency requires individual efficiency when external effects are all detrimental. A producer's net output vector (including negative entries for inputs) is "individually efficient" if no dominating net output vector is available to him without changing some other producer's net output choice. Finally, we show that when each individual producer's choice set is convex *any* socially efficient net output vector can be sustained by profit-maximizing production with externality-offsetting taxes.

Let $x^i = x^i_1, \ldots, x^i_n$ be the net output vector of the ith producer, where negative entries represent net inputs. We assume that x^i is chosen from a *feasible set* S^i. For the usual reasons (see, e.g. Quirk and Saposnik [6]) we assume that S^i always includes the origin and the negative orthant of Euclidian n-space (free disposal), and that S^i has no elements in the strictly positive orthant (no outputs without inputs).

The size and shape of S^i depend in general on the net output choices of *other producers.* Let X stand for the matrix of net output choices of the m producers in the economy, whose k, jth element, x_j^k, is the net output of commodity j by producer k. We shall represent the dependence of S^i on the choices of other producers as a functional relationship, mapping X into subsets of n-space, and denote this relationship by $S^i = S^i(X)$. Thus $S^i(X)$ is defined as the set $\{x^n | x^1, \ldots, x^{i-1}, x^{i+1}, \ldots, x^m\}$. We shall say that *these relationships embody detrimental external effects if for any two different producers k and i*

$$\lambda > 1 \Rightarrow S^i(\ldots \lambda x_j^k \ldots) \subset S^i(X),$$

where $(\ldots \lambda x_j^k \ldots)$ *denotes the matrix obtained from X by replacing x_j^k with λx_j^k*. This definition implies that a producer tends to hurt other producers by increasing the intensity of any net output *or* net input. Obviously a definition of beneficial external effects would be obtained by reversing the set inclusion sign in this definition. The definition could also be made specific to a particular element x_j^k so that some variables could exhibit detrimental, some beneficial, external effects. We confine our attention to relationships involving only detrimental externalities, which by the definition, include the case of zero externality.

Under this definition it follows trivially that socially efficient production (total net output of some one item maximized for any set of given totals of the other net outputs) requires efficiency on the part of each individual producer. For if a producer has chosen an individually inefficient net output vector he can alter his choice in a way which preserves his net output but reduces his net input usage. Since the external effect of the latter action must in our case *enhance* the productive opportunities of the other producers, such a choice is clearly required by social efficiency.

Since, by assumption, the feasible set of each producer, considering only variations in the vector under his control, is convex, any point in that set which is efficient from the producer's point of view will be a profit-maximizing choice for some vector of prices. And since, as we have just shown, any socially efficient point will be composed of a sum of points efficient for each producer, it follows that any socially efficient net output vector can be sustained as a profit-maximizing point for all producers if the prices are appropriately adjusted for each producer by a set of taxes. It remains only to show that the "appropriate" taxes are precisely equal to the marginal external damages arising from changes in output and input choices.

For this demonstration it will be convenient to assume that the feasible set of the ith producer is defined by the inequality $f^i(X) \leq 0$, where f^i is a differentiable function. Recall that X is a *matrix* of which the typical element, x_j^i, specifies the net output (negative for inputs) of the jth commodity by the ith producer. Fixing net output vectors x^j for $j \neq i$, $f^i(X) = 0$ defines the "private production possibility frontier" constraining x^i, the net output vector choice of the ith producer. By assumption, the set of vectors x^i satisfying $f^i(X) \leq 0$ in this case is convex.

If the rows of the matrix $\bar{X}$ sum to a point on the social production possibility frontier, then it is a solution to the non-linear programming problem

$$\underset{X}{\text{maximize}} \sum_{i=1}^{m} x_1^i,$$

subject to

$$\sum_{i=1}^{m} \bar{x}_k^i - \sum_{i=1}^{m} x_k^i \leq 0 \quad (k=2, \ldots, n)$$

$$f^j(X) \leq 0 \quad (j=1, \ldots, m).$$

By a simple extension of the Kuhn–Tucker theorem on optimization with inequality constraints, necessary conditions for a solution to this problem are that there exist non-negative multipliers $\lambda_2, \ldots, \lambda_n$, corresponding to the constraints requiring minimum amounts of commodities other than commodity 1, and $\gamma_1, \ldots, \gamma_m$, corresponding to the individual production constraints, such that

$$1 - \sum_{j=1}^{m} \gamma_j f_{i1}^j = 0 \quad (i=1, \ldots, m)$$

$$\lambda_h - \sum_{j=1}^{m} \gamma_j f_{kh}^j = 0 \quad (k=1, \ldots, m)$$

$$(h=2, \ldots, n).$$

(The notation f_{kh}^j stands for the partial derivative of f^j with respect to x_h^k.) By the usual interpretation, λ_j equals the amount of commodity 1 (in effect here the *numeraire* commodity) obtained by a unit reduction in the amount of commodity j produced. The multiplier γ_k is the value (in commodity 1 terms) of the extra output which could be obtained if firm k's production constraint were relaxed by requiring $f^k(X) \leq 1$ instead of $f^k(X) \leq 0$.

Consider next the profit-maximizing problem faced by producer i faced with a vector p of prices *and* a vector t^i of taxes:

$$\underset{x^i}{\text{maximize}} \sum_{j=1}^{m} (p_j - t_j^i) x_j^i,$$

subject to

$$f^i(X) \leq 0,$$

where all variables other than the "own", vector, x^i, are treated as exogenously fixed in the constraint. By the Kuhn–Tucker theorem, if x^i is a solution there necessarily exists a non-negative multiplier, δ^i, such that

$$p_j - t_j^i - \delta^i f_{ij}^i (X) = 0 \quad (j=1, \ldots, m).$$

Furthermore, since the constraint set is convex, these conditions, together with the constraint, are sufficient as well as necessary for a constrained maximum. The multiplier δ^i indicates the profit which would be *lost* to the ith producer if his production constraint were "tightened" by one unit.

Now we need only put the two problems together. If $\bar{X}$ is a set of individual producer vectors summing to a point on the social production possibility function, use the Lagrange multipliers from the associated non-linear programming problem and set

$$p_1 = 1, \quad p_2 = \lambda_2, \quad \ldots, \quad p_n = \lambda_n \quad (i=1, \ldots, m)$$

$$t_{ik} = \sum_{\substack{j \neq 1 \\ j=1}}^{m} \gamma_j f_{ik}^j, \quad (k=1, \ldots, n).$$

We see that $\bar{x}^i$ satisfies the necessary and sufficient conditions for a profit maximum for producer i, with the multiplier δ^i of his problem equal to γ_i in the economy-wide problem.

To interpret this result note that, for $j \neq i$, f_{ik}^j is, in effect, the constriction in the jth production constraint per unit increase in the ith producer's net output of the kth good. Hence $\gamma_j f_{ik}^j$ is the external cost, in *numeraire* units, of such an increase, imposed by the ith producer on the jth per unit increase in x_k^i, and $\sum\limits_{j \neq i}^{m} \gamma_j f_{ik}^j$ is the total external social cost per unit increase in x_k^i. Furthermore, since $\delta^i = \gamma_i$, the external social cost will also exactly equal the marginal external profit loss per unit increase in output of X_k by firm i observed when the proper corrective taxes are applied.

REFERENCES

[1] Baumol, W. J., *Welfare Economics and The Theory of the State*, 2nd ed., Cambridge, Mass., 1965.

[2] ——, "External Economies and Second-Order Optimality Conditions", *American Economic Review*, vol. 54 (1964), pp. 358–72.

[3] Kolm, S. C., "Les Non-Convexités d'Externalité", CEPREMAP Rapport No. 11, mimeograph, 1971.

[4] Pigou, A. C., *The Economics of Welfare*, 4th ed., 1932.

[5] Portes, R. D., "The Search for Efficiency in the Presence of Externalities", in Paul Streeten (ed.), *Unfashionable Economics: Essays in Honour of Lord Balogh*, 1970, pp. 348–61.

[6] Quirk, J. and R. Saposnik, *Introduction to General Equilibrium Theory and Welfare Economics*, New York, 1968.

[7] Samuelson, P. A., "Social Indifference Curves", *Quarterly Journal of Economics*, vol. 70 (1956), pp. 1–22.

[8] Starett, D., "On a Fundamental Non-Convexity in the Theory of Externalities", Harvard Institute of Economic Research, Discussion Paper 115, 1970.

[3]

The symmetry between controlling pollution by price and controlling it by quantity

JOHN PEZZEY University of Bristol and UK CEED

Abstract. Under ideally competitive conditions, both controlling pollution by price (using a combined charge-subsidy scheme) and controlling it by quantity (using a marketable permit scheme) can achieve short- and long-run efficiency and also political acceptability, provided that both schemes embody the same degree of environmental ownership. The resulting full symmetry between control by price and control by quantity, a symmetry overlooked in the literature because of the entry-exit assumptions automatically made for most subsidy schemes, allows a useful practical choice to be made between the two control systems.

La symétrie entre le contrôle de la pollution par les prix et les quantités. Sous des conditions concurrentielles idéales, le contrôle de la pollution soit par les prix (en utilisant un système de taxes et subventions) soit par les quantités (en utilisant un système de mise en marché de permis) peut également réussir à assurer l'efficacité à court et à long terme et être politiquement acceptable, pour autant que ces mécanismes incorporent le même degré de propriété environnementale. La symétrie qui en résulte entre le contrôle par les prix et par les quantités, une symétrie qu'ignore la littérature spécialisée à cause des postulats d'entrée et de sortie que la plupart des mécanismes de subventions entérinent automatiquement, permet de faire des choix pratiques et utiles entre les deux systèmes de contrôle.

I. INTRODUCTION

The essential result of this paper is simple. Under ideal conditions, controlling excessive pollution or congestion of a scarce public or common property resource by using a price-based instrument such as a fee or charge can be made symmetrical,

I am grateful for research support from the Harkness Fellowship, the UK Department of the Environment and the UK Centre for Economic and Environmental Development. I also thank seminar participants at the University of Colorado, Chuck Howe, Gene Mumy, Paul Downing, and three anonymous referees for helpful comments on earlier drafts. All views expressed and remaining errors are mine.

Canadian Journal of Economics Revue canadienne d'Economique, XXV, No. 4
November novembre 1992. Printed in Canada Imprimé au Canada

984   John Pezzey

in terms of short-run efficiency, long-run efficiency,[1] and political acceptability, to using a quantity-based instrument such as a marketable licence or permit. The symmetry between pure charges and sold or auctioned marketable permits has already been shown by Spulber (1985). Here we show that there is also symmetry between charge-subsidies and corresponding marketable permit schemes where some or all of the permits are freely granted rather than sold. There is thus no *fundamental* reason, as long as the decision has been taken to use some market instrument rather than direct regulation, for choosing control by price instead of control by quantity, or vice versa.

The key condition for attaining this useful freedom of choice is that, in any given application, both types of instrument embody the same degree of 'environmental ownership' in the form of symmetrical, private property rights in the resource. However, the relevant literature implicitly or explicitly, but in either case rather inconsistently, rules out this property rights condition for control by price but does not rule it out for control by quantity. As a result, it often happens that efficient and acceptable instruments are rejected by economists; efficient but unacceptable instruments are proposed instead; while inefficient but acceptable instruments are the ones actually used by policy makers. The aim of this paper is to encourage the use of control instruments that are both efficient and acceptable.

In keeping with the existing literature, the argument below uses the language of pollution control, specifically the control of water pollution. However, it can also apply to a range of natural and man-made resources which are not yet privately owned, such as the atmosphere, land for waste dumping, wilderness and wildlife, road space or airport landing slots, where either one-way or mutual (congestion) externalities may arise.

The 'ideal' conditions that are assumed to hold here constitute perfect competition in its fullest sense. We consider a perfectly competitive industry comprising many small firms, each of which is a rational profit-maximizer producing a single output and discharging a single effluent, emission, or waste stream. The effluent is neither storable on the factory site, nor cumulative in the environment, but is continuously assimilated into a well-mixed but finite environmental reservoir. Firms face perfectly competitive markets for their outputs and for their capital and labour inputs, but they own different sets of fixed factors like enterprise and therefore have different marginal cost schedules for effluent control. Time-dependent phenomena such as uncertainty and technical innovation in pollution control are ignored. Perfect information is freely available to all firms and to the pollution control authority (hereafter just 'the authority'), and transaction costs are zero. Last, but by no means least, a perfect authority, whose sole objective is to maximize public welfare, is assumed.

Contrary to normal practice, in section II and III we first state the case for the equivalence of control by price and control by quantity, and then, in section IV, we

---

1 As usual in the pollution control literature, 'short run' takes as given the firms that exist in the industry, while 'long run' allows for the entry and exit of firms.

relate the ideas thus raised to the existing literature. Finally, in section v, we draw out some implications for policy.

II. CONTROL BY PRICE: THE CASE FOR THE CHARGE-SUBSIDY

Controlling pollution by price implies the use of charges[2] per unit of effluent added and/or subsidies[3] per unit of effluent reduced. The way in which charges and subsidies can be combined into a 'charge-subsidy' scheme, which achieves short-run efficiency, long-run efficiency, and political acceptability, has been fully spelt out in a neglected paper by Mumy (1980).[4] His scheme is effectively that each polluting firm pays

$$V(E - E_b) \text{ (in, say, dollars per month)} \qquad (1)$$

to the authority, where

V = the charge rate (in, say, dollars per ton of effluent) set by the authority, which does not vary from firm to firm or with time.

E = the effluent level (in, say, tons per month). This is under the firm's control and so may vary from firm to firm and over time.[5]

E_b = the baseline effluent right (in tons per month) which is initially given *as a property right* to each existing firm by the authority. E_b may vary from firm to firm but does not vary over time.

If a firm has a positive baseline, and its effluent is less than its baseline ($E < E_b$), it receives a subsidy from the authority. If $E_b = 0$ for all firms, the scheme reduces to a pure Pigovian pollution charge. V (which will of course equal the industry's marginal cost of effluent control in equilibrium) is chosen so that the marginal damage cost of the resulting total effluent ΣE is equal to V, thus achieving *short-run efficiency*, given the ideal conditions assumed.[6] ΣE is thus determined on economic grounds and is not necessarily the same as total baseline effluent ΣE_b, which is determined on political grounds (see below). The scheme therefore may not be revenue-neutral for the authority.

Long-run efficiency is achieved because E_b is a full property right. New firms entering the industry are therefore *not* given effluent rights (so for them, $E_b = 0$), while existing firms exiting from the industry *keep* their effluent rights and receive a subsidy of VE_b in perpetuity. Under these entry-exit rules, the opportunity cost to

2 Also known as fees or taxes.
3 Also known as bribes, payments, or compensation.
4 The name 'charge-subsidy' is mine; Mumy himself referred to 'efficient property rights sharing,' to emphasize the property rights involved in the scheme.
5 Mumy actually considered the more restricted case where effluent is strictly proportional to output, and output itself is taxed.
6 Because each firm remains small in relation to the environmental reservoir, the marginal damage cost curve of each firm's effluent is constant. See Burrows (1979) and Collinge and Oates (1982) for the modifications required to the charge scheme if marginal damages increase as the firm's effluent increases.

986   John Pezzey

any firm of producing output Q and effluent E rather than closing down production (or not starting production in the first place, in the case of a new firm) is the sum of $C(Q, E)$, the firm's ordinary cost function excluding effluent charges and subsidies; $V(E - E_b)$, the effluent charge-subsidy; and VE_b, the cost of not receiving the perpetual subsidy for closing down. The net opportunity cost to the firm is then

$$C(Q, E) + V(E - E_b) + VE_b = C(Q, E) + VE, \tag{2}$$

and since $C(Q, E) + VE$ is the social opportunity cost of production, long-run efficiency is achieved. The baseline effluent right E_b disappears from formula (2), so it has no effect on production costs or resource allocation; the invariance proposition of Coase (1960) is thus recovered. Owning E_b effluent rights simply increases the wealth of the firm's owners, and there are no wealth effects, because firms are small. Holderness (1979) observed how Coase invariance exists only 'when rights are assigned to closed classes of individuals or entities,' and the above entry-exit assumptions do indeed close the class of owners of effluent rights.

In a charge-subsidy scheme, baseline effluent rights E_b for each firm should be chosen entirely on political grounds (which is why ΣE_b and ΣE may differ). The choice is unlikely to be easy. In many cases de facto effluent rights clearly exist in the form of existing effluent standards (Buchanan and Tullock 1975, 142; Pezzey 1988, 207). However, both environmental and industrial interests often fear, if for quite opposite and incompatible reasons, that formally recognizing effluent rights will be disadvantageous to them in the long-term struggle that usually precedes the establishment of any property rights over unowned resources. Whichever is the case, the more quickly and firmly that a formula can be found to settle disagreements between environmental and industrial interests, the sooner and greater will be the economic gain which can then be shared between these interest groups and also taxpayers and consumers.

III.   CONTROL BY QUANTITY, AND SYMMETRY WITH
       CONTROL BY PRICE

The authority can achieve effluent control by a quantity instrument, in a way that is formally symmetrical to the above scheme of control by price, as follows. As with charge-subsidies, the control authority starts by knowing the optimal total effluent ΣE. The authority gives ('grandfathers') each existing firm a free baseline amount E_b of marketable effluent permits (MEPS),[7] and takes such steps as are necessary to create an efficient market to bring together potential buyers and sellers of MEPS. If $\Sigma E_b > \Sigma E$, the authority must then rent back $(\Sigma E_b - \Sigma E)$ permits from the lowest offerer; if $\Sigma E_b < \Sigma E$, it must create an extra $(\Sigma E - \Sigma E_b)$ permits and offer these out for rental to the highest bidder. In either case, the equilibrium rental price of an

---

7 Also known as transferable discharge permits (TDPS), tradeable emission licences, tradeable effluent rights, marketable pollution consents, etc., etc.

TABLE 1
Categorization of market instruments for effluent control by method of control, and by effluent rights embodied

Control by price or by quantity?	Effluent rights owned by firm		
	Zero	Intermediate	Free market level of effluent
Price	P1. Pure charge	P2. *Charge-subsidy*[a]	P3. Pure subsidy
Quantity	Q1. Sold or auctioned MEPs	Q2. Freely granted (grandfathered) MEPs	Q3. *Granted and bought back MEPs*[a]

a Instruments in *italics* are frequently ignored in the literature (for example, by Milliman and Prince 1989).

MEP becomes V, the optimal effluent price.[8] If a firm's effluent $E > E_b$, it legally must rent $(E - E_b)$ permits at a rental price V, whereas if $E < E_b$ it will wish to lease out $(E_b - E)$ spare permits. If the firm closes down $(E = 0)$, it can lease out all E_b spare permits and receive a permanent income of VE_b. As with the charge-subsidy scheme, firms entering the industry do not receive effluent rights (i.e., $E_b = 0$). In all cases a firm producing output Q and effluent E therefore ends up paying $V(E - E_b)$ to the authority but faces opportunity costs of production equal to $C(Q, E) + VE$. These are the same formulae as (1) and (2) for the charge-subsidy scheme, so the MEPs achieve the same short- and long-run efficiency, and, as before, baseline effluent permits can be distributed according to political criteria without impairing efficiency. The whole scheme is presumably similar to that envisaged in a comment on Mumy by Beavis and Walker (1981), though with the important difference that here the total ΣE_b of the effluent baselines does *not* need to be exactly equal to the 'total amount of acceptable discharge,' that is, the optimal total effluent ΣE; if it does, political and economic considerations become entangled again.

The available schemes for control by price and control by quantity are summarized in table 1, and under our ideal conditions we have shown that the two types of control are fully symmetrical in terms of efficiency and acceptability. Our key conclusion is therefore that the best control scheme is to formalize the de facto effluent rights of each firm into precise baselines, and then incorporate these baselines as property rights into either charge-subsidy or MEP schemes, with the choice between charge-subsidies and MEPs being determined by practical departures from the ideal conditions.

IV. ARGUMENTS AGAINST SUBSIDIES

There is little in sections II and III that is technically new, as already noted. However, the symmetry we have established and depicted in table 1 is widely rejected in the

8 The talk is of renting rather than selling permits in order to make the symmetry between marketable permits and charge-subsidies more obvious. If the interest rate is r and the permit is permanent, the selling price would be V/r. Other details of this market, for example, whether it uses quoted prices or auctions, are not discussed here.

literature. It is therefore important to examine this rejection here, before briefly reviewing in section v why a fundamentally free choice between control by price and control by quantity is desirable and how it should be made, and suggesting how the debate can move forward.

The literature on effluent charges and subsidies stretches from Kamien, Schwartz, and Dolbear (1966) to modern textbooks like Baumol and Oates (1988, chap. 14).[9] Its essential conclusion is that subsidies are undesirable, for three reasons: one economic, one administrative, and one political. The economic reason given is that, in the long run, subsidies encourage excessive entry into a polluting industry, and avoiding this would require the practically and politically impossible task of tracking down potential polluters and subsidizing them to stay out of the industry. However, this conclusion entirely depends on the (usually implicit) assumptions that subsidy payments are available to all firms that enter, and terminated for all firms that exit. The case for these standard 'open-class' entry-exit assumptions, which differ crucially from our 'closed-class' assumptions above, is rarely given. While the standard assumptions may represent the way in which real subsidy schemes generally operate, as noted by Baumol and Oates (1988, 214), there is no theoretical reason why a new firm should not have to buy or rent its effluent rights from existing owners of the environment, just as it must buy or rent its new factory site from existing owners of land.

The administrative reason given, for example, by Baumol and Oates (1988, 216), is that it would be infeasible to pay subsidies *indefinitely* to firms which have exited. If so, the solution would be the suggestion in Dewees and Sims (1976, 330) that the authority buys out exiting firms' effluent rights by offering lump-sum subsidies in compensation (although this could make big demands on the authority's cashflow). The political reason is that given by writers such as Spulber (1985, 106), who object to firms' owning effluent rights, on the grounds that society owns the environment, and recommend pure charging instead. As argued above, this ignores the political reality that many firms have de facto effluent rights and the clout to defend them.

Despite the formal symmetry that we have shown to exist between freely granted MEPS and charge-subsidies under ideal conditions, the former are both much better known and much less likely to be criticized in the literature than the latter; see, for example, the approval given to granted MEPS in Baumol and Oates (1988, 179). Such writers are much more prepared to accept the notion of environmental property rights with control by quantity than they are with control by price. As a result, they explicitly or implicitly accept the closed-class entry-exit assumptions for control by quantity, and thus ensure that the long-run economic objections of excessive entry to the industry do not arise with freely granted MEPS. Also, MEPS do not get tainted with criticism of related instruments, because of the asymmetric choices of instruments that are made when comparing control by price and control by quantity. For example, Milliman and Prince (1989), in an otherwise comprehensive study of how instrument choice affects technical innovation, choose pure charges

---

9 An earlier version of this paper (Pezzey 1990) contains a more detailed review of this and related literature.

and pure subsidies (P1 and P3 in table 1) as instruments which control by price, but sold MEPs and freely granted MEPs (Q1 and Q2) as instruments that control by quantity. Choosing to study pure subsidies instead of the charge-subsidy option (P2) tends to associate control by price in general with the specific moral hazard of pure subsidies, which arises when the level of effluent that firms initially (or hypothetically) discharge in the absence of all regulation is used as the starting point for subsidies. The equivalent objection to MEPs does not arise because no one thinks it sensible even to consider option Q3, whereby firms are given permits equal to what their free-market, unregulated discharges would be.

V. IMPLICATIONS FOR POLICY

Conventional economic wisdom thus unnecessarily excludes a rights-based charge-subsidy scheme (option P2 in table 1) from serious consideration as a policy instrument. This may have expensive consequences in real cases where pure charging (option P1) is politically unacceptable because of well-established de facto effluent rights, but control by price is more cost-effective than control by quantity. In any given case, practical choices between control by price and control by quantity, and about how much regulation should be retained as a backstop to market instruments, should be based on how well each instrument copes with the way the real world departs from the ideal conditions set out in section I. These departures include uncertainty; monitoring and enforcement costs, and how they are distributed among firms and the control authority; storage or accumulation of pollutants; changes over time due to economic growth and technical progress; and vulnerability to monopoly power (see Rose-Ackerman 1977 and Pezzey 1988 for surveys of many of these points). Because of the variety of practical circumstances that can occur, there can be *no general presumption that control by quantity is superior to control by price*.

Uncertainty is worth a special mention. It is well established, following a seminal contribution of Weitzman (1974) and a recent summary by Baumol and Oates et al. (1988, chap. 5), that if the authority has good information on the marginal benefits of effluent control, is uncertain about the absolute level of control costs, but is reasonably sure that marginal benefits decrease less steeply than marginal costs increase as effluent is reduced, then control by prices will give greater expected social welfare than control by quantities. Harrison (1983) and Oates et al. (1989, fn4) record cases (concerning aircraft landing noise and urban air pollution, respectively) where these conditions are met, and control by price is economically preferable. In the context of global warming, the choice between carbon taxes and tradeable carbon emission permits may be one where, if effective progress is to be made, using control by price to avoid excessive costs to industry is more important than using control by quantity to achieve precise control over carbon dioxide emissions.

How then can the charge-subsidy idea be added to the menu of instruments considered by policy makers? One way to overcome resistance to the idea may be to change the language used. Kelman's (1981) survey showed that attitudes to effluent

990   John Pezzey

charging are greatly influenced by the choice of particular words, such as 'fees,' 'charges,' or 'taxes.' Clearly, there is also a vast difference in political perception between 'a bribe,' 'a subsidy,' and 'compensation,' even if all three are financially identical; which word many writers have chosen to use can hardly be accidental. However, it is also clear from other policy studies, such as the analysis of the u.s. emissions trading scheme in Hahn (1989, 101), that a fundamental message of economic analysis – that once a resource has become scarce, it needs to be owned, and priced, if it is to avoid becoming even scarcer – is one that many people do not want to hear, particularly when it is applied to the natural environment. The implications of an economic need for the deep oceans and the stratosphere to be 'owned' can indeed be disturbing, both practically and psychologically, and may provoke second thoughts about how far the physical demands of continued economic growth can be allowed to proceed. However, while they do proceed, there is an urgent need to find ways of controlling resource use that are both efficient and acceptable. The delicate task of promoting schemes that contain the necessary elements of subsidy and effluent rights, while trying to avoid direct use of such emotive words, is therefore one that economists should not duck.

REFERENCES

Baumol, W.J., and W.E. Oates (1988) *The Theory of Environmental Policy*, 2nd edition (Cambridge: Cambridge University Press)
Beavis, B., and M. Walker (1981) 'Long-run efficiency and property rights sharing for pollution control: a comment.' *Public Choice* 37, 607–8
Buchanan, J.M., and G. Tullock (1975) 'Polluters' profits and political response: direct control versus taxes.' *American Economic Review* 65, 139–47
Burrows, P. (1979) 'Pigovian taxes, polluter subsidies, regulation, and the size of a polluting industry.' This JOURNAL 12, 494–501
Coase, R.H. (1960) 'The problem of social cost.' *Journal of Law and Economics* 3, 1–44
Collinge, R.A., and W.E. Oates (1982) 'Efficiency in pollution control in the short and long runs: a system of rental emission permits.' This JOURNAL 15, 346–54
Dewees, D.N., and W.A. Sims (1976) 'The symmetry of effluent charges and subsidies for pollution control.' This JOURNAL 9, 323–31
Hahn, R.W. (1989) 'Economic prescriptions for environmental problems: how the patient followed the doctor's orders.' *Journal of Economic Perspectives* 3, 95–114
Harrison, D.J. (1983) 'The regulation of aircraft noise.' In *Incentives for Environmental Protection*, ed. T.C. Schelling (London, England: MIT Press)
Holderness, C.G. (1989) 'The assignment of rights, entry effects, and the allocation of resources.' *Journal of Legal Studies* 18, 181–9
Kamien, M.I., N.L. Schwartz, and F.T. Dolbear (1966) 'Asymmetry between bribes and charges.' *Water Resources Research* 2, 147–57
Kelman, S. (1981) *What Price Incentives? Economists and the Environment* (Boston: Auburn House)
Milliman, S.R., and R. Prince (1989) 'Firm incentives to promote technological change in pollution control.' *Journal of Environmental Economics and Management* 17, 247–65
Mumy, G.E. (1980) 'Long-run efficiency and property rights sharing for pollution control.' *Public Choice* 35, 59–74
Oates, W.E., P.R. Portney, and A.M. McGartland (1989) 'The *net* benefits of incentive-

based regulation: a case study of environmental standard setting.' *American Economic Review* 79, 1233–42

Pezzey, J.C.V. (1988) 'Market mechanisms of pollution control: "polluter pays," economic and practical aspects.' In *Sustainable Environmental Management – Principles and Practice*, ed. R.K. Turner (London: Belhaven Press)

— (1990) 'Charge-subsidies versus marketable permits as efficient and acceptable methods of effluent control: a property rights synthesis.' *Department of Economics Discussion Papers* 90/271 (Bristol: University of Bristol)

Rose-Ackerman, S. (1977) 'Market models for water pollution control: their strengths and weaknesses.' *Public Policy* 25, 383–406

Spulber, D.F. (1985) 'Effluent regulation and long-run optimality.' *Journal of Environmental Economics and Management* 12, 103–16

Weitzman, M.L. (1974) 'Prices vs. quantities.' *Review of Economic Studies* 41, 477–92

Part II
Instruments

A   Fees and Charges

[4]

THE
QUARTERLY JOURNAL
OF ECONOMICS

Vol. LXXXVII   November 1973   No. 4

SPECIFIC TAXES AND THE CONTROL OF POLLUTION:
A GENERAL EQUILIBRIUM ANALYSIS *

Thomas H. Tietenberg

I. INTRODUCTION

Statement of Purpose

This paper will imbed industrial pollution in a simple general equilibrium model and then use this model as a basis for addressing the issue of whether or not specific taxes on effluent are a sufficient means for a government to control pollution. The model views waste outputs as joint products. Any one production process is capable of producing a vector of these residuals. The firm can either purify these products, thereby rendering them ecologically neutral, or it can dispose of them in a common property resource (e.g., a lake) where they become pollutants. Pollution can thus be controlled by either (1) purification of residuals or (2) reduction of the level of intensity of the process yielding the offending residuals. Pollutants cause disutility to the consumers (1) by forcing the otherwise unnecessary consumption of some pollution-avoiding commodities (e.g., air conditioners), (2) by forcing a reduction in the consumption of other desirable commodities (e.g., swimming in a particular lake), and (3) directly by its very presence in the environment.

*An earlier version of this paper was delivered at the Winter Meetings of the Econometric Society, New Orleans, Louisiana, December 27, 1971. The financial support of this research from Grant #T01–MH–12117 from the National Institute of Mental Health is gratefully acknowledged. I also wish to acknowledge the benefit I have derived from comments on previous drafts by Theodore Groves, Eugene Smolensky, Donald Hester, Stephen Henderson, and William Hogan, Jr. The reader is asked to grant them immunity for the shortcomings of this paper.

With this characterization of pollution we then proceed to use social welfare functions with unspecified weights to derive a set of socially efficient allocations of resources including pollution. These allocations represent the targets for government policy. The actors in the model are classified into three categories — consumers, producers, and the government. Each actor is assumed to have a specific objective. By assumption, consumers maximize utility, producers maximize profits, and the government seeks a socially efficient allocation of resources.

The mechanism for controlling pollution, a competitive market augmented by a system of specific taxes on effluent, is then examined to determine whether the government, armed with these instruments, can sustain a socially efficient allocation of resources when all other agents are acting noncooperatively in their own self-interest. This capability is examined by (1) defining and characterizing decentralized equilibria when these instruments are applied and (2) establishing whether there exists a particular set of values for these instruments such that every decentralized equilibrium is socially efficient and every socially efficient state can be sustained as a decentralized equilibrium. The results are shown to have important implications for pollution control policy.

Historical Perspective

Arrow was one of the first economists to question the prevailing notion that such a system clearly existed:

> If any component of x entered as a variable into the utility functions of more than one individual, the whole analysis will be vitiated as it stands. . . . The general feeling is that in these cases, optimal allocation can be achieved by a price system, accompanied by a suitable system of taxes and bounties. However, the problem has only been discussed in simple cases; and no system has been shown to have, in the general case, the important property possessed by the price system. . . ; not only can optimal distributions (usually) be achieved by the price system but any distribution so achieved is optimal.[1]

Other economists have put the argument more strongly. Bohm, for instance, argues "that the unit taxation approach may be used to fulfill the necessary optimum conditions . . . but that it cannot satisfy the sufficiency conditions of positive net total benefits."[2] This latter argument, which is presented in partial equilibrium

---

1. Kenneth J. Arrow, "An Extension of the Basic Theorems of Classical Welfare Economics," *Second Berkeley Symposium on Mathematical Statistics and Probability*, J. Neyman, ed. (Berkeley: University of California Press, 1951), pp. 527–29.
2. Peter Bohm, "Pollution, Purification and the Theory of External Effects," *Swedish Journal of Economics*, LXXII (June 1970), 156–57.

terms, is based upon the notion that, in general, profits will not be a perfect signal to the firm as to when to shut down. This notion is developed from a theoretical construct known as the marginal damage function that describes the marginal loss in welfare to consumers from increasing amounts of pollution.

This paper will present sufficient conditions (in terms of more conventional concepts such as utility and production functions) for the existence of a tax and transfer system that can satisfy modified versions of the two fundamental theorems of classical welfare economics in spite of the fact that the externality in question (pollution) cannot directly be taxed. We can then examine the nature of this efficient tax.

II. The Model

Introduction

The model [3] described in this paper will be a highly simplified view of reality in several respects. It will be static in that there is no intertemporal component to either the consumption or the production decision. As a result, the very important problems associated with the intergenerational transfer of externalities will be ignored. The model will be concerned only with industrially generated pollution that inflicts technological externalities on both consumers and producers. The costs of implementing the system of effluent charges will be assumed to be zero.[4]

In our analytical economy there exist J normal commodities, A waste products, B pollutants, and A types of waste purification.[5] With respect to waste purification, a technology is posited that is capable of rendering waste products ecologically neutral. To produce these commodities and services, there are K firms. Some of these firms create waste products as by-products of a production process that supplies the J normal commodities or the A types of waste purification. Some of these production processes are adversely affected by the presence of pollutants in the common property resources.[6] To consume these commodities there are I consumers.

3. This model is a modified version of a model suggested by Takashi Negishi, "Optimal Policies with Externality," a paper delivered at the Second World Congress of the Econometric Society, Summer 1970.

4. This is not to deny the importance of transactions costs, but rather to emphasize other dimensions of choice that are logically distinct from costs of implementation, enforcement, etc.

5. The difference between waste products and pollutants in this model will be explained below.

6. A common property resource for our purposes refers to any body of water or airshed that is controlled by no unified directing power. For a theo-

Utility functions (U_i) and production functions (F_k) are assumed to be functions of (1) the activities controlled by that economic agent (i.e., the ith consumer or the kth firm) and (2) the stocks of pollutants, which are frequently out of the control of these agents. Symbolically we can represent the utility function for the ith consumer as

(2.1) $U_i = U_i(R,X_i)$ $i = 1, \ldots, I,$

where R is a B dimensional vector of each of B types of pollution in the common property resource and X_i is a J dimensional vector of consumption levels of the J normal goods for the ith consumer. The production possibilities for the $J+A$ commodities (the J private goods and A waste purification services) of the kth firm are those satisfying

(2.2) $F_k(R,Y_k,Z_k) \geq 0$ $k = 1, \ldots, K,$

where Y_k is a $J+A$ dimensional vector of outputs of the J normal commodities and the A purification services and Z_k is a J dimensional vector of inputs. Associated with each input and output combination for each firm is an A dimensional vector of waste products (W_k), which is determined by

(2.3) $W_k = f_k(Y_k,Z_k)$ $k = 1, \ldots, K.$

Equation set (2.3) maps the relationship between the combination of inputs and outputs chosen and the level of waste products created. Note that it is possible for a firm to produce only one of the $J+A$ commodities defined by (2.2) and still produce a vector of waste products. The level and composition of waste products is dependent on both the composition of the output bundle of the $J+A$ commodities and the input mix used to produce them. If the kth firm produces no waste products, then f_k is the zero function for the particular values of Y_k and Z_k being used.

We still have to define the relationship between pollutants and waste products (i.e., between the B dimensional vector R and the A dimensional vector W_k). Pollutants, by definition, exist in the common property resource, whereas waste products are merely the residuals in the production process. Waste products can either be purified or expelled untreated into the common property resource where they become pollutants. Any particular element of the R vector (say the bth) can be defined as

retical exposition of the theory of common property resources, see H. Scott Gordon, "The Economic Theory of a Common Property Resource: The Fishery," *Journal of Political Economy*, LXII (April 1954), 124–42; and Steven N. S. Cheung, "The Structure of a Contract and the Theory of a Non-Exclusion Resource," *Journal of Law and Economics*, XII (April 1970), 49–70.

SPECIFIC TAXES AND POLLUTION CONTROL 507

(2.4) $R_b = g(V,M)$ $b = 1, \ldots, B,$

where V is the set of $K \times A$ unpurified waste products and M is a vector of exogenous variables that govern the production of pollutants from emissions.[7] M may contain, for example, wind velocity and direction, temperature, amount of sunlight, etc. — natural elements that are largely immune to human control. Thus the production function g is rather unique in that only some of the inputs are subject to control.

Before proceeding, we consider the analytical treatment of pollution as reflected in (2.1) and (2.2). There are two aspects of pollution that are captured by these equations. First, pollution is a vector of commodities that enters the utility or production functions of more than one economic agent. This formulation suggests that pollution reduces social welfare not only by raising the costs or reducing the output of other commodities, but also by its mere presence in the common property resource. Second, we mean that the *same* set of arguments, the vector of pollutants, enters the utility or production function of more than one economic agent.[8] This maintained hypothesis is quite common in nonquantitative discussions of pollution, but it rarely appears in published formal models so that its impact on policy alternatives has not been assessed.[9] It can be justified by the realization that pollutants enter the utility

7. Any particular member of the set V (say the ka^{th} can be expressed as $V_{ka} = W_{ka} - Y_{kJ+a}$, where W_{ka} is the amount of the ath waste product produced by the kth firm and Y_{kJ+a} is the amount of the ath waste product purified by the kth firm.

8. It should be noted at this point that because more than one person is affected by the stock of pollution does not mean that their marginal utilities have the same magnitude or even that they have the same sign. It is perfectly possible, for instance, that distance from the center of pollution is also an argument in the utility function. For an examination of spatial adaptation to pollution, see Robert H. Strotz and Colin Wright, "Spatial Adaptation to Urban Air Pollution" (mimeo, 1970).

Although we assume that the relevant vector is defined over a specific geographic area, establishing the boundaries of this geographic area would be a difficult proposition. It is possible, for instance, that consumers have a global concern for pollution. Firms, however, are presumably only concerned with the stocks that affect their productive process that are probably local stocks. In this case the vector of pollutants in (2.2) could be different from the vector of pollutants in (2.1), since the geographic dimension of their measurement would not be the same.

9. For an example of the nonquantitative discussions, see Donald A. Nichols and Clark W. Reynolds, *Principles of Economics* (New York: Holt, Rinehart and Winston, Inc., 1971), p. 123; Martin Shubik, "A Curmudgeon's Guide to Microeconomics," *Journal of Economic Literature*, VIII (June 1970), 421–22; and Edwin G. Dolan, *TANSTAAFL: The Economic Strategy for Environmental Crisis* (New York: Holt, Rinehart and Winston, Inc., 1971), pp. 44–54. One example of a quantitative treatment can be found in Allen V. Kneese, Robert U. Ayers, and Ralph C. d'Arge, *Economics and the Environment: A Materials Balance Approach* (Washington: Resources for the Future, Inc., 1970), pp. 95–96.

functions as negative proxies for resources generally consumed out-side the market process (e.g., air breathed, scenery viewed, sun-shine absorbed, etc.) as well as for reasons of their adverse effects on health.[1]

The next section of this paper will address the problem of de-fining and characterizing efficient allocations of resources. These allocations will be characterized in terms of a version of the now familiar Kuhn-Tucker conditions for the optimal solution of a non-linear programming problem.[2]

Socially Efficient Allocations of Resources

The region of feasible allocations in our economy is bounded by several different kinds of constraints. Production relationships govern the technological feasibility of transforming factors into final products. These relationships have appeared above as (2.2) and (2.3). The second type of relationship is the market balance equation that states that the aggregate level of consumption of any good by firms or consumers cannot exceed the inventory of that good plus its produced output. This can be expressed as

$$(2.5) \quad \sum_{k=1}^{K} Z_{kj} + \sum_{i=1}^{I} X_{ij} \leqq \sum_{k=1}^{K} Y_{kj} + \sum_{i=1}^{I} X''_{ij} \quad j=1, \ldots, J,$$

where X_{ij} is the amount of the jth good consumed by the ith consumer, Z_{kj} is the amount of the jth good used as a factor input by the kth firm, Y_{kj} is the amount of output of the jth good by the kth firm, and X''_{ij} is the inventory of the jth good held by the ith consumer prior to production and exchange activities. These inventories exist be-cause either the resources were not all used previously or they are renewable resources.

The final type of constraint limits the amount of feasible puri-fication to the current produced output of that particular waste product by all firms. Once the waste product enters the common property resource it is irretrievable. This can be expressed symbol-ically as

$$(2.6) \quad \sum_{k=1}^{K} [Y_{kJ+a} - f_{ka}(Y_k, Z_k)] \leqq 0 \quad a=1, \ldots, A.$$

We can now define a technically feasible allocation.

DEFINITION. *An allocation of resources* {X,Y,Z,R} *is technically*

1. For some statistical support for the proposition that pollution ad-versely affects health, see Lester B. Lave and Eugene P. Seskin, "Health and Air Pollution," *Swedish Journal of Economics*, LXXIII (March 1971), 76–95.
    2. H. W. Kuhn and A. W. Tucker, "Nonlinear Programming," in J. Ney-man, *op. cit.*, pp. 481–92.

SPECIFIC TAXES AND POLLUTION CONTROL   509

feasible *if and only if it satisfies equation sets* (2.2), (2.4), (2.5), *and* (2.6).

We can also define socially optimal allocations of resources.

DEFINITION. *A technically feasible allocation of resources is* socially efficient $\{X^*, Y^*, Z^*, R^*\}$ *if and only if*

$$\sum_{i=1}^{I} a_i U_i(X^*_i, R^*) \geqslant \sum_{i=1}^{I} a_i U_i(X_i, R),$$

for some vector $\{a \mid 0 < a_i < \infty\}$ *for all* $\{X, R\}$, *which are technically feasible.*

Thus, an allocation of resources is socially efficient if it maximizes some social welfare function composed of weighted individual preferences. This is, of course, only a partial ordering of states until such time as a particular vector a is selected. Our purpose will be to show that regardless of what a vector is selected, the resulting allocation of resources can be sustained as a decentralized equilibrium.

The final step in this stage of the analysis is to derive necessary and sufficient conditions for an allocation to be socially efficient. This can be accomplished by appealing to the Kuhn-Tucker conditions for the optimal solution of a nonlinear programming problem using the following assumptions:

ASSUMPTION 1. U_i, f_k, *and* F_k *are concave, and all functions are twice differentiable.*

This assumption implies diminishing returns to scale for all arguments except the pollutants and a nondecreasing effect of the pollutants as their magnitude increases.[3] Assumption 1 also implies that the indirect effect of pollutants (the effect on the marginal utility and marginal productivity of other commodities) is not strong enough to counteract the direct effects.[4] In addition, we need

ASSUMPTION 2. *There exists at least one technically feasible allocation of resources* $\{X, Y, Z, R\}$ *such that* (2.2), (2.5), *and* (2.6) *hold as strict inequalities, and* $X_{ij} > 0$, $Y_{kj} > 0$, $Z_{kj} > 0$, *and* $R_b > 0$ *for all* i, j, k, *and* b.

---

3. It should be noted that this assumption is possibly incompatible with the experimental psychology literature, which argues that equal step increases of a stimulus are less easily perceived at higher degrees of intensity of the stimulus. See R. Duncan Luce *et al.*, eds., *Handbook of Mathematical Psychology* (New York: John Wiley & Sons, Inc., 1963), pp. 195–240.

4. In other words, the off-diagonal terms in the second-degree matrix cannot dominate the diagonal terms. For further discussion and explanation of this point, see William J. Baumol, "External Economies and the Second Order Optimality Conditions," *American Economic Review*, LIV (June 1964), 358–72.

This assumption merely assumes the existence of technically feasible allocation in which no constraint is binding. This allocation, of course, will not in general be optimal.

From the theory of nonlinear programming we can derive the optimality conditions for a socially efficient allocation of resources for any given vector a by forming the Lagrangean expression,

$$(2.7) \quad \sum_{i=1}^{I} a_i U_i(R,X_i) + \sum_{k=1}^{K} \gamma_k F_k(R,Y_k,Z_k)$$

$$- \sum_{j=1}^{J} P_j \left[\sum_{i=1}^{I} X_{ij} + \sum_{k=1}^{K} Z_{kj} - \sum_{k=1}^{K} Y_{kj} - \sum_{i=1}^{I} X''_{ij} \right]$$

$$+ \sum_{a=1}^{A} \psi_a \sum_{k=1}^{K} [f_{ka}(Y_k,Z_k) - Y_{kJ+a}],$$

and maximizing this function over the nonnegative orthant with respect to the decision variables for society as a whole.[5] This yields[6]

LEMMA 2.1. *Given Assumptions 1 and 2 and the constrained objective function (2.7), necessary and sufficient conditions for an allocation of resources $\{X^*,Y^*,Z^*,R^*\}$ to be socially efficient are*

$$(2.8) \quad F_k(R^*,Y^*_k,Z^*_k) = 0 \qquad (\geq 0 \text{ if } \gamma_k = 0) \qquad k = 1, \ldots, K;$$

$$(2.9) \quad \sum_{i=1}^{I} X^*_{ij} + \sum_{k=1}^{K} Z^*_{kj} - \sum_{k=1}^{K} Y^*_{kj} - \sum_{i=1}^{I} X''_{ij} = 0 \qquad (\leq 0 \text{ if } P_j = 0),$$
$$j = 1, \ldots, J;$$

$$(2.10) \quad \sum_{k=1}^{K} [Y^*_{kJ+a} - f_{ka}(Y^*_k,Z^*_k)] = 0 \qquad (\leq 0 \text{ if } \psi_a = 0),$$
$$a = 1, \ldots, A;$$

$$(2.11) \quad a_i \frac{\partial U_i}{\partial X_{ij}} - P_j = 0 (\leq 0 \text{ if } X^*_{ij} = 0), \qquad i = 1, \ldots, I,$$
$$j = 1, \ldots, J;$$

$$(2.12) \quad \sum_{i=1}^{I} \sum_{b=1}^{B} \sum_{a=1}^{A} a_i \frac{\partial U_i}{\partial R_b} \cdot \frac{\partial R_b}{\partial f_{ka}} \cdot \frac{\partial f_{ka}}{\partial Z_{kj}} + \sum_{s=1}^{K} \sum_{b=1}^{B} \sum_{a=1}^{A} \gamma_s$$

$$\frac{\partial F_s}{\partial R_b} \cdot \frac{\partial R_b}{\partial f_{ka}} \cdot \frac{\partial f_{ka}}{\partial Z_{kj}} + \gamma_k \frac{\partial F_k}{\partial Z_{kj}} - P_j$$

$$+ \sum_{a=1}^{A} \psi_a \frac{\partial f_{ka}}{\partial Z_{kj}} = 0 \qquad (\leq 0 \text{ if } Z^*_{kj} = 0),$$
$$k = 1, \ldots, K; \quad j = 1, \ldots, J;$$

---

5. Note that society has to choose only $\{X,Y,Z\}$, because once these variables have been determined, they completely define R. Thus before the maximization we substitute for R using (2.4).

6. This set of conditions is a special case of the more general Kuhn-Tucker conditions in which all variables are restricted to the nonnegative

SPECIFIC TAXES AND POLLUTION CONTROL 511

$$(2.13) \quad \sum_{i=1}^{I} \sum_{b=1}^{B} \sum_{a=1}^{A} a_i \frac{\partial U_i}{\partial R_b} \cdot \frac{\partial R_b}{\partial f_{ka}} \cdot \frac{\partial f_{ka}}{\partial Y_{kj}}$$

$$+ \sum_{s=1}^{K} \sum_{b=1}^{B} \sum_{a=1}^{A} \gamma_s \frac{\partial F_s}{\partial R_b} \cdot \frac{\partial R_b}{\partial f_{ka}} \cdot \frac{\partial f_{ka}}{\partial Y_{kj}}$$

$$+ P_j + \gamma_k \frac{\partial F_k}{\partial Y_{kj}} + \sum_{a=1}^{A} \psi_a \frac{\partial f_{ka}}{\partial Y_{kj}} = 0 \quad (\leqslant 0 \text{ if } Y^*{}_{kj}=0),$$

$$k=1, \ldots, K; \quad j=1, \ldots, J;$$

$$(2.14) \quad \sum_{i=1}^{I} \sum_{b=1}^{B} \sum_{a=1}^{A} a_i \frac{\partial U_i}{\partial R_b} \cdot \frac{\partial R_b}{\partial f_{ka}} \cdot \frac{\partial f_{ka}}{\partial Y_{kj}} + \sum_{i=1}^{I} \sum_{b=1}^{B} a_i \frac{\partial U_i}{\partial R_b} \cdot \frac{\partial R_b}{\partial Y_{kj}}$$

$$+ \sum_{s=1}^{K} \sum_{b=1}^{B} \sum_{a=1}^{A} \gamma_s \frac{\partial F_s}{\partial R_b} \cdot \frac{\partial R_b}{\partial f_{ka}} \cdot \frac{\partial f_{ka}}{\partial Y_{kj}} + \gamma_k \frac{\partial F_k}{\partial Y_{kj}}$$

$$+ \sum_{s=1}^{K} \sum_{b=1}^{B} \gamma_s \frac{\partial F_s}{\partial R_b} \cdot \frac{\partial R_b}{\partial Y_{kj}} - \psi_a + \sum_{a=1}^{A} \psi_a \frac{\partial f_{ka}}{\partial Y_{kj}} = 0$$

$$(\leqslant 0 \text{ if } Y^*{}_{kj}=0),$$

$$k=1, \ldots, K; \quad j=J+1, \ldots, J+A;$$

$$(2.15) \quad \gamma_k, P_j, \psi_a, X_{ij}, Y_{kj}, Z_{kj} \geqq 0,$$

for all i, j, k and a. All partial derivatives are evaluated at $\{X^*, Y^*, Z, {}^*R^*\}$.

Proof. Let $\{X+, Y+, Z+, R+, a, \gamma, P, \psi\}$ be a solution to Lemma 2.1. To establish that the Kuhn-Tucker conditions correctly characterize the maximization of our social welfare function, we must show (1) that the Kuhn-Tucker conditions are necessary conditions (i.e., that $\{X^*, Y^*, Z^*, R^*, a, \gamma, P, \psi\}$ is a solution to Lemma 2.1), and (2) that the Kuhn-Tucker conditions are sufficient (i.e., that $\{X+, Y+, Z+, R+\}$ is socially efficient).

(a) Necessity. The Kuhn-Tucker conditions are necessary if the constraints satisfy one of the constraint qualifications.[7] The particular one we will use is Slater's constraint qualification.

Slater's constraint qualification. Let X^0 be a convex set in R^n. The m-dimensional vector function g on X^0, which defines the convex feasible region $X = \{X \mid x \epsilon X^0, g(x) \leq 0\}$ is

---

orthant. For the unconstrained case see Olvi L. Mangasarian, *Nonlinear Programming* (New York: McGraw-Hill Book Company, 1969), p. 111. The constrained case is a simple extension of this.

7. We could do away with the constraint qualification by using a notion of stability that implies that there is no resource or combination of resources for which the optimal solution will improve infinitely fast as we move away from some arbitrary allocation. Some interesting theorems in this regard are presented in A. M. Geoffrion, "Duality in Nonlinear Programming: A Simplified Applications-Oriented Development," *SIAM Review,* XIII (Jan. 1971), 1–37.

said to satisfy Slater's constraint qualification (on X^0) if there exists an $x' \, \epsilon \, X^0$ such that $g(x') < 0$.[8]

The constraints in our problem are arranged for the maximization problem while the theorem is stated in terms of a minimization problem. We can test our constraints by showing that there exists an $x' \, \epsilon \, X^0$ such that $g(x') > 0$. This is guaranteed directly by Assumption 2. Therefore necessity is established.

(b) *Sufficiency.* This proof is in two parts. We have to show that $\{X^+, Y^+, Z^+, R^+\}$ is feasible and that the Kuhn-Tucker conditions are sufficient. Feasibility of $\{X^+, Y^+, Z^+, R^+\}$ is insured by equation sets (2.8), (2.9), and (2.10). It is well-known that the Kuhn-Tucker conditions are sufficient if the objective function and all constraints are concave.[9] F_k, f_{ka}, and U_i are all concave by Assumption 1. The other constraint is concave because it is linear. Therefore sufficiency is established.

The value of Lemma 2.1 to the theorist as a social norm lies in the fact that its merit is independent of the economic system that prevails in the hypothetical economy. It presupposes the existence of neither markets nor planners in its choice of a socially optimal allocation of resources. We now turn to a private ownership market economy that is supplemented by a government with limited authority, and we ask whether such an economy is able to sustain a socially efficient allocation of resources.

III. Decentralized Equilibrium with Effluent Charges

Agents and Their Behavior

Our economy is composed of three kinds of economic agents — consumers, firms, and the government. We can characterize their decision process through the use of several behavioral assumptions that facilitate the transition from the mathematics to the economics.

ASSUMPTION 3. *Consumers maximize their utility subject to a budget constraint, and firms maximize their profits subject to a production constraint. All agents in the economy are price takers.*

ASSUMPTION 4. *Consumers and producers are assumed to have perfect knowledge of prices, their own utility or production functions, and the available commodities (including the amount of unpurified waste produced by all other agents).*

8. Mangasarian, *op. cit.*, p. 78.
9. See, for example, Kenneth J. Arrow and Alain C. Enthoven, "Quasi-Concave Programming," *Econometrica*, XXIX (Oct. 1961), 779.

SPECIFIC TAXES AND POLLUTION CONTROL   513

ASSUMPTION 5. *A government exists in our hypothetical economy that has the power* (1) *to set the* K × A *dimensional matrix of specific taxes on* A *unpurified wastes for* K *firms and* (2) *to choose the* I *dimensional vector* L, *which contains the lump sum transfers to and from each of the* I *consumers. It costlessly administers these programs. Its goal is to sustain a socially efficient allocation of resources. It has at its disposal all the information needed to solve Lemma 2.1.*

In the decentralized choice of resource allocations, certain factors are given.

DEFINITION: *A private ownership economy is characterized by a* K × I *matrix* H (*which has as its* kith *element* η_{ki}) *of profit shares, an* I × J *matrix* X" *of resource endowments of* J *goods among* I *consumers,* K *production possibility functions (the* F$_k$'s) K × A *waste production relationships (the* f$_{ka}$'s*), the* B *pollution production functions (the* g$_b$'s*) and the* I *utility functions (the* U$_i$'s*).*

In a private ownership economy with a government imposing effluent charges and lump sum transfers, the production choice can be expressed as maximizing the following constrained objective function with respect to the variables under the firm's control:

(3.1) $\pi_k + \mu_k \, F_k\,(R_k, Y_k, Z_k)$,

where R_k is the vector of pollutants formed by holding the waste products of all other firms fixed and adding the level of unpurified waste produced by the kth firm, and where

$$\pi_k \equiv \sum_{j=1}^{J} P_j(Y_{kj} - Z_{kj}) - \sum_{a=1}^{A} T_{ka}(f_{ka}(Y_k, Z_k) - Y_{kJ+a})$$
$$k = 1, \ldots, K$$

is the level of profit to the kth firm and T_{ka} is the specific tax on the amount of unpurified waste disposal.

The consumer's budget constraint is affected by the imposition of a tax on waste discharge because the profits of firms in which he owns a share will be affected by the implementation of this augmented price system. Symbolically we can represent the consumer's choice as the maximization of the following constrained function:

(3.2) $U_i(X_i, R) - \lambda_i[\sum_{j=1}^{J} P_j(X_{ij} - X''_{ij}) - L_i - \sum_{k=1}^{K} \eta_{ki}\,\pi_k]$
$$i = 1, \ldots, I,$$

where π_k is the profit of the kth firm (defined above in (3.1)) and η_{ki} is the percentage of the kth firm's profits distributed to the ith consumer. The parameter π_{ki} is treated as if it were given to the consumer and is not an element of choice.

There are, of course, many possible allocations of resources that could result from this noncooperative decentralized decision making. We now wish to characterize the subset of these allocations, which imply a lack of incentives on the part of any agent to make any other choice given the choices of all other agents.

DEFINITION. *Given a private ownership economy, a state of this economy* $\{X',Y',Z',R',X'',H',P',T',L'\}$ *is called a* decentralized equilibrium with taxes *if the associated allocation of resources* $\{X',Y',Z',R'\}$ *satisfies the following conditions:*

(i) $U_i(X'_i,R') \geqslant U_i(X_i,R')$ $i=1, \ldots, I$ for all X_i,

which satisfy

$$\sum_{j=1}^{J} P'_j(X_{ij}-X''_{ij}) - \sum_{k=1}^{K} \eta_{ki}\,\pi'_k - L'_i \leqq 0.$$

(ii) $\pi'_k \equiv \sum\limits_{j=1}^{J} P'_j(Y'_{kj}-Z'_{kj}) - \sum\limits_{a=1}^{A} T'_{ka}\,[f_{ka}(Y'_k,Z'_k) - Y'_{kJ+a}]$

$$\geqq \sum_{j=1}^{J} P'_j(Y_{kj}-Z_{kj}) - \sum_{a=1}^{A} T'_{ka}\,[f_{ka}(Y_k,Z_k) - Y_{kJ+a}]$$

$k=1, \ldots, K$ for all $\{Y_k,Z_k\}$ such that $F_k(R'_k,Y_k,Z_k) \geqq 0$,

where $R'_{ka} = g(V'_K/_k,V_k,M)$

and $V'_K/_k$ *is the set of* $K-1 \times A$ *levels of unpurified waste discharged by all firms other than the* k^{th} *firm in a decentralized equilibrium:*

(iii) $\sum\limits_{i=1}^{I} X'_{ij} + \sum\limits_{k=1}^{K} Z'_{kj} \leqq \sum\limits_{i=1}^{I} X''_{ij} + \sum\limits_{k=1}^{K} Y'_{kj}$

$j=1, \ldots, J$ and

(iv) $\sum\limits_{k=1}^{K} [\,Y'_{kJ+a} - f_{ka}(Y'_k,Z'_k)\,] \leqq 0$ $a=1, \ldots, A.$

We can quite readily interpret this definition. Given profit shares and resource endowments, which have been exogenously determined, a decentralized equilibrium with taxes is a state of the economy in which, when confronted by a government-imposed set of taxes, lump sum transfers, and a market clearing set of prices, consumers maximize their utility by selecting their equilibrium consumption bundle, and firms maximize their profits (assuming that all other agents remain at their equilibrium bundle) by selecting their equilibrium bundle of inputs and outputs.

Before characterizing any decentralized equilibrium with taxes we need to modify Assumption 2.

SPECIFIC TAXES AND POLLUTION CONTROL 515

ASSUMPTION 2'. *There exists at least one technically feasible allocation of resources* $\{X,Y,Z,R\}$ *such that Assumption 2 holds and*

$$\sum_{j=1}^{J} P'_j(X_{ij}-X''_{ij}) - L'_i - \sum_{k=1}^{K} \eta_{ki}\,\pi_k < 0 \qquad i=1, \ldots , I.$$

This assumption is needed because Assumption 2 is not sufficient to guarantee that given a set of prices, profits, and lump sum transfers, every consumer has an internal point in his budget set. With this assumption we can characterize this decentralized equilibrium for a given pattern of factor ownership as follows:

LEMMA 3.1. *If* f_{ka} *is linear for all k and a, and Assumptions 1 and 2' hold, necessary and sufficient conditions for a particular allocation of resources* $\{X',Y',Z',R'\}$ *to be sustained as a decentralized equilibrium with taxes* $\{X',Y',Z',R',X'',H',P',T',L'\}$ *given (3.1) and (3.2) are*

(3.3) $\displaystyle \sum_{j=1}^{J} P'_j(X'_{ij}-X''_{ij}) - L'_i - \sum_{k=1}^{K} \eta_{ki}\,\pi'_k = 0\,(\leq \text{ if } \lambda_i = 0)$
$i=1, \ldots , I;$

(3.4) $F_k(R',Y'_k,Z'_k) = 0 \qquad (\geqq 0 \text{ if } \mu_k = 0) \qquad k=1, \ldots , K;$

(3.5) $\displaystyle \frac{\partial U_i}{\partial X_{ij}} - \lambda_i P'_j = 0 \qquad (\leqq 0 \text{ if } X'_{ij}=0) \qquad i=1, \ldots , I;$
$j=1, \ldots , J;$

(3.6) $\displaystyle -P'_j - \sum_{a=1}^{A} T_{ka}\frac{\partial f_{ka}}{\partial Z_{kj}} + \mu_k$

$\displaystyle \left[\frac{\partial F_k}{\partial Z_{kj}} + \sum_{b=1}^{B}\sum_{a=1}^{A} \frac{\partial F_k}{\partial R_b}\cdot\frac{\partial R_b}{\partial f_{ka}}\cdot\frac{\partial f_{ka}}{\partial Z_{kj}} \right] = 0$

$(\leqq 0 \text{ if } Z'_{kj}=0) \qquad k=1, \ldots , K; \qquad j=1, \ldots , J;$

(3.7) $\displaystyle T_{ka} - \sum_{t=1}^{A} T_{kt}\frac{\partial f_{kt}}{\partial Y_{kJ+a}} + \mu_k$

$\displaystyle \left[\frac{\partial F_k}{\partial Y_{kJ+a}} + \sum_{b=1}^{B}\sum_{t=1}^{A} \frac{\partial F_k}{\partial R_b}\cdot\frac{\partial R_b}{\partial f_{kt}}\cdot\frac{\partial f_{kt}}{\partial Y_{kJ+a}} \right]$

$\displaystyle +\mu_k \left[\sum_{b=1}^{B} \frac{\partial F_k}{\partial R_b}\cdot\frac{\partial R_b}{\partial Y_{kJ+a}} \right] = 0 \qquad (\leqq 0 \text{ if } Y'_{kJ+a}=0)$

$k=1, \ldots , K; \qquad a=1, \ldots , A;$

(3.8) $\displaystyle P'_j - \sum_{a=1}^{A} T_{ka}\frac{\partial f_{ka}}{\partial Y_{kj}} + \mu_k$

$\displaystyle \left[\sum_{a=1}^{A}\sum_{b=1}^{B} \frac{\partial F_k}{\partial R_b}\cdot\frac{\partial R_b}{\partial f_{ka}}\cdot\frac{\partial f_{ka}}{\partial Y_{kj}} + \frac{\partial F_k}{\partial Y_{kj}} \right] = 0$

$(\leqq 0 \text{ if } Y'_{kj}=0) \qquad k=1, \ldots , K; \qquad j=1, \ldots , J;$

(3.9) $\lambda_i, \mu_k, X_{ij}, Z_{kj}, Y_{kj} \geq 0$ for all i, j, k, a.

Proof. The necessity portion of the proof is established by showing that Slater's constraint qualification is fulfilled. This follows directly from Assumption 2′. Sufficiency follows directly from Assumption 1 and the linearity of f_{ka}.

IV. Two Classical Welfare Theorems for an Economy Containing Pollution

The Theorems [1]

We can now use this Lemma and the previous Lemma to derive the theorems of interest. The first theorem concerns whether every socially efficient allocation of resources can be sustained as a decentralized equilibrium with taxes.

THEOREM 1. *Given any distribution of factor endowments X″, any distribution of profit shares H′, and any socially efficient allocation of resources {X*,Y*,Z*,R*}, a sufficient condition for the existence of a J dimensional vector P′, an I dimensional vector L′, and a K × A dimensional matrix T′ such that {X*,Y*, Z*,R*,X″,H′,P′,T′,L′} is a decentralized equilibrium with taxes is for f_{ka} to be linear for all k and a, and Assumptions 1 and 2′ to hold.*[2]

The proof will also establish

COROLLARY 1. *The set of taxes that will accomplish this objective is given by*

$$T_{ka} = - \sum_{\substack{s=1 \\ s \neq k}}^{K} \sum_{b=1}^{B} \gamma_s \frac{\partial F_s}{\partial R_b} \cdot \frac{\partial R_b}{\partial f_{ka}} - \sum_{i=1}^{I} \sum_{b=1}^{B} a_i \frac{\partial U_i}{\partial R_b} \cdot \frac{\partial R_b}{\partial f_{ka}} - \psi_a,$$

where this expression is evaluated at a particular socially efficient allocation.

Proof. Using the trial solution that $\mu'_k = \gamma^*_k$ plus the fact that $\partial R_a/\partial f_{ka} = -\partial R_{j-J}/\partial Y_{kj}$ for $j=J+a$, and collecting terms yields the fact that any allocation of resources $\{X^*,Y^*,Z^*,R^*\}$ that satisfies (2.12), (2.13), and (2.14) will also satisfy (3.6), (3.8), and (3.7), respectively. Since (2.8) holds for any solution to Lemma

1. Variables or parameters superscripted by "′" represent a solution to Lemma 3.1, whereas variables or parameters superscripted by "*" represent a solution to Lemma 2.1.
2. Neither Assumption 1 nor the linearity of f_{ka} is necessary, but without them the equivalence in the theorem depends on a more detailed knowledge of where the solution vector to Lemma 3.1 lies.

SPECIFIC TAXES AND POLLUTION CONTROL   517

2.1, then $\{X^*,Y^*,Z^*,R^*\}$ will satisfy (3.4) in Lemma 3.1. The P' must be chosen so that (2.5) is satisfied and clearly from (2.9) P^* satisfies this criterion. Finally since ψ^*_{ka} in Lemma 2.1 is identical to ψ^*_{ka} in Lemma 3.1 (by the construction of T), (2.6) will be satisfied so that $\{X^*,Y^*,Z^*,R^*\}$ is technically feasible.

The only remaining part of the proof is to show that there exists a selection of L_i such that (3.3) is satisfied and (2.11) is identical to (3.5). To be compatible with $\{X^*,Y^*,Z^*,R^*\}$, the budget constraint must satisfy

$$(3.14) \quad \sum_{j=1}^{J} P'_j X^*_{ij} - \sum_{k=1}^{K} \eta'_{ki}$$

$$\left[\sum_{j=1}^{J} P'_j (Y^*_{kj} - Z^*_{kj}) - \sum_{a=1}^{A} T'_{ka} (f_{ka}(Y^*_k, Z^*_k) - Y^*_{kJ+a}) \right]$$

$$-L'_i \leqslant \sum_{j=1}^{J} P'_j X''_{ij} \qquad i=1, \ldots, I.$$

Therefore we can satisfy (3.3) by choosing L_i so that

$$(3.15) \quad L'_i = \sum_{j=1}^{J} P'_j (X^*_{ij} - X''_{ij}) + \sum_{k=1}^{K} \eta'_{ki}$$

$$\left[\sum_{a=1}^{A} T'_{ka} (f_{ka}(Y^*_k, Z^*_k) - Y^*_{kJ+a}) - \sum_{j=1}^{J} P'_j (Y^*_{kj} - Z^*_{kj}) \right]$$

$$i=1, \ldots, I.$$

With this choice λ'_1 will equal $\dfrac{1}{a^*_i}$ and $\{X^*,Y^*,Z^*,R^*,X'',H',P',T',L'\}$ will be a decentralized equilibrium with taxes.

The next issue concerns whether or not every decentralized equilibrium with taxes will be socially efficient. Clearly since we have not yet placed any limitations on the admissible set of taxes, not all decentralized equilibria with taxes will be efficient. It is possible to impose inefficient taxes. It should also be clear that not all decentralized equilibria with efficient taxes will be efficient in the absence of lump sum transfers. This can be inferred from Corollary 1 since the efficient tax, in general, depends on the particular socially efficient allocation of resources chosen as the target. It is, of course, possible to choose a target that is not sustainable with the given endowment of factors and profit shares in the absence of lump sum transfers. If the government computed the tax rates on one of these allocations and then imposed these taxes, the resulting equilibrium would not, in general, be socially efficient. Therefore in the next theorem we turn our attention to socially efficient tax-transfer pairs.

DEFINITION. *Let S be a collection of all socially efficient states of the economy with an arbitrary member*

$S_1 = \{X^*, Y^*, Z^*, R^*, a^*, P^*, \gamma^*, \psi^*\}$.

A socially efficient tax-transfer pair $\{T^*, L^*\}$ *is any tax-transfer pair that can satisfy all of the following conditions for at least one* $S_1 \in S$:

(i) $T^*_{ka} = - \sum\limits_{t=1}^{K} \sum\limits_{b=1}^{B} \gamma^*_t \dfrac{\partial F_t}{\partial R^*_b} \cdot \dfrac{\partial R^*_b}{\partial f_{ka}}$

$\qquad\qquad t \neq k$

$\qquad - \sum\limits_{i=1}^{I} \sum\limits_{b=1}^{B} a^*_i \dfrac{\partial U_i}{\partial R^*_b} \cdot \dfrac{\partial R^*_b}{\partial f_{ka}} - \psi^*_a$

$\qquad\qquad k = 1, \ldots, K; \qquad a = 1, \ldots, A;$

(ii) $L^*_i = \sum\limits_{j=1}^{J} P^*_j (X^*_{ij} - X''_{ij}) + \sum\limits_{k=1}^{K} \eta_{ki}$

$\left[\sum\limits_{a=1}^{A} T^*_{ka} (f_{ka}(Y^*_k, Z^*_t) - Y^*_{kJ+a}) - \sum\limits_{j=1}^{J} P^*_j (Y^*_{kj} - Z^*_{kj}) \right]$

$i = 1, \ldots, I.$

This definition is motivated by the previous proof in which these particular values were found to be efficient. With the aid of two more definitions and one more assumption, we can state and prove our second theorem.

DEFINITION. *Decentralized equilibria with taxes* $\{X', Y', Z', R', X'', H',$
$P', T^*, L^*\}$ *are said to be* unique *if for each socially efficient tax-transfer pair* $\{T^*, L^*\}$ *there exists one and only one state of the economy that is a decentralized equilibrium with taxes relative to this tax-transfer pair.*

DEFINITION. *A consumer is said to be* satiated *if* $\dfrac{\partial U_i}{\partial X_{ij}} = 0$ *for* $j = 1, \ldots, J.$

ASSUMPTION 6. *No consumer is satiated in any decentralized equilibrium with taxes.*

THEOREM II. *Given any distribution of factor endowments* X'', *any distribution of profit shares* H', *and any socially efficient tax-transfer pair* $\{T^*, L^*\}$, *if* $\{X', Y', Z', R', X'', H', P', T^*, L^*\}$ *is unique relative to* $\{T^*, L^*\}$, *Assumptions 1, 2, 6, and the linearity of* f_{ka} *are sufficient for the allocation of resources* $\{X', Y', Z', R'\}$ *associated with this decentralized equilibrium with taxes to be socially efficient.*

Sketch of the Proof. Since this proof is similar to the proof of the previous theorem, in the interest of brevity we will only sketch the proof while highlighting the differences. We have to show that given a particular X'', H', and any socially efficient tax-transfer pair $\{T^*,L^*\}$, the allocation of resources that is achieved in a decentralized equilibrium with taxes is socially efficient or that a solution to Lemma 3.1 will also satisfy Lemma 2.1 given that a tax-transfer pair that is compatible with Lemma 2.1 is chosen. As a trial solution assume that $P^* = P'$, $\gamma^* = \mu'$, and T^* and L^* take on the values described in the definition of a socially efficient tax-transfer pair. It is easily established using the uniqueness property that the single allocation that solves Lemma 3.1 will also satisfy (2.13), (2.14), and (2.15).[3]

Since the allocation $\{X',Y',Z',R'\}$ satisfies (3.4), then it will also satisfy (2.8). Equation sets (2.9) and (2.10) are satisfied from the definition of a decentralized equilibrium. We can show that (2.11) holds if we can find an a vector such that any allocation that satisfies (3.5) will also satisfy (2.12). Since $\lambda'_i \neq 0$ for $i = 1, \ldots, I$ from Assumption 6, then choosing $a^* = 1/\lambda'$ yields the desired compatibility. Since the decentralized equilibrium is unique, this choice of a^* is the same as the a^* used to define the tax-transfer pair. Therefore $\{X',Y',Z',R'\}$ will be socially efficient.

The Nature of Socially Efficient Taxes

An examination of Corollary 1 yields some aspects of efficient effluent charges that have important policy consequences. One traditional reason why specific taxes on effluent have been preferred to effluent standards by some economists is due to the belief that the government can achieve an efficient allocation of resources by establishing only one tax rate per waste product (or a total of A tax rates) whereas efficiency requires a separate standard for each firm (or a total of $K \times A$ standards). If this belief is valid, the benefits to be derived from choosing the tax system over effluent standards are (1) it is easier and less costly to administer and (2) unlike the standards system it can achieve both efficiency and political equity at the same time.[4]

---

3. As is well-known, there are two kinds of situations when nonuniqueness arises: (1) when certain production or utility functions are linear so that a single price can yield more than one allocation of resources, and (2) when a single endowment of factors and profit shares can sustain more than one equilibrium price and hence more than one allocation of resources. Of these two the latter is most troublesome in Theorem II because the tax transfer is defined with respect to and is compatible with only one of these allocations. Thus, our definition of uniqueness is stronger than needed.

4. Political equity is a phrase that refers to the desire of legislative or regulatory bodies to impose the same legal constraints on every agent. As stated above, imposing the same effluent standard on all those firms discharging the same type of effluent will not, in general, lead to an efficient allocation of resources.

520   *QUARTERLY JOURNAL OF ECONOMICS*

Contrary to this belief, it can be seen from Corollary 1 that in a socially efficient decentralized equilibrium not all firms will face the same per unit tax rate in general. In what follows, we shall examine the two mathematical cases in which a single tax rate will not suffice and then relate those conditions to pollution control policy.

The first case and empirically the one of lesser importance occurs if some firm is adversely affected by its own pollution and adjusts its production process accordingly. It can be seen from Corollary 1 that the efficient tax rate for any firm is one that is equal to the marginal external damage caused by that firm's waste products. For firms that are unaffected by their own pollution, the efficient tax they face is equal to total marginal social damage since all damage is external. However, for firms that are adversely affected by their own pollution, the efficient tax rate is equal to total marginal social damage minus the marginal social damage already internalized. Accordingly, any firm that has already internalized part of the damage caused by its waste products should, in an efficient allocation of resources, face a lower tax rate than firms that have not internalized any of the damage their pollution is causing. Hence the use of a single tax rate in this case would not lead to an efficient allocation of resources.

More important empirically is the case in which $\partial R_b / \partial f_{ka}$ is not the same for all polluting firms. When this occurs, it means that the marginal physical product (in terms of pollution) of one unit of waste product is not the same for all firms discharging that waste product. This nonuniformity has both temporal and spatial dimensions. We shall consider each of these in turn, but first a short discussion of the environmental processes that govern this nonuniformity is in order.

Pollutants, the end products of these processes and the targets of pollution control policy, are defined in terms of concentrations (e.g., parts per million) rather than on a weight basis. They have been so defined because most observed pollution damage is related to concentrations rather than weights. Waste products, on the other hand, (the source of control) are expressed on a weight per unit time basis (e.g., grams per minute). It is not correct to model the process of translating waste products into pollutants as if the waste products could simply be added together to obtain pollutants.[5]

In fact, the relationship is a very complex one that depends on

5. If it were possible to model the process this way, then $\partial R_b / \partial f_{ka}$ would be the same for all firms and a uniform tax would be efficient.

SPECIFIC TAXES AND POLLUTION CONTROL 521

climatic conditions, geography, and chemical reactions. The impact of modeling the world to account for these factors, as the above development has, is that although a single per unit tax for each type of waste product will minimize the resource cost of reducing total waste products at the source, such a tax system will not minimize the cost of achieving the socially efficient pollutant concentration, which is the real policy objective.

In the temporal dimension of pollution control policy, knowledge of this process immediately leads to the conclusion that the efficient set of taxes depends on factors that are not under the control of environmental managers and that change over time. Therefore, the efficient set of taxes will change over time not only because the benefits of a unit reduction in pollutant concentrations change over time and the number of sources changes, but also because the relationship among pollutants and waste products is governed by varying and basically uncontrollable elements in nature. Hence, social efficiency will not in general be achieved by temporally uniform tax rates.

There is also a spatial dimension of pollution control policy in which $\partial R_b / \partial f_{ka}$ may not be the same for all firms. On a national scale it is obvious that the inputs to this production process are not the same. Wind, climate, and geography vary markedly from region to region, and these factors affect $\partial R_b / \partial f_{ka}$.

For example, the marginal contribution to a pollutant concentration of a unit of waste product will be lower, ceteris paribus, in a windy city than in a city covered by a stagnant air mass. Similarly a unit of nitrogen dioxide will cause more photochemical smog in a warm region with lots of sunlight, ceteris paribus, than in an overcast, cold region. Therefore, the efficient set of taxes on waste products will not in general be spatially uniform, not only because there are different tolerances for pollution in different regions, but also because the production process by which waste products are transformed into pollutants depends on the location of the sources and the location of the receptors.[6] Hence, social efficiency will not, in general, be achieved by spatially uniform tax rates.

One other aspect of an efficient set of taxes, which has policy relevance and should be mentioned, is that taxes are calculated at a particular socially efficient allocation of resources. This fact is

6. It should be noted that, although we have discussed regional differences in tax rates, the argument holds for much smaller areas (such as cities) as well. There are good reasons for believing that even within as small a geographic area as a city it is unreasonable to expect $\partial R_b / \partial f_{ka}$ to be the same for all firms.

important for a couple of reasons. Attempts to derive empirically the marginal social cost of pollution in the absence of taxes may be of little help in determining the appropriate tax rates. It is the marginal social cost *at the efficient allocation of resources* that is relevant, and it seems unreasonable to expect that this will be the same before and after the imposition of the tax.

V. Concluding Comments

We have confronted the argument that effluent charges may fulfill the necessary but not the sufficient conditions for efficiency by deriving a set of sufficient conditions for the competitive mechanism augmented by a system of specific taxes on effluent to sustain an efficient allocation of resources. The only unusual assumptions in terms of conventional general equilibrium and welfare economics were (1) that waste products were linear functions of inputs and outputs and (2) that for each socially efficient tax-transfer pair the allocation of resources associated with the decentralized equilibrium relative to that tax-transfer pair was unique.

We have also shown that modifying the simple externality models to account for the environmental production process that transforms waste products into pollutants does not alter the conclusion that a specific tax on industrially generated waste products can achieve social efficiency, but it does change the nature of the efficient tax rather drastically. The tax rates that will achieve efficiency will, in general, not be the same for all firms. This nonuniformity of efficient taxes erases the two major benefits cited for the effluent charge — ease of administration and the capability of achieving efficiency and political equity simultaneously. In short, choosing policy instruments to control pollution is a much more difficult task than our early externality models would lead us to believe.

Williams College

[5]

EFFLUENT CHARGES: A CRITIQUE°

SUSAN ROSE-ACKERMAN   *The Wharton School, University of Pennsylvania*

Effluent Charges: A Critique. This paper argues that the case for effluent charges as a solution to environmental problems has been substantially overstated mainly because of a general failure to confront the numerous complexities generated by any real world application of the charge technique.

The paper considers problems raised by costs of information, regulatory lag, capital intensity of pollution investment, collusive behaviour by dischargers, discontinuities in marginal cost and benefit schedules, as well as the difficulty of setting optimal charges when polluters' marginal damage functions differ and when joint treatment among dischargers is an efficient outcome.

Les droits de pollution: une critique. L'auteur prétend ici que l'on a grandement exagéré l'efficacité des droits de pollution comme solution au problème de l'environnement, principalement parce que, de façon générale, on ne tient pas compte du grand nombre de difficultés pratiques rencontrées dans l'application de ce principe. Il énumère comme suit les principales parmi les difficultés pratiques. Des entreprises peuvent être forcées de fermer leur porte même si en principe elles pourraient rester en affaire, si on applique le principe suivant lequel chaque firme ne paye que les « coûts sociaux » reliés à la partie non-traitée de ses déchets. Des installations communes de traitement peuvent être localisées au « mauvais » endroit. Si les échelles de droits de pollution ne sont pas révisées fréquemment, la qualité de l'eau peut se détériorer à un rythme plus rapide que celui qui serait optimal. Les coûts de la fixation des droits à des niveaux « inadéquats » peuvent être élevés si on tient compte du « problème des points-seuils ». Les adjustements sur les marchés prennent plusieurs années et, entre-temps, les niveaux des droits conduisent à un traitement qui est soit trop élaboré, soit insuffisant, suivant la structure de marché et l'existence de collusion. Les discontinuités des fonctions de coûts et de bénéfices peuvent conduire à l'inutilité du critère marginal; dans ces conditions, le besoin d'une information précise sur les coûts de traitement ou de nettoyage dans chaque cas particulier est encore plus important que d'habitude, mais cette information est tout aussi difficile à obtenir. En conclusion, l'auteur prétend que, dans le domaine concret, les droits de pollution ne représentent probablement que « le meilleur moindre mal », et peut être moins que cela dans plusieurs cas.

I / Introduction

Since the model of the price system is the one with which economists work most comfortably, it should be no surprise that it is readily invoked by the profession as a solution to the problem posed by environmental degradation. An effluent charge converts pollution into a good for which firms and municipalities must pay, thereby seeming a simple approximation of the invisible hand of the marketplace.

Unfortunately, the argument for effluent charges generally proceeds with models which do not reflect the complexity of the problem that confronts a decision-maker who attempts to use the device in the real world. When these complexities are recognized, it becomes apparent that, in any concrete application, an operational effluent charge may be a less efficient means of controlling pollution than a comparable non-market regulatory device.

°This paper was funded by a grant from the Council on Law-Related Studies.

Canadian Journal of Economics/Revue canadienne d'Economique, VI, no. 4 November/novembre 1973. Printed in Canada/Imprimé au Canada.

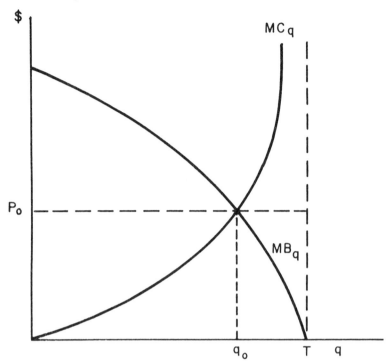

FIGURE 1

II / A simple effluent charge model

Before considering models which incorporate real world constraints, it is
important to recognize that the case for an effluent charge is not straightforward even when perfect competition prevails, and there are no externalities
except those created by the discharge of effluents into society's single lake.[1]
To make the situation as simple as possible, assume that (a) each discharger's
impact upon water quality is independent of its location; (b) the marginal
costs of pollution abatement for each discharger can be added together with
no cross effects; (c) marginal benefits (MB) intersect marginal costs (MC)
only once when MB of reducing waste discharged is falling and MC is rising.[2]

[1]While this paper focuses on the problems of devising a charge system to control water
pollution, many of the points made are applicable to the public control of air pollution and
solid wastes as well.
[2]Marginal costs and marginal benefits are expressed as functions of the quantity of waste
emitted by dischargers (measured from right to left in Figure 1). In order to relate marginal benefits to waste discharge, three separate models are needed: one relating discharges
to water quality in the lake; a second relating lake quality to physical measures of recreational activity; and a third assigning dollar values to these physical benefits. For a full
discussion of the serious problems that can arise in constructing these models see Ackerman
et al. (1973).
   The marginal cost curve also requires interpretation. Since firms will respond to pollution

In such a world, Kneese and Bower (1968, 98–101) argue that a uniform effluent charge, set at the intersection of MB and MC, would generate an optimal result (see Figure 1 where q is waste treated and T is raw waste load). Unfortunately, however, even this conclusion will not be valid in every case.

Under an effluent charge regime, a firm not only must spend money on a treatment facility, but must also pay a tax on the waste it continues to discharge. This tax is often justified as compensation for the damage caused by the untreated waste, but if MB is falling, the polluter's payment is, in fact, more than the value of the damage done (see Figure 1). Thus, if a fee is used, a firm might shut down when, in fact, the damage it caused did not warrant this action.[3]

Contrast this result with a perfectly efficient non-market scheme of control. Under an "allocation" regime, each polluter's conduct is not regulated by pecuniary incentives; instead, the discharger receives a legal command ordering it to emit no more than a specified quantity of waste. If each allocation is "optimally" set so that each firm's marginal cost equals MB, no firm will be required to shut down when it is inefficient to do so. On the other hand, since no payment is required for damage caused, some firms may remain in operation when they could not compensate the victims of their pollution from their profits.[4] Thus, even at this level of generality, neither an effluent charge nor an "allocation" approach appears necessarily superior.[5]

---

charges or regulations not only by building treatment facilities and making process changes but also by cutting back on output, the MC curve for any firm does not represent the marginal cost of treating a *given* raw waste load to a certain level. Rather it shows the level of treatment, process change, and raw load cut back for which each firm's profits are maximized for any given fee. For a related discussion, see Baumol and Oates (1971, 54).

[3]This is just one example of the problems that can arise when marginal and total criteria contradict each other. For related discussions see, Meyer, Kain, and Wohl (1965, 344–5), Coase (1960), Patinkin (1963), and Turvey (1963).

[4]It is possible to put the comparison between charges and allocations in mathematical terms for the general case in which dischargers can reduce wasteloads through a combination of treatment and reductions in the production of final goods. (Compare Baumol and Oates (1971, 63–5) who restrict their proof to the case in which output of final goods is unchanged.)

Under an effluent charge scheme a polluter's profits (π_1) are:

(i) $\pi_1 = \bar{P}_Z Z - \bar{P}_q M - K(Z) - C[q,T(Z)]$,

where: $\bar{P}_Z$ = price per unit of final good (assumed to be fixed); Z = quantity of final good; M = load discharged in pounds; $\bar{P}_q$ = price per pound of waste dumped (assumed to be fixed); $K(Z)$ = total direct production costs of producing Z; $C[q,T(Z)]$ = total costs of removing q pounds of waste by treatment given raw load produced, $T(Z)$; $q = T(Z) - M$ = load treated in pounds.

The first order conditions for profit maximization are:

(ii) $\partial \pi_1 / \partial Z = 0 = \bar{P}_Z - \bar{P}_q T'(Z) - K'(Z) - C_T \cdot T'(Z)$,

(iii) $\partial \pi_1 / \partial q = 0 = \bar{P}_q - C_q$.

This implies:

(iv) $\bar{P}_Z = K'(Z) + (C_q + C_T) \cdot T'(Z)$

If, instead, the authority assigns the firm an allocation $\bar{M}$, where $MB = \bar{P}_q$, profits (π_2) are:

(v) $\pi_2 = \bar{P}_Z Z - K(Z) - C[T(Z) - \bar{M}, T(Z)]$,

(vi) $\partial \pi_2 / \partial Z = 0 = \bar{P}_Z - K'(Z) - (C_q + C_T) \cdot T'(Z)$.

III / Beyond the simple model: dispersed pollution sources, co-operation, and growth

A. THE RIVER CASE

Now let us consider a more realistic pollution "market" – an estuary with polluters and water users located at different points along its length. We shall demonstrate that in this case only a complex charge scheme is certain to be more efficient than a primitive non-market mode of allocation. Moreover, under an efficient charging scheme, the central authority will require extensive information before it can set the optimal fee schedule.

The essential complication introduced here is that different polluters affect different parts of the river in different ways. Using contemporary engineering models to chart the differential impact of polluters in different locations on water quality, it is possible to assess the degree of complexity required of a charge system under our modified assumptions. The sophisticated engineering model, devised by Robert Thomann, divides an estuary's length into a large number of sections, and reports the extent to which a discharge in any particular river section affects water quality (measured in dissolved oxygen units) in any other.[6] The Thomann model develops a matrix of coefficients, a_{ij}, that sets forth the impact of a pound of pollution dumped into section i upon the dissolved oxygen level of section j. Therefore, because of the different conditions in the different sections of the river, different effluent charges, x_j, must be set for each section. The actual fee per pound paid by a polluter in section i is not the fee in that section but instead is a weighted average of the fees in all the sections where the weights are the a_{ij}s, i.e., $\Sigma_j a_{ij} x_j$.[7] This proliferation of data is likely to make the process of finding the equilibrium point very time consuming. Given n polluters, one in each of the n sections of the river, the problem for any one polluter is equivalent to that of a producer with a production function that leads to the output of products in fixed proportions, except that the "products" are negative goods that he must pay

Since (iv) is equivalent to (vi), we have shown that the marginal conditions for profit maximization are identical under both pollution control schemes. Of course total profits will be lower with a fee by the amount of the total charge $(\bar{P}_q \cdot \bar{M})$: $\pi_1 = \pi_2 - \bar{P}_q \cdot \bar{M}$, and π_1 might be negative even though π_2 was positive.

[5]If there is only a single polluter, the first best alternative is a total charge set equal to residual damage i.e., the area under MB to the right of q in Fig. 1. With a large number of dischargers this solution is not possible since there is no unambiguous way of allocating total damage among the polluters.

Further difficulties arise if we drop the assumption of perfect competition. For example, in the absence of externalities, a monopolist produces too little and sells for too high a price from the point of view of over-all efficiency. If he is earning excess profits, these funds could theoretically be tapped for pollution control without worsening the situation. However, if a pollution control policy is instituted that raises the monopolist's marginal costs, he will cut back output still further and raise price still higher. See Buchanan (1969).

[6]Thomann's model is critically discussed in Ackerman and Sawyer (1972).

[7]This procedure assumes that the impact of a pound of pollutant dumped in i on water quality in j is independent of the total loads discharged by other polluters in other locations, i.e., the level of marginal damage in a section is affected, but the rate at which waste decays is not. This is an oversimplification of the physical processes that actually occur. Kneese and Bower (1968, 110–12) raise this issue and Ackerman and Sawyer (1972, 447–78) discuss the problem in detail.

for, rather than ones for which he receives revenue. Given the fees, $x_1, ..., x_n$, each polluter will minimize costs by setting $\Sigma_j a_{ij} x_j = MC_i(q_i)$, where q_i is the load removed[8] by the discharger in section i.

Using the matrix of a_{ij}s generated by the Thomann model, the equilibrium effluent charges, $x_1, ..., x_n$, can be found if we specify the marginal benefit curves for each section and the marginal cost curves for each polluter.

We then have in equilibrium:

(1) $x_j = MB_j(\Sigma_i a_{ij} q_i)$, $j = 1, ..., n$,

(2) $\Sigma_j a_{ij} x_j = MC_i(q_i)$, $i = 1, ..., n$.

Equation 1 states that marginal benefit in any section j equals x_j and that marginal benefit is a function of the loads, $q_1, ..., q_n$, removed from each section, while equation 2 stipulates that the fee per pound paid by any polluter i equals the marginal cost of treating an extra pound. This is a system of $2n$ independent equations in $2n$ unknowns. The system can thus be solved for the x_j and q_i so long as certain non-negativity conditions are fulfilled.[9]

Since the Thomann model and similar work indicate that the location of a polluter along the river will substantially affect the impact its discharge will have upon water quality levels,[10] this more sophisticated model is likely to generate equilibrium fees which vary considerably from polluter to polluter. If the central authority chooses to ignore this fact, and sets a single effluent charge for the river, the results may be even less efficient than a highly simplistic non-market allocation scheme. Assume, for example, that the perceived alternative to a uniform charge is a primitive allocation scheme which assigns all polluters in the region quotas equal to a uniform percentage of their raw wasteload. The relative efficiency of these alternatives may be most easily seen if, for a moment, we assume that each polluter regulated by the central agency is located on a different lake. In this case a uniform fee will generally be more efficient than an allocation plan requiring equal percentage removal when polluters' marginal costs are close together at the "true" efficient solution. In contrast, the "allocation" approach will be best when the efficient solution requires similar percentage cutbacks at widely different marginal costs. This situation can arise when the marginal benefits of cleaning up one lake are much higher than those resulting from improving water quality in the other. If there are two polluters, one and two, and if polluter one would have to treat at a higher percentage treatment and marginal cost in the

[8]The problem is analogous to that of maximizing profits when a firm's production function leads to the output of several products. See Henderson and Quandt (1958, 67–72).
[9]A boundary condition must be added stating that if q_i represents the load removed by discharger i, then $0 \leq q_i \leq T_i$ when T_i is the total raw load of i. Johnson (1967) includes a similar constraint in his study comparing effluent charges with other allocation schemes, but defines q_i as the additional load removed by discharger i over and above current treatment. Therefore, his constraint is $0 \leq q_i \leq Q_i$ where Q_i is the total load currently discharged by i. Johnson's requirement is a political rather than an economic or technical constraint and is one a rational pollution control authority would not impose.
[10]See the Thomann model's coefficients for the Delaware Estuary which are presented in Graves, Whinston, and Hatfield (1970, Table 22, 97–9).

efficient solution, we can show[11] that a uniform fee will be cheapest so long as:

(3) $c_2 T_2 > c_1 T_1$,

where c_i = constant slope of MC_i and T_i = raw load of i.

When polluters in one section affect river conditions in another section, in the manner portrayed by the Thomann model, the result is basically similar, except that the precise formula shown in equation 3 does not hold. Consider the case of two polluters and two sections. Let W_1 and W_2 be water quality standards in the two sections expressed in terms of load removed and let q_1, q_2 be the amount treated by each of the polluters. For river standards to be met:

(4) $q_1 + a_{21} q_2 \geq W_1$

(5) $a_{12} q_1 + q_2 \geq W_2$.

The solution must be on or above both the lines W_1 and W_2 in Figure 2. The cost of compliance is $(\frac{1}{2}) (c_1 q_1^2 + c_2 q_2^2)$ i.e., the sum of the areas under the marginal cost curves for any levels of clean-up q_1 and q_2, where c_i is the slope of marginal cost curve i. Lines of equal cost are thus shown in the diagram as a family of curves concave to the origin.

We can represent the "allocation" approach, requiring uniform percentage treatment, by a line with the equation: $q_1/T_1 = q_2/T_2$, and a uniform fee by the line $c_1 q_1 = c_2 q_2$. In Figure 2 uniform treatment is clearly the cheaper solution. However, other values of a_{ij}, T_i, and c_i could be chosen so that an

---

[11]Let W_1 and W_2 be the desired water quality objectives in water bodies one and two expressed in terms of load treated. Let q_1 and q_2 be the amounts each polluter treats. Then $q_1 \geq W_1$, $q_2 \geq W_2$ for the objectives to be met. Let T_1 and T_2 be the raw loads of the two polluters and c_1 and c_2 be the constant rates of increase of marginal costs. Thus $MC_1 = c_1 q_1$, $MC_2 = c_1 q_2$. It follows that the total treatment costs for both dischargers taken together are:

$$TC = (\tfrac{1}{2}) (c_1 q_1) q_1 + (\tfrac{1}{2}) (c_2 q_2) q_2$$
$$TC = (\tfrac{1}{2}) c_1 q_1^2 + (\tfrac{1}{2}) c_2 q_2^2.$$

Assume further that $c_1 W_1 > c_2 W_2$ and $W_1/T_1 > W_2/T_2$. Thus in order for the water quality objectives to be met with a uniform fee, $q_1 = W_1$ and $c_1 W_1 = c_2 q_2$ or $q_2 = (c_1/c_2)W_1$. Therefore, total costs, excluding the costs of paying the fee, are: $(\tfrac{1}{2}) c_1 W_1^2 + (\tfrac{1}{2}) c_2 (c_1 W_1/c_2)^2$.

In contrast, under an allocation scheme requiring uniform percentage removal, we must have $q_1 = W_1$ and $W_1/T_1 = q_2/T_2$ or $q_2 = W_1 T_2/T_1$. Therefore, total costs are:

$$(\tfrac{1}{2}) c_1 W_1^2 + (\tfrac{1}{2}) c_2 (T_2 W_1/T_1)^2.$$

It follows that a uniform fee is cheaper than a uniform percentage treatment scheme if and only if:

$$(\tfrac{1}{2}) c_1 W_1^2 + (\tfrac{1}{2}) (c_1^2 W_1^2/c_2) < (\tfrac{1}{2}) c_1 W_1^2 + (\tfrac{1}{2})c_2(T_2^2 W_1^2/T_1^2). \text{ Simplifying, we obtain:}$$

$c_1^2/c_2 < c_2 T_2^2/T_1^2$, or:

$(c_1 T_1)^2 < (c_2 T_2)^2$.

Since $c_i > 0$ and $T_i > 0$, $i = 1, 2$: $c_1 T_1 < c_2 T_2$.

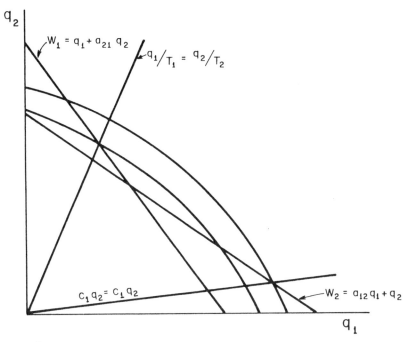

FIGURE 2

effluent charge would be cheapest. One simple means of reversing the results is to change the a_{ij}. If the relationships between raw loads and marginal costs are as shown in Figure 2, uniform treatment is cheaper when a_{12} is relatively large (close to one) and a_{21} is relatively small. Thus, it is only a sophisticated effluent charge which is certain to be more efficient than a primitive non-market mode of allocation, even under the relatively simple conditions we have thus far hypothesized.[12]

B. GROWTH IN EXISTING FIRMS

To complicate the problem further, consider the difficulties presented by the growth of existing firms. Let us assume, realistically, that once the pollution authority has set an effluent charge, it will find it difficult to change the schedule frequently, and compare such a "sticky" charge with an equally "sticky" non-market control mechanism. Two alternative non-market strategies

[12]Thus, while leading authorities have often recommended uniform effluent charges as more efficient than uniform percentage treatment, this policy preference is grounded on a number of assumptions which are not generally made explicit and which do not necessarily hold in every case. See Baumol and Oates (1971, 57–8), Johnson (1967, 296–7), and Kneese (1971b). Although Johnson's empirical study of the Delaware Estuary, using data criticized in Ackerman *et al.* (1973), suggests the superiority of uniform effluent charges in this single case, he does not note the critical factors which make this result possible. On the other hand, Kneese (1971a, 158, n. 11), does recognize that a uniform fee might be inferior to uniform treatment, but this observation is passed over lightly.

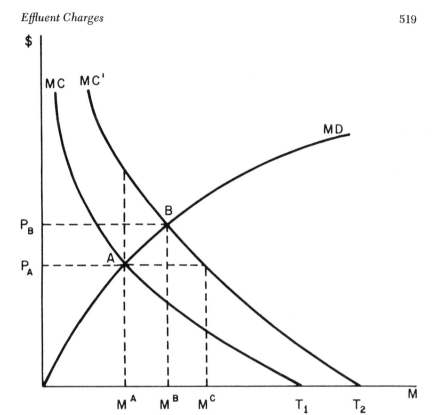

FIGURE 3

must be considered. Under the "allocation" approach, the quantity of a polluter's permissible discharge is held constant over time. Under the "fixed
percentage treatment" scheme, however, percentage removal is held constant
but the actual quantity discharged is permitted to rise with the growth in raw
waste load.

Consider the consequences of a rise in demand for the final product manufactured by a polluter. In a competitive industry, as the market price rises, the
firm will wish to increase its output. Using Figure 3 we shall first consider the
way such a firm will respond to a "sticky" charge compared to a "sticky"
allocation scheme. For ease of exposition the units on the horizontal axis in
Figure 3 are reversed so that the load discharged, M, rather than the load
treated, q, increases as one moves to the right. This means that a single
marginal damage curve, MD, can be drawn that is zero when nothing is
emitted and rises at a decreasing rate as more waste enters the stream. Thus
the curve does not shift when raw load increases. MC is the initial marginal
cost curve and A is the equilibrium point where discharge equals M^A and
marginal damages equal P_A. In response to increased production of both final

product and waste, the marginal cost of reducing discharge shifts outward to
MC' since it is now more expensive to maintain any given level of emission.
The new optimal level of pollution is M^B, but if a fee of P_A persists, the firm
will increase dumping beyond M^B to M^C and raise its output by more than the
optimal amount indicated by B. With a fixed allocation, M^A, however, the
firm will expand output of the final good it produces by less than the optimal
amount. In short, under an "allocation scheme," a growing firm will dump
"too little," while under an "effluent charge scheme" it will dump "too much."
In contrast, a "fixed percentage treatment" scheme promises to handle the
growth problem somewhat better since each firm is permitted to increase its
total waste discharge so long as the percentage of raw waste load remains the
same.[13] The firm would dump more than M^A but the amount would still not
equal the optimum, M^B, unless the geometry of the problem were very
simple.[14]

Of course, under either an allocation scheme or an effluent charge plan, the
problem of growth can be partially resolved by setting the allocation or fee so
that the discounted present value of treatment costs plus residual damages is
minimized over the period during which regulations or prices will not be
changed. This means that, with a fixed MD curve, if the discharger is expected
to expand over time, both allocations and fees should be set at levels that are
initially too high. Under an allocation plan the growing firm would treat "too
little" in the early years and "too much" in the later years, while with a fee it
would treat "too much" at first and "too little" later. Unfortunately, however,
reliable estimates of future growth and pollution control costs, which are
needed for this alternative approach, are very difficult to obtain; indeed
relatively sophisticated pollution control agencies cannot obtain good data on
even current costs and pollution loads (Ackerman *et al.*, 1973).

C. JOINT TREATMENT

There are often substantial economies of scale in water pollution control, and
polluters may, therefore, find costs minimized by piping their wastes to a
joint treatment facility. This section will consider the use of an effluent charge
as a device to induce a group of polluters to build a joint treatment plant, and
demonstrate that a charge system will generally be unsatisfactory when used
as the sole allocation strategy.

A single tool, the effluent charge, cannot be expected to resolve two distinct
allocation problems – that of plant location and that of treatment level – in an
efficient manner.[15] Since the marginal benefits obtained from different levels of

[13]It is true, of course, that a uniform percentage treatment scheme has problems of its own,
most notably the difficulty of ascertaining reliable measurements of raw waste loads which
would be generated by firms if they had engaged in no treatment whatever. In practice,
such raw load calculations often contain significant inaccuracies. See Kneese (1971b), and
Ackerman *et al.* (1974 forthcoming, chap. 17).

[14]In one simple case, constant percentage treatment will in fact lead to an optimal amount
of discharge. If MD is a straight line and if MC and MC' are parallel straight lines, then by
similar triangles OM^A/OT_1 (the percentage dumped at point A) equals OM^B/OT_2 (the percentage dumped at B). The most extreme assumption here is the parallelism of MC and
MC' which implies that no economies of scale exist.

[15]An analogous point has been developed by macroeconomic and international trade theor-

cleanup will vary depending upon the location of the regional plant, the fee should vary with plant location.[16] In particular, fees should generally be higher for upstream locations than downstream sites because at each level of discharge, marginal damages are higher the greater the length of the river the waste must traverse. Given a fee schedule which recognizes the importance of the location of the regional plant, dischargers will choose the site that minimizes the sum of treatment costs, piping costs, and *effluent fees*. From a global point of view, however, the actual cost-minimizing location is the site that minimizes the sum of treatment costs, piping costs, and *residual damages*. If, at each location, marginal damages fall as treatment levels increase, then, as noted in section II, the total fee paid will exceed marginal damages for all sites, and we can show that in the case in which the total fee is always some fixed multiple, K, of residual damages, polluters acting together will never choose the optimal plant location. Instead, they will choose the location, d, that minimizes the sum of treatment costs, $C(d)$, plus piping costs, $S(d)$, plus effluent charge payment, $P(d) \cdot M(d)$, where $P(d)$ is fee per pound, and $M(d)$ is load discharged at d. For each location, $P(d)$ is set by the authority at the level where marginal costs equal marginal damages. If $B(d)$ equals residual damages at d, then $P(d) \cdot M(d) = KB(d)$. The dischargers will choose the site where the sum of marginal treatment and piping costs and marginal fee costs from a move in either direction equals zero – i.e., where:

(6) $C'(d) + S'(d) + KB'(d) = 0.$

From the point of view of society, however, d should be chosen such that:

(7) $C'(d) + S'(d) + B'(d) = 0.$

Therefore, unless the total fee collected just equals total residual damage at each d, the co-operating dischargers will not choose the optimal site. In order to induce the polluters to locate their plant at the place defined in equation 7, the central authority must instead adopt one of two policy options. On the one hand, it may specify the plant location on the basis of its own cost-benefit analysis and then set an effluent charge where marginal cost equals marginal benefit at that site. As long as side payments are permitted, however, the authority need not require participation but can rely on the polluters to allocate costs in such a way as to induce those polluters to join who would contribute to over-all efficiency. On the other hand, the authority can design a complex scheme under which it first establishes a schedule of effluent fees set where MC equals MB at each location d and next announces that it will pay lump sum subsidies at every d calculated to equal the difference between total fees paid and total residual damages actually caused by the dischargers. If this combination charge and subsidy procedure were followed, the actual

---

ists who have argued the necessity of having at least as many policy instruments as policy goals. See Meade (1952, 99–124), Tinbergen (1952).

[16]We assume in the text that only one joint plant will be built to treat all discharges. Finding the solution becomes much more complex if we allow numerous joint treatment plants and permit some polluters to remain independent of co-operative pollution control arrangements. Graves, Whinston, and Hatfield (1970) have developed an algorithm to solve this problem and have applied it to data from the Delaware Estuary.

net fee paid in equilibrium at any location d would equal the value of residual damages, and hence the polluters would be certain to locate their joint treatment plant at the optimal location. But, of course, the information required to undertake such a regulatory scheme would be immense, and it is thus unlikely that the charge and subsidy plan should be preferred to one in which the central authority simply chooses the location of the regional plant.

IV / The dynamics of adjustment

In a world in which pollution loads are predetermined and unchanging, the central authority which has *perfect information* about each polluter's marginal treatment costs can choose either an effluent charge system or an allocation scheme to achieve its water quality goal at minimum cost. Of course, we do not live in a world of perfect information, and it is commonly suggested that this fact weighs heavily in favour of an effluent charge (Kneese and Bower, 1968, 135–9). This section will demonstrate, however, that using a pricing system does not in itself solve the difficulties posed by extensive information requirements.

A. THE COST OF MAKING A MISTAKE

The first important source of trouble arises because the pollution control board must set its effluent charge at the point where the marginal benefit curve intersects the *estimated* aggregate marginal cost curve. Since the MC curve is based upon a guess, it is necessary to consider the cost of the guess being wrong. Moreover, guesses can be wrong in two directions: costs may be underestimated or overestimated. Nevertheless, for purposes of discussion, we shall consider the case that is more common in the real world, in which clean-up costs are seriously underestimated although an analogous discussion may easily be developed for the case in which costs are overestimated.

Figure 4 illustrates the social loss caused by an underestimate of costs. If MC_E and MC_A are estimated and actual marginal costs respectively, under an erroneous effluent charge, P_o, the loss is area ABC; under a non-market scheme which orders each polluter to limit his discharge so that total pollution load is $T - q_o$ (i.e., the quantity at which MB = the *erroneous* MC), the loss is measured by the area DBE. Obviously, one area may be greater or smaller than another, depending upon the relative slopes and positions of the *actual* MC and MB curves.[17] Furthermore, as we shall see, there are many situations in the environmental context in which the MB schedule contains large discontinuities. When this is the case, an effluent charge based on an underestimate of costs will result in a *somewhat* greater level of pollution than expected, which in turn (given the discontinuity in marginal benefits) will mean that the cost of making a mistake is very large. In contrast, a non-market scheme which fixes the quantity discharged will not run this risk.

To reduce the costs of making a mistake, the pollution control authority

[17]If both *actual* MC and MB are straight lines, with MC originating at the origin, the loss will be equal under a charge scheme or a non-market allocation scheme if the slope of MC equals the negative slope of MB.

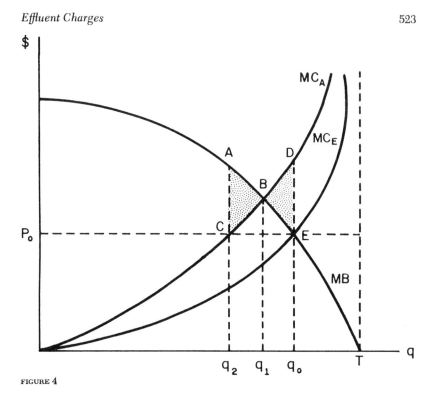

FIGURE 4

can be expected to reserve the right to change its prices as new information is obtained concerning the polluters' response. This ability to experiment, however, is limited by the capital intensity of pollution control. Since there is a time lag between the decision to invest and the completion of the project, the river quailty resulting from any given charge will not be known for several years. Moreover, the possibility that fees will change over time raises new problems to which we shall now turn.

B. REACHING EQUILIBRIUM
If polluters believe that changes in the effluent charge are likely, this will itself have an effect on the process by which equilibrium will be attained.

(i) No Collusion
If collusion is effectively prevented but the number of polluters is small, every polluter has an incentive to provide overstated cost information to other polluters in the hope that they will over-invest. However, if every polluter suspects that others are overstating their costs, all dischargers will be conservative and only build small scale plants until they can assess the accuracy of the others' cost data.

If there is a large number of dischargers, none of whom acting alone can

substantially affect water quality, then – in the absence of collusion – each
polluter will have little basis for guessing the equilibrium fee if all he knows
is the authority's marginal benefit schedule. In addition, each polluter must
have some sense of the marginal cost curves of all the other dischargers. In
the typical case, no particular discharger will have better information on this
point than the pollution controllers; hence they will set their *MC* equal to
the announced fee.

(ii) Collusion

Polluters will have an incentive to collude if they recognize that the agency
has poor information concerning marginal treatment costs. Given the authority's lack of reliable data, it is only reasonable to expect that the controllers
will judge the optimality of the charge by observing polluter behaviour: if
polluters clean up more than the *MB* curve suggests is optimal, the charge
will be lowered, and *vice versa*. This control strategy, however, may be exploited by polluters who collude to present the controllers with misleading
behavioural responses. It makes sense to form a cartel and agree to treat to
the level of water quality that minimizes the sum of treatment costs plus fee
costs (the shaded area in Figure 5). Instead of choosing the optimal amount
of treatment, q_o, they will choose some higher level of treatment, q_1, since the
marginal cost of more extensive treatment is balanced by the fact that the fee
paid on all units discharged will fall.[18] The amount of over-treatment will
depend positively upon the steepness of the marginal benefit schedule and
negatively upon the rapidity with which marginal costs increase. In this
situation every polluter will be treating at a point where his *MC* is above the
fee charged per pound and thus will have an incentive to cheat his fellow
dischargers by treating at lower levels and dumping more into the river. Only
if the joint cost-minimization point can be enforced by the cartel can the
non-optimal solution persist.

Clearly, the central agency can eliminate the incentive to collude by refusing
to alter the fee if treatment rises above q_o. But this response, of course, violates
the assumptions of the analysis which presupposes (realistically) that, within
a range, the agency does not know what q_o ought to be, and it is precisely the
treatment cost information which would permit the determination of q_o that
the cartel is interested in keeping secret.

---

[18]In more formal terms, suppose $C_i(q_i)$ is the total cost for polluter i of treating q_i of its
total raw load T_i. Then if the discharger is a perfect competitor in the waste market, he
seeks to minimize $C_i(q_i) + P(T_i - q_i)$ where P is the fee per pound dumped. This minimum occurs where $C_i'(q_i) = P$. However, if polluters collude, they know that total benefits
B are a function of $Q = \Sigma_i q_i$. Thus their choice of treatment level will influence the level
of the fee or $B'(Q)$. Converting the sum of the individual total cost curves into a cartel-
wide total cost function $TC(Q) = \Sigma_i C_i(q_i)$, the expression that the cartel wishes to minimize is:

$TC(Q) + B'(Q)[T - Q]$, where $T = \Sigma_i T_i$.

The minimum occurs at: $TC'(Q) = B'(Q) - B''(Q)(T - Q)$ and if $B''(Q) < 0$, this implies
$TC'(Q) > B'(Q)$. In order for the distribution of loads among polluters to be efficient,
$TC'(Q) = C_i'(q_i)$ for all i. Therefore, $C_i'(q_i) > B'(Q)$ for all i.

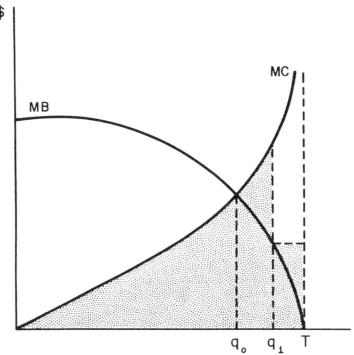

FIGURE 5

V / Adding realism to benefit and cost calculations

While it is conventional to portray marginal costs as continuously rising and marginal benefits as continuously falling over the relevant range, this is only sometimes true in problems posed by environmental regulation. Indeed, in water pollution control, it is impossible to make coherent policy without recognizing that these assumptions are unrealistic.

A. BENEFITS

In order to understand the benefits of "clean water" it is necessary to consider "threshold effects." Below a certain quality level it is unsafe to swim; below another, it is impossible to fish successfully since the fish are all dead; below another, it is unpleasant to boat since the water is malodorous. After each of these thresholds is attained, the marginal benefit to recreationists of a particular kind is relatively small or non-existent.[19]

Therefore, the marginal benefit schedule may alternate between very rapid increases over some ranges of water quality and very small increases over

---

[19]The benefits of water pollution control in a particular case – the Delaware Estuary – are discussed in Ackerman *et al.* (1973).

others (Scott, 1972, 270–3). Thus even a conventional marginal cost curve might intersect *MB* in several places where *MC* is rising and *MB* is falling. Total costs and total benefits must be calculated at each potential solution in order to determine the most efficient one, and a stable outcome, arrived at by trial and error on the basis of ignorance about costs, might be only a local optimum rather than the over-all efficient solution.

Even more serious problems for an effluent charge will arise if there are actual discontinuities in the marginal benefit schedule – a situation which will commonly occur in water pollution control, given the threshold phenomena we have described. In this situation, the central authority could, in theory, set the effluent charge so that polluters will clean up to any desired level, but if the optimal level occurs at one of the discontinuities in the marginal benefit schedule, the charge could not equal marginal benefits but should be equal to *MC* at the desired treatment level. Consequently the central authority must possess relatively accurate cost information if equilibrium is to be attained within a reasonable time. It is possible to avoid this information requirement, however, by charging a polluter not the social harm caused by the marginal pound discharged but the value of all the benefits not realized. This strategy will lead to the efficient amount of treatment no matter how irregular the *MB* curve. Unfortunately, it can be used only when one polluter is involved or when all polluters can collude successfully, since when several polluters together cause the remaining damage there is no obvious way to divide total damage costs among them to insure the efficient solution. We do not know how to measure the incremental damage caused by a single polluter unless other polluters' discharges are held constant.

B. COSTS

Analogous problems may arise as a result of discontinuities in the water pollution control cost curve. If only a finite number of treatment plant scales are available to the polluter, then the long-run marginal cost curve will also have discontinuities. More capital investment becomes worthwhile when the marginal operating cost of the previous scale becomes so high that it equals annual fixed costs required for the next level.[20] If the marginal cost schedule is in fact of this form, an effluent charge may not correspond to a unique plant scale. Once again, the desired level of treatment will only be obtained if each individual polluter is told how its *total* payment to the central authority shifts as per cent removal rises.[21] A different, but analogous, difficulty arose in one empirical study of industrial polluters. The firms examined were found to have constant average and marginal costs (Dorcey, 1971, 62 and Figure 4.8). Thus a fee set equal to the level of marginal costs prevailing amongst the polluters could not determine a unique treatment level.

[20]If process changes and output cutbacks are possible, the cost information needed is more complex but still may exhibit discontinuities.
[21]Of course an allocation scheme will face its own problems because of discontinuities. A rigidly enforced system of waste load allocations might require a polluter to make a huge capital investment in a tertiary treatment plant even though the additional waste load such a plant must remove is very small.

VI / Conclusion

This essay has attempted to suggest the difficulties a pollution control agency would have if it chose to adopt a charge strategy to carry out its policies. To recall only the most important: firms may be forced to close, even when they could have remained in business if they were required to pay only the social cost imposed by the untreated portion of their waste; joint treatment facilities may be located in the "wrong" places; if charge schedules are not readily altered over time, water quality will deteriorate at a rate more rapid than optimal; the costs of setting an "incorrect" charge may be high, given the "threshold" problem; market adjustments will take many years and in the meantime, charges will induce overtreatment or undertreatment, depending upon the market structure and the existence of collusion; discontinuities in cost and benefit functions may mean that marginal criteria are unusable, and accurate information regarding each polluter's cost of cleanup will then be even more important than ordinarily, but equally difficult to obtain. Doubtless, non-market schemes encounter substantial, and in many cases, greater obstacles than those which we have canvassed.[22] We have emphasized the weaknesses of a "market solution" in our analysis of the "effluent charge" scheme only to counteract the common professional tendency to criticize non-market methods of resolving the externalities problem without undertaking a similar careful scrutiny of the economist's favoured strategy.

References

Ackerman, B., S. Rose-Ackerman, and D. Henderson. "The Uncertain Search for Environmental Policy: The Costs and Benefits of Controlling Pollution Along the Delaware River," *University of Pennsylvania Law Review* 121 (June 1973), 1225–308.

Ackerman, B., S. Rose-Ackerman, D. Henderson, and J. W. Sawyer Jr. *The Uncertain Search for Environmental Quality* (Free Press, New York, forthcoming, 1974).

Ackerman, B. and J. W. Sawyer Jr. "The Uncertain Search for Environmental Policy: Scientific Factfinding and Rational Decisionmaking along the Delaware River." *University of Pennsylvania Law Review* 120 (Jan. 1972), 419–503.

Baumol, W. J. and W. E. Oates. "The Use of Standards and Prices for Protection of the Environment." In P. Bohm and A. V. Kneese (ed.), *The Economics of Environment* (London, 1971), 53–65.

Buchanan, J. "External Diseconomies, Corrective Taxes, and Market Structure." *American Economic Review* 59 (March 1969), 174–7.

Coase, R. H. "The Problem of Social Cost." *Journal of Law and Economics* 3 (Dec. 1960), 1–44.

Dales, J. H. *Pollution, Property and Prices* (Toronto, 1968).

Dorcey, A. "The Economic and Fiscal Aspects of a Regional Water Quality Management System." Vol. VIII of I. Fox. *Institutional Design for Water Quality Management: A Case Study of the Wisconsin River Basin* (University of Wisconsin, Water Resources Center, Technical Report OWRR C-1228, 1970).

[22]Moreover, alternative market schemes have been proposed which overcome some of the difficulties of an effluent charge system. Indeed, a forthcoming book – Ackerman *et al.* (1974, chap. 18) – will argue that a market scheme first proposed by Dales (1968, 77–100), under which a fixed quantity of pollution rights is auctioned off to the highest bidders, will often provide a far more promising approach to pollution control than either charges or polluter-by-polluter wasteload allocations.

Graves, G., A. Whinston, and G. Hatfield. *Mathematical Programming for Regional Water Quality Management*, Water Pollution Control Research Series, 16110 FPX 08/70, US Department of the Interior, Federal Water Quality Administration (Washington, DC, 1970).

Henderson, J. and R. Quandt. *Microeconomic Theory* (New York, 1958).

Johnson, E. "A Study in the Economics of Water Quality Management." *Water Resources Research* 3 (2nd quarter, 1967), 291–305.

Kneese, A. V. "Environmental Pollution: Economics and Policy." *American Economic Review Papers and Proceedings* 61 (May 1971a), 154–66.

Kneese, A. V. "The Political Economy of Water Quality Management." In J. S. Bain and W. F. Ilchman (ed.), *The Political Economy of Environmental Control* (Berkeley, 1971b), 35–66.

Kneese, A. V. and B. T. Bower. *Managing Water Quality: Economics, Technology, Institutions* (Baltimore, 1968).

Meade, J. *The Balance of Payments* (London, 1952).

Meyer, J., J. Kain, and M. Wohl. *The Urban Transportation Problem* (Cambridge, Mass., 1965).

Patinkin, D. "Demand Curves and Consumer's Surplus." In C. Christ *et al.*, *Measurement in Economics* (Stanford, Cal., 1963), 83–112.

Scott, A. D. "The Economics of International Transmission of Pollution." In Organisation for Economic Co-operation and Development, *Problems of Environmental Economies* (Paris, 1972), 255–73.

Tinbergen, J. "Four Alternative Policies to Restore Balance of Payments Equilibrium." *Econometrica* 20 (July 1952), 372–90.

Turvey, R. "On Divergences between Social Cost and Private Cost." *Economica* 30 (Aug., 1963), 309–13.

[6]

Journal of Public Economics 24 (1984), 29–46. North-Holland

EFFLUENT FEES AND MARKET STRUCTURE

Wallace E. OATES

Bureau of Business and Economic Research, University of Maryland, College Park, MD 20742, USA

Diana L. STRASSMANN*

Rice University, Houston, TX 77001, USA

Received May 1982, revised version received May 1983

This paper explores the efficiency properties of a system of effluent fees in a mixed economy in which polluting agents take a variety of organizational forms: private monopoly, the managerial firm, regulated firms, and public bureaus. The analysis, including some crude empirical estimates, suggests that the welfare gains from pollution control are likely to dwarf in magnitude the potential losses from the various imperfections in the economy. The tentative conclusion is that the case for a system of fees that is invariant with respect to organizational form is not seriously compromised by likely deviations from competitive behavior.

1. Introduction

The formal analysis of a Pigouvian tax on polluting activities typically proceeds in terms of perfectly competitive firms whose productive pursuits impose external costs on other agents in the economy. Moreover, the optimality properties of the Pigouvian measure depend upon this assumption of perfect competition. A cursory inspection of the real world, however, reveals that the major sources of pollution encompass a wide variety of institutional structure. The public sector, for example, is itself a major polluter [see Oates and Strassmann (1978)]. Municipal waste-treatment plants dump enormous quantities of wastes into our waterways, and publicly-owned power plants are heavy contributors to air pollution. The largest single sulfur polluter in the United States is the Tennessee Valley Authority (TVA), which accounts for 16 percent of sulfur emissions in the nation [see King (1977)]. In addition, private but publicly-regulated firms (including utilities that provide electrical power) are among the very largest of polluters. Finally, many of the large factories that emit massive quantities of wastes are owned and operated by huge firms in highly concentrated

*We are grateful for many helpful comments on earlier drafts of this paper to Peter Altroggen, William Baumol, Richard Caves, Robert Dorfman, Joseph Kalt, Margaret Lewis, Robert Mackay, Albert McGartland, Lee Preston, Eugene Seskin, Jeffery Smisek, and anonymous referees. We are also indebted for the support of parts of this work to the National Science Foundation and the Sloan Foundation.

industries like steel, chemicals, and automobile manufacturing. The application of the competitive model with its myriad of small firms acting as price-takers is thus suspect for many classes of polluters.

The economic analysis of market incentives for pollution control must, therefore, push beyond the simple competitive model. We must ask how polluters with widely varying sets of objectives are likely to respond to these incentives. In this paper, we seek to determine what some standard models of organizational behavior tell us about how decision-makers in different institutions would respond to the introduction of a set of effluent charges.[1] We then use these results to evaluate the implications of these responses for efficient resource allocation, taking explicit account of the distortions that market imperfections, bureaucratic behavior, and public regulation of private firms themselves introduce.

2. The problem of 'allocative efficiency'

2.1. The conceptual issue

We begin the analysis with the standard monopoly model under which a profit-maximizing firm has some discretion over the price it charges for its output. We assume that the firm has a production function of the form:

$$Q = Q(L, E), \tag{1}$$

where L (which we shall call 'labor') represents a vector of all inputs other than E, the firm's level of waste emissions. We thus treat the source of pollution, namely waste emissions, as a productive input from the perspective of the firm. If factor markets are perfectly competitive and if the environmental authority confronts the firm with a Pigouvian charge on its emissions equal to marginal social damage (MSD), it follows that the firm, in the process of minimizing its costs, will select what from society's point of view is the cost-minimizing combination of factor inputs for whatever level of output it chooses. In short, cost-minimizing behavior ensures 'technical efficiency' in the use of all inputs including the services of the environment. A corollary is that, for the case where units of emissions from all sources are equally damaging, a uniform effluent fee will lead cost-minimizing polluters to equate their marginal abatement costs and hence to achieve the desired level of environmental quality at the minimum aggregate abatement cost (our 'least-cost theorem').

---

[1]The analysis also has relevance for systems of marketable pollution permits. However, it would need to be extended to account for any imperfections in the permit market itself [Hahn (1981)]. The assumption here is that, under a system of effluent fees, individual polluters take the fee structure as given.

The problem in this case concerns allocative distortions in the pattern of final outputs. As Buchanan (1969) has pointed out, the monopolist's sub-optimal level of output is the source of a basic dilemma for the formulation of policy to regulate externalities. An effluent fee provides an incentive for needed pollution abatement, but, at the same time, raises the firm's marginal cost and thereby induces a reduction in output. The result is some gain in efficient resource allocation from reduced waste emissions, but some loss in efficiency from the contraction in output; the *net* effect on social welfare is uncertain. In short, an effluent fee (Pigouvian or otherwise) may represent too much of a good thing.

The analysis must, therefore, take explicit account of Buchanan's tradeoff between pollution abatement and monopolistic output restriction. Following Baumol and Oates (1975, ch. 6), we depict the nature of this tradeoff in fig. 1. Let DD' represent the industry demand curve confronting a monopolist, with DMR being the corresponding marginal-revenue curve. We assume that the monopoly can produce at constant cost (PMC=private marginal cost), but that its production activities impose costs on others. In particular, in the absence of any fees, the monopolist's (private) cost-minimizing technique of production generates pollution costs per unit equal to AB so that the SMC_0 (social marginal cost) curve indicates the true cost to society of each unit of output. To maximize profits, the monopolist would produce OQ_m.

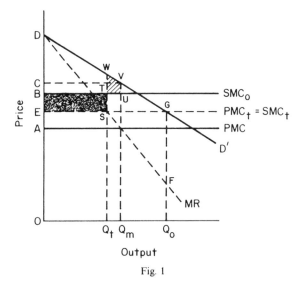

Fig. 1

Suppose next that we subject the monopolist to a pollution tax, a fee per unit of waste emissions. This will provide an incentive to alter the production process in a way that yields lower emissions per unit of output. In fig. 1 this would have two effects: it would raise the PMC curve and, over some range, would tend to lower SMC. This second effect results from the choice of what, from society's standpoint, is a lower-cost method of production (taking into account the costs of pollution). The minimum social cost of production will be reached when the pollution costs are wholly internalized so that $PMC_t = SMC_t$ (where the subscript t refers to costs in the presence of a Pigouvian tax). At this point, the firm's selection of a production process will be based upon a set of input prices (including a price of waste emissions) that reflects true social opportunity costs.

Since there exist two distinct sources of allocative distortions, a full resolution of the problem will, in general, require two policy actions: a Pigouvian tax on waste emissions equal to MSD and a subsidy per unit of output equal to GF (the difference between marginal cost and marginal revenue at the Pareto-efficient level of output). A typical environmental agency, however, will have neither the authority nor the inclination to offer subsidies to monopolists.

In this constrained setting, the problem of the environmental authority takes on a second-best character: the determination of the effluent fee which balances, at the margin, the social gain from increased abatement against the social loss from reduced output from the monopolist. Lee (1975), and more recently Barnett (1980), have derived formally the first-order conditions for this second-best optimal fee. Although the Lee–Barnett results take a slightly different form, they can be expressed as:

$$t_j = t^* - \frac{C_j}{|\eta_j|}, \tag{2}$$

where

t_j = optimal tax per unit of waste emissions for the jth polluter,
t^* = Pigouvian tax on the competitive firm,
C_j = marginal abatement cost, and
η_j = price elasticity of demand.

The second term on the RHS of eq. (2) reflects the marginal welfare loss from reduced output associated with a unit increase in the tax.[2] For a perfectly competitive firm, this term is zero so that $t_j = t^*$. But for a firm with some

[2]It is easy to see, incidentally, that in accord with intuition this term is equal to the marginal welfare loss per unit of reduced output (i.e. price minus marginal cost) multiplied by the reduction in output associated with an additional unit of abatement. More formally:

$$\frac{C_j}{|\eta_j|} = \frac{P_j}{|\eta_j|} \cdot \frac{\partial Q_j}{\partial a_j} = (P_j - MR_j)\frac{\partial Q_j}{\partial a_j} = (P_j - MC_j)\frac{\partial Q_j}{\partial a_j},$$

where a_j is the level of abatement activity and where profit maximization implies that $MR_j = MC_j$. The marginal cost of abatement, $P_j(\partial Q_j/\partial a_j)$ is expressed in terms of the value of forgone output.

control over market price, the optimal unit tax on emissions will vary inversely with marginal abatement cost and directly with the price elasticity of demand for the firm's output.

In principle, therefore, we can determine the optimal set of effluent fees on all polluters, be they competitive firms or monopolists. However, this is not, in fact, very comforting. First, such a determination would require an enormous amount of information encompassing both the price elasticities of demand *and* the abatement costs for each polluter. And second, even if the environmental authority were able to assemble all these data, it is difficult to envision a legal and political setting in which such a discriminatory set of fees would be acceptable.

At the policy level, the real choice may well be that between a single fee applicable both to perfect and imperfect competitors or the abandonment of a system of pricing incentives for environmental protection. From this perspective, the important issue is the *extent* of the welfare loss associated with the pattern of reductions in output induced by the charge on waste emissions. There is a substantial empirical literature suggesting that the magnitude of the overall allocative losses in the economy attributable to monopolistic distortions is quite small.[3] Since the large estimated welfare gains from pollution abatement would seem to dwarf the apparently small welfare losses from effects on the pattern of industry outputs, it is tempting simply to conclude that concern over monopolistic distortions represents, in this case, a theoretical nicety that we can safely ignore in the design of environmental policy.

This, however, will not quite do. The proper question is: *Given the existing pattern of monopolistic distortions* (i.e. existing divergences between price and marginal cost), do the *additional* reductions in monopoly outputs generate efficiency losses of a substantial magnitude?

2.2. A rough estimate

In order to get some feel for just how damaging the existence of monopolistic elements in the economy is to the case for an effluent fee that is invariant with respect to industry structure, we have undertaken some admittedly quite crude, partial-equilibrium calculations making use of a representative polluter.[4] Our procedure involves a comparison of two

[3]The seminal paper presenting this result is Harberger (1954). Several later studies support Harberger's general finding [e.g. Schwartzman (1960)]. However, for a dissenting view, see Cowling and Mueller (1978) who have criticized the earlier work on methodological grounds and have calculated their own estimates of allocative losses from monopoly for both the United States and the United Kingdom. Their estimates for these losses are much more sizable.

[4]The shortcoming of a partial-equilibrium approach is clear. A full, general-equilibrium treatment of the problem would take into account the interaction between markets; the outcome in this setting would depend not only on the magnitude of the initial distortion and the own price elasticity of demand, but also on cross elasticities.

equilibrium positions: the first involves no control over the externality (and hence no abatement), and the second is the outcome under a system of effluent fees where the fee does not vary with industry structure. In moving from the former to the latter, we compare the welfare gains from reduced pollution *net* of abatement costs to the *increment* of new allocative losses associated with monopoly elements in the existing market structure. We use existing environmental programs in the United States as a (rough) benchmark for overall abatement efforts.

Proceeding in terms of a 'representative polluter', we assume that the social marginal cost of production associated with each level of abatement activity is approximately constant over the relevant range. We can then approximate the welfare gain from reduced pollution net of abatement costs by:

$$W_g = Q[SMC_0 - SMC_t], \tag{3}$$

where SMC_0 and SMC_t are, respectively, social marginal cost before and after the introduction of a set of Pigouvian taxes. The welfare gain is simply the reduced cost (private plus external) per unit times the level of output (area $EBTS$ in fig. 1). Likewise, we can approximate the welfare loss from reduced output by:

$$W_l = \Delta Q(P - SMC_0). \tag{4}$$

Welfare loss is the loss per unit (equal to the difference between price and *social* marginal cost) times the change in output (an approximation to area $TWVU$ in fig. 1). To determine the relative sizes of these two effects, we divide (4) by (3) to obtain:

$$W_r = \frac{\Delta Q}{Q} \frac{[P - SMC_0]}{[SMC_0 - SMC_t]}. \tag{5}$$

It will facilitate the numerical comparisons to divide both numerator and denominator by P:

$$W_r = \frac{\Delta Q}{Q} \frac{[P - SMC_0]/P}{[SMC_0 - SMC_t]/P}. \tag{6}$$

The next step is to try to make some educated guesses as to the orders of magnitude of the various terms in eq. (6). To do this, we construct a profile of a representative polluter that incorporates reasonable estimates of the parameters. In each instance we lean in the direction of magnitudes that are favorable to the finding of a relatively large welfare loss associated with monopolistic distortions.

W.E. Oates and D.L. Strassmann, Effluent fees and market structure   35

The first term in eq. (6) is the percentage change in quantity, which we can express as the percentage change in price times the price elasticity of demand (η). There are available for the United States fairly detailed data on abatement expenditures by sector, and we make use of a careful ongoing study by H. David Robison (1983) in which he uses these data and a large input–output model to estimate existing abatement costs per dollar of output for 78 sectors in the U.S. economy. Using 1977 data, Robison finds, for example, the following percentage increases in costs attributable to abatement expenditures: 2.0 for the paper industry, 2.1 for ferrous metals, 2.3 for copper, and 3.3 for electric utilities. He assumes that these costs are passed forward in terms of higher prices. Deriving a 'representative' increase in prices under a fee system from these estimates is problematic, since the effect of a set of fees on costs would differ in two important respects from the existing command-and-control system. First, as existing studies indicate, a fee system would tend to reduce abatement costs significantly through more cost-efficient patterns of abatement and technology, but, second, such savings in abatement costs must be balanced against the effluent fees that sources would have to pay. We take as a 'typical' increase in price for our representative polluter a figure of 5 percent, where we assume that this cost increase to our polluter (and the consequent rise in price) is constituted in equal parts of control costs (2.5 percent) and effluent fees (2.5 percent).[5] A best estimate for a representative price elasticity of demand is also uncertain, but for the major industries of concern, including power generation, chemicals, pulp and paper, etc., a typical value of two is probably a generous assumption. This gives us a value of $\Delta Q/Q$ for our representative polluter of:

$$\frac{\Delta Q}{Q}=|\eta|\frac{\Delta P}{P}=2(0.05)=0.1. \tag{7}$$

We thus assume a 10 percent reduction in output for our representative polluter attributable to the adoption of the fee program.

Turning next to the numerator of the second term in (6), we note that $[P-SMC_0]$ is equal to that portion of the difference between price and private marginal cost that is not offset by the marginal social damage of waste emissions; that is:

$$\frac{[P-SMC_0]}{P}=\frac{[P-(PMC+MSD)]}{P}, \tag{8}$$

where MSD is the marginal external cost associated with polluting emissions. Finding appropriate magnitudes for these variables is somewhat

---

[5]This distinction is important, because control costs represent actual social costs, while fee payments are, from the perspective of society, a transfer payment.

more conjectural. However, we can make a very rough guess by noting that existing estimates of the benefits from air pollution control in the United States are about twice the level of abatement costs; Lave and Seskin (1977, p. 230), for example, offer a conservative estimate of health benefits alone from meeting standards for ambient air quality of about $16 billion as compared to EPA's estimate of $9.5 billion for abatement costs.[6] If we double our representative estimate for abatement costs of 2.5 percent and add on another 2.5 percent for residual damages, we reach a figure of 7.5 percent of marginal cost for our estimate of marginal social damages in the absence of any control program.[7]

Next we need a figure for monopolistic markup over marginal cost. There exists an empirical literature that has estimated the relationship between price–cost margins and industry concentration [e.g. Shepard (1972)]; these studies find that the margin of price over cost rises by about one percentage point for every increase of ten percentage points in the four-firm industry concentration ratio (C_4). If we assume that 'competition' involves a C_4 of about 20 and that 'monopolistic' industries have a typical C_4 of 70, we would have a level of monopolistic prices that exceeds competitive prices by about 5 percent. Leaning in the direction of a more generous estimate, we take 10 percent as our representative monopolistic markup over private marginal cost. This leaves us with an estimate of 2.5 for the difference between price and social marginal cost as a percentage of price.[8]

The denominator of the second term in eq. (6) follows directly from the preceding profile of our representative polluter. The representative reduction in marginal social damages from the control program is 5 percent of price (since we assume benefits equal to twice the level of abatement costs); from this, we subtract 2.5 percent of price for abatement costs leaving us with a net reduction in social marginal cost of 2.5 percent of price.

Pulling together our results for the various terms in (6), we arrive at:

$$W_r = (0.1)\frac{0.025}{0.025} = 0.1. \qquad (9)$$

[6]These figures admittedly refer to total, rather than marginal, benefits and costs. However, since we are considering the entire increment from the introduction of a program of effluent fees, we take them as a reasonable approximation.

[7]Our estimate thus implies that the fee program reduces the representative marginal damages from pollution by about two-thirds. This seems to us a relatively conservative figure, since abatement efforts have typically led sources to reduce emissions by well over 50 percent (in excess of 90 percent in several cases); such reductions in the presence of increasing marginal damages (the typical case) would suggest reductions in damages from pollution far in excess of one-half.

[8]Note that if the markup were only 5 percent, then social marginal cost for our representative polluter would actually exceed the monopoly price so that a contraction in output would raise, rather than lower, social welfare. Such may well be the case in some instances.

We thus estimate the monopolistic welfare loss from a program of effluent fees to be roughly an order of magnitude smaller than the welfare gain from reduced pollution (net of abatement costs). While this estimate obviously depends on our choice of values for the various parameters, we believe that the 'representative calculation' is relatively generous to the magnitude of the potential welfare loss from reduced monopoly outputs and that 'reasonable' parameter values are unlikely to suggest that this loss can rival in size the gains from improved environmental quality. In view of the range of policy options available to the environmental authority, our conclusion is that it is probably safe to ignore the issue of incremental output distortions associated with a system of effluent fees.

3. The problem of technical efficiency

In the preceding section, the assumption of simple profit-maximizing behavior with its corollary of cost minimization allowed us to ignore the issue of technical efficiency: faced with a Pigouvian fee, cost-minimizing agents will select the socially least-cost combination of inputs and will operate along their minimum cost curves. In this section of the paper we drop this assumption and explore the implications of technical inefficiencies for the efficacy of a system of effluent fees. We examine a series of models in which the failure to minimize costs comes from either of two sources: a more complex objective function that incorporates variables other than (or in addition to) the level of profits, or some type of regulatory constraint that provides the decision-maker with an incentive to choose something other than the cost-minimizing pattern of factor inputs.

3.1. Managerial models of maximizing firms

For our purposes, the Williamson (1963) model captures the spirit of the results that emerge when a firm's managers maximize an objective function that contains variables other than simply short-run profits. In particular, Williamson formulates a managerial utility function that incorporates 'expense preferences' for expenditures on staff (S), managerial emoluments (M) (extra salary and perquisites), and 'discretionary profits' consisting of the difference between actual profits and the minimum profits demanded. The firm thus maximizes:

$$U = U[S, M, \pi_R - \pi_0 - T] \quad \text{subject to} \quad \pi_R > \pi_0 + T \tag{10}$$

or

$$U = U\{S, M, (1-t)[R(X) - C(X) - S - M] - \pi_0\}, \tag{11}$$

where

$$R = \text{revenue} = P \cdot X; \ \partial^2 R / \partial X \partial S \geq 0,$$
$$P = \text{price} = P(X, S; \varepsilon); \ \partial P / \partial X < 0; \ \partial P / \partial S \geq 0; \ \partial P / \partial \varepsilon > 0,$$
$$X = \text{output},$$
$$S = \text{staff (in money terms) or (approximately) general}$$
$$\text{administrative and selling expense;}$$
$$\varepsilon = \text{a demand shift parameter},$$
$$C = \text{costs of production} = C(X),$$
$$M = \text{managerial emoluments},$$
$$\pi = \text{actual profits} = R - C - S,$$
$$\pi_R = \text{reported profits} = \pi - M,$$
$$\pi_0 = \text{minimum (after-tax) profits demanded},$$
$$T = \text{taxes, where } t = \text{tax rate, and}$$
$$\pi_R - \pi_0 - T = \text{discretionary profits}.$$

To treat waste emissions and the effluent fee explicitly, we amend the Williamson model to distinguish between these emissions (E) and all other inputs (L). Eq. (11) thus becomes:

$$U = U\{S, M, (1-t)[R(g(L, E)) - P_l L - f E - S - M] - \pi_0\}, \qquad (12)$$

where P_l is the price of other inputs and f denotes the effluent fee. Maximization of this utility function yields as one of the first-order conditions the familiar result:

$$\frac{P_l}{f} = \frac{\partial X / \partial L}{\partial X / \partial E}. \qquad (13)$$

This is the usual condition for the cost-minimizing combination of factor inputs: marginal products proportional to factor prices. It may seem surprising at first glance to find that firms that are technically inefficient (i.e. do not produce at minimum cost *overall*) are effectively cost-minimizers with regard to pollution abatement. However, the rationale is quite straightforward: since abatement activities contribute nothing to staff or emoluments and reduce discretionary profits, the firm's managers have an incentive to minimize the expenditure on abatement (consisting of effluent fees plus pollution-control costs) by extending abatement activity to the point where marginal abatement cost equals the effluent fee. We can thus extend our cost-minimization theorem to encompass certain managerial models of maximizing firms: a world of such firms subject to an effluent fee can, in principle, achieve the desired standard of environmental quality at the minimum aggregate abatement cost.[9]

[9]As one potential qualification to this result, we note that abatement activities could, under certain circumstances, enter directly into the managerial utility function. The firm's managers might perceive, for example, that activities to curtail pollution produce some valuable 'good-will' for the firm. In such instances, the firm might well extend abatement activities beyond the point at which marginal abatement cost equals the effluent fee.

3.2. Organizational models of firm behavior

Organizational models of firm behavior treat managerial decisions in the context of the firm's internal structure and environment. Firm behavior in these models cannot be characterized by an explicit objective function. These models include both the Carnegie Tech type that emphasizes internal coalitions, information costs, limited time, and bounded rationality [e.g. Cohen and Cyert (1962)] and the Harvard Business School variety that focus on the internal dynamics and structure of the firm [e.g. Chandler (1962) and Bower (1970)]. Without probing in detail into these alternative views of managerial decision-making, we wish to note in passing that both of these approaches imply the possible presence of a degree of 'managerial slack'.

Such slack may well have some implications for the effectiveness of a system of effluent fees. In particular, the case for effluent fees rests on the presumption that an increase in the price of effluents will induce firms to use less effluents relative to other inputs; over the longer haul, fees will induce firms to engage in R&D that will allow them to develop cheaper abatement technologies. However, the incentives to change production policies quickly in response to relative price changes may be quite weak in a managerial context. Changes in production methods (particularly to less pollution-intensive methods) may involve major changes in equipment. If an important perquisite is 'ease of management', the firm might conceivably employ some of the fat in its budget to avoid the effort and possible complications associated with the adoption and development of new abatement techniques. There are, in fact, some investigations of the diffusion of knowledge and new technologies that have found that firms in concentrated industries do not respond as quickly to price changes and the availability of new innovations as do firms in more competitive industries [see Kamien and Schwartz (1975)]. The evidence on this issue, however, is not conclusive; yet the lack of consensus certainly provides some justification for skepticism about the belief that effluent fees will work as well in highly concentrated industries as they might in more competitive cases.

We shall return to this issue again in our discussion of bureaucratic leanings toward 'ease of management'. What we can say is that all of the models of imperfectly competitive firm behavior that we have discussed establish some presumption that effluent fees will induce firms to pollute less; how closely the outcomes approach a cost-minimizing solution is less clear.

3.3. Public bureaus

Since models of bureaucratic behavior typically posit neither profit maximization nor cost minimization, the response of public agencies to pricing incentives is problematic [see Oates and Strassmann (1978)]. To explore the impact of effluent fees on public decision-makers, we first examine a variant of the Niskanen (1977) model of bureaucratic behavior. As we shall see, the model and its implications for abatement activities bear a strong resemblance to our analysis in the Williamson framework of private managerial maximization. We then discuss the implications of Wilson et al.'s (1980) richer study of public agencies.

In our varient of the Niskanen model, we postulate that the bureau's decision-makers seek to maximize an objective function that contains as arguments the bureau's output (Q) and its level of perquisites (P):

$$U = U(Q, P).\tag{14}$$

Bureaucrats desire an increased output (or 'size'), for this enhances the bureau's power and prestige and with these its capacity to influence the course of events. Migue and Belanger (1974) have contended that agency officials also place a premium on the bureau's 'discretionary budget', the excess of the bureau's funding above its necessary costs. This 'fat' in the budget can be employed for a variety of perquisites ranging from higher salaries and expanded staff to additional facilities or, perhaps, reduced effort.

As earlier, we assume that the production function for the bureau's output depends on waste emissions, E, and a vector of other outputs, L:

$$Q = Q(L, E).\tag{15}$$

Moreover, the bureau is subject to a budget constraint:

$$B = wL + fE + cP,\tag{16}$$

where B is the bureau's budget, w is the price of 'other' inputs (given to the bureau), f is the effluent fee per unit of waste emissions, and c is the (constant) marginal cost of perquisites. Note that L is defined to include only the minimally necessary quantity of other inputs such as labor to provide a given output; likewise, w can be thought of as the lowest wage that will keep employees. Extra salary and labor are viewed as perquisites.

The budget-determination process is the remaining issue. Here we follow Niskanen and assume that the bureau possesses a kind of monopoly power in its dealings with the legislative agency that provides its funding. In particular, the bureau submits (and obtains) a budget of an all-or-nothing character that extracts the entire area under the legislature's demand curve up to the bureau's proposed level of output. In short, the bureau behaves much like a perfectly discriminating monopolist; for whatever level of output it selects, say $\hat{Q}$, the bureau's budget equals:

$$B = \int_0^{\hat{Q}} D(Q)\,dQ,\tag{17}$$

where $D(Q)$ is the legislature's inverse demand function for the bureau's output.[10]

In this framework the bureaucrat's problem becomes that of maximizing the utility function in eq. (14) subject to its budget constraint:

$$M = U[Q(L, E), P] + \left[wL + fE + cP - \int_0^Q D(Q)\,dQ \right].$$ (18)

Solving for the stationary values of (18) yields (among other results):

$$\frac{\partial Q/\partial E}{\partial Q/\partial L} = \frac{f}{w}.$$ (19)

This result is essentially the same as that obtained from our analysis of the Williamson model. Eq. (19), like that for the managerial model of the maximizing firm, implies cost minimization in only a limited sense: the bureau minimizes pollution abatement and other costs that do not generate perquisites. Effluent fees are effectively lost dollars; they provide no utility to the bureaucrat. By minimizing pollution-abatement costs, the bureau maximizes the remaining budget for the procurement of perquisites. Like cost-minimizing firms, a bureau behaving according to this model has an incentive to extend pollution-abatement activities to the point where marginal abatement cost equals the effluent fee.

An important qualification to this result introduces an indeterminacy similar to that in the satisficing models of firm behavior. Bureaucrats, like the employees of firms, are likely to have some preferences for the perquisite 'ease of management'. Just how pervasive such behavior is in public agencies is unclear; however, it could introduce some inefficiencies in the allocation of abatement quotas among polluters. Although, as the analysis suggests, bureaucrats are likely to have some incentive to economize on abatement costs, the discussion in the previous section on diffusion of innovations is also likely to apply to bureaus. Bureaus, like firms protected by managerial slack, do not need to respond as quickly to price changes and to the availability of new technologies as do competitive firms, since the survival of a bureau does not, in general, depend on an aggressively tight management.

The model is admittedly a very simplistic one that cannot begin to encompass the diversity in circumstances and particular objectives of different public agencies. In a recent series of case studies of bureaucratic behavior, Wilson et al. (1980) criticize such simplistic approaches to the characterization of public agencies. Their studies find that the behavior of these agencies is 'complex and changing' (p. 373); it is not subject, for

---

[10]Alternatively, we might simply assume that the budget is, for our purposes, some predetermined sum [see Oates and Strassmann (1978)]. This would not alter the results.

example, to the broad generalization of the 'captive theory', that, as a rule, 'Regulation is acquired by the industry and is designed and operated primarily for its benefit' [Stigler (1971, p. 3)]. Instead, Wilson et al. suggest that 'We view these agencies as coalitions of diverse participants who have somewhat different motives' (p. 373). Wilson et al. find that the studies reveal public agencies to '... prefer security to rapid growth, autonomy to competition, stability to change... Government agencies are more risk averse than imperialistic' (p. 376). This serves to underline our earlier observations on the potential for public agencies to respond sluggishly to incentives for change.

We note of particular interest the general similarity of the findings and qualifications of the behavior of managers of public and (imperfectly competitive) private enterprises. In a set of case studies of several electric utilities, Roberts and Bluhm (1981) can likewise find no systematic differences in behavior between publicly and privately owned concerns; they conclude that 'The mere fact of public or private ownership by itself does not tell us very much about the kind of behavior to expect' (p. 335).

In summary, although the formal model in this section predicts minimization of abatement costs by public agencies, the interpretation of this result and the broadening of our perspective on bureaucratic behavior suggest some basic reservations. We surely cannot, in a simplistic way, extend the umbrella of our cost-minimization theorem to encompass public agencies. At the same time, the analysis does suggest that an explicit price on pollution activities will present managers (public or private) with a real incentive for abatement. Managers are obviously not entirely oblivious to the costs of alternatives; Roberts and Bluhm, for example, in describing the history of the huge publicly-owned utility, TVA, found that 'Despite the agency's broad responsibility for conservation and regional development, most of its engineering decisions have reflected a continuing attempt to minimize the cost of power' (p. 63).

3.4. Regulated firms

The presence of regulated firms introduces, in principle, another source of technical inefficiency: a regulatory constraint that effectively distorts the relative prices of inputs to the firm. By setting some maximum 'fair' rate of return to capital inputs, the regulatory authority creates an incentive for the firm to extend its use of capital, for by using more capital, the firm is able to enlarge the base upon which its profit constraint is determined. All this is well known and has been described in terms of the Averch and Johnson (1962) (A–J) model; the distortion in factor inputs involving an excessive use of capital is the A–J effect.[11]

[11]For excellent, comprehensive treatments of the analytics of the A–J model, see Baumol and Klevorick (1970) and Bailey (1973).

It is fairly straightforward exercise to take the standard version of the A–J model (in which the regulated firm maximizes profits subject to a rate-of-return constraint), to introduce waste emissions into the production function as earlier, and then to examine the first-order conditions for profit maximization in the presence of an effluent fee. We shall not go through the mechanics here, but wish simply to note that the results of this exercise indicate that the regulated firm will not, in general, be a cost minimizer with respect to abatement activity.[12] It would be quite plausible, for example, for such a firm to extend abatement activity beyond the level at which marginal abatement cost equals the effluent fee, if by doing so the firm could expand its capital stock through the use of pollution-control equipment. The rationale from the perspective of the firm is the higher level of absolute profits that the expanded capital stock would allow. But from society's vantage point, this represents, of course, an excessive level of abatement. Under these circumstances, a system of effluent fees could not be expected to generate the least-cost set of pollution-abatement quotas among sources.

More generally, we cannot even conclude that an increase in the effluent fee will lead to a reduction in the waste emissions of the regulated firm. Bailey (1973, pp. 135–137) shows that for the two-factor case, a rise in the wage rate need not lead to reduced labor input; although there must surely be a strong presumption of this result, the outcome is formally indeterminate. Likewise, in his two-factor formulation of the A–J model, Cowing (1976) finds that the sign of the derivative of waste emissions with respect to the effluent fee is ambiguous.

While these results are, *in principle*, disturbing, there remains the important practical question of the actual magnitude of the A–J effect. The literature has not produced any compelling evidence of a widespread A–J effect [e.g. Baron and Taggart (1977), Spann (1974)]. Not only is there an absence of empirical support for 'over-capitalization' by regulated firms, but there is evidence that the rate-of-return constraint may often not even be binding.

In addition to the empirical studies, recent theoretical work casts further doubt on the applicability of the A–J theorem. Peles and Stein (1976) show that the theorem is 'highly sensitive' to the treatment of uncertainty; if uncertainty is multiplicative in form, the A–J effect is reversed! Perhaps even more basic is the issue of the interaction between the regulated firm and the regulating authority [e.g. Stigler (1971), Peltzman (1976), Joskow (1972, 1973)]. In the A–J model, the firm simply takes the rate-of-return constraint as exogenous. In the real world, however, the regulated firm must typically make application for any price changes, and, as Joskow (1972) shows, the determination of the allowed rate of return appears to involve a fairly complicated process of interaction between the regulator and the regulated firm.

---

[12]We would be happy to provide the formal analysis to any interested readers.

The industrial-organization literature thus establishes some compelling reasons for being quite skeptical about the A–J description of the behavior of the regulated firm. However, it does not, at this juncture, provide a straightforward alternative framework for our analysis of the regulated firm's response to an effluent fee. Some of the literature notes (as does our treatment of managerial models of the firm and of public agencies) the presence of other variables in the objective function in addition to profit maximization; for example, Roberts and Bluhm (1981) conclude that 'Our studies have shown that regulated firms are not pure incentive-oriented profit maximizers. To varying extents, managers are sensitive — albeit within financial limits — to not harming the public and to doing "the right thing"' (p. 384). Moreover, to the extent that the costs associated with effluent charges can be passed along through viable requests for higher prices to the regulatory authority, there may be grounds for questioning a highly 'cost-conscious' response to a system of fees.

We emerge with the sense that our expectations for the regulated firm are much in the spirit of those for the 'managerial' firm and the public bureau. In all these cases, we envision a managerial utility function which contains a multiplicity of objectives as a consequence of which cost-minimization plays an important, but not a singular, role in the determination of behavior.

4. Conclusion

While the analysis in this paper cannot yield any firmly grounded conclusions about the effectiveness of a system of effluent fees in a mixed economy, it does, we think, suggest some tentative results. First, our admittedly crude calculations in section 2 suggest that any distortions in the vector of final outputs resulting from a system of effluent fees (or probably from other forms of pollution control) are unlikely to be the source of substantial welfare loss. The 'allocative' issue that has troubled Buchanan and others in the design of systems to regulate externalities appears to be relatively unimportant in terms of its magnitude.

Second, although it is more difficult to get a sense of the extent of the 'technical inefficiencies' associated with deviations from cost minimization, the analysis does indicate that even where profit maximization is not the sole or dominant objective of the source, there are other considerations that can make it in the interest of polluters to engage in (relatively) efficient levels of abatement activities in response to effluent charges. We found, for example, that in our versions of both the Williamson model of utility-maximizing managers of private firms and the Niskanen model of bureaucracy, there exist incentives promoting cost-minimizing behavior with respect to abatement activities. While these incentives are no doubt blunted somewhat by a certain amount of 'managerial slack', there is at least a real rationale for managers and bureaucrats to seek out cost-saving abatement techniques.

More generally, the central point is that the case for relying on pricing incentives for pollution control in a mixed economy is really little different from that for using prices to guide the allocation of other inputs. While there obviously exists some degree of technical inefficiency in the economy associated with departures from cost-minimizing behavior, there is little reason to believe that the extent of such inefficiencies in pollution control under a system of effluent fees (or marketable emission permits) will be any greater (or any less) than in the use of labor, capital, and other factor inputs.

The importance of introducing pricing incentives for pollution control is underscored by an emerging empirical literature that indicates that existing programs of direct controls are generating enormous waste: abatement costs on the order of two to ten times as large as needed to attain the designated standards of environmental quality [see, for example, Atkinson and Lewis (1974), Palmer et al. (1980), Seskin et al. (1983), and Kneese et al. (1971, appendix C)]. These estimates, moreover, refer only to savings based upon existing abatement technology; they do not address the important long-run issues of the stimulus such pricing incentives would provide for research and development of new techniques for the curtailment of emissions. Our judgement is that a least-cost solution over time to the achievment of our objectives for cleaner air and water would probably involve aggregate costs no larger than 20–25 percent of those under the command-and-control programs that have been adopted in the United States.

Even in the presence of a substantial amount of 'slippage' in the form of technical inefficiencies, it is our view that the likely gains from a system of pricing incentives are quite large. We doubt that the complications arising from the existence of a mixed economy compromise significantly the case for a system of pricing incentives for environmental management.

References

Atkinson, S. and D. Lewis, 1974, A cost-effectiveness analysis of alternative air quality control strategies, Journal of Environmental Economics and Management 1, 237–250.

Averch, H. and L. Johnson, 1962, Behavior of the firm under regulatory constraint, American Economic Review 52, 1052–1069.

Bailey, E., 1973, Economic theory of regulatory constraint (Heath, Lexington, Mass.).

Barnett, A., 1980, The Pigouvian tax rule under monopoly, American Economic Review 70, 1037–1041.

Baron, D. and R. Taggart, Jr., 1977, A model of regulation under uncertainty and a test of regulatory bias, Bell Journal of Economics 8, 151–167.

Baumol, W. and A. Klevorick, 1970, Input choices and rate of return regulation: An overview of the discussion, Bell Journal of Economics and Management 1, 162–190.

Baumol, W. and W. Oates, 1975, The theory of environmental policy (Prentice-Hall, Englewood Cliffs, N.J.).

Bower, J., 1970, Managing the resource allocation process (Harvard Business School, Cambridge, Mass.).

Buchanan, J., 1969, External diseconomies, corrective taxes, and market structure, American Economic Review 59, 174–177.

Chandler, Jr., A., 1962, Strategy and structure (MIT Press, Cambridge, Mass.).

46 *W.E. Oates and D.L. Strassmann, Effluent fees and market structure*

Cohen, K. and R. Cyert, 1962, Theory of the firm: Resource allocation in a market economy (Prentice-Hall, Englewood Cliffs, N.J.).

Cowing, T., 1976, The environmental implications of monopoly regulation: A process analysis approach, Journal of Environmental Economics and Management 2, 207–223.

Cowling, K. and D. Mueller, 1978, The social costs of monopoly power, Economic Journal 88, 727–748.

Hahn, R., 1981, Market power and transferable property rights, unpublished paper.

Harberger, A., 1954, Monopoly and resource allocation, American Economic Review 44, 77–87.

Joskow, P., 1972, The determination of the allowed rate of return in a formal regulatory hearing, Bell Journal of Economics and Management Science 3, 632–644.

Joskow, P., 1973, Pricing decisions of regulated firms: A behavioral approach, Bell Journal of Economics and Management Science 4, 118–140.

Kamien, M. and N. Schwartz, 1975, Market structure and innovation: A survey, Journal of Economic Literature 13, 1–37.

King, W., 1977, T.V.A., a major polluter, faces suit to cut sulfur dioxide fumes, New York Times 76 (July 4).

Kneese, A., S. Rolfe and J. Harned, eds., 1971, Managing the environment (Praeger, New York).

Lave, L. and E. Seskin, 1977, Air pollution and human health (Johns Hopkins Press, Baltimore).

Lee, D., 1975, Efficiency of pollution taxation and market structure, Journal of Environmental Economics and Management 2, 69–72.

Migue, J. and G. Belanger, 1974, Toward a general theory of managerial discretion, Public Choice 17, 27–43.

Niskanen, Jr., W., 1977, Bureaucracy and representative government (Aldine, Chicago).

Oates, W. and D. Strassmann, 1978, The use of effluent fees to regulate public-sector sources of pollution: An application of the Niskanen model, Journal of Environmental Economics and Management 5, 283–291.

Palmer, A., et al., 1980, Economic implications of regulating chlorofluorocarbon emissions from nonaerosol applications (Rand, Santa Monica, California).

Peles, Y. and J. Stein, 1976, The effect of rate of return regulation is highly sensitive to the nature of uncertainty, American Economic Review 66, 278–289.

Peltzman, S., 1976, Toward a more general theory of regulation, Journal of Law and Economics 19, 211–240.

Roberts, M. and J. Bluhm, 1981, The choices of power: Utilities face the environmental challenge (Harvard University Press, Cambridge, Mass.).

Robison, H.D., 1983, Three essays on input–output analysis, unpublished Ph.D. dissertation, University of Maryland, College Park, Md.

Schwartzman, D., 1960, The burden of monopoly, Journal of Political Economy 68, 627–630.

Seskin, E., R. Anderson, Jr. and R. Reid, 1983, An empirical analysis of economic strategies for controlling air pollution, Journal of Environmental Economics and Management 10, 112–124.

Shepard, W., 1972, Elements of market structure: An inter-industry analysis, Southern Economic Journal 38, 531–537.

Spann, R., 1974, Rate of return regulation and efficiency in production: An empirical test of the Averch–Johnson thesis, Bell Journal of Economics and Management Science 5, 38–52.

Stigler, G., 1971, The theory of economic regulation, Bell Journal of Economics and Management Science 2, 3–21.

Williamson, O., 1963, Managerial discretion and business behavior, American Economic Review 53, 1032–1057.

Wilson, J., et al., 1980, The politics of regulation (Basic Books, New York).

[7]

GARDNER M. BROWN, JR.* and RALPH W. JOHNSON*

Pollution Control by Effluent Charges: It Works in the Federal Republic of Germany, Why Not in the U.S.†

INTRODUCTION

This article describes the recent Federal Republic of Germany effluent charge law and the political and legal background that permitted this law to be enacted. The impact of that law is assessed, although the assessment is necessarily tentative in view of the short experience with the law to date.

The economic and legal implications of enacting an effluent charge law in the United States also are analyzed. Included in this discussion are the advantages and disadvantages of state vs. federal enactment, the constitutional objections that might be raised to such a law, and how it might be coordinated with existing water pollution control laws in the United States.

THE POLITICAL BACKGROUND OF THE 1976 FEDERAL WATER ACT AND THE EFFLUENT CHARGE LAW

Water management historically has been controlled locally in the Federal Republic of Germany (FRG). Even after the 1871 unification of the nation's water management, as well as many other areas of domestic policy, water management remained under local control. In 1937, the

*Gardner M. Brown, Jr., Department of Economics and Ralph W. Johnson, School of Law, University of Washington.

†Research for this article was funded by the Environmental Protection Agency, the German Marshall Fund of the United States and the Federal Republic of Germany Ministry of Interior. Inquiries about research on other market oriented approaches to environmental management can be addressed to Mr. Mahesh Podar, EPA. We are grateful for the translation services of Casey O'Rourke and research assistance and translation services of Barbara Fritzemeier. Extension comments on various drafts were provided by Blair Bower, Will Irwin, Marvin Kosters, Allen Kneese, J. Salzwedel, E. Rehbinder, H. Massing, Dr. F. Schendel, M. Uppenbrink, H. Luhr, F. Schröder, G. Gedlitschka, Dr. W. Kitschler, Dr. H. Roth, Dr. J. Gilles, C. D. Malloch, Michael DeBusschere, Malte Faber, Lutz Wicke and W. Dorau. The editors also wish to thank and acknowledge the assistance with translations received from Ms. Ulrike von Juene, Department of Modern and Classical Languages, University of New Mexico, Albuquerque.

National Socialist Government tried to centralize governmental power at the national level, including water management. This trend, however, was reversed in 1949 with adoption of the postwar Basic Law which encouraged decentralization again. The long tradition of local control over water management in the FRG played a critical role in shaping the debates on the 1976 Federal Water Act and Effluent Charge Law.[1]

By the start of the last decade environmental quality in the FRG was badly in need of improvement. Rapid industrialization had placed excessive demands on the self-purifying capabilities of receiving waters. In many regions traditional uses of water bodies, for example as a source of drinking water, had been precluded by the deterioration of water quality. The need for new water legislation in particular and new environmental law in general was recognized widely.

The heightened concern for environmental quality led to the creation of a Cabinet Committee for Environmental Problems to coordinate environmental actions of all federal ministries.[2] This Committee produced a document entitled "A Program for the Protection of the Human Environment" (PPHE), published in 1971, which strongly advocated a market-oriented approach to environmental control. The report also recommended a constitutional amendment to give the federal government preemptive power over the Länder[3] to enact appropriate environmental legislation. Constitutional amendments were proposed in 1973, 1974, and 1975 but failed to pass. Failure of these proposed amendments meant that the federal government would have to share legal authority over this subject with the Länder, as required by the 1949 Basic Law. Under the Basic Law the federal government could only enact framework legislation, leaving all implementation and enforcement to the Länder.[4] ·

---

1. Federal Water Act, Wasserhaushaltsgesetz (1976) [hereinafter cited as FWA]; Effluent Charge Law, Abwasserabgabengesetz (1976) [hereinafter cited as ECL].

2. *See* Hans-Dietrich Genscher, INTRODUCTION TO A PROGRAMME FOR THE PROTECTION OF THE HUMAN ENVIRONMENT (1971) (adopted by the Government of the Federal Republic of Germany).

3. "Länder" are the states in the Federal Republic of Germany.

4. *See* GOVERNMENT OF THE FEDERAL REPUBLIC OF GERMANY, A PROGRAM FOR THE PROTECTION OF THE HUMAN ENVIRONMENT 72, 74, 75 (1971) [hereinafter cited as PPHE]; L. DINKLOH, VORTRAG JAHRESTAGUNG 1982, FACHGRUPPE WASSERCHEMI, STAND DES GEWÄSSERSCHUTZES 1982 IN DER BUNDESREPUBLIK DEUTSCHLAND AUS DER SICHT DER GESETZGEBUNG (Conditions of Water Protection Law in Germany in 1982 in View of the Legislation) (May 1982). Everyone interviewed agreed that water quality in the early 1970s was unacceptable. Disagreement occurs over the solution.

The PPHE recommended expansion of federal legislative competence in the environmental field in view of the limitations placed on those powers in the Basic Law of 1949. The FRG's constitution, the Basic Law (Grundgesetz), apportions legislative competence between the federal and Länder governments on the basis of four classifications. In the first category, the federal government has exclusive competence regarding foreign affairs, national citizenship, commerce with foreign nations, postal affairs, the national railroads and air transportation, currency. In the second category, the federal government and the Länder have concurrent legislative competence over civil, criminal, and

Internal and external pressures built and forced the improvement of water quality. Switzerland and the Netherlands, downstream from the FRG, expressed increased concern about the deterioration of water quality in mutually shared water bodies, such as Lake Constance and the Rhine. International organizations, such as the Organization for Economic Cooperation and Development (OECD) and the European Community, called for stronger environmental and water pollution control laws, as did the 1972 World Environmental Conference held under United Nations auspices in Stockholm, Sweden. These international organizations urged adoption of the market-oriented "polluter pays" principle as the best means for implementing environmental and water pollution control programs.[5]

Political support in the FRG increased for a market approach to environmental management during the early 1970s. Initial proposals for water pollution control legislation looked very much like the "ideal" systems urged by economists. Charges would be levied on waste dischargers in direct proportion to the damage caused by their use of public waters.[6] Some Länder, however, especially Bavaria and Baden-Wuerttemberg in the south, opposed these radical innovations and recommended a more moderate charge system which would operate in tandem with the traditional standards/regulatory system.[7] By 1976, the idea of a combined

---

real estate law, health and welfare, local commerce and natural resource development, road construction and maintenance, and land use regulation. In the third category, the federal government's competence is limited to the enactment of "framework" legislation in the areas of water management, press and film industries, land distribution, regional planning, and public services. The fourth category consists of matters that are reserved wholly to the Länder.

Of significance here is the fact that in water management matters the federal government has neither exclusive nor concurrent legislative competence: if it had exclusive competence, it would have total control over water quality management; if it had concurrent competence, it could preempt Länder laws to the extent that a uniform national approach is appropriate. The Basic Law, however, provides only for federal "framework" competence and, thus, leaves all implementation and enforcement to the Länder where those powers have traditionally been found.

5. UNITED NATIONS, FOOD AND AGRICULTURAL ORGANIZATION, LEGAL AND INSTITUTIONAL RESPONSES TO GROWING WATER DEMAND (Rome, 1977). ORGANIZATION FOR ECONOMIC COOPERATION AND DEVELOPMENT [hereinafter cited as OECD], ENVIRONMENT DIRECTORATE, WATER MANAGEMENT SECTOR GROUP, WATER MANAGEMENT POLICIES AND INSTRUMENTS (Final Report, Paris, 1975). OECD, ENVIRONMENT DIRECTORATE, THE POLLUTER PAYS PRINCIPLE: NOTE ON THE IMPLEMENTATION OF THE POLLUTER PAYS PRINCIPLE (May, 1974). OECD, RECOMMENDATIONS OF THE COUNCIL ON THE IMPLEMENTATION OF THE POLLUTER PAYS PRINCIPLE, (March, 1974). BULLINGER, G. RINCKE, OBERHAUSER, & SCHMIDT, THE POLLUTER-PAYS PRINCIPLE AND ITS INSTRUMENTS (1972); and E. REHBINDER, THE POLITICS OF THE POLLUTER-PAYS PRINCIPLE (1972). Both studies were published by the FRG as background papers for the United Nations' 1972 World Environmental Conference, Stockholm, Sweden.

6. Interview with Dr. H. Massing, Deputy of President, Head of Department of Water Resources in Düsseldorf (Mar. 16, 1983); interview with Dr. M. Uppenbrink, Director, Department of Environmental Planning, Environmental Protochon Agency, FRG, Dr. H. P. Luhr, Ministerialrat, Dr. Kanowski, Ministerialrat, Umweltbundesamt (the FRG's EPA), in Berlin (Mar. 17, 1983).

7. For the broad powers sought, *see* PPHE, *supra* note 4, at 72–73. Some discussion of the

system of regulations plus charges had become dominant. This system would levy charges high enough to create market-like incentives to abate pollution but, at the same time, would continue an administrative management regime for pollution control.

Industry initially opposed the idea of any effluent charge system. As political support for the system gained momentum, however, opposition shifted to implementation issues, such as the criteria for setting charges, the level of charges, and the dates when the system would go into effect.[8]

Some industries actually supported the effluent charge concept. The two main sources of support were the newer plants with new waste-saving production processes and the latest pollution control technology, and those older plants with recently installed new pollution control equipment. They believed their charges would be relatively smaller thus giving them a competitive edge over industrial facilities with less up-to-date equipment. A few other industries supported the idea because they believed that the levying of charges would even out the serious inequities caused by variations in the water quality regulatory systems among the Länder.[9]

THE 1976 FEDERAL WATER ACT, WASSERHAUSHALTSGESETZ

The Federal Water Act (FWA) continues the operation of the permit systems that were in effect in the Länder under the 1957 law.[10] The FWA sets forth the conditions governing the granting of permits to use public waters for the discharge of effluents. The FWA empowers the federal government to establish uniform discharge standards for certain major pollutants and to determine the level of technology that must be achieved by municipalities and industries. In addition, the FWA grants the federal government authority to establish a minimum national water quality goal for receiving waters, and it did so by setting this goal at quality level II (Guetezustand II).[11] Quality level II is moderately polluted water with a

---

problems arising from limited federal competency is found in R. JOHNSON & G. BROWN, JR., CLEANING UP EUROPE'S WATERS (1976) [hereinafter cited as JOHNSON & BROWN].

8. There is a feeling in the FRG that once a consensus on the need for legislation is achieved, the various parties are more inclined to work in cooperation toward the common goal in contrast to the United States where a more adversarial philosophy seems to operate. Interview with Drs. Uppenbrink, Luhr, and Kanowski, *supra* note 6.

9. Interview with Dr. W. Kitschler, Ministerialrat, Dr. H. Roth, Ministerialrat, and Dr. J. Gilles, Ministerialrat, Ministry of Interior, in Bonn (Mar. 15, 1983); interview with J. Salzwedel, Director, Institut für das Recht der Wasserwirtschaft an der Universität Bonn, in Bonn (Mar. 15, 1983).

10. *See* Malle, Sind Abgaben ein geeignetes Instrument der Umweltspolitik? (Are Charges an Appropriate Environmental Policy Tool?), 1982 UMWELT 35 (1982).

11. GOVERNMENT OF THE FEDERAL REPUBLIC OF GERMANY, GEWÄSSERGÜTE KARTE DER BUNDESREPUBLIK DEUTSCHLAND, AUSGABE 1976 (Water Quality Map of the Federal Republic of Germany, 1976 Edition). There are four classes of water. Class I is oxygen saturated, low in nutrients, and supports high quality fish; Class II is defined in the text; Class III is heavily polluted; Class IV is excessively polluted. The FRG uses a method of classifying water quality developed by Kolkwitz and Marssond and revised by Liebman. *See* H. Liebmann, *Die Notwendigkeit*

good oxygen supply capable of supporting a large variety of algae, cray-fish and insect larvae, and fish.[12]

The FWA made an important change in existing law by banning future issuance of any "licenses" by the Länder.[13] Licenses, which were used extensively by some Länder, created vested rights for 20 years or longer and required compensation when revoked. Now all waste dischargers must operate under permits which are issued for shorter periods of time and are subject to change and even revocation as water quality demands change over time.[14] The FWA also subjects present license holders to reasonable regulations to conform the licensed discharges with the federal minimum standards under Art. 7a(2). More importantly, the effluent charge law subjects license holders to the same charges as permit holders.[15] While the federal government establishes the overall national water quality level (i.e., Level II), the Länder establish definite water quality targets and programs for achieving those targets.

The most important provision in the FWA is article 7(a), which authorizes the federal government to establish technology-based standards (allgemein anerkannte Regeln der Technik) such as *best practicable,* or *commonly accepted* technology. These standards form one of the basic measurements used in the Effluent Charge Law, discussed later. The standards vary depending on whether the waste water originates with a municipality or industry and, if the latter, the standards vary by industry. The federal government appointed some 50 task forces to establish these technological standards for different industries and for cities. In addition to the basic regulatory system provided for in the FWA the law provides that dischargers causing harm or injury to others are liable for damages.[16] Those who violate the provisions of the Act are also liable for fines of as much as $100,000 Deutschemark (DM).[17]

THE EFFLUENT CHARGE LAW, ABWASSERABGABENGESETZ

After years of extensive public discussion, the Effluent Charge Law (ECL) passed overwhelmingly in September 1976.[18] It calls for the Länder

---

einer Revision des Saprobiensystems und deren Bedentung für die Wasserbeurteilung, GESUND-HEITS-INGENIEUR 68 (1947) (The Necessity of a Revision of the Saprob Systems and Its Importance to the Classification of Waters).

12. FWA, art. 2(1).

13. FWA, art. 8(2).

14. FWA, art. 4(1), (2); art. 18.

15. ECL, art. 4(1).

16. FWA, art. 22(1).

17. FWA, art. 41.

18. Menke-Glueckert reports that there were only seven dissenting votes in the Bundestag and they wanted a more strict effluent charge law. *See* Menke-Glueckert, *Stand der Vorbereitungen zum Inkrafttreten des Abwasserabgabengesetzes,* BERICHTE DER ABWASSERTECHNISCHEN VEREINIGUNG E.V. (Status of the Preparations for the Implementation of the Effluent Charge Law);

to levy charges (Article I) on direct dischargers for specified effluents into public waters.[19] Firms and households discharging into municipal sewerage facilities are not charged directly. The effluent charge policy reflects the polluter-pays principle, which broadly states that the parties discharging waste should pay for the abatement costs actually or implicitly imposed on society.

The discharge permit issued by the Länder is divided into two parts. The first, a legal part, establishes the discharge right, and contains all the physical, chemical and biological data and monitoring procedures pertaining to waste water quality (pH, temperature, biochemical oxygen demand (BOD_5), other concentrations) and establishes the maximum amount of waste water in specified time periods. The specified waste water quality levels must be equal or higher in quality than the minimum requirements of the federal administrative regulation. This part of the discharge permit is subject to the water laws of the FRG and the Länder.[20]

The second part of the discharge permit contains all the data necessary to calculate the waste water discharge bill. The pollutants considered for purposes of the effluent charge are settleable solids, chemical oxygen demand (COD), cadmium (Cd), mercury (Hg) and toxicity for fish. The permits also specify the annual volume of water that can be discharged. The standard may be specified in terms of concentration per cubic meter of discharge volume or per ton of product produced.

The permit specifies a maximum concentration of each pollutant and volume of waste water a discharger expects to produce (Höechstwert). The average (standard) amount of the waste to be discharged and the expected concentrations (Regelwert) are provided by each discharger and are reflected in the permit. Under normal circumstances the figure or reference value (Bezugswert) on which the charge is based is the volume and concentration the entity expects to discharge. Notice that the charge normally is based on the expected rather than the actual level of discharge.

Table 1 illustrates these ideas. The hypothetical firm discharges only settleable solids and COD whose reference (expected average) and maximum values have been specified. Under normal circumstances, the waste discharge bill is calculated easily. The data in Table 1 are converted to damage units using the coefficients provided in an appendix to the ECL and exhibited in Table 2 below.[21]

ATV-JAHRESHAUPTVERSAMMLUNG, Tr. 32 (1980) (Reports of the Waste Water Technical Association, Proceedings of the Annual Conference).

19. ECL, art. 3(1).

20. ECL, art. 4(4).

21. Table reproduced from B. BOWER, R. BARRE, J. KUCHNER & C. RUSSELL, INCENTIVES IN WATER QUALITY MANAGEMENT: FRANCE AND THE RUHR AREA 301 (1981) [hereinafter cited as BOWER]. Other pollutants for which minimum requirements may be established for some industries include: biological oxygen demand (BOD), hydrocarbons, phenols, cyanide,

TABLE 1

SELECTED POLLUTION PARAMETER VALUES
FOR A HYPOTHETICAL FIRM

	WATER LAW Component Values	WATER CHARGE Component Values
Total Discharge (cubic meter/yr)	12,000,000	10,755,000
Specific Amount of Waste Water Per Ton of Product (cubic meter/ton)	190	160
Settleable Substances (ml/l)	.18	.15 (Ref. Val.) .30 (Max. Val.)
COD (O_2-kg/ton of product)	140	112 (Ref. Val.)
COD (O_2 mg/liter)	740	700 (Max. Val.)

The total damage units of pollution, based on the data in Table 1, and the conversion factors in Table 2 are summarized as follows:[22]

Damage Units (DU)

Settleable Solids	1,600
COD	165,600
	167,200

The charge per damage unit is 12 DM in 1981 and rises to 40 DM per damage unit in 1986.[23] Thus the initial bill for this hypothetical firm in 1981 is 2,006,400 DM (about \$722,300—1 DM is about \$.36 in round numbers, as of May 1984).

The ECL contains an economic incentive for polluters to meet the federal minimum standards. Dischargers in compliance with the federal minimum standards will have the charge liability halved by the unit charge. In the event that the Länder have imposed stricter standards than

heavy metals, halogenated hydrocarbons, sulfide, ammonia, fluoride, phosphorus, and total suspended solids. *See* Hornef & Kanowski, *New Federal Waste Water Discharge Standards in Germany,* 1981 EFFLUENT AND WATER TREATMENT JOURNAL 513 (Nov. 1981) [hereinafter cited as Hornef & Kanowski].

22. For settleable solids:
10,755,000 cubic meter/yr × .15 ml/liter (.1 damage unit/cubic meter = 1613 damage units.
For COD:
10,755,000 cubic meter/yr × (700 mg/l) × 2.2 damage units/100 kg = 165,627 damage units.
Help in understanding the computations was obtained during an interview with Dr. W. Dorau, Umweltbundesamt (EPA for FRG), in Berlin (Sept. 30, 1982), and in a letter from Dr. Dorau to the authors (Dec. 15, 1982).
23. ECL, art. 9(4).

TABLE 2

CRITERIA TO BE USED FOR ASSESSMENT OF DAMAGE
OF DISCHARGES, NATIONAL EFFLUENT CHARGE SYSTEM
OF THE FEDERAL REPUBLIC OF GERMANY

Criteria	Unit of Measurement, Quantity/Yr.	Damage Units Per Unit of Measurement
Settleable Substances for Which Organic Content $\geqslant 10\%$[a]	1 cubic meter settled	1.0
Settleable Substance for Which Organic Content $\geqslant 10\%$	1 cubic meter settled	0.1
Oxidizable Substance, as Measured by COD[b]	100 kg	2.2
Mercury & Compounds[c]	100 g Hg	5.0
Cadmium & Compounds[c]	100 g Cd	1.0
Toxicity Toward Fish	1000 cubic meter wastewater	$0.3\ G_F$[d]

[a]Measurement procedure: reduce amount by 0.1 ml/liter waste water beforehand.
[b]Measurement procedure: reduce amount by 16 mg per liter waste water beforehand. Silver sulfate is the catalyst in the dichromate method specified.
[c]Measurement procedure for Hg and Cd: atomic absorption spectrometer.
[d]G_F is the dilution factor, e.g., down or up to nontoxicity. If waste water is discharged in coastal waters, toxicity is not considered for those substances whose content is based on salts which are comparable to those in ocean water.

those set by the federal government, the standard of the Länder must be met in order to qualify for the 50 percent discount.[24]

The normal (expected) value will ordinarily not exceed the federal minimum standard. In the example above, the bill would be halved to 1,003,200 DM (about $361,150) if the firm met the federal minimum. If actual waste discharge is above the federal minimum, using the average (monitor value) of the last five observations, the polluter faces legal consequences under the FWA[25] and loses the 50 percent reduction in the charge obligation.[26]

The ECL and FWA are keyed primarily to expected performance. Seasonal and other variations in discharge, however, are important considerations. Damage generally is a function of actual, not average, discharge. In recognition of this, maximum concentration values and volume are defined and the maximum cannot be more than twice the expected values stipulated by the discharger. The physical basis for the charge

24. ECL, art. 9(5).
25. FWA, art. 12 (withdrawal of license).
26. ECL, art. 9(5).

therefore is at least one-half the maximum value. If the maximum value is exceeded more than once, then the value on which the charge typically is computed (Regelwert) is increased. Thereafter the new basis for the charge increases by the amount the maximum actually is exceeded.[27]

The Hardship Clause

The ECL contains a hardship clause that permits temporary exemptions where imposition of the effluent charge would result in significant, detrimental economic consequences.[28] An exemption may be for the whole charge or a part of it. Eleven industries, several counties, and a number of cities have petitioned the Minister of the Interior to date for exemptions. Although the hardship clause has yet to be used, many who were interviewed thought the clause was important in gaining political acceptance of the legislation.[29]

Magnitude of the Charge and the Minimum Standard

Because few revenues have been collected and too little time has elapsed to make a representative study of actual impacts, it is not possible to report the actual economic effect of the new water quality laws on municipalities and industries. One study, however, appraised the likely impact of an effluent charge on 26 of the major water polluting industries in the country. The cost of the charge and avoidance measures was less than two percent of sales for the most serious polluters except in the pulp, yeast, and leather industrial branches.[30] Sales for the last two industries rank in the lowest twenty percent of the group surveyed.[31] Only in the pulp sector does the charge component loom large. Should the new water quality laws put some of the pulp plants out of business, it would be seen as a modest advancement of the anticipated date of demise of old,

---

27. The new standard value is the old level increased by one-half the amount by which the maximum exceeded the old minimum. *See* ECL, art. 4(4). The basis for computing the charge is reduced if the discharger anticipates that his actual volume and concentration will be below his previously stipulated expected value (or standard value) by at least 25 percent for at least one-fourth of a year. *See* ECL, art. 4(5). In this case, the charge is based either on the actual performance or on the downward revised expected value.

28. ECL, art. 9(6).

29. Interviews with Dr. W. Kitschler, Ministerialrat, Ministry of Interior, in Bonn (Mar. 15, 1983); Dr. Martin Uppenbrink, Director, Department of Environmental Planning, Umweltbandesamt (EPA for FRG) in Berlin (Mar. 17, 1983); Dr. Hans-Peter Luhr, Umweltbundesamt (EPA for FRG), in Berlin (Mar. 17, 1983); Dr. Herbert Massing, *supra* note 6; Dr. Jurgen Salzwedel, *supra* note 9; Dr. E. Rehbinder, Professor, School of Law, J. W. Goethe University in Frankfurt/Main (Mar. 14, 1983); Dr. F. Schröder, Department of Interior in Munich, Bavaria (Mar. 18, 1983).

30. G. RINCKE, UNTERSUCHUNG ÜBER WIRTSCHAFTLICHE AUSWIRKUNGEN DER VORGESEHENEN ABWASSERABGABE AUF ABWASSERINTENSIVE PRODUKTIONSZWEIGE (Study of the Economic Effects of the Expected Effluent Charge on Effluent Intensive Branches of Production) (Feb. 1976).

31. *Id.*

technologically dated plants. Short of a full-scale study of each sector's domestic and international competitive position, a one or two percent increase in the cost of products is not necessarily innocuous. This increase, however, is small compared to variations in advertising budgets and annual changes in interest and wage rates, and probably is small compared to annual changes in raw material costs.

The charge for waste treatment by municipalities depends on the size of the municipality, desired level of waste treatment, and the age of equipment.[32] It is high when new facilities are built and tapers off as the financing obligations are met because the charge varies with financial costs rather than real costs.[33] One study found that sewerage charge rates varied from .60 DM per cubic meter to 3 DM per cubic meter, but the charge in large municipalities did not exceed 1 DM per cubic meter.[34]

The effluent charge component of the new laws increased the unit cost by .03 DM per cubic meter in 1981. The increase will amount to .11 DM per cubic meter in 1986. The effluent charge component in 1986 will amount to about 3.26 DM or $1.30 per year per inhabitant. The cost of adding facilities to meet the minimum standards expressed on a volume basis was estimated to be about .33 DM per cubic meter for the municipalities surveyed or perhaps 10 DM ($4.00) per year if per capita annual consumption is 30 cubic meters. Adding together the cost of the charge (about $1.30) and the necessary new facilities (about $4.00) the estimated total cost is under about $6 per year per inhabitant to meet the requirements of the new water quality laws.

TASK FORCE GROUPS TO ESTABLISH
MINIMUM REQUIREMENTS

The Federal Ministry of Interior initially established 60 task forces, one for each major polluting activity. The Minister of the Interior appointed the task forces and the members were drawn from the federal and Länder governments and representatives from the relevant industries. Technical expertise was brought in from universities, technical institutes, and consulting firms. The purpose of the task force was to establish minimum standards compatible with generally accepted standards of tech-

---

32. Interview with H. Massing, *supra* note 6.
33. *See* Ewringmann, Hansmeyer, Hoffmann, & Kibat, *Auswirkungen des Abwasserabgaben-gesetzes auf Industrielle Indirekeinleiter* (Effects of the Effluent Charge Law on Industrial Indirect Dischargers), 2/81 UMWELTBUNDESAMT BERICHTE 14 (Feb. 1981) [hereinafter cited as Ewringmann].
34. *Id*.

TABLE 3

FEDERAL MINIMUM STANDARDS FOR MUNICIPALITIES

Samples According to Load Category of Discharger	Settleable Solids ml/l	Chemical Oxygen Demand (COD) mg/l	Biochemical Oxygen Demand After 5 Days (BPD₅) mg/l
Load Cat. 1: Less than 60 kg per day BOD₅ (Untreated)			
Grab Sample	0.2	—	—
2-Hr. Mixed Sample	—	180	45
24-Hr. Mixed Sample	—	120	30
Load Cat. 2: 60–600 kg per day BOD₅ (Untreated)			
Grab Sample	0.3	—	—
2-Hr. Mixed Sample	—	160	35
24-Hr. Mixed Sample	—	110	25
Load Cat. 3: More than 600 kg per day BOD₅ (Untreated)			
Grab Sample	0.3	—	—
2-Hr. Mixed Sample	—	140	30
24-Hr. Mixed Sample	—	100	20

nology.[35] Volumes and concentrations regularly issued with new effluent discharge permits and standards, acceptable to a majority of experts in the field, describe the minimum standard level desired. Table 3 illustrates the standards for municipalities of three different sizes.[36]

The idea of a task force to establish minimum standards and the com-

35. INSTITUTE FÜR WASSER-BODEN-UND LUFTHYGIENE DES BUNDESGESUN-DHEITSAMTES, HINWEISE ZUR ERARBEITUNG DER MINDESTANFORDERUNGEN NACH ARTIKEL 7aWHG DURCH DIE ARBEITGRUPPEN FÜR EINZELNE INDUSTRIEBEREICHE (Berlin, Sept. 9, 1977) (Suggestions for the Establishment of Minimum Standards Under the FWA, art. 7(a) Through the Task Forces for Specific Industries).

36. The minimum standards for municipalities are found in the relevant task force report, SCHMUTZWASSER VWV, ERSTE ALLGEMEINE VERWALTUNGSVORSCHRIFT ÜBER MIN-DESTANFORDERUNGEN AN DAS EINLEITEN VON SCHMUTZWASSER AUS GEMEINDEN IN GEWÄSSER-1 (January 24, 1979) (First Comprehensive Administrative Regulations on the Flow of Polluted Waters of Our Municipalities in Flood Area-1). It was reported that these standards are equivalent to 93 percent removal of BOD₅ for small communities and 94.5 percent for large cities. Interview with F. Schafhausen, Umweltbundesamt (EPA for FRG), in Berlin (Nov. 25, 1982).

position of that task force form the crucial ingredients of the new laws in the FRG and in the laws' implementation. The Länder, by voting for the new FWA, gave up their control to set minimum water quality standards before they knew what the new minimum effluent standards were. Basically, they were being asked to give up an unspecified amount of power.

An effective safeguard against too much loss of control is to provide a role for the Länder and the polluters in the standard-setting process. The task force provided the institutional vehicle for this protection. However strong the appetite for improving water quality the Ministry of Interior may have had, the appointment process had to recognize the bare fact that each state had to enact implementing legislation and carry out the attendant enforcement responsibilities. Moreover, the Bundesrat must pass the regulations recommended by each task force. The task force created the practical means for postponing the debate over technical minutia which would have mired the legislative process and extended the date of enactment further into the future.

THE POLLUTER-PAYS PRINCIPLE AND EFFLUENT CHARGES

A first step in evaluating the actual effluent charge system is to consider the characteristics of an ideally efficient system. From this comparison, it will be seen that the actual effluent charge system bears little resemblance to an idealized one.[37] This unsurprising finding means, however, that the search for merit and deficiencies must be made in the murky realm of second-best analysis where judgment and partial analysis play a more prominent role than rigorous proofs in a general equilibrium context.

---

37. An effluent charge is a financial obligation that must be borne by some entity discharging waste, treated or untreated, into a natural water course. The entity can be a firm or municipality or even an individual household. The size of the bill for the effluent discharge varies with the amount of pollution produced, at least in principle. the effluent charge is imposed even if the authority does not treat the particular effluent. JOHNSON & BROWN, *supra* note 7, at 14.

Under an effluent charge system the person who benefits from depositing wastes into public waters is charged in proportion to the benefits received. This tends to induce polluters to reduce the amount of wastes discharged, and promotes economic efficiency.

For example, a polluter who is charged $25 for each ton of suspended solids discharged will pay only if that is the cheapest way for the enterprise to dispose of the waste. If $25 reflects the cost of treatment others would willingly spend to remove each ton of suspended solids, the value of resources used by one party to remove waste is just balanced by the value other parties gain by producing the waste. In contrast, if the polluter does not pay an effluent charge for disposing of his waste, there is no reason for him to economize on waste production. The polluter understandably will act as though his cost of discharging waste is zero, but society will bear an expenditure of $25 to treat or endure the waste. This is inefficient, as the additional expenditure of resources to treat waste is not matched by a corresponding positive value derived from discharging the waste.

No significant differences exist between an idealized effluent charge system and an idealized standards system. For each, public managers, blessed with adequate information, calculate just that level of water quality in a river for which the benefits of extra quality are matched by the cost society necessarily must bear to preserve that extra quality.[38] Discovering this magic point (or vector with multiple qualitative characteristics) requires enough knowledge to permit the rule maker to calculate what each polluter would be willing to pay to discharge an extra unit of waste. The water quality manager can achieve the desired outcome either by posting a common effluent charge or by issuing individual (optimum) standards to each "consumer" of water quality. The charges or the standards change through time in keeping with changing circumstances.

Although the idealized system is of little practical interest its attributes have considerable merit. First, under these circumstances each polluter places the same value on an extra unit of pollution. Thus no discharger pays any more than another for an additional unit of effluent discharge. Veiled behind the single characterization, yet nevertheless of crucial significance, is the second attribute. There is no cheaper way to achieve the desired quality level because the least cost technology has been adopted by all. Those who can treat their effluent cheaply will trade this service for a price to others whose cost of treatment is high. Of course, the incentive to discover low cost measures to reduce effluent discharge diminishes as the level of the effluent charge decreases.[39]

The third attribute, which is less important for the present study, is the marginal cost to dischargers which is just matched by the benefits to those from marginally improved water quality. If this condition is not met there is economic waste. A charge or standard set too high results in polluters paying more than the beneficiaries gain from the last bit of water quality achieved. For many reasons, not the least of which is the difficulty of measuring the benefits of water quality improvement, no one seriously

---

Following any absolute interpretation of the beneficiary-pays principle, polluters should pay the full cost of the pollution as measured by opportunity cost. In this instance, opportunity cost means either the cost of restoring the water quality to its desired level or the value (to members of society) given up because water quality is now less than the desired level, whichever is less. The extra costs of treatment borne by downstream users of water (to achieve the previous level of quality) or the value of foregone days of swimming and fishing on a particular stream (as measured pragmatically, perhaps by the extra cost of obtaining the same quality of recreation elsewhere) are illustrative opportunity costs.
JOHNSON & BROWN, *supra* note 7, at 10.

38. A more complete treatment of water quality management systems is available in A. KNEESE & B. BOWER, MANAGING WATER QUALITY: ECONOMICS, TECHNOLOGY, INSTITUTIONS (1968); A. FREEMAN III, R. HASEMAN, & A. KNEESE, THE ECONOMICS OF ENVIRONMENTAL POLICY (1973).

39. Strategies for avoiding effluent charges include better waste treatment technologies, different production techniques, different inputs and an altered output mix or level.

has argued that the federal minimum requirements or effluent charges in the FRG will result in this condition. When the three conditions are met, there are no further gains from trade among polluters, among beneficiaries or between polluters and beneficiaries of clean water.[40]

These three attributes fall within the realm of efficiency. In addition, when an effluent charge is adopted it satisfies the equity criterion known as the polluter-pays principle in the case of water quality. Those who pollute are those who pay. Standards fall short of this equity goal because they permit the free discharge of a given amount of pollution.

The ECL and the FWA at the federal level do not satisfy the efficiency criteria set for the above ideals because each producer of a given product faces the same minimum standard and each must meet the same discharge concentration levels whether the cost of treatment is high or low. Even by paying a charge the uniform minimum standard cannot be avoided. The marginal cost of treatment in one branch of industry is not equal to the marginal cost in another, except fortuitously, because the task force groups were not charged with that responsibility. The next two sections discuss the degree to which a policy of minimum standards leads to resource inefficiency in the municipal and industrial sectors.

UNIFORM STANDARDS ARE COSTLY UNLESS REQUIRED WASTE TREATMENT LEVELS ARE HIGH

Economists have long argued strenuously that uniform standards are inefficient. A uniform standard refers to a policy in which all dischargers of a type, such as municipalities, are required to achieve the same level of purification or waste removal or to adopt the same technology. Whatever its practical or equitable merits, the policy is costly and inefficient whenever individual waste dischargers differ in ways substantially effecting the cost of waste treatment, for example, when there are economies of size in waste treatment costs.

According to the Council of Experts for Environmental Questions, the effluent charge policy is about one-third cheaper than a uniform standards policy.[41] A charge level of 40 DM (1974 prices) would have achieved a 73 percent removal for a cost of 1.2 billion DM per year whereas a uniform standard achieving the same level of purification would have cost just under 1.8 billion DM per year.[42] Inflation and technical progress

---

40. Additionally, the marginal value of a given water quality characteristic is equated across all beneficiaries when the quality characteristic is not a collective good.

41. RAT VON SACHVERSTÄNDIGEN FÜR UMWELTFRAGEN (The Council of Experts for Environmental Questions), DIE ABWASSERABGABF, WASSERGUTWIRTSCHAFTLICHE UND GESAMTÖKONOMISCHE WIRKUNGEN, SONDERGUTACHTEN 70 (1974) (The Effluent Charge: Effects on Water Quality Management and the General Economy).

42. Interview with Professor Dr. G. Rincke, formerly of the Technische Hochschule, in Darmstadt (Sept. 23, 1982).

have occurred since 1974 when these data were assembled. Increasing the DM values by 50 percent or more would produce estimates more appropriate for the present. By 1986, the expected real value of the 40 DM charge will be around 22 DM.[43] It would be over 80 DM per damage unit if the charge was indexed for inflation.

The potential economic advantages of an effluent charge over a uniform standard apply to the industrial sector as well. Using data from one widely quoted study, at a uniform standard of 80 percent removal of chemical oxygen demand, some pollution-intensive industries, such as chemicals, have (marginal) treatment costs more than twice as high as other pollution-intensive industries, such as food processing.[44]

The potential cost savings from eschewing uniform standard policies are greatest when there is a big difference in treatment cost opportunities among polluters. As required levels of treatment or the effluent charge increase, opportunities for substituting low cost for high cost treatment diminish, and the economic advantage of the effluent charge over uniform standards is eroded.[45] A charge high enough to achieve 100 percent removal for all is the same as a uniform standard. At the required levels of purification cited above, the efficiency gains of a charge over a uniform standard are modest.[46]

LOW EFFLUENT CHARGE LEVELS REDUCE BUT DO NOT ELIMINATE INCENTIVES TO ECONOMIZE

A charge of 12 DM in 1981 rising to 40 DM per damage unit in 1986 was, and is, too small to achieve the desired water quality objectives for the country and it cannot be a very great incentive to discover low cost abatement strategies.[47] But there are important exceptions worth citing even if the frequency is unknown.

In response to the new water quality legislation, a giant chemical firm, BASF, has made a serious effort to manage water quality. BASF treats

---

43. Interview with L. Wicke, Scientific Director, Umweltbundesamt (EPA for FRG), in Berlin (Sept. 29, 1982); and interview with Dr. Klaus Zimmermann, International Institute for Environment and Society, in Berlin (Sept. 29, 1982). The 1986 estimate was provided in a letter from Dr. Zimmermann to the authors (Dec. 3, 1982).

44. G. Rincke, *The German Federal Law on Wastewater Charges*, 95–102, PROG. WAT. TECH., 10 (1978).

45. *See* comparison of charges and standards in A. KNEESE & B. BOWER, *supra* note 38, at 131–42.

46. *See* lecture by Dr. Lutz Wicke, *The Experience with the German Effluent Charge System in the Light of Irish Considerations in That Field*, Dublin University (Apr. 15, 1983).

47. This view was held almost universally by those interviewed: *e.g.*, interviews with J. Salzwedel, *supra* note 9, H. Massing, *supra* note 6, M. Uppenbrink, *supra* note 9, M. Faber, Professor, University of Heidelburg, L. Wicke, *supra* note 43. *See also* M. FABER & H. NIEMES, DAS ABWASSERABGABENGESETZ: RICHTUNGSWEISEND FÜR DIE UMWELTPOLITIK, 1982 UMWELT 1 (1982) (The ECL: A New Direction for Environmental Policy).

its own waste as well as the waste of two large municipalities and three smaller ones with populations of over 300,000, and achieves low unit abatement costs by large-scale integrated treatment processes. BASF achieves purification levels greater than what is required presumably because it is cheaper than paying the effluent charge. There are numerous other large industries with comparable performance records.

The second feature of the BASF system is of substantial economic interest. BASF has practiced the polluter-pays principle within its plant since 1975.[48] Individual branches basically face shadow or implicit prices for the volume and concentration of COD.[49] The response to the introduction of an internal liability system has been a 20 percent decrease in discharge. Rather than mandate physical decreases the intra-firm charge elicited a "voluntary" decrease in effluent discharge achieved through process change, recycling of solvents, improved pretreatment facilities, and replacement of old facilities.[50] Even if the charge is modest, it induces cost savings.

The charge also provides an incentive for municipalities and industries to operate treatment plants and operate them efficiently. Inefficient treatment is incompatible with minimum requirements and inefficient operation will prevent qualification for the 50 percent discount on the effluent charge. The charge, by encouraging increased operating and maintenance expenditures, partially offsets the efficiency distortion created by existing subsidy programs where only capital costs are subsidized.[51]

One consequence of the ECL (and the FWA) is the remarkable level of investment in waste treatment plants and equipment during the announcement phase, generally, 1974–1979. One study reported the industrial responses to the new water laws[52] while another investigated the response to the new laws by municipalities.[53] Slightly more than one-third of the towns or cities interviewed cited the effluent charge law as the primary reason for undertaking more extensive waste treatment mea-

---

48. In a letter from Blair Bower to the authors (Apr. 9, 1983), Bower stated that Dow Chemical Co. began an intra-firm effluent charge policy in the U.S. in 1958.

49. BASF calculates the effluent charge bill for each branch of the company. The bill is based on an accounting price per unit of effluent and the amount of effluent for that branch.

50. Letter from W. Haltrich Prokurist, BASF, Ludwigshafen, to the authors (Dec. 7, 1982).

51. A. GIWER, WAS DARF AUS DER ABGABE FINANZIERT WERDEN? (What May Be Financed with the Effluent Charges?) (1980).

52. R-U. Sprenger and M. Pupeter, *Evaluierung von gesetzlichen Massnahmen mit Auswirkungen im Unternehmensbereich* (*Evaluation of Legal Measures with Consequences in the Business Sector*), IFO-Institut für Wirtschaftsforschung (Munich, May 1980). *See also* Ewringmann, *supra* note 33, and H. Hoffman & D. Ewringmann, *Auswirkungen des Abwasserabgabengesetzes auf Investitionsplanung und Abwicklung in Unternehmen, Gemeinden und Abwasserverbänden* (1977) (Effects of the Effluent Charge Law on Investment Planning and Arrangements in Firms, Municipalities and Effluent Associations, Study prepared for the Umweltbundesamt (EPA for FRG)).

53. *See supra*, note 33.

sures, while an additional 14 percent declared that the minimum requirements alone were responsible for increased expenditures. Another 20
percent stated that they had accelerated their construction plans due to
the effluent charge law. When the planned construction phase of their
sample municipalities is completed, 80 percent of the inhabitants will
receive full secondary treatment.[54] This is compared to a national goal
of 90 percent in 1985 established in 1971 and estimated levels of under
40 percent and 53 percent in 1963 and in 1978, respectively.[55] As a result
of dedicated efforts to manage waste discharge more efficiently in 1981,
more than one-half the waste dischargers met the minimum requirements
and qualified for the halving of the charge in general and, in Baden-
Wuerttemburg, 90 percent qualified for the charge reduction.[56]

The new laws necessarily improved ambient water quality. No quantitative estimate of the change in water quality has been made but there
has been an improvement in the biological quality, judging from a comparison of water quality maps between 1975–1980.[57] Other actual or likely
consequences, some of them good and others not beneficial, are discussed
below. The subsequent evaluation is largely qualitative because the laws
are so new. There has been too little time to have practical experience
with administering or enforcing the law, or spending the revenues collected.

EFFLUENT CHARGE REVENUES: A POTENTIAL SUBSTITUTE
SOURCE OF SUBSIDIES

The effluent charge amassed revenue amounting to about 350 million
DM in 1981.[58] The Länder use the revenues for water quality management
administration expenses associated with the ECL,[59] and for projects or
purposes which maintain or improve water quality, including industrial
production processes which are pollution-saving.[60] The fraction devoted

---

54. *Id.*

55. *See* PPHE, *supra* note 4, and P. Menke-Glueckert, *supra* note 18.

56. Estimates obtained during interviews with F. A. Schendel, Bayer A. G., in Leverkusen (Mar.
14, 1983); M. Schell, Ministerium Für Ernährung, Landwirtschaft und Forsten, Schleswig-Holstein,
in Kiel (Mar. 16, 1983); and W. Baumgärtner, Ministerium für Ernährung, Landwirtschaft und
Umwelt, Baden-Württemberg, in Stuttgart (Mar. 18, 1983).

57. *See* GOVERNMENT OF THE FEDERAL REPUBLIC OF GERMANY, GEWÄSSERGÜ-
TEKARTE DER BUNDESREPUBLIK DEUTSCHLAND, AUSGABEN 1976 UND 1980 (Water
Quality Map of the Federal Republic of Germany, 1976 and 1980 Editions).

58. Actual collections not completed as of the writing of this paper. Estimate provided by F. A.
Schendel in interview, *supra* note 56. Excerpts of document provided by one interviewee contained
an estimate of 650 million DM for the two years 1981 and 1982.

59. The revenues can be used only for that portion of the administrative expenditure associated
with enforcement of the Act and for the Länder's own supplementary regulations. ECL, Art. 13.

60. The application to support investment in special production processes is described in F.
Boelam, INTERIM REPORT ON THE DRAFT OF AN EFFLUENT CHARGE LAW, 1976 WAS-
TEWATER CORRESPONDENCE 23 (June 1976).

to administration varies among the Länder. One Länder used about 50 percent in the first year but this is expected to fall to 20 to 25 percent in future years.[61] Effluent charge revenues are an obvious and important source of subsidies for waste-treatment investments. This inevitably raises concern that the new source of subsidy may be substituted for the old source, general fund moneys.[62] At the present time, Länder governments offer investment subsidies in the neighborhood of 40 percent or more.[63]

It takes little political acumen to imagine that the Länder government will decrease subsidies for waste treatment from the general fund once effluent charge revenues roll in. This would be a particularly attractive substitution in times of fiscal conservancy. The polluter-pays principle can be invoked in defense of the reallocation. It will be hard to argue against the proposition that the dischargers who benefit from waste treatment facilities (which meet the minimum standards or reduce the bill for discharge) ought to pay for the facilities. Those who approve of shifting fiscal responsibilities from higher to lower echelons in the political hierarchy can see the merit of effluent charge as a new source of subsidies.

MORE POLICY INSTRUMENTS ARE BETTER THAN LESS

Some have argued that an effluent charge is a more flexible policy tool because it can be changed more readily than an effluent standard.[64] Others have argued just the opposite. For example, "one serious practical liability" of the effluent charge is the inability to change it as quickly as may be desired.[65] The truth probably rests between the two extremes.

The benefit of having both a system of standards and charges is that the water quality regulations can each be adjusted through time to produce a result more harmonious with the desired water quality objectives. The objectives will change through time as a result of changing environmental and economic conditions.[66]

---

61. Data obtained during interviews with F. Schendel, *supra* note 56, C. A. Conrad, Dr. Jur, in Kiel, March 1983, and Baumgärtner, *supra* note 56.
62. Interview with H. Massing, Deputy of President, Head of Department of Water Resources in Düsseldorf (Sept. 21, 1983).
63. JOHNSON & BROWN, *supra* note 7, at 126–27; BOWER, *supra* note 21, at 237–40, 270–71; and interview with P. Michaelis, Justitiar, Ruhrverband, in Essen (Sept. 22, 1982); interview with F. Schröder, Department of Interior, in Munich, Bavaria (Sept. 28, 1982). In extraordinary circumstances, the subsidy for waste treatment plants has been as high as 80 percent. Letter from P. Michaelis, Justitiar, Ruhrverband, in Essen (Dec. 7, 1982).
64. Kneese and Bower, *supra* note 38.
65. W. BAUMOL & W. OATES, THE THEORY OF ENVIRONMENTAL POLICY: THE EXTERNALITIES, PUBLIC OUTLAYS & THE QUALITY OF LIFE 154 (1975). To support their view they refer the reader to the history of tax changes.
66. The added flexibility provided by multiple regulatory instruments is a further point in favor of adopting a charge policy in the United States.

A combined charge and standards system is advantageous in a decentralized decisionmaking framework where the control from above is circumscribed. For example, Länder can regulate the aggregate discharge level of a municipality but they are powerless to establish charges for the firms and households in the municipalities. The Länder may charge the municipalities for their waste discharge, but they cannot force municipalities to adopt pricing policies for water quantity or quality which make the indirect dischargers, the customers of the municipalities, see the marginal economic consequences of their waste discharge decisions. Introducing an effluent charge typically increases the costs of a continued average or nonmarginal pricing policy for all customers. Customers who are not the cause of the increased price, because they do not pollute or their pollution is more benign, now have an economic incentive to pressure the municipality to adopt a more rational charge policy.

Evidence of the inducement to change customer pricing policies created by the effluent charge is provided in a survey of 52 municipalities.[67] Nearly one-fourth of the municipalities had decided to change the structure of their water and sewerage fees in response to the effluent charge prior to the policy actually taking effect. More can be expected to change their fee structure with time.[68]

There is a further advantage of a combined charge and standard regime in a decentralized system. It is difficult for an authority like the state to use an effluent charge alone to achieve a desired ambient water quality when the state has no control over the pricing policy of municipalities or other pubic agencies with their own pricing policies. One reasonable strategy is for the state to set standards to achieve desired water quality goals and then introduce a charge system which satisfies non-water quality efficiency criteria such as equity considerations.

EFFLUENT CHARGES CHANGE THE COSTS OF ENFORCEMENT

There is no reason why enforcement costs should be different with an ideal effluent charge compared to an ideal standard.[69]

In the absence of an effluent charge, the reward for violating a standard is the expected gross profit of the actions less the expected costs associated

---

67. *See Zur Bedeutung von Awasserabgabe und Entwässerungsgebühren für die Effizienz der rommunalen Entwässerung* 39 FINANZARCHIV 101 (1981) (Concerning the Significance of the Effluent Charge and Sewerage Fees for the Efficiency of Municipal Sewerage); and H. NIEMES, UMWELT ALS SCHADSTOFFEMPFÄNGER (1981) (Environment as Receiver of Waste Materials).

68. *Id.*

69. We can only speculate on the truth of this assertion at present, since it is too soon to obtain qualitative or quantitative evidence on the new German program.

with being caught.[70] If it is reasonable to assume that in a combined charge and standard system those caught accidentally or intentionally exceeding the legal standard would have to pay fines plus charges which vary with the unreported quantities discharged, then the charge system reduces the expected net benefit of violating the standard. Thus, a given level of compliance can be achieved at a lower enforcement cost in the presence of a charge. Alternatively, a higher level of compliance can be achieved (with a standard and charge) than was obtained at the old cost of enforcement, when there only was a standard. In short, noncompliance should decrease when it is less rewarding so enforcement can be cut back accordingly.

Enforcement costs also will be lower if there is some trade-off between "justice" and economic sanctions in the world of practical affairs. Polluters might argue successfully that because they are paying, the frequency of punitive proceedings or level of punishment should be mitigated. This argument is unavailable in a pure standards system because no effluent charges are levied.

The arguments for decreased enforcement costs focus on the (net) benefits of evasion to the evader. The outcome, when viewed from the supply side, is different. Prior to an effluent charge the reward to the Länder water quality management authority for enforcing water quality standards is improved water quality. Because effluent charge revenues cover the Länder's costs of administering the effluent charge law, the water quality management agencies in the Länder will be encouraged to increase enforcement activities. The rewards are improved water quality and a larger agency, with the expansion automatically financed by effluent charge revenues. The net result of these qualitative arguments is: (1) there will be a greater resemblance of actual discharge with legally mandated standards (in this sense, one can say that the quality of water law has improved); (2) the reduction in the discrepancy between the actual result and the certain legal requirement, in effect, reduces the uncertainty about enforcement to polluters; (3) the cost of a given level of compliance has decreased. It is not possible, however, to conclude that total enforcement costs will increase or decrease, unless the agency aggrandizement effect can be assumed to outweigh the diminished value of compliance averting behavior for firms and municipalities.

The new legal and economic instruments are more precisely stated than before. The Länder have had to develop a more precise measurement and

---

70. For an interesting discussion of compliance averting behavior, *see* paper presented by D. Lee, *Protecting Our Environment: Some Public Choice Considerations*, Conference on Market Perspectives in Natural Resources Economics, Political Economy Research Center, Montana State University (Bozeman, June 10–14, 1982). Also of interest is Viscusi & Zeckhauser, *Optimal Standards with Incomplete Enforcement*, 27 PUBLIC POLICY 437–56 (1979).

monitoring system and to sharpen their enforcement practices. The increased quality of data removes ambiguity and reduces the costs of enforcement.[71]

THE COSTS OF INSTITUTIONAL CHANGE

The idea of a nationwide effluent charge was a dramatic new idea. Integrating it with standards and permits made it a complex undertaking. It required education of legislators, Länder officials, and municipal and industrial administrators. As a result of the long gestation period, all parties had ample opportunity to present their interests and it therefore can be argued that the resulting policies accurately reflect the relative weights of all interested parties to the decision.[72]

Changes in management and administration at the local level in response to the new policies were costly. On the other hand, the difficulties of implementing a charges system were greatly overestimated. One of three Länder to strongly oppose the effluent charge law was Schleswig-Holstein. As a predominantly rural region, their concern focused on the cost and the ominous task of acquiring sufficient technical capability for administering the new legislation. After a few years of experience, several experts with substantial responsibility for administering the ECL have found it to be a far easier task than they had imagined, to their great surprise. Simple practical ways have been devised to implement the "economic point of view." Illustratively, the need for increased analysis of samples has been handled, in part, by contracting with private labs. These former foes are now staunch supporters of the effluent charge system.[73]

All practical and effective water quality management programs require the specification of variables, parameter, and threshold values. An effluent

---

71. Interviews with Schröder, *supra* note 63, and Schell, *supra* note 56.

72. The idea of adequate representation by all parties in the political process contrasts with the manner in which principal water quality legislation in the United States allegedly occurred. In a remarkable and little known piece of public policy analysis, Marc Roberts explains persuasively how environmentalists played a disproportionate role in the passage of the Clean Water Act of 1972. Roberts, *The Political Economy of the Clean Water Act of 1972: Why No One Listened to the Economists,* UTILIZATION OF SOCIAL SCIENCE IN POLICY MAKING IN THE UNITED STATES (OECD 1974). The U.S. Senate version, calling for a standard of zero discharge by 1983, passed by a vote of 80-0. Only in the final version was the standard compromised to the best available technology standard. Roberts argues that the technical complexity of the issue gave great authority to the subcommittees of the Public Works Committee, which handled the water quality legislation. The ranking members of the committee and the technical staff had a special position and played a substantial role in the final outcome. According to Roberts, a strong environmental influence on the staff was evident: one staff member was married to an environmental lobbyist, some staff members were persuaded that any discharge was hazardous, and the environmental lobby groups were well organized and effective.

73. Interviews with Conrad, *supra* note 61; O. Behrend, Ltd. Ministerialrat, in Kiel (Mar. 16, 1983); and T-W. Krahl, Deutscher Städtetag, Landesverband Schleswig-Holstein, in Kiel (Mar. 16, 1983).

charge system has the greatest chance of meeting the criterion of political feasibility if it is kept simple—few pollutants, strictly limited number of threshold values, uncomplicated rate schedules, etc. The bane of naive marginal efficiency is simplicity. Simplifying eventually involves making charges and standards and other debatable components of policy more uniform by aggregating and averaging. It saves transactions and political costs, ultimately at the expense of efficiency.[74]

One benefit ascribed to a policy, which applies to all, viz. meeting minimum requirements, is that it greatly reduces the incentive of any one firm or industry to curry special favor.[75] To do so singles one out for public scrutiny much more than if there is a distribution of policies subject to interpretation, adjustment, or reclassification, where the administrator has broad discretion in enforcement. If bargaining for a narrow interest is discouraged by announcements that policies will be uniform, then it can be argued that policy decisions will be made more quickly. The duration of the uncertainty about the date and content of new legislation also is reduced, thus creating a further source of benefit. If these arguments have merit then the resulting benefits must be weighed against the costs of uniformity. Only path-breaking empirical research will tell us when the net benefits of simplifying rules actually are positive.

IMPLEMENTATION PITFALLS

One of the largest stumbling blocks remaining in the way of practical implementation is devising an acceptable and an effective policy for charging indirect discharges. About 90 percent of all firms in the FRG discharge their effluent into the sewerage systems of municipalities and are not directly liable for the effluent charge.[76] Three elements of the indirect discharger problem warrant discussion. First, how are those dischargers whose waste enters municipal systems to be charged; second, do their costs resemble the costs of direct dischargers; and third, can there be relief for a firm whose economic viability is threatened by charges a municipality levies for that firm's discharge?

---

74. For example: Charges under the ECL are based on measurments of cadmium and mercury discharges. Other heavy metals, such as lead, are not measured or used as the basis for charges. Interview with Dr. H. Luhr, *supra* note 6, in Berlin (Mar. 17, 1983). Dr. Luhr reported that a "somewhat proportional relationship" seems to exist between the quantities of the two measured heavy metals and others found in industrial effluents, but it is by no means exact. A company whose effluent contains a disproportionately high quantity of lead in relation to cadmium and mercury will have no incentive to remove the lead because the charge is unrelated to it. This reduces the efficiency of the charge system.

75. Interview with J. Salzwedel, Director, Institut für das Recht der Wasserwirtschaft an der Universität Bonn, in Bonn (Sept. 22, 1982).

76. Ewringmann, *supra* note 33.

An important criterion for a municipal charge system is administrative simplicity. This feature is sacrificed to the degree that a second desirable characteristic, the polluter-pays principle, is achieved. Ideally, each firm faces a (marginal) charge that reflects the (marginal) cost of discharge imposed on the municipality. For example, firms with high concentrations of cadmium, mercury, COD, settleable solids or toxicity would pay more than those with lower concentrations in their expected waste. In this manner, the polluter-pays principle is passed back to the entity making the marginal pollution decision.

In practice, municipalities in the FRG have charge systems so rudimentary that the cost of waste treatment is embedded in the charge for fresh water withdrawals. Clearly their charge policy is used primarily as a financial instrument by the municipalities and not as an allocative device. Thus, finding a solution to the practical pricing policy problem has wide ramifications in terms of efficiency.

When all firms are homogeneous in their residuals discharge, municipalities can continue to practice undifferentiated charge systems. When individual discharge varies greatly in volume and concentrations, a pricing policy which does not distinguish differences in volume, concentration, or pollutants will greatly favor the big pollution-intensive industries and discriminate against the mild polluters. A uniform pricing policy acts as a wet blanket on incentives to reduce discharge, which would be undertaken by an estimated 80 percent of the firms for a cost lower than the municipalities to which they are hooked up.[77]

Fairness between the direct and indirect discharger with regard to the federal water quality laws is a consideration which should be raised. Inadequate data, however, preclude reaching definitive conclusions. Even qualitative answers are not possible because of the presence of two major counterforces.

Subsidies to municipalities and non-fee revenues such as ad valorem taxes tend to decrease the cost of effluent treatment to indirect dischargers.[78] The advantage will decrease to the extent that subsidies from the effluent charge revenues will be made available to firms. On the other hand, indirect dischargers pay for treatment of storm water runoff which is not of their making but can amount to as much as 50 percent of the total cost in some communities.

---

77. G. Rincke, *Die Abwasserabgabe in der kommunalen Gebuehrensatzung*, Berichte der abwassertechnischen Vereinigung E.V., Nr. 32, ATV-Jahreshauptversammlung, Mainz (1980) (The Effluent Charge in the Municipal Fee-Regulation).

78. Less than one-half of the communities covered their costs, net of subsidies, with fees in the 1976–1978 period. *See* Ewringmann, *supra* note 33, at 14.

INCORPORATING EFFLUENT CHARGES INTO
U.S. WATER POLLUTION CONTROL LAW[79]

The existing U.S. system of water pollution control is dominated by a legalistic approach in two ways. First, it emphasizes as its goal the total ban of discharges of wastes into public waters instead of applying cost-benefit principles which would proscribe only those discharges of waste which are not cost-justified for a particular body of water, considering the alternative uses for those waters and their assimilative capacity.[80] Second, the U.S. system relies heavily on the threat of punishment, i.e., fines and/or imprisonment, rather than on economic incentives to induce industries, municipalities, and other waste dischargers to reduce the pollutants they discharge into public waters.

The first of the above two concepts, the ban-the-discharge approach, was explicitly incorporated into the Federal Water Pollution Control Act Amendments (FWPCA) of 1972.[81] This concept was subjected to heavy criticism by the National Water Commission,[82] the National Commission on Water Quality,[83] and independent economists, who considered the concept to be too costly.

It was not surprising that the 1977 Amendments to the FWPCA altered the emphasis of the federal program in the direction of the receiving water standards approach and away from the no-waste-discharge principle. This change is important to our consideration of effluent charges as a supplement to the existing pollution control system. While effluent charges are consistent with a receiving water standards approach, they tend to conflict with the ban-the-pollution approach. Effluent charges are based on the assumption that some wastes will continue to be deposited into public waters and this use is not, *per se,* legally wrong or inherently evil. An effluent charge system is a legitimate means of allocating the use-opportunities for this resource among competitors. In addition, this system will create a pool of revenues that can be used for the construction of treatment facilities, research, and pollution control administration.

There are three major options for enacting an effluent charge law. The

---

79. F. ANDERSON, A. KNEESE, P. REED, R. STEVENSON, & S. TAYLOR, ENVIRONMENTAL IMPROVEMENT THROUGH ECONOMIC INCENTIVES, (1977) [hereinafter cited as F. ANDERSON] contains an excellent analysis of the use of money charges to discourage environmental harm and the practical problems posed in the United States by different implementation strategies. We refer the reader to this work for a fuller analysis of some of the problems discussed here.

80. "[I]t is the national goal that the discharge of pollutants into the . . . waters [of the nation] be eliminated by 1985[.]" 33 U.S.C. § 1251(a)(1) (1982).

81. Pub. L. No. 92-500, 86 Stat. 816 (codified as amended at 33 U.S.C. § 1251-1376 (1982)).

82. NATIONAL WATER COMMISSION, FINAL REPORT, WATER POLICIES FOR THE FUTURE 69, 74-76 (1976).

83. NATIONAL COMMISSION ON WATER QUALITY, REPORT TO CONGRESS 5 (1976).

advantages and disadvantages of each are noted as follows: (1) the federal government could enact an effluent charge law for the entire nation and could collect the charges and disburse them as it saw fit. Under this plan, Congress might carry forward the same federal-state relationship that is used in administering the Clean Water Act.[84] Thus a state would be permitted to implement the charge system under continuing federal supervision, so long as the state met federal standards.[85] Alternatively,if a state decided not to implement the federal charges program, EPA would itself carry out the implementation in that state; (2) the states could enact effluent charge systems of their own choosing, so long as their choices were not preempted by the Clean Water Act; (3) the federal government could enact a law that would set minimum requirements for any state effluent charge law. States could then enact such laws as they saw fit, so long as those laws met federal standards. If a state chose not to have an effluent charge law, then none would exist in that state, e.g., EPA would not implement any federal charge system in that state.

CONGRESS' POWER UNDER THE FEDERAL CONSTITUTION TO ENACT AN EFFLUENT CHARGE LAW

Congress doubtless has the constitutional power to enact an effluent charge law applicable throughout the United States if it chooses to do so.[86]

Until the mid 1960s, water pollution control had always been dominated by state regulation. By then, however, it was apparent that state regulation was failing to achieve the kind of water pollution control desired by the public. At first, federal intervention was gradual. In the Water Quality Act of 1965, Congress sought simply to oversee state regulation and made no attempt to regulate waste discharges directly. With the rediscovery of the Rivers and Harbors Act of 1899, the federal government undertook, in 1969, through the Corps of Engineers permit system to regulate directly the discharge of wastes into public waters by industries.[87] In 1972, the federal government changed the rules of the game entirely and took over the field of water pollution control from the states, essentially reversing

---

84. *See* 33 U.S.C. § 1342(b) (1982).

85. *Id.*

86. *See* J. NOWAK, R. ROTUNDA, & J. YOUNG, CONSTITUTIONAL LAW 150–56 (1977) [hereinafter cited as NOWAK]; and F. ANDERSON, *supra* note 79. *See also, e.g.,* Wickard v. Filburn, 317 U.S. 111 (1942).

87. *See* Comment, *Discharging New Wine into Old Wineskins: The Metamorphosis of the Rivers and Harbors Act of 1899,* 33 U. PITT. L. REV. 483 (1972); Barry, *The Evolution of the Enforcement Provisions of the Federal Water Pollution Control Act: A Study of the Difficulty in Developing Effective Legislation,* 68 MICH. L. REV. 1103 (1970).

the federal/state roles and thereafter allowing state regulation only under strict federal supervision.[88]

The courts have supported this expansion of the federal government's role in the environmental law field as well as in other areas of social and economic regulation,[89] and have done so via an increasingly broad interpretation of Article I, Sec. 8(3) of the federal Constitution, the so-called "commerce clause."[90] This clause says that Congress shall have the power "to regulate commerce with foreign nations, and among the several States. . . ."

While early cases suggested that Congress' legislative power under this clause might be limited to navigable waters because that is where commerce occurs, in recent years, the court has made it clear Congressional power is much broader.[91] In 1942, the Court said that Congress' power extends to any activity that "affects" interstate commerce. In *Wickard v. Filburn*[92] the Court held that Congress constitutionally could enact a law regulating the acreage of wheat a farmer could plant even though the wheat was destined solely for use on his own farm. The cumulative effect of private wheat growing by many farmers would affect the price of the grain and would "affect" interstate commerce.

Subsequent cases have established the applicability of this principle to the environmental law field. In *United States v. Ashland Oil & Transportation Co.*[93] the court held that Congress had the constitutional authority to enact the Federal Water Pollution Control Act Amendments of 1972, by which the federal government took over much of the direct regulation of water pollution. On the impact of water pollution on interstate commerce, the court said:

> Obviously water pollution is a health threat to the water supply of the nation. It endangers our agriculture by rendering water unfit for irrigation. It can end the public use and enjoyment of our magnificent rivers and lakes for fishing, for boating, and for swimming. These health and welfare concerns are, of course, proper subjects for Congressional attention because of their many impacts upon interstate commerce generally. But water pollution is also a direct threat to navigation—the first interstate commerce system in this country's history and still a very important one.[94]

88. Pub. L. 92-500, 86 Stat. 880 (codified as amended at 33 U.S.C. § 1342(b) (1982)).
89. *See* L. TRIBE, AMERICAN CONSTITUTIONAL LAW 236–38 (1978).
90. *See, e.g.,* Wickard v. Filburn, 317 U.S. 111 (1942); Katzenbach v. McClung, 379 U.S. 294 (1964); Perez v. United States, 402 U.S. 146 (1971).
91. *See* United States v. Appalachian Electric Power Co., 311 U.S. 377 (1940); Oklahoma v. Guy F. Atkinson Co., 313 U.S. 508 (1941); Arizona v. California, 283 U.S. 423 (1931).
92. 317 U.S. 111 (1942).
93. 504 F.2d 1317 (6th Cir. 1974).
94. *Id.*

Other potential impacts of water pollution that "affect" interstate commerce easily can be identified, any one of which would justify Congressional legislation on this subject. It seems clear therefore that Congress has the constitutional power under the commerce clause to enact an effluent charge law to control water pollution, if it chooses to do so.

CONSTRAINTS ON CONGRESS' POWER TO LEGISLATE

The Bill of Rights of the United States Constitution contains two concepts that might be the basis for challenges to a federal effluent charge law: the due process and equal protection concepts.[95] Assuming that law is carefully written and reasonably related to the pollution goals to be achieved, however, challenges under either of these two concepts should fail.

The due process clause has two separate aspects, one called "substantive" due process and the other "procedural" due process. Under the substantive due process requirement, private property cannot be taken by the government without payment of just compensation. The courts, however, have held that waste dischargers, even those who have been depositing wastes into public waters for many years, have no vested property right to continue doing so, and cannot demand compensation when their activities are regulated or prohibited.[96]

Procedural due process requires that fair procedures be followed in applying any regulatory scheme, such as notice of hearings and orders, and the opportunity to present one's own arguments before a proper forum. Defects in procedural due process can be corrected ordinarily by modifying the process to one that meets judicially approved tests of fairness.

The principal tenet of the equal protection doctrine is that persons

---

95. "No person shall be . . . deprived of life, liberty, or property, without due process of law. . . ." U.S. CONST. amend. V. In Bolling v. Sharpe, 347 U.S. 497 (1954), the Supreme Court ruled that the fifth amendment's due process clause includes the concept of equal protection. *See* Buckley v. Valeo, 424 U.S. 1 (1976); Weinberger v. Weisenfield, 420 U.S. 636, 638 n. 2 (1975); Schlesinger v. Ballard, 419 U.S. 498 (1975); Frontiero v. Richardson, 411 U.S. 677 (1973).

96. *See* W. RODGERS, ENVIRONMENTAL LAW 357–58 (1977), and the numerous cases upholding sec. 13 of the Rivers & Harbors Act of 1899, 33 U.S.C. § 407 (1982). *E.g.,* United States v. Standard Oil Co., 384 U.S. 224 (1966). *See also* the cases upholding the Federal Water Pollution Control Act Amendments of 1972. *E.g.,* City of Milwaukee v. Illinois, 451 U.S. 304 (1981); United States v. Ashland Oil & Transportation Co., 504 F.2d 1317 (6th Cir. 1974); Leslie Salt Co. v. Froehlke, 403 F. Supp. 1292 (N.D. Ca.), *aff'd in part, rev'd in part, modified in part, and remanded* 578 F.2d 742 (9th Cir. 1978).

> Today the due process and equal protection guarantees are not significant restraints on the government's ability to act in matters of economics or social welfare. . . . as long as there is any conceivable basis for finding . . . a rational relationship [to any legitimate end of government] the law will be upheld. Only when a law is a totally arbitrary deprivation of liberty will it violate the substantive due process guarantee.

NOWAK, *supra* note 86, at 409–10.

similarly situated must be treated alike under the law. Conceivably, an industrial waste discharger might complain that his charges were higher than another who was similarly situated. The courts, however, have almost uniformly rejected those claims, where the classification is "rationally" based,[97] i.e., based "upon a state of facts that reasonably can be conceived to constitute a distinction, or difference in state policy. . . ."[98] The Supreme Court recognizes a strong presumption of constitutionality under the rational basis test.[99] If the differential treatment were to be based on race, gender, or some other suspect classification then the standard of judicial review would be "strict scrutiny," which is usually "fatal" to the legislation.[100] But carefully drafted effluent charge legislation would classify persons and firms on the basis of the amount and quality of effluent they discharged into public waters, not on the basis of any suspect classification, and thus should satisfy the constitutional equal protection requirement.

STATE AUTHORITY TO ENACT EFFLUENT CHARGE LAWS

The states also have the legal power to enact effluent charge laws if they choose to do so.[101] As noted above, states traditionally have enacted most of the legislation in the health and environmental fields. This power generally is referred to as the state's "police power" and is the basis for

---

97.   In applying the rationality requirement, the Court has ordinarily been willing to uphold any classification based "upon a state of facts that reasonably can be conceived to constitute a distinction, or difference in state policy. . . ."
Allied Stores of Ohio v. Bowers, 358 U.S. 522, 530 (1959).
    This remarkable deference to state objectives has operated in the sphere of economic regulation quite apart from whether the conceivable "state of facts" (1) actually exists, (2) would convincingly justify the classification if it did exist, or (3) has ever been urged in the classification's defense by those who either promulgated it or have argued in its support. Often only the Court's imagination has limited the allowable purposes ascribed to government.
TRIBE, *supra* note 89, at 956 (1978).
    The first standard of review is the rational relationship test which we saw developed for use in both equal protection and substantive due process issues in the post 1937 decisions of the Court. The Court will not grant any significant review of legislative decisions to classify persons in terms of general economic legislation.
NOWAK, *supra* note 86, at 524.
    98. Allied Stores of Ohio v. Bowers, 358 U.S. 522, 530 (1959).
    99. McGowan v. Maryland, 366 U.S. 420 (1961).
    100. Gunther, *The Supreme Court, 1971 Term—Foreword: In Search of Evolving Doctrine on a Changing Court: A Model for a Newer Equal Protection,* 86 HARV. L. REV. 1 (1972).
    101. Such legislation would be upheld as an exercise of the state's "police powers," which encompasses the inherent right of state and local governments to enact legislation protecting the health, safety, morals or general welfare of the people within their jurisdiction. *See* Charles River Bridge v. Warren Bridge Co., 36 U.S. (11 Pet.) 420 (1847). "Police power' is the name given to one agent of a state's sovereign power of government. The principal limitation on this power, relevant here, arises from the due process and equal protection guarantees. *See supra* text accompanying notes 95–100.

regulations protecting health, morals, aesthetic appearance, environmental quality, recreation, fish and wildlife, and economic welfare.

The more serious challenge to state effluent charge laws arises from two other sources: (1) the law might violate the federal Constitution's "dormant" commerce clause requirement that guarantees free interstate commerce and (2) the law might be preempted by existing federal statutes in the field of pollution control. Under the first challenge the courts have held that legislation may be suspect if it places a greater burden on out-of-state enterprises than on those operating within the state.[102] The typical case of an invalid state law under this concept is the law that places special requirements on the length of trains[103] or requirements on trucks' mud flaps[104] that pass through the state on interstate travel.

A state-enacted effluent charge system should not violate the "dormant" commerce clause because it should be drawn to apply equally to in-state and out-of-state waste dischargers. The dormant commerce clause is "not important to the charges approach," because "most charges plans can function effectively without unreasonable impacts on interstate commerce."[105]

The question of federal preemption of state water pollution control laws is more complex. Under the supremacy clause of the federal Constitution,[106] if Congress enacts a law that conflicts with a state law, or that occupies the field so completely that no room is left for state legislation, or where the congressional intent to preempt the field is manifest, then the state law is preempted and cannot stand.[107] Clearly Congress could

---

102. Although the criteria for determining the validity of state statutes affecting interstate commerce have been variously stated, the general rule that emerges can be phrased as follows: Where the statute regulates even-handedly to effectuate a legitimate local public interest, and its effects on interstate commerce are only incidental, it will be upheld unless the burden imposed on such commerce is clearly excessive in relation to the putative local benefits. If a legitimate local purpose is found, then the question becomes one of degree. And the extent of the burden that will be tolerated will of course depend on the nature of the local interest involved, and on whether it could be promoted as well with a lesser impact on interstate activities. Occasionally the Court has candidly undertaken a balancing approach in resolving these issues, but more frequently it has spoken in terms of "direct" and "indirect" effects and burdens.

Pike v. Bruce Church, Inc. 397 U.S. 137, 142 (1970) (case citations within quotation omitted). Huron Cement Co. v. City of Detroit, 362 U.S. 440 (1960).

103. Southern Pacific Co. v. Arizona, 325 U.S. 761 (1945).

104. Bibb v. Navajo Freight Lines, Inc., 359 U.S. 520 (1959).

105. ANDERSON, *supra* note 79, at 130–31.

106. U.S. CONST. art. VI, § 2.

107. The principle to be derived from [the Supreme Court's] decisions is that federal regulation of a field of commerce should not be deemed preemptive of state regulatory power in the absence of persuasive reasons—either that the nature of the regulated subject matter permits no othe concusion, or that the Congress has unmistakably so ordained.

Florida Lime & Avocado Growers v. Paul, 373 U.S. 132, 142 (1963). However, where Congress

enact a comprehensive effluent charge law that would preempt state laws in the field. Most federal legislation in the environmental field provides that state laws on the same subject are not preempted if they are more strict than the federal act.[108] If a provision of this type were included in the federal effluent charge law, then a state could levy charges, if it chose to do so, that would be added on to those levied under the federal law.

A related question is whether existing federal water pollution control laws preempt the field so there is no room left for state effluent charge laws. The Clean Water Act expressly reserves to the states the power to enact water quality control laws with stricter standards than those promulgated under the federal act.[109] In theory, one can argue that after 1985, the states could not possibly have stricter standards, because by then the nation will have achieved the no-discharge goal. The 1977 amendments, however, make it clear that the government intends to perpetuate a technology-based program of pollution control into the foreseeable future.[110] This program allows sufficient leeway for implementation of state effluent charge programs.

A technology-based federal program, however, might raise the question whether state effluent charges could be levied on waste dischargers who were already meeting the federal technology-based standards. An argument of this nature would probably fail, because add-on state effluent charges would necessarily reflect stricter standards than those required by the federal law. In addition they would come within the provision of the federal act allowing stricter state laws.[111] A credible counterargument, however, can be made to the effect that the federal disclaimer allows stricter state standards only in terms of the quantities of chemicals or other substances discharged into public waters and not in terms of charges assessed against polluters. In view of the uncertainty raised by this argument, Congress ought to enact an amendment to the Clean Water Act making it clear that state effluent charge laws would not be preempted by existing federal pollution control laws.

Under an amended Clean Water Act, the states could continue to im-

legislates "in a field which the States have traditionally occupied . . . we start with the assumption that was the clear and manifest purpose of Congress." Rice v. Santa Fe Elevator Corp., 331 U.S. 218, 230 (1947). *See also* Huron Cement Co. v. City of Detroit, 362 U.S. 440 (1960) and Askew v. American Waterways Operators, Inc., 411 U.S. 325 (1973).

108. *See, e.g.*, Clean Water Act, 33 U.S.C. § 1370 (1982); *see also*, 33 U.S.C. § 1314(i), 1316(d), 1318(c) (1982). Resources Conservation and Recovery Act of 1976, 42 U.S.C. § 6948, 6947 (1982).

109. Clean Water Act, 33 U.S.C. §§ 1314(l), 1316(c), 1318(c) (1982).

110. The 1977 amendments contain several important changes authorizing extensions of timetables for achieving standards and objectives. *See*, Pub. L. No. 96-217, 91 Stat. 1582–86 (codified as amended at 33 U.S.C. § 1311 (1982)). Significantly, *no* timetable is set for achieving the 1972 goal of eliminating "the discharge of pollutants into the . . . waters by 1985." 33 U.S.C. § 125(a)(1) (1900).

111. *See* ANDERSON, *supra* note 79, at 131 (same conclusion).

plement their own standards-oriented water pollution control systems as they do now, so long as they meet minimum federal standards. (Thirty-six states have met these federal standards and carry out their own programs under the supervision of the Environmental Protection Agency.[112]) Alternatively, the states could add a charge system to their bag of tools for controlling pollution.

As noted above, congressional legislation in this area might take one of two basic approaches. The legislation could provide the states with authority to enact whatever charge systems they deem appropriate and those systems would not be preempted by existing federal water pollution control laws. Secondly, the federal act could set minimum physical standards and minimum charge standards for state effluent charge laws. Obviously a major concern in making the choice between these two, or among other variables, will be to assure that the nation does not return to the era when industries bargained one state's pollution control laws against another and threatened to move from states with strict laws to those with more lenient programs.

VARIATIONS IN EFFLUENT CHARGE LEVELS
BY STATE OR REGION

If Congress enacted an effluent charge law, the question arises whether that law should establish uniform charges for waste dischargers all across the nation, or should it vary those charges by state or region. If the charges are uniform everywhere, then the states or regions with cleaner waters may complain they are being penalized because their charges are higher than necessary to achieve the desired water quality levels.

One important argument in favor of uniform national charges is that, if variations were permitted, some states might set charges low for their less developed regions with less pollution, thus inviting industries to move to those places and discharge their pollution there. This possibility could well raise the ire of both environmentalists who want to keep the clean areas clean and of larger cities and industrialized areas who want to keep jobs.

Probably the most important reason for applying a uniform charge across the nation is the political difficulty of deciding on the level of charges that should be applied in different regions. No acceptable formula

---

112. States with approved National Pollution Damage Elimination System programs are: Alabama, California, Colorado, Connecticut, Delaware, Georgia, Hawaii, Illinois, Indiana, Iowa, Kansas, Kentucky, Maryland, Michigan, Minnesota, Mississippi, Missouri, Montana, Nebraska, Nevada, New Jersey, New York, North Carolina, North Dakota, Ohio, Oregon, Pennsylvania, South Carolina, Tennessee, Vermont, Virgin Islands, Virginia, Washington, West Virginia, Wisconsin, Wyoming. Letter from David A. Greenburg, Attorney with Office of Water Enforcement and Permits, U.S. Environmental Protection Agency, to Ralph W. Johnson (May 17, 1984).

exists for regionalizing charges without creating great political controversy. Nonetheless some modifications in the uniform charge theme might prove feasible and politically desirable.[113] The modifications include special surcharges on new plants, allowing states some variation but with a federally established minimum charge below which the states cannot go, recognition of assimilative capacity as a basis for modest charge adjustment, and application of a varying time schedule for phasing in charges reflecting the different amortization needs of diverse industries. These approaches, as well as others that might be conceived, would tend to discourage migration of industries from one state or region to another and, at the same time, would speak to the complaints of environmentalists who want to discourage degradation of the more pristine areas.

If states enacted their own effluent charge systems, without the constraint of any federal minimum, then significant variations are likely to exist between the states, and the political problem of threatened industrial migration can be expected.

COULD AN EFFLUENT CHARGE SYSTEM BE GRAFTED ONTO THE PRESENT TECHNOLOGY-BASED STANDARDS SYSTEM?

No insurmountable legal problems should arise by enacting an effluent charge system on top of the current standards system. As noted above, the current water pollution control system in the United States, while professing a no-discharge-by-1985 goal, is in fact a technology-based system, applying criteria such as best available technology or best conventional technology.[114] Under this system, the states have set ambient water quality standards for receiving waters.[115] For many bodies of water, these standards are being met, or can be met, by application of the technology-based standards. For other waters, however, these technology-based standards are deemed too lax to assure compliance with existing standards. For these waters, Sec. 303(d) of the CWA requires that they be classed as water quality "limited segments;"[116] special procedures are then established to encourage achievement of the desired ambient water quality level.

Obviously an effluent charge system appropriately could be applied to waste dischargers on the "limited segments" of water where technology-based standards will not achieve the desired ambient water quality stan-

---

113. ANDERSON, *supra* note 79, at 166–72.

114. *See* 33 U.S.C. § 1311 (1982).

115. *See* Water Quality Standards and Implementation Plans, 33 U.S.C. § 1313 (1982), and 33 U.S.C. § 1342 (1982). All of the state programs that have been approved by the EPA under § 1342 necessarily have met the requirements for setting § 1313's water quality standards. *See supra* note 112 for states that have complied with federal standards.

116. *See* 40 C.F.R. § 130.20 (1984). *See generally* 33 U.S.C. 1313(d)(1)(A) (1982).

dards. An effluent charge system would be an appropriate means to encourage industries and municipalities to improve their technology, or to consider alternative disposal systems.

A different situation arguably could exist regarding those waters where the ambient water quality standards have been achieved. On these waters the waste dischargers might claim that charges are inappropriate because the desired water quality has already been achieved. A somewhat similar argument might be made by an industry on any body of water that is using the legally required level of technology. The answer to these arguments is that both the ambient water quality standards and the technology-based standards are simply waypoints; they are not final resting places. Because of continuing population and industrial growth and the need to dispose of an ever-increasing quantity of waste material, the U.S. needs to continue developing better waste control technology and alternative methods and locations for disposing of wastes. An effluent charge system provides a built-in incentive for encouraging these continuing efforts.

The data presently generated by the National Pollutant Discharge Elimination System (NPDES) make it quite feasible to adopt an effluent charge system with only modest additional effort. The applicant for an NPDES permit must provide EPA or the relevant state agency extensive and precise data on the quantity and content of the wastes to be discharged under the permit. These data include chemical parameters, metal content, physical and biological parameters, and radioactive parameters, and cover a total of some 68 different items. The permits identify the permissible discharges of each of these substances, and require appropriate self-monitoring to assure that the permissible limits are not exceeded or, if exceeded, are reported. EPA and the relevant state agencies have a well established spot-monitoring system of their own to assure the validity of the self-monitoring system. With this body of data already available it would not be technically difficult to graft an effluent charge system onto the present regime. The principal decisions to be made concern the choice of the wastes that would be the basis of the charges.

One of the surprising consequences of the enactment of the FRG charge system was the degree to which more complete and more precise data were developed because money changed hands on the basis of that data. While the United States is further along now than the Germans were when they enacted the effluent charge law—and it has considerably more data than the Germans did then—it nonetheless seems likely that implementation of an effluent charge system here also would generate important new information, for the same reason—money changes hands on the basis of that information.

SYNOPSIS

The Federal Republic of Germany's 1976 Effluent Charge Law was produced at the crest of that nation's environmental movement and against a background of broad support for the application of market economics to resolve the country's water pollution problems. The ECL was designed to operate in tandem with the existing standards/permit system established in a 1957 law and modified by 1976 Amendments.

The federal act established a minimum national water quality goal and authorized federally created task forces to set technology-based standards for all industries and municipalities. The federal act also determined the pollutants on which the charges are to be based and set the annual charges for each pollutant. The Länder carry out all enforcement of the Act, including timing of implementation, collection, and disbursement of charge revenues. The technology-based standards established uniform thresholds for individual discharge levels across the country. The Länder can set higher minimums if they are necessary for achieving particular quality goals in given water bodies.

The effluent charge system enacted in 1976 is tied to five pollutants: settleable solids, COD (chemical oxygen demand), mercury, cadmium, and toxicity for fish. The charge level started at 12 DM (about $5.00) per damage unit in 1981 and rises to 40 DM (about $16.00) per damage unit in 1986. A damage unit is a specified amount of effluent such as 45.45 Kg of COD. The charge per damage unit is uniform across regions and polluters.

Each discharger pays for the expected amount of pollution stipulated in the effluent charge portion of the individual permit. The charge liability is lower under two circumstances. If the expected discharge level meets the federal minimum standards, the unit charge is reduced by one-half (e.g., from 12 DM to 6 DM in the first year). Second, if the actual discharge level is substantially below the expected level, the bill is based on the actual level of discharge. When maximum levels of discharge stipulated in the permits are exceeded, polluters are penalized by having to pay more in the future.

Revenues from the charges can be used by the Länder for the costs of administering the ECL and for supporting pollution abatement activities.

The short experience of the FRG teaches the following lessons about an effluent charge system: it is most likely politically viable and administratively attractive if

(1) It covers a small number of pollutants;
(2) It is combined with permit systems;
(3) The charges begin at some specified level and escalate during a transition period;

    (4) The charge levels result from a process involving the participation of interested parties including those benefitted and harmed by waste discharge;

    (5) Measures and levels of volumes and pollution concentrations are simplified;

    (6) Effluent charge revenues are made available for abatement related expenditures[117]—see below;

    (7) Hardship clauses are provided to protect dischargers or industrial sectors under exceptional circumstances;[118]

    (8) Care is taken to demonstrate how the effluent charge program actually can be implemented.

If an effluent charge system meeting the above constraints is implemented, then the U.S. can expect the following:

(1) Charges to increase the incentive for firms to find treatment technologies, substitute production processes, and substitute input and output combinations which diminish residuals discharge. The qualitative evidence is that firms whose discharge licenses did not change generally found ways to reduce their charge obligation. An intrafirm effluent charge system resulted in a 20 percent decline in waste discharge in the seven years since its introduction.[119]

(2) Charges increase the incentives for municipalities to adopt customer sewage pricing policies which not only are acceptable financial instruments but also offer incentives for the indirect dischargers to economize on waste production.[120]

(3) Charges encourage, if not require, municipalities to find satisfactory procedures for better monitoring the intake and outflow of effluent. This will help public authorities to reduce the average cost of their sewage services and will aid them in executing an effluent charge policy which better reflects the marginal cost of treating a given customer's effluent.[121]

(4) The present system of subsidies in the U.S. for waste treatment rewards capital intensive municipal waste treatment technologies by subsidizing capital expenditures. By encouraging municipalities to use more operation and maintenance expenses to reduce waste discharge, the effluent charge system helps to correct the resource allocation distortions the subsidies created.[122]

(5) If revenues generated from charges are made available for ex-

---

117. The small number of pollutants, the escalation of prices and the availability of revenues for abatement expenditures are qualities of the French effluent charge system. *See* JOHNSON & BROWN, *supra* note 7 and BOWER, *supra* note 21.

118. *See supra* text accompanying notes 28–29.

119. *See supra* text accompanying notes 47–57.

120. *Id.*

121. *Id.*

122. *Id.*

penditures for water quality improvement, some portion of these funds will be available for use by industry. The present subsidy system in the U.S., by excluding firms directly, distorts the marginal cost of waste treatment between private (firms) and public (municipality) dischargers. There may be equity considerations which justify the present policy, but such goals are achieved at the cost of a loss in efficiency. These losses will be mitigated, in part, if firms qualify for subventions. Final discussions prior to the passage of the ECL defined the uses of charge revenues to include the industrial expenditures for effluent conserving production processes, in addition to more straightforward pollution abatement expenditures.[123]

(6) If a charge system is generating billions of dollars per year in revenues, it is likely that this source increasingly would look attractive as a substitute to the U.S. Treasury for pollution abatement subsidies. Since revenues have yet to be collected in the FRG, there is no evidence to support the concern of several water quality experts interviewed that this substitution would take place. A decreased dependence on the Treasury redistributes the cost of pollution from the general taxpayer to the consumer of pollution intensive products and to the owners of factories specializing in the production of those products. This shift in the source of subsidy would further emphasize the acceptance of the polluter-pays-principle.[124]

(7) If the cost of administering and enforcing the effluent charge system is covered in part or totally by revenues created, as it is in the FRG, then we can expect greater availability of enforcement services and more compliance compared to the precharge period. There is too little empirical evidence regarding enforcement levels in the U.S. to know whether and to what extent the present situation is optimal. There is the danger of excessive enthusiasm for enforcement when the budget for enforcement comes from charge revenues. Representation of heterogeneous interests on the board in charge of revenue disbursement, is one way to reduce the chance for this resource misallocation—admixtures are effective antidotes for excessive zeal.[125]

(8) Introducing an actual effluent charge system on top of a standards system, in all likelihood, increases the total cost of managing water quality. The fixed cost of educating legislators and others unfamiliar with such a policy so they can vote intelligently should not be overlooked. In return, water quality is improved and the flexibility, quality, and recision

---

123. *See supra* text accompanying notes 76–78.
124. *See supra* text accompanying notes 58–63.
125. *See supra* text accompanying notes 69–71.

of the management program is improved when more policy options are available.[126]

(9) An effluent charge system combined with a permit system creates a more flexible bag of policy tools capable of better responding to changing circumstances than either system alone.[127]

(10) In recent years residuals producers have been permitted to trade environmental quality permits. The bubble concept introduces greater flexibility into the system by enabling exchange, in effect, to remove constraints on some firms' behavior. Is a charge system unnecessary if a bubble policy is in place? Other things being equal, the introduction of charges results in a loss to polluters because the implicit value of discharge permits is depreciated by the introduction of an effluent charge. In contrast, since introducing the bubble removes some constraints, the value of tradeable permits increases. Thus, the distributive consequences of these two policies is quite different. In practice, the efficiency aspects seem to be different. To date, the number of air pollution offsets consummated is modest and the number of water quality trades is miniscule. This suggests that there are practical impediments to the development of offset markets in water. These may or may not be of a short-run nature. On the other hand, the effluent charge impinges on all municipalities, direct dischargers, and on some indirect dischargers. Thus the effluent charge system generally is superior to a bubble system. The extent of the resource savings created by an effluent charge depends largely on its magnitude. What is clear from this discussion is that in a pollution offset program, effluent charges are complementary, not competitive programs, when the criterion is economic efficiency. Finally, if the experience in the Federal Republic of Germany is a guide, the introduction of an effluent charge will improve water quality.[128]

In the United States it is clear that Congress has the constitutional power to enact an effluent charge law for the nation as a whole. Alternatively Congress could enact a framework law establishing minimum standards for state effluent charge laws, and then allow the states to enact such laws as they saw fit. Objections might be raised to federal or state effluent charge laws on the basis of constitutional equal protection and due process grounds, but these objections would fail.

States have authority to enact effluent charge laws under their own constitutions, and under the federal constitution; however, state water pollution control laws might be preempted by the existing federal laws in the field. The Clean Water Act (CWA) explicitly provides that state

---

126. *See supra* text accompanying notes 64–68.
127. *Id.*
128. *Id.*

laws are not preempted by the federal Act if they are "stricter" than the federal law, and state effluent charge laws might conceivably meet this criterion.[129] Sufficient uncertainty, however, surrounds this question that we recommend enactment of federal legislation explicitly authorizing state effluent charge laws.

An effluent charge law could be enacted in the U.S. to operate in tandem with the existing CWA standards/permit system. In spite of the much publicized no-discharge "goal" of the Clean Water Act, the system actually is technology based. It would be quite feasible to coordinate an effluent charge law with the existing NPDES system. The data generated by the NPDES process provides the technical information that would be required for establishing effluent charges. An effluent charge system should not excuse waste dischargers merely because they are meeting the technology-based standards.

Caution should be exercised about considering variations in charge levels if they are initiated by a state or a region. Not only is it exceedingly difficult to determine the technically proper and politically acceptable variance, but those variances might also encourage industries to bargain among states for the lowest charges.

129. 33 U.S.C. § 1370 (1982).

[8]

Transpn Res.-E (Logistics and Transpn Rev.), Vol. 33, No. 1, pp. 1–13, 1997
© 1997 Elsevier Science Ltd
All rights reserved. Printed in Great Britain
1366-5545/97 $17.00 + 0.00

Pergamon

FEES AND REBATES ON NEW VEHICLES: IMPACTS ON FUEL EFFICIENCY, CARBON DIOXIDE EMISSIONS, AND CONSUMER SURPLUS

KENNETH E. TRAIN,*
Department of Economics, University of California, Berkeley, CA 94720-3880, U.S.A.

WILLIAM B. DAVIS and MARK D. LEVINE
Energy Analysis Program, Lawrence Berkeley Laboratory, 1 Cyclotron Road, Berkeley, CA 94720, U.S.A.

(Received 20 February 1996; revised version received 15 July 1996)

Abstract—Several incentive systems are examined that provide rebates on vehicles with higher-than-average fuel efficiency and levy fees on vehicles with less efficiency. The rebates and fees are applied to new vehicles at the time of purchase, and the rates are set such that the total outlay for rebates equals the revenues from fees. We find that moderately-sized rebates and fees result in a substantial increase in average fuel efficiency. Most of the effect is due to manufacturers' incorporating more fuel-efficiency technologies into the vehicles that they offer, since the rebates and fees effectively lower the price to manufacturers of these technologies. Consumer surplus is found to rise, and the profits of domestic manufacturers are estimated to drop only slightly under most systems and actually to rise under one system. © 1997 Elsevier Science Ltd

1. INTRODUCTION

The term 'feebates' denotes a system of fees and rebates applied to new cars and trucks based on their fuel efficiency, emissions rates, or other attributes. A fee is added to the sales price of vehicles with, for example, low fuel efficiency, and a rebate is subtracted from the price of vehicles with high fuel efficiency. The feebate 'schedule' gives the amount of fee or rebate as a function of specified attributes. For example, the schedule might be linear in miles per gallon (mpg), with each unit of higher mpg being rewarded with a given increase in rebate or decrease in fee; or it might be linear in the gallons of fuel that the vehicle consumes in travelling a mile (gpm), in which case it is NONLINEAR in mpg. The feebate 'rate' is the slope of the schedule, that is, the change in the fee or rebate for a one-unit change in the specified attribute(s). The 'zero-point' is the level of the attribute(s) at which neither a fee or rebate is applied. The zero-point can be set such that the feebates are revenue neutral, in that the revenues obtained from the fees just off-set the payments of rebates. This aspect of feebates makes them attractive politically, since the feebates need not be considered a new tax.

The legislative bodies in Arizona, California, Connecticut, Maine, Massachusetts, New York, Oregon and Wisconsin are considering or have considered state-level feebates. Maryland passed a law that establishes a state feebates program; however, it has been not been implemented. The Chief Counsel of the National Highway Traffic Safety Administration under then-President Bush wrote to the Maryland Attorney General arguing that Maryland's law violated the federal law that established the Corporate Average Fuel Efficiency (CAFE) standards, which preempts state-level regulation 'relating to fuel economy standards.' The state Attorney General investigated the issue and concluded that some, but not all, provisions in the law seemed to violate the federal preemption. In particular, the Maryland law established a labeling requirement as well as a feebate system, and the Attorney General stated that the labeling requirement seemed to be preempted but that the feebates themselves, without a labeling requirement, were not.

Five bills have been proposed in Congress (three in the House and two in the Senate) to establish feebates at the federal level. The concept was fairly active during the early days of the Clinton administration but has since waned. Given the fluctuating political climate regarding fuel

---

*Author for correspondence.

efficiency, it is likely that interest will return eventually, and it is important that we have an understanding of feebates to use at that time.*

Several studies have examined the potential impacts of feebates. Gordon and Levenson (1989) and Davis (1991) estimate the effects of California's proposed feebates, called DRIVE +. They examine only the demand response (i.e. they do not investigate the effect of feebates on manufacturers' design decisions), and find the effects to be very small. Their result is consistent with Greene (1991a) who shows that large changes in prices—far beyond the feebate levels being proposed in California—are needed to induce meaningful improvements in average fuel efficiency. DRI (1991) and Charles River Associates (1991) determine the level of incentives that would be necessary to meet a target level of emissions or average fuel efficiency. These studies consider the demand and supply response of incentives applied nationally. CRA set a target of 39.6 mpg for the average new car fuel-efficiency for the period 2001–10 and found that a feebate rate between $32 and $50 per mpg would achieve this target. Meeting this target represents an improvement in average new car fuel efficiency ranging from 6 to 20%, depending on the year, over the fuel efficiency that CRA forecast would occur without feebates or other intervention. DRI specified a penalty schedule that levied fees on vehicles that achieved less than 50 mpg, with the rate per mpg decreasing as fuel efficiency increases. No rebates were paid. It found that a schedule with rates that ranged from $52 to $105 per mpg, with an average of about $75, would stabilize CO_2 emissions at their 1989 levels. Given their baseline forecast of emissions without feebates or other intervention, the fees are forecast to reduce emissions and fuel consumption by 13% in 2010.

Our analysis builds upon these studies, confirming some results and clarifying others. We find, consistent with Gordon/Levenson, Davis, and Greene, that the demand response is very small for the level of feebates that are being considered. However, we find a substantial manufacturer response. Our findings on the combined demand and supply effect are consistent with CRA's and DRI's. In particular, we find that a feebate rate of $70 per mpg increases average new car fuel efficiency by 18% by 2010 and that there are decreasing returns from higher feebates such that a rate that is $35 per mpg can be expected to obtain about 14% improvement in new car fuel efficiency—which is in the middle of the range obtained by CRA. In comparison to DRI, we find that a feebate rate of $70 per mpg reduces CO_2 emissions by nearly 8% in the year 2010, which is less than the 13% reduction that DRI predicts for a penalty schedule whose average rate is about $75 per mpg. However, part, and perhaps all, of this difference is attributable to the ways the two incentive schemes are specified. Our feebate schedules are constructed to be revenue-neutral, while DRI's schedule is revenue-generating, applying fees but no rebates. For comparison, the DRI scheme can be considered a revenue-neutral feebate schedule plus an excise tax on all vehicles with less than 50 mpg. The feebate portion accounts for only part of DRI's predicted 13% reduction in emissions.

We extend the previous studies in several ways. (1) We examine a variety of feebate schemes that differ in the size of feebate, the factor on which the feebate is applied (gpm, mpg, or size-adjusted gpm), and whether cars and trucks are treated differently in the calculation of feebates. The analysis provides information for the most appropriate design of feebates. (2) We examine the impact of these feebates on a wider set of outcome variables. Most importantly, we estimate the impact on consumer surplus and sales of domestic manufacturers, in addition to fuel efficiency, fuel consumption, and CO_2 emissions. (3) The analysis is more detailed than previous analyses of the demand and supply sides. For example, we disaggregate the vehicle market into 95 'subclasses' of vehicles (where a subclass is a subset of a class), compared to the ten classes used by DRI and the aggregate demand approach of CRA which did not distinguish classes. This disaggregation is important for several reasons. Consumers are more likely to switch among subclasses than classes; with the demand analysis conducted on classes or at the aggregate level, a small estimated demand response could be an artifact of the method which masks possibly substantial switching at the subclass level. Our analysis, which examines subclass switching, verifies that the demand response is small even when subclass switching is included. The disaggregation into subclasses also allows the supply model to more realistically represent the design decisions of manufacturers, whose response is estimated to be critical to the effectiveness of feebates.

---

*General discussions of the purpose, characteristics, and history of feebates are provided by DeCicco *et al.* (1993), and Davis *et al.* (1994).

Section 2 describes the demand and supply models. The succeeding sections present the baseline forecast, the feebate schedules, and the forecasted impacts. The final section concludes. Details on all aspects of the analysis are provided in Davis *et al.* (1994).

2. METHOD

A supply model predicts the vehicle offerings of manufacturers, and a demand model predicts the purchase decisions of consumers. A 'baseline forecast' is produced under the assumption of no feebates or other changes in regulation. For each feebate scheme, a 'scenario forecast' is produced by changing the inputs to the models to represent the feebates. The difference between the baseline and scenario forecasts is the estimated impact of the feebates.

2.1. Supply

The supply model developed by Energy and Environmental Analysis (EEA; Duleep, 1992) is used to forecast the characteristics of vehicle offerings. All makes and models of vehicles are grouped into 95 subclasses, which represent further disaggregation of the standard 14 classes developed by the U.S. Environmental Protection Agency (EPA). In particular, the 14 EPA classes are broken into 19 classes, with the correspondence designated in Table 1. Each of the classes is further subdivided into subclasses by three criteria: performance (three levels, as measured by horsepower), technology (two levels representing the degree of more advanced technology implemented in the vehicle), and import status (foreign or domestic). Many subclasses are empty because no vehicles were produced with the necessary characteristics; of the possible 228 subclasses, 95 are non-empty and hence are used in the ananlysis.

In EEA's model, each of the 95 subclasses is considered a 'platform' that has observed characteristics in the base year (1990). The base-year description of each platform includes its price, fuel efficiency, size (interior space, luggage space), weight, horsepower, and the design of the vehicle that provides these characteristics. Table 2 gives the average characteristics for each of 19 classes in the base year (sales-weighted averages over the subclasses in each class.)

EEA's model considers 55 technologies that can be incorporated into vehicles to increase their fuel efficiency. Table 3 enumerates these technologies, the year in which they become available, their cost, and their effect on the vehicle's fuel efficiency and horsepower. It is important to note that EEA's model does not consider the possibility of increasing fuel efficiency by simply reducing the size, weight, and/or power of the vehicle. The intent is for safety and performance to be unimpaired by the adoption of any of the technologies. For example, material substitution (technologies 3–6) provides lighter but stronger materials: while the weight of the vehicle decreases, the level of safety it provides does not necessarily decrease. Given these constraints on the technologies that we consider, the forecasts of feebate impacts should be viewed as the impacts that would occur if safety and performance are not traded-off for higher fuel efficiency.

Table 1. Correspondence of EEA classes with EPA classes

	EEA class		EPA class
1	Minicompact	1	Minicompact
2	Subcompact	2	Subcompact
3	Sports	6	Sports
4	Compact	3	Compact
5	Intermediate	4	Midsize
6	Large	5	Large
7	Luxury	7	Luxury
8	Near luxury	2	1/2 Compact, 1/2 subcompact
9	Midsize wagon	4	Midsize
10	Large wagon	5	Large
11	Near truck	2	1/2 Compact, 1/2 subcompact
12	Minivan	10	Compact van
13	Mini utility	14	Mini utility
14	Compact pickup	8	Compact pickup
15	Compact van	10	Compact van
16	Compact utility	12	Compact utility
17	Standard pickup	9	Standard pickup
18	Standard van	11	Standard van
19	Standard utility	13	Standard utility

4   Kenneth E. Train *et al.*

Table 2. Base-year (1990) vehicle characteristics

	Class	No. of sub-classes	No. of makes (imp.)	No. of makes (dom.)	Sales (1000)	Price (1990$)	CAFE (mpg)	Horsepower
Minicompact	1	5	26	0	259	11 236	37.2	97
Subcompact	2	9	28	6	1547	10 618	34.0	94
Sports	3	8	24	6	663	15 439	26.9	142
Compact	4	7	13	28	2095	13 433	29.4	115
Midsize	5	8	5	39	1934	15 738	26.6	137
Large	6	2	0	9	599	18 331	24.5	162
Luxury	7	6	57	12	941	32 904	22.7	180
Near luxury	8	7	19	8	477	18 851	25.8	144
Midsize wagon	9	5	2	10	149	16 469	26.0	141
Large wagon	10	2	0	5	31	18 746	22.7	143
Near truck	11	4	10	3	77	15 101	26.5	121
Minivan	12	2	3	0	20	13 569	33.8	95
Mini utility	13	2	4	0	55	12 625	30.8	79
Compact Pickup	14	6	16	11	890	11 799	24.7	122
Compact van	15	9	8	17	968	17 579	22.8	140
Compact Utility	16	5	7	8	478	19 209	20.6	150
Standard pickup	17	3	0	27	945	15 339	18.0	177
Standard van	18	3	0	22	292	15 770	17.4	175
Standard utility	19	2	0	14	175	20 860	16.4	200
All	1–19	95	219	222	12 596	16 173	25.2	136

The supply model determines the extent to which each technology is adopted by comparing the costs and the benefits of the technology. The benefits are calculated as the present value of the fuel savings that result from the technology, summed over a four year period discounted at 8% annually (labeled, PVFS), plus an 'adder' to represent the fact that the adoption of some technologies increases performance. This adder (labeled A) is set at $15 per percent improvement in horsepower for all platforms except those representing sports and luxury vehicles, for which the adder is twice as large (to represent the fact that performance-enhancing technologies are more desired in these subclasses.) The 'cost-effectiveness' of the technology is $CE = [(PVFS + A)/C] - 1$, where C is the cost of the technology. The proportion of vehicles in the platform for which the technology is adopted is a logistic function of CE: $M = MAX * [1/(1 + \exp(-2CE))]$, where MAX represents a production constraint that rises over time.* The price of the platform is raised to cover the cost of the technology in the share of vehicles in which the technology is adopted.[†]

As an example, consider variable valve timing 1 (technology 32 in Table 3). This technology costs $140 to incorporate into vehicles and raises fuel-efficiency by 5% and horsepower by 10%. For a vehicle whose original fuel efficiency is 26 mpg, adoption of this technology would raise it to 27.3 mpg, such that fuel costs drop by 0.214 cents per mile at a fuel price of $1.17/gallon. For this technology, PVFS = $85.17 (with 12 000 miles driven annually and no increase in fuel prices) and A = $150 (10% improvement in fuel efficiency times 15) such that the decision rule implies that the technology would be adopted in 80% of the vehicles in which it is possible to do so (i.e. 80% of MAX.)

The introduction of feebates changes manufacturers' evaluation of technologies by providing an extra benefit. If a technology is adopted, the increased fuel efficiency of the vehicle translates into a change in the feebate that is applied to the vehicle (a decrease in the fee or an increase in the rebate.) This change in feebate is added to the numerator in calculating CE, thereby increasing the adoption of fuel-efficiency technologies. In the above example, if the feebate changed by $91.50 (as it would under the GPM LOW scenario described below), then adoption of the technology would rise to 94% of MAX.

2.2. Demand
We use the demand model of Train (1986) to represent consumers' choices among the vehicles that manufacturers offer. The model is a customer-level discrete/continuous model of the household's

---

*MAX is a step-function of previous values of M, such that greater penetration leads to a higher production constraint.

[†]With a list of technologies, the savings of each depends on which other technologies have been adopted. Therefore, the sequence in which technologies are considered can affect which ones are adopted. EEA has developed a sequencing procedure based on its analysis of auto industry practices. To minimize the effect of such sequencing on the results, the same sequencing is used for the baseline and all scenario forecasts.

Environmental Instruments and Institutions

Table 3. Fuel-efficiency technologies

	Technology	Change in fuel economy (%)	Cost ($)	Change in horsepower (%)	Year available
1	Front wheel drive	6.0	160	0	80 (85)
2	Unit body	4.0 (6.0)	80	0	80 (95)
3	Material substitution II	3.3	0.6/lb	0	87 (96)
4	Material substitution III	6.6	0.8/lb	0	97 (06)
5	Material substitution IV	9.9	1/lb	0	07 (16*)
6	Material substitution V	13.2	1.5/lb	0	17* (26*)
7	Drag reduction II	2.3	32	0	85 (90)
8	Drag reduction III	2.3	32	0	91 (97)
9	Drag reduction IV	2.3	48	0	04 (07)
10	Drag reduction V	2.3	64	0	14* (17*)
11	Torque converter lockup	3.0	40	0	80
12	4-Speed automatic	4.5	225	5	80
13	5-Speed automatic	6.5	325	7	95 (97)
14	Continuously variable transmission	7.0		250	795 (05)
15	6-Speed manual	2.0	100	5	91 (97)
16	Electronic transmission I	0.5	20	0	88 (91)
17	Electronic transmission II	1.0	20	0	98 (06)
18	Roller cam	2.0	16	0	87 (86)
19	Overhead cam 4	3.0	100	20 (15)	80
20	Overhead cam 6	3.0	140	20 (15)	80 (85)
21	Overhead cam 8	3.0	170	20 (15)	80 (95)
22	4 (3) Valves per cylinder 4	8.0 (6.0)	240	45 (30)	88 (90)
23	4 (3) Valves per cylinder 6	8.0 (6.0)	320	45 (30)	91 (90)
24	4 (3) Valves per cylinder 8	8.0 (6.0)	400	45 (30)	91 (02)
25	Cylinder reduction	3.0	(100)	−10	88 (90)
26	5 (4) Valves per cylinder 4	10.0 (8.0)	300	55	98 (97)
27	Turbocharger	5.0	500	45	80
28	Friction reduction I	2.0	20	0	87 (91)
29	Friction reduction II	2.0	30	0	96 (02)
30	Friction reduction III	2.0	40	0	06 (12*)
31	Friction reduction IV	2.0	50	0	16* (22*)
32	Variable valve timing I	5.0	140	10	98 (06)
33	Variable valve timing II	3.0	40	15	08 (16*)
34	Lean burn	10.0	150	0	12* (18*)
35	Two stroke	15.0	150	0	04 (08)
36	Throttle-body injection	2.0	40	5	82 (85)
37	Multi-point injection	3.5	80	10	87 (85)
38	Air pump	1.0	0	0	82 (85)
39	Idle off	1.5	15	10	87 (85)
40	Oil SW-30	0.5	2	0	87
41	Oil synthetic	1.5	5	0	97
42	Tires I	1.0	16	0	92
43	Tires II	1.0	16	0	02
44	Tires III	1.0	16	0	12*
45	Tires IV	1.0	16	0	18*
46	Accessory improvements I	0.5	15	0	92 (97)
47	Accessory improvements II	0.5	15	0	97 (07)
48	Electric power steering	1.5	40	0	02
49	4WD improvements	3.0	100	0	02
50	Air bags	−1.0	300	0	87 (92)
51	Emissions tier I	−1.0	150	0	94 (96)
52	Emissions tier II	0.0	150	0	03 (04)
53	Anti-lock brakes	−1.0	300	0	87 (90)
54	Side impact	−1.0	100	0	96
	Roof crush	−1.0	100	0	01

Notes: When data differ for trucks and cars, the truck data are given in parentheses. Costs are given in 1990 $ or 1990 $/lb (for material substitution II–V).

*Indicates that the fuel-economy technology was not available in the time horizon of the feebates forecast, and thus not included in this analysis.

When a fuel-economy technology has more than one stage (for example, Drag reduction II–V) then the latter stages are marginal, in addition to the earlier stages, and thus do not supersede the previous stages, but do require them. For example, variable valve timing II will save a total of 8% on fuel economy over the baseline, and cost $180. It cannot, however, be evaluated for cost-effectiveness independent of variable valve timing I). The last 6 options are not fuel economy technologies *per se*, but are regulatory requirements on vehicle technology that affect fuel economy.

6   Kenneth E. Train *et al.*

choice of how many vehicles to own, which subclasses and vintages of vehicles to own, and how much to drive each vehicle annually. Let i denote a particular portfolio of vehicle holdings, where a portfolio represents a number of vehicles (perhaps zero), the subclass of each vehicle, and the vintage (new, one year old, and so on) of each vehicle. Let $I(n)$ denote the set of all portfolios with n vehicles. $I(0)$ contains only one portfolio (denote $i = 0$ for no vehicle holdings), $I(1)$ contains each possible subclass/vintage combination, and $I(2)$ contains each possible pair of subclass/vintage combinations. The utility that household h obtains from portfolio i is denoted U_{hi}, consisting of a portion V that depends on observed characteristics of the portfolio and the household, and a portion that is unobserved: $U_{hi} = V_{hi} + \epsilon_{hi}$. Assuming that ϵ_{hi} for all i is distributed Generalized Extreme Value with correlation over i within each set $I(n)$ but not across these sets, then the probability that the household owns portfolio i is nested logit (McFadden, 1978):

$$P_{hi} = \frac{e^{V_{hi}/\lambda}\left[\sum_{j\in I(n_i)} e^{V_{hj}/\lambda}\right]^{\lambda-1}}{e^{V_{h0}} + \left[\sum_{j\in I(1)} e^{V_{hj}/\lambda}\right]^{\lambda} + \left[\sum_{j\in I(2)} e^{V_{hj}/\lambda}\right]^{\lambda}}$$

where n_i is the number of vehicles in portfolio i. The parameter $(1-\lambda)$ is a measure of the correlation of $_{hi}$ within each nest. V_{hi} is linear in parameters with explanatory variables that include: the price, fuel cost (i.e. the price of fuel divided by mpg), shoulder room, luggage space, weight and horsepower of each vehicle in the portfolio, as well as the income, number of workers, and number of members of the household. The parameters were estimated on a sample of 1095 households whose vehicle holdings were observed. The estimated parameters are provided in Train (1986, Ch. 8) and for brevity are not repeated here. The parameters that are most critical to the results are those associated with price and operating cost. The estimated parameters imply that consumers are willing to pay at least $830 in higher vehicle price in return for a one-cent per mile reduction operating cost, with higher income households willing to pay even more. For comparison, $830 constitutes the discounted present value of savings from a one cent per mile reduction in fuel cost over eight years at about 12 000 miles driven per year with a real discount rate of 3% and no anticipated growth in real fuel prices.*

Conditional on its portfolio choice, the amount the household drives each vehicle is given by a regression equation with explanatory variables that include the operating cost of the vehicle. Since the household chooses its portfolio, the operating cost of its chosen vehicle(s) is endogenous. In estimation, actual operating cost was replaced with the expected operating cost, calculated as the probability-weighted average of the operating cost of each possible subclass and vintage of vehicle. Other variables explaining vehicle mileage include: household income, location, number of workers, number of members, and a dummy indicating whether the vehicle is the newer one in a two-vehicle portfolio. The expected fuel consumption of the household is calculated as the sum over portfolios of the probability that the household chooses the portfolio times the miles that each vehicle in the portfolio are predicted to be driven divided by the mpg of the vehicle.

By Small and Rosen (1981), the expected consumer surplus that household h obtains in the vehicle market (that is, in its choice of portfolio and its driving of the chosen vehicles) is linear in the log of the denominator of P_{hi}:

$$CS_h = W + \mu \ln D_h,$$

---

*It is important to note that consumers' willingness to pay was not determined by calculating the present value of the operating cost savings under assumptions about the discount rate, future gas prices, and annual vehicle-miles driven. Rather, willingness to pay was represented by parameters within the discrete choice model, whose estimated values were those that best fit the data. That is, willingness to pay was inferred from consumers' actual choices. In the text above, the estimated willingness to pay of $830 is compared to the present value of operating cost savings under particular assumptions as a means of determining whether the estimate is reasonable and the extent to which the estimate implies that customers behave rationally. The fact that the $830 estimate corresponds to the present value of savings under reasonable assumptions implies that the estimate is reasonable and that consumers do indeed behave fairly rationally in their choice of fuel efficiency in vehicles. The estimate of $830 is for households whose income is below average. For households with higher-than-average income, Train's model obtains an estimated willingness to pay that is higher than $830. This result implies that either (i) the estimate for these households is too high (i.e., above the true value), (ii) higher income households apply a lower discount rate than 3%, expect gas prices to rise, and/or expect to drive more than 12,000 miles annually (any of which would provide a higher present value of fuel efficiency improvements), or (iii) irrationally have a higher willingness to pay for fuel efficiency improvements than can be justified financially by the present value of the savings that the improvements provide. The estimates of willingness to pay obtained by Train's model are in the midrange of the findings of other researchers' models of automobile demand (Train, 1985.).

where W is the constant of integration, μ is the inverse of the marginal utility of income, and D_h is the denominator of P_{hi}. The change in consumer surplus from the baseline forecast to a scenario does not depend on W, only on μ and the difference between D_h in the baseline forecast and that in the scenario (that is, the difference in the characteristics of the vehicles that are offered by manufacturers in the baseline forecast vs the scenario). The value of μ is obtained by using the fact from utility theory that a marginal change in fuel prices changes consumer surplus by the amount of fuel that is consumed: $dCS/dGP = -X$, where GP is the price of fuel and X is the quantity of fuel consumed. Specifically, the value of μ was found that caused this equality to hold for a one-tenth cent change in the price of fuel.

Aggregate forecasts are obtained from the household-level demand model through sample enumeration, that is, by taking the weighted sum of the household-level predictions, with the weights set such that the weighted sample is representative of the population. Over time, the weight associated with each household and the income of the household are adjusted to represent population changes over time (such that the weighted sample in each forecast year looks like the population is forecast to look in that year.)

3. BASELINE FORECAST

The baseline forecast was generated using EIA's (EIA, 1992) fuel price projection, which represents an average growth rate in real prices of 1.1% per year over the period 1995 to 2010. The number of US households was assumed to grow at a 1.2% average annual rate, with income per household growing 1.3% per year on average and household size dropping from 2.815 in 1990 to 2.534 in 2010. Under these inputs, new car CAFE (the average fuel efficiency rating of all new cars sold) is forecast to rise to 31.7 mpg by the year 2000 and 37.0 mpg by 2010. For trucks, the comparable figures are 23.3 and 25.6 mpg. These forecasts are similar to those of EIA (1992), though slightly higher in 2010. (This difference with the EIA forecast is largely due to our inclusion of the two-stroke engine technology which is expected to come on line around 2005.) On-road fuel efficiency is calculated by adjusting the CAFE ratings downward by 15% to account for the difference between CAFE ratings and actual on-road fuel economy (Westbrook and Patterson, 1989.) The average on-road fuel efficiency of the stock of cars and trucks (both new and used) is projected to rise to 22.6 mpg in 2005 and 25.6 mpg in 2010. Vehicle ownership increases steadily, from a total of 165 million in 1990 to 215 million in 2010. Nearly all of this 30% increase is due to the growth in number of households: vehicles per household rises only slightly, from 1.75 in 1990 to 1.76 in 2010. (The rise in income, which would lead to more vehicles per household, is essentially off-set by decreased household size.) Vehicle miles travelled (VMT) rises from 1.8 trillion in 1990 to 2.6 trillion in 2010. Again, this growth is mostly due to increased number of households: VMT per household rises by only 8% over the entire period. Despite the 42% rise in VMT, fuel consumption and CO_2 emissions are projected to increase only 11% by 2010, since on-road fuel efficiency rises by 28%.

4. FEEBATE SCENARIOS

Six feebate schedules are examined. Each schedule is revenue-neutral, in that the fees collected equal (or nearly equal) the rebates disbursed. Revenue neutrality is accomplished by the appropriate choice of the 'zero point' in the feebate schedule—the level of fuel efficiency at which neither a fee or a rebate is applied. The six schedules are:

GPM LOW is based on the gallons per mile (gpm) that the vehicle consumes. The feebate rate is $50000 per gpm. Thus, a vehicle whose fuel efficiency rating is 25 mpg, which translates to 0.04 gpm, would receive a $500 greater rebate or lower fee than a vehicle that obtains 20 mpg, which is 0.05 gpm. A different zero point is applied for cars and trucks. For most vehicles, the feebate is less than 1% of the vehicle's price.

GPM HIGH is the same as GPM LOW except that the rate is twice as high: $100000 per gpm.

ONE ZERO POINT applies the same rate as GPM LOW but uses the same zero point for both cars and trucks.

MPG LOW is based on the fuel efficiency (mpg) rating of the vehicle. The rate is $70 per mpg, and different zero points are applied for cars and trucks. At 26.7 mpg, this rate translates into $50000 per gpm; and at 37.8 mpg, $100000 per gpm.

8   Kenneth E. Train *et al.*

NONLINEAR LOW is based on gpm but with a rate that decreases as the distance to the zero point gpm increases. The intent of this schedule was to apply a high rate for the vast majority of vehicles that are close to the average fuel efficiency, without imposing extreme feebates on vehicles with very high or low fuel efficiency. The feebate is calculated by comparing the vehicle's gpm with the zero point, taking the square root of the difference, and multiplying by $8000. For most vehicles, the feebate rate per gpm is close to that applied in GPM HIGH; however, the range of feebates (from the highest fee to the largest rebate) is comparable to that in GPM LOW.

SIZE-BASED applies a feebate to cars that accounts for the size of the car. The feebate is proportional to gpm per unit of interior volume, with a rate of $3 750 000 per gpm/cubic foot. For an average sized car of 92 cubic feet, this rate translates into $40 760 per gpm, and is higher for smaller cars and lower for larger cars. Feebates on trucks are the same as in GPM LOW.

5. RESULTS

Table 4 gives the estimated impacts of these six feebate schemes. The impacts are expressed as changes from the base-line forecast to the scenario forecast. Under GPM LOW, new car CAFE improves by 14% compared to the base by 2010, and new truck CAFE improves by 11%. These are fairly substantial improvements given that the feebates are generally less than 1% of the price of the vehicle. Nearly all of the improvement comes from supply-side response, with very little attributable to consumers shifting among the 95 subclasses. Of the 14% improvement in new car CAFE, 13% is due to manufacturers' response and only 1% is due to consumers' response. A similar pattern exists for trucks, and the pattern is evidenced in all the other feebate schemes. This result confirms the findings of Gordon/Levenson, Davis, and Greene who estimated small demand response. The result also emphasizes that the effectiveness of feebates depends critically on manufacturers' response, a topic that we return to in Section 6. One policy implication of this finding is that, for most states, state-level feebates are unlikely to be effective, since manufacturers would probably not adjust the design of vehicles in response to feebates that affect only a small portion of their national sales. For large states, such as California (which captures over 10% of the vehicle market), manufacturers might respond to state feebates, particularly since vehicles are already specially equipped to meet California's emissions standards. Also, if coalitions of states initiated state-level feebates at about the same time, manufacturers might respond. Generally, however, federal feebates will be needed if the critical manufacturer response is to be obtained.

On-road fuel efficiency is forecast to improve by 10% for the stock of all vehicles by the year 2010 under GPM LOW. Since increased fuel efficiency means lower operating costs, vehicle miles travelled increases. This increased driving provides extra surplus for consumers (which is captured

Table 4. Estimated impact of feebates

Schedule	GPM LOW	GPM HIGH	ONE ZERO POINT	MPG LOW	NON-LIN LOW	SIZE-BASED
Increase in CAFE in 2010 (%)						
New cars	14	18	14	18	16	12
New trucks	11	13	11	11	12	10
Entire on-road vehicle stock	10	13	11	12	12	9
Fuel savings (billion gallons)						
Annually in 2010	6.9	8.2	7.1	7.7	7.7	6.3
Cumulative 1995–2010	75.6	89.9	77.8	80.8	83.8	66.7
Total CO_2 emissions reduction (million tons)						
Annually in 2010	68.3	81.2	70.3	76.2	76.2	62.4
Cumulative 1995–2010	750	890	770	800	830	660
New vehicle sales (% change in cumulative discounted sales 1995–2010)						
Total	−0.7	−0.2	0.2	0.0	0.1	−0.8
Domestic vehicles	−1.3	−1.0	−0.1	−0.7	−0.5	0.2
Foreign vehicles	0.4	1.4	0.8	1.3	1.2	−3.2
Benefits to consumers (in 1990 $)						
Annually in 2010						
Per household ($)	82	91	75	87	90	70
All households (billion $)	10.0	11.1	9.1	10.6	11.0	8.5
Cumulative discounted total 1995–2010						
All households (billion $)	51	56	47	52	55	43

in the consumer surplus measure discussed below), but translates into less fuel savings than would be obtained if the fuel efficiency improvement were not accompanied by more driving. About 25% of the potential savings are lost, which is within the range found by previous researchers for the 'take-back' effect. (Greene, 1991*b*, reviews and critiques these previous studies and argues that the take-back effect is actually lower, in the range of 10–15% or less. If Greene's estimates are correct, the fuel savings and emissions reductions from feebates are larger than those indicated in our analysis.) Even with this 25% take-back, the estimated reductions in fuel consumption and CO_2 emissions are fairly large, constituting 6.9 billion gallons of fuel and 68.3 million tons of CO_2 emissions saved per year by 2010 under GPM LOW compared to the base forecast. These figures represent a 7% reduction in fuel use and emissions from personal vehicles.

Consumers benefit from the feebates. Under GPM LOW, consumer surplus rises by $82 per household per year by 2010 (expressed in 1990 $). Summed over all households, the benefits to consumers approach $10 billion annually by 2005, and stays nearly constant thereafter. Consumer surplus increases for all income groups. The consumer surplus incorporates the cost of fuel-efficiency technologies that manufacturers adopt, since these costs are represented as higher prices for the vehicles. Consumer surplus also incorporates the benefits of greater travel due to lower operating costs. Essentially, the feebates induce manufacturers to supply vehicles whose attributes consumers like more. We return to this issue, and its plausibility, in Section 6.

Feebates induce a rise in new vehicle sales at first followed by a drop. Because consumers favor the more efficient vehicles produced under the feebates, new vehicle sales are stimulated in the early years of the feebates. For example, new U.S. vehicle sales in 1995 are forecast to increase by 1.9% under GPM LOW compared to the baseline. This increase in sales diminishes, and eventually reverses, as the more efficient vehicles penetrate the entire stock and used vehicles become more attractive. By 2010, new vehicle sales under GPM LOW are forecast to drop 3.5% below baseline sales. Over the entire period 1995–2010, total sales are forecast to decrease by 0.75%. For evaluating the effect of feebates on manufacturers, the discounted sales stream is most relevant. Using an 8% discount rate, cumulative discounted sales over 1995–2010 drop by 0.70% under GPM LOW compared to the baseline. The percent decrease in the discounted total is somewhat smaller than when not discounted because annual sales rise and then fall, and discounting reduces the impact of more distant events.

When feebates are imposed, the fees tend to fall largely on the purchasers of domestic vehicles, while the rebates are obtained by the buyers of foreign vehicles. The GPM LOW feebates result in an average fee in 1995 of $80 on domestic vehicles, which in effect subsidizes an average rebate of $150 on foreign vehicles. Over time, domestic manufacturers are predicted to make up some of the difference by capturing the larger untapped fuel efficiency potential in their vehicles. However, the disparity is forecast to remain through 2010. As a result of this difference, foreign manufacturers gain market share. Over the entire period 1995–2010, discounted cumulative sales rise by 0.4% for foreign manufacturers and drop by 1.3% for domestic manufacturers.

We now turn to the other feebate schedules, each of which provides information about the appropriate design of feebates. Under GPM HIGH, the feebate rate is twice as high as under GPM LOW. However, the impact is far less than double: new car CAFE improves by 18% under GPM HIGH compared to 14% under GPM LOW, and fuel savings and CO_2 reductions are only one-fifth greater. These results suggest that there are highly decreasing returns to raising the feebate rate.

With different zero points for cars and trucks, a fee could be levied on a car that achieves the same gpm as a truck that receives a rebate, and a consumer could be induced to buy a truck that receives a rebate instead of a more efficient car that has a fee. ONE ZERO POINT applies the same zero point to cars and trucks. The impacts are nearly exactly the same as GPM LOW, for easily identifiable reasons. Manufacturer response depends only on the feebate rate (i.e.the change in feebate that comes from an improvement in a vehicle's gpm.) Since the feebate rate is not affected by the choice of zero point, manufacturer response is the same in ONE ZERO POINT as in GPM LOW. Consumers shift from trucks to cars, since ONE ZERO POINT feebates raise the effective price (including feebate) of trucks relative to cars. However, since consumer response to feebates is very small, the shift to cars is small. Only about 0.2 billion gallons of fuel and 2 million tons of CO_2 emissions are saved beyond that obtained under the GPM LOW feebates.

The MPG LOW feebates are linear in mpg; their effective rate per gpm therefore rises with efficiency. The effective rate per gpm is equivalent to GPM LOW at 26.7 mpg and equivalent to

GPM HIGH at 37.8 mpg. Since cars in the forecast period are generally closer to 37.8 mpg, the impact of MPG LOW on cars is about the same as in GPM HIGH: an 18% improvement in fuel efficiency. For trucks, the impact is about the same as in GPM LOW. The overall impacts are therefore about midway between the impacts of GPM HIGH and GPM LOW.

From a policy perspective, mpg-based feebates are perhaps useful because the public is accustomed to thinking in terms of mpg. However, if the purpose of feebates is to account for externalities in fuel use, mpg-based feebates are hard to justify since they apply a non-constant value to a gallon of fuel saved. For example, an increase from 10 to 11 mpg receives the same change in feebates as an increase from 60 to 61 mpg, despite the fact that the former saves 33 times as much fuel. The forecasts indicate that gpm-based feebates can be designed that attain the same total impacts as mpg-based feebates, such that the choice between the two forms hinges on the tradeoff between the advantages of public familiarity vs the advantages of having a constant value assigned to a gallon saved.

The NONLINEAR LOW feebates apply a higher rate for vehicles near the zero point and lower rates for vehicles that are further from the zero point. For most vehicles, the rate is close to GPM HIGH; however the range of feebates is closer to GPM LOW. The impacts, as expected, are about midway between GPM HIGH and GPM LOW. The one exception is consumer surplus, which is very near the GPM HIGH level. The reason is that the very large fees that linear feebates impose on vehicles with very low fuel efficiency (which reduce consumer surplus) are mitigated under the NONLINEAR feebate formula. The question from a policy perspective is whether this benefit compensates for the complexity of the NONLINEAR feebates.

Domestic manufacturers are generally hurt, though only very slightly, by the feebate schemes discussed so far. A proposal that has been advanced by manufacturers and others (e.g. DeCicco, 1994) is to base feebates on the fuel efficiency of the car relative to its size, thereby recognizing that larger cars provide more 'service' and hence are expected to consume more fuel than smaller cars. For cars, our SIZE-BASED scenario applies feebates that are linear in gpm per unit of interior space with a rate that, for an average sized car, is somewhat lower than GPM LOW. (For an average-sized car, SIZE-BASED feebates translate into a rate of $40 760 per gpm while GPM LOW applies a rate of $50 000 per gpm.) For trucks, an appropriate measure of size is not so apparent, and the GPM LOW feebates are applied in the SIZE-BASED scenario.

The SIZE-BASED feebacks are predicted to have less impact on fuel efficiency and CO_2 emissions than any of the other feebate schedules that we consider. It seems, however, that the lower impact on efficiency and emissions is primarily due to the feebate rate being lower on average in this scenario, rather than the size-basing *per se*. For example, total fuel savings and emissions reductions by 2010 are estimated to be 8.7% lower under size-based than under GPM LOW (e.g. 6.3 billion gallons per year under size-based compared to 6.9 billion gallons under GPM LOW.) However, as stated above, the feebate rate for cars is about 20% lower under our SIZE-BASED scenario than GPM LOW. Though we have not estimated the effects, we expect that SIZE-BASED feebates with a comparable average rate would achieve about the same results as GPM LOW, and that SIZE-BASED feebates with higher average rates would obtain greater savings.* The reason that size-basing *per se* does not reduce energy savings is similar to that for ONE ZERO POINT. Manufacturers respond to the feebate rate, not the absolute level of feebates. Size-basing causes some vehicles to have a higher rate per gpm and others to have a lower rate; however, the average rate can be set to be the same as under consumption-based feebates. With the average rate the same, the manufacturer response aggregated over all vehicles is about the same.† Consumers respond to the level of feebate as well as the rate. However, consumer response is extremely small under all the feebate schemes. Therefore, whatever size-basing does to consumer response is inconsequential since consumer response is small no matter what.

---

*We specified the SIZE-BASED feebates such that they would have the same range of feebates (i.e. the distance between the largest fee and the largest rebate) as under GPM LOW. If the rate per gpm/cubic-foot was raised for the size-based feebates such that the average rate per gpm matched that of GPM LOW, then the range under SIZE-BASED would be much larger than that under GPM LOW. If the range of feebates is an issue for policymakers, then our results indicate that size-basing provides slightly less impact on fuel efficiency than consumption-based feebates that have a comparable range.

†Given the non-linearity of the technology penetration curve and the fact that different costs and benefits apply for each technology for each vehicle subclass, manufacturer response is not necessarily the same for any feebates that have the same average rate. It is an outcome of the analysis that, in the case of the SIZE-BASED scenario, the smaller manufacturer response to the lower rates for large cars approximately balances out the larger response to the higher rates for small cars.

As expected, domestic manufacturers are indeed helped by size-basing. The discounted cumulative sales of domestic vehicles rise by 0.2% under SIZE-BASED compared to a drop of 1.3% under GPM LOW. Sales of foreign vehicles drop by 3.2% under SIZE-BASED, whereas they rose slightly under the non-size based scenarios; whether this is a benefit or loss depends, of course, on one's perspective. Consumer surplus is lower under SIZE-BASED than GPM LOW, but is still positive. The results indicate that SIZE-BASED feebates can benefit domestic manufacturers and consumers (relative to no feebates) while attaining close to the same improvements in fuel efficiency and reductions in fuel consumption and CO_2 emissions as under consumption-based feebates.

One caveat is important in this regard. Under SIZE-BASED feebates, manufacturers have an incentive to increase the size of their vehicles, even if doing so decreases the fuel efficiency of the vehicle—as long as the vehicle's gpm increases by a smaller proportion than its size. The supply component of our analysis assumes that the size of vehicles is held constant, and thereby does not incorporate the effect of this incentive. Insofar as manufacturers would increase the size of vehicles, the impact of SIZE-BASED feebates on fuel efficiency, fuel consumption, and CO_2 emissions will be lower than predicted. In this case, size-basing would produce benefits for domestic manufacturers at the cost of smaller fuel savings and CO_2 reductions.

6. DISCUSSION

Perhaps the most interesting, and controversial, aspect of the findings is that feebates can induce manufacturers to produce vehicles that increase consumer surplus. The question arises: if consumers like the extra fuel efficiency sufficiently to pay for it through higher prices, then why don't manufacturers improve the fuel efficiency of their vehicles without the inducement of feebates? Stated more succinctly: why don't free market forces result in the amount of fuel efficiency that consumers are willing to pay for?*

A reviewer and others have suggested that one answer perhaps lies in the market risk associated with investing in new technologies. Feebates provide incentives for the new technologies that can overcome this market barrier. Of course, risk is a market barrier (in the sense of resulting in sub-optimal outcomes) only if each individual firm bears more risk than society does, such that investment in the technologies is desirable from a social perspective but not from each individual manufacturer's perspective.

Theoretical analyses have indicated that market forces will result in a non-optimal amount of product variety (Spence, 1976a,b; Dixit and Stiglitz, 1977, 1979; Scherer, 1979), a non-optimal level of product durability (Schmalensee, 1974; Su, 1975; Liebowitz, 1982; Rust, 1986), and too few low-quality products (Anderson *et al.*, 1992). Each of these studies specifies situations that differ from ours in important ways. (For example, Anderson *et al.* assume no fixed cost associated with higher quality and that firms are randomly assigned a level of quality and choose to either produce or not produce.) However, the analyses of product durability identify a phenomenon that is very relevant for fuel efficiency. In particular, these studies point out that the manufacturer of a durable good is constrained in its market power by the scrappage decision of the customer. If the customer can decide how long to hold the durable good before scrapping it, then the manufacturer cannot raise price as far as it could otherwise since doing so will induce customers to hold onto the product longer—which means that they will buy the product less often. Essentially, the manufacturer competes with its own previously sold products. A similar phenomenon is evident in our forecasts of new vehicle sales. At first, sales increase because consumers prefer the more fuel-efficient vehicles (i.e. consumers are more-than-willing to pay for the cost of the fuel efficiency improvements). However, the new vehicles later enter the stock of used vehicles. Since used vehicles compete with new vehicles in the purchase decisions of customers, the improved used vehicles cause the sales of new vehicles to decrease. Essentially, by improving new vehicles, manufacturers are creating greater competition for themselves when these new vehicles become used. This observation does not imply that manufacturers will provide less of all attributes than

---

*To be sure that the effects on consumer surplus and new vehicle sales were attributable to the higher fuel efficiency within the vehicle and not to the change in prices represented by the feebates *per se*, we calculated consumer surplus and new vehicle sales under the vehicle offerings that manufacturers were predicted to provide under GPM LOW but without the feebates being added/subtracted from the price of the vehicles. (The cost of the technologies that improved the vehicle's fuel efficiency was included in the price, but the feebates were not.) Consumer surplus was higher in this scenario than in the baseline.

consumers are willing to pay for. As Rust (1986) demonstrates for product durability, competition with the scrappage market can induce manufacturers to produce either more or less durability than they would without such competition, depending on a variety of factors. The observation simply implies that manufacturers, because of the competition with used vehicles that they themselves can affect, cannot be expected to produce the level of each attribute that customers are willing to pay for. In the case of fuel efficiency, our empirical analysis suggests that manufacturers are producing less fuel efficiency than consumers are willing to pay for and, since cumulative discounted sales are forecast to decrease when fuel efficiency is raised, that the under-provision of fuel efficiency is profit-maximizing for manufacturers. However, this direction of suboptimality need not be the case necessarily or for all attributes. Theoretical analysis is needed to identify the conditions under which an attribute is under- or over-provided.

Our analysis indicates that foreign manufacturers benefit from non-size-based feebates, and domestic manufacturers benefit from size-based feebates. The question arises: why don't manufacturers who would benefit from a particular type of feebate introduce the changes that it would make under that feebate schedule without the feebates? That is: why are the feebates needed to induce changes by manufacturers who would benefit from the changes? If the decision variables are considered to be vehicle attributes, Nash equilibrium in a static game requires that no manufacturer be able to increase its profits by changing the attributes of its products assuming that all other manufacturers keep their vehicle attributes the same. Our results do not, in themselves, imply that any one manufacturer could increase its profits by unilaterally increasing the fuel efficiency of its vehicles. Rather, our results indicate that some manufacturers benefit when all manufacturers increase the fuel efficiency of their vehicles. Our model does not distinguish manufacturers, and so an analysis of the profit impacts of unilateral changes by individual manufacturers cannot be conducted on our model. Nevertheless, we suspect that if our vehicle classes were further disaggregated by manufacturer, the model would indeed predict that some manufacturers could unilaterally benefit from increasing the fuel efficiency of their vehicles. This would indicate that the model implies that the market is not currently in a Nash equilibrium of a static game where the decision variables are the attributes of the vehicles. The non-cooperative game that the manufacturers are playing might be different. Tirole (1992) discusses a variety of theories of dynamic oligopoly behavior that have the property that in equilibrium an individual firm could enhance its profits by making a marginal change in price or quality if other firms did not change their prices or qualities; the situation is in equilibrium because each firm realizes that its competitors will indeed change their prices and/or qualities in response and that its own profits will fall after all the changes. The issue of equilibrium in the automobile market has only recently been examined. Berry *et al.* (1992), Pakes *et al.* (1993), and Goldberg (1994) estimate and apply models of oligopoly behavior in the automobile industry, assuming Nash equilibrium with the decision variable being price given the other attributes of the vehicles. These authors discuss the importance of incorporating the endogeneity of non-price attributes in future work.

Our analysis suggests that the effectiveness of feebates is critically dependent on manufacturers' increasing the fuel efficiency of their vehicles in response to the feebates. However, little is known about the determination of product attributes, either in a free market or with intervention. Perhaps the most important implication of the study, therefore, is in highlighting the need for far more research on manufacturer behavior.

REFERENCES

Anderson, S., dePalma, A. and Thisse, J.-F. (1992) *Discrete Choice Theory of Product Differentiation*. MIT Press, Cambridge, MA.

Berry, S., Levinsohn, J. and Pakes, A. (1992) Automobile prices in market equilibrium. Working paper, Department of Economics, Yale University.

Charles River Associates (1991) *Policy Alternatives for Reducing Petroleum Use and Greenhouse Gas Emissions*. CRA Report Number 766.00.

Davis, W. B. (1991) Economic incentives to improve fuel economy and reduce carbon dioxide emissions in California automobiles. Testimony before the California Energy Commission, Conservation Report Hearings.

Davis, W. B., Levine, M. and Train, K. (1994) *Effects of Feebates on Vehicle Fuel Economy, Carbon Dioxide Emissions, and Related Factors*. Technical report 2, Energy Efficiency in the U.S. Economy series, Office of Policy, Planning, and Program Evaluation, U.S. Department of Energy.

DeCicco, J. (1994) Size-based fees and feebates for reducing light vehicle CO_2 emissions. Working paper, American Council for an Energy Efficient Economy, Washington, D.C.

DeCicco, J., Geller, H. and Morrill, J. (1993) *Feebates for Fuel Economy: Market Incentives for Encouraging Production and Sales of Efficient Vehicles.* Report of the American Council for an Energy Efficient Economy, Washington, D.C.

Dixit, A. and Stiglitz, J. (1977) Monopolistic competition and optimum product diversity. *American Economic Review* **67**, 297–308.

Dixit, A. and Stiglitz, J. (1979) Monopolistic competition and optimum product diversity: a reply. *American Economic Review* **69**, 961–963.

DRI, (1991) *An Analysis of Public Policy Measures to Reduce Carbon Dioxide Emissions from the U.S. Transportation Sector.* Report to the Energy Policy Branch, Office of Policy, Planning, and Evaluation, U.S. Environmental Protection Agency.

Duleep, K. G. (1992) *NEMS Transportation Sector Model.* Energy and Environmental Analysis Inc., Report.

EIA, (1992) *Annual Energy Outlook.* U.S. Energy Information Administration.

Goldberg, P. (1994) Product differentiation and oligopoly in international markets: the case of the U.S. automobile market. Working paper, Department of Economics, Princeton University.

Gordon, D. and Levenson, L. (1989) *DRIVE + : A Proposal for California to Use Consumer Fees and Rebates to Reduce New Motor Vehicle Emissions and Fuel Consumption.* Lawrence Berkeley Laboratory Report.

Greene, D. L. (1991) Short-run pricing strategies to increase corporate average fuel economy. *Economics Inquiry* **29**, 101–114.

Greene, D. L. (1991b) Vehicle use and fuel economy: how big is the 'rebound' effect? Working paper, Center for Transportation Analysis, Energy Division, Oak Ridge National Laboratory.

Liebowitz, S. (1982) Durability, market structure, and new-used goods models. *American Economic Review* **72**, 816–824.

McFadden, D. (1978) Modeling the choice of residential location. In *Spatial Interaction Theory and Planning Models*, eds A. Karquist *et al.* North-Holland, Amsterdam.

Pakes, A., Berry, S. and Levinsohn, J. (1993) Applications and limitations of some recent advances in empirical industrial organization: price indexes and the analysis of environmental change. *American Economic Review* **83**, 241–246.

Rust, J. (1986) When is it optimal to kill off the market for used durable goods?. *Econometrica* **54**, 65–86.

Scherer, F. M. (1979) The welfare economics of product variety: an application to the ready-to-eat cereals industry. *Journal of Industrial Economics* **28**, 113–134.

Schmalensee, R. (1974) Market structure, durability, and maintenance effort. *Review of Economics Studies* **41**, 277–287.

Small, K. and Rosen, H. (1981) Applied welfare economics with discrete choice models. *Econometrica* **49**, 105–130.

Spence, M. (1976) Product selection, fixed costs, and monopolistic competition. *Review of Economics Studies* **43**, 217–235.

Spence, M. (1976) Product differentiation and welfare. *American Economics Review* **66**, 407–414.

Su, T. (1975) Durability of consumption goods reconsidered. *American Economics Review* **65**, 148–157.

Tirole, J. (1992) *The Theory of Industrial Organization.* MIT Press, Cambridge, MA.

Train, K. (1985) Discount rates in consumers' energy-related decisions: a review of the literature. *Energy* **10**, 1243–1253.

Train, K. (1986) *Qualitative Choice Analysis: Theory, Econometrics, and an Application to Automobile Demand.* MIT Press, Cambridge, MA.

Westbrook, F. and Patterson, P. (1989) Changing patterns and their effect on fuel economy. Paper presented at the 1989 Government/Industry Meeting of the Society of Automotive Engineers, Washington, D.C.

B   Tradable Permits

[9]

THE USE OF STANDARDS AND PRICES FOR PROTECTION OF THE ENVIRONMENT

*William J. Baumol and Wallace E. Oates**

Princeton University, Princeton, N.J., USA

Summary

In the Pigouvian tradition, economists have frequently proposed the adoption of a system of unit taxes (or subsidies) to control externalities, where the tax on a particular activity is equal to the marginal social damage it generates. In practice, however, such an approach has rarely proved feasible because of our inability to measure marginal social damage.

This paper proposes that we establish a set of admittedly somewhat arbitrary standards of environmental quality (e.g., the dissolved oxygen content of a waterway will be above x per cent at least 99 per cent of the time) and then impose a set of charges on waste emissions sufficient to attain these standards. While such *resource-use prices* clearly will not in general produce a Pareto-efficient allocation of resources, it is shown that they nevertheless do possess some important optimality properties and other practical advantages. In particular, it is proved that, for any given vector of final outputs such prices can achieve a specified reduction in pollution levels at minimum cost to the economy, even in the presence of firms with objectives other than that of simple profit maximization.

In the technicalities of the theoretical discussion of the tax-subsidy approach to the regulation of externalities, one of the issues most critical for its application tends to get the short end of the discussion. Virtually every author points out that we do not know how to calculate the ideal Pigouvian tax or subsidy levels in practice, but because the point is rather obvious rarely is much made of it.

This paper reviews the nature of the difficulties and then proposes a substitute approach to the externalities problem. This alternative, which we shall call the environmental pricing and standards procedure, represents what we consider to be as close an approximation as one can generally achieve in practice to the spirit of the Pigouvian tradition. Moreover, while this method does not aspire to anything like an optimal allocation of resources, it will be shown to possess some important optimality properties.

---

* The authors are members of the faculty at Princeton University. They are grateful to the Ford Foundation whose support greatly facilitated the completion of this paper.

Swed. J. of Economics 1971

1. Difficulties in Determining the Optimal Structure of Taxes and Subsidies

The proper level of the Pigouvian tax (subsidy) upon the activities of the generator of an externality is equal to the marginal net damage (benefit) produced by that activity.[1] The difficulty is that it is usually not easy to obtain a reasonable estimate of the money value of this marginal damage. Kneese & Bower report some extremely promising work constituting a first step toward the estimation of the damage caused by pollution of waterways including even some quantitative evaluation of the loss in recreational benefits. However, it is hard to be sanguine about the availability in the foreseeable future of a comprehensive body of statistics reporting the marginal net damage of the various externality-generating activities in the economy. The number of activities involved and the number of persons affected by them are so great that on this score alone the task assumes Herculean proportions. Add to this the intangible nature of many of the most important consequences—the damage to health, the aesthetic costs—and the difficulty of determining a money equivalent for marginal net damage becomes even more apparent.

This, however, is not the end of the story. The optimal tax level on an externality generating activity is not equal to the marginal net damage it generates *initially*, but rather to the damage it would cause if the level of the activity had been adjusted to its *optimal* level. To make the point more specifically, suppose that each additional unit of output of a factory now causes 50 cents worth of damage, but that after the installation of the appropriate smoke-control devices and other optimal adjustments, the marginal social damage would be reduced to 20 cents. Then a little thought will confirm what the appropriate mathematics show: the correct value of the Pigouvian tax is 20 cents per unit of output, that is, the marginal cost of the smoke damage *corresponding to an optimal situation*. A tax of 50 cents per unit of output corresponding to the current smoke damage cost would lead to an excessive reduction in the smoke-producing activity, a reduction beyond the range over which the marginal benefit of decreasing smoke emission exceeds its marginal cost.

The relevance of this point for our present discussion is that it compounds enormously the difficulty of determining the optimal tax and benefit levels. If there is little hope of estimating the damage that is currently generated, how much less likely it is that we can evaluate the damage that would occur in an optimal world which we have never experienced or even described in quantitative terms.

There is an alternative possibility. Instead of trying to go directly to the optimal tax policy, one could instead, as a first approximation, base a set of

---

[1] We will use the term marginal *net* damage to mean the difference between marginal social and private damage (or cost).

taxes and subsidies on the current net damage (benefit) levels. Then as outputs and damage levels were modified in response to the present level of taxes, the taxes themselves would in turn be readjusted to correspond to the new damage levels. It can be hoped that this will constitute a convergent, iterative process with tax levels affecting outputs and damages, these in turn leading to modifications in taxes, and so on. It is not clear, however, even in theory, whether this sequence will in fact converge toward the optimal taxes and resource allocation patterns. An extension of the argument underlying some of Coase's illustrations can be used to show that convergence cannot always be expected. But even if the iterative process were stable and were in principle capable of yielding an optimal result, its practicality is clearly limited. The notion that tax and subsidy rates can be readjusted quickly and easily on the basis of a fairly esoteric marginal net damage calculation does not seem very plausible. The difficulty of these calculations has already been suggested, and it is not easy to look forward with equanimity to their periodic revision, as an iterative process would require.

In sum, the basic trouble with the Pigouvian cure for the externalities problem does not lie primarily in the technicalities that have been raised against it in the theoretical literature but in the fact that we do not know how to determine the dosages that it calls for. Though there may be some special cases in which one will be able to form reasonable estimates of the social damages, in general we simply do not know how to set the required levels of taxes and subsidies.

2. The Environmental Pricing and Standards Approach

The economist's predilection for the use of the price mechanism makes him reluctant to give up the Pigouvian solution without a struggle. The inefficiencies of a system of direct controls, including the high real enforcement costs that generally accompany it, have been discussed often enough; they require no repetition here.

There is a fairly obvious way, however, in which one can avoid recourse to direct controls and retain the use of the price system as a means to control externalities Simply speaking, it involves the selection of a set of somewhat arbitrary standards for an acceptable environment. On the basis of evidence concerning the effects of unclean air on health or of polluted water on fish life, one may, for example, decide that the sulfur-dioxide content of the atmosphere in the city should not exceed x percent, or that the oxygen demand of the foreign matter contained in a waterway should not exceed level y, or that the decibel (noise) level in residential neighborhoods should not exceed z at least 99 % of the time. These acceptability standards, x, y and z, then amount to a set of constraints that society places on its activities. They represent the decision-maker's subjective evaluation of the minimum standards that must be met in order to achieve what may be described in persuasive terms as "a reasonable quality of life". The defects of the concept will immediately be clear to the reader, and, since we do not want to minimize them, we shall examine this problem explicitly in a later section of the paper.

For the moment, however, we want to emphasize the role of the price system in the implementation of these standards. The point here is simply that the public authority can levy a uniform set of taxes which would in effect constitute a set of prices for the private use of social resources such as air and water. The taxes (or prices) would be selected so as to achieve specific acceptability standards rather than attempting to base them on the unknown value of marginal net damages. Thus, one might tax all installations emitting wastes into a river at a rate of $t(b)$ cents per gallon, where the tax rate, t, paid by a particular polluter, would, for example, depend on b, the BOD value of the effluent, according to some fixed schedule.[1] Each polluter would then be given a financial incentive to reduce the amount of effluent he discharges and to improve the quality of the discharge (i.e., reduce its BOD value). By setting the tax rates sufficiently high, the community would presumably be able to achieve whatever level of purification of the river it desired. It might even be able to eliminate at least some types of industrial pollution altogether.[2]

Here, if necessary, the information needed for iterative adjustments in tax rates would be easy to obtain: if the initial taxes did not reduce the pollution of the river sufficiently to satisfy the preset acceptability standards, one would simply raise the tax rates. Experience would soon permit the authorities to estimate the tax levels appropriate for the achievement of a target reduction in pollution.

One might even be able to extend such adjustments beyond the setting of the tax rates to the determination of the acceptability standards themselves. If, for example, attainment of the initial targets were to prove unexpectedly inexpensive, the community might well wish to consider making the standards stricter.[3] Of course, such an iterative process is not costless. It means that at least some of the polluting firms and municipalities will have to adapt their

---

[1] BOD, biochemical oxygen demand, is a measure of the organic waste load of an emission. It measures the amount of oxygen used during decomposition of the waste materials. BOD is used widely as an index of the quality of effluents. However, it is only an approximation at best. Discharges whose BOD value is low may nevertheless be considered serious pollutants because they contain inorganic chemical poisons whose oxygen requirement is nil because the poisons do not decompose. See Kneese and Bower on this matter.
[2] Here it is appropriate to recall the words of Chief Justice Marshall, when he wrote that "The power to tax involves the power to destroy" (McCulloch vs. Maryland, 1819). In terms of reversing the process of environmental decay, we can see, however, that the power to tax can also be the power to restore.
[3] In this way the pricing and standards approach might be adapted to approximate the Pigouvian ideal. If the standards were revised upward whenever there was reason to believe that the marginal benefits exceeded the marginal costs, and if these judgments were reasonably accurate, the two would arrive at the same end product, at least if the optimal solution were unique.

46   *W. J. Baumol and W. E. Oates*

operations as tax rates are readjusted. At the very least they should be warned in advance of the likelihood of such changes so that they can build flexibility into their plant design, something which is not costless (See Hart). But, at any rate, it is clear that, through the adjustment of tax rates, the public authority can realize whatever standards of environmental quality it has selected.

3. Optimality Properties of the Pricing and Standards Technique

While the pricing and standards procedure will not, in general, lead to Pareto-efficient levels of the relevant activities, it is nevertheless true that the use of unit taxes (or subsidies) to achieve the specified quality standards does possess one important optimality property: it is the least-cost method to realize these targets.[1] A simple example may serve to clarify this point. Suppose that it is decided in some metropolitan area that the sulfur-dioxide content of the atmosphere should be reduced by 50 %. An obvious approach to this matter, and the one that often recommends itself to the regulator, is to require each smoke-producer in the area to reduce his emissions of sulfur dioxide by the same 50 %. However, a moment's thought suggests that this may constitute a very expensive way to achieve the desired result. If, at existing levels of output, the marginal cost of reducing sulfur-dioxide emissions for Factory *A* is only one-tenth of the marginal cost for Factory *B*, we would expect that it would be much cheaper for the economy as a whole to assign *A* a much greater decrease in smoke emissions than *B*. Just how the least-cost set of relative quotas could be arrived at in practice by the regulator is not clear, since this obviously would require calculations involving simultaneous relationships and extensive information on each polluter's marginal-cost function.

It is easy to see, however, that the unit-tax approach can *automatically* produce the least-cost assignment of smoke-reduction quotas without the need for any complicated calculations by the enforcement authority. In terms of our preceding example, suppose that the public authority placed a unit tax on smoke emissions and raised the level of the tax until sulfur-dioxide emissions were in fact reduced by 50 %. In response to a tax on its smoke emissions, a cost-minimizing firm will cut back on such emissions until the marginal cost of further reductions in smoke output is equal to the tax. But, since all economic units in the area are subject to the same tax, it follows that the marginal cost of reducing smoke output will be equalized across all activities. This implies that it is impossible to reduce the aggregate cost of the specified decrease in smoke emissions by re-arranging smoke-reduction quotas: any alteration in this pattern of smoke emissions would involve an increase in

---

[1] This proposition is not new. While we have been unable to find an explicit statement of this result anywhere in the literature, it or a very similar proposition has been suggested in a number of places. See, for example, Kneese & Bower, Chapter 6, and Ruff, p. 79.

smoke output by one firm the value of which to the firm would be less than the cost of the corresponding reduction in smoke emissions by some other firm. For the interested reader, a formal proof of this least-cost property of unit taxes for the realization of a specified target level of environmental quality is provided in an appendix to this paper. We might point out that the validity of this least-cost theorem does not require the assumption that firms are profit-maximizers. All that is necessary is that they minimize costs for whatever output levels they should select, as would be done, for example, by a firm that seeks to maximize its growth or its sales.

The cost saving that can be achieved through the use of taxes and subsidies in the attainment of acceptability standards may by no means be negligible. In one case for which comparable cost figures have been calculated, Kneese & Bower (p. 162) report that, with a system of uniform unit taxes, the cost of achieving a specified level of water quality would have been only about half as high as that resulting from a system of direct controls. If these figures are at all representative, then the potential waste of resources in the choice between tax measures and direct controls may obviously be of a large order. Unit taxes thus appear to represent a very attractive method for the realization of specified standards of environmental quality. Not only do they require relatively little in the way of detailed information on the cost structures of different industries, but they lead automatically to the least-cost pattern of modification of externality-generating activities.

4. Where the Pricing and Standards Approach is Appropriate

As we have emphasized, the most disturbing aspect of the pricing and standards procedure is the somewhat arbitrary character of the criteria selected. There does presumably exist some optimal level of pollution (i.e., quality of the air or a waterway), but in the absence of a pricing mechanism to indicate the value of the damages generated by polluting activities, one knows no way to determine accurately the set of taxes necessary to induce the optimal activity levels.

While this difficulty certainly should not be minimized, it is important at the outset to recognize that the problem is by no means unique to the selection of acceptability standards. In fact, as is well known, it is a difficulty common to the provision of nearly all public goods. In general, the market will not generate appropriate levels of outputs where market prices fail to reflect the social damages (or benefits) associated with particular activities. As a result, in the absence of the proper set of signals from the market, it is typically necessary to utilize a political process (i.e., a method of collective choice) to determine the level of the activity.[1] From this perspective, the selec-

---

[1] As Coase and others have argued, voluntary bargains struck among the interested parties may in some instances yield an efficient set of activity levels in the presence of externalities. However, such coordinated, voluntary action is typically possible only in small groups. One can hardly imagine, for example, a voluntary bargaining process involving all the persons in a metropolitan area and resulting in a set of payments that would generate efficient levels of activities affecting the smog content of the atmosphere.

48   *W. J. Baumol and W. E. Oates*

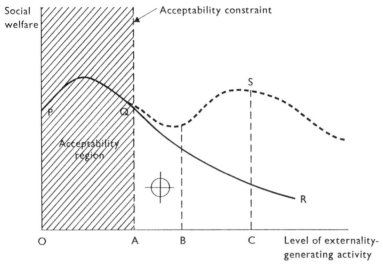

Fig. 1

tion of environmental standards can be viewed as a particular device utilized in a process of collective decision-making to determine the appropriate level of an activity involving external effects.

Since methods of collective choice, such as simple-majority rule or decisions by an elected representative, can at best be expected to provide only very rough approximations to optimal results, the general problem becomes one of deciding whether or not the malfunction of the market in a certain case is sufficiently serious to warrant public intervention. In particular, it would seem to us that such a blunt instrument as acceptability standards should be used only sparingly, because the very ignorance that serves as the rationale for the adoption of such standards implies that we can hardly be sure of their consequences.

In general, it would seem that intervention in the form of acceptability standards can be utilized with any degree of confidence only where there is clear reason to believe that the existing situation imposes a high level of social costs *and* that these costs can be significantly reduced by feasible decreases in the levels of certain externality-generating activities. If, for example, we were to examine the functional relationship between the level of social welfare and the levels of particular activities which impose marginal net damages, the argument would be that the use of acceptability standards is justified only in those cases where the curve, over the bulk of the relevant range, is both decreasing and steep. Such a case is illustrated in Fig. 1 by the curve *PQR*. In a case of this kind, although we obviously will not have an accurate knowledge of the relevant position of the curve, we can at least have some assurance that the selection of an acceptability standard and the imposition of a unit tax sufficient to realize that standard will lead to an increase in social welfare. For example, in terms of the curve PQR in Fig. 1, the levying of a tax sufficient to reduce smoke outputs from level OC to OA to ensure that the quality of the air meets the specified environmental standards would obviously increase social welfare.[1]

On the other hand, if the relationship between social welfare and the level of the externality-generating activity is not monotonically decreasing, the changes resulting from the imposition of an acceptability standard (e.g., a move from S to Q in Fig. 1) clearly may lead to a reduction in welfare. Moreover, even if the function were monotonic but fairly flat, the benefits achieved might not be worth the cost of additional intervention machinery that new legislation requires, and it would almost certainly not be worth the risk of acting with highly imperfect, inconclusive information.

In some cases, notably in the field of public utility regulation, some economists have criticized the employment of acceptability standards on both these grounds; they have asserted that the social costs of monopolistic misallocation of resources are probably not very high (i.e., the relevant portion of the social-welfare curve in Fig. 1 is not steep) and that the regulation can itself introduce inefficiencies in the operations of the regulated industries.

Advocacy of environmental pricing and standards procedures for the control of externalities must therefore rest on the belief that in this area we do have a clear notion of the general shape of the social welfare curve. This will presumably hold true where the evidence indicates, first that a particular externality really does have a substantial and unambiguous effect on the quality of life, if, for example, it makes existence very unpleasant for everyone or constitutes a serious hazard to health; and second that reductions in the levels of these activities do not themselves entail huge resource costs. On the first point, there

---

[1] The relationship depicted in Fig. 1 is to be regarded as an intuitive device employed for pedagogical purposes, not in any sense as a rigorous analysis. However, some further explanation may be helpful. The curve itself is not a social-welfare function in the usual sense; rather it measures in terms of a numeraire (kronor or dollars) the value, summed over all individuals, of the benefits from the output of the activity minus the private *and* net social costs. Thus, for each level of the activity, the height of the curve indicates the *net* benefits (possibly negative) that the activity confers on society. The acceptability constraint indicates that level of the activity which is consistent with the specified minimum standard of environmental quality (e.g., that level of smoke emissions from factories which is sufficiently low to maintain the quality of the air in a particular metropolitan area). There is an ambiguity here in that the levels of several different activities may jointly determine a particular dimension of environmental quality, e.g., the smoke emissions of a number of different industries will determine the quality of the air. In this case, the acceptable level of pollutive emissions for the firm or industry will clearly depend on the levels of emissions of others. If, as we discussed earlier, unit taxes are used to realize the acceptability standards, there will result a least-cost pattern of levels of the relevant externality-generating activities. If we understand the constraint in Fig. 1 to refer to the activity level indicated by this particular solution, then this ambiguity disappears.

50   *W. J. Baumol and W. E. Oates*

is growing evidence that various types of pollutants do in fact have such un-
fortunate consequences, particularly in areas where they are highly concentrated. [On this see, for instance, Lave & Seskin]. Second, what experience
we have had with, for example, the reduction of waste discharges into waterways suggests that processes involving the recycling and reuse of waste materials can frequently be achieved at surprisingly modest cost.[1] In such cases
the rationale for the imposition of environmental standards is clear, and it
seems to us that the rejection of such crude measures on the grounds that they
will probably violate the requirements of optimality may well be considered
a kind of perverse perfectionism.

It is interesting in this connection that the pricing and standards approach
is not too different in spirit from a number of economic policy measures that
are already in operation in other areas. This is significant for our discussion,
because it suggests that regulators know how to work with this sort of approach
and have managed to live with it elsewhere. Probably the most noteworthy
example is the use of fiscal and monetary policy for the realization of macro-
economic objectives. Here, the regulation of the stock of money and the
availability of credit along with adjustments in public expenditures and tax
rates are often aimed at the achievement of a selected target level of employment or rate of inflation. Wherever prices rise too rapidly or unemployment
exceeds an "acceptable" level, monetary and fiscal variables are readjusted in
an attempt to "correct" the difficulty. It is noteworthy that this procedure is
also similar to the pricing and standards approach in its avoidance of direct
controls.

Other examples of this general approach to policy are not hard to find.
Policies for the regulation of public-utilities, for instance, typically utilize a
variety of standards such as profit-rate ceilings (i.e., "fair rates of return")
to judge the acceptability of the behavior of the regulated firm. In the area
of public education, one frequently encounters state-imposed standards (e.g.,
subjects to be taught) for local school districts which are often accompanied
by grants of funds to the localities to help insure that public-school programs
meet the designated standards. What this suggests is that public administrators
are familiar with this general approach to policy and that the implementation
of the pricing and standards technique should not involve insurmountable
administrative difficulties. For these reasons, the achievement of specified
environmental standards through the use of unit taxes (or subsidies) seems to
us to possess great promise as a workable method for the control of the quality
of the environment.

---

[1] Some interesting discussions of the feasibility of the control of waste emissions into
waterways often at low cost are contained in Kneese & Bower. In particular, see their
description of the control of water quality in the Ruhr River in Germany.

5. Concluding Remarks

It may be useful in concluding our discussion simply to review the ways in which the pricing and standards approach differs from the standard Pigouvian-prescription for the control of externalities.

(1) Under the Pigouvian technique, unit taxes (or subsidies) are placed on externality-generating activities, with the level of the tax on a particular activity being set equal to the marginal net damage it generates. Such taxes (if they could be determined) would, it is presumed, lead to Pareto-efficient levels of the activities.

(2) In contrast, the pricing and standards approach begins with a predetermined set of standards for environmental quality and then imposes unit taxes (or subsidies) sufficient to achieve these standards. This will not, in general, result in an optimal allocation of resources, but (as is proved formally in the appendix) the procedure does at least represent the least-cost method of realizing the specified standards.

(3) The basic appeal of the pricing and standards approach relative to the Pigouvian prescription lies in its workability. We simply do not, in general, have the information needed to determine the appropriate set of Pigouvian taxes and subsidies. Such information is not, however, necessary for our suggested procedure.

(4) While it makes no pretense of promising anything like an optimal allocation of resources, the pricing and standards technique can, in cases where external effects impose high costs (or benefits), at least offer some assurance of reducing the level of these damages. Moreover, the administrative procedures—the selection of standards and the use of fiscal incentives to realize these standards—implied by this approach are in many ways quite similar to those used in a number of current public programs. This, we think, offers some grounds for optimism as to the practicality of the pricing and standards technique for the control of the quality of the environment.

References

1. Bohm, P.: Pollution, Purification, and the Theory of External Effects. *Swedish Journal of Economics 72*, no. 2, 153-66, 1970.
2. Coase, R.: The Problem of Social Cost. *Journal of Law and Economics 3*, 1-44, 1960.
3. Hart, A.: Anticipations, Business Planning, and the Cycle. *Quarterly Journal of Economics 51*, 273-97, Feb. 1937.
4. Kneese, A. & Bower, B.: *Managing Water Quality: Economics, Technology, Institutions*. Baltimore, 1968.
5. Lave, L. & Seskin, E.: Air Pollution and Human Health. *Science 21*, 723-33 Aug. 1970.
6. Portes, R.: The Search for Efficiency in the Presence of Externalities. *Unfashionable Economics: Essays in Honor of Lord Balogh* (ed. P. Streeten), pp. 348-61. London, 1970.
7. Ruff. L.: The Economic Common Sense of Pollution. *The Public Interest*, Spring 1970, 69-85.

APPENDIX

In the text, we argued on a somewhat intuitive level that the appropriate use of unit taxes and subsidies represents the least-cost method of achieving a set of specified standards for environmental quality. In the case of smoke-abatement, for instance, the tax-subsidy approach will automatically generate the cost-minimizing assignment of "reduction quotas" without recourse to involved calculations or enforcement.

The purpose of this appendix is to provide a formal proof of this proposition. More precisely, we will show that, to achieve *any* given vector of final outputs along with the attainment of the specified quality of the environment, the use of unit taxes (or, where appropriate, subsidies) to induce the necessary modification in the market-determined pattern of output will permit the realization of the specified output vector at minimum cost to society.

While this theorem may seem rather obvious (as the intuitive discussion in the text suggests), its proof does point up several interesting properties which are noteworthy. In particular, unlike many of the propositions about prices in welfare analysis, the theorem does not require a world of perfect competition. It applies to pure competitors, monopolists, or oligopolists alike so long as each of the firms involved seeks to minimize the private cost of producing whatever vector of outputs it selects and has no monopsony power (i.e., no influence on the prices of inputs). The firms need not be simple profit-maximizers; they may choose to maximize growth, sales (total revenues), their share of the market, or any combination of these goals (or a variety of other objectives). Since the effective pursuit of these goals typically entails minimizing the cost of whatever outputs are produced, the theorem is still applicable. Finally, we want simply to emphasize that the theorem applies to whatever set of final outputs society should select (either by direction or through the operation of the market). It does not judge the desirability of that particular vector of outputs; it only tells us how to make the necessary adjustments at minimum cost.

We shall proceed initially to derive the first-order conditions for the minimization of the cost of a specified overall reduction in the emission of wastes. We will then show that the independent decisions of cost-minimizing firms subject to the appropriate unit tax on waste emissions will, in fact, satisfy the first-order conditions for overall cost minimization.

Let

x_{iv} represent the quantity of input i used by plant v ($i = 1, ..., n$), ($v = 1, ..., m$),

z_v be the quantities of waste it discharges,

y_v be its output level,

$f_v(x_{1v}, ..., x_{nv}, z_v, y_v) = 0$ be its production function,

p_i be the price of input i, and

k the desired level of $\sum z_v$, the maximum permitted daily discharge of waste.

In this formulation, the value of k is determined by the administrative authority in a manner designed to hold waste emissions in the aggregate to a level consistent with the specified environmental standard (e.g., the sulphuric content of the atmosphere). Note that the level of the firm's waste emissions is treated here as an argument in its production function; to reduce waste discharges while maintaining its level of output, the firm will presumably require the use of additional units of some other inputs (e.g., more labor or capital to recycle the wastes or to dispose of them in an alternative manner).

The problem now becomes that of determining the value of the x's and z's that minimize input cost

$$c = \sum_i \sum_v p_i(x_{iv})$$

subject to the output constraints

$$y_v = y_v^* = \text{constant} \qquad (v = 1, ..., m)$$

and the constraint on the total output of pollutants

$$\sum_v z_v = k.$$

It may appear odd to include as a constraint a vector of given outputs of the firms, since the firms will presumably adjust output levels as well as the pattern of inputs in response to taxes or other restrictions on waste discharges. This vector, however, can be *any* vector of outputs (including that which emerges as a result of independent decisions by the firms). What we determine are first-order conditions for cost-minimization which apply to *any* given vector of outputs no matter how they are arrived at. Using $\lambda_v(v=1, ..., m)$ and λ as our $(m+1)$ Lagrange multipliers, we obtain the first-order conditions:

$$\left.\begin{aligned}
&\lambda_v f_{vz} + \lambda = 0 &&(v = 1, ..., m)\\
&p_i + \lambda_v f_{vi} = 0 &&(v = 1, ..., m)(i = 1, ..., n)\\
&y_v = y_v^* &&(v = 1, ..., m)
\end{aligned}\right\} \qquad (1)$$

where we use the notation $f_{vz} = \partial f_v / \partial z_v$, $f_{vi} = \partial f_v / \partial x_{iv}$.

Now let us see what will happen if the m plants are run by independent managements whose objective is to minimize the cost of whatever outputs their firm produces, and if, instead of the imposition of a fixed ceiling on the emission of pollutants, this emission is taxed at a fixed rate per unit, t. So long as its input prices are fixed, firm v will wish to minimize

$$c = tz_v + \sum_i p_i x_{iv}$$

subject to

$$y_v = y_v^*.$$

Swed. J. of Economics 1971

54   *W. J. Baumol and W. E. Oates*

Direct differentiation of the m Lagrangian functions for our m firms immediately yields the first-order conditions (1)—the same conditions as before, provided t is set equal to λ. Thus, if we impose a tax rate that achieves the desired reduction in the total emission of pollutants, we have proved that this reduction will satisfy the necessary conditions for the minimization of the program's cost to society.[1]

---

[1] In this case, λ (and hence t) is the shadow price of the pollution constraint. In addition to satisfying these necessary first-order conditions, cost-minimization requires that the production functions possess the usual second-order properties. An interesting treatment of this issue is available in Portes. We should also point out that our proof assumes that the firm takes t as given and beyond its control. Bohm discusses some of the problems that can arise where the firm takes into account the effects of its behavior on the value of t.

[10]

JOURNAL OF ECONOMIC THEORY 5, 395–418 (1972)

Markets in Licenses and Efficient Pollution Control Programs*

W. DAVID MONTGOMERY

Division of the Humanities and Social Sciences, California Institute of Technology, Pasadena, California 91109

Received May 19, 1972

1. INTRODUCTION

Artificial markets have received some attention as a means of remedying market failure and, in particular, dealing with pollution from various sources. Arrow [1] has demonstrated that when externalities are present in a general equilibrium system, a suitable expansion of the commodity space would lead to Pareto optimality by bringing externalities under the control of the price system. Since his procedure is to define new commodities, each of which is identified by the type of externality, the person who produces it and the person who suffers it, his conclusion is pessimistic. Each market in the newly defined commodities involves but one buyer and one seller, and no forces exist to compel the behavior which would bring about a competitive equilibrium.

On the other hand, many forms of pollution are perfect substitutes for each other. Sulfur oxide emissions from one power plant trade off in the preferences of any sufferer with sulfur oxide emissions from some other power plant at a constant rate. This fact leads to the possibility of establishing markets in rights (or "licenses") which will bring together many buyers and sellers. Dales [2] has discussed a wide variety of such arrangements.

Unfortunately, because of the elements of public goods present in most environmental improvements, it appears unlikely that markets in rights, containing many sufferers from pollution as participants, will lead to overall Pareto optimality. They can only serve the more limited, but still

---

* Parts of this article appeared in my Ph.D. dissertation "Market Systems for the Control of Air Pollution," submitted to the Department of Economics at Harvard University in May, 1971. Research on this thesis was partly supported under Grant No. AP-00842 from the Environmental Protection Agency to Walter Isard. I am also indebted to Kenneth Arrow and James Quirk for valuable advice. Needless to say, all errors are solely the responsibility of the author.

395

valuable, function of achieving specified levels of environmental quality in an efficient manner. An example of this function is found in a proposal by Jacoby and Schaumburg [6] to establish a market in licenses (or "BOD bonds") to control water pollution from industrial sources in the Delaware estuary. The purpose of the present article is to provide a solid theoretical foundation for such proposals. Markets such as those proposed by Jacoby and Schaumburg will be characterized in a general fashion, and it will be proved that even in quite complex circumstances the market in licenses has an equilibrium which achieves externally given standards of environmental quality at least cost to the regulated industries.

Two types of license are discussed: a "pollution license," and an "emission license." The emission license directly confers a right to emit pollutants up to a certain rate. The pollution license confers the right to emit pollutants at a rate which will cause no more than a specified increase in the level of pollution at a certain point. Since a polluter will in general affect air or water quality at a number of points as a result of his emissions, he will be required to hold a portfolio of licenses covering all relevant monitoring points. All such licenses are free transferable. A main thesis of this article is that the market in pollution licenses will be more widely applicable than the market in emission licenses.

1.1. *The Applicable Pollution Control Problem*

Consider the following problem of pollution control: In a certain region there is a set of n industrial sources of pollution, each of which is fixed in location and owned by an independent, profit-maximizing firm. The prices of the inputs and outputs of these firms are fixed, because the region is small relative to the entire economy. Therefore any change in the level of output of a firm or industry in the region will have only a negligible impact on the output of the economy as a whole, and prices will be unaffected by output changes in the region. These firms are represented by a set of integers $I = \{1,..., n\}$.

Some regional standard of environmental quality in terms of a single pollutant has been chosen as a goal by a resource management agency. This standard is denoted by a vector $Q^* = (q_1^*,..., q_m^*)$. If air pollution is the particular area of interest, q_j^* might be an annual average concentration of sulfur dioxide at point j in an air basin. If water pollution is involved, q_j^* might be a measure of dissolved oxygen deficit at point j on a river. Since there is only one pollutant present in the region, the elements of Q^* represent concentrations of the one pollutant at various locations. The development of a decentralized system for achieving environmental goals at a number of different locations is the most important contribution of this article.

All pollution in the region arises from the industrial sources, each of which emits a single pollutant at the rate e_i. The emission vector $E = (e_1, ..., e_n)$ is mapped into concentrations by a semipositive matrix H, so that $E \cdot H = Q$. The standard Q^* imposes constraints on allowable emission rates of the form $E \cdot H \leqslant Q^*$. The problem of pollution control is to achieve Q^* at least cost to the polluters.

Some discussion of the limitations which the model places on the results presented is in order. The assumption that concentrations are a linear function of emissions is the only part of this problem which does not generalize easily. Therefore, the market in licenses to be described must be construed as applicable only in situations in which the assumption is approximately true. Fortunately, there are at least two important problems of pollution control in which it is true. One such is the management of dissolved oxygen deficit in a river. The DOD at a point downstream of a source releasing BOD (bacteriological oxygen demand) effluent is proportional to the BOD released [6, 8].

Management of concentrations of nonreactive atmospheric pollutants is another problem in which linearity is approximately true, as long as the variables to be related are average emission rates and average concentrations [5]. In this case, which will be used in this article as the source of illustrative examples, a meteorological diffusion model provides the means of relating long-run average concentrations to average rates of emission. As formulated by Martin and Tikvart [11], the model is based on an equation describing the shape of a smoke plume from an elevated source emitting at a constant rate with a wind of constant direction and speed. From this equation, the contribution of any source to concentration at any receptor can be calculated for given wind direction and speed. By taking the frequency distribution of wind direction and speed and appropriately modifying the predicted concentrations, one arrives at a theoretical relationship between average rates of emission and average concentrations [14].

The results of the diffusion model can be conveniently represented as an $m \times n$ matrix of unit diffusion coefficients, denoted

$$H = \begin{pmatrix} & \vdots & \\ \cdots & h_{ij} & \cdots \\ & \vdots & \end{pmatrix}.$$

The typical element states the contribution which one unit of emission by firm i makes to average pollutant concentration at point j.

The assumption that only one pollutant is present in the region can be justified by appeal to the external decision on desired air quality. If desired air quality in terms of one pollutant is independent of desired air quality in terms of any other so that, for example, the decision on the desirability of a certain concentration of sulfur dioxide is independent of the concentration of particulates permitted in the region, then nothing is lost. The management problem can be generalized by adding constraints representing emission vectors which achieve desired levels of many pollutants and joint production of pollution. In principle, it is solved in the same way as the one-pollutant system developed here.

The assumption that all prices (except those associated with pollution) are unaffected by measures undertaken to control pollution is a common one in economic analysis of environmental problems. It is necessary to allow consideration of problems in isolation, and to avoid full-sized (and nonoperational) general-equilibrium models [9].

When this assumption is made, it is possible to define for each firm a single-valued function which associates a cost with any emission rate adopted by the firm.

1.2. The Cost Function

The purpose of this section is to construct a function relating each level of emission which might be adopted by the firm to its cost and to establish that the profit-maximizing firm will minimize this function. Moreover, it will be argued that no firm will ever choose a level of emission greater than that which is observed in the complete absence of regulation.

Consider the typical multiproduct firm i. Let

$$G_i(y_{i1}, ..., y_{iR}, e_i)$$

represent the minimum total cost of producing a vector of output $(y_{i1}, ..., y_{iR})$ and emissions e_i. This is the cost incurred when inputs are optimally adjusted for that output and emission level. For the static analysis with which we deal, we can assume that both operating costs and an annual capital cost are included. Profit then will be

$$\pi_i = \sum_r p_r y_{ir} - G_i(y_{i1}, ..., y_{iR}, e_i).$$

Assume that G_i is convex and twice differentiable and that its domain is the positive orthant of the $r + 1$-dimensional space of real numbers. Define $(\bar{y}_{i1}, ..., \bar{y}_{ir}, \bar{e}_i)$ by

$$\sum_r p_r \bar{y}_{ir} - G_i(\bar{y}_{i1}, ..., \bar{y}_{iR}, \bar{e}_i) = \max_{y_{ir}, e_i} \left[\sum_r p_r y_{ir} - G_i(y_{i1}, ..., y_{iR}, e_i) \right].$$

Now consider the case in which the firm must adopt an emission level e_i and adjusts its output in order to obtain maximum profit for the fixed level of emission. Define $\tilde{y}_{ir}$ by

$$\sum_r p_r \tilde{y}_{ir} - G_i(\tilde{y}_{i1}, ..., \tilde{y}_{iR}, e_i) = \max_{y_{ir}} \left[\sum_r p_r y_{ir} - G_i(y_{i1}, ..., y_{iR}, e_i) \right].$$

The cost to firm i of adopting emission level e_i is defined as the difference between its unconstrained maximum of profit and its maximum of profit when emissions equal e_i. That is,

$$F_i(e_i) = \sum_r p_r(\bar{y}_{ir} - \tilde{y}_{ir}) - [G_i(\bar{y}_{i1}, ..., \bar{y}_{iR}, \bar{e}_i) - G_i(\tilde{y}_{i1}, ..., \tilde{y}_{iR}, e_i)]. \tag{1.1}$$

This cost is composed of two terms: the change in gross income from altering the output vector and the change in costs from setting emissions at a nonoptimal level (with an optimal adjustment of output).[1]

Consider the variation in $F_i(e_i)$ when a small change is made in e_i. Differentiating totally with respect to e_i, we find

$$dF_i(e_i) = -\sum_r \left(p_r - \frac{\partial G_i}{\partial y_{ir}} \right) \frac{dy_{ir}}{de_i} de_i + \frac{\partial G_i}{\partial e_i} de_i. \tag{1.2}$$

We have assumed that output levels adjust to maximize profit for a given level of e_i. That is, y_{ir} adjusts so that

$$p_r - \partial G_i / \partial y_{ir} = 0$$

for $j = 1, ..., r$. Therefore [13],

$$dF_i(e_i)/de_i = \partial G_i / \partial e_i. \tag{1.3}$$

It can further be shown that the convexity of $G_i(y_{i1}, ..., y_{iR}, e_i)$ implies the convexity of $F_i(e_i)$.

THEOREM 1.1.   *If $G_i(y_{i1}, ..., y_{iR}, e_i)$ is convex, $F_i(e_i)$ is also convex.*

Proof.   The proof is immediate from the definition of convexity.

It is convenient to be able to use a single-valued function $F_i(e_i)$ to associate with any emissions level its cost. The properties of $F_i(e_i)$ proved above allow us to conclude that any relevant conditions which are satisfied

---

[1] Three general classes of techniques of emission reduction are available. First, emissions can be reduced by reducing the scale of output, or by altering the product mix of the firm. Second, the production process or the inputs used, such as fuels, can be altered. Finally, "tail-end" cleaning equipment can be installed to remove pollutants from effluent streams before they are released into the environment. All three of these techniques will commonly be found in combination.

400   MONTGOMERY

by the partial derivative of G_i with respect to e_i will be satisfied by the derivative of F_i. In particular, we can conclude that if the profit-maximizing firm has any choice of e_i, it will minimize $F_i(e_i)$ subject to whatever costs or constraints we impose on it. Moreover, if G_i is convex, it follows that the conditions under which $\sum_i F_i(e_i)$ is minimized are the same as the conditions under which the total economic cost to firms of emissions control is minimized.

2. The Characterization of an Efficient Emission Vector

The goal of management is limited to bringing about an emission vector which will result in air of quality Q^* at least total cost to the region. Such an emission vector is called efficient, and designated E^{**}. The concept of least cost to the community is also given a specific meaning: it is the minimum of the sum $\sum_i F_i(e_i)$. With some risk of ambiguity, this sum is called "joint total cost."

To provide a reference to which later results can be compared, a general solution for the efficient emission vector can now be derived. The problem is to choose the vector $E = (e_1,..., e_n)$ to minimize $\sum_i F_i(e_i)$ subject to the constraints

$$E \geqslant 0 \quad \text{and} \quad EH \leqslant Q^*,$$

where $Q^* \geqslant 0$, $h_{ij} \geqslant 0$ for all i, j. We will label this the "total joint cost minimum problem." Our exploration will proceed throughout this article on the assumption that G_i is convex. This implies that $F_i(e_i)$ is convex, and therefore that $\sum_i F_i(e_i)$ is also convex. It is also assumed that H is semipositive. The typical shape of $F_i(e_i)$ is illustrated in Fig. 1.

Minimizing a convex function subject to linear constraints and non-negativity constraints is equivalent to finding the saddle point of an associated Lagrangean. Formally, $(E^{**}, U^{**}) = (e_1^{**},..., e_n^{**}, u_1^{**},..., u_m^{**})$ will be a saddle point of the expression

$$-\sum_i F_i(e_i) + \sum_j u_j \left(q_j{}^* - \sum_i h_{ij} e_i \right)$$

with $E^{**} \geqslant 0$, $U^{**} \geqslant 0$. The differential Kuhn–Tucker conditions for this saddle point are

$$F_i'(e_i) + \sum_j u_j h_{ij} \geqslant 0, \qquad \sum_i \left[e_i \left(F_i'(e_i) + \sum_j u_j h_{ij} \right) \right] = 0, \quad (2.1)$$

$$q_j{}^* - \sum_i h_{ij} e_i \geqslant 0, \qquad \sum_j \left[u_j \left(q_j - \sum_i h_{ij} e_i \right) \right] = 0. \qquad (2.2)$$

These conditions are necessary and sufficient [7]. Moreover, it is easy to show that the minimum does in fact exist.

THEOREM 2.1. E^{**} and U^{**} satisfying (2.1) and (2.2) exist.

Proof. Since $\sum_i F_i(\bar{e}_i) = 0$ and $\sum_i F_i(e_i) \geqslant \sum_i F_i(\bar{e}_i)$, for $e_i \geqslant 0$, $\sum_i F_i(e_i)$ is bounded from below. By hypothesis, the set

$$\Psi = \{E \mid EH \leqslant Q^*, E \geqslant 0\}$$

is not empty. Therefore, $\sum_i F_i(e_i)$ is defined on a nonempty closed set and bounded from below; therefore, it attains a minimum over the set Ψ for some element of Ψ.

If $\sum_i F_i(e_i)$ is not strictly convex, then E^{**} need not be unique. Since, however, $\min_{E \in \Psi} \sum_i F_i(e_i)$ is unique, it does not matter what particular minimizer is chosen. Therefore, I shall refer to the vector which minimizes costs; the reader may interpret this reference as meaning "any element of the set of E which minimizes $\sum_i F_i(e_i)$."

The following theorem is true if $\sum_i F_i(e_i)$ is strictly convex.

THEOREM 2.2.   If E^{**} minimizes $\sum_i F_i(e_i)$ subject to $EH \leqslant Q^*$ and $E \geqslant 0$, then $E^{**} \leqslant \bar{E}$.

Proof. Assume *per contra* that $e_i^{**} > \bar{e}_i$ for some $i = i'$. Then $F_{i'}(e_{i'}^{**}) > F_{i'}(\bar{e}_{i'})$ and $h_{ij}e_{i'}^{**} > h_{ij}\bar{e}_{i'}$. Therefore

$$\sum_{i \neq i'} F_i(e_i^{**}) + F_i(\bar{e}_{i'}) < \sum_i F_i(e_i^{**}) \tag{2.3a}$$

and

$$\sum_{i \neq i'} h_{ij}e_i^{**} + h_{i'j}\bar{e}_i < \sum_i h_{ij}e_i^{**}. \tag{2.3b}$$

By (2.3b) the vector $(e_1^{**},..., \bar{e}_{i'} ,..., e_n^{**})$ satisfies $EH \leqslant Q^*$ and by (2.3a) E^{**} does not minimize $\sum_i F_i(e_i)$.

3. MARKETS IN LICENSES

We can now proceed to the construction of markets which, in equilibrium, lead to emission rates which satisfy the conditions of Theorem 2.1. A set of licenses are defined, such that the possession of licenses confers the right to carry out a certain average rate of emission.

Consider the function

$$\Lambda(H_i , L_i),$$

where H_i is the i-th row of the matrix H and $L_i = (l_{i1}, ..., l_{ik})$. We define l_{ik} as the number of licenses of type k held by firm i. This function defines the right to emit which is generated by holding a portfolio of licenses L_i. Then firm i can maximize profits by minimizing direct emission costs plus the cost of purchasing licenses, subject to the constraint that emissions not exceed $\Lambda(H_i, L_i)$. We assume throughout that some initial allocation of licenses l_{ik}^0, is made. Then the firm's problem is to minimize

$$F_i(e_i) + \sum_k p_k(l_{ik} - l_{ik}^0)$$

subject to $e_i \leqslant \Lambda(H_i, L_i)$.

A market equilibrium will exist if there exist nonnegative prices P^* such that when e_i^*, L_i solve the firm's minimization problem for p_k^*, the following market clearing conditions hold:

$$\sum_i (l_{ik}^* - l_{ik}^0) \leqslant 0, \qquad \sum_k p_k^* \left[\sum_i (l_{ik}^* - l_{ik}^0)\right] = 0. \qquad (3.1a)$$

That is, there is some set of prices of licenses such that when each firm minimizes the sum of the cost of reducing emissions and the net cost of buying licenses, excess demand for licenses is nonpositive, and excess supply of a license drives its price to zero.

The market equilibrium is efficient if e_i^* represents equilibrium emissions and in any equilibrium

$$\sum_i F_i(e_i^*) = \sum_i F_i(e_i^{**}). \qquad (3.1b)$$

Note that when all licenses are allocated to firms, (3.1a) implies that any expenditure on licenses by one firm is a revenue to another firm. Therefore total expenditure among all firms, associated with the control of pollution, just equals the total cost of emission control. That is,

$$\sum_i \left[F_i(e_i^*) + \sum_k p_k(l_{ik}^* - l_{ik}^0)\right] = \sum_i F_i(e_i^*) + \sum_k p_k \left[\sum_i (l_{ik}^* - l_{ik}^0)\right]$$
$$= \sum_i F_i(e_i^*).$$

These three properties do not exhaust the set of desirable properties of a market system. It might be that an equilibrium exists, or is efficient, only under strong conditions on the initial allocations of licenses which can be adopted. The more variation which is possible in the choice of initial allocations, the more freedom the management agency will have to pursue such goals as equity of the treatment, subsidization of "deserving" industries, and so on. We begin by defining a licensing system which has an efficient equilibrium for all distributions of a fixed total of licenses.

Analogously to the distinction between ambient standards and emission standards, we must differentiate between emission licenses and pollution licenses. Emission licenses are perhaps the most natural to think of trading, but there are great problems in using them when quality at many locations is a matter of concern. In particular, it is not possible to allow the licenses to be traded on a one-for-one basis [12].

Suppose there are two sources of pollution and one monitoring point, that each source is assigned licenses which allow it to emit 5 units of pollutant, and that $h_{11} = 1$ and $h_{21} = 2$. Under these circumstances, there will be $5 \cdot 1 + 5 \cdot 2 = 15$ units of pollution at the monitoring point. The marginal rate of substitution between emissions at source 2 and emissions at source 1 which keeps air quality constant is 2. If licenses are exchanged on a one-for-one basis, the transfer of one license from firm 1 to firm 2 will result in air quality being degraded to 16 units of pollution. If there is a second monitoring point, and $h_{22}/h_{12} \neq 2$, the marginal rate of substitution between emissions at sources 1 and 2 will change, depending on which monitoring point imposes the *operative* constraint on emissions. By defining rights to *cause* pollution at each of the monitoring points, we can avoid these problems completely, although they can be resolved with emission licenses if certain restrictions on trades are observed.

3.1. *The Market in Licenses to Pollute*

In this section we establish the existence and efficiency of equilibrium in a system of transferable licenses to pollute. Let l_{ij} represent the quantity of licenses allowing pollution at point j held by firm i, and let

$$L_i = (l_{i1}, ..., l_{im})$$

be the "portfolio" of licenses held by firm i. The licensing function can have the form

$$\Lambda(H_i, L_i) = \min_j \frac{l_{ij}}{h_{ij}},$$

which implies that each firm faces the constraints

$$h_{ij}e_i \leqslant l_{ij} \qquad j = 1, ..., m.$$

That is, the relevant element of the diffusion matrix is taken to be a correct predictor of the amount which an average rate of emission at point i contributes to pollutant concentration at point j. Each firm is allowed to have an average rate of emission which produces no more pollution at any point than the amount which the firm is licensed to cause at that point. The firm will minimize $F_i(e_i) + \sum_j p_j(l_{ij} - l^0_{ij})$ subject to the licensing constraint.

In the theorems which follow we use the convention that l_j is a scalar, a total number of licenses allowing pollution at point j. Thus $\sum_i l_{ij} = l_j$. When $L_i = (l_{i1}, ..., l_{im})$ and $L = (l_1, ..., l_m)$, $\sum_i L_i = L$.

The strategy of proof is to define a market equilibrium relative to an initial allocation of licenses and to derive necessary and sufficient conditions for its existence. A subsidiary construction, called a "license-constrained joint cost minimum," is defined and shown to exist. It generates a second set of necessary and sufficient conditions. It is shown that the emission vector and shadow prices which satisfy the conditions of a license-constrained joint cost minimum for given totals of licenses also satisfy the conditions of competitive equilibrium relative to any initial allocation of licenses in which the given totals are completely distributed among firms. An equilibrium license portfolio for each firm is constructed, and shadow prices on each firm's licensing constraints are identified. To prove that a competitive equilibrium achieves the joint cost minimum defined in Section 2, we show that when license totals equal desired air qualities any emission vector and price vector which satisfy the equilibrium conditions also satisfy the conditions for efficiency. In the course of the proof the efficient emission vector is identified as the equilibrium emission vector and shadow prices on the air quality constraints in the overall joint cost minimum are identified as the prices of licenses. The equilibrium license portfolio has each element just equal to the pollution caused by the efficient rate of emission for the corresponding firm. The proof itself is rigorous and abstract.

DEFINITION. A market equilibrium is an $n + 2$ tuple of vectors $L_i^* \geqslant 0$, $E^* \geqslant 0$, and $P^* \geqslant 0$ such that L_i^* and E^* minimize

$$F_i(e_i) + \sum_j p_j^*(l_{ij} - l^0_{ij})$$

subject to $l_{ij} - h_{ij}e_i \geqslant 0, j = 1, ..., m$, for all i and which also satisfy the market clearing conditions

$$\sum_i (l^*_{ij} - l^0_{ij}) \leqslant 0, \qquad \sum_j p_j^* \left[\sum_i (l^*_{ij} - l^0_{ij}) \right] = 0. \qquad (3.1a)$$

LEMMA 3.1. *A market equilibrium exists if and only if there exist vectors*

$$(u_{i1}^*,..., u_{im}^*) \geqslant 0 \qquad i = 1,..., n,$$

$$(p_1^*,..., p_m^*) \geqslant 0$$

such that

$$F_i'(e_i^*) + \sum_j u_{ij}^* h_{ij} \geqslant 0, \qquad e_i^* \left[F_i'(e_i^*) + \sum_j u_{ij}^* h_{ij} \right] = 0, \quad (3.2a)$$

$$p_j^* - u_{ij}^* \geqslant 0, \qquad \sum_j l_{ij}^*[p_j^* - u_{ij}^*] = 0, \quad (3.2b)$$

$$l_{ij}^* - h_{ij}e_i^* \geqslant 0, \qquad \sum_j u_{ij}^*[l_{ij}^* - h_{ij}e_i^*] = 0, \quad (3.2c)$$

for all i and

$$\sum_i (l_{ij}^* - l_{ij}^0) \leqslant 0, \qquad \sum_j p_j^* \left[\sum_i (l_{ij}^* - l_{ij}^0) \right] = 0. \quad (3.2d)$$

Proof. First we characterize the vectors L_i^* and e_i^* which minimize cost for the firm. Minimizing a function is equivalent to maximizing its negative; and the negative of a convex function is concave. Therefore, we can state the problem of the firm as one of maximizing the concave function

$$-F_i(e_i) - \sum_j p_j^*(l_{ij} - l_{ij}^0).$$

Form the Lagrangean

$$\phi_i(l_{i1},..., l_{im}, e_i, u_{i1},..., u_{in})$$

$$= -F_i(e_i) - \sum_j p_j^*(l_{ij} - l_{ij}^0) + \sum_j u_{ij}(l_{ij} - h_{ij}e_i).$$

From the Kuhn–Tucker theorem the following conditions are necessary and sufficient for the constrained maximum; where $\phi_i(l_{i1}^*,..., l_{in}^*, e_i^*, u_{i1}^*,..., u_{in}^*) = \phi_i^*$.

$$\partial\phi_i^*/\partial e_i \leqslant 0, \qquad e_i^* \cdot (\partial\phi_i^*/\partial e_i) = 0,$$

$$\partial\phi_i^*/\partial l_{ij} \leqslant 0, \qquad \sum_j l_{ij}^* \cdot (\partial\phi_i^*/\partial l_{ij}) = 0,$$

$$\partial\phi_i^*/\partial u_{ij} \geqslant 0, \qquad \sum_j u_{ij}^* \cdot (\partial\phi_i^*/\partial u_{ij}) = 0.$$

Performing the indicated differentiation gives 3.2a to 3.2c which must be satisfied for all *i*. Equation (3.2d) repeats the market clearing condition.

DEFINITION.   A license-constrained joint cost minimum is a vector E^{**} which minimizes

$$\sum_i F_i(e_i)$$

subject to $EH \leqslant L^0$ and $E \geqslant 0$.

In making this definition we assume that some arbitrary vector of licenses L^0 is issued by the management agency. We must assume that the set $\{E \mid EH \leqslant L^0 \text{ and } E \geqslant 0\}$ is not empty. Then the same argument used in Section 2 to establish the existence of a joint cost minimum will establish the existence of a license-constrained minimum. We now can use the following lemma to prove existence of an equilibrium on the pollution license market.

LEMMA 3.2.   *An emission vector E^{**} is a license-constrained joint cost minimum if and only if there exists a vector $(u_1^{**},..., u_m^{**}) \geqslant 0$ such that*

$$F_i'(e_i^{**}) + \sum_j u_j^{**} h_{ij} \geqslant 0, \qquad \sum_i e_i^{**}\left[F_i'(e_i^{**}) + \sum_j u_j^{**} h_{ij}\right] = 0, \quad (3.3a)$$

$$l_j^0 - \sum_i h_{ij} e_i^{**} \geqslant 0, \qquad \sum_j u_j^{**}\left[l_j^0 - \sum_i h_{ij} e_i^{**}\right] = 0. \quad (3.3b)$$

Proof.   The proof is as in Lemma 3.1.

The market equilibrium will exist for any distribution of licenses such that $l_{ij}^0 \geqslant 0$ and $\sum_i l_{ij}^0 = l_j^0$.

THEOREM 3.1.   *A market equilibrium of the pollution license system exists for $\sum_i l_{ij}^0 = l_j^0$.*

Proof.   We proceed constructively by using (3.3a) and (3.3b) to show that $e_i^* = e_i^{**}$, $l_{ij}^* = h_{ij} e_i^{**}$, $p_j^* = u_j^{**}$ and $u_{ij}^* = u_j^{**}$ for all i satisfy (3.2a)–(3.2d).

Equation (3.2a).   Since $F_i'(e_i^{**}) + \sum_j u_j^{**} h_{ij} \geqslant 0$ for all i, and $e_i^{**} \geqslant 0$, it follows from

$$\sum_i e_i^{**}\left[F_i'(e_i^{**}) + \sum_j u_j^{**} h_{ij}\right] = 0$$

that

$$e_i^{**}\left[F_i'(e_i^{**}) + \sum_j u_j^{**} h_{ij}\right] = 0$$

for all i. Therefore, e_i^{**} and u_j^{**} satisfy 3.2a for all i.

Equation (3.2b). If $p_j{}^* = u_j{}^{**}$ and $u_{ij}^* = u_j{}^{**}$, $p_j{}^* - u_{ij}^* = 0$ for all i and j, and (3.2b) is satisfied by any l_{ij}^*.

Equation (3.2c). If $l_{ij}^* = h_{ij}e_i^{**}$ for all i and j, clearly l_{ij}^* and e_i^{**} satisfy (3.2c) for any u_{ij}^*, and in particular for $u_j{}^{**}$.

Equation (3.2d). Let $\sum_i l_{ij}^0 = l_j{}^0$ and $l_{ij}^* = h_{ij}e_i^{**}$. Then, (3.3b) gives by substitution

$$\sum_i l_{ij}^0 - \sum_i l_{ij}^* \geqslant 0 \qquad \text{and} \qquad \sum_j u_j{}^{**}\left[\sum_i l_{ij}^0 - \sum_i l_{ij}^*\right] = 0.$$

Therefore, $p_j{}^* = u_j{}^{**}$ and l_{ij}^* satisfy (3.2d).

Thus we conclude that for any choice of license totals which imply a feasible air quality vector, a market equilibrium exists. If we choose the license totals correctly, we can show that any market equilibrium is a joint cost minimum. The joint cost minimum was defined in Section 2 as a vector E^{**} which minimizes $\sum_i F_i(e_i)$ subject to $EH \leqslant Q^*$ and $E \geqslant 0$. First we prove that any emission vector which results from a market equilibrium with $\sum_i l_{ij}^0 = l_j{}^0$ minimizes $\sum_i F_i(e_i)$ subject to $EH \leqslant L^0$ and $E \geqslant 0$.

THEOREM 3.2. *Any emission vector which satisfies the conditions of a market equilibrium with* $\sum_i L_i{}^0 = L^0$ *is a license-constrained joint cost minimum.*

Proof. We show that any e_i^* which satisfies (3.2a)–(3.2d) satisfies (3.3a) and (3.3b) with $u_j{}^{**} = p_j{}^*$.

Equation (3.3a). By (3.2b), either $u_{ij}^* = p_j{}^*$ or $l_{ij}^* = 0$. By (3.2c), $l_{ij}^* \geqslant h_{ij}e_i^*$; therefore, whenever $p_j{}^* \neq u_{ij}^*$, $l_{ij}^* = 0$, and it follows that $e_i^* = 0$, or $h_{ij} = 0$. Whenever $h_{ij} = 0$, $p_j{}^*h_{ij} = u_{ij}^*h_{ij} = 0$. Therefore, $e_i^*[F_i'(e_i^*) + \sum_j p_j{}^*h_{ij}] = 0$ holds whether or not $p_j{}^* = u_{ij}^*$.

Since $u_{ij}^* \leqslant p_j{}^*$, $\sum_j u_{ij}^*h_{ij} \leqslant \sum_j p_j{}^*h_{ij}$ and $F_i'(e_i^*) + \sum_j u_{ij}^*h_{ij} \geqslant 0$ imply $F_i'(e_i^*) + \sum_j p_j{}^*h_{ij} \geqslant 0$. Therefore, e_i^* and $p_j{}^*$ satisfy the inequality in (3.3a).

Equation (3.3b). Since $\sum_i l_{ij}^0 = l_j{}^0$, (3.2d) implies that $l_j{}^0 = \sum_i l_{ij}^0 \geqslant \sum_i l_{ij}^*$. Since, by (3.2c), $l_{ij}^* - h_{ij}e_i^* \geqslant 0$, $\sum_i l_{ij}^* - \sum_i h_{ij}e_i^* \geqslant 0$. If $l_j{}^0 \geqslant \sum_i l_{ij}^*$, then it must be true that $l_j{}^0 - \sum_i h_{ij}e_i^* \geqslant 0$ and the inequality in (3.3b) is satisfied by e_i^*.

Substitute $l_j{}^0$ for $\sum_i l_{ij}^0$ in (3.2d) giving $\sum_j p_j{}^*[l_j{}^0 - \sum_i l_{ij}^*] = 0$. If $l_{ij}^* = h_{ij}e_i^*$ for all i and all j, clearly $\sum_j p_j{}^*[l_j{}^0 - \sum_i h_{ij}e_i^*] = 0$. Assume that $l_{ij}^* - h_{ij}e_i^* > 0$ for some i and j. Then, by (3.2c), $u_{ij}^* = 0$. If $p_j{}^* \neq 0$,

(3.2b) implies that $l_{ij}^* = 0$ for that i and j, and since $l_{ij}^* > h_{ij}e_i^*$ for that i and j, e_i^* must be negative. Since this is impossible, we must have either $l_{ij}^* = h_{ij}e_i^*$ for all i or $p_j^* = 0$; and the alternative holds for each j. Therefore,

$$p_j^* \left[\sum_i h_{ij}e_i^* - l_j^0 \right] = 0,$$

and p_j^*, e_i^* satisfy (3.3b).

Thus, if we take the totals of each type of license distributed to firms we will find that firms exchange licenses so as to minimize joint total cost subject to the constraint that concentrations of pollutants at each monitoring point be no greater than the total of licenses issued for that point. The following corollary is immediate.

COROLLARY. *If* $L^0 = Q^*$, *the equilibrium emission vector is a joint cost minimum.*

Proof. If $L^0 = Q^*$, by Theorem 3.2 an equilibrium emission vector minimizes $\sum_i F_i(e_i)$ subject to $EH \leqslant Q^*$ and $E \geqslant 0$.

Theorem 3.1 can now be restated as "an efficient emission vector can be achieved as a competitive equilibrium" and 3.2 as "any competitive equilibrium with appropriate license distribution achieves an efficient vector." We can also prove an interesting theorem on the initial allocation of licenses.

THEOREM 3.3. *If* $l_{ij}^0 \geqslant 0$ *and* $\sum_i l_{ij}^0 = q_j^*$, *then* E^*, P^*, *and* L_i^* *are independent of* L_i^0.

Proof. Equations (3.2a)–(3.2c) depend in no way on L_i^0. In (3.2d) L_i^0 appears, but only in the form of the sum $\sum_i L_i^0$.

This result is somewhat unusual, in that the particular equilibrium achieved in a system usually depends on the initial allocations. The reason that this system is independent of the initial allocation is that the firm's behavior is independent of its asset position. Any redistribution which preserves totals of each type of license does not change the equilibrium. A graphical depiction of the equilibrium of the firm when a system of pollution licenses is imposed reveals the independence of initial allocations. The equilibrium is depicted in Fig. 1.

In the course of proving Theorem 3.1 it was noted that $p_j^*(h_{ij}e_i^* - l_{ij}^*) = 0$, so that $\sum_j p_j^*(h_{ij}e_i^* - l_{ij}^*) = 0$. Any emission level chosen by the firm implies that the firm purchases certain quantities of licenses, so that we can associate with any emission rate a cost equal to $\sum_j p_j^* h_{ij}e_i^*$. The minimization of cost (of emission control plus net purchases of licenses) can then be represented as the minimization of the sum $F_i(e_i) + \sum_j h_{ij} p_j^* e_i$. The emission rate e_i^* in Fig. 1 is the minimizer of this sum. Theorem 3.1 states that there exist prices which clear markets for licenses when each firm chooses license holdings, and emission rates e_i^*, to minimize cost.

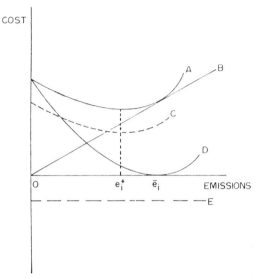

FIG. 1. A—$F_i(e_i) + \sum_j p_j h_{ij} e_i$; B—$\sum_j p_j h_{ij} e_i$; C—$F_i(e_i) + \sum_j p_j h_{ij} e_i - \sum_j p_j l_{ij}^0$; D—$F_i(e_i)$; E—$\sum_j p_j l_{ij}^0$.

The initial allocation of licenses is equivalent to a lump sum subsidy, and is independent of emission level. Therefore, this subsidy can be represented as a horizontal line, $\sum_j p_j^* l_{ij}^0$, in Fig. 1. The curve $F_i(e_i) + \sum_j p_j^* h_{ij} e_i - \sum_j p_j^* l_{ij}^0$ is the net cost function which represents the actual cost of emission control and licenses. Note that e_i^* is independent of the size of the subsidy. Because of this result, the management agency can distribute licenses as it pleases. Considerations of equity, of administrative convenience, or of political expediency can determine the allocation. The same efficient equilibrium will be achieved.

It should, however, be noted that in assuming the convexity of $F_i(e_i)$ we impose certain conditions relating to nonnegative profits. Let $\bar{\pi}_i$ be the (maximal) profit earned before the introduction of a licensing system, and let $\tilde{\pi}_i(e_i)$ be the profit earned when emissions are set at rate e_i . Then by (1.1) we have $\tilde{\pi}_i(e_i) = \bar{\pi}_i - F_i(e_i)$. In the long run the firm will only stay in business if $\tilde{\pi}_i(e_i) \geqslant 0$. In this case the cost function will have the form $F_i(e_i) = \min(\bar{\pi}_i - \tilde{\pi}_i(e_i), \bar{\pi}_i)$. An upper bound, equal to $\bar{\pi}_i$, is placed on costs incurred by the firm. This upper bound destroys the convexity of the cost function unless $F_i(0) \leqslant \bar{\pi}_i$. Such an assumption is implicit in the assumption that $F_i(e_i)$ is convex.

It would appear that the need to purchase licenses imposes a cost on the firm additional to the cost of emission control $F_i(e_i)$. Even though this cost sums to zero for all firms taken together, it may be positive for some individual firms and negative for others. Fortunately, we can prove the following theorem, namely, that if $F_i(0) \leqslant \bar{\pi}_i$, then even if a firm is allocated no licences initially (i.e., $l^0_{ij} = 0$ for some i and all j), it can still earn nonnegative profits at any levels of emissions and license holding.

THEOREM 3.4.   *If $F_i(0) \leqslant \bar{\pi}_i$, $F_i(e_i^*) + \sum_j p_j^* l^*_{ij} \leqslant \bar{\pi}_i$.*

Proof.   We have proved that

$$F_i(e_i^*) + \sum_j p_j^* l^*_{ij} = F_i(e_i^*) + \sum_j p_j^* h_{ij} e_i^*.$$

If $F_i(e_i^*) + \sum_j p_j^* h_{ij} e_i^* > \bar{\pi}_i$, then $F_i(e_i^*) + \sum_j p_j^* h_{ij} e_i^* > F_i(0)$. But $e_i = 0$ and $l_{ij} = 0$ satisfy $e_i \leqslant \Lambda(H_i, L_i)$, so that e_i^* does not minimize cost subject to the licensing constraint. This contradiction establishes the theorem.

This demonstration completes the discussion of pollution licenses. We began by showing that for any vector of licenses L^0 which implies feasible concentration levels at each monitoring point there exists a competitive equilibrium in the license market. We then showed that the concentrations which result from the equilibrium will be less than or equal to the levels permitted by the vector of licenses and that joint total costs are minimized subject to this constraint. Finally we showed that when $L^0 = Q^*$, the problem of achieving desired air quality standards at minimum cost is solved by the market in pollution licenses.

The major generalization provided by this theorem is that it establishes the possibility of achieving environmental goals at a *number* of geographic points while maintaining the advantages of a market system. Thus one important objection to the use of economic incentives, that they could lead to change in the pattern of emissions such that although air quality improvements at one point are achieved, it is at the expense of deteriorating air quality elsewhere, is laid to rest. Moreover, we discovered that the fixed totals could be allocated arbitrarily among firms.

Overall convexity and the possibility, for each firm, of absorbing all costs of abatement in profits were assumptions necessary for the operation of the system when no information on cost functions is available.

We turn now to an alternative licensing system. It will turn out that this system of emission licenses is interesting because it provides a means of linking up the proposal to issue transferable licenses with other proposals for achieving efficient solutions in a decentralized manner.

3.2. *The Market in Emission Licenses*

The effluent charge is a tax which a firm must pay on each unit of pollutant which it emits into a water course. A corresponding charge for air pollution control might be called an emission charge. In order to calculate a charge which will lead to efficiency in air pollution control, the manager must solve in advance the overall cost-minimization problem. It is not difficult to show that the correct tax on emission by firm i is equal to the shadow price on its emissions determined by the minimization of joint total cost. The tax is $\sum_j u_j{}^* h_{ij}$, where $u_j{}^*$ is the value of the Lagrange multiplier on the j-th quality constraint evaluated at the optimum. But in order to calculate such a tax the manager must know the cost functions of each firm. It is, of course, possible to obtain that information in an iterative process by varying the tax. This is a cumbersome and politically unattractive procedure, and it has been shown by Marglin [10] that the information transferred to the regulatory authority by such a procedure is as great as the information needed to set quantity standards for each firm. That is, whenever it is possible to calculate the correct tax it is possible to achieve E^{**} in the initial allocation.

A licensing scheme does not require such prior or iterative gathering of information. The market makes the necessary calculations independently in the course of reaching equilibrium. For this reason we are led to consider licensing schemes as superior to taxation. The natural correlate of emission charges is a system of emission licenses.

An emission license confers on the firm holding it the right to emit pollutants at a certain rate. It is not always desirable to allow such rights to be transferred on a one-for-one basis: the desirable rule governing exchange of emission rights is that a firm may be allowed to emit up to a level which causes pollution equal to that which would be caused if each firm from which it obtained rights emitted to the maximum extent permitted by the rights which it has given up. We must differentiate rights to emit by the location at which they permit emissions to take place. Then l_k , $k = 1,..., n$, is a quantity of licenses to emit at location k. It is sufficient to allow k to run over the set of firms I since each firm is in a fixed location. Let l_{ik} represent the quantity of licenses allowing emissions at location k held by firm i.

If the exchange of such licenses between polluters at different locations is to be permitted, some rule must be stated regarding the right to emit which a license to emit at location k confers on a firm at location i. Consider a firm i which emits at a rate $e_i = h_{kj}l_{ik}/h_{ij}$. Then the pollution which firm i causes at point j is precisely the pollution which firm k would cause if it emitted at the rate $e_k = l_{ik}$, since $h_{ij}e_i = h_{kj}l_{ik} = h_{kj}e_k$. The licensing function can have the form

$$\Lambda(H_i, L_i) = \min_j \left(\sum_k h_{kj}l_{ik}/h_{ij} \right),$$

which implies that each firm faces the constraints

$$h_{ij}e_i \leqslant \sum_k h_{kj}l_{ik} \qquad j = 1,..., m,$$

and will minimize

$$F_i(e_i) + \sum_k p_k(l_{ik} - l_{ik}^0)$$

subject to those constraints.

A restriction on the initial allocation of licenses is needed if the market equilibrium with emission licenses is to be efficient. It is that $\sum_i l_{ik}^0 \geqslant 0$ and $\sum_k h_{kj} \sum_i l_{ik}^0 = q_j^*$ for all j. Note that this assumption is equivalent to the assumption that there exists a nonnegative emission vector E^0 such that

$$E^0 \cdot H = Q^*.$$

This is quite a strong condition, since even if the matrix H is of full rank, for arbitrary semipositive H and Q^* the equations

$$E \cdot H = Q^*$$

will not in general have nonnegative solutions.

If $\sum_k h_{kj}l_k^0 < q_j^*$ for some j it may not be possible to achieve the minimum of joint total cost without prior knowledge of cost functions. Suppose that the joint cost minimum vector E^{**} satisfies

$$\sum_i h_{ij}e_i^{**} = q_j^*$$

for some j, and that $\sum_k h_{kj}l_k^0 < q_j^*$ for the same j. Then since the equilibrium emission vector E^* satisfies

$$\sum_i h_{ij}e_i^* \leqslant \sum_k h_{kj}l_k^0,$$

it follows that

$$\sum_i h_{ij} e_i^* < \sum_i h_{ij} e_i^{**}$$

and $e_i^* \neq e_i^{**}$, so that the market equilibrium is inefficient.

DEFINITION. A market equilibrium in emission licenses is an $n + 2$-tuple of vectors $L_i^* \geqslant 0$, $E^* \geqslant 0$ and $P^* \geqslant 0$ such that L_i^* and E^* minimizes

$$F_i(e_i) + \sum_k p_k^*(l_{ik} - l_{ik}^0)$$

subject to

$$\sum_k h_{kj} l_{ik} - h_{ij} e_i \geqslant 0 \qquad j = 1,..., m$$

and

$$e_i \geqslant 0; \qquad l_{ik} \geqslant 0$$

for all i and which also satisfy the market clearing conditions

$$\sum_i (l_{ik}^* - l_{ik}^0) \leqslant 0, \qquad \sum_k p_k^* \left[\sum_i (l_{ik}^* - l_{ik}^0) \right] = 0.$$

LEMMA 3.3.   *A market equilibrium exists if there exist vectors*

$$\begin{aligned} (u_{i1}^*,..., u_{im}^*) &\geqslant 0 \qquad i = 1,..., n, \\ (p_1^*,..., p_n^*) &\geqslant 0, \end{aligned} \tag{3.4}$$

such that

$$F_i'(e_i^*) + \sum_j u_{ij}^* h_{ij} \geqslant 0, \qquad e_i^* \left[F_i'(e_i^*) + \sum_j u_{ij}^* h_{ij} \right] = 0, \quad (3.5a)$$

$$p_k^* - \sum_j u_{ij}^* h_{kj} \geqslant 0, \qquad \sum_k \left[l_{ik}^* \left(p_k^* - \sum_j u_{ij}^* h_{kj} \right) \right] = 0, \quad (3.5b)$$

$$\sum_k h_{kj} l_{ik}^* - h_{ij} e_i^* \geqslant 0, \qquad \sum_j \left[u_{ij}^* \left(\sum_k h_{kj} l_{ik}^* - h_{ij} e_i^* \right) \right] = 0, \quad (3.5c)$$

for all i and

$$\sum_i (l_{ik}^* - l_{ik}^0) \leqslant 0, \qquad \sum_k p_k^* \left[\sum_i (l_{ik}^* - l_{ik}^0) \right] = 0. \quad (3.5d)$$

Proof.   The proof is as in Lemma 3.1.

THEOREM 3.5. *A market equilibrium in emission licenses exists.*

Proof.   In Theorem 2.2 it was shown that an emission vector minimizing joint total costs subject to the air quality constraints exists, and that in consequence E^{**} and U^{**} satisfying (2.1) and (2.2) exist. Let licenses be issued initially so that

$$\sum_k h_{kj} \sum_i l_{ik}^0 = q_j^*$$

for all j. Then we show that E^{**} is an emission vector and U^{**} a price vector satisfying (3.5a)–(3.5d).

We begin by proving the following proposition:

P.1.   *If $\sum_k h_{kj} l_k^0 \geqslant \sum_i h_{ij} e_i^{**}$, then there exist l_{ik}^* such that $\sum_i l_{ik}^* \leqslant l_k^0$, $l_{ik}^* \geqslant 0$, and $\sum_k h_{kj} l_{ik}^* \geqslant h_{ij} e_i^{**}$ for all i and k. Letting $L_i = (l_{i1},..., l_{in})$ and H_i be the i-th row of the matrix H we may write the inequalities which must have a nonnegative solution in matrix form as*

$$(L_1^*,..., L_n^*) \begin{bmatrix} -H & & I \\ & \ddots & \vdots \\ & & -HI \end{bmatrix} \leqslant (-H_1 e_1^{**},..., -H_n e_n^{**}, L^0).$$

It is a theorem [3] that either these inequalities or the following inequalities have a nonnegative solution:

$$\begin{bmatrix} -H & & I \\ & \ddots & \vdots \\ & & -HI \end{bmatrix} \begin{bmatrix} X_1 \\ \vdots \\ X_n \\ X_{n+1} \end{bmatrix} \geqslant 0, \tag{1}$$

$$(-H_1 e_1^{**},..., -H_n e_n^{**}, L) \begin{bmatrix} X_1 \\ \vdots \\ X_n \\ X_{n+1} \end{bmatrix} < 0. \tag{2}$$

We can write (1) as

$$-\sum_{j=1}^m h_{ij} x_{1j} + x_{n+1i} \geqslant 0 \qquad (i = 1,..., n)$$

$$\vdots$$

$$-\sum_{j=1}^m h_{ij} x_{nj} + x_{n+1i} \geqslant 0 \qquad (i = 1,..., n)$$

and (2) as

$$- \sum_{i=1}^{n} \sum_{j=1}^{m} h_{ij} e_i^* x_{ij} + \sum_i l_i^0 x_{n+1i} < 0.$$

We assume that there do exist nonnegative solutions denoted with superscript 0's to (1) and (2). Let us multiply each line of (1) through by l_i^0 and sum the result over i, giving

$$- \sum_{j=1}^{m} \sum_{i=1}^{n} h_{ij} l_i^0 x_{kj} + \sum_i l_i^0 x_{n+1i} \geq 0$$

for all k.

Comparing this inequality with (2) we find

$$- \sum_{i=1}^{n} \sum_{j=1}^{m} h_{ij} e_i^* x_{ij} < - \sum_{i=1}^{n} \sum_{j=1}^{m} h_{ij} l_i^0 x_{kj} .$$

Since $\sum_i h_{ij} e_i^* \leq \sum_i h_{ij} l_i^0$ by hypothesis,

$$- \sum_j x_{kj}^0 \sum_i h_{ij} l_i^0 \leq - \sum_j x_{kj}^0 \sum_i h_{ij} e_i^*,$$

and

$$- \sum_{i=1}^{n} \sum_{j=1}^{m} h_{ij} e_i^* x_{ij}^0 < - \sum_{i=j}^{n} \sum_{j=1}^{m} h_{ij} e_i^* x_{kj}^0$$

for all k. We remove the minus signs and reverse the inequality, giving

$$\sum_i e_i^* \sum_j h_{ij} x_{ij}^0 > \sum_i e_i^* \sum_j h_{ij} x_{kj}^0$$

for all k. Therefore it must be true for some i that

$$\sum_j h_{ij} x_{ij}^0 > \sum_j h_{ij} x_{kj}^0$$

for all k. Therefore, it must be true for $k = i$, which implies

$$h_{i1} x_{i1}^0 + \cdots + h_{in} x_{in}^0 > h_{i1} x_{i1}^0 + \cdots + h_{in} x_{in}^0 .$$

This contradiction establishes that there is no nonnegative solution to inequalities (1) and (2) and P.1 is proved.

We can now proceed line by line to show that E^{**} and U^{**} satisfying (2.1) and (2.2) also satisfy (3.5a)–(3.5d).

Equation (3.5a). From (2.1), e_i^{**} and u_j^{**} satisfy (3.5a) for all i.

Equation (3.5b). Let $p_k{}^* = \sum_j u_j^{**} h_{kj}$ and $u_{ij}^* = u_j^{**}$ for all i. Then they satisfy (3.5b) since $p_k{}^* - \sum_j u_{ij}^* h_{kj} = 0$ for all i and k.

Equation (3.5c). Let $\sum_k h_{kj} \sum_i l_{ik}^0 = q_j{}^*$. Then by (2.2),

$$0 \leqslant \sum_k h_{kj} \sum_i l_{ik}^0 - \sum_i h_{ij} e_i^{**},$$

and by P.1 there exist $l_{ik}^* \geqslant 0$ such that

$$\sum_i l_{ik}^0 \geqslant \sum_i l_{ik}^* \quad \text{and} \quad \sum_k h_{kj} l_{ik}^* - h_{ij} e_i^{**} \geqslant 0$$

for all i. If $>$ holds for some i and j,

$$q_j{}^* = \sum_k h_{kj} \sum_i l_{ik}^0 \geqslant \sum_k h_{kj} \sum_i l_{ik}^* > \sum_i h_{ij} e_i^{**},$$

and $u_j^{**} = 0$. Therefore, (3.5c) is satisfied with $u_{ij}^* = u_j^{**}$.

Equation (3.5d). If $\sum_i l_{ik}^0 > \sum_i l_{ik}^*$ for some k and $h_{kj} > 0$,

$$q_j{}^* = \sum_k h_{kj} \sum_i l_{ik}^0 > \sum_k h_{kj} \sum_i l_{ik}^* \geqslant \sum_i h_{ij} e_i^{**},$$

and $u_j^{**} = 0$ for all j. If $h_{kj} = 0$ for that k and some j, then for the corresponding j, $u_j^{**} h_{kj} = 0$. In either case $p_k{}^* = \sum_j u_j^{**} h_{kj} = 0$ and (3.5d) is satisfied.

We reverse the direction of inference to prove that if $\sum_i h_{ij} l_i^0 = q_j{}^*$, the competitive equilibrium emission vector is efficient. We assume in addition that the rank of H is m: this involves no significant loss of generality since any constraint matrix can be made to satisfy the condition by striking out redundant constraints. The operation of eliminating redundant constraints does not change the set Ψ of emission vectors which satisfy the constraints.

THEOREM 3.6. *If $\sum_k h_{kj} l_k^0 = q_j{}^*$, E^* minimizes $\sum_i F_i(e_i)$ subject to $EH \leqslant Q^*$, $E \geqslant 0$.*

Proof. First we note that in proving Theorem 3.4 we established that (3.5a)–(3.5d) are satisfied, for all i, by $u_{ij}^* = u_j^{**}$, and that the rank of H

equals *m*. Therefore, the matrix of partial derivatives of the licensing constraints for each firm also has rank *m*, and the multipliers on those constraints are unique [4]. Since the Kuhn–Tucker conditions are satisfied by identical multipliers for each firm, they are *only* satisfied by identical multipliers. Let u_j^{**} be equal to any of the u_{ij}^*, identical for all *i*. Then,

Equation (2.1). e_i^* and u_j^{**} as defined satisfy (2.1) whenever they satisfy (3.5a).

Equation (2.2). By (3.5d)

$$\sum_i l_{ik}^0 \geqslant \sum_i l_{ik}^*, \quad \text{and} \quad q_j^* = \sum_k h_{kj} \sum_i l_{ik}^0 \geqslant \sum_k h_{kj} \sum_i l_{ik}^*.$$

By (3.5c)

$$\sum_k h_{kj} \sum_i l_{ik}^* \geqslant \sum_i h_{ij} e_i^*.$$

Therefore $q_j^* - \sum_i h_{ij} e_i^* \geqslant 0$. If $q_j^* > \sum_i h_{ij} e_i^*$, either $\sum_i l_{ik}^0 > \sum_i l_{ik}^*$ for some *k* with $h_{kj} \neq 0$, or $\sum_k h_{kj} l_{ik}^* > h_{ij} e_i^*$ for some *i* and that *j*. If the latter, $u_{ij}^* = 0$ and $u_j^{**} = 0$. If the former, $p_k^* = 0$ and since by (3.5b) $p_k^* - \sum_j u_{ij}^* h_{kj} \geqslant 0$, $\sum_j u_{ij}^* h_{kj} = 0$ and $u_{ij}^* = 0$, so that $u_j^{**} = 0$. Therefore, $u_j^{**}(q_j^* - \sum_i h_{ij} e_i^*) = 0$ when $u_j^{**} = u_{ij}^*$.

This completes the proof that a competitive equilibrium, satisfying the conditions of joint cost minimization, exists in the market for emission licenses. An integral part of the proof was the assumption that the total of each type of license is determined by solving the *equations*

$$\sum_k h_{kj} l_k^0 = q_j^*.$$

If the management agency is restricted to assigning all licenses of type *i* which it issues to firm *i*, then its ability to redistribute costs will be severely limited by the necessity of choosing l_i^0 to satisfy the air quality constraints with equality if indeed such a solution exists in the problem at hand.

REFERENCES

1. K. J. ARROW, The organization of economic activity: Issues pertinent to the choice of market versus non-market allocation, *in* "The Analysis and Evaluation of Public Expenditures: The PPB System," pp. 47–64, U.S. Congress, Joint Economic Committee, U.S. Government Printing Office, Washington, DC, 1969.
2. J. H. DALES, "Pollution, Property, and Prices," University of Toronto Press, Toronto, Canada, 1968.

3. D. GALE, "The Theory of Linear Economic Models," p. 47, McGraw-Hill, New York, 1960.

4. G. HADLEY, "Nonlinear and Dynamic Programming," p. 190, Addison-Wesley Publ. Co., Inc., Reading, MA, 1964.

5. H. W. HERZOG, The air diffusion model as an urban planning tool, *Socio-Econ. Plan. Sci.* **3** (1969), 329–349.

6. H. JACOBY AND G. SCHAUMBURG, "Administered Markets in Water Quality Control: A Proposal for the Delaware Estuary," unpublished.

7. S. KARLIN, "Mathematical Methods and Theory in Games, Programming and Economics," Vol. 1, pp. 203–204, Addison-Wesley Publ. Co., Inc., Reading Mass., 1962.

8. A. KNEESE AND B. BOWER, "Managing Water Quality: Economics, Technology, Institutions," Chapter 2, Johns Hopkins Univ. Press, Baltimore, 1968.

9. S. MARGLIN, Objectives of water resources development: A general statement, *in* "Design of Water Resource Systems" (A. Maass *et al.*, Eds.), p. 23, Harvard Univ. Press, Cambridge, MA, 1962.

10. S. MARGLIN, Information in price and command systems of planning, *in* "Public Economics" (J. Margolis, Ed.), St. Martin's Press, New York, 1969.

11. D. O. MARTIN AND J. A. TIKVART, A general atmospheric diffusion model for estimating the effects on air quality of one or more sources, APCA Paper No. 68-148, presented at the Annual Meeting of the Air Pollution Control Association, June 1968.

12. W. D. MONTGOMERY, Artificial markets and the theory of games, Social Science Working Paper No. 8, California Inst. of Technol., March 1972.

13. P. SAMUELSON, "Foundations of Economic Analysis," pp. 34–35, Harvard Univ. Press, Cambridge, MA, 1966.

14. TRW Systems Group, "Regional Air Pollution Analysis, Phase I: Status Report," pp. 4.3–4.15, U.S. Department of Health, Education and Welfare and National Air Pollution Control Administration, Washington, DC, 1969.

[11]

Auctions and Alternative Procedures for Allocating Pollution Rights

Randolph M. Lyon

Systems of transferable pollution rights are currently of interest to both researchers and policymakers because of their potential as efficient means of controlling pollution.[1] Despite this potential, there has been relatively little investigation of the actual mechanisms through which rights would be allocated. This paper examines a range of alternative procedures for allocating pollution rights since the approach used can have substantial impact upon the implementation and efficacy of a rights program.

Using simulations based upon realistic data, this paper investigates rights programs based upon (1) sales of rights by the government under two distinct types of auctions and (2) free initial distribution of rights by the government to dischargers on some basis, followed by exchange among dischargers of rights and funds. It is shown that under programs based on rights sales by the government, dischargers' payments for rights can result in substantial financial burdens upon the dischargers. While these payments for rights are transfers—as opposed to real economic costs—they may be a significant concern. In one case study considered here, for example, dischargers' aggregate payments for rights alone would be larger than their treatment expenses under an inefficient regulatory program requiring uniform percentage removal of waste by all dischargers.

Because rights programs may involve small numbers of dischargers—as poli-cies may be designed for regions such as river basins—this paper considers both potential strategic behavior of dischargers bidding for rights and price-taking behavior. Particular attention is given to examining the types of auction procedures that could be used to implement rights sales by the government, if the approach of government sales is adopted. A special type of auction—a type of Groves mechanism, as defined by Green and Laffont (1977)—that discourages strategic bidding by dischargers is contrasted with auctions that use a single clearing price as a method for government sales. A number of properties of the Groves mechanism, including its efficiency and disincentives for manipula-

The author is an assistant professor of economics at the University of Texas, Austin. The suggestions of R. Brandis, E. D. Brill, J. W. Eheart, R. Engelbrecht-Wiggans, J. H. Leuthold, S. A. Matthews, F. Schoumaker, and two anonymous referees are deeply appreciated. J. W. Eheart also provided data and market simulations which were invaluable in calculating the simulation results presented here. The author is also very grateful for support provided by a University of Illinois Fellowship in Economics and by the National Science Foundation, under grant PRA 79-13131 to J. W. Eheart and E. D. Brill. Responsibility for shortcomings of this work is, of course, solely the author's.

[1] The U.S. Environmental Protection Agency and several states have expressed interest in systems of transferable rights to control air pollution (see, for example, U.S. EPA, 1981; *Wall Street Journal*, Dec. 15, 1978; *Washington Post*, Aug. 10, 1980), and in Wisconsin the state's Department of Natural Resources has considered the possibility of using rights to help attain water quality standards in the Fox River (David 1980). Tietenberg (1980) and U.S. EPA (1980) provide recent surveys of research on pollution rights and related policies.

Land Economics, Vol. 58, No. 1, February 1982
0023-7639/79/0003-0299 $1.00/0

tion, suggest that it should receive serious consideration if the approach of rights sales by the government is adopted.[2]

Rights programs based on free initial distribution relieve the financial burden of rights purchases by reducing dischargers' payments for rights. These programs, however, may be affected by dischargers' strategic manipulation of both the initial distribution and the exchange of rights for funds following the initial distribution. The case studies demonstrate that even with fairly large numbers of dischargers (e.g., one case study has 53 dischargers), there may be relatively small numbers of dischargers that dominate either the sale or purchase of rights. These small numbers would encourage strategic (price-making) behavior under the free initial distribution approach. Unfortunately, the strategy-free aspects of the Groves auction mechanism cannot be adapted to the approach where rights are initially distributed by the government free of charge and then exchanged.

The structure of this paper is as follows. Section I describes the alternative rights allocation procedures and uniform treatment regulations, which are assumed to represent contemporary regulatory policies. Dischargers' expenses under the alternative programs are then examined using a simple model. Section II presents results of simulations of the alternative programs based upon data from two water pollution-control case studies: a phosphorus removal problem for 53 municipal treatment plants in the Lake Michigan watershed in Wisconsin and a simplified biochemical oxygen demand (BOD) removal problem for 11 firms and municipalities on the Willamette River in Oregon.[3] In section III results of the simulations are analyzed and the procedures are examined with

respect to their properties of strategic manipulability, efficiency, and equity and financial burden. Implications of the findings for the design of pollution rights programs are discussed in the concluding section.

I. THE ALTERNATIVE PROCEDURES

The goal of this paper is to examine procedures that might be used to allocate

---

[2] Groves mechanisms are typically considered as a class of public-goods preference revelation mechanisms. While the applicability of this class of mechanisms to auctions has been apparent since Vickrey's (1961) seminal work (and is also noted by Monash (1980)), to this author's knowledge the mechanisms have not previously been extensively applied and analyzed as auctions for multiple objects in realistic settings.

[3] BOD consists primarily of organic wastes, whose decay by micro-organisms removes dissolved oxygen (DO) from the water. In contrast to phosphorus, BOD is generally modeled as being a nonconservative pollutant. In this case the dual processes of oxygen depletion and reaeration are considered. Optimal environmental management in the case of a nonconservative pollutant is more complex than for a conservative pollutant if one desires to attain a specific level of ambient environmental quality. In the case of BOD discharge, ambient environmental quality is measured in terms of DO. The DO concentration is not a simple function of the pounds of BOD (or sewage) discharged by all the sources, however, because the amount of BOD in a unit of sewage changes over time, or equivalently, depending on the location of the discharges along a river.

Despite the fact that BOD is more accurately modeled as a nonconservative pollutant, in the simulations of allocation procedures here BOD is modeled conservatively. That is, a zero reaeration rate is assumed. The simulations, therefore, actually examine the least costly removal of an aggregate amount of BOD, as opposed to the least costly attainment of a dissolved oxygen standard. This procedure has previously been used by Eheart (1980) to investigate permit systems to control BOD in the Willamette River. Using a water quality model, Eheart determined that there is little loss of optimality of the rights program under the conservative assumption in this case because of the specific physical characteristics of the river and dischargers. Moreover, the conservative modeling is not problematic for this study because the emphasis here is on examining general properties of alternative allocation procedures, as opposed to making detailed investigations of specific programs to control nonconservative pollutants. Several alternative programs specifically for controlling such pollutants are examined by David (1980) and Eheart (1980).

pollution rights by contrasting these procedures both with each other and with present regulatory policies. Regulatory procedures are considered in this study because they represent a standard against which programs involving transferable rights can be compared. Current environmental management policy—particularly in the water pollution realm —is often modeled as being of the uniform regulation type (Thomann 1972, p. 244). Such regulations require all dischargers to remove equal percentages of their wastes.

In contrast to uniform treatment, policies yielding cost-efficient treatment generally imply that dischargers operate at different percentage removal levels because of their different treatment costs. In particular, the well-known first-order condition for minimizing aggregate treatment costs implies that dischargers would generally have equal marginal treatment costs, where marginal costs are increasing.[4] Pollution rights programs and effluent charges are among the ways to reach a decentralized cost-efficient distribution of treatment effort, given an environmental standard.[5]

This paper considers two different types of auctions which the government could use to sell discharge rights. These are (1) an auction where all rights are sold for a single clearing price and (2) a special type of discriminating auction that discourages strategic manipulation of bids by individual dischargers. Under both of these auctions, dischargers would submit sealed bid schedules for rights to the governmental pollution-control authority. Dischargers would base these bids upon their marginal treatment costs, because purchase of each right (allowing, for example, discharge of one pound of a pollutant) would entitle the discharger to forgo removal of the corresponding amount of the pollutant from its effluent.[6]

Under the single-price auction, rights would be sold to the highest bidders for a price that could represent either the lowest accepted bid or the highest rejected bid. Under either pricing scheme, there is an incentive for bidders to manipulate their bids. The bidder who submits the marginal bid, for example, could have an incentive to reduce his bid under either version.[7]

The second type of auction considered is designed specifically to eliminate dischargers' incentives for strategic bidding. This auction is a member of the class of procedures termed Groves mechanisms and has the property of being individually incentive-compatible. This property implies that under this mechanism it is always an individual discharger's dominant strategy to reveal truthfully its valuations for rights.[8]

Under the incentive-compatible auc-

---

[4] See, for example, Kneese and Bower (1968). Where locational effects are important, dischargers theoretically should have equal marginal costs of reducing their impacts on ambient environmental quality at the relevant location or locations. As observed in footnote 4, however, examination of rights systems that address locational aspects of pollution is beyond the scope of this paper.

[5] Note that globally least-costly attainment of an environmental standard might involve centralized approaches such as regional treatment plants and environmental modifications. Consideration of centralized options, however, is beyond the scope of this paper.

[6] The rights sold might be valid for some limited life span, such as five or ten years, or in perpetuity. This paper does not address the trade-offs between choices among different life spans.

[7] Before bids are submitted, the dischargers are generally uncertain about which of their bids will be the marginal bid. Thus, it is possible that all of their bids will be reduced somewhat to reflect the possibility that any one bid might be marginal in the single-price auction case. Alternatively, if there are many bidders, it is possible that the small probability that any one bid is marginal may imply that there is relatively little strategic manipulation on the part of bidders. More rigorous analysis of equilibrium bidding under multiple-object auctions is quite complex. Some initial results are presented in Lyon (1980; 1981).

[8] See, for example, Hurwicz [(1972; 1973)] and Dasgupta, Hammond, and Maskin (1979).

tion, rights are allocated to the highest bidders, as under a single-price auction. Where there are m identical rights, a discharger (denoted by i) winning k_i rights ($k_i \leq m$) would be assessed the k_i highest rejected bids of all other dischargers except itself. Thus, a discharger's own bids never directly affect its payment under the incentive-compatible procedure.[9]

The final procedure considered is based upon free initial distribution of the rights by the government, followed by trading. Under this approach, rights are initially distributed on a basis such as property value, or influent flow or load.[10] Following their initial distribution, rights could be exchanged among dischargers for funds to allow attainment of an efficient allocation of rights. Under the centralized exchange suggested by David et al. (1980), dischargers would submit supply and demand schedules and a single market-clearing price would be determined by the governmental pollution-control authority. The centralized exchange would facilitate large transactions and would prevent exchanges of permits just ahead of inspections (David et al. 1980). Alternatively, rights might also be exchanged via an unregulated market following their initial distribution.

Dischargers' expenses under the alternative rights and regulation policies may be investigated using the following model. Figure 1 presents typically shaped continuous marginal treatment cost curves for two dischargers. Discharger 1 represents a discharger with high marginal treatment costs, and discharger 2 represents one with low marginal costs. For simplicity, assume both generate equal amounts of waste, measured in pounds.

Suppose uniform treatment regulations imply 85% removal for both dischargers. Total treatment costs under this policy

are given by the areas under the marginal cost curves (IAK and IFK) plus constants (which are assumed to be zero).

The first-order condition for cost-efficient treatment generally implies that both dischargers operate at equal marginal treatment costs. Pollution rights selling for a price of P will implement this efficient solution and provide aggregate removal equal to that of uniform treatment.

Under a government sale of rights via a single-price auction, the high-cost discharger will pay IBJ for treatment plus

---

[9] The procedure can also be described as follows. Define R as the sum of the winning bids under the efficient assignment of rights to bidders. Define R_{-i} as the sum of winning bids if discharger i's bids are omitted from consideration. (Thus, if discharger i is assigned some rights under the efficient assignment, then R_{-i} would be less than R because to calculate R_{-i} i's winning bids would be replaced by bids which previously would have been rejected.) Define the extra value created by discharger i as C_i, where $C_i = R - R_{-i}$. Let i's payment (P_i) equal the sum of i's winning bids (B_i) minus C_i. Where discharger i bids truthfully (as is its dominant strategy), its profit will equal C_i because the profit also equals $B_i - P_i$. (In the case of an auction for a single right this procedure implies that the right would be sold to the highest bidder for the price bid by the second highest bidder. Thus, in the single right case this procedure is a second-price auction (Vickrey 1961).)

Under the incentive-compatible procedure, dischargers have no incentive to lower their bids, because by lowering their bids they risk losing a right which they value more than it would cost them, yet they do not directly lower their assessment if they still win the right. There is also no incentive for dischargers to raise their bids because in doing so they only risk winning a right for which they may be assessed more than its worth to them.

The mechanism is guaranteed to be incentive-compatible, however, only if participants have additively separable preferences (Green and Laffont 1977). (The expected profit or expected cost functions typically assumed for dischargers, however, meet the criterion of additive separability.) The mechanism is also theoretically susceptible to manipulation by coalitions of bidders (Green and Laffont 1979).

[10] See, for example, David et al. (1980) and Eheart et al. (1980). Influent load equals influent flow times the average concentration of the pollutant.

The EPA's bubble and emissions offset policies are actually special versions of the free initial distribution approach. For more on these policies see, for example, Maloney and Yandle (1980), and U.S. EPA (1980; 1981).

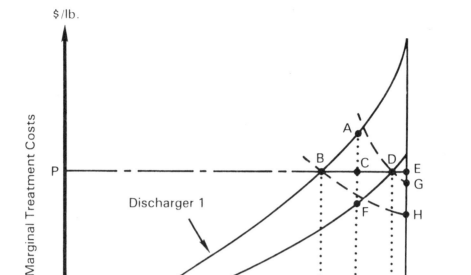

FIGURE 1
COST CURVES AND ALTERNATIVE POLICIES

JBEM for rights. This total expense may be more or less then its uniform treatment expense, depending on whether *JBEM* is greater or less than *JBAK*. Under the same sale, the low-cost discharger will pay *IDL* for treatment plus *LDEM* for rights, which is always greater than its uniform treatment costs by the area *KFDEM*.

Under the free initial distribution procedure based on load, both dischargers are allotted rights allowing discharge of 15% of their waste (*KM*). Following

efficient exchange, discharger 1 will pay *IBJ* for treatment and *JBCK* for rights, which implies lower expenses than both uniform treatment and the government single-price sale for this discharger. Discharger 2 pays *IDL* for treatment and receives payments for rights from discharger 1 worth *KCDL* (which equals *JBCK*). The net expense for discharger 2 is *IFK* minus *FCD*, which also yields lower expenses than uniform treatment and the government single-price sale.

Under the government sale of rights

via the incentive-compatible auction, both dischargers face a price of rights approximately of P on the margin, but the prices of the rights purchased are less than or equal to P. The discounts that the dischargers receive may be different, so the dashed lines representing rights prices under this auction are not parallel in general. If dischargers behave as price takers under the single-price auction to the extent that the clearing price (P) is not affected, then dischargers would have lower expenses under an incentive-compatible auction than under the single-price auction when P is the clearing price in the latter because of the possible discount under the incentive-compatible procedure. High-cost dischargers may have lower expenses under the incentive-compatible auction than under uniform treatment, even when their expenses are not lower under the single-price auction than under uniform treatment, if the area *JBHM* is less than *JBAK*. If the area *JBHM* is less than *JBCK*, then the high-cost discharger would have lower expenses under the incentive-compatible auction than under the free initial distribution procedure. The low-cost discharger would always have lower expenses under the free initial distribution procedure—under which it would sell rights—than under the government sale via the incentive-compatible auction. The next section presents simulation results that suggest insights into the possible magnitudes of the dischargers' expenses considered above.

II. SIMULATION RESULTS

Two case studies based on engineering estimates of treatment costs are used to investigate the alternative pollution-control programs. Because the case studies both share features and have unique characteristics, a number of insights can be gained by considering the two examples together.

One case study uses treatment cost data for the control of phosphorus for 53 municipal dischargers in the Lake Michigan basin. These cost data were originally collected by Braasch and Joeres (1975) and were subsequently used by David et al. (1980) and Eheart et al. (1980) to investigate rights programs using free initial distribution. The policy considered is one approach to implementing a 1969 order for Wisconsin to have 80% of all phosphorus removed from point-source discharges in the Lake Michigan basin (Braasch and Joeres 1975). Wisconsin's Department of Natural Resources decided to implement this order by calling for a uniform 85% reduction in phosphorus discharges by the 53 largest municipal dischargers in the basin. This reduction implies daily removal of approximately 17,700 pounds of phosphorus and allows daily discharge of 3,120 pounds by the dischargers.

The second case study is a biochemical oxygen demand (BOD) management problem. This study is based upon treatment cost data originally compiled by Liebman (1965) for 11 dischargers on the Willamette River in Oregon. Five of these dischargers are municipalities while six are firms associated with pulp and paper or related industries. Within the BOD case study, the problem considered here is that of attaining the least costly decentralized removal of an aggregate amount of BOD by the dischargers. The 11 dischargers generate a total of approximately 411,000 pounds of BOD per day. It is assumed that only 168,000 pounds per day can be discharged if a given dissolved oxygen standard is to be met. These figures imply an overall removal efficiency of 59%.[11]

To simulate the outcomes of the allocation procedures, it is assumed in all cases that participants truthfully reveal treatment cost data and other information. Importantly, only one of the procedures (the Groves mechanism) is theoretically individually incentive-compatible; under all the other procedures, participants would generally have incentives to misrepresent preferences. Straightforward bidding is assumed here for the simulations, however, because the case studies are too complex (i.e., there are too many objects and bidders) for equilibrium bidding strategies to be calculated for the nonincentive-compatible procedures.[12] The extent to which the straightforward bidding assumption is justified for the different procedures is discussed later in this paper.

Tables 1 and 2 present dischargers' expenses for the two case studies under (1) uniform treatment, (2) least-cost operation,[13] (3) a single-price auction for rights, (4) an incentive-compatible auction for rights, and (5) free initial distribution of rights by the government on the basis of influent load followed by a centralized exchange. For both case studies, all five policies yield the same levels of environmental quality. In addition, because of the straightforward bidding assumption, all three procedures involving transferable rights yield a distribution of treatment effort among dischargers identical to the least-cost operation.

The efficiency gains for the rights programs over the uniform regulations are given by the differences in total treatment costs under the uniform treatment and least-cost programs. Note that because of differences in the plants' treatment efforts under uniform treatment and cost-efficient programs, however, some

individual plants have lower treatment costs under uniform treatment than under least-cost operation.

Under the single-price auction, rights are assumed to sell for $1.899/pound and $.0139/pound in the respective phosphorus and BOD cases. These figures are calculated under the assumption that the allocating agency sells the rights for a price equal to the highest rejected bid.[14]

The incentive-compatible allocation procedure is the Groves mechanism under which rights are allocated to the highest bidders, and a discharger i win-

---

[11] Recall from footnote 3 that BOD is modeled in this paper as a conservative pollutant, and hence, locational issues are ignored.

[12] A further complication in determining equilibrium outcomes for these case studies is that dischargers may be asymmetric because their sizes and demands for rights may vary by a few orders of magnitude. In the phosphorus case, for example, one may contrast the influent load of Milwaukee's Jones Island plant (7473 lbs/day) with that of Oconto Falls (12 lbs/day). Without symmetry different probability distributions for costs would have to be assumed for different plants, leading to extremely complicated expressions for the equilibrium bidding functions. A simplified model of equilibrium bidding for rights, however, is presented in Lyon (1980; 1981).

[13] The uniform treatment and least-cost treatment costs are approximated from the cost data under the assumption that only discrete levels of treatment are feasible. (This approximation procedure has previously been used by Eheart (1980).) Thus, if a plant reported costs only for 35% and 65% removal, and if uniform treatment required 59% removal, then the plant is assumed to operate at the 65% level. In the phosphorus case some dischargers did not report cost data for removal levels as high as 85%. Under uniform treatment, these dischargers are assumed to operate at the highest feasible level they reported. Milwaukee's costs are relatively low because the city currently receives an industrial waste free of charge which is useful in treating phosphorus.

[14] An advantage of using the highest rejected bid as the clearing price is that the purchaser of the last unit still makes a profit on its purchase—if it bids straightforwardly—which would not be the case if the lowest accepted bid became the clearing price. In these two case studies, the lower and higher marginal bids are in fact nearly identical. In the phosphorus case the highest rejected bid is 1.899 and the lowest accepted bid is 1.903, while in the BOD case the respective bids are .0139 and .0145. At other points of the aggregate demand curves, however, there are more significant differences between the marginal values.

TABLE 1

DISCHARGERS' EXPENSES UNDER ALTERNATIVE PROGRAMS FOR PHOSPHORUS CONTROL[a]

Discharger		Treatment Costs ($/day)		Total Expense of Rights Plus Treatment ($/day)		
Number	Name	Uniform Treatment	Least-Cost	Single-Price[b]	Incentive-Compatible[c]	Free Distribution[d,e]
1	Milwaukee-Jones Is.	32.0	32.0	1,590	1,090	-536
2	Milwaukee SS	183	183	474	455	-396
3	Racine	19.1	19.1	201	194	-45.4
4	Green Bay	337	337	633	614	337
5	Appleton	274	274	608	582	223
6	Manitowoc[f]	552	552	899	870	639
7	Kenosha	199	328	433	432	171
8	Sheboygan[f]	675	3.75	953	624	775
9	Oshkosh	199	199	334	331	199
10	Menom. Falls	60.9	60.9	118	118	60.9
11	De Pere	120	85.2	154	153	98.5
12	Fond du Lac[f]	307	307	526	515	415
13	Grafton	39.3	44.9	75.3	75.2	29.7
14	West Bend	88.0	88.0	136	135	63.3
15	Kaukauna	60.3	60.3	98.3	98.1	52.7
16	Cedarburg	83.3	83.0	121	121	83.0
17	South Milwaukee	127	55.4	131	131	93.4
18	Shawano S.D.	40.5	40.5	65.2	65.1	40.5
19	Port Washington	33.0	52.3	63.7	63.7	25.7
20	New London	52.1	39.4	81.2	81.0	48.9
21	Kiel	85.5	69.0	113	113	80.4
22	Plymouth[f]	56.1	56.1	94.1	93.5	71.3
23	North Park S.D.	26.6	33.2	57.9	57.9	25.6
24	Two Rivers	94.8	94.8	113	113	94.8
25	Chilton	47.0	54.2	71.3	71.3	46.6
26	Sturgeon Bay	71.2	71.2	97.8	97.8	73.1
27	Ripon[f]	46.3	46.3	78.6	78.5	53.9
28	Germantown	27.0	32.2	43.6	43.6	20.8
29	Sheboygan Falls[f]	46.8	46.8	69.6	69.6	52.5
30	Mariette	74.4	66.0	92.5	92.5	69.8
31	Kewaskum	44.7	33.3	58.0	57.9	37.1
32	Kimberly	33.2	33.2	46.5	46.5	35.1
33	Oconto	89.9	4.19	82.0	81.7	66.9

Continued overleaf

TABLE 1 *Continued*

Discharger		Treatment Costs ($/day)		Total Expense of Rights Plus Treatment ($/day)		
Number	Name	Uniform Treatment	Least-Cost	Single-Price[b]	Incentive-Compatible[c]	Free Distribution[d,e]
34	Sturtevant	31.0	31.0	42.3	42.3	31.0
35	Clintonville	58.9	58.9	72.2	72.2	58.9
36	Portage[f]	73.8	73.8	98.5	98.2	79.5
37	Thiensville	36.2	28.3	47.3	46.9	35.9
38	Menasha S.D. (East)	28.4	28.4	41.6	41.6	22.6
39	New Holstein	33.7	33.7	43.2	43.2	33.7
40	Little Chute	27.7	22.0	35.2	35.2	25.8
41	Union Grove	35.8	35.8	45.3	45.3	35.8
42	Algoma	43.1	5.96	51.5	51.4	40.1
43	Waupaca	47.2	.96	38.9	38.8	29.4
44	Kewaunee	28.2	23.1	32.6	32.6	25.0
45	Peshtigo[f]	48.3	.54	30.9	30.8	25.2
46	Butte Des Morts	23.0	23.0	34.5	34.5	23.0
47	Weyauwega	21.4	21.4	30.9	30.9	21.4
48	Kohler	19.4	20.5	24.3	24.3	18.6
49	Berlin[f]	45.0	1.45	73.6	73.3	60.3
50	North Fond du Lac[f]	29.7	29.7	41.1	41.1	33.5
51	Oconto Falls[f]	.23	23.9	19.2	19.2	15.4
52	Neenah-Menasha[f]	.00	.00	281	264	169
53	Holland S.D.[f]	.00	.00	51.3	51.1	36.1
	Total	4,880	3,920	9,860	8,880	3,930[g]

[a] Results are presented to no more than three significant figures.
[b] Single-price auction.
[c] Incentive-compatible auction.
[d] Free initial distribution on the basis of influent load followed by single-price exchange.
[e] Negative figures indicate receipts.
[f] Plants for which estimated treatment costs were unavailable for treatment levels above 85% (See footnote 13).
[g] Value differs from treatment costs under least cost operation only because of round-off error.

TABLE 2

DISCHARGERS' EXPENSES UNDER ALTERNATIVE PROGRAMS FOR BOD CONTROL[a]

Discharger		Treatment Costs ($/day)		Total Expense of Rights plus Treatment ($/day)		
Number	Name	Uniform Treatment	Least-Cost	Single-Price[b]	Incentive-Compatible[c]	Free Distribution[d,e]
1	Springfield	0	0	45.8	45.8	-11.0
2	Eugene	1,620	1,620	2,030	1,840	1,610
3	Evans Prods.	80.1	80.1	90.4	90.4	67.0
4	Corvallis	849	518	636	636	561
5	Albany	677	677	737	737	675
6	West. Kraft P. & P.	713	548	700	700	604
7	Columbia R. P. & P.	1,270	929	998	998	955
8	Salem	1,770	1,860	2,300	2,190	1,580
9	Spaulding P. & P.	396	396	563	563	289
10	Oregon Flax	29.8	33.6	46.1	46.1	12.0
11	Crown-Zeller & Publishers P. & P.	2,440	1,840	2,690	2,220	2,150
	Total	9,840	8,500	10,800	10,100	8,500

[a] Results are presented to no more than three significant figures.
[b] Single-price auction.
[c] Incentive-compatible auction.
[d] Free initial distribution on the basis of influent load followed by single-price exchange.
[e] Negative figures indicate receipts.

ning k_i rights pays the k_i highest rejected bids of dischargers other than itself. Because the k_i highest rejected bids are by definition less than or equal to the highest rejected bid, the incentive-compatible procedure will always imply payments for rights that are less than or equal to those under the single-price auction, if it can be assumed that the clearing price in the single-price procedure is not affected by strategic bidding.[15] Tables 1 and 2 show that under this assumption, dischargers' total payments for rights under the incentive-compatible auction are 16% and 33% less than their payments under the single-price auction, in the respective phosphorus and BOD cases. Dischargers purchasing the largest numbers of rights make the largest savings compared to the single-price auction.

Importantly, the payments for rights under both the single-price and incentive-compatible auctions are substantial for both case studies. In particular, in the phosphorus case dischargers' payments as a group for the rights are greater than the costs of uniform treatment, regardless of which of the two auctions is considered. (The phosphorus dischargers' total payments for rights alone are $5,940/day ($9,860/day − $3,920/day) and $4,960/day ($8,880/day − $3,920/day), under the single-price and incentive-compatible auctions, respectively.) Such large financial burdens are a major reason for considering rights programs involving transfers from the government to dischargers.

The free initial distribution procedure examined here is based upon an approach considered by David et al. (1980) and Eheart et al. (1980).[16] Rights are initially distributed here on the basis of pollutant load, measured in pounds of phosphorus or BOD per day. By definition, this approach implies that the uniform treatment configuration is duplicated prior to exchanges of rights.[17] The total numbers of rights distributed (3,120 pounds of phosphorus/day and 168,000 pounds of BOD/day) are identical to those sold in the single-price and incentive-compatible auction procedures above.

Following their initial free distribution, the rights may be transferred among dischargers via a governmentally operated single-price exchange. Following a procedure considered by Eheart et al. (1980), buyers would submit demand and supply schedules to the government. If participants are straightforward, the outcome of this procedure would be the same final allocation of rights as in the previously discussed efficient single-price and incentive-compatible auctions. Furthermore, the marginal clearing prices would also be the same as in the single-price auction case. In the simulations reported in tables 1 and 2, it has been assumed that the same price would result as in the previous single-price auctions.[18]

---

[15] Recall, however, that the single-price procedure is theoretically not incentive-compatible.

[16] The free distribution results presented for the phosphorus case are based on one of the simulations considered in Eheart et al. (1980).

[17] All of the dischargers in Table 2 have lower expenses under the free distribution program than under uniform treatment. In the phosphorus case (Table 1) 14 dischargers (prior to rounding off data for presentation) have higher total costs than under uniform treatment because of the manner in which uniform treatment costs were approximated for that case study. 12 of these dischargers are among those plants assumed to operate at their highest reported levels of feasible treatment under the uniform policy, even though those levels are below the 85% standard. Two other dischargers, Sturgeon Bay and Kimberly, have slightly higher costs under the rights distribution program due to round-off error in the simulation.

[18] Note that in this case (where dischargers, as opposed to the government, sell rights), dischargers who sell rights would prefer that the higher of the two marginal bids be used (e.g., $1.903/lb phosphorus and $.0145/lb BOD), so perhaps some intermediate figure might be chosen in practice. In the case where the government sells

Among the key results of the simulation of the free initial distribution approach in the phosphorus case is that, despite the fact that there are 53 dischargers, the two Milwaukee plants sell 604 out of the 806 rights exchanged, and with a third plant, Kenosha, they account for 687 of the total.[19] Sheboygan buys about half the rights transferred. In the BOD case one discharger, Crown-Zellerbach, purchases over 70% of the 32,000 rights exchanged, while another, Salem, sells just under two-thirds of the total rights exchanged. More detailed results of the free initial distribution approach are presented in Eheart et al. (1980) and Lyon (1980).

III. ANALYSIS

In this section the alternative procedures are examined with respect to several properties that are fundamental to the implementation and efficacy of the programs. Because of the importance of incentive compatibility in affecting the outcomes of the allocation procedures, the discussion begins with this property.

Incentive Compatibility

Incentive compatibility is important because a procedure designed to operate on information such as costs, bids, or pollution flows submitted voluntarily by participants may not meet its objectives if this information is manipulated. If the procedure is not incentive-compatible it is possible, for example, that certain dischargers, who would not receive rights under an efficient allocation, will end up with rights.

Strategic behavior may also have a range of distributional effects. Strategic behavior by individuals and coalitions

may lower the receipts of the government if it sells rights via auctions and may either raise sellers' receipts or reduce buyers' payments for rights under the free initial distribution approach.

Incentive compatibility is also closely related to the objective of ease of administration, because a procedure that encourages voluntary truthful revelation of information will require less government investigation and litigation to operate. Furthermore, an incentive-compatible procedure may have outcomes that are more certain and stable, because it will be implemented on the basis of true information, which may be less likely to change than strategic revelations.

As well as being complementary with respect to other properties, however, incentive compatibility is also conflicting. Hurwicz (1972) has shown that no informationally decentralized allocation procedure can be (1) individually incentive-compatible, (2) Pareto efficient, and (3) individually rational in all economic environments. (Individual rationality implies that no participant is made worse-off than his initial condition by participating in the procedure.) Thus, there are fundamental trade-offs with respect to other desirable properties which must be faced if an incentive-compatible procedure is implemented. While an efficient and individually incentive-compatible procedure can be implemented for government rights sales through the use of a Groves mechanism, transfers to dischargers—of either rights or funds—must be of a

---

the rights, it is implicitly assumed that revenue generation is not a primary purpose of the auction, so that there is no major disadvantage in using the lower of the two marginal values as the price.

[19] Following Eheart et al. (1980), the two Milwaukee plants are assumed to be separate and competing entities for the simulations.

lump-sum nature to avoid introducing strategic incentives.[20]

While only one of the approaches simulated here—the Groves mechanism—is individually incentive-compatible in theory, the single-price auction of rights by the government may tend toward being individually incentive-compatible when there are large numbers of dischargers and rights. In this case any one individual has a small probability of affecting the marginal bid, and by lowering bids in order to attempt to lower the marginal bid, dischargers risk losing rights they would otherwise win. It must be emphasized, however, that in many pollution rights cases there will not be large numbers of bidders. In the BOD case study, for example, there are only 11 major dischargers.

Manipulation may also affect the reallocation of rights under the free initial distribution procedure. Under the governmentally operated single-price exchange, dischargers would have incentives to misrepresent both demand and supply schedules submitted to the government agency. An alternative approach to reallocation, an unregulated—or free-market—exchange, would presumably function efficiently if there were large numbers of participants who act as price takers. It is important to consider, however, whether there are sufficient numbers of buyers and sellers in the program to get price-taking behavior. In the phosphorus case, for example, although there are 53 plants and 3,120 rights, two plants would sell 75% of the rights traded at the market-clearing price, and three plants would sell 85% of the total. Similarly, in the BOD case, one discharger sells and another buys roughly two-thirds of the rights exchanged. Thus, even with relatively large numbers of participants and objects, imperfectly competitive behavior cannot be dismissed as unlikely.

Under uniform treatment, dischargers could have incentives to misrepresent the quantities of waste that they must treat in order to be permitted to discharge greater amounts. An additional problem with regulations is that they may provide insufficient incentives for firms to change production processes that might cut down on wastes.[21]

A final incentives issue for all the programs considered is the problem of enforcing dischargers' removal of wastes. One may assume that the enforcing agency can attempt to gather information about actual discharges in order to ensure compliance.[22] Enforcement is a problem under all pollution-control schemes, and there are unlikely to be major differences in this context among alternative rights and regulatory programs.

Efficiency

In this study, efficiency is defined as the least costly decentralized attainment of an environmental standard. Assuming straightforward bidding, all of the rights

[20] All of the procedures considered here, including the Groves mechanism, are susceptible to manipulation by coalitions. In all the potential coalitions, however, individual dischargers have incentives to deviate from collusive behavior. Thus, the coalitions would tend to be inherently unstable, unless agreements can be enforced among the colluding parties. The problem of collusion is difficult, if not impossible, to eliminate via any useful mechanism which is informationally decentralized and which uses bidders' messages to determine allocations in a nontrivial manner (Green and Laffont 1979). Some effects of collusion under the government sale might be reduced if a reservation (i.e., minimum) price is set for the rights by the government.

[21] The costs of rights—or the revenues obtainable through sales of rights possessed—provide continuing incentives for implementation of such changes, whereas regulations provide only the incentives of savings in treatment costs.

[22] Note that obtaining of information by an enforcement agency to ensure straightforward *bidding*, however, might be possible only to a limited extent because bidding reflects costs that may be specific to dischargers.

programs would lead to this solution. Without the assumption of straightforward bidding, efficiency or near-efficiency would also generally follow under most of the rights programs, given large numbers of bidders and no collusion.

Uniform treatment regulations are clearly not efficient in general. By requiring equal percentage removal by all dischargers, there is no attempt to equalize marginal treatment costs across dischargers. This inefficiency is one of the strongest arguments against uniform regulations and is a major reason why investigators have looked favorably upon programs such as rights or taxes. In addition, rights or taxes provide continuing economic incentives in a dynamic sense for the reduction of discharges.

Equity and Financial Burden

Uniform treatment can imply lower expenses for dischargers with relatively low marginal treatment costs than some types of efficient pollution-control programs. Under efficient programs, the low-cost dischargers would find themselves treating at higher percentages than under uniform treatment. Only if the low-cost dischargers received some transfer from the government—such as free initial pollution rights or refunds from rights sales—would they benefit from an efficient program. An important advantage of uniform treatment for all dischargers over government sales of rights via either the single-price or incentive-compatible auctions is that under the regulations there is no financial burden from the payments for rights.[23]

In the phosphorus case (Table 1), only three dischargers out of 53 (numbers 33, 43, and 45) are estimated to have lower total expenses under the single-price auction than under uniform treatment, while those three and a fourth (number 8)

have lower expenses under the incentive-compatible procedure than under uniform treatment.[24] Three and four dischargers have lower expenses under the respective auctions than under uniform treatment in the BOD case.

The free initial distribution approach implies that no payments are made by the dischargers as a group to the government, yet the approach ideally results in an efficient allocation of rights among dischargers. Because of the reduced financial burden, all dischargers in both cases—with the exception of Sheboygan (Table 1)—would have lower costs under the free initial distribution procedure followed by a single-price exchange than under the government sale of rights via either the single-price or the incentive-compatible auctions.[25] All of the dischargers should have expenses at least as low under the free initial distribution approach based on influent load as under uniform treatment, because initial distribution on the basis of load by definition

[23] Buchanan and Tullock (1975) observe that the imposition of financial burdens by programs such as pollution taxes may explain why legislation implementing such efficient programs has not been widely adopted.

[24] The number of dischargers having lower expenses under rights sales than under regulations may be somewhat underestimated because using maximum reported treatment levels to approximate uniform treatment costs for some plants underestimates these costs, where the maximum levels are less than the levels actually required by uniform treatment. For some other plants, however, uniform treatment costs are overestimated because of the use of discrete treatment levels to approximate costs (see note 13).

[25] Sheboygan would have lower expenses under a government sale of rights via the Groves mechanism than under the free initial distribution procedure followed by a single-price exchange. Sheboygan is large and relatively inefficient and would buy many rights under either the free initial distribution procedure or the government sale. (It requires the second largest number of rights of all plants in the least-cost outcome.) Under the Groves procedure Sheboygan realizes a bigger savings via the procedure's quality discount than the city gains under the initial allocation proportional to load. In addition, Sheboygan's uniform treatment costs are underestimated because of the approximation procedure used (see note 13).

implies granting all dischargers sufficient initial amounts of transferable rights to permit them to operate at the uniform treatment level.[26]

An important issue surrounding the free initial distribution approach is that it requires transfers from the government to dischargers, and hence, determination of "shares." Whether such shares can be determined equitably and without manipulation may depend upon the specific application. It is not clear, for example, whether the distributional bases of this procedure would or should be updated to reflect entry of new dischargers into the region or growth of existing firms.

One attractive equity aspect of the incentive-compatible auction may be precisely its encouragement of straightforward revelation. This procedure also tends to give quantity discounts to purchasers of large numbers of rights because rights are sold for the k_i highest rejected prices as opposed to the single highest rejected price. A related issue, however, is that the incentive-compatible procedure is likely to sell rights at different prices to different dischargers because each discharger's payment depends on the bids of the others. The price discrimination involved is impersonal and does not involve, for example, extracting maximum willingness-to-pay. It is possible, however, that dischargers may view the payments under the incentive-compatible procedure as being arbitrary or even unfair. The price discrimination property cannot be altered, however, without losing the property of incentive compatibility.

A final question is whether equity in this policy problem is best considered at the level of dischargers or individual citizens. This issue is noted because if the real goal is equitable treatment of individuals, then government sales of rights to dischargers might be acceptable if coupled with cycling of the revenues into the general revenue fund.[27]

IV. CONCLUSIONS

Based on both theoretical and practical considerations, this paper suggests that there is a fundamental trade-off between reducing the financial burden of rights purchases and eliminating the incentives for strategic behavior. Very few dischargers in either case study would have lower expenses under government sales of rights than under uniform treatment regardless of whether sales are by a single-price or incentive-compatible auction. The simulations also suggest that there may be opportunities for strategic behavior under nonincentive-compatible procedures because of small numbers of dischargers.

If one accepts views such as those of Dales (1968, p. 93) and Rose (1973) that the government should sell rights to dischargers for a positive price, however, then the incentive-compatible auction deserves serious consideration. Its property of individual incentive compatibility would discourage noncollusive forms of strategic behavior, and it would encourage efficient and stable allocations.

The alternative approach of free initial

---

[26] Though see note 17 with respect to the figures in Table 1.

[27] Other equity questions may be whether the dischargers should have to buy rights at all, and whether environmentalists or the public could enter the allocation procedures. Whether the public or the dischargers should pay for pollution rights or receive compensation for their sale is largely a question of property rights. Coase (1960), of course, suggests the symmetry between the parties bearing or causing externalities. Examination of the efficiency, distributive, and incentives arguments behind these questions is beyond the scope of this study.

distribution on some basis followed by exchange may be affected by strategic manipulation. With small numbers of dischargers the exchange following the initial distribution could be manipulated by buyers or sellers. To invoke a large numbers argument of incentive compatibility for this procedure, there must be both many buyers and sellers who act as though they expect to have little effect upon the clearing price. The case studies suggest that even with relatively many dischargers, there may be sufficient variations in size and treatment costs to weaken this assumption. In addition, strategic behavior could affect the initial distribution of rights unless this distribution is based upon information that is extremely difficult to manipulate.

Despite its problems with respect to incentive compatibility (which has implications for its properties of efficiency, equity, ease of administration, and certainty of outcome), the rights allocation approach of free distribution followed by exchange may present the most desirable package of features. Importantly, this approach can yield dischargers savings over the expenses of present uniform regulations, as well as provide incentives for efficient allocation of treatment efforts. In contrast, the incentive-compatible procedure may be optimal with respect to objectives including individual incentive compatibility and efficiency. If dischargers' payments for rights are not viewed as a problem—or are viewed as a policy goal—then the incentive-compatible procedure deserves consideration. Where environmental policy is primarily aimed at minimizing dischargers' expenses while maintaining or improving environmental quality, however, rights programs based upon free initial distribution by the government appear to merit serious consideration.

References

Braasch, Daryl A., and Joeres, Erhard F. 1975. *Analysis of the Lake Michigan Basin Phosphorus Removal Policy in Wisconsin.* Sea Grant College Technical Report, WIS–SG–75–224, Madison: University of Wisconsin.

Buchanan, James M., and Tullock, Gordon. 1975. "Polluters' Profits and Political Response: Direct Controls versus Taxes." *American Economic Review* 65 (Mar.): 139–47.

Coase, Ronald H. 1960. "The Problem of Social Cost." *Journal of Law and Economics* 3 (Oct.): 1–44.

Dales, J. H. 1968. *Pollution, Property and Prices.* Toronto: University of Toronto Press.

Dasgupta, Partha; Hammond, Peter; and Maskin, Eric. 1979. "The Implementation of Social Choice Rules: Some General Results on Incentive Compatibility." *Review of Economic Studies* 46 (Apr.): 185–216.

David, Elizabeth. 1980. "Cost Effective Management Options for Attaining Water Quality." Madison: Bureau of Planning, Wisconsin Department of Natural Resources (Oct.).

David, M.; Eheart, W.; Joeres, E.; and David, E. 1980. "Marketable Permits for the Control of Phosphorus Effluent into Lake Michigan." *Water Resources Research* 16 (Apr.): 263–70.

Eheart, J. Wayland. 1980. "Cost-Efficiency of Transferable Discharge Permits for the Control of BOD Discharges." *Water Resources Research* 16 (Dec.): 980–86.

Eheart, J. Wayland; Joeres, Erhard F.; and David, Martin H. 1980. "Distribution Methods for Transferable Discharge Permits." *Water Resources Research* 16 (Oct.): 833–43.

Green, Jerry, and Laffont, Jean-Jacques. 1977. "Characterization of Satisfactory Mechanism for the Relevation of Preferences for Public Goods." *Econometrica* 45 (Mar.): 427–38.

———. 1979. "On Coalition Incentive Compatibility." *Review of Economic Studies* 46 (Apr.): 243–54.

Hurwicz, Leonid. 1972. "On Informationally Decentralized Systems." In *Decision and Organization,* eds. C. B. McGuire and R. Radner. Amsterdam: North-Holland.

———. 1973. "The Design of Mechanisms for Resource Allocation." *American Economic Review* 63 (May): 1–30.

Kneese, Allen V., and Bower, Blair T. 1968. *Managing Water Quality: Economics, Technology, and Institutions.* Baltimore: The Johns Hopkins University Press.

Liebman, Jon C. 1965. "The Optimal Allocation of Stream Dissolved Oxygen Resources." Ph.D. dissertation, Cornell University, Ithaca, N.Y.

Lyon, Randolph M. 1980. "Auctions and Alternative Procedures for Public Allocation: With Applications to the Distribution of Pollution Rights." Ph.D. dissertation, University of Illinois, Urbana.

—————. 1981. "Dischargers' Expenses under Pollution Rights Programs and Regulations." mimeo., Department of Economics, University of Texas, Austin (Jan.).

Maloney, M. T., and Yandle, Bruce. 1980. "Bubbles and Efficiency." *Regulation* 4 (May/June): 49–52.

Monash, Curt Alfred. 1980. "Efficient Allocation of an Uncertain Supply." mimeo., Kennedy School of Government, Harvard University (Apr.).

Rose, Marshall. 1973. "Market Problems in the Distribution of Emission Rights." *Water Resources Research* 9 (Oct.): 1132–44.

Thomann, Robert V. 1972. *Systems Analysis and Water Quality Management.* New York: McGraw-Hill.

Tietenberg, Thomas H. 1980. "Transferable Discharge Permits and the Control of Stationary Source Air Pollution: A Survey and Synthesis." *Land Economics* 56 (Nov.): 391–416.

U.S. Environmental Protection Agency. 1980. "Emission Reduction Banking and Trading Project: Annotated Bibliography." Fifth edition. Office of Planning and Evaluation, U.S. EPA, Washington, D.C. (Oct.).

—————. 1981. "An Analysis of Economic Incentives to Control Emissions of Nitrogen Oxides from Stationary Sources." Office of Planning and Management, U.S. EPA, Washington, D.C. (Jan.).

Vickrey, William. 1961. "Counterspeculation, Auctions, and Competitive Sealed Tenders." *Journal of Finance* 16 (May): 8–37.

[12]

MARKET POWER AND TRANSFERABLE PROPERTY RIGHTS*

ROBERT W. HAHN

The appeal of using markets as a means of allocating scarce resources stems in large part from the assumption that a market will approximate the competitive ideal. When competition is not a foregone conclusion, the question naturally arises as to how a firm might manipulate the market to its own advantage. This paper analyzes the issue of market power in the context of markets for transferable property rights. First, a model is developed that explains how a single firm with market power might exercise its influence. This is followed by an examination of the model in the context of a particular policy problem—the control of particulate sulfates in the Los Angeles region.

I. INTRODUCTION

The idea of implementing a market to ration a given quantity of resources is by no means novel. Working examples include markets for taxi medallions and liquor licenses. Suggested applications for the use of a market approach abound in the economics literature, especially in the fields of air and water pollution.[1] Why has the idea of setting up a market in transferable property rights received so much attention? One key reason, and the reason which motivates this paper, is that such markets have the potential to achieve a given objective in a cost-effective manner. Whether this potential is realized depends, among other things, on the design of the market and the extent to which individual firms can exert a significant influence on the market.

The purpose of this paper will be to explore how the initial distribution of property rights can lead to inefficiencies. Section II develops the basic model for the case in which one firm can influence the market. Section III considers a potential application of the model. The results of the theoretical analysis are then

*The work reported here was supported by the Environmental Quality Laboratory at Caltech and the California Air Resources Board. I would like to thank Jim Quirk, Roger Noll, Jennifer Reinganum, and Robert Dorfman for providing helpful comments. The views expressed herein, including any remaining errors, are solely the responsibility of the author.

1. Tietenberg [1980] provides a comprehensive survey of the application of marketable permits to the control of stationary source air pollution. A general list of references to potential applications in air and water pollution is provided in the study by Anderson *et al.* [1979].

compared with the conventional wisdom, and directions for future research are discussed in Section IV.

For analytical purposes, firms are divided into two categories. A firm will be said to have market power if it realizes it has an influence on price. A firm will not have market power if it acts as a price taker. The question for analysis, then, is how a single firm with market power might influence the market by affecting the price at which a commodity sells. More precisely, this essay examines how the price strategy of a firm with market power varies with changes in the initial distribution of property rights.

In the static models developed below, all transactions take place at a single price. Restricting the model in this way permits analysis of a range of inefficient outcomes. This is in contrast to the approach taken by Coase [1960] in his seminal article, who does not restrict the bargaining space and, consequently, emphasizes the range of efficient outcomes that can result, irrespective of the initial endowment of property rights.

The principal result is that the degree of inefficiency observed in the market is systematically related to the distribution of permits. For the case of one firm with market power, the results have some intuitive appeal. If a firm with market power would elect to buy permits in a competitive market (i.e., where all firms act as if they were price takers), then it follows a strategy resembling that of a monopsonist. If it would choose to sell permits in a competitive market, then the firm with market power follows a strategy resembling that of a monopolist. These results are formalized in the next section.

II. THE BASIC MODEL

A critical assumption underlying the competitive model is that firms act as if they were price takers. In the model developed below, it will be assumed that all firms except one are price takers. The basic question to be answered is how (and whether) the equilibrium price and quantities will vary as a function of the initial distribution of permits among firms.

Consider the case of m firms with firm 1 designated as the firm with market power. A total of L permits are distributed to the firms, with the ith firm receiving Q_i^0 permits. Firms are allowed to trade permits in a market that lasts for one period. The number of permits that the ith firm has after trading will be denoted by Q_i. All firms except the market power firm are assumed to have downward sloping inverse demand functions for permits of the form $P_i(Q_i)$ over the region $[0,L]$. P_i represents firm i's willingness to pay. All trades in the market are constrained to take place at a single equilibrium price P. For concreteness, we shall consider the case of a classical pollution externality. All price-taking firms attempt to minimize the sum of abatement costs and permit costs. For the case of pollution, the assumption of downward sloping demand curves is equivalent to the assumption that marginal abatement costs are increasing. Let $C_i(Q_i)$ be the abatement cost associated with emitting Q_i units. Marginal abatement costs, $-C_i'$, are assumed to be positive and increasing, which implies that $C_i' < 0$ and $C_i'' > 0$ for $i = 2, \ldots, m$. Price takers solve the following optimization problem:

(1) $\underset{Q_i}{\text{minimize}} \ C_i(Q_i) + P(Q_i - Q_i^0) \quad (i = 2, \ldots, m)$.

The first-order condition for an interior solution is

(2) $C_i'(Q_i) + P = 0$.

This merely says that price takers will adjust the quantity used Q_i until the marginal abatement cost equals the equilibrium price P.[2] Equation (2) implicitly defines a demand function $Q_i(P)$, which is downward sloping on $[0,L]$ for $i = 2, \ldots, m$. Furthermore, note that the number of permits the ith price-taking firm will use is *independent* of its initial allocation of permits.

The analysis of the firm with market power is less straightforward. Begin by defining an abatement cost function $C_1(Q_i)$, where $C_1' < 0$ and $C_1'' > 0$. This says that the firm with market power faces increasing marginal abatement costs. Firm 1 has the power to pick a price that will minimze its expenditure on abatement costs and permits subject to the constraint that the market clear. Formally, the problem is to

(3) $\underset{P}{\text{minimize}} \ C_1(Q_1) + P(Q_1 - Q_1^0)$

        subject to $Q_1 = L - \sum_{i=2}^{m} Q_i(P)$.

2. The assumption of increasing marginal abatement cost implies that the firm attains a regular minimum in solving the problem.

Substituting the constraint into the objective function and differentiating yield the following first-order condition for an interior minimum:

$$(4) \quad (- C_1' - P) \sum_{i=2}^{m} Q_i' + \left(L - \sum_{i=2}^{m} Q_i(P) - Q_1^0 \right) = 0.$$

Equation (4) reveals that the only case in which the marginal cost of abatement $- C_1'$ will equal the equilibrium price is when firm 1's distribution of permits just equals the amount it chooses to use. In effect, this says that the only way to achieve a cost-effective solution, where marginal abatement costs are equal for all firms, is to pick an initial distribution of permits for firm 1 which coincides with the cost-minimizing solution.

This gives rise to the following result:

PROPOSITION 1. Suppose that there is one firm with market power. If it does not receive an amount of permits equal to the number that it holds in equilibrium, then the total expenditure on abatement will exceed the cost-minimizing solution.

The key point to be gleaned from the analysis is that the distribution of permits matters, with regard not only to equity considerations but also to cost. Traditional models of such markets view problems of permit distribution as being strictly an equity issue.[3] With the introduction of market power, it was shown that the distribution of permits may also impinge on efficiency considerations.

The next logical question to explore is how the market equilibrium will vary as a function of firm 1's initial distribution of permits. Doing the necessary comparative statics yields

$$(5) \quad \left. \frac{\partial P}{\partial Q_1^0} \right|_{L = \text{constant}} = \left((- C_1' - P) \sum_{i=2}^{m} Q_i'' \right.$$
$$\left. + \sum_{i=2}^{m} Q_i^2 C_i'' - 2 \sum_{i=2}^{m} Q_i' \right)^{-1}.$$

3. The analysis by Montgomery [1972] is one such example. In this analysis firms are assumed to be price takers. For the case of one pollutant, one market, and a linear relationship between source emissions and environmental quality, Montgomery finds that the distribution of permits will have no effect on achieving the target in a cost-effective manner.

MARKET POWER AND TRANSFERABLE PROPERTY RIGHTS 757

The expression for the denominator is the second-order condition for the cost minimization and will be positive if the second-order sufficiency condition for a minimum obtains. For example, in the case of linear demand curves (i.e., $Q_i'' = 0$), the expression will be positive. Thus, for the case when a regular interior minimum exists, a transfer of permits from any of the price takers to the firm with market power will result in an increase in the equilibrium price. An immediate corollary to this result is that the number of permits that the firm with market power uses will increase as its initial allocation of permits is increased. Formally, the problem is to show $(\partial Q_1/\partial Q_1^0) > 0$. By the chain rule,

$$(6) \qquad \frac{\partial Q_1}{\partial Q_1^0} = \frac{\partial Q_1}{\partial P}\frac{\partial P}{\partial Q_1^0}.$$

It suffices to show that $(\partial Q_1/\partial P)$ is positive. By direct substitution for Q_i,

$$(7) \qquad \frac{\partial Q_1}{\partial P} = \frac{\partial\left(L - \sum_{i=2}^{m}Q_i(P)\right)}{\partial P}.$$

The expression on the right-hand side of (7) equals $-\sum_{i=2}^{m}Q_i'(P)$, which is positive, because demand curves are presumed to be negatively sloped.

One question that arises in this model is whether there is any systematic relationship between the distribution of permits to the firm with market power and the degree of inefficiency. If inefficiency is measured by the extent to which abatement costs exceed the minimum required to reach a stated target, then it is possible to show the following result:

PROPOSITION 2. Let Q_1^* denote the distribution of permits for the case when permit distribution equals permit use for the firm with market power. Then inefficiency increases both as Q_1^0 increases above Q_1^* and as Q_1^0 decreases below Q_1^*.

The proposition is verified by determining how total cost TC varies as a function of Q_1^0.

The efficient solution is derived from the following minimization:

$$(8) \qquad \underset{Q_1, \ldots, Q_m}{\text{minimize } TC} = C_1(Q_1) + \sum_{i=2}^{m} C_i(Q_i)$$

subject to

$$Q_1 + \sum_{i=2}^{m} Q_i = L.$$

First-order conditions imply that

(9) $-C_i'(Q_i) = P_i(Q_i) = P \quad (i = 2, \ldots, m).$

Differentiation of total cost with respect to Q_1^0 yields

$$\frac{\partial TC}{\partial Q_1^0} = C_1' \frac{\partial Q_1}{\partial Q_1^0} + \sum_{i=2}^{m} C_i' \frac{\partial Q_i}{\partial Q_1^0}$$

(10)
$$= C_1' \sum_{i=2}^{m} \frac{\partial Q_i}{\partial Q_1^0} + \sum_{i=2}^{m} C_i' \frac{\partial Q_i}{\partial Q_1^0}$$

$$= \sum_{i=2}^{m} (C_i' - C_1') \frac{\partial Q_i}{\partial Q_1^0}.$$

The above expression can be simplified by noting that

(11) $$\frac{\partial Q_i}{\partial Q_1^0} = -\frac{\partial P}{\partial Q_1^0} \bigg/ C_i''.$$

Equation (11) is obtained by differentiating (9) with respect to Q_1^0. Substituting equation (11) into (10) yields

$$\frac{\partial TC}{\partial Q_1^0} = -\frac{\partial P}{\partial Q_1^0} \sum_{i=2}^{m} \frac{(C_i' - C_1')}{C_i''}$$

(12)
$$= -\frac{\partial P}{\partial Q_1^0} \sum_{i=2}^{m} \frac{(-P - C_1')}{C_i''}$$

$$= \frac{\partial P}{\partial Q_1^0} (P + C_1') \sum_{i=2}^{m} \frac{1}{C_i''}.$$

Equation (12) implies that

(13) $$\frac{\partial TC}{\partial Q_1^0} > (<) \, 0 \quad \text{as } (P + C_1') > (<) \, 0.$$

Combining (13) with equation (4) yields the result that total cost achieves a minimum at Q_1^* and will increase as the permit distribution deviates from Q_1^* in either direction.

III. A POTENTIAL APPLICATION

In order to apply the basic model described in the previous section, it is necessary to develop an operational test for identifying a firm with market power. How this might be done is beyond the scope of this paper. In the application discussed below, the firm holding the largest share of permits under a competitive market simulation is designated as the market power firm.

To demonstrate how the basic model can be applied, the problem of controlling particulate sulfates in the Los Angeles region was selected. This problem was chosen because it appeared to be a likely candidate for a transferable property rights scheme, and because the problem of market power could conceivably arise. Market simulations based on the assumption that firms are price takers indicate that the largest emitter of sulfur oxides, an electric utility, could account for as much as half of the total emissions, and an even higher proportion of emissions for which abatement technologies are known—i.e., controllable emissions.[4]

The extent of market power will in general, vary with the level of allowable emissions, the shape of the marginal abatement cost schedule for the market power firm, and the marginal abatement costs faced by all other firms. For this particular example, a permit will be defined as the right to emit one ton of sulfur oxides emissions per day for one day. Based on this definition, Figure I shows the marginal costs of abatement for the firm designated as the market power firm.[5] Two curves are drawn in Figure I, a discrete step function (based on the data in Hahn [1981b]), and a continuous approximation that has the following functional form:

$$(14) \qquad\qquad - C_1' = 88,300 Q_1^{-0.87}.$$

Actually, for the case of the market power firm, a continuous approximation is probably more reasonable because the abatement strategy under consideration is the desulfurization of fuel oil or the purchase of lower sulfur residual fuel oil.

A similar graph for all other firms is shown in Figure II, which illustrates the derived demand for permits at any given

4. A more detailed discussion of the market power question can be found in Hahn [1981a], and Hahn and Noll [1982].

5. Further assumptions underlying the development of these data, such as the availability of natural gas, are discussed in Hahn [1981a].

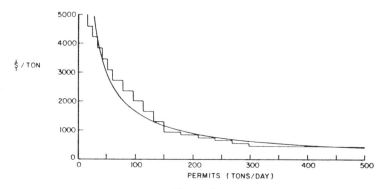

FIGURE I
Marginal Abatement Costs for Market Power Firm

price. The continuous approximation to the discrete case takes
the following form:

$$(15) \qquad \sum_{i=2}^{m} Q_i(P) = 73 + \frac{154{,}000}{P}.$$

The demand curve is based on some discrete technologies such as
scrubbers as well as some continuous abatement strategies such
as the one mentioned above. The continuous approximation will
be used for purposes of illustration. Note that the particular form
used in (15) implies that emissions by others will be at least 73
tons per day for all positive permit prices.

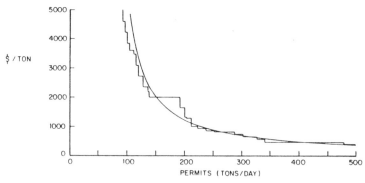

FIGURE II
Derived Demand for Permits by All Other Firms

MARKET POWER AND TRANSFERABLE PROPERTY RIGHTS 761

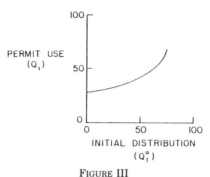

FIGURE III
Permit Use versus Permit Distribution—Market Power

To compute how the initial distribution of permits affects prices, quantities, and overall abatement, it is first necessary to select a value for the total number of permits. In this example the parameter L was set equal to 149 tons/day, an amount which will ensure that both state and federal standards related to sulfur oxides emissions and particulate sulfates will be met. Having chosen a value for L, we find it possible to examine how permit use varies with initial distribution by substituting equations (14) and (15) into equation (4) and solving. The graphical solution to the problem is shown in Figure III. Note that Q_1 increases as a function of Q_1^0 until a corner solution is approached. This point corresponds to a permit distribution where all other firms receive an amount of permits that just equals their uncontrollable emissions. If all other firms receive an amount of permits that falls short of their uncontrollable emissions, then the relationship between Q_1 and Q_1^0 is not unique. In this latter case, the market power firm can reap infinite rewards by exploiting the perfectly inelastic part of the demand curve.[6]

Prices vary widely as a function of the initial distribution of permits. The monopsony price is approximately $3,200/ton, while the competitive price, associated with $Q_1^0 = 36$, is about $3,900/ ton.[7] When all other firms receive permits corresponding to their uncontrollable emissions, the price of a permit jumps to approximately $21,000/ton. The monopoly price, i.e., when $Q_1^0 = L$, is not well defined both in theory and in practice: in theory, because

---

6. In practice, such rewards would be limited by the decision of other firms to shut down operations.
7. All prices and costs are given in 1977 dollars.

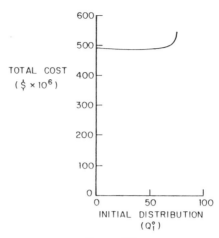

FIGURE IV
Total Annual Abatement Cost versus Initial Distribution

(15) is a hyperbola with an asymptote; and in practice, because of insufficient information on the value of firms and possible technologies that might be available for controlling so-called uncontrollable emissions.

Given permit use as a function of the initial distribution of permits, it is then possible to estimate the total annual costs of abatement by integrating equations (14) and (15). The relationship between total annual abatement expenditures and the initial distribution of permits is shown in Figure IV. Note that abatement expenditures remain relatively constant (in the neighborhood of 490 million dollars annually) until the market power firm is able to exert some monopoly power when it receives permits in excess of 60 tons per day.

The relative importance of monopolistic and monopsonistic behavior may be quite sensitive to parameter changes. In this case, monopsonistic behavior does not appear to present a problem. The reason is that in the range of the competitive equilibrium for emissions limits around this stringent standard, the supply of permits from other firms to the largest source is very sensitive to price changes. This undermines the opportunity of the firm with market power to take advantage of its high market share. As the emissions ceiling is relaxed, inefficiencies resulting from monopsonistic behavior tend to increase.

If the primary objective in setting up a market is to minimize total abatement costs, Figure IV indicates that the policymaker should try to avoid a situation where the firm with market power can act as a monopolist. However, because of the uncertainty associated with the cost data, it makes sense to try to minimize the likelihood that a firm or group of firms will be able to induce a price-quantity equilibrium which departs from the competitive result in either direction. Alternatives for dealing with this issue are discussed in Hahn and Noll [1982]. The theory developed in this paper indicates that the expected excess demand of each firm may be a critical variable over which the policymaker can exercise control.

IV. Conclusions

This paper has explored the issue of market power in the context of markets in transferable property rights. The simple model developed in Section II reveals two essential points. First, just because a firm may have a large share of the permits, this does not necessarily mean it can influence the outcome in the permit market. Second, if a firm does have market power in the permit market, its effect on price (assuming that there is one firm with market power) varies with its excess demand for permits. That is to say, once the potential for market power has been ascertained, it is a flow—excess demand of the firm with market power—which determines the equilibrium.

The importance of the flow has immediate implications for market design. In particular, with full knowledge of demand functions, a central authority could effectively pick the quantity of permits it wanted the market power firm to use through a suitable initial allocation. The limits to the discretion of the authority would be dictated by two extreme cases: pure monopsony in which all permits are distributed to the price takers, and pure monopoly in which all permits are distributed to the firm with market power.

Of course, the more realistic situation is one in which the authority has, at most, only a crude estimate of the demand functions. In this case, the basic model can be applied to assess the possibilities for exerting market influence. The sensitivity of the results could be checked by varying the demand functions and the initial distribution of permits. This would allow the policymaker to determine whether the type of market influence considered here is likely to pose a problem in a given application.

The formal analysis in Sections II and III indicates the range of potential outcomes that might arise when firms can exert rather specific types of influence in markets which ration a fixed supply of intermediate or final goods. There are clearly other strategies that large firms might pursue, particularly when the market is just getting under way. For example, it is quite likely that the total number of permits issued and the pattern of distribution could be affected by the behavior of such firms. In the case of pollution rights, some firms might refuse to play the game if they do not care for the new set of rules. Such actions are difficult to model explicitly, which is why the focus here has been on the potential for gain within a well-defined set of rules. Even within this setting, further research is warranted.

One avenue for further research would be to extend the basic model to the case where two or more firms have market power. Hahn [1981a] has examined this issue for the case of two firms with market power. The result on cost minimization and permit distribution (Proposition 1) was shown to generalize. A second potentially fruitful area of research would be to extend the model to more than one period along the lines of Stokey [1981], who considers a durable goods monopolist. Finally, it might be useful to test the theory of the basic model in a small-group experimental setting and determine when, and under what types of institutions, it is supported.

The key result obtained here, that it is the pattern of excess demands that ultimately determines the extent to which any firm can influence the market, does not appear to be widely recognized. One reason is that many people feel that manipulation of such markets will not be a problem. For example, Tietenberg, in surveying the literature on air rights markets, expresses the view that "the anticompetitive effects of a TDP (transferable discharge permit) system are not likely to be very important in general" [1980, p. 414]. For several applications such as the one considered by DeLucia [1974] and the one considered by Hahn [1981a], the assumption that the market will approximate the competitive solution would appear to depend critically on how the institutions are designed. Because there is a very real possibility that several markets in transferable property rights could be subject to different kinds of systematic manipulation, there is a need to explore further the ramifications of such problems in theory and applications.

CARNEGIE-MELLON UNIVERSITY

MARKET POWER AND TRANSFERABLE PROPERTY RIGHTS 765

REFERENCES

Anderson, R. S., Jr., R. O. Reid, E. P. Seskin, *et al.*, "An Analysis of Alternative Policies for Attaining and Maintaining a Short Term NO_2 Standard," A report to the Council on Environmental Quality prepared by MATHTECH, Inc., Princeton, New Jersey, 1979.

Coase, R. H., "The Problem of Social Cost," *Journal of Law and Economics*, III (1960), 1–44.

DeLucia, R. J., *An Evaluation of Marketable Effluent Permit Systems*, U.S. Environmental Protection Agency, EPA-600/5-74-030, Washington, D.C., 1974.

Hahn, R. W., "An Assessment of the Viability of Marketable Permits," Ph.D. thesis, California Institute of Technology, Pasadena, California, 1981a.

——, "Data Base and Programming Methodology for Marketable Permits Study," Open File Report No. 80-8, Environmental Quality Laboratory, California Institute of Technology, Pasadena, California, 1981b.

——, and R. G. Noll, "Designing a Market for Tradable Emissions Permits," in *Reform of Environmental Regulation*, W. A. Magat, ed. (Cambridge, MA: Ballinger, 1982).

Montgomery, D. W., "Markets in Licenses and Efficient Pollution Control Programs," *Journal of Economic Theory* (1972), 395–418.

Stokey, N. L., "Rational Expectations and Durable Goods Pricing," *Bell Journal of Economics* (1981), 112–128.

Tietenberg, T. H., "Transferable Discharge Permits and the Control of Stationary Source Air Pollution: A Survey and Synthesis," *Land Economics*, LVI (1980), 391–416.

[13]

JOURNAL OF ENVIRONMENTAL ECONOMICS AND MANAGEMENT **29**, 133–148 (1995)

Transaction Costs and Tradeable Permits*

ROBERT N. STAVINS

John F. Kennedy School of Government, Harvard University, Cambridge, Massachusetts 02138, and Resources for the Future, Washington, DC 20036

Received January 12, 1994; revised April 5, 1994

Tradeable-permit systems are at the center of current interest and activity in market-based reforms of environmental policy, because these systems can offer significant advantages over conventional approaches to pollution control. Unfortunately, claims made for their relative cost-effectiveness have often been exaggerated. Transaction costs, which may be significant in these markets, reduce trading levels and increase abatement costs. In some cases, equilibrium permit allocations and hence aggregate control costs are sensitive to initial permit distributions, providing an efficiency justification for politicians' typical focus on initial allocations. © 1995 Academic Press, Inc.

1. INTRODUCTION

The past five years have witnessed a dramatic increase in the attention given by policy makers to market-based environmental policy instruments as supplements to the conventional command-and-control standards that dominated the previous two decades of environmental law and regulation. One market-based instrument— tradeable emission permits—has been the center of much of this activity. The enthusiasm for this new approach has been so great that policy action and implementation has, in some cases, advanced beyond the understanding of some fundamental design issues. This paper seeks to illuminate an area that has received little attention, the effects of transaction costs on the performance of markets for pollution control.

The claims made for the cost-effectiveness of tradeable-permit systems have often exceeded what can reasonably be anticipated. Tietenberg [39] assimilated the results from 10 analyses of the costs of air pollution control, and in a frequently cited table, indicated the ratio of cost of actual command-and-control programs to least-cost benchmarks. Unfortunately, the resulting ratios (which ranged from 22.0 to 1.1) have sometimes been taken by others to be directly indicative of the potential gains from adopting specific ("cost effective") mechanisms such as tradeable emission permits. A more realistic and appropriate comparison would be between actual command-and-control policies and either actual trading programs

*Helpful comments were provided by Dallas Burtraw, Robert Dorfman, Lawrence Goulder, Robert Hahn, James Hines, Adam Jaffe, Joseph Kalt, Richard Morgenstern, Richard Newell, Wallace Oates, Thomas Tietenberg, Martin Weitzman, Richard Zeckhauser, seminar participants at Harvard University, two anonymous referees, and an associate editor. The author alone is responsible for remaining errors.

133

134   ROBERT N. STAVINS

(such as the EPA's bubble policy) or *reasonably constrained* theoretical permit
programs [22].

1.1. *Markets for Pollution Control and the Potential Role of Transaction Costs*

More than two decades ago, Crocker [5] and Dales [8] developed the idea of
using transferable discharge permits to allocate the pollution-control burden among
firms or individuals; and Montgomery [29] provided a rigorous proof that a
tradeable-permit system could, in theory, provide a cost-effective policy instrument
for pollution control. A sizeable literature on tradeable permits has followed.[1]

A number of factors can adversely affect the performance of tradeable-permit
systems, concentration in the permit market [15, 28], concentration in the output
market [27], non-profit-maximizing behavior, such as sales or staff maximization
[42], the preexisting regulatory environment [3], and the degree of monitoring and
enforcement [23]. Additionally, several authors have commented on the potential
importance of transaction costs in tradeable permit markets [18, 41, 2], although
there has been only one attempt to allow for transaction costs within a model of
tradeable-permit activity [42].[2]

In general, transaction costs are ubiquitous in market economies and can arise
from the transfer of any property right because parties to exchanges must find one
another, communicate, and exchange information. There may be a necessity to
inspect and measure goods to be transferred, draw up contracts, consult with
lawyers or other experts, and transfer title. Depending upon who provides these
services, transaction costs can take one of two forms, inputs of resources—including time—by a buyer and/or a seller or a margin between the buying and selling
price of a commodity in a given market. This paper focuses on the latter characterization of transactions costs. Hence, transaction costs in our analytical model can be
thought of as the direct financial costs of brokerage services. This analytical
approach is also empirically relevant, given the important role that brokers have
played in the operation of actual permit systems [12, 19, 38].

We can identify three potential sources of transaction costs in tradable-permit
markets: (1) search and information; (2) bargaining and decision; and (3) monitoring and enforcement.[3] The first source, search and information, may be the most
obvious. Due to the public-good nature of some information, it can be underpro-
vided by markets. Brokers step in, provide information about firms' pollution-con-

---

[1] Extensive surveys of the literature are found in Tietenberg [39, 40]. A more recent, though less
comprehensive, survey is provided by Cropper and Oates [6].

[2] In this case, the author allowed for a dispersion between a constant selling price and a constant
purchase price of permits, but did not pursue the implications of this and other forms of transaction
cost functions for the performance of the respective markets. Kohn [24] examined the effect of
transaction costs on the optimal (efficient) level of pollution control by assuming an arbitrary magnitude
for transaction costs and comparing the consequent efficient level of control with that predicted when a
pollution tax is employed. The model of transaction costs is itself problematic, since Kohn assumes that
the magnitude of these costs increases with the level of control, as opposed to the level of exchange.

[3] All three categories can be interpreted as representing cost due to lack of information [7]. One
alternative taxonomy is direct financial costs of engaging in a trade, costs of regulatory delay, and
indirect costs associated with uncertainty of completing a trade [14]. In the context of this taxonomy, our
analytical approach focuses on the first category—direct financial costs of trade.

trol options and potential trading partners, and thus reduce transaction costs, while absorbing some as fees.[4] Although less obvious, the second source of transaction costs, bargaining and decision, is potentially as important. There are real resource costs to a firm involved in entering into negotiations [24], including time and/or fees for brokerage, legal, and insurance services [20, 13]. The third source of transactions costs—monitoring and enforcement—can also be significant, but these costs are typically borne by the responsible governmental authority and not by trading partners, and hence do not fall within our notion of transaction costs incurred by firms.

There are two sets of circumstances in which transaction costs might be particularly high: (1) transfer is expensive for technological reasons; and (2) institutions are designed to impede trade. Both apply in the tradeable-permit context [13, 16], and in either case, transaction costs in the form of direct, financial costs of trading are frequently the result.

1.2. *Empirical Evidence of Transaction Costs in Permit Markets*

There is abundant anecdotal evidence indicating the prevalence of significant transaction costs in tradeable permit markets. Atkinson and Tietenberg [1] survey six empirical studies that found trading levels—and hence cost savings—in permit markets to be lower than anticipated by theoretical models. Liroff [26, p. 2] suggests that this experience with permit systems "demonstrates the need for . . . recognition of the administrative and related transaction costs associated with transfer systems."[5] More specifically, Hahn and Hester [18] suggest that the Fox River water-pollutant trading program failed due to high transaction costs in the form of administrative requirements that essentially eliminated potential gains from trade. Likewise, under the EPA's Emissions Trading Program for criteria air pollutants, there is no ready means for buyers and sellers to identify one another, and—as a result—buyers frequently pay substantial fees to consultants who assist in the search for available permits [19, 16].

At the other extreme, the high level of trading that took place under the program of lead-rights trading among refineries as part of the EPA's leaded gasoline phasedown has been attributed to the program's minimal administrative requirements and the fact that the potential trading partners (refineries) were already experienced at striking deals with one another [18]. Hence, transaction costs were kept to a minimum and there was little need for intermediaries.

---

[4]In the newly established sulfur dioxide (SO_2) trading program under the Clean Air Act amendments of 1990, there is a substantial role for brokers for consulting with electrical utilities to help them understand their options. Brokerage firms maintain computer models used to predict the supply and demand for permits to provide forecasting services for utilities (Thomas Brooks, AER* X, personal communication, September 17, 1992). In local programs, such as the EPA's Emissions Trading Program, the broker may also carry out air-quality modeling required for trades between noncontiguous sources of nonuniformly mixed pollutants [25].

[5]Alternative explanations of low observed trading levels have also been advanced: lumpy investment in pollution-control technology, concentration in permit or product markets, the sequential and bilateral nature of the trading process (in the context of a nonuniformly mixed pollutant) leading to some initial trades that then preclude better trades from being carried out subsequently [1], and the regulatory environment [21, 3]. Some but not all of these "alternative explanations" of low trading levels can be viewed as special cases of transaction costs, broadly defined.

Similarly, Tripp and Dudek [41] claim that the success of the New Jersey Pinelands transferable development rights program was due to its design which minimized transaction costs (by the government taking on a feeless brokerage role).

Another source of indirect evidence of the prevalence of transaction costs in permit markets comes form the well-known bias in actual trading toward "internal trading" within firms, as opposed to "external trading" among firms. It has been hypothesized that the crucial difference favoring the internal trades and discouraging the external trades is the existence of significant transaction costs that arise once trades are between one firm and another [43, 19]. Finally, the existence of commercial brokers charging significant fees to facilitate transactions is another body of evidence.

2. TRADEABLE EMISSION PERMITS IN A MARKET WITH TRANSACTION COSTS

Consider first a cost-minimizing pollution control program for a uniformly-mixed, flow pollutant. For such a problem, we can focus on aggregate emissions per unit of time, where ággregate emissions, E, are simply the sum of emissions, e_i, from N individual firms (or sources), where emissions from each source are the difference between unconstrained[6] emissions, u_i, and emission reductions, r_i. A cost-effective emission-control program is one that controls aggregate emissions from all sources at minimum total cost. It is well known that if the control cost functions are convex in their relevant ranges, then the necessary and sufficient conditions for cost minimization yield the result that the marginal cost of control will be the same among all sources that carry out positive levels of control.

To achieve this cost-effective allocation of the pollution-control burden, the government could conceivably establish a nonuniform (source-specific) standard to ensure that all firms would control emissions at the same marginal cost of control, but this would require detailed information about the costs faced by each source, which could be obtained by the authority only at very great cost, if at all. One way out of this impasse is a system of marketable emission permits. Consider a system under which the responsible authority allocates a total of $\overline{E}$ emission permits, q_{0i} to each firm ($i = 1 \ldots N$).[7] Firms are free to trade permits among themselves and may meet government standards by exercising control and/or by possessing permits for their residual emissions. Under these conditions, the permit system achieves the cost-effective allocation of emissions control among sources, but without the government needing to acquire information about control costs; the final, equilibrium allocation of the control burden will be the same for any initial allocation of permits [29].

---

[6]The phrase "unconstrained emissions" is conditional upon prior pollution-control efforts. Hence, u_i can be though of alternatively as *status quo* or *ex ante* emissions.

[7]To use the taxonomy of Tietenberg [40], we are considering an undifferentiated discharge permit, which gives any holder the same emission privileges; transfers among firms are on a one-for-one basis. There are three reasons for considering this simplest type of system: first, it is analytically the most convenient; second, the results generalize to the case of ambient permits for nonuniformly mixed pollutants; and third, simple emission permits have been the system used in nearly all applications. A new and potentially important exception is the RECLAIM trading program in the Los Angeles area, which imposes additional constraints when partners to trade are from different geographically defined regions; see South Coast Air Quality Management District [35].

2.1. *A Model of Permit Trading with Transaction Costs*

We now consider a market for emission permits in which costs are associated with the exchange of permits. Let t_i denote the quantity of permits traded by source i,

$$t_i = |u_1 - r_i - q_{0i}|. \tag{1}$$

We define a common transaction cost function, $T(t_i)$, for which $T'(t_i) > 0$ and for which $T''(t_i)$ may be positive, negative, or zero-valued.[8] Each firm faces the problem,

$$\min_{\{r_i\}} \left[c_i(r_i) + p \cdot (u_i - r_i - q_{0i}) + T(t_i) \right] \tag{2}$$

$$\text{subject to:} \quad r_i \geq 0, \tag{3}$$

where $c_i(r_i)$ is pollution abatement (control) cost and p is the price of permits. This problem yields the following solution:

$$\frac{\partial c_i(r_i)}{\partial r_i} + \frac{\partial T(t_i)}{\partial r_i} - p \geq 0 \tag{4}$$

$$r_i \cdot \left[\frac{\partial c_i(r_i)}{\partial r_i} + \frac{\partial T(t_i)}{\partial r_i} - p \right] = 0 \tag{5}$$

$$r_i \geq 0. \tag{6}$$

The environmental constraint is satisfied, but in contrast to the cost-effective and tradeable-permit solutions without transaction costs, we now find that rather than equilibrating marginal control costs among sources, the result of trading—for situations in which positive levels of control occur—is equilibration of the *sum* of marginal control costs and marginal transaction costs. Also, the total cost incurred by all regulated firms is no longer the simple sum of control costs but rather this amount plus total transaction costs.

2.2. *Consequences of Transaction Costs*

How should we think about the condition that the sum of marginal control costs and marginal transaction costs be equilibrated across sources? First, by the chain rule, we know that

$$\frac{\partial T(t_i)}{\partial r_i} = \left[\frac{\partial T(t_i)}{\partial t_i} \right] \cdot \left[\frac{\partial t_i}{\partial r_i} \right]. \tag{7}$$

---

[8] We assume that $T(t_i)$ is known with certainty. This is not unreasonable, but it is restrictive. Still, the function $T(t_i)$ is admittedly a simple characterization of the transaction cost function; later we add some structure to this to approximate empirical realities, but further work could lead to representations linked to other aspects of the taxonomy of transaction costs. For example, transaction costs are likely to be a function not only of the size of trades but of other attributes as well, since these costs should be affected by the relationship between trading partners. In particular, the anecdotal evidence summarized in the text suggests that transaction costs will be less for intrafirm (internal) than for interfirm (external) trades. This implies an avenue for further analytical work, allowing one component of transaction costs themselves to be endogeneous to a trader's decision problem.

If a source is a purchaser of permits $(u_i - r_i - q_{0i} > 0)$, then $t_i = u_i - r_i - q_{0i}$, and so in this case

$$\frac{\partial t_i}{\partial r_i} = -1 \quad \text{and} \quad \frac{\partial T(t_i)}{\partial r_i} = -\frac{\partial T(t_i)}{\partial t_i}. \tag{8}$$

If a source is a seller of permits $(u_i - r_i - q_{0i} < 0)$, then $t_i = -u_i + r_i + q_{0i}$, and we have

$$\frac{\partial t_i}{\partial r_i} = 1 \quad \text{and} \quad \frac{\partial T(t_i)}{\partial r_i} = \frac{\partial T(t_i)}{\partial t_i}. \tag{9}$$

By substituting the results from Eq. (8) or Eq. (9) into Eq. (5), it is clear that if marginal transaction costs are nonzero, the original "cost-effective equilibrium," where marginal control costs are equated across all sources, will not be achieved.[9] The marginal control costs experienced by the permit buyer exceed those experienced by the permit seller by the amount of marginal transaction costs that the trading partners bear.

This can be perceived most readily in the context of a two-source scenario. Assuming for the time being that marginal transaction costs are constant (i.e., that $T''(t_i) = 0$) and that these costs ($T' = \alpha$) are paid directly by the seller of permits (as with most brokerage fees), we can view the marginal transaction costs and the new equilibrium condition in Fig. 1,[10] where the outcome of trading is the pollution-control allocation r_A^*, as different from the equilibrium without transaction costs, r^*. If the initial allocation of control responsibility is located to the right of the posttrading equilibrium with transaction costs, r_B^*, then the locus of points representing the sum of marginal control costs and marginal transaction costs is found above source 2's control cost function (achieving outcome r_B^* in Fig. 1). The equilibria differ because the identities of buyer and seller have switched.[11]

[9]Since some or all transaction costs may be real resource costs, we should really refer to the original equilibrium as "the cost-effective equilibrium in the absence of transaction costs." Furthermore, to whatever degree transaction costs reflect real resource costs, their existence also affects the optimal level of control [11, 24]. We use the abbreviated description of the equilibria simply for convenience, but the distinction should not be overlooked. To whatever degree transaction costs are real resource costs, they will result in a different equilibrium, but one that is still cost-effective, although it will necessarily involve greater aggregate costs than the cost-effective equilibrium in the absence of transaction costs. It might be said that this implies that the message here is not about transaction costs per se, but rather about correctly measuring the true costs of control. The response, as will become clear below, is that the crucial issue is not whether we call this category of typically omitted costs "transaction costs" or "other control costs," rather, it is whether these costs are a function of the degree of control—as with control or abatement costs—or a fixed and/or variable function of trading activity itself.

[10]In Fig. 1, the vertical axis is in monetary terms; control for source 1, r_1, increases to the right; control for source 2, r_2, increases to the left; all points on the horizontal axis represent compliance with the aggregate emission constraint. An interior solution is depicted where marginal costs are equilibrated at a positive level of control for both sources. If the initial allocation of emission permits is q_{01} and q_{02}, respectively, then in the absence of trades the emission control *required* by each of the two sources is implicitly $r_i = u_i - q_{01}$ for $i = 1, 2$. By taking the initial allocation as a new vertical axis, the two truncated marginal control cost functions yield permit supply and demand functions, and by horizontally summing all such relationships, permit market supply and demand can be examined.

[11]This result that the trading equilibrium is sensitive to the initial allocation in the presence of transaction costs is, of course, fully consistent with the Coase Theorem, which states that in the presence of transaction costs, the anticipated outcome from a process of bilateral negotiation is variant with respect to the initial assignment of property rights [4]. In Fig. 1, if the initial allocation of permits is between r_A^* and r_B^*, then no trading will take place and the initial allocation *is* the final equilibrium.

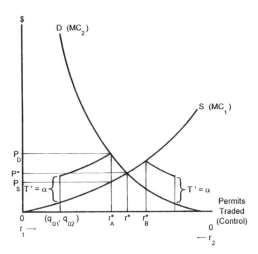

FIG. 1.   Constant marginal transaction costs.

These effects are not dependent upon the marginal transaction costs being constant. The effect of transaction costs is unambiguously to decrease the volume of permit trading, regardless of the specific forms that the marginal control cost functions and transaction cost functions take, as long as the marginal control cost functions are nondecreasing over the relevant ranges.[12] As transaction costs increase, the price received by sellers is depressed and the price paid by purchasers is driven upward. Analogous with the well-known tax-incidence result, the degree of these price effects is dependent upon the relative elasticity values. The change in price will be greater for relatively high-cost controllers, essentially because high-cost (inelastic) controllers have less flexibility in decision making.[13]

This leads naturally to the question of how the burden of transaction costs is shared between permit suppliers and demanders. Following the analogy with tax incidence, it is not surprising to find that the gains from trade accruing to both sides of the market decrease as a consequence of transaction costs, and that the distribution of the burden is a function of the relative elasticities of the underlying cost-of-control functions. The burden (loss of potential gains from trade) of transaction costs will fall most heavily on the higher cost controllers (*steeper*

[12] A proof of this is found in [36]. In the case of constant marginal transaction costs, whether trading will take place depends upon whether the difference between the marginal control costs of the two sources at the initial allocation is greater than the transaction costs of trading the minimal size permit. In those situations in which trading takes place, the magnitude of individual trades will be less than in the absence of transaction costs. Hence, constant marginal transaction costs may reduce the total number of trades (particularly if combined with significant fixed transaction costs) and will certainly reduce aggregate trading volume.

[13] Because these results are analogous to well-known tax-incidence results, derivations are not provided in this paper. Instead, the reader is referred to [36]. The effect of transaction costs is also parallel to the effect of transportation costs, which result in a single market price being replaced by a cluster of prices. By "high-cost controller," we mean the source with the steeper marginal control cost function.

marginal control-cost functions), regardless of who may actually pay direct transaction costs, such as brokerage fees.[14]

In general, in the presence of transaction costs, if the initial allocation deviates from what would be the equilibrium allocation in the absence of transaction costs, total expenditures on pollution control (even putting aside transaction costs themselves) will exceed the cost-minimizing solution. Thus, transaction costs reduce welfare—partly by absorbing resources directly and partly by suppressing exchanges that otherwise would have been mutually beneficial.

This brings us to the central question, can the initial allocation of permits affect the outcome of trading (putting aside the most obvious effect of the initial allocation being located on one side or the other of the cost-effective equilibrium)? In other words, does the frequently restated finding of Montgomery [29], that the equilibrium allocation of control and hence the aggregate costs of control are independent from the initial permit allocation, still hold in the presence of transaction costs? The answer is "it depends." To see this, we examine again a two-source model in which source 1 is a potential permit seller and source 2 is a potential buyer. We assume, without loss of generality, that transaction costs are paid by sellers.

For positive levels of control by two sources, Eqs. (5) and (9) yield the market equilibrium condition

$$c_1'(r_1) + T'(t_1) = c_2'(r_2), \qquad (10)$$

where $t_1 = -u_1 + r_1 + q_{01}$. In order to investigate the effect of the initial allocation on the equilibrium outcome, we differentiate both sides of Eq. (10) with respect to the primary variables, r_1, r_2, and q_{01},

$$c_1''(r_1) \cdot dr_1 + T''(t_1) \cdot \frac{\partial t_1}{\partial r_1} \cdot dr_1 + T''(t_1) \cdot \frac{\partial t_1}{\partial q_{01}} \cdot dq_{01} = c_2''(r_2) \cdot dr_2. \quad (11)$$

Any change in t_1 is necessarily equal to a corresponding change in t_2. Since u_1 and u_2 do not—by definition—change and since any change in q_{01} must of necessity be equal to -1 times the change in q_{02}, it must be the case that any change in r_1 is equal to -1 times the change in r_2. In other words, holding constant the total quantity of permits, aggregate emissions reductions must be unchanged. Hence, we can substitute $-dr_1$ for dr_2 in Eq. (11). Also, note that from the definition of t_1, we know that

$$\frac{\partial t_1}{\partial r_1} = \frac{\partial t_1}{\partial q_{01}} = 1. \qquad (12)$$

---

[14] If transaction costs are paid by permit buyers, then as Eq. (8) illustrates, the locus of points representing the sum of marginal control costs and marginal transaction costs is found *below* the respective control cost functions. The respective equilibria, however, are identical in the two cases. This suggests an alternative approach to modeling transaction costs in tradeable permit markets, namely a game theoretic approach, allowing for the presence of bilateral monopoly. In that context, the division of the gains from trade—with or without transaction costs—will depend upon possible information asymmetries [30].

Thus, rearranging terms, we rewrite Eq. (11) as

$$[c_1''(r_1) + T''(t_1) + c_2''(r_2)] \cdot dr_1 + T''(t_1) \cdot dq_{01} = 0. \tag{13}$$

Dividing through by dq_{01}, we have

$$\frac{dr_1}{dq_{01}} = \frac{-T''(t_1)}{[T''(t_1) + c_1''(r_1) + c_2''(r_2)]}, \tag{14}$$

enabling us to examine the impact of the initial allocation on the equilibrium control level,

$$\text{if } T''(t_1) = 0 \quad \text{then } \frac{dr_1}{dq_{01}} = 0 \tag{15}$$

$$\text{if } T''(t_1) > 0 \quad \text{then } \frac{dr_1}{dq_{01}} < 0 \tag{16}$$

$$\text{if } T''(t_1) < 0 \quad \text{then } \frac{dr_1}{dq_{01}} > 0. \tag{17}$$

If marginal transaction costs are constant $(T''(t_1) = 0)$, the usual result in the absence of transaction costs still holds; the initial allocation of permits has no effect on the equilibrium allocation of control responsibility and aggregate control costs.[15]

On the other hand, if marginal transactions costs are increasing $(T''(t_1) > 0)$, then the initial allocation would seem to affect the posttrading outcome (Fig. 2). In particular, Eq. (16) implies that as we increase the allocation of emission permits to a source (reduce its initial control responsibility), its equilibrium control level will be reduced, thus increasing the departure of the post-trading equilibrium outcome from the "cost-effective equilibrium," driving up aggregate control costs in the process.[16] Although increasing marginal transaction costs, on their own, are not sustainable (since parties would simply split their transactions into smaller trades to economize), increasing marginal transaction costs *are* sustainable if combined with sufficient fixed transaction costs.[17]

---

[15]Although direct abatement costs are therefore unaffected, aggregate transaction costs themselves change as the initial allocation of control responsibility becomes more or less remote from the cost-effective equilibrium allocation.

[16]As we shift the initial allocation of control responsibility in Fig. 2 from r_{0A} to r_{0B}, the trading equilibrium also changes (from r_A^* to r_B^*), because $T''(t_i) \geq 0$. As a result of employing the initial allocation r_{0B} instead of r_{0A}, both transaction costs *and* deadweight loss have increased.

[17]A further comment on the case of fixed transaction costs is merited. If $T(t_i) = \alpha$ and $T'(t_i) = 0$, then trading will occur, *ceteris paribus*, in the simple $N = 2$ model if potential gains from trade exceed α at the initial allocation of permits (and control responsibility); otherwise it will not. In this case, sources may have incentives to reduce the number of separate trades but the equilibrium level of each trade is not affected. In the aggregate, the result is decreased trading volume. On its own, the case of fixed transaction costs is analytically identical to Hahn's [17] examination of the EPA's "20% rule" on criteria air-pollutant trading; if all sources engage in trading, the cost-effective equilibrium allocation of the control burden among sources will be achieved, but in the presence of fixed transaction costs, there may be fewer trades than would otherwise occur, in which case the cost-effective allocation will not be achieved.

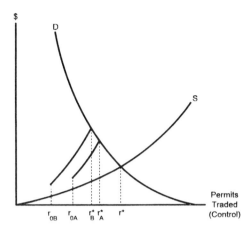

FIG. 2.   Increasing marginal transaction costs.

Finally, decreasing marginal transaction costs might occur where brokers offer quantity discounts[18] on their services. In this case $(T''(t_1) < 0)$, a shift in the initial permit allocation *away* from the cost-effective equilibrium leads to a posttrading outcome that is *closer* than otherwise to the cost-effective equilibrium[19] (Fig. 3). What may appear to be a counterintuitive result is simply due to the fact that decreasing marginal transaction costs mean that there are scale economies of trading of which firms can take advantage.[20]

Thus, in the presence of transaction costs, the initial distribution of permits can matter in terms of efficiency, not only in terms of equity (Table I).[21] In his 1972 paper, Montgomery [29] observed that because of the independence of the equilib-

---

[18] Typical volume discounts are expressed in terms of average, not marginal costs. If fixed costs are present, there could be declining average costs without declining marginal costs, but if there are no fixed costs, then declining average and declining marginal costs obviously imply one another. A referee points out that at the market level, decreasing marginal transaction costs could also be due to positive information externalities (associated with larger trading volumes) that systematically lower transaction costs. See Section 3.1.

[19] When $T''(t) < 0$, the denominator in Eq. (14) is of ambiguous sign, depending upon the relative magnitudes of the slopes of the marginal transaction cost function and the marginal control cost functions. Since we assume that the second-order condition for the permit-trading problem is satisfied, however, the denominator must be positive; thus the apparent ambiguity is removed.

[20] This happens unless marginal transaction costs are reduced to zero before the cost-effective equilibrium is reached, in which case costly trading is followed by costless trading, and the cost-effective equilibrium is achieved. Hence, moving the initial allocation away from the cost-effective one either will cause the final equilibrium to be closer to the cost-effective equilibrium or will have no effect, in which case the final equilibrium is the cost-effective one.

[21] One other potentially important type of transaction cost is associated with percentage brokerage fees. Analogous to the *ad valorem* tax, such constant percentage brokerage fees appear as constant marginal transaction costs to individual permit buyers or sellers (who take the permit price to be exogenously determined), but at the level of market supply and demand for permits, this type of brokerage fee structure looks something like declining marginal transaction costs. However, in this case, a change in the initial permit allocation has no effect on the final equilibrium, since the transaction cost function does not itself shift.

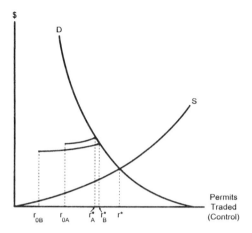

FIG. 3. Decreasing marginal transaction costs.

rium from the initial allocation, "the management agency can distribute licenses as it pleases. Considerations of equity, of administrative convenience, or of political expediency can determine the allocation. The same efficient equilibrium will be achieved." Not so, potentially, in the presence of transaction costs. This can reduce the discretion of the environmental agency and the legislature, and may thereby reduce the political attractiveness and feasibility of a tradeable permit system.

3. IMPLICATIONS FOR PUBLIC POLICY

3.1. *Instrument Choice*

At the most basic level, the message of this analysis is obvious; choices between conventional, command-and-control environmental policies, and market-based in-

TABLE I

Consequences of Transaction Costs

Form transaction costs	Example	Effect on number of trades	Effect on size of individual trades	Effect on aggregate quantity traded	Sensitivity of posttrading equilibrium to initial allocation
Fixed cost only	$1000 per trade	Decrease or no change	No change	Decrease or no change	$dr / dq_0 > 0$
Constant MTC	$10 per ton of SO_2 traded	Decrease or no change	Decrease	Decrease	$dr / dq_0 = 0$
Increasing MTC	$10 \times (\text{ton traded})^2$	Decrease or no change	Decrease	Decrease	$dr / dq_0 < 0$
Decreasing MTC	$10 \times (\text{ton}) - .001 \times (\text{ton})^2$	Decrease or no change	Decrease or no change	Decrease or no change	$dr / dq_0 > 0$
Value of trade	$0.10 \times$ value of trade	Decrease or no change	Decrease	Decrease	$dr / dq_0 = 0$

struments ought to reflect the imperfect world in which these instruments are applied.[22] But even if transaction costs prevent significant levels of trade from occurring, aggregate costs of control will most likely not exceed those of a conventional command-and-control approach. A trading system *with no trading* taking place will likely be less costly than a technology standard (because the trading system provides flexibility to firms regarding their chosen means of control) and no more costly than a uniform performance standard. On the other hand, it is possible that in some circumstances the total cost of compliance (including transaction costs) of a tradeable permit system could exceed (depending upon the initial allocation of permits) the costs of a uniform performance standard (which exhibited small administrative costs). There is no simple answer, no policy panacea; case-by-case examinations are required.

Nevertheless, some general implications of the analysis do emerge. Transaction costs increase the aggregate costs of control indirectly by reducing total trading volume and directly by adding to total costs of control. However, these effects should be ameliorated somewhat in markets with relatively large number of potential trading sources. As the pool of potential trading partners increases, it should be easier for sources to identify potential trading partners, thereby lowering transaction costs. A larger number of firms can mean more frequent transactions, generating more information, and thereby reducing uncertainty[23] [31]. Hence, due to possible transaction-cost effects (and due to the likely effects of market concentration and strategic behavior), we ought to be least confident of relying on tradeable-permit system in situations of thin markets.

Economists have tended to give greater emphasis to the symmetry between tradeable permits and pollution charges[24] than to their differences, although the two approaches are not symmetric under conditions of uncertainty [44], in the presence of transaction costs, or under a number of other conditions [37]. Analyses that have compared taxes and permits have assumed zero transaction costs, which is troubling considering the evidence that these costs are common in permit markets. Systems of pollution taxes, of course, can also involve substantial administrative costs, both fixed (per firm) and variable [32]. Hence, these instruments should also be compared only on a case-by-case basis.

An interesting implication of the analysis in regard to the choice between taxes and permits comes from our finding that the initial allocation of permits affects the final equilibrium when marginal transaction costs are nonconstant. Recall that the major advantage of tradeable permits over command-and-control *and* over emission taxes is that the government can achieve a given aggregate target cost effectively *without* knowing anything about individual firms' costs of pollution control. However, when transaction costs exist, operating an efficient permit

---

[22] The point that it is wrong to compare an actual, imperfect institution with a theoretically ideal one goes back at least to Coase [4]. A referee notes that Demsetz [9] objected to comparisons of actual imperfect markets with idealized governments; our point may be said to be the reverse—actual imperfect government should not be compared with idealized markets. Since transactions costs are essentially the market counterpart of administrative costs in command-and-control systems, both need to be considered.

[23] In a model with many sources, transaction costs would be a function of the size of trades and a function of total trading volume in the market, if trades produce positive informational externalities.

[24] The assumed symmetry is between taxes and auctioned permits, and between freely allocated permits and taxes with specific redistribution of revenues to selected firms [10].

market means distributing permits initially so that the sum of control costs and transaction costs is minimized. Clearly, the initial allocation that minimizes this sum is the solution to the least-cost pollution control problem in the absence of transaction costs (since if permits where thus distributed, no trading would occur, and transaction costs would also be minimized). But to solve this problem, the government must know firms' cost-of-abatement functions. Unless transaction costs are prohibitive, tradeable permits will retain an information advantage over command-and-control approaches and emission taxes, but that advantage will be less in the presence of significant transaction costs than otherwise.

3.2. *Instrument Design*

Three sets of tradeable-permit design issues stand out as being particularly affected by the presence of transaction costs. First, there is the issue of what point in the "product cycle" to regulate pollution. The simplest systems (whether tradeable permits or other instruments) focus on *inputs* to the production process, such as the lead content of gasoline or the carbon content of fossil fuels. One step toward greater sophistication but also substantially greater administrative complexity and transaction costs is represented by *emissions* permit trading. Further in the same direction is *ambient* or *concentration* permit trading. And further still would be *exposure* trading [34], and finally *risk* trading [33]. As we move along this path, each system may come closer to a theoretical ideal, but each system is also likely to bring greater public costs associated with monitoring and enforcement and greater private transaction costs. Indeed, these practical considerations provide one explanation of why, contrary to the models going back to Montgomery [29], only input and emissions trading have actually be adopted by public authorities.[25]

Second, given the close linkage between information problems and transaction costs, another obvious implication is that programs should be designed to provide needed information. In principal, there are three ways this could be done, government can take actions that directly reduce regulatory uncertainty, barriers to private brokerage services can be reduced, and allowance can be made for the development of futures markets. In the first case, at a minimum, the government authority can avoid creating regulatory barriers (such as requirements for government preapproval of trades) that drive up transaction costs and discourage trading.[26] More actively, the government can seek to reduce market uncertainty by taking on a brokerage role—supplying information about potential buyers and sellers, and thus helping sources identify one another [41].

Private provision of brokerage services can also play an important role in information provision. Thus, although commercial brokers can certainly be recipients of transaction costs, their activities reduce transaction costs below what they would otherwise be. Intermediaries, in general, can contribute to social welfare by

---

[25]As noted above, the new RECLAIM trading program for SO$_x$ and NO$_x$ in the Los Angeles area provides a partial exception, in the sense that due to concerns about the nonuniformly mixed nature of the pollutants in the airshed, additional constraints have been placed upon emissions trading.

[26]Regulatory uncertainty may also include "general program uncertainty" about whether a particular program will be maintained in the future, and uncertainty about how a new trading program fits into the preexisting regulatory environment (as in the case of the SO$_2$ allowance trading system among regulated electrical utilities).

helping parties economize on transaction costs. Brokers can play the role of consultants, adding value by understanding the regulatory process and by maintaining information about prospective suppliers and demanders of permits. Under the more conventional function of bringing together buyers and sellers ("brokering deals" by matching buy orders and sell orders), these firms both absorb and reduce transaction costs. Finally, brokers may assume risk by buying, holding, and selling permits.

A third and final set of design issues is associated with determining the initial allocation of permits, a subject that takes on added significance in the presence of transaction costs. For any tradeable permit system, political feasibility can be established or destroyed over this single aspect of design. Because of the necessity of establishing a constituency for a proposed system, the route that has inevitably been chosen for distributing permits has been free distribution (endowment).[27] This politically attractive route enables all sorts of initial allocations to be devised in order to win support for a program.

Typically, economists have been quite agnostic about these alternative allocations, because they have considered them to have only distributional implications, since it has been assumed that aggregate abatement costs would be unaffected. As we have seen, however, this may not be true when transaction costs are present. Thus, a successful attempt to establish a politically viable program through specific permit allocations can actually result in a program that will be far more costly than promised. This may argue for the economist's favorite permit-allocation mechanism—auctions, an approach that becomes even more attractive in the presence of transaction costs. We are left, however, with the reality that political barriers against permit auctions and political incentives in favor of all sorts of free distributions are likely to remain in place for the foreseeable future.

The general message for public policy that arises when we begin to consider the presence of transaction costs in markets for tradeable permits is that "the devil is likely to be in the details." Although the existence of transaction costs may make the choice between ambient and emission permits more obvious, it may well make the choice between conventional approaches and permits more difficult because of the ambiguities that are introduced. Likewise, the supposed symmetry of taxes and permits becomes questionable, and the need to compare these instruments on a case-by-case basis becomes more compelling. Finally, with transaction costs as with other departures from frictionless markets, greater attention should be paid to the details of design of specific systems, in order to lessen the risk of overselling these policy ideas and in order to create systems that stand a chance of being implemented successfully.

REFERENCES

1. S. Atkinson and T. Tietenberg, Market failure in incentive-based regulation: The case of emissions trading, *J. Environ. Econom. Management* **21**, 17–31 (1991).
2. W. J. Baumol and W. E. Oates, "The Theory of Environmental Policy," Second Edition, Cambridge Univ. Press, Cambridge, UK (1988).

[27]As Zeckhauser [45] has noted, the distribution of gains and losses arising from a policy is likely to have greater effects on whether that policy is adopted (in a democratic society) than the magnitude (or even the sign) of net benefits.

3. D. R. Bohi and D. Burtraw, Utility investment behavior and the emission trading market, *Resour. Eng.* **14**, 129–153 (1992).
4. R. H. Coase, The problem of social cost, *J. Law Econom.* **3**, 1–44 (1960).
5. T. D. Crocker, The structuring of atmospheric pollution control systems, in "The Economics of Air Pollution" (Harold Wolozin, Ed.), Norton, New York (1966).
6. M. L. Cropper and W. E. Oates, Environmental economics: A survey, *J. Econom. Literature* **30**, 675–740 (1992).
7. C. J. Dahlman, The problem of externality, *J. Law Econom.* **22**, 141–162 (1979).
8. J. Dales, "Pollution, Property and Prices," University Press, Toronto (1968).
9. H. Demsetz, Information and efficiency: Another viewpoint, *J. Law Econom.* **12**, 1–22 (1969).
10. D. N. Dewees, Instrument choice in environmental policy, *Econom. Inq.* **21**, 53–71 (1983).
11. P. B. Downing and W. D. Watson, Jr., The economics of enforcing air pollution controls, *J. Environ. Econom. Management* **1**, 219–236 (1974).
12. D. J. Dudek and J. Palmisano, Emissions trading: Why is this thoroughbred hobbled?, *Columbia J. Environ. Law* **13**, 217–256 (1988).
13. J. P. Dwyer, A free market in tradeable emissions is slow growing, *Public Affairs Rep.* **1** (January), 6–7 (1992).
14. V. Foster and R. W. Hahn, "ET in LA: Looking Back to the Future," working paper, American Enterprise Institute, Washington, DC (1993).
15. R. W. Hahn, Market power and transferable property rights, *Quart. J. Econom.* **99**, 753–765 (1984).
16. R. W. Hahn, Economic prescriptions for environmental problems: How the patient followed the doctor's orders, *J. Econom. Perspect.* **3**, 95–114 (1989).
17. R. W. Hahn, Regulatory constraints on environmental markets, *J. Public Econom.* **42**, 149–175 (1990).
18. R. W. Hahn and G. L. Hester, Marketable permits: Lessons for theory and practice, *Ecol. Law Quart.* **16**, 361–406 (1989).
19. R. W. Hahn and G. L. Hester, Where did all the markets go? An analysis of EPA's emissions trading program, *Yale J. Reg.* **6**, 109–153 (1989).
20. R. W. Hahn and R. G. Noll, Designing a market for tradable emissions permits, *in* "Reform of Environmental Regulation," (W. A. Magat, Ed.), Ballinger Publishing Company, Cambridge, MA (1982).
21. R. W. Hahn and R. G. Noll, Barriers to implementing tradeable air pollution permits: Problems of regulatory interactions, *Yale J. Reg.* **1**, 63–91 (1983).
22. W. H. Hahn and R. N. Stavins, Economic incentives for environmental protection: Integrating theory and practice, *Amer. Econom. Rev.* **82**, 464–468 (1992).
23. A. G. Keeler, Noncompliant firms in transferable discharge permit markets: Some extensions, *J. Environ. Econom. Management* **21**, 180–189 (1991).
24. R. E. Kohn, Transactions costs and the optimal instrument and intensity of air pollution control, *Policy Sci.* **24**, 315–332 (1991).
25. A. J. Krupnick, W. E. Oates, and E. Van De Verg, On marketable air-pollution permits: The case for a system of pollution offsets. *J. Environ. Econom. Management* **10**, 233–247 (1983).
26. R. A. Liroff, "The Evolution of Transferrable Emission Privileges in the United States," paper presented at the Workshop on Economic Mechanisms for Environmental Protection, Jelenia Gora, Poland (1989).
27. D. A. Malueg, Welfare consequences of emission credit trading programs, *J. Environ. Econom. Management* **18**, 66–77 (1990).
28. W. S. Misolek and H. W. Elder, Exclusionary manipulation of markets for pollution rights, *J. Environ. Econom. Management* **16**, 156–166 (1989).
29. W. D. Montgomery, Markets in licenses and efficient pollution control programs, *J. Econom. Theory* **5**, 395–418 (1972).
30. R. B. Myerson and M. A. Satterthwaite, Efficient mechanisms for bilateral trading, *J. Econom. Theory* **29**, 265–281 (1983).
31. R. G. Noll, Implementing marketable emissions permits, *Amer. Econom. Rev.* **72**, 120–124 (1982).
32. A. M. Polinsky and S. Shavell, Pigouvian taxation with administrative costs, *J. Public Econom.* **19**, 385–394 (1982).
33. P. R. Portney, Reforming environmental regulation: Three modest proposals, *Issues Sci. Technol.* **4**, 74–81 (1988).
34. J. A. Roumasset and K. R. Smith, Exposure trading: An approach to more efficient air pollution control, *J. Environ. Econom. Management* **18** 276–291 (1990).

35. South Coast Air Quality Management District, "The Regional Clean Air Incentives Market, Final Volume I," Los Angeles, CA (1993).

36. R. N. Stavins, "Transaction Costs and the Performance of Markets for Pollution Control," paper presented at American Economic Association Annual Meeting, Boston, MA, Jan. (1994).

37. R. N. Stavins and B. W. Whitehead, Pollution charges for environmental protection: A policy link between energy and environment, *Annual Rev. Energy Environ.* **17**, 187–210 (1992).

38. J. Taylor, CBOT plan for pollution-rights market is encountering plenty of competition, *Wall Street J.* Aug. 24, C-1, C-16 (1993).

39. T. H. Tietenberg, "Emissions Trading: An Exercise in Reforming Pollution Policy," Resources for the Future, Washington, DC (1985).

40. T. H. Tietenberg, Transferable discharge permits and the control of stationary source air pollution: A survey and synthesis, *Land Econom.* **56**, 391–416 (1980).

41. J. T. B. Tripp and D. J. Dudek, Institutional guidelines for designing successful transferable rights programs, *Yale J. Reg.* **6**, 369–391 (1989).

42. J. T. Tschirhart, Transferable discharge permits and profit-maximizing behavior, *in* "Economic Perspectives on Acid Deposition Control" (Thomas D. Crocker, Ed.), Butterworth, Boston (1984).

43. U.S. General Accounting Office, "A Market Approach to Air Pollution Control Could Reduce Compliance Costs Without Jeopardizing Clean Air Goals," U.S. Government Printing Office, Washington, DC (1982).

44. M. L. Weitzman, Prices vs quantities, *Rev. Econom. Stud.* **41**, 477–491 (1974).

45. R. Zeckhauser, Preferred policies when there is a concern for probability of adoption, *J. Environ. Econom. Management* **8**, 215–237 (1981).

[14]

JOURNAL OF ENVIRONMENTAL ECONOMICS AND MANAGEMENT 31, 269–286 (1996)
ARTICLE NO. 0044

A Model of Intertemporal Emission Trading, Banking, and Borrowing*

JONATHAN D. RUBIN

Department of Economics and Energy, Environment and Resources Center, University of Tennessee, Knoxville, Tennessee 37996

Received October 30, 1995; revised November 13, 1995

This paper provides a general treatment of emission trading, banking, and borrowing in an intertemporal, continuous-time model. Using optimal-control theory, the decentralized behavior of firms is shown to lead to the least-cost solution attainable under joint-cost minimization. Explicit solutions for the time paths of emissions and permit prices are derived when firms are allowed to both bank and borrow and when firms are only allowed to bank emission permits. The policy implications of emission banking and borrowing are discussed. © 1996 Academic Press, Inc.

1. INTRODUCTION

Marketable emission permits are an economic instrument used to attain a predetermined level of environmental quality. The basic concept behind emission trading, originated by Dales [5], is a simple one. Initially, a regulatory agency limits the overall level of emissions, either by setting a standard or allocating emissions, and then it allows firms to trade their emission allocations or surplus permits. It is now widely agreed among economists that marketable emission permits can be a cost-effective strategy for controlling environmental pollutants, and an extensive literature on their properties has developed (see Tietenberg [12] and Cropper and Oates [4] for thorough reviews). This work has identified three sources of potential cost savings: emission trading between firms, emission averaging between sources within a firm, and emission trading through time. Despite common reference to these three components, previous theoretical and empirical research has focused almost exclusively on the first two items, trading and averaging.

The small amount of research into emission banking and borrowing is regrettable since public policymakers have already begun to incorporate banking and borrowing rules into law. For example, the banking (and trading) of sulfur dioxide is now authorized by the Clean Air Act Amendments of 1990, and California allows manufacturers of passenger cars to bank (and trade) hydrocarbon emissions. In the context of fuel efficiency, rather than emissions, the borrowing and banking of

---

*The author appreciates helpful comments on earlier versions of this paper from Michael Caputo, Amy Farmer, Gloria Helfand, Catherine Kling and two anonymous referees.

269

CAFE (corporate average fuel economy) credits for up to three years has been allowed since 1980 (Congressional Quarterly 1980 Almanac [2, pp. 487, 488]).[1]

However, the concept of marketable permits, and intertemporal trading in particular, goes far beyond the realm of environmental cleanup or fuel economy standards. It applies to the trade of goods whose existence is statutorily generated but privately transferable. For example, the concept applies to the allocation of residential or commercial property development rights or to the buying and selling of federal funds between banks who must meet reserve requirements. In the latter case, conservative banks who choose to carry surplus reserve requirements act as sellers of overnight loans (e.g., trading) and, in addition, may choose to hold additional excess reserves speculatively over several time periods (e.g., banking) in the hope that a lucrative loan opportunity arises. Throughout this paper, however, the discussion of banking will be couched in terms of standard usage within the environmental field, which refers to emission of pollutants by firms.

Within the environmental literature this paper fills the theoretical gap by providing a comprehensive treatment of intertemporal emission trading.[2] In particular, it explores the problem of minimizing the cost of intertemporal emission control by N heterogeneous firms in the presence of pollution standards and emission permits that are tradable across firms and through time. In such a setting, firms may directly reduce emissions, and they may purchase, sell, bank and borrow emission permits in order to meet applicable standards or to take advantage of any speculative opportunities that may arise. To generate permits, a firm may choose to pollute less than the current standard. These generated permits may be sold to a different firm or deposited in an emission bank to be used later by the firm or sold at a later time to another firm. The borrowing of permits occurs when a firm pollutes more than its current standard, but the cumulative deficit must be repaid by the end of the planning horizon. Banking and borrowing lower the cost of compliance with emission standards by allowing firms to adjust their emission stream more flexibly through time.

The model used here builds on the seminal static paper by Montgomery [7] on marketable permits and extends the work of Tietenberg [12] and Cronshaw and Kruse [3], who examine some aspects of permit markets with banking in discrete time. Tietenberg characterizes the joint least-cost allocation of pollution for a cumulative pollutant given a constraint on the total amount of pollution over time. In this cumulative emission-permit system it appears that all permits are issued at the beginning of the time horizon. Some permits are therefore always banked until used up. Before the permits are used up their price must rise at the rate of interest. Cronshaw and Kruse examine a permit system with banking where permits are allocated to firms in each of T periods. They show that in a market for transfer-

---

[1]With the borrowing of emission credits there is the obvious concern about firms borrowing and then declaring bankruptcy. In the CAFE program, enforcement of the CAFE standards takes the form of fines and a review of company product planning to determine the likelihood of future compliance (Shaffer [10]). The enforcement problem is inversely related to the cost of entry and exit. For capital-intensive industries such as electrical power producers or automobile manufacturers, enforcement should not be a problem. For industries characterized by a rapid turnover of firms, enforcement of all environmental regulations is a more serious issue.

[2]See Rubin and Kling [8] for an empirical examination of emission banking and borrowing for light-duty vehicle manufacturers.

able and bankable permits, a competitive equilibrium exists and achieves aggregate emission targets at least system cost. Cronshaw and Kruse also show that if firms do not desire to bank, permit prices will rise by less than the rate of interest.

This paper extends the work of Tietenberg and Cronshaw and Kruse by providing a more general treatment of permit trading in continuous time through the use of optimal-control theory. Instead of limiting intertemporal trading to banking, this paper starts from the context of allowing both borrowing and banking with restrictions on borrowing as a special case. The formulation allows the derivation and extension of the results of Tietenberg and Cronshaw and Kruse, but interpretations are necessarily somewhat different. Importantly, the continuous-time framework allows for the explicit solution for the time path of emissions, which, cost savings notwithstanding, may be intertemporal emission trading's most significant aspect. Neither Tietenberg's nor Cronshaw and Kruse's paper examines this issue in detail.[3] In addition, this analysis extends the model of Cronshaw and Kruse by considering the qualitative impacts of emission borrowing on permit prices, the emission stream, and total social damages.

2. THE FORMAL MODEL

Consider a firm, i, that has two control variables, $e_i(t)$, the quantity of emissions each period, and $y_i(t)$, the quantity of emission permits bought ($y_i > 0$) or sold ($y_i < 0$) at price $P(t)$, which will be determined by the equilibrium conditions of all firms in the permit market over the planning horizon of length T. The level of emissions that are in the bank, $B_i(t)$, is a state variable.

Following Montgomery [7], the abatement-cost function for firm i, $C_i(e_i)$, is equal to the difference between the unconstrained profits and profits in which the firm adopts an emission level S (where S is less than the unconstrained emission level, e) and adjusts its output in order to obtain maximum profits for the constrained level of emissions. It is assumed that C_i is twice continuously differentiable and convex in e_i and that marginal abatement costs, $-C_i'(e_i)$, are positive and strictly increasing, that is, $C_i'(e_i) < 0$ and $C_i''(e) > 0$.

Firms choose abatement levels and permit purchases and sales given that permit prices, $P(t)$, might change over time. This makes the formal model nonautonomous. In addition, firms are assumed to be making decisions over a finite time horizon of length T. The length of T can, of course, be relatively short or very long.

3. THE JOINT-COST PROBLEM

In the joint-cost problem, the regulator's goal is to minimize the total costs of pollution abatement of N heterogeneous firms subject to reaching a regional emission standard, $S(t)$. The regulator assigns each firm a standard or endowment of emissions, $S_i(t)$, so that $\sum_{i=1}^{N} S_i(t) = S(t)$. Note that emission banking and

---

[3]Tietenberg [12, p. 29] notes that his cumulative-emission permit system does not regulate emission rates but only total emissions. Cronshaw and Kruse [3, p. 6] note that banking could lessen environmental damages by delaying emissions or could allow firms to possibly emit large amounts of emission using previously banked permits.

borrowing, as opposed to trading, change the nature of standards since firms can emit above the standard at some points in time. What is required is that firms have to fulfill the cumulative standard over the entire planning horizon. In the joint-cost problem, the regulator is able to achieve cost savings by optimally adjusting the emission stream. Defining the optimal-value function as J^{**}, $B = \Sigma B_i$ as the aggregate stock of banked emissions, $\dot{B}$ as the rate of change of aggregate banked emissions, and T as the terminal time period, the problem can be written as[4]

$$J^{**} \equiv \underset{e_i}{\text{Min}} \int_0^T e^{-rt} \sum_i^N C_i(e_i(t))\, dt \tag{1}$$

$$\text{s.t.} \quad \dot{B} = \sum_{i=1}^N (S_i(t) - e_i(t)) \tag{2}$$

$$B(0) = 0, \qquad B(t) \geq 0 \tag{3}$$

$$e_i(t) \geq 0, \qquad \forall i. \tag{4}$$

The objective functional (1) minimizes the sum of the N firms' present discounted emission-abatement costs. Equation (2) is the state equation that shows that the changes in the aggregate stock of banked emissions equal the aggregate differences between all firms' standards and emissions. The standards, S_i, can be interpreted either as firm-specific emission standards or as initial endowments of permits where the emission standards are understood to be equal to zero. Equation (3) says that the initial aggregate stock of banked emissions is zero, the aggregate stock of permits may never have negative balances (this constraint is later relaxed to allow borrowing), and at the terminal period the aggregate stock of emission permits is non-negative. Equation (4) requires that firms cannot have negative emission levels.

The joint-cost problem can be solved through use of optimal-control theory. Using this framework, defining $\Lambda(t)$ as the costate variable on the state equation and Φ as the multiplier function for the bank non-negativity constraint, the present value Hamiltonian is

$$H \equiv e^{-rt}\left[\sum_{i=1}^N C_i(e(t))\right] + \Lambda\left[\sum_{i=1}^N (S_i(t) - e_i(t))\right], \tag{5}$$

and the Lagrangian is

$$L \equiv e^{-rt}\left[\sum_{i=1}^N C_i(e(t))\right] + \Lambda\left[\sum_{i=1}^N (S_i(t) - e_i(t))\right] - \Phi B. \tag{6}$$

This problem yields the following necessary conditions for an optimal solution (Kamien and Schwartz [6, Part II, Section 17]):

$$\dot{B} = \frac{\partial L}{\partial \Lambda} = \sum_{i=1}^N (S_i(t) - e_i(t)) \tag{7}$$

$$\dot{\Lambda} = -\frac{\partial L}{\partial B} = \Phi; \qquad B \geq 0,\ \Phi \geq 0,\ \Phi B = 0 \tag{8}$$

---

[4]If both banking and borrowing are always allowed, then (3) and (13) should be replaced with the simpler constraint $B(0) = 0$ and $B(T) \geq 0$, rather than $B(0) = 0$ and $B(t) \geq 0 \in [0, T]$.

$$\frac{\partial L}{\partial e_i} = e^{-rt}C_i'(e_i) - \Lambda \geq 0; \quad e_i \geq 0, \quad e_i\frac{\partial L}{\partial e_i} = 0, \quad \forall i \qquad (9)$$

$$B(T) \geq 0, \quad -\Lambda(T) \geq 0, \quad B(T)\Lambda(T) = 0. \qquad (10)$$

Economic Interpretation of Necessary Conditions for Optimization
of the Joint-Cost Problem

Equation (7) simply restates the state equation. Next, (9) says that all firms that discharge some emissions should have present discounted marginal abatement costs $(-e^{-rt}C_i'(e_i))$ equal to the marginal cost of an additional unit of banked emissions $(-\Lambda)$, and that all firms that discharge emissions should have equal marginal abatement costs. This equation also shows that $\Lambda(t)$, the present value costate variable, which is the marginal or shadow value of a unit of emissions in the bank, is negative. This reflects the fact that if there were an additional unit of emissions in the systemwide bank, systemwide total abatement cost would be lower because firms could then discharge more emissions, which would lower their costs. For later use denote the vector of optimal systemwide emissions from this problem as $\mathbf{E}^{**}(t) = (e_1^{**}(t),\ldots,e_N^{**}(t))$.

Equation (8) says that if, in total, firms bank a positive quantity of emissions over some time interval, then the marginal value of an additional unit of emissions in the systemwide bank is constant for that time interval. This implies that the marginal value of a unit of emission in the bank would always be a negative constant if firms were allowed to borrow emissions (i.e., if $B(t)$ were unrestricted in sign, implying that $\Phi \equiv 0$). If firms, in total, would like to borrow emissions, but are not allowed to do so, the marginal value of an additional unit of emissions in the bank would be decreasing (i.e., going to zero from the left since the multiplier, $\Phi(t)$, is positive.

The above conditions tell us what necessary conditions must hold for the minimization of the present value of total costs. The sufficient conditions needed to guarantee an optimum to this problem are that H be convex in e_i and B and that the above necessary conditions hold; additional continuity and regularity conditions that are met for this problem are given in full in Theorem 1 of Seierstad and Sydsæter [9, pp. 317, 318].[5] In particular, the sufficient conditions require that each firm's cost function be convex in emissions. Moreover, existence follows from Theorems 2.1 and 3.1 of Steinberg and Stalford [11], since the state equation (7) is linear in the e_i, and the integrand is strictly convex in the e_i.

Next, it is necessary to look at how individual firms will make their decisions to optimally control abatement costs and permit purchases and sales, given that firms take permit prices as exogenous.

[5]According to Seierstad and Sydsæter, for optimal-control problems with pure state constraints (e.g., $f(B(t),t) \geq 0$, or in this case $B(t) \geq 0$), the costate variable may be discontinuous at the terminal period. Given appropriate regularity conditions an optimal solution to the problem still exists, but this possibility will not be considered in the analysis.

274   JONATHAN D. RUBIN

4. THE FIRM'S PROBLEM

Formally, defining the optimal value function as J_i^*, and letting $P(t)$ be the instantaneous price of permits $y_i(t)$ purchased or sold by firm i, and $A_i(t)$ and $D_i(t)$ be the bounds on the instantaneous purchase and sale of permits, firm i's problem is

$$J_i^* \equiv \underset{y_i, e_i}{\text{Min}} \int_0^T e^{-rt}\left[C_i(e_i(t)) + P(t)y_i(t)\right] dt \tag{11}$$

$$\text{s.t.} \quad \dot{B}_i = S_i(t) - e_i(t) + y_i(t) \tag{12}$$

$$B_i(0) = 0, \qquad B_i(t) \geq 0 \tag{13}$$

$$e_i(t) \geq 0 \tag{14}$$

$$-A_i(t) \leq y_i(t) \leq D_i(t), \qquad A_i(t) > 0, \qquad D_i(t) > 0. \tag{15}$$

The objective functional (11) says that firm i chooses the level of emissions and purchases or sales of permits to minimize its present discounted costs. Permits may be instantaneously purchased or sold with no transaction costs. Equation (12) is the state equation, which says that changes in the bank of emissions equal the difference between the standard and the level of emission plus any permits bought or sold. Equation (13) says that the initial stock of the bank is zero, firms may never have a negative bank balance (this constraint is later relaxed), and at the terminal period the firm is free to hold any non-negative stocks of emission permits. Equation (14) requires that firms cannot have negative emission levels. Finally, Eq. (15) bounds the instantaneous purchase or sales of permits.

This last constraint, Eq. (15), is a necessary technical requirement that arises if one wants to consider arbitrary price paths, since the objective function is linear in y_i. The solution to this type of control problem is straightforward but unnecessarily complicated to get at the economic issues at hand.[6] An alternative approach is to consider price paths for which a solution to the firms problem, without bounds, exists. This is equivalent to assuming that the firm has an internal solution. The economic intuition of this assumption is discussed below. However, there are two cases in which a firm would not have an internal solution. One situation is when the least-cost strategy for the firm is to sell a bounded maximum number of permits at any point in time. The other situation is when a firm's marginal abatement costs are so high that it would rather take no abatement activities, but simply buy as many permits as possible. The maximum number of permits a firm could buy at any point in time is likely to be controlled by the regulatory agency to prevent localized hot spots of concentrated emissions. Thus, in both these cases, a firm could potentially desire to sell, borrow, or buy some maximum number of permits at any point in time.

By assuming that firms do not buy or sell permits at a bounded rate, this problem can be solved through the use of an optimal-control-theory framework by defining $\lambda_i(t)$ as the present-value costate variable on the state equation and $\phi_i(t)$ as the multiplier function for the bank non-negativity constraint for firm i,

---

[6]See Kamien and Schwartz [6, Part II, Section 12] for the solution technique suitable to this problem.

$i = 1, \ldots, N.$[7] Then, the present-value Hamiltonian is

$$H_i = e^{-rt}[C_i(e_i) + Py_i] + \lambda_i[(S_i - e_i) + y_i].$$ (16)

The Hamiltonian yields the following necessary conditions:[8]

$$\dot{B}_i = \frac{\partial H_i}{\partial \lambda_i} = (S_i - e_i) + y_i$$ (17)

$$\dot{\lambda}_i = -\frac{\partial H_i}{\partial B_i} = \phi_i; \qquad B_i \geq 0, \; \phi_i \geq 0, \; \phi_i B_i = 0$$ (18)

$$\frac{\partial H_i}{\partial e_i} = e^{-rt}C_i'(e_i) - \lambda_i \geq 0; \qquad e_i \geq 0, \; e_i \frac{\partial H_i}{\partial e_i} = 0$$ (19)

$$\frac{\partial H_i}{\partial y_i} = e^{-rt}P + \lambda_i = 0$$ (20)

$$B_i(T) \geq 0, \; -\lambda_i(T) \geq 0, \; B_i(T)\lambda(T) = 0.$$ (21)

Economic Interpretation of the Firm's Necessary Conditions

Equation (17) simply restates the state equation. Next, looking at Eq. (19), if $\partial H/\partial e_i > 0$, then $e_i = 0$ and if $e_i > 0$, then $-e^{-rt}C'(e_i) = -\lambda_i$. Thus, if the firm emits some pollution, then the present discounted marginal abatement costs equal the marginal value of an additional unit of emissions in the firm's emission bank account. If present discounted marginal abatement costs (evaluated at the complete abatement level of emissions) are less than the present value of an additional unit of emissions in the bank, then the firm discharges no emissions. Furthermore, λ_i, the marginal value of a unit of banked emissions for firm i, is negative. This reflects the fact that if the firm had an additional unit of emissions in the bank, abatement costs would be lower, as would the present value of total costs.

Equation (18) says that the marginal value of a unit of emissions in the bank is increasing (becoming less negative, or smaller in absolute value) if the non-negativity constraint on the bank is binding.[9] That is, the value to the firm of an additional unit of emissions in the bank is decreasing if the firm desires to borrow a unit of emissions but it cannot. If the constraint against borrowing were not present, then the marginal value of a unit of banked emissions would be constant whenever the firm was not selling permits at its maximum level.

Looking now at (20), we see that the firm will purchase or sell y_i permits so that the discounted marginal cost of a unit in the bank is equal to its discounted price,

---

[7]The more general case requires the introduction of additional multiplier variables for the permit bounds given in (15).

[8]See Seierstad and Sydsæter [9, pp. 332–33] for a complete list of the necessary conditions. "Almost necessary" conditions basically require that the Lagrangian multiplier, ϕ_i, only has jumps at a finite number of points and has piecewise continuous derivatives elsewhere.

[9]It is also possible for the bank constraint to be binding just enough that the value of a unit of banked emissions is constant and the stock in the bank is equal to zero. This case is not fundamentally different and will not be explicitly addressed.

$e^{-rt}P = -\lambda_i$. Permit prices are determined by equilibrium conditions of all firms in the market for permits. For firm i (as well as all other firms) to have a solution without bounds on permit purchases or sales over the entire time horizon, permit prices must follow along a singular path defined by $\alpha \equiv e^{-rt}P + \lambda_i = 0$. Then, along the singular path, $\dot{\alpha}_i \equiv 0$, so the following must hold: $\dot{\alpha}_i = -re^{-rt}P + e^{-rt}\dot{P} + \dot{\lambda}_i = 0$. Substituting in for $\dot{\lambda}_i$ from Eq. (18), the following equation results: $0 = -re^{-rt}P + e^{-rt}\dot{P} + \phi_i$. After some manipulation, the following restrictions on the percentage rate of growth in the price of emission permits for the firm to remain on the singular path are derived:

$$\frac{\dot{P}}{P} = \left\{ \begin{array}{c} r \\ r - \dfrac{e^{rt}\phi_i}{P} \end{array} \right\} \text{ whenever } \left\{ \begin{array}{c} B_i > 0 \\ \phi_i > 0 \end{array} \right\}. \tag{22}$$

If firm i either wants to bank emissions or does not face a restriction on the borrowing of emission credits, then a nonbounded solution requires that the price path of permits follow Hotelling's rule and grow at the rate of return earned from holding any other asset. Equivalently, the present value price of permits, $e^{-rt}P(t)$, is constant. If a firm faces a binding constraint on the borrowing of emission credits, then $\phi_i > 0$, and the rate of growth in prices must be less than the comparable interest rate. In this case the present-value price of permits is decreasing through time since $e^{-rt}P(t) = -\lambda_i(t)$, $\dot{\lambda}_i(t) = \phi_i(t) > 0$, and $\dot{\lambda}_i(t) < 0$. Summarizing, for a particular firm to have a non-bounded solution, the present-value price of permits must be constant when each firm can bank and borrow emissions. If the bank is required to be non-negative, and the firm desires to borrow, then a non-bounded solution requires that permit prices be decreasing. The transversality condition, (21), implies that if a firm holds a permit at the terminal period, its value must be zero.

Combining conditions (19) and (20), under the condition that a firm emits a positive quantity of pollution, then $-C_i'(e_i) = P$. That is, under normal conditions a firm will equate the marginal cost of pollution abatement with the price of a permit. Therefore, present-value marginal abatement costs are constant when each firm can freely bank and borrow emissions, and may decline if borrowing is restricted and firms desire to borrow.

5. EQUILIBRIUM CONDITIONS

As discussed above, the regulator's goal is to minimize the joint total costs of pollution abatement of N heterogeneous firms subject to reaching a regional emission standard, S, over the T-period horizon so that $B(T) \geq 0$. It is necessary to show that an equilibrium in a market for permits exists and is efficient in the sense that it is equal to the system-cost minimization by regulators who know the cost functions of all the firms. Montgomery's [7] definition of a market equilibrium for pollution permits is used, but modified to account for the dynamic problem at hand.

Especially, but not exclusively, in problems with linear-control variables, the optimal path of the linear control $y_i^*(t)$ can be discontinuous in t. Therefore in Lemma 1, which appears in the Appendix, it is necessary to assume that the optimal control $y_i^*(t)$ is continuous in the price of permits, P, even at particular instances, $t = \tau$, where $y_i^*(t)$ is discontinuous in t. This assumption is not too restrictive; its economic content is simply that at a certain point in time if a firm suddenly (e.g., discontinuously) decides to buy or sell a different quantity of permits, the decision is based on a continuous function of the permit price. An intertemporal equilibrium in the permit market also requires that firms have an internal solution.

DEFINITION. An intertemporal market equilibrium in emission permits over a T period horizon consists of the vectors $\mathbf{Y}^*(t) = (y_1^*(t),\ldots,y_N^*(t))$ and $\mathbf{E}^*(t) = (e_1^*(t),\ldots,e_N^*(t))$ and the scalar $P^*(t) \geq 0$ such that $\mathbf{Y}^*(t)$ and $\mathbf{E}^*(t)$ minimize each firm's costs given $P^*(t)$,

$$\int_0^T e^{-rt}\left[C_i(e_i(t)) + P^*(t)y_i(t)\right] dt \tag{23}$$

subject to each firm's constraints (12)–(15) for each $t \in [0,T]$, and the market clearing condition on permits

$$\sum_{i=1}^N y_i^*(t) = 0. \tag{24}$$

The market-clearing condition is a flow condition that simply requires that at any point in time a permit purchased by one firm be sold by some other firm. In addition, cost minimization will also impose the following terminal stock condition:

$$P^*(T)\left[\sum_{i=1}^N B_i^*(T)\right] = 0. \tag{25}$$

Cost minimization requires that the price of a unit of pollution in the bank be zero at the terminal period or that the stock of permits from all firms be zero.

THEOREM 1. *An intertemporal market equilibrium in emission permits over a T-period horizon exists when firms are not buying and selling permits at their maximum instantaneous rate.*

Proof. See the Appendix.

The Market Equilibrium is the Least-Cost Solution

Before proving that the market solution is at least as inexpensive as that from the social planner, it is interesting to think about what banking and borrowing mean in the context of a market. As shown above, the price path of permits must be either constant or decreasing. Thus, for all firms that have not shut down or are not emitting at their maximum rate, $e^{-rt}P(t) = \lambda_i(t)$. Thus λ_i must be equal for all firms and be of either constant or declining value. This says that the value of a unit of banked emission is the same for all firms that have an internal solution for permit purchases, even though it is possible for some firms to borrow while others bank and yet others do neither. Furthermore, if $e^{-rt}P$ is falling over some period, then the result of all firms buying and selling permits is an overall net desire to borrow.

Now it is necessary to show that the market solution is as least as good as the solution from the social planner. In general, permit prices are some function of all firms' cost functions of emissions abatement, $P = g(C_1(e_1), \ldots, C_N(e_N))$. Given the assumption of interior solutions, (19) and (20) can be equated to yield $C_i'(e_i) + P(t) = 0$. Since $C_i'' > 0$, the implicit function theorem implies that $e_i = e_i(P(t))$ is the solution to this marginal condition, where $\partial E / \partial P(t) \equiv -1/C_i'' < 0$. Over the T-period horizon, the total quantity of emissions must be

$$\int_0^T \sum_{i=1}^N e_i(P^*(z)) \, dz = \int_0^T \sum_{i=1}^N S_i(z) \, dz \qquad (26)$$

since the stock of emissions in the bank is defined to be zero at $t = 0$ and will be chosen to be zero at $t = T$.

An implication of the equilibrium conditions is a relationship between the price path of permits and equilibrium banking. At any time t the aggregate change in the bank can be expressed as

$$\dot{B} = \sum_{i=1}^N S_i(t) - e_i(P(t)) + y_i(P(t)), \qquad (27)$$

and the stock in the bank at any time t equals

$$B(t) = \int_0^t \sum_{i=1}^N (S_i(z) - e_i(P(z)) + y_i(P(z))) \, dz. \qquad (28)$$

Given that the permit market clears in each period, this reduces to

$$B(t) = \int_0^t \sum_{i=1}^N (S_i(z) - e_i(P(z))) \, dz. \qquad (29)$$

This says that the stock of all firms' permits equals the cumulative number of permits that have been banked. If borrowing is allowed, then $B(t)$ can be negative, indicating that, cumulatively, firms are borrowing emissions. When borrowing is not allowed, then it must be non-negative. In either case, over the whole time horizon the sum of all firms' stocks must be non-negative

$$B(T) = \int_0^T \sum_{i=1}^N (S_i(z) - e_i(P(z))) \, dz \geq 0. \qquad (30)$$

The last result will be used in the proof that the market-equilibrium solution from firms' individually buying and selling permits is efficient in the sense that it is at least as good as the systemwide cost minimization by regulators who know the cost functions of all firms. This proof is based on a proof by Cronshaw and Kruse [3], who show that a discrete-time permit system with banking attains the minimum joint-cost solution.

THEOREM 2. *The market-equilibrium solution $(e_1^*(t), \ldots, e_N^*(t))$, and $(y_1^*(t), \ldots, y_N^*(t))$ is at least as inexpensive as the system-cost minimization, so that $\sum_{i=1}^N J_i^* \leq J^{**}$, where J^* and J^{**} are the optimal-value functions defined in (11) and (1), respectively.*

Proof. See the Appendix.

6. CHARACTERIZING THE PATH OF EMISSIONS

From an environmental perspective, perhaps the most important result of allowing firms to intertemporally move permits, as opposed to simply trading them instantaneously among themselves to attain the least-cost allocation, is the effect that banking and borrowing can have on the stream of emissions through time and the resulting health and environmental impacts. If firms use banking to smooth decreasing emission standards over time and if marginal damages from pollution are increasing, then banking generates lower total pollution damages when considering the integral over the whole time horizon of each period's damage. From the formal model we can derive some insight into how firms adjust their emission streams when they can take advantage of the ability to bank and borrow.

For all firms not buying or selling permits at their maximum rate, but able to equate marginal abatement costs with the marginal value of banked emissions, it was shown earlier that $e^{-rt}C_i'(e_i^*) - \lambda_i = 0$. If this can be sustained for some nonzero time interval, then along this interval the following condition on the time path of $e_i(t)$ is derived by differentiating (25) with respect to t and collecting terms:

$$\dot{e}_i = \frac{rC_i'(e_i)}{C_i''(e_i)} + \frac{e^{rt}\dot{\lambda}_i}{C_i''(e_i)}, \qquad (31)$$

which after substituting from Eq. (18) gives

$$\dot{e}_i = \frac{rC_i'(e_i)}{C_i''(e_i)} + \frac{e^{rt}\phi_i}{C_i''(e_i)}. \qquad (32)$$

Given the sign conditions described earlier, the first term on the right in (32) is negative and the second term is positive or zero. The second term is zero whenever all the firms desire only to bank emissions or are allowed to borrow emissions. This means that if firms never desire to borrow emissions or desire to borrow and are allowed to do so, so that $\phi_i(t) = 0$, then the firms' emission streams will decline through time. When borrowing is not allowed and firms desire to borrow, the second term on the right is positive and could be larger in magnitude than the first term, which is negative. Thus, in principle, aggregate emissions could increase through time when borrowing is not allowed. This would occur, however, only when emission standards are becoming less stringent through time.

The discounting of abatement costs causes firms to delay abatement and abatement expenditures beyond the time firms would have engaged in them had costs not been discounted. When standards are constant through time, $\sum_{i=1}^{N}\dot{S}_i = 0$, then the least-cost behavior of firms is to borrow emissions, $\sum_{i=1}^{N}B_i < 0$, in early years and pay them back in later years. This is true since, as was shown above, the present-value marginal cost of abatement must be constant when banking and borrowing are allowed. When firms choose to bank emissions, $\sum_{i=1}^{N}B_i > 0$, it must be because they expect future (undiscounted) abatement costs to rise by more than the rate of discount; they would not otherwise bear the additional earlier abatement expenditures. The only way for undiscounted abatement costs to rise through time is for the level of abatement to rise through time, since $-C_i'(e_i) > 0$, $-C_i''(e_i) > 0$. Since constant standards engender borrowing, it must be that banking arises from standards that tighten, $\sum_{i=1}^{N} \dot{S}_i < 0$, over some interval of time.[10]

If damages from emissions are a convex function and standards are tightening through time, then allowing firms to bank lowers social damages. Banking allows firms to smooth their emission stream through time. For a convex damage function, where cumulative damage is the integral of damages in all time periods, total damages are reduced if the level of emissions is more constant through time. This result is one of the qualitative differences between a static and a dynamic marketable permit system. By giving firms an incentive (cost savings) to more than meet current standards, firms will intertemporally move emissions to lower social damages. This occurs when firms are allowed to bank emissions and when standards are becoming stricter through time. Emission standards' becoming stricter through time is, however, often the norm. If pollution damages are convex, then allowing firms to shift emissions into the present will increase environmental damages. Whether or not firms should be allowed to borrow, at the expense of increasing environmental damages, depends on balancing the cost savings to firms with the additional harm caused by borrowing.

Since firms discount future costs, they will want to borrow when permit prices grow at less than the rate of interest, are constant, or actually decrease over time. In equilibrium, permit prices will equal marginal abatement costs. Thus firms will want to borrow emissions when standards are constant or not becoming more stringent at a sufficiently high rate. Hence, if standards are constant through time so that firms desire to borrow, but are not allowed to do so, then the rate of emissions will remain constant through time. In this case we can give an economically interesting interpretation to ϕ_i, the multiplier on the non-negativity constraint on the bank for firm i. Setting $e = 0$ in Eq. (32) and rearranging yields $C_i'(e_i) = -e^{rt}\phi_i/r$. Substituting in $-P(t)$ for $C_i'(e_i)$ and rearranging, it is seen that $\phi_i(t) = re^{-rt}P(t)$. Therefore, $\phi_i(t)$, the multiplier on the bank's non-negativity constraint, can be interpreted as the periodic payment that firm i would be willing to make on a perpetual annuity whose purchase price is equal to the present discounted cost of an emission permit.

A special case arises when $r = 0$. In this case firms have no incentive to put off costs until the future, and given the assumption of strict convexity of the emission-control function, firms want to emit pollution at the same rate over the entire time horizon, regardless of whether emission standards are becoming stricter. If emission rates are becoming stricter through time and the discount rate is zero, firms will want to save or buy permits in the beginning time periods for later use. Higher discount rates lower the value of future cost savings and decrease the incentive for firms to bank emissions.

This analysis has been framed in a world of certainty. In actuality, marginal abatement costs and marginal damages are uncertain. In particular, a firm might use banked permits to cover a period of particularly high emission levels because of

---

[10] This conclusion needs to be tempered when considering a more general setting than is considered in this model. For example, a firm may wish to bank emissions if investments in abatement equipment are coordinated with other nonabatement investments. Additionally, firms may want to bank emissions in anticipation of future growth, especially when there exists considerable uncertainty about the future price of permits.

a sudden increase in the demand for output. If marginal damages are increasing, this excess burst of emissions could cause unacceptably large damages. An environmental regulator, therefore, may wish to put a ceiling on the total allowable discharge of emissions in any one given period.

7. FINAL REMARKS

This paper characterizes the optimal behavior of firms that face emission standards set by regulators, but are allowed to purchase, sell, bank, and borrow emission permits given a finite planning horizon of length T. It was shown that an equilibrium solution to the firms' problems exists and is efficient in the sense of achieving the least-cost solution (the solution attained by a social planner who knows the cost functions of all firms). The equilibrium permit-price path was shown to be constant or decreasing depending on whether firms desired to bank or were allowed to borrow permits and whether firms faced a binding constraint on borrowing. Existence of an equilibrium permit-price path also requires that firms not buy and sell permits at their maximum instantaneous rate. Strategic behavior and price and cost uncertainty were not taken into consideration.

As argued in the body of the paper, perhaps the most important consequence of emission banking and borrowing is the ability of firms to shift their emission stream through time. In particular, when social damages are an increasing function of the level of pollution emitted at any one time, then it is good public policy to allow firms to bank emissions when standards are becoming stricter through time. In contrast, when standards are constant or easing ($\dot{S} \geq 0$), then allowing firms to borrow will raise social damages while lowering firms' costs.

At least for now our society is becoming increasingly concerned with environmental quality, and emission standards are becoming more stringent. This paper has shown that in these circumstances, banking of emission permits can lower social damages. Banking also provides cost savings to firms by allowing them to adjust their own internal rates of emission reductions to an externally set standard. Along with averaging and trading, then, banking can also lower the monetary costs of compliance. Banking, therefore, along with averaging and trading, should be considered by public policymakers in charge of ensuring the safety of our environment.

APPENDIX

Proof of Theorem 1

The definitions, lemmas, and proofs that follow are structured after those of Montgomery [7], who proves the existence of an equilibrium in a permit market for a static problem. The strategy is to derive the sufficient conditions that guarantee a market equilibrium given the market-clearing price $P^*(t)$. Next, the sufficient conditions for the existence of a joint-cost minimum are derived and shown to exist. It is then demonstrated that the conditions that satisfy the joint-cost minimum also satisfy the conditions for the market equilibrium where $P^*(t)$ is not identified as the negative of the present-value costate variable from the joint-cost problem, i.e., $-e^{rt}\Lambda^{**}(t) = P^*(t)$.

DEFINITION. An intertemporal market equilibrium in emission permits over a T-period horizon consists of the vectors $\mathbf{Y}^*(t) = (y_1^*(t), \ldots, y_N^*(t))$ and $\mathbf{E}^*(t) = (e_1^*(t), \ldots, e_N^*(t))$ and the scalar $P^*(t) \geq 0$ so that $\mathbf{Y}^*(t)$ and $\mathbf{E}^*(t)$ minimize each firm's costs given $P^*(t)$

$$\int_0^T e^{-rt} \left[C_i(e_i(t)) + P^*(t) y_i(t) \right] dt \tag{A1}$$

subject to each firm's constraints (11)–(15) for each $t \in [0, T]$, and the market-clearing condition on permits

$$\sum_{i=1}^N y_i^*(t) = 0. \tag{A2}$$

The market clearing condition is a flow condition that simply requires that at any point in time a permit purchased by one firm must be sold by some other firm. In addition, cost minimization will also impose the following terminal stock condition:

$$P^*(T) \left[\sum_{i=1}^N B_i^*(T) \right] = 0. \tag{A3}$$

Cost minimization requires that the price of a permit be zero at the terminal period or that the stock of permits from all firms be zero.

LEMMA 1. *An intertemporal market equilibrium in emission permits over a T-period horizon exists if there exist optimal shadow values for the emissions for each firm,* $\lambda^* = (\lambda_1^*(t), \ldots, \lambda_N^*(t)) \geq 0$; *non-negativity multipliers for the permit bank,* $\phi^* = (\phi_1^*, \ldots, \phi_N^*)$; *an optimal permit price* $P^*(t) \geq 0$ *for all* $t \in [0, T]$, *and if, in addition, firms are not buying and selling permits at their maximum instantaneous rate so that*

$$\dot{B}_i^* = \frac{\partial L_i}{\partial \lambda_i^*} = (S_i - e_i^*) + y_i^* \tag{A4}$$

$$\dot{\lambda}_i^* = -\frac{\partial L_i}{\partial B_i^*} = \phi_i^*; \qquad B_i^* \geq 0, \ \phi_i^* \geq 0, \ \phi_i^* B_i^* = 0 \tag{A5}$$

$$\frac{\partial L_i}{\partial e_i^*} = e^{-rt} C_i'(e_i^*) - \lambda_i^* \geq 0; \qquad e_i^* \geq 0, \ e_i^* \frac{\partial L_i}{\partial e_i^*} = 0 \tag{A6}$$

$$\frac{\partial L_i}{\partial y_i} = e^{-rt} P^* + \lambda_i^* = 0 \tag{A7}$$

$$B_i^*(T) \geq 0, \ -\lambda_i^*(T) \geq 0, \ B_i^*(T) \lambda_i^*(T) = 0 \tag{A8}$$

and

$$\sum_{i=1}^N y_i^*(t) = 0, \qquad P^*(T) \left[\sum_{i=1}^N B_i^*(T) \right]. \tag{A9}$$

Proof. This strategy of this proof is to show that firms that minimize costs given $P^*(t) \geq 0$ do indeed generate the conditions listed above and that the conditions listed above are sufficient for each firm to minimize costs.

Form the Lagrangian

$$L_i = e^{-rt}[C_i(e_i) + P^*y_i] + \lambda_i[(S_i - e_i) + y_i] - \phi_i B_i. \tag{A10}$$

For all i, the solution (λ_i^*, ϕ_i^*, e_i^*, and y_i^*) to (A1) subject to (12)–(15) exists by Theorem 2.1 and 3.1 of Steinberg and Stalford [11]. Moreover, the sufficient conditions of Theorem 3.1 of Seierstad and Sydsæter [9, Chapter 5] are met since the Hamiltonian is convex in (e_i, y_i, B_i). Differentiation yields the conditions listed above.

DEFINITION. An intertemporal emission-constrained joint-cost minimum over a T-period horizon is the vector of emissions $\mathbf{E}^{**}(t) = (e_i^{**}(t),\ldots,e_N^{**}(t))$, which minimizes the joint costs of all firms

$$J^{**}(\beta) \equiv \min_{e_i} \int_0^T e^{-rt} \sum_i^N C_i(e_i(t))\, dt \tag{A11}$$

subjet to the constraints for the joint cost problem (2)–(4).

This definition requires the same aggregate level of emission abatement and the same restrictions on the banking of emissions as required by the sum of the abatement activities by each firm in the intertemporal market equilibrium defined above.

The following lemma gives sufficient conditions for the existence of an intertemporal joint-cost minimum to exist. This lemma is then used to prove the existence of an equilibrium in the pollution-permit market.

LEMMA 2. *An emission vector* $\mathbf{E}^{**}(t)$ *is an intertemporal emission-constrained joint-cost minimum over a T-period horizon if there exist optimal shadow values for the systemwide emissions,* $\Lambda^{**} \geq 0$, *and a non-negativity multiplier for the emission bank,* $\Phi^{**} \geq 0$, *so that*

$$\dot{B}^{**} = \frac{\partial L}{\partial \Lambda^{**}} = \sum_{i=1}^N (S_i - e_i^{**}) \tag{A12}$$

$$\dot{\Lambda}^{**} = -\frac{\partial L}{\partial B^{**}} = \Phi^{**}; \qquad B^{**} \geq 0, \Phi^{**} \geq 0, \Phi^{**}B^{**} = 0 \tag{A13}$$

$$\frac{\partial L}{\partial e_i} = e^{-rt}C_i'(e_i^{**}) - \Lambda^{**} \geq 0; \qquad e_i^{**} \geq 0, e_i^{**}\frac{\partial L}{\partial e_i} = 0, \forall i \tag{A14}$$

$$B^{**}(T) \geq 0, -\Lambda^{**}(T) \geq 0, B^{**}(T)\Lambda^{**}(T) = 0. \tag{A15}$$

Proof. Defining Λ as the costate variable on the state equation and Φ as the multiplier function for the emission bank non-negativity constraint, the Lagrangian for the joint cost minimum is

$$L \equiv e^{-rt}\left[\sum_{i=1}^N C_i(e(t))\right] + \Lambda\left[\sum_{i=1}^N (S_i(t) - e_i(t))\right] - \Phi B. \tag{A16}$$

Using the same arguments as in the proof for Lemma 1, the solution (Λ^{**}, Φ^{**}, e_i^{**}) to (A11) subject to (1)–(4) exists.

THEOREM 1. *An intertemporal market equilibrium in emission permits over a T-period horizon exists when firms are not buying and selling permits at their maximum instantaneous rate.*

Proof. This proof proceeds by using the sufficient conditions for the joint-cost problem, (A12)–(A15), to show that $e_i^* = e_i^{**}$, $y_i^* = \dot{B}_i^{**} - (S_i - e_i^{**})$, $-e^{-rt}P^* = \Lambda^{**}$, and $\lambda_i^* = \Lambda^{**}$ will also satisfy the sufficient conditions for decentralized problem (A4)–(A9) for all $i = 1, \ldots, N$ and $t \in [0, T]$.

Equation (A5). (A5) is met by setting $\lambda_i^* = \Lambda^{**}$ and $\phi_i^* = \Phi^{**}$ given in (A13).

Equation (A6). From (A14)

$$\frac{\partial L}{\partial e_i} = e^{-rt}C_i'(e_i^{**}) - \Lambda^{**} \geq 0; \qquad e_i^{**} \geq 0, e_i^{**}\frac{\partial L}{\partial e_i} = 0, \forall i.$$

It is true, therefore, that $e_i^* = e_i^{**}$ and $\lambda_i^* = \Lambda^{**}$ also satisfy (A6) for all i and t.

Equation (A7). Let $\lambda_i^* = \Lambda^{**} = -e^{-rt}P^*$ for all i and t; then (A7) is satisfied.

Equation (A4). From (A12) we have

$$\dot{B}^{**} = \sum_{i=1}^{N} \dot{B}_i^{**} = \sum_{i=1}^{N}(S_i - e_i^{**}).$$

Let $\dot{B}_i^* = \dot{B}_i^{**}$, and since at $\lambda_i^* = \Lambda^{**} = -e^{-rt}P^*$, $e_i^* = e_i^{**}$, then in addition we know, using (A4), that

$$\dot{B}_i^{**} = S_i - e_i^{**} + y_i^*.$$

Summing over all i, this gives

$$\sum_{i=1}^{N} \dot{B}_i^{**} = \dot{B}^{**} + \sum_{i=1}^{N} y_i^*.$$

Therefore, $\sum_{i=1}^{N} y_i^* = 0$, which is necessary and feasible. It is also the market-clearing condition.

Equation (A8). Let $B_i^*(T) = B_i^{**}(T)$ and $\lambda_i^*(T) = \Lambda^{**}(T)$. Let no firm hold any stock of pollutants in the joint-cost problem at $t = T$; then $\sum_{i=1}^{N} B_i^{**}(T) = B^{**}(T) = 0$. By setting $B_i^*(T) = B_i^{**}(T) = 0$ we have a feasible solution that satisfies the equality requirement of (A8). Now let only one firm, firm i, hold a stock of permits at $t = T$. Then $\sum_{i=1}^{N} B_i^{**}(T) > 0$. Letting $B_i^*(T) = B_i^{**}(T)$, we have a feasible solution that satisfies the inequality part of (A8). In addition, let $P^*(T) = \Lambda^{**}(T)$, and we get the cost-minimizing condition on the terminal-period stock, which requires that $P^*(T)[\sum_{i=1}^{N} B_i^*(T)] = 0$. Therefore, we have found a feasible solution that exists and satisfies all the conditions of Lemma 1.

Proof of Theorem 2.[11] Let $e_i^*(t)$ and $y_i^*(t)$ be the cost-minimizing emission and permit paths for firm i. Suppose, however, that $e_i^*(t)$ and $y_i^*(t)$ do not minimize the system costs. Then there exist some other $\tilde{e}_i(t)$ and $\tilde{y}_i(t)$ that are feasible and

[11]This proof is a continuous-time adaptation of a proof by Cronshaw and Kruse.

produce a lower system cost. That is,

$$\int_0^T \sum_{i=1}^N \bar{e}_i(z) \, dz \le \int_0^T \sum_{i=1}^N S_i(z) \, dz \quad \text{(feasibility)}, \qquad \text{(A17)}$$

$$\int_0^T \sum_{i=1}^N e^{-rt} C_i(\bar{e}_i(z)) \, dz < \int_0^T \sum_{i=1}^N e^{-rt} C_i(e_i^*(z)) \, dz \quad \text{(lower cost)}. \quad \text{(A18)}$$

Now, defining alternative permit purchases and sales so that the first $N - 1$ firms have enough permits for their emissions and so that the permit market clears

$$\tilde{y}_i = \begin{cases} \bar{e}_i - S_i & \text{for } i = 1, \ldots, N-1 \\ -\sum_{j=1}^{N-1} \tilde{y}_j & \text{for } i = N \end{cases}. \qquad \text{(A19)}$$

In addition, from (A19), firm N also has a sufficient initial endowment and permits to cover its emissions

$$\int_0^T (\tilde{y}_N(z) + S_N - \bar{e}_N(z)) \, dz = \int_0^T \sum_{i=1}^N \left(\bar{S}_i - \bar{e}_i(z) \right) dz \ge 0. \qquad \text{(A20)}$$

Since $\sum_{i=1}^N y_i = 0$, then $\sum_{i=1}^N Py_i = 0$ because $P(t)$ is the same for all firms. Therefore, upon using the last observations in (A18) and (A19), the cost of the alternative program is less than that of the equilibrium,

$$\int_0^T \sum_{i=1}^N \left(e^{-rt} [C_i(\bar{e}_i(z)) + P\tilde{y}_i(z)] \right) dz < \int_0^T \sum_{i=1}^N \left(e^{-rt} [C_i(e_i^*(z)) + Py_i^*(z)] \right) dz.$$

$$\text{(A21)}$$

So for some firm i, it must be true that

$$\int_0^T \left(e^{-rt} [C_i(\bar{e}_i(z) + P\tilde{y}_i(z))] \right) < \int_0^T \left(e^{-rt} [C_i(e_i^*(z)) + Py_i^*(z)] \right) dz, \quad \text{(A22)}$$

but this contradicts the optimality of $e_i^*(t)$ and $y_i^*(t)$ for the firm. This proves that the market equilibrium is efficient.

REFERENCES

1. California Air Resources Board, "Proposed Regulations for Low-Emission Vehicles and Clean Fuels," staff report, August 1990.
2. "Congressional Quarterly 1980 Almanac," Congressional Quarterly Inc., Washington, D.C. (1981).
3. Mark B. Cronshaw and Jamie B. Kruse, "Permit Markets with Banking," working paper, Dept. of Economics, University of Colorado, Boulder (1993).
4. Maureen Cropper and Wallace Oates (1992), Environmental economics: A survey, *J. Econ. Lit.* **30**, 675–740 (1992).
5. J. H. Dales, "Pollution, Property, and Prices," University of Toronto Press, Toronto, Canada (1968).
6. Morton I. Kamien, and Nancy L. Schwartz, "Dynamic Optimization: The Calculus of Variations and Optimal Control in Economics and Management," Second Edition, Elsevier, New York (1991).
7. David W. Montgomery, Markets in licenses and efficient pollution control programs, *J. Econ. Theory* **5**, 395–418 (1972).

8. Jonathan Rubin and Catherine Kling, An emission saved is an emission earned: An empirical study of emission banking for light-duty vehicle manufacturers, *J. Environ. Econom. Manag.* **25**(3), 257–74 (1993).

9. Atle Seierstad and Knut Sydsæter, "Optimal Control Theory with Economic Applications," (Advanced Textbooks in Economics, 24) Elsevier, New York (1987).

10. Brian Shaffer, Regulation, competition, and strategy: The case of automobile fuel economy standards, 1974–1991, *Markets, Politics and Social Perform.*, **13**, 191–218 (1992).

11. A. M. Steinberg and H. L. Stalford, On existence of optimal controls, *J. Optim. Th. Appl.* **11**(3) 267–273 (1973).

12. Thomas Tietenberg, "Emission Trading, An Exercise in Reforming Pollution Policy," Resources for the Future, Washington, D.C. (1985).

C Command-and-Control

[15]

Journal of Public Economics 34 (1987) 19–44. North-Holland

PERFORMANCE VERSUS DESIGN STANDARDS IN THE REGULATION OF POLLUTION

David BESANKO*

School of Business, Indiana University, Bloomington, IN 47405, USA

Received May 1986, revised version received May 1987

This paper presents a model of pollution regulation in an oligopolistic market. Two forms of regulation are considered: performance standards which regulate pollution directly by an upper limit on emissions, and design standards which regulate pollution indirectly by a minimum usage requirement of an emissions control input. Equilibria under each regulatory regime are characterized. A welfare analysis reveals that performance standards are preferred to design standards if the objective is minimization of emissions plus pollution damage costs. However, the comparison is indeterminate if the regulator's objective is total surplus less pollution damage. An equivalence between emissions taxes and performance standards is established, so the above welfare comparisons also apply to emissions taxes versus design standards.

1. Introduction

A common form of regulation in the presence of externalities or information problems is standard-setting. Standard-setting is widely used in environmental regulation, regulation of product and occupational safety, and regulation of product efficacy and durability.

An important issue in devising a standard is whether the standard should specify performance goals (a performance standard) or should specify the procedures used to meet the performance goal (a design standard). In the context of air quality regulation, a performance standard would place a limit on a factory's emission of sulfur dioxide per million parts of air, while a design standard would specify the minimum efficiency of a stack gas scrubber designed to reduce SO_2 emissions.

Although a performance standard and a design standard may be directed toward the same goal (cleaner air, fewer workplace accidents), their economic consequences may be quite different. Breyer (1982) notes that although design standards are easier to enforce than performance standards:

*This work was begun while I was visiting the Economics Research Group at Bell Communications in 1985. I would like to thank David Baron, David D. Martin, Martin Perry, and David Sappington for helpful discussions on this paper, and two referees for their useful comments and suggestions.

... design standards limit the firm's flexibility. A firm that finds a cheaper or more effective way of achieving the regulation's objective must undertake the heavy burden of forcing a change in the standard before it can use its new method. For the same reason, a design standard tends to freeze existing technology and to favor those firms already equipped with that technology over potentially innovative new competitors.

A performance standard permits flexibility and change. It is directly addressed to the problem that must be solved. And since the agency must, in any event, consider the comparative performance of different machines in order to write a design standard, it may be easy for the agency to write its standard directly in terms of performance goals such as cleaner air or fewer injuries [Breyer (1982, p. 105)].

The purpose of this paper is to analyze the economic consequences of performance and design standards in a stylized model of regulation. To be concrete, the model is described in the context of pollution regulation, though the model could apply to occupational safety and health regulation or product safety with the appropriate redesignation of variables. The model focuses on the regulation of an oligopolistic industry consisting of identical firms each of which emits pollution. The government can set either a performance standard or a design standard. The industry equilibrium under each regulatory regime is analyzed, and the economic consequences of each type of standard are discussed. As the comments of Breyer suggest, we find that a performance standard allows the regulator to achieve a given pollution target at a lower cost of emissions control than does a design standard. It is therefore easy to show that a regulator who is concerned with minimizing the costs of emissions control plus pollution damage (subject to a minimum profit constraint) will prefer an optimally chosen performance standard to an optimally chosen design standard. However, there is a cost to the efficiency gain from a performance standard: for any given level of pollution, a performance standard results in a smaller aggregate industry output than does a design standard. If the regulator's objective is social welfare maximization there may be circumstances under which the regulator will prefer an optimally chosen design standard to an optimally chosen performance standard. It has been argued by some authors [e.g. Roberts and Farrell (1978)] that environmental regulators care about other factors besides emissions control and pollution damage costs (e.g. local employment impacts of environmental regulations). To the extent that social welfare maximization adequately represents regulatory objectives in pollution control, our analysis provides a positive rationale (that does not rest on differences in enforcement costs) why environmental regulators might prefer design standards to performance standards.

Models of pollution standards and optimal pollution standards have been studied by many authors including Kwerel (1977), Spence and Weitzman (1978), Dasgupta, Hammond and Maskin (1979), Dasgupta (1982), and Baron (1985a, 1985b). Our paper differs from these papers in its emphasis upon the comparative advantages of different *types* of standards. In addition, the focus of many of these papers is on the impact of uncertainty or asymmetric information on optimal pollution regulation. This paper demonstrates that even under certainty and symmetric information there are nontrivial tradeoffs to be considered in the choice of the method of regulating pollution. The paper most closely related to ours is a recent study by Farber and Martin (1986) of the relationship between market structure and pollution control activity. Farber and Martin's paper focuses on performance standards in a setting in which the regulator can monitor only imperfectly. A principal theoretical finding of their study is that under plausible circumstances pollution control activity by a firm will decrease as rivalry in the market increases. The empirical analysis presented by the authors supports this result.

Our paper is organized in the following manner. Section 2 gives an overview of the model. Section 3 characterizes the industry equilibrium under a performance standard, and section 4 characterizes the industry equilibrium under a design standard. Section 5 compares the welfare effects of performance standards and design standards. Section 6 elaborates on the analysis of section 5 in the context of a specific example. Section 7 contains a discussion of the issue of pollution taxes versus standards. Section 8 discusses the limitations and possible extensions of the model. Section 9 summarizes and concludes.

2. Overview of the model

2.1. The regulated industry

We consider an oligopoly consisting of a fixed number n of identical firms where $1 \leqq n < \infty$. The output x_i of a typical firm i in conjunction with a quantity e_i of a composite emissions-control input (hereafter e.c.i.) determines a quantity q_i of pollution emitted by that firm according to:

$$q_i = f(x_i, e_i), \tag{1}$$

where f is a twice-continuously differentiable quasi-convex function, with $f_x > 0$ and $f_e < 0$. The function $f(\cdot)$ is a reduced-form pollution production function. To see how this reduced form is derived, suppose that each firm in the industry manufactures a joint product (x, q) according to production functions:

$$x = g(z, e), \qquad q = h(z, e),$$

where $z \in R^l$ is the vector of productive inputs. To develop the specification in (1), we assume that $h(\cdot)$ has the specific form:

$$h(z, e) = f(g(z, e), e),$$

where $f(\cdot)$ is the function defined in (1). This specification assumes that pollution is a direct byproduct of a firm's output. An example where this specification is plausible would be the case of a producer of electrical capacitors that have PCBs as an insulating fluid. In the process of manufacturing capacitors, a certain proportion of defectives are produced, and the disposal of these defectives creates pollution. Consequently, the amount of pollution will be an increasing function of the number of nondefective capacitors produced and sold.[1]

The assumption that $f(\cdot)$ is quasi-convex in (x, e) implies that the marginal rate of substitution of output for e.c.i. is an increasing function of output. Thus, the more output the firm produces, the greater is the required increase in the e.c.i. in order to keep pollution constant.

The cost function of a typical firm i takes the form:

$$c(e_i)x_i + k(e_i),$$

where $c(e_i)$ is the marginal cost of output, and $k(e_i)$ is the fixed cost of installing e_i units of the e.c.i. The marginal cost function $c(\cdot)$ is assumed to be twice-continuously differentiable, nondecreasing and (weakly) convex in e; i.e. $c' \geq 0$, $c'' \geq 0$.[2] The emissions control cost function $k(\cdot)$ is assumed to be twice-continuously differentiable, increasing, and (weakly) convex in e; i.e. $k' > 0$; $k'' \geq 0$. In addition, $k(0) = 0$.

Firms are assumed to behave as Cournot competitors and thus set outputs. The profit of a typical firm i is given by:

$$\pi_i = P(X)x_i - c(e_i)x_i - k(e_i),$$

---

[1]Another set of assumptions applicable to air pollution will also generate the specification in (1). Suppose z denotes the tons of coal used by a manufacturer and let $x = g(z)$ denote a firm's production function. Let $q = s(1 - e)z$, where e represents the percentage of sulfur removed by a flue gas desulfurization system (scrubber), and s is the sulfur content of a ton of coal. Solving out z one obtains:

$$q = s(1 - e)g^{-1}(x) \equiv f(e, x),$$

where $g^{-1}(\cdot)$ is the inverse of g.

[2]Under the interpretation of the underlying technology discussed above,

$$c(e_i)x_i = \min w \cdot z;$$

$$\text{s.t. } x_i = g(z_i, e_i),$$

where w is a vector of input prices. We are thus implicitly assuming that g is linearly homogeneous in z. The assumption that $c' \geq 0$ would follow if $g_e \leq 0$, i.e. if the e.c.i. had a nonpositive marginal product of output.

where $P(X)$ denotes the industry inverse demand function, and $X \equiv \sum x_j$ denotes industry output. The inverse demand function is assumed to be a twice-continuously differentiable decreasing function of X over the interval $[0, \bar{X}]$, where $\bar{X} < \infty$ is such that $P(\bar{X}) = 0$. We also assume that for any $X \leqq \bar{X}$, $P''(X)X + P'(X) \leqq 0$. This assumption will be satisfied if the inverse demand is concave in X or not 'too convex'.[3] The role of this assumption and the assumption that $P(X)$ goes to zero at a finite X is to ensure the existence of an industry equilibrium.

2.2. The regulatory constraints

The regulator is presumed able to specify either a *performance* standard (a maximum allowed pollution per firm) or a *design* standard (a minimum level of the e.c.i. per firm). Under a performance standard each firm faces a regulatory constraint of the form:

$$f(x_i, e_i) \leqq q_P,$$

where q_P is the maximum level of pollution allowed per firm.[4] Each firm then chooses output x_i and a quantity e_i of e.c.i. to satisfy the constraint.

Under a design standard, the regulatory constraint for each firm is of the form:

$$e_i \geqq e_D.$$

Each firm is then free to select its output.

The chief administrative difference between a performance standard and a

---

[3]It is straightforward to prove that the assumption that $P''(X)X + P'(X) \leqq 0$ is a sufficient condition for a firm's revenue function $P(x + X_0)x$ to be concave in x for $X_0 \in [0, \bar{X}]$ and $x \in [0, \bar{X}]$. The concavity of the firm's revenue function is used in the proof of the existence of an industry equilibrium. (See footnote 7 below.) The assumption that $P''(X)X + P'(X) \leqq 0$ is also a sufficient condition for a firm's Cournot reaction function $x(X_0)$ to be decreasing in the output X_0 of all other firms in the industry.

[4]The performance standard in this model can be thought of as arising in one of two ways. First, we can think of the regulator as imposing a maximum level of pollution on each firm by fiat. Alternatively, the upper limit q_P can be thought of as arising because each firm in the industry has purchased q_P 'marketable rights' to pollute. For this second interpretation to be convincing, we need to include in our model the decisions by firms to trade these marketable rights among themselves and the determination of an equilibrium price of rights. However, in our setting a detailed theoretical treatment of these issues is unnecessary. To see this, suppose that before the product market opens each firm in the industry is awarded a certain number of marketable rights (which may or may not equal q_P but which add up to nq_P). In an equilibrium in marketable rights, firms will trade the rights among themselves until the marginal benefit of a right is equated across all firms (where the calculation of marginal benefit takes into account profits in the subsequent product market equilibrium). Since all firms are symmetric, by assumption, marginal benefits will be equated when each firm acquires an equal number q_P of the rights.

design standard is that under a performance standard the regulator would monitor the actual pollution emitted by a firm while under a design standard the regulator would monitor the efficiency of the system the firm puts in place to control emissions. For example, suppose a firm manufactures railroad ties and generates hazardous creosote waste as a byproduct. A performance standard would involve monitoring the quantity of creosote waste generated by the firm while a design standard would involve monitoring the procedures followed by the firm to dispose of or store the hazardous creosote waste. A firm may do a poor job of disposing of the waste (low e) but may not generate much pollution because its output x is low. Intuitively, a performance standard gives a firm more freedom because the standard can be met either by increasing the efficiency of the emissions-control system or by cutting back on output, whereas a design standard can only be met by increasing the efficiency of the emissions-control system. In practice, the distinction between a performance standard and a design standard may be quite subtle. Consider the example of a manufacturer of electrical capacitors that uses PCBs as an insulating fluid. A design standard might be a specification of procedures for disposing of defective capacitors that contain PCBs. To monitor compliance, the regulator might send an official to observe whether these disposal procedures are being followed. However, if this type of monitoring is costly, the regulator may end up monitoring compliance by measuring the amount of PCBs in the soil where the firm dumps its wastes. If the latter enforcement procedure is used, the 'effective' standard is a performance standard because compliance depends on the amount of pollution created by the firm, even though the 'nominal' standard is a design standard.

2.3. Welfare analysis and regulatory objectives

In analyzing optimal performance and design standards, two regulatory objectives will be considered. The first objective is social welfare maximization (SWM) subject to a minimum profit constraint. Social welfare is defined as the sum of consumers' and producers' surplus minus the social costs $B(\sum q_j)$ of pollution damage. The social cost function is assumed to be increasing and (weakly) convex, i.e. $B' > 0$, and $B'' \geq 0$. Under SWM the regulator's objective can be written as:

$$V(X) - \sum c(e_j)x_j - \sum k(e_j) - B(\sum q_j),$$

where $V(X) = \int_0^X P(t)\, dt$ is the gross willingness to pay for the industry output.

The second regulatory objective studied is cost minimization (CM) subject to a minimum profit constraint. The regulator's objective function under CM is the sum $\sum k(e_j) + B(\sum q_j)$ of emissions control and pollution costs.

The regulator is assumed to have full information about the technological, cost, and demand conditions. The regulator will be assumed to precommit to a standard, and the firms will then respond and reach an equilibrium in the product market. With precommitment the appropriate game-theoretic structure is to treat the regulator as a Stackelberg leader who takes into account how the standard affects the equilibrium in the product market.[5]

3. Industry equilibrium under a performance standard

Under a performance standard a firm's profit-maximization problem can be stated as:[6]

$$\max P(X)x - c(e)x - k(e),$$

s.t.

$$f(x, e) \leqq q_P,$$ (2)

$$e \geqq 0,$$

$$x \geqq 0.$$

Given our assumption on technology and demand, an application of Friedman's (1977) Theorem 7.8 can be used to prove that an equilibrium exists.[7] Throughout the analysis we will assume that the equilibrium is symmetric and we will focus on the case in which the regulatory constraint (2) is binding and in which industry output X^P and the quantity e^P of the e.c.i. are positive. The conditions characterizing such an equilibrium are:

$$P'(X^P)x^P + P(X^P) - c(e^P) = \lambda f_x(x^P, e^P),$$ (3)

$$-c'(e^P)x^P - k'(e^P) = \lambda f_e(x^P, e^P),$$ (4)

---

[5]The precommitment assumption is important. In the absence of precommitment an alternative formulation would be to have the firms and the regulator move simultaneously and to analyze the Nash equilibrium in outputs and a standard. To see the implications of this setup, notice that in our model the regulator's objective function under either SWM or CM does not depend *directly* on the standard (either q_P or e_D). Instead, it depends on the choices made by the firms (x, e, and q). This implies that there will be many Nash equilibria in the game in which the firms and regulator move simultaneously. One Nash equilibrium is where each firm conjectures that the regulator will set a standard which is sufficiently lax so that the unregulated Cournot output and a zero level of e.c.i. are feasible. Given such a conjecture each firm will produce its unregulated Cournot output. In turn, if the regulator conjectures that the firms will produce this output and choose zero e.c.i., the regulator (being indifferent across standards, given conjectures about output, e.c.i., and pollution levels) will have no incentive not to choose the lax standard.

[6]Where there is no ambiguity, the subscript designating a firm will be dropped.

[7]A proof of the existence of an equilibrium can be found in Besanko (1987).

$$f(x^P, e^P) = q_P,$$ (5)

where $x^P \equiv X^P n^{-1}$, and $\lambda > 0$ is the Lagrange multiplier for the regulatory constraint (2).

Analysis of the first-order conditions yields the following result about the equilibrium industry output under a performance standard.

Proposition 1. Under a performance standard, the industry equilibrium output X^P is less than the unregulated Cournot industry equilibrium output X^C.

Proof. In the absence of regulation, each firm would spend nothing on emissions control, and the industry equilibrium output X^C would satisfy:

$$P'(X^C)x^C + P(X^C) - c(0) = 0,$$ (6)

where $x^C \equiv X^C n^{-1}$. Because $c(e^P) \geq c(0)$, (6) and (3) imply:

$$P'(X^P)x^P + P(X^P) > P'(X^C)x^C + P(X^C).$$ (7)

The assumption that $P''(X)X + P'(X) \leq 0$ can be shown to imply that $P'(X)Xn^{-1} + P(X)$ is strictly decreasing in X.[8] Thus, (7) implies $X^P < X^C$.

Q.E.D.

The result holds for two reasons. First, the regulatory constraint induces each firm to use a positive quantity of the e.c.i. which raises marginal production cost. Second, the regulatory constraint creates an implicit marginal cost λf_x of output. Both effects work to decrease the industry output below the unregulated Cournot output X^C.

4. Industry equilibrium under a design standard

Under a design standard e_D the profit-maximization problem of a typical firm can be stated as:

[8]The proof of this assertion is follows:
$$d[P'(X)Xn^{-1} + P(X)]/dX = P'(X)(1 + n^{-1}) + P''(X)Xn^{-1}$$
$$< P'(X) + P''(X)Xn^{-1}.$$ (F1)
If $P''(X) \leq 0$, the right-hand side of (F1) is negative, which proves the assertion for the case in which $P''(X) \leq 0$. If $P''(X) > 0$, then
$$P'(X) + P''(X)Xn^{-1} \leq P'(X) + P''(X)X \quad \text{(since } Xn^{-1} \leq X),$$
$$\leq 0,$$
which proves the assertion for the case in which $P''(X) > 0$.

$$\max P(X)x - c(e)x - k(e)$$

s.t.

$$e \geq e_D, \tag{8}$$

$$x \geq 0.$$

Again using Friedman's (1977) Theorem 7.8 an equilibrium can be shown to exist.[9] In equilibrium, the regulatory constraint (8) will be binding.

The condition for the symmetric industry equilibrium output X^D is:

$$P'(X^D)x^D + P(X^D) = c(e_D), \tag{9}$$

where $x^D \equiv X^D n^{-1}$.

This yields the following characterization of the industry equilibrium.

Proposition 2. (a) *If* $c' > 0$ *and* $e_D > 0$, *the industry equilibrium output* X^D *is less than the unregulated Cournot industry output* X^C; (b) *if* $c' = 0$, *the industry equilibrium output equals the Cournot industry output.*

Proof. (a) If $c' > 0$ and $e_D > 0$, then $c(e_D) > c(0)$. Thus:

$$P'(X^D)x^D + P(X^D) > P'(X^C)x^C + P(X^C)$$

which implies $X^D < X^C$.

(b) If $c' = 0$ the condition for the unregulated Cournot equilibrium is precisely (9). Q.E.D.

Proposition 3. *Under a design standard* (a) *industry output is nonincreasing in the required amount of the e.c.i., i.e.* $dX^D/de_D \leq 0$; (b) *equilibrium pollution per firm decreases in the required amount of the e.c.i., i.e.* $dq^D/de_D < 0$.

Proof. (a) Differentiating each side of (9) with respect to e_D and noting that $x^D = X^D/n$, yields:

$$[P''x^D + (1 + n^{-1})P'] dX^D/de_D = c'(e_D).$$

The term in the square brackets is negative (see footnote 8) and $c'(e_D) \geq 0$, so the result follows immediately.

(b) Using (1):

$$dq^D/de_D = f_x dx^D/de_D + f_e < 0. \quad \text{Q.E.D.}$$

[9]See Besanko (1987) for a proof.

Because the equilibrium pollution per firm, $q^D(e_D)$, is a decreasing function, we can invert this function to obtain the design standard $e^D(q_D)$ needed to implement a target level q_D of pollution per firm. This way of thinking about design standard-setting will prove useful in developing comparisons between performance and design standards.

5. Comparison of performance and design standards

This section compares the equilibria that arise under a performance standard and a design standard. Two types of comparisons are made. In the first subsection we compare a performance standard q to a design standard which induces an equilibrium pollution per firm q. In the next subsection, we compare an optimal performance standard to an optimal design standard.

5.1. Comparison of industry equilibria for a fixed level of pollution per firm

The first result we present confirms the conventional wisdom that performance standards are more cost-effective than design standards.

Proposition 4. For any target pollution level q, emissions costs are lower under a performance standard than under a design standard.

Proof. To prove the result it is sufficient to demonstrate that $e^P(q) < e^D(q)$. The proof is by contradiction. Suppose, to the contrary, that $e^D(q) \leq e^P(q)$.

Let $X^D(q)$ denote the equilibrium industry output under a design standard that induces a pollution level per firm of q, and $x^D(q) \equiv X^D(q)n^{-1}$. From (9):

$$P'(X^D(q))x^D(q) + P(X^D(q)) - c(e^D(q)) = 0. \tag{10}$$

By contrast, from (3):

$$P'(X^P(q))x^P(q) + P(X^P(q)) - c(e^P(q)) > 0. \tag{11}$$

By assumption, $e^D(q) \leq e^P(q)$, so because $c' \geq 0$, (10) and (11) imply:

$$P'(X^P(q))x^P(q) + P(X^P(q)) > P'(X^D(q))x^D(q) + P(X^D(q)). \tag{12}$$

The inequality (12) implies:

$$X^P(q) < X^D(q). \tag{13}$$

But,

$$q = f(x^P(q), e^P(q)), \tag{14a}$$

$$q = f(x^D(q), e^D(q)).$$ (14b)

Because $e^P(q) \geqq e^D(q)$ and $X^P(q) < X^D(q)$, (14a) and (14b) cannot simultaneously hold. Thus, $e^P(q) < e^D(q)$.   Q.E.D.

An immediate consequence of Proposition 4 is:

Proposition 5. For any target pollution level q, the industry output $X^P(q)$ under a performance standard is less than the industry output $X^D(q)$ under a design standard.

Proof. Because

$$q = f(x^P(q), e^P(q)) = f(x^D(q), e^D(q)),$$

and $e^P(q) < e^D(q)$, it follows that $X^P(q) < X^D(q)$.   Q.E.D.

Propositions 4 and 5 are depicted in fig. 1. The upward-sloping line 00′ is the set of (x, e) pairs that generate a constant pollution per firm. The equilibrium under a performance standard q puts a firm at point R, where a firm's isoprofit contour is tangent to 00′, while the equilibrium under a design standard puts a firm at point S where a firm's isoprofit contour reaches its peak. The intuition behind Propositions 4 and 5 is that under a performance standard firms substitute output for the e.c.i. to achieve the pollution target q. Under a design standard, no such substitution takes place.

Propositions 4 and 5 indicate the policy tradeoff the regulator faces under SWM. If a performance standard is used, the regulator sacrifices output for low emissions control costs while under a design standard, the regulator sacrifices emissions control costs for higher output. For a *fixed* pollution target q the social welfare under a performance standard may or may not exceed the social welfare under a design standard.[10] It has been argued by some authors [e.g. Roberts and Farrell (1978)] that the objective functions of environmental regulators typically include other factors besides emissions control and pollution costs. For example, state environmental regulators in the United States may be sensitive to the employment implications of pollution control regulations. To the extent that social welfare maximization is a reasonable approximation of regulatory objectives in pollution control, our analysis provides a rationale unrelated to differences in enforcement costs why environmental regulators may prefer design standards to performance standards.

We next compare a firm's equilibrium profit under both types of standards.

[10]In the example in section 6 we show that the same ambiguity arises when the standards are chosen optimally.

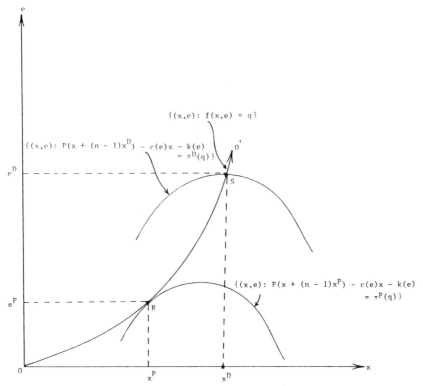

Fig. 1. Comparison of equilibrium under performance and design standards for a fixed pollution target.

Proposition 6. For any target pollution level q, a firm's equilibrium profit $\pi^P(q)$ under a performance standard exceeds a firm's equilibrium profit $\pi^D(q)$ under a design standard.

Proof. Again recall that $x^D \equiv X^D[n]^{-1}$ and $x^P \equiv X^P[n]^{-1}$. By assumption, (x^D, e^D) and (x^P, e^P) both yield pollution per firm q, and therefore (x^D, e^D) is a *feasible* choice for a firm under a performance standard $q_P = q$. Because (x^P, e^P) maximizes a firm's profits in equilibrium, this implies:

$$\pi^P(q) = P(X^P)x^P - c(e^P)x^P - k(e^P)$$

$$> P([1 - n^{-1}]X^P + x^D)x^D - c(e^D)x^D - k(e^D). \tag{15}$$

Now, from Proposition 5, $X^P < X^D$. Thus:

$$P([1 - n^{-1}]X^P + x^D)x^D - c(e^D)x^D - k(e^D)$$

$$> P(X^D)x^D - c(e^D)x^D - k(e^D) = \pi^D(q). \tag{16}$$

Combining (15) and (16) yields the desired result. Q.E.D.

A firm prefers that a given amount of pollution be achieved by a performance standard rather than a design standard because a performance standard gives the firm more flexibility in adjusting to the pollution target. [A firm can choose (x, e) instead of just x.]

5.2. Comparison of optimal standards

As indicated in the previous section, a regulator whose objective is SWM faces a tradeoff in the use of performance versus design standards. By contrast a regulator whose objective is CM has an unambiguous preference for performance standards over design standards.

Proposition 7. When the regulator's objective is CM, the optimal performance standard results in a lower cost than an optimal design standard, i.e. $C^P(\pi)$, where:

$$C^P(\pi) \equiv \min_{q_P} nk(e^P(q_P)) + B(nq_P), \, .$$

$$\text{s.t. } \pi^P(q_P) \geqq \pi. \tag{CM-P}$$

$$C^D(\pi) \equiv \min_{q_D} nk(e^D(q_D)) + B(nq_D),$$

$$\text{s.t. } \pi^D(q_D) \geqq \pi. \tag{CM-D}$$

Proof. Let q_D^* denote the optimal solution to [CM–D]. Thus,

$$\pi^D(q_D^*) \geqq \pi. \tag{17}$$

From Proposition 6, it follows that

$$\pi^P(q_D^*) > \pi^D(q_D^*) \geqq \pi. \tag{18}$$

Thus, q_D^* is a *feasible* solution to [CM–P]. Moreover, from Proposition 4, $e^P(q_D^*) < e^D(q_D^*)$, so that

$$nk(e^P(q_D^*)) + B(nq_D^*) < nk(e^D(q_D^*)) + B(nq_D^*). \tag{19}$$

Expression (19) implies that *a feasible* solution to [CM–P] results in lower costs than the optimal solution to [CM–D]. This implies that $C^P(\pi) < C^D(\pi)$.   Q.E.D.

The result follows because performance standards induce a lower equilibrium quantity of emissions input per firm and higher profit per firm than does a design standard that results in the same equilibrium level of pollution. The implication of this proposition is that a regulator concerned only with costs would support performance standards over design standards. The Congressional Budget Office (CBO) has advocated performance standards over design standards in air pollution precisely on grounds of cost efficiency. According to Crandall (1983), the CBO has estimated that the cost of using a scrubbing standard relative to the performance standards that had been in place prior to 1977 would be about \$2,400 per ton of SO_2 abated by the year 2000.

6. An example

To further compare performance and design standards, an example will be presented.

The specification of the example is:

$$f(x, e) = x/e,$$

$$P(X) = a - X,$$

$$c(e) = 0,$$

$$k(e) = ke,$$

$$B(nq) = bnq.$$

6.1. Industry equilibrium under a performance standard

The equilibrium quantity of e.c.i. per firm and the equilibrium industry output under a performance standard q_P are given by:

$$e^P(q_P) = \begin{cases} [a[q_P]^{-1} - k[q_P]^{-2}][n+1]^{-1}, & q_P \geq ka^{-1}, \\ 0, & q_P < ka^{-1}; \end{cases}$$

$$X^P(q_P) = \begin{cases} [a - k[q_P]^{-1}]n[n+1]^{-1}, & q_P \geq ka^{-1}, \\ 0, & q_P < ka^{-1}. \end{cases}$$

These functions are depicted in fig. 2. Fig. 2 indicates that if the performance standard is sufficiently stringent ($q_P < ka^{-1}$) firms produce zero

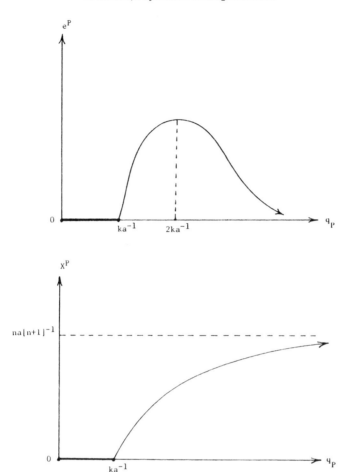

Fig. 2. Equilibrium under performance standard for example.

output and use no e.c.i. As the standard is relaxed (q_P is increased), each firm initially uses a higher quantity of the e.c.i., but beyond a certain point ($q_P = 2ka^{-1}$), further relaxation of the standard induces each firm to reduce e. The equilibrium industry output is a strictly increasing function of the allowed pollution q_P per firm over the range where $q_P > ka^{-1}$. As the standard is relaxed industry output approaches $na[n+1]^{-1}$ – the unregulated Cournot industry output.

6.2. Industry equilibrium under a design standard

The equilibrium under a design standard that results in pollution q_D per firm is given by

$$X^{D}(q_{D}) = na[n+1]^{-1},$$

$$e^{D}(q_{D}) = a[[n+1]q_{D}]^{-1}.$$

The e.c.i. quantity that generates an equilibrium pollution q_{D} per firm is a strictly decreasing function, while the equilibrium industry output is independent of the pollution target q_{D}.

6.3. Optimal performance standard under SWM

To investigate the tradeoffs present under SWM, we will investigate the optimal performance standard under SWM in this section. In the next section, we will characterize the optimal design standard under SWM.

Under SWM the regulator solves:

$$\max_{q_{P}} W^{P}(q_{P}) \equiv aX^{P}(q_{P}) - (1/2)[X^{P}(q_{P})]^{2} - nke^{P}(q_{P}) - bnq_{P}$$

s.t.

$$\pi^{P}(q_{P}) \geqq 0,$$

where, for simplicity, the minimum allowed profit is assumed to equal zero.

Fig. 3 depicts social welfare as a function of q_{P}. The function is defined at $q_{P} = 0$ and for $q_{P} \in [ka^{-1}, \infty)$, and the minimum profit constraint can be shown to be satisfied for all values of q_{P} in the domain of $W^{P}(\cdot)$.[11] The function $W^{P}(\cdot)$ can be shown to have no more than one local maximum.[12] The global maximum for $W^{P}(\cdot)$ will thus either be at this unique local maximum if W^{P} is positive at this point, or it will be at $q_{P} = 0$ if W^{P} is negative at the local maximum.

Assuming an interior optimum, the optimal performance standard can be shown to satisfy the first-order condition:[13]

$$k[n+2][n+1]^{-2}[q_{P}]^{-3}\{aq_{P}-k\} = b. \tag{20}$$

Eq. (20) defines the optimal standard as a function $q_{P}(n, a, b, k)$ of the

---

[11]Social welfare is not 'defined' at $q_{P} \in (0, ka^{-1})$ because if the regulator sets a standard in this interval, firms will produce zero output and use no emissions control input. The regulator constraint will thus be slack over this interval and the actual level of pollution per firm will be zero.

[12]However, for some parameter values it may not have a local maximum.

[13]The second-order condition can be shown to imply that $q_{P} \geqq (3/2)ka^{-1}$. It can be shown that the cubic equation in (20) has at most one root such that $q_{P} \geqq (3/2)ka^{-1}$. If such a root exists and if the solution to (20) is such that $W^{P}(q_{P}) > 0$, then (20) describes the global maximum. If (20) has no root that exceeds $(3/2)ka^{-1}$ or if it does and that root is such that $W^{P}(q_{P}) < 0$, the global maximum is $q_{P} = 0$.

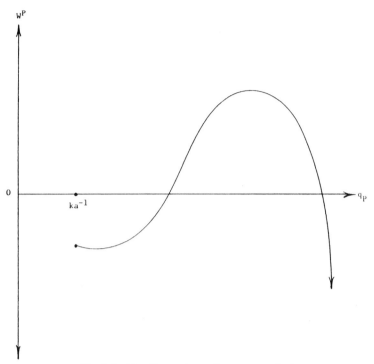

Fig. 3. Social welfare under performance standard.

number of firms and of demand and cost parameters. Proposition 8 indicates how the optimal standard varies with the number of firms in the industry.

Proposition 8. *For the example, as the number of firms in the industry increases the optimal performance standard becomes more stringent, i.e.* $\mathrm{d}q_P/\mathrm{d}n < 0$.[14]

Proof. Differentiation of (20) with respect to n yields:

$$\{k(n+2)(n+1)^{-2}q_P^{-3}[3kq_P^{-1}-2a]\}\, \mathrm{d}q_P\,/\mathrm{d}n$$

$$-kq_P^{-3}(aq_P-k)(5n+1)(n+1)^{-4}=0.$$

The term in the curly brackets is negative by the second-order condition of the regulator's problem. Thus, $\mathrm{d}q_P/\mathrm{d}n < 0$.   Q.E.D.

This result is driven by two effects. First, a benefit from relaxing the

[14]As can be seen from the numerical example below, this proposition does *not* imply that total industry pollution $Q^P \equiv nq^P$ decreases in n.

pollution standard is that a relaxed standard induces more output. As n increases, however, the marginal benefit of output (i.e. the industry price), decreases. Thus, for any q_P, the marginal benefit of relaxing the standard decreases in n. Second, as n increases the marginal social cost nb of relaxing the standard increases. Together, these effects work to decrease q_P as n increases.

To further characterize the optimal performance standard some numerical examples are presented in table 1. The examples indicate the optimal standard becomes less stringent as demand shifts outward (a increases) and as the marginal emissions control cost k increases. The standard becomes more stringent as the marginal social cost b of pollution damage increases. Table 1 also indicates that even though pollution per firm declines with n, total industry pollution $Q^P \equiv nq_P$ rises with n: a tighter standard is not sufficient to offset the pollution of an additional firm.

The numerical examples also indicate that under an optimal performance standard, social welfare does not unambiguously increase with the number of

Table 1

Optimal performance standards.

$a = 100, k = 10, b = 10$				$a = 100, k = 10, b = 15$			
n	q_P	Q^P	$W^P(q_P)$	n	q_P	Q^P	$W^P(q_P)$
1	8.61	8.61	3,577	1	7.02	7.02	3,539
2	6.62	13.24	4,179	2	5.39	10.78	4,119
5	4.36	21.80	4,423	5	3.55	17.75	4,325
10	3.09	30.90	4,194	10	2.52	25.20	4,195
20	2.18	43.60	4,105	20	1.77	35.40	3,910
100	0.95	95.00	3,052	100	0.76	76.00	2,630

$a = 100, k = 20, b = 10$				$a = 100, k = 20, b = 15$			
n	q_P	Q^P	$W^P(q_P)$	n	q_P	Q^P	$W^P(q_P)$
1	12.14	12.14	3,506	1	9.90	9.90	3,452
2	9.32	18.64	4,069	2	7.60	15.20	3,986
5	6.13	30.75	4,243	5	4.99	24.95	4,105
10	4.35	43.50	4,078	10	3.53	35.30	3,883
20	3.05	61.00	3,746	20	2.47	49.40	3,472
100	1.30	130.00	2,280	100	1.04	104.00	1,702

$a = 200, k = 10, b = 2$				$a = 300, k = 10, b = 2$			
n	q_P	Q^P	$W^P(q_P)$	n	q_P	Q^P	$W^P(q_P)$
1	27.36	27.36	14,890	1	33.52	33.52	33,616
2	21.05	42.10	17,609	2	25.80	51.60	39,793
5	13.92	69.60	19,166	5	17.06	85.30	43,409
10	9.93	99.30	19,436	10	12.18	121.80	44,141
20	7.04	140.80	19,390	20	8.63	172.60	44,207
100	3.14	314.00	18,738	100	3.85	385.00	43,450

producers in the market as it would in an unregulated Cournot equilibrium. This occurs for two reasons. First, emissions control costs are fixed so the regulatory constraint induces economies of scale. Secondly, as indicated previously, total industry pollution rises as n increases, so the social costs of pollution damage rise as more producers are added to the market. Together, these effects reduce welfare as n increases and counterbalance the pro-competitive effects on the product market equilibrium from having a large number of producers.[15]

6.4. Optimal design standards under SWM

The regulator's problem is

$$\max_{q_D} W^D(q_D) \equiv aX^D(q_D) - (1/2)[X^D(q_D)]^2 - nke^D(q_D) - nbq_D$$

s.t.

$$q_D \geq k[a/[n+1]]^{-1} \text{ or } q_D = 0.$$

The constraint indicates the pollution targets under which the equilibrium profit of each firm is non-negative. The function $W^D(q_D)$, sketched in fig. 4, can be shown to be strictly concave over the range $[k[a/[n+1]]^{-1}, \infty)$. The global maximum will thus either be the local maximum q_D^* in the figure or at $q_D = k[a/[n+1]]^{-1}$.

Assuming an interior solution, the optimal pollution target per firm can be shown to equal

$$q_D = \{akb^{-1}[n+1]^{-1}\}^{1/2}, \tag{21}$$

which implies that the optimal design standard is equal to

$$e_D = \{ab[[n+1]k]^{-1}\}^{1/2}. \tag{22}$$

The expressions in (21) and (22) allow us to establish the following analogue of Proposition 8.

Proposition 9. For the example, as the number of firms in the industry increases: (a) the original pollution target becomes more stringent, i.e. $dq_D/dn < 0$; (b) the design standard that implements the optimal pollution target is relaxed, i.e. $de_D/dn < 0$; (c) total industry pollution increases, i.e. $dQ^D/dn > 0$.

---

[15]An analogous tradeoff between economics of scale and imperfect competition has been analyzed by Perry (1984).

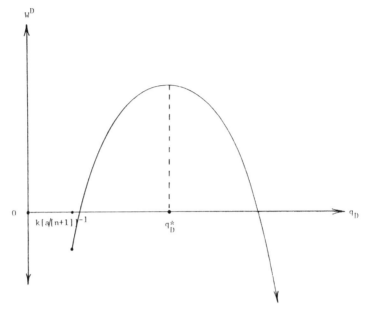

Fig. 4. Social welfare under a design standard.

Proof. Parts (a) and (b) follow immediately from (21) and (22). Total industry pollution Q^D under the optimal design standard is given by

$$Q^D(n) \equiv nq_D = n[n+1]^{-1/2}[akb^{-1}]^{1/2}. \qquad (23)$$

Differentiation of (23) with respect to n yields:

$$dQ^D/dn = (1/2)[n+2][n+1]^{-3/2}[akb^{-1}]^{1/2} > 0. \quad \text{Q.E.D.}$$

The e.c.i. quantity decreases as the pollution target decreases because the output of each firm decreases in n. Thus, with more producers in the market the regulator can relax the design standard and still achieve a lower quantity of pollution per firm.

Part (a) of Proposition 9 pertains only to an interior solution to the regulator's problem. If n is sufficiently large, a corner solution in which $q_D = k[n+1]/a$ can be shown to be optimal. In this case the optimal pollution target increases in n.

6.5. Comparison of performance and design standards

The tradeoff between performance and design standards under SWM will now be considered. As a first step, the expressions for the equilibrium industry output and e.c.i. can be shown to imply that for $q \in [ka^{-1}, \infty)$, the difference $W^P(q) - W^D(q)$ between total welfare under a performance standard and a design standard for any pollution target q is given by:

$$W^P(q) - W^D(q) = nk[[n+1]q]^{-2}[(1/2)[n+2]k - aq].$$ (24)

This expression can be used to derive a sufficient condition for a performance standard to dominate a design standard.

Proposition 10. *If, in the example,*

$$(n+2)(n+1)^{1/2} \geq 2a^{3/2}k^{-1/2}b^{-1/2},$$ (25)

the regulator will prefer a performance standard to a design standard.

Proof. Define

$$\Delta(q) \equiv nk[[n+1]q]^{-2}[(1/2)[n+2]k - aq],$$ (26)

and let q_D^* and q_P^* be the optimal pollution targets under a design standard and performance standard, respectively. We now show that if $\Delta(q_D^*) > 0$, then maximum welfare $W^P(q_P^*)$ under a performance standard exceeds maximum welfare $W^D(q_D^*)$ under a design standard. Because $\Delta(q_D^*) > 0$,

$$W^D(q_D^*) < W^D(q_D^*) + \Delta(q_D^*).$$

$$= W^P(q_D^*) \quad \text{[from (24)]}$$

$$\leq W^P(q_P^*).$$

Now, if $q_D^* = k[a/[n+1]]^{-1}$, $\Delta(q_D^*) > 0$. If $q_D^* = \{akb^{-1}[n+1]^{-1}\}^{1/2}$, $\Delta(q_D^*) > 0$ if and only if (25) holds. Q.E.D.

The sufficient condition (25) is more likely to hold when b, k, and n are large and when a is small. Thus, a performance standard is likely to dominate a design standard when marginal emissions-control and marginal pollution costs are large and when demand is low. Finally, when the number of firms is large, the regulator is more likely to prefer a performance standard. The relationship between market structure and the choice of standard stems from the interaction between the pollution externality and imperfect competition. The pollution externality distorts output upward while imperfect competition distorts output downward. The performance standard counteracts the upward distortion of output, but the downward distortion remains. The design standard deals inefficiently with the pollution externality because it does not induce firms to take into account the marginal effects of output on pollution. However, it leaves open the possibility that the downward output distortion due to imperfect competition will be offset by the upward output distortion. When the downward distortions are less serious (high n) this possibility is less likely, and a performance standard is more likely to be preferred to a design standard.

A comparison between optimal performance and design standards is presented in table 2. As the preceding analysis suggests, performance standards dominate design standards when the number of producers in the industry is sufficiently large, while design standards dominate when the number of producers is small.

7. Performance standards and emissions taxes

Suppose that instead of setting a performance standard or a design standard, the regulator levied an emissions tax t on each firm. Each firm would then face the problem:

$$\max P(X)x - c(e)x - k(e) - tf(x, e)$$

s.t.

$$x \geq 0,$$

$$e \geq 0.$$

Table 2

Comparison of optimal performance standards and optimal design standards.

	$a = 100, k = 20, b = 15$		
n	W^{P*}	W^{D*}	sign$[W^{P*} - W^{D*}]$
1	3,452	3,505	−
5	4,105	4,154	−
10	3,883	3,914	−
20	3,472	3,275	+
60	2,413	−6,143	+
80	2,032	−14,563	+
100	1,702	−25,399	+

W^{P*} = maximum social welfare, performance standard.
W^{D} = maximum social welfare, design standard.

Let x^T, e^T denote the equilibrium output and e.c.i. per firm and let $q^T = f(x^T, e^T)$. It is well known that in the absence of uncertainty or asymmetric information, the regulator can implement the equilibrium under an emissions tax by setting an appropriately chosen performance standard [see, for example, Dasgupta and Heal (1979)]. In this case, the performance standard $q_P = q^T$ will implement the equilibrium under the emissions tax.

Because a design standard may dominate a performance standard if the regulator's objective is social welfare, this result implies that the regulator may prefer a design standard to a tax, even in the absence of considerations such as uncertainty or asymmetric information.

8. Discussion and extensions

Because the formal model presented in the previous section is fairly simple, it understates the richness of the standard-setting problem in the real world. The objective of this section is to discuss a number of important issues in the standard-setting problem that our model ignores and to suggest possible directions for future research into the standard-setting problem.

By modelling regulation as a simple constraint, we are implicitly assuming that the regulator is able to continuously monitor the firm's compliance with the standard and that the regulator can impose on the firm sufficiently stiff penalties for noncompliance with the standard. An extension of our analysis would be to assume that pollution cannot be monitored perfectly and that monitoring is costly. The regulator would then be faced with the problem of how frequently to monitor compliance. In a model with positive monitoring costs, it seems plausible to expect that continuous monitoring is not optimal. This will weaken the effects of standards whose compliance is difficult to monitor. A very stringent pollution performance standard whose compliance is monitored only infrequently could well induce the firm to invest less in an e.c.i. than it would under a very lax design standard which can be monitored much more frequently. Thus, an important consideration in choosing a performance or a design standard would be the relative costs of monitoring. In pollution regulation it may be relatively easy to monitor certain forms of pollution (e.g. air pollution by an electric utility) but relatively difficult to monitor other forms of pollution (e.g. contamination of soil by PCBs). In many cases it is conceivable that differences in monitoring costs could dominate the economic tradeoffs identified in our theoretical model.

Another important aspect of the enforcement problem in standard-setting is the existence of limitations on penalties for violating standards. In most cases the regulatory agency bears the burden of proof in establishing a record of noncompliance. In addition, the magnitude of the penalty is often left to the discretion of the regulator or the courts. For example, under the Clean Air Act Amendments of 1977, Congress left unspecified the penalties for noncompliance but directed the Environmental Protection Agency to consider a variety of factors (including severity of the pollution damage and the impact of the penalty on the violating firm) in assessing penalties [Farber and Martin (1986)]. Typically, the agency must go through extensive administrative and legal proceedings before a fine can be assessed. Sometimes the outcome of these proceedings is a revision of the standard and not the assessment of a penalty [Breyer (1982)]. For first-time violators of standards, penalties are usually not assessed [Roberts and Farrell (1978)]. When penalties are assessed they are often small in relation to the assets of the violating firm. Because penalties are often small, firms will find it advantageous to violate the standard, and the regulator must take this into account in the formulation of the standard. With respect to the choice between performance and design standards, the size of the penalty may be an important consideration. Courts will generally not impose penalties for violations of standards when those penalties do not bear a reasonable relationship to the expected damage resulting from the violation. Thus, one might expect a tendency for the courts to impose a higher penalty for a violation of a pollution performance standard than for a violation of a design standard because in the former case the perceived damages from 'over-polluting' may seem far more substantial than the perceived damages from installing the 'wrong' type of pollution control equipment.

Another issue ignored in our model is that firms may be able to successfully circumvent the standards without legally violating them. For example, under a design standard, a firm might install pollution control equipment and then let it sit idle if the use of the equipment adds to the marginal cost of the final product. What prevents a firm from doing this in our model is our implicit assumption that the regulator not only monitors the installation of the e.c.i. but verifies that it is used and maintained to its fullest level of efficiency. In practice, if inspection resources are limited and if the pollution control inputs are sufficiently complex, verification efforts by the regulator may turn out to be quite superficial. Indeed, the regulator may end up making inferences about the efficiency of the emissions control procedures by observing actual emissions. But in this case, what is nominally a design standard in fact becomes a performance standard.

The firms may also be able to legally circumvent a performance standard. If, for example, the performance standard is viewed as an endowment of marketable rights to pollute (see footnote 4) then two or more firms in the industry may have an incentive to merge in order to relax the performance standard.[16]

---

[16]Salant, Switzer and Reynolds (1983) have shown that in a symmetric Cournot model with linear demand and linear marginal cost horizontal mergers are generally unprofitable unless all firms in the industry merge together to form a monopoly. An open question is whether the presence of performance standards alters this result, at least in some cases.

A final important issue that we have ignored in our model is the effect of pollution standards on entry. The imposition of standards raises the costs of entering the regulated industry. Because performance standards give firms more flexibility than design standards, it would seem that performance standards would have less harmful effects on entry than design standards. Of course, in the spirit of Salop and Scheffman (1983) and Rogerson (1984), incumbent firms may have a preference for design standards for this very reason. An important and worthwhile extension of our analysis would be to examine the implications of alternative regulatory regimes for entry and to study the preferences of both the regulator and the firms for regulatory regimes when the full impacts of alternative types of standards are considered.

9. Summary and conclusions

The conventional wisdom on standard-setting in pollution seems to be that, absent considerations of enforcement costs, performance standards should be preferred to design standards because performance standards give firms more flexibility to adjust to the standard and thus pollution targets can be achieved at a lower cost. The principal policy implication of this analysis is that the conventional wisdom is correct if one counts only emissions and damage costs. If, however, one also counts the impact of pollution regulation on the output supply decisions of firms, then it is no longer the case that performance standards dominate design standards, and the choice between the two (in the absence of enforcement cost differences) will depend on parameters of the demand and cost functions and on the market structure of the regulated industry.

References

Baron, D.P., 1985a, Noncooperative regulation of a non-localized externality, Rand Journal of Economics 16, 553–568.
Baron, D.P., 1985b, Regulation of prices and pollution under incomplete information, Journal of Public Economics 28, 211–231.
Besanko, D., 1987, Standard setting in pollution regulation: A comparison of regulatory regimes, unpublished working paper.
Breyer, S., 1982, Regulation and its reform (Harvard University Press, Cambridge, MA).
Crandall, R.W., 1983, Air pollution, environmentalists, and the coal lobby, in: Roger G. Noll and Bruce M. Owen, eds., The political economy of deregulation: Interest groups in the regulatory process (American Enterprise Institute, Washington, D.C.).
Dasgupta, P.S., 1982, The control of resources (Harvard University Press, Cambridge, MA).
Dasgupta, P.S. and G.M. Heal, 1979, Economic theory and exhaustible resources (Cambridge University Press, Cambridge, UK).
Dasgupta, P.S., P.J. Hammond and E.S. Maskin, 1979, On imperfect information and optimal pollution control, Review of Economic Studies 46, 185–216.
Farber, S.C. and R.E. Martin, 1986, Market structure and pollution control under imperfect surveillance, The Journal of Industrial Economics 35, 147–160.

Friedman, J.W., 1977, Oligopoly and the theory of games (North-Holland, Amsterdam).

Kwerel, E., 1977, To tell the truth: Imperfect information and optimal pollution control, Review of Economic Studies 44, 595–601.

Perry, M.K., 1984, Scale economies, imperfect competition, and public policy, Journal of Industrial Economics 32, 313–333.

Roberts, M.J. and S.O. Farrell, 1978, The political economy of implementation: The Clean Air Act and stationary sources, in: A. Friedlander, ed., Approaches to controlling air pollution (MIT Press, Cambridge, MA).

Rogerson, W.P., 1984, A note on the incentive for a monopolist to increase fixed costs as a barrier to entry, Quarterly Journal of Economics 99, 399–402.

Salant, S.W., S. Switzer, and R.J. Reynolds, 1983, Losses from horizontal merger: The effects of an exogenous change in industry structure on Cournot Nash equilibrium, Quarterly Journal of Economics 98, 185–199.

Salop, S.C. and D.T. Scheffman, 1983, Raising rivals' costs, American Economics Review 73, 267–271.

Spence, A.M. and M.L. Weitzman, 1978, Regulatory strategies for pollution control, in: A. Friedlander, ed., Approaches to controlling air pollution (MIT Press, Cambridge, MA).

[16]

JOURNAL OF ENVIRONMENTAL ECONOMICS AND MANAGEMENT **14**, 112–127 (1987)

Firm Behaviour under Regulatory Control of Stochastic Environmental Wastes by Probabilistic Constraints

Brian Beavis and Ian Dobbs

Department of Economics, University of Newcastle Upon Tyne, England

Received October 17, 1983; revised January 16, 1986

Environmental regulation by deterministic constraints is not really meaningful when, as is common in practice, effluent generation is inherently stochastic. In such a case, "probabilistic constraints" are more appropriate. This paper examines the regulatory control of firm behaviour under mean and percentile probabilistic constraints. © 1987 Academic Press, Inc.

INTRODUCTION

As a consequence of the inherent difficulty of measuring environmental damage resulting from waste discharges, environmental control authorities have directed attention toward a policy of attaining "satisfactory" environmental quality levels. The choice of appropriate policy instruments for the efficient achievement of such environmental quality levels has been widely debated in the literature in recent years.[1] Virtually all of this discussion has been conducted on the implicit premise that waste discharges can be accurately controlled by those responsible for them. Thus uncertainty exists only in relation to the environmental control authority's monitoring problem; (see, e.g., Harford [7] and Yohe [15]). In the real world, however, waste dischargers are often unable to control with any great degree of accuracy the quantity and quality of wastes associated with specific levels of their productive activities; the discharger is also faced with a stochastic control problem. The inherent stochastic nature of waste discharges (as distinct from "monitoring uncertainty"), is recognised by environmental control authorities, who for this reason, acknowledge that, realistically, environmental quality objectives and constraints must be formulated in probabilistic terms. For example, in a major policy statement in 1977 the National Water Council for England and Wales [11] recommended that water quality objectives be determined by Water Authorities for all rivers, canals, and major streams in their region and that the water quality objectives "have regard to uses of the waters and environmental considerations, based on values of quality parameters which can be expected to be achieved by 95 per cent of samples taken (95 percentile values)." Having determined appropriate water quality objectives, the Water Authorities were to review consent conditions attached to existing discharges. Although discharge consent conditions (specifying maximum

---

[1] There is considerable literature on the comparison of control instruments. In the general planning context, see, e.g., Weitzman [13], Laffont [9], and Ireland [8]. In the environmental pollution context, see, e.g., Roberts and Spence [12], Yohe [15], Adar and Griffin [1], Dewees [5], and Baumol and Oates [2].

permissible concentrations of pollutants in discharges) were to remain as "fixed figures which could not legally be exceeded," it was "not expected that Water Authorities would prosecute if occasional samples exceeded the consent conditions (provided these were not outside the range of variation to be expected from well-run treatment plants)." The recommendations of the National Water Council were subsequently accepted by the Water Authorities, so that while consent conditions are regarded as absolute limits for legal purposes, in judging compliance with such conditions most Water Authorities, as a rule, have deemed the conditions to have been met if 80% of discharge samples have complied with the consent conditions. In effect, from the viewpoint of the discharger, the consent conditions require that discharges in excess of a specific level can only appear with a probability less than some given number, i.e., the discharger faces a percentile constraint.

The purpose of this article is to extend the analysis of enforcement of discharge constraints to the case where the discharger is faced with a constraint specifying allowable upper bounds on either the mean level or some particular percentile of its waste discharge. At the same time the process by which dischargers are "controlled" by such constraints is given more detailed consideration than has hitherto been the case (cf. Downing and Watson [6], Harford [7], Yohe [15]). That is, we explicitly model the monitoring, adjudication, and conviction (court) process. This allows consideration of the importance and impact of various information flows (as between court authorities, environmental authorities, and dischargers).

Our concern is thus with a firm's choice of production and waste-treatment activities in an environment where the firm's waste discharges are stochastic and the firm is faced with an environmental restriction on percentile or mean waste discharges to the environment. Should the firm be deemed by the environmental control authority to be violating its environmental restrictions, it is prosecuted in the courts, and if found guilty, incurs a penalty for such a violation.[2]

When a firm's waste discharges are stochastic, the control authority will experience difficulty in proving to the courts that any observed excess discharges are not due to random fluctuations but to the firm's choice of production and waste treatment activities. This is because the authority's monitoring activities will only produce an imperfect estimate of the probability distribution of the firm's waste discharges. Typically, the firm itself will be aware of the authority's difficulties and will consequently take these into consideration when formulating its production and waste treatment plans. It will thus not regard the imposed environmental restriction as a constraint which it must automatically satisfy. In Section 2 of the paper the firm's production and waste treatment technology is described. Section 3 examines the nature of the regulatory constraint, and Section 4, the authority's monitoring and enforcement problem. Section 5 is devoted to an examination of the firm's behavior, in light of its recognition of the authority's monitoring and enforcement problems. Section 6 concludes the paper with a discussion of the factors likely to influence the environmental control authority's choice of environmental restriction.

---

[2] Maler [10] and Beavis and Walker [3] are the only contributions, to our knowledge, that explicitly recognise the need for such a form of regulatory constraint. Maler explicitly assumes an infinite fine, in that he assumes that the firm *must* obey the probabilistic standard. Thus, he bypasses the information issues and the role of the authority vis-a-vis the courts which arise in this paper. Beavis and Walker do not discuss any of the issues raised in the present paper.

2. FIRM TECHNOLOGY

We assume that the firm's production technology can be described by a conventional, twice-continuously differentiable, strictly concave production function $q = q(x_1)$, where q denotes output and x_1 input used in production. In addition to producing output the firm also generates wastes as a by-product of its production activities. The firm has the option of either discharging the wastes it generates directly to the environment, or of treating the wastes before they are discharged. The wastes, w, discharged by the firm to the environment depend on the level of input used in production, x_1, the level of input used in waste treatment, x_2, and a continuous random variable, θ, which has probability density function $\psi(\theta)$. In a general formulation, where $w = H(x_1, x_2, \theta)$, the firm would have the two instruments, x_1, x_2, by which it could influence the mean, variance, and structure of the waste-distribution. For analytical convenience, we shall assume the additively separable form

$$w = G(x_1, x_2) + \theta, \tag{1}$$

where $G(\cdot)$ is twice differentiable and $G_1 > 0$, $G_2 < 0$.[3] $G(\cdot)$ is assumed non-negative for all $x_1, x_2 > 0$. $G(\cdot)$ is further assumed to be strictly convex in (x_1, x_2); this implies $G_{11} > 0$, $G_{22} > 0$ and[4]

$$G_{11}G_{22} - G_{12}^2 > 0 \tag{2}$$

$$G_2^2 G_{11} + G_1^2 G_{22} - 2G_1 G_2 G_{12} \geq 0. \tag{3}$$

$G_{11} > 0$ may be justified on various grounds; for example, if wastes are essentially residuals from production and there are diminishing returns in production (as assumed above). $G_{22} > 0$ corresponds to an assumption of diminishing returns to abatement. Although abatement–treatment can be subject to increasing returns over some region, the assumption that clean-up gets progressively more difficult is not unreasonable.

The sign of the cross-partial, G_{12}, is of significance in the comparative statics section. $G_{12} = 0$ corresponds to the "separable" case where the efficiency with which x_2 abates effluent is unaffected by throughput. $G_{12} > 0$ corresponds to the case where increasing product output decreases the marginal efficiency of abatement, and $G_{12} < 0$ to the case where it increases the marginal efficiency of abatement. The comparative statics results turn out to be unambiguous if $G_{12} > 0$ and also if $G_{12} < 0$ so long as, in the latter case, $|G_{12}|$ is relatively small; that is, where the interaction effect is not too strong.

The only assumptions required on the p.d.f. are that $E(\theta) = k$, constant ($k > 0$), and that $\psi(\theta) > 0$ for $\theta > 0$, the domain of definition.[5] Thus, the random variable

---

[3] Subscripts denote the appropriate partial derivatives.

[4] Strict convexity implies convexity of the level sets and (3) is a standard result for the latter when $G_2 < 0$.

[5] The additive specification (1) is assumed to be a good approximation in the region of the optimum. Clearly, if $G = 0$, say because $x_1 = 0$, there are still positive wastes. Thus with no production, discharges continue; this may be plausible in many cases. The multiplicative specification $w = G(x_1, x_2)\theta$ implies that when $G = 0$, $w = 0$ and so zero production implies zero wastes. Unfortunately, the multiplicative specification renders the comparative statics much less tractable.

w has an associated p.d.f. which may be defined as

$$g(w) \equiv \psi(w - G(\cdot))$$

3. THE REGULATORY CONSTRAINTS

The firm is required to obey an environmental discharge restriction of one of two types.

(i) *The percentile constraint*

$$\text{Prob}(w > w^*) \leq 1 - \alpha, \tag{4}$$

where w^* and α are parameters whose values are set by the environmental control authority. w^* is the critical level above which the firm must not transgress more than a fraction $1 - \alpha$ of the time. Denoting by w_α the α100th percentile of the distribution of w, (4) may be equivalently expressed as

$$w_\alpha \leq w^*, \tag{5}$$

where, given $w = G(x_1, x_2) + \theta$, w_α is defined by

$$I \equiv \int^{w_\alpha - G} \psi(\theta) \, d\theta = \alpha. \tag{6}$$

With $\alpha = 0.5$, w_α denotes the median; with $\alpha = 0.95$, w_α denotes the 95th percentile.

(ii) *The mean constraint*

$$\bar{w} \leq \bar{w}^*. \tag{7}$$

Equation (7) parallels (5). $\bar{w}$ denotes the waste discharge mean.

The percentile and mean constraints under a general specification of the form $w = H(x_1, x_2, \theta)$, give rise to different forms of firm behaviour; it is *not* possible, in general, to translate a mean constraint into an equivalent percentile constraint, or vice versa. However, under an additive specification as in (1), or the multiplicative specification ($w = G(x_1, x_2)\theta$), it *is* possible to translate a mean constraint into a percentile constraint and vice versa. This is so because, in the latter cases, there is a unique relationship between $\bar{w}$ and w_α, whilst under the general specification, this is not necessarily true; in the general case, the firm can use x_1, x_2 to independently vary w_α and $\bar{w}$.

The possibility that w_α and $\bar{w}$ can be varied independently when $w = H(x_1, x_2, \theta)$ is straightforward to establish. Thus

$$\bar{w} = \int H(x_1, x_2, \theta) \psi(\theta) \, d\theta$$

$$w_\alpha = H(x_1, x_2, \theta_\alpha),$$

116   BEAVIS AND DOBBS

where θ_α is defined by

$$\int^{\theta_\alpha} \psi(\theta) \, d\theta = \alpha.$$

A simple example now establishes the point; let $\psi(\theta) = 1/(b - a)$ over the domain (a, b) and let $w = Ax_1 + Bx_2\theta$. Then

$$\bar{w} = Ax_1 + Bx_2(b + a)/2$$
$$w_\alpha = Ax_1 + Bx_2(\alpha(b - a) + a).$$

Thus, except if $\alpha = 0.5$, $\bar{w}$ and w_α may be varied independently via x_1, x_2. Under the additive specification, since $E(\theta) = k$,

$$\bar{w} = G + k$$

whilst, from (6),

$$\frac{\partial I}{\partial G} = \psi(w_\alpha - G)\frac{\partial}{\partial G}(w_\alpha - G) = 0$$

so, given $\psi(w_\alpha - G) \neq 0$, $\partial/\partial G(w_\alpha - G) = 0$ so

$$w_\alpha = G + \theta_\alpha$$

(since, when $G = 0$, $w_\alpha = \theta_\alpha$). Hence we have

$$w_\alpha = \bar{w} + (\theta_\alpha - k).$$

It follows that a constraint of the form $w_\alpha \leq w_1^*$ may be translated into an

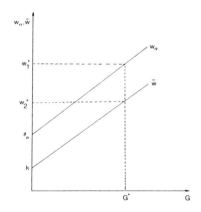

FIG. 1.   Relation between Percentile and Mean Constraints.

equivalent constraint on the mean, $\bar{w} \leq w_2^*$ and vice versa, as illustrated in Fig. 1. A similar translation is possible in the multiplicative case.[6]

The additive specification (1) is adopted since neither the general, nor the multiplicative specification proves particularly tractable. Given the above observations, it is a matter of taste as to whether we pursue the analysis of mean or percentile constraint. Since the sampling distribution of the mean is, analytically, the more straightforward, we shall focus the exposition on this case. Points of difference pertaining to the percentile constraint are then noted as we proceed.

4. THE MONITORING AND ENFORCEMENT PROBLEM

If the firm is deemed by the environmental control authority to be in violation of the environmental restriction, it is prosecuted in the courts, and if found guilty, it incurs a fine, F. If the legal costs are borne by the unsuccessful party, then such costs to the firm may be included in F. One would expect therefore F to contain a substantial fixed component although there might also be a sliding scale fine associated with the extent of the violation. However, in view of the substantial fixed costs involved, it seems an appropriate and reasonable approximation to assume a fixed fine, F.

The control authority monitors the firm's waste discharge activities by taking periodic random samples of effluent discharges. On the basis of its sample observations the authority has to decide whether the firm is in violation of the environmental discharge restriction, and, if so, whether to take legal action against the firm for such violation.

Let $\tilde{\bar{w}}$ denote the sample mean, based on a set of sample observations taken by the Authority. If legal proceedings were costless to the authority then the authority could be expected to prosecute the firm for violation of the environmental discharge restriction whenever $\tilde{\bar{w}} > w^*$. However, the court is aware that $\tilde{\bar{w}}$ is a random variable, and will not regard $\tilde{\bar{w}} > w^*$ as necessarily establishing "beyond reasonable doubt" that in fact $\bar{w} > w^*$. Specifically, the court may be expected to convict the firm when the observed $\tilde{\bar{w}}$ implies that

$$\text{Prob}(\bar{w} > w^*) \geq \beta, \tag{8}$$

where β reflects the court's view of what constitutes "beyond reasonable doubt." For simplicity, we assume that the authority knows β, and thus what will constitute "beyond reasonable doubt."[7] Assuming that the unsuccessful party bears the costs, then the authority will only institute proceedings if (8) holds (otherwise it would be involved in unsuccessful and costly proceedings).

Given that effluent discharges are not normally distributed, the sample mean has an asymptotically normal distribution. We shall take its distribution to be normal,

---

[6] The same is true for the multiplicative specification $w = G(x_1, x_2)\theta$. Here, choice of G (through x_1, x_2) affects both mean and variance of w but the relationships between w_α and $\bar{w}$ remains unique: Assuming $E(\theta) = 1$ in this case, $\bar{w} = G$ whilst the percentile is defined by $\int^{w_\alpha/G} \psi(\theta) \, d\theta = \alpha$ from which $\partial(w_\alpha/G)/\partial G = 0$ so $w_\alpha = kG$ where k is a constant ($k > 1$): Thus $w_\alpha = k\bar{w}$; a constraint on the percentile may be translated into a constraint on the mean.

[7] In principle, β could be estimated from past court records.

with mean $\bar{w}$ and variance γ^2 given by[8]

$$\gamma^2 = \sigma^2/n, \tag{9}$$

where n is the sample size and σ^2 (constant) denotes the variance of the underlying waste distribution ψ; $\mathrm{var}(\theta) \equiv \sigma^2$ so $\mathrm{var}(w) = \sigma^2$.

The random variable $(\tilde{\bar{w}} - \bar{w})/\gamma$ is therefore a standard normal random variable, so that letting ϕ_β denote the β100th percentile of the standard normal distribution, it follows that

$$\mathrm{Prob}\left\{ (\tilde{\bar{w}} - \bar{w})/\gamma > \phi_\beta \right\} = 1 - \beta \qquad \text{which implies Prob}\left\{ \bar{w} > \tilde{\bar{w}} - \gamma\phi_\beta \right\} = \beta$$

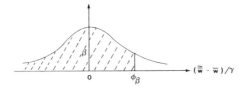

Thus, an observed sample mean, $\tilde{\bar{w}}$, in excess of $w^* + \gamma\phi_\beta$ implies that $\mathrm{Prob}(\bar{w} > w^*) \geq \beta$. The court and authority do not know γ but can estimate it; we assume this is done with minimal error.[9] Thus, the authority will invoke a successful prosecution if and only if

$$\tilde{\bar{w}} > w^* + \gamma\phi_\beta. \tag{10}$$

Given the above description, there is a clear separation of control instruments; the courts choose the fine F and β, whilst the authority chooses w^* and n, the sample size.

5. THE FIRM'S OPTIMISATION PROBLEM

The firm itself will be aware of the difficulty facing the control authority in determining whether or not the environmental restriction is being met and of obtaining a successful conviction in the courts. The firm will therefore take these matters into consideration when selecting its input levels. The firm's objective we take to be that of maximization of expected profits, so that the expected fine which the firm envisages it will face, consequent upon its choice of inputs is of importance.

---

[8] This amounts to assuming a sufficiently large sample to validate the approximation. In the case of the percentile constraint, the sample gives an estimate $\hat{w}_\alpha$ of w_α. This has a p.d.f., $h(\hat{w}_\alpha)$, which can be shown to be asymptotically normal with mean w_α and variance $\gamma^2 = (1/[g(w_\alpha)]^2)(\alpha(1 - \alpha)/n)$. (Cramer [4, p. 367, ff] cf. (8). The analysis proceeds in identical fashion to the text.

[9] In fact, in the additive case, all samples provide information on the variance (which is invariant to the firm's choice of input levels).

The expected fine is given by

$$EF = \int_{w_0}^{\infty} h(\tau)\, d\tau, \tag{11}$$

where $w_0 \equiv w^* + \gamma\phi_\beta$ and $h(\cdot)$ is the p.d.f. of the random variable $\tilde{w}$. That is, $\tilde{\tilde{w}} \sim N(\bar{w}, \gamma^2)$.

In Appendix 1 the above expected fine function is shown to be characterised by[10]

$$EF_i = FG_i h(w_0), \qquad i = 1, 2 \tag{12}$$

$$EF_{ij} = Fh(w_0)\big[G_{ij} + (w_0 - \bar{w})G_i G_j/\gamma^2\big] \tag{13}$$

with

$$EF_{11}EF_{22} - EF_{12}^2 = F^2\big[(h(w_0))\big]^2\big[G_{11}G_{22} - G_{12}^2\big]$$
$$+ F^2\big[h(w_0)\big]^2\big[(w_0 - \bar{w})/\gamma^2\big]\big[G_{11}G_2^2 + G_{22}G_1^2 - 2G_1 G_2 G_{12}\big]. \tag{14}$$

The firm's problem is one of selecting its productive and waste treatment inputs in order to maximise profits. In doing so, it will take account of the authority's choice of sampling policy and standard (w^*) and the court's conviction policy (β) and the fine (F) it incurs if successfully prosecuted.

The firm is assumed to operate in competitive output and input markets, so that letting p denote the price of output and r_i the price of input i, $i = 1, 2$, its expected profit function is given by

$$E\pi(x_1, x_2) = pq(x_1) - r_1 x_1 - r_2 x_2 - F\int_{w_0}^{\infty} h(\tau)\, d\tau. \tag{15}$$

Necessary conditions for an interior solution to the firm's problem require that

$$pq'(x_1) = r_1 + FG_1 h(w_0) \tag{16}$$

$$r_2 = -FG_2 h(w_0). \tag{17}$$

Equations (16) and (17) are only necessary conditions for expected profits to be locally maximised; they are not sufficient conditions. However, it seems reasonable to suppose that the firm will select input levels which ensure that $\bar{w} < w_0$. For should $\bar{w} \geq w_0$ then, on at least 50% of the occasions when the firm's discharge activities are monitored, the firm will expect to subsequently appear in court and be found guilty of violation of the environmental restriction. Intuitively, this seems overly frequent—something which the courts in any event could be expected to respond to by increasing F. We therefore proceed on the assumption that F is set realistically so that the firm's choice of inputs results in $\bar{w} < w_0$. It then follows that $\bar{w} < w_0$ together with the restrictions imposed on $G(\cdot)$ (Eqs. (1) and (2)) are sufficient to ensure that the firm's expected fine function is strictly convex and

---

[10] $EF_i \equiv \partial EF/\partial x_i$, $EF_{ij} = \partial^2 EF/\partial x_i \partial x_j$.

hence its expected profit function is strictly concave over the relevant domain.[11] Conditions (16) and (17) then become sufficient, as well as necessary, conditions for expected profits to be globally maximised over this domain.

COMPARATIVE STATICS

The firm, in its choice of x_1, x_2 takes w^*, n (the authority's control variables) and F, β (the court's control variables) as parametric. In Appendix 2, the comparative statics qualitative responses depend upon:

(1) the nature of the effluent generation and treatment technology, $G(\cdot)$

(2) the way parametric shifts affect the expected fine function—in particular, how changes affect $h(w_0)$.

Appendix 2 provides the basis for the analysis

$$\frac{dx_1^*}{dF} = Y\frac{h(w_0)}{F}, \qquad \frac{dx_2^*}{dF} = Z\frac{h(w_0)}{F}$$

$$\frac{dx_1^*}{d\eta} = Y\frac{\partial h(w_0)}{\partial \eta}, \qquad \frac{dx_2^*}{d\eta} = Z\frac{\partial h(w_0)}{\partial \eta},$$

where $\eta = w^*, n, \beta$ and where

$$Y \equiv \frac{F^2 h(w_0)}{D}[G_2 G_{12} - G_1 G_{22}] \tag{18}$$

$$Z \equiv \frac{1}{D}\{pq''FG_2 + F^2 h(w_0)[G_1 G_{12} - G_2 G_{11}]\}, \tag{19}$$

and D denotes the determinant of the hessian matrix of $E\pi$ evaluated at the optimum ($D > 0$ is assumed). Assuming $q'' \leq 0$; $pq''FG_2 \geq 0$. Clearly, if $G_{12} \geq 0$, then $Y < 0, Z > 0$. This will also be true if $G_{12} < 0$ so long as $|G_{12}|$ is not too large; that is, the interaction effect is not too strong in this case. As discussed in Section 2, this is assumed to be the case, and so $Y < 0, Z > 0$.

To complete the analysis we require sign $[\partial h(w_0)/\partial \eta]$ for each parameter η. The equations for these are given in Appendix 2 ((xii)–(xv)). Referring to these, we obtain

$$\text{sign}\left(\frac{\partial h(w_0)}{\partial w^*}\right) = \text{sign}(w_0 - \bar{w}) \tag{20}$$

$$\text{sign}\left(\frac{\partial h(w_0)}{\partial \beta}\right) = \text{sign}(-(w_0 - \bar{w})\phi_\beta'), \tag{21}$$

$\phi_\beta' > 0$ (recall Fig. 2) and we assumed, $w_0 - \bar{w} > 0$. Hence

$$\frac{\partial h(w_0)}{\partial w^*} < 0, \qquad \frac{\partial h(w_0)}{\partial \beta} < 0. \tag{22}$$

---

[11] The importance of the curvature of the expected fine function is discussed in some detail in a related context by Harford [7, pp. 32–33]. The expected profit function could still be concave if $\bar{w} > w_0$, but we do not pursue this case.

Now, from Appendix 2(xiii),

$$\text{sign}\left(\frac{\partial h(w_0)}{\partial n}\right) = \text{sign}\left[\gamma^2 + (w_0 - \overline{w})(\overline{w} - w^*)\right]. \tag{23}$$

Since $\gamma^2 > 0$, $w_0 - \overline{w} > 0$ so if $\overline{w} - w^* > 0$ then $\partial h(w_0)/\partial n > 0$. Although we have argued for an upper bound on $\overline{w}$ (viz. $\overline{w} < w_0$) there is no necessary reason why $w^* < \overline{w} < w_0$ should hold. If $\overline{w} < w^*$ the ambiguity remains. We defer the intuition for this.

The comparative statics responses of $\overline{w}$ are then given by

$$\frac{d\overline{w}}{d\eta} = \frac{\partial \overline{w}}{\partial x_1} \cdot \frac{dx_1}{d\eta} + \frac{\partial \overline{w}}{\partial x_2} \cdot \frac{dx_2}{d\eta} = G_1 \frac{dx_1}{d\eta} + G_2 \frac{dx_2}{d\eta}.$$

Using the results for $dx_i/d\eta$,

$$\frac{d\overline{w}}{d\eta} = \frac{\partial h(w_0)}{\partial \eta} A, \qquad \text{where } \eta = \beta, w^*, n$$

and

$$\frac{d\overline{w}}{dF} = \frac{Ah(w_0)}{F},$$

where $A \equiv [F^2 h(w_0)(2G_1 G_2 G_{12} - G_1^2 G_{22} - G_2^2 G_{11}) + Pq''FG_2^2]/D$. Since $D > 0$ and $2G_1 G_2 G_{12} - G_1^2 G_{22} - G_2^2 G_{11} < 0$ from (2), $A < 0$ so

$$\frac{d\overline{w}}{dF} < 0 \quad \text{and} \quad \text{sign}\left(\frac{d\overline{w}}{d\eta}\right) = -\text{sign}\left(\frac{\partial h(w_0)}{\partial \eta}\right),$$

where $\eta = \beta, w^*, n$. We now summarise these results in Table 1 and then give some intuition for the ambiguous results for n.

Possible ambiguity in responses to n arise because changes in n affect the expected fine function in two ways—by affecting w_0 and by affecting $h(\ )$ the p.d.f. of $\overline{w}$. An increase in n reduces γ and hence w_0 ($\partial w_0/\partial n = -\gamma \phi_\beta/2n < 0$), providing the firm with an incentive to reduce $\overline{w}$. However, increases in n also reduce the variance of $\tilde{w}$, thereby providing the authority with a more precise estimate of $\overline{w}$. If $\overline{w} > w^*$, the latter effect will induce the firm to respond by

TABLE I

Comparative Statics Responses

Parameter Variable	Courts control F	β	Authority controls w*	n
x_1	−	+	+	?[−]^a
x_2	+	−	−	?[+]
$\overline{w}$	−	+	+	?[−]

^{a} Responses in [ ], if $w^* < \overline{w} < w_0$ holds.

reducing $\bar{w}$, so that, in this case, both effects work in the same direction. However, if $\bar{w} < w^*$, the effect of greater precision in the estimation of $\bar{w}$ provides the firm with an incentive to increase $\bar{w}$—so the effects have opposite signs and the overall result is qualitatively ambiguous. This latter case may occur if the firm is trying to keep $\bar{w}$ below the prosecution level since even if it does so, the stochastic nature of $\bar{w}$ means that it could still incur a fine ("unfairly") on some occasions; an increase in n would reduce this possibility, so making it worthwhile to increase $\bar{w}$ to some extent.

To sum up, the firm in general responds to all the parameters F, β, w^*, n. The responses to a change in fine (F), to what constitutes "reasonable doubt" (β) and to the announced environmental standard (w^*) are fairly intuitive.

6. ENVIRONMENTAL QUALITY CHOICE

Consider first the average level of environmental quality which will result from the application of the above form of regulation. Section 5 derived the comparative static responses of $\bar{w}$ to parametric changes in w^*, n, β, F. The firm may choose $\bar{w} \gtrless w^*$ (depending on parameter values) so little can be said a priori about the resulting level of environmental quality. For example, suppose the fine is small—ceteris paribus, this would tend to imply the environmental standard will be violated frequently and environmental quality relatively poor. If the authority has in mind some particular level of actual environmental quality which it considers acceptable, then it can set about achieving this by altering one of its controls, n, w^*. It could, for example, decrease the announced standard, w^*. Firms would respond by cutting back $\bar{w}$ (and hence also expected wastes). The announced standard would still be frequently violated, but the resulting actual environmental quality would then be satisfactory. Of course, unless the underlying desired *actual* level of environmental quality is announced in some way, regulation *appears* to be ineffective—since the apparently desired constraint ($\bar{w} < w^*$) is not being satisfied. The authority could try to bring together its announced standard, w^*, and the firm's choice of $\bar{w}$ by revising its choice of n. Pressure for revision of fine levels in the courts would also help to bring about a $\bar{w}$ less than w^*. In practice, "reasonable doubt," β, probably cannot be regarded as a smoothly variable choice parameter: it is probably more correctly viewed as exogenous. Separation of control leaves β and F to be chosen by the courts and w^* and n by the authority. It seems reasonable to suggest that there will be optimal choices of w^* and n by the authority conditional on the given exogenous F and β ("chosen" by the courts) and the fact that firms respond to choices of w^* and n.

Suppose, for example, we take the view that environmental control authorities seek to minimise conflict and criticism appearing as "signals" from the economic and social environment in which they operate.[12] A typical form of objective function

---

[12] The statutes establishing most environmental control authorities are usually quite vague, giving the authorities a great deal of freedom for choosing how the vague statutory mandates should be implemented. Environmental control authorities therefore typically operate within an environment which gives them a certain amount of flexibility in procedure. Statutes and court decisions limit the kind of things that environmental control authorities can do, but no reasonable interpretation of statutes, court decisions, etc., could lead to the implication that the objective of environmental control authorities is to locate and enforce the optimal level of environmental quality.

might be $U[E\pi, w^*, n]$. n, the sample size, implies sampling costs which would be of concern (positively or negatively) to the authority. w^*, the announced environmental standard, is the focus of attention for environmental pressure groups; an improvement in announced standards could be expected to reduce conflict with such groups.[13] $E\pi$ denotes firms expected profits (perhaps net of environmental fines); these enter $U(\cdot)$ to proxy the interests of the industrial pressure groups; conflict with such groups increases as the impact of environmental standards increases. The authorities choice of w^*, n thus enter directly into $U(\ )$ and also indirectly since $E\pi = E\pi(x_1(w^*, n), x_2(w^*, n))$. The result of optimising $U(\cdot)$ would be that authority's optimal choices of w^* and n. We do not pursue this formally as the problem proves fairly complex and intractable (requiring knowledge of third order derivatives etc.). The above discussion however articulates the idea that there may be optimal choices of w^*, n.

The observation that the authority has two control variables, w^* and n, leads to some interesting possibilities. In practice, "conflict" with firms will tend to vary with macroeconomic activity and firms' profitability. In a recession, industrial pressure on the authority to reduce the burden on the firms increases; one might then expect environmental standards (w^*) to fall in such periods. However, the control variable n provides the authority with the possibility of maintaining apparent environmental standards (w^*) whilst reducing the burden on firms by changing n (so allowing actual environmental standards to fall). If environmentalists focus only upon announced standards (as discussed above) this would be fairly attractive to the authority. This suggests that environmentalists would need to be able to track all control variables if they are to monitor the situation and accord "blame" appropriately—a virtual impossibility.

APPENDIX 1 : PROPERTIES OF THE EXPECTED FINE FUNCTION

The expected fine is

$$EF = \int_{w_0}^{\infty} Fh(\tau) \, d\tau \qquad\qquad (i)$$

where

$$h(\tau) \equiv 1/\gamma\sqrt{2\pi} \exp\left(-\tfrac{1}{2}((\tau - \overline{w})/\gamma)^2\right). \qquad\qquad (ii)$$

Waste discharge is defined as

$$w = G(x_1, x_2) + \theta \qquad\qquad (iii)$$

---

[13] If such groups recognise the separation of control between the setting of the standard and the setting of the fine, it seems reasonable that they should focus on w^* (in practice this seems to be the case). Their concern is presumably with *actual* environmental quality and it is possible for a high quality standard to be associated with low actual environmental quality. This would be likely if the fine involved were small. However, this outcome could be seen to be the "fault" of the courts and one would expect such pressure groups to direct themselves to the courts "failings" in such a case. Hence, as far as the authority is concerned, it is the announced standard, rather than the actual environmental quality that is of concern.

where θ has p.d.f. $\psi(\theta)$. The successful prosecution level, w_0, is defined by

$$w_0 = w^* + \gamma\phi_\beta. \tag{iv}$$

Now, $\bar{w} = G(x_1, x_2) + k$ since $E(\theta) = k$. Hence

$$\frac{\partial \bar{w}}{\partial x_i} = G_i. \tag{v}$$

From Eq. (9),

$$\frac{\partial \gamma^2}{\partial x_i} = \frac{\partial \gamma}{\partial x_i} = 0, \tag{vi}$$

so, from (iv),

$$\frac{\partial w_0}{\partial x_i} = 0. \tag{vii}$$

Equation (vi) is the result that the firm is powerless to alter the variance of the sample mean, although it can alter the mean through (v). Given the above, the properties of the expected fine function (i) are

$$EF_i = -F\frac{\partial w_0}{\partial x_i}h(w_0) + F\int_{w_0}^{\infty}\left\{\frac{\partial h(\tau)}{\partial w}\frac{\partial \bar{w}}{\partial x_i} + \frac{\partial h(\tau)}{\partial \gamma}\frac{\partial \gamma}{\partial x_i}\right\}d\tau, \tag{viii}$$

which, from (v), (vi), (vii), reduces to

$$EF_i = FG_i\int_{w_0}^{\infty}\frac{\partial h(\tau)}{\partial \bar{w}}d\tau. \tag{ix}$$

Now $h(\cdot)$ is simply a normal density function and

$$\frac{\partial h(\tau)}{\partial \bar{w}} = \frac{1}{\gamma^2}h(\tau)(\tau - \bar{w}),$$

so if we write $s \equiv (\tau - \bar{w})/\gamma$, (ix) may be written as

$$EF_i = \frac{FG_i}{\sqrt{2\pi}}\int_{(w_0 - \bar{w})/\gamma}^{\infty}se^{-1/2s^2}ds.$$

The latter integral may be directly integrated to give

$$EF_i = FG_ih(w_0). \tag{x}$$

The second order derivatives are given by

$$EF_{ij} = FG_{ij}h(w_0) + FG_i\frac{\partial h(w_0)}{dx_j}.$$

Now, since

$$\frac{\partial w_0}{\partial x_j} = \frac{\partial \gamma}{\partial x_j} = 0, \quad \frac{\partial h(w_0)}{\partial x_j} = \frac{\partial h(w_0)}{\partial \bar{w}} \frac{\partial \bar{w}}{\partial x_j}.$$

Also,

$$\frac{\partial h(w_0)}{\partial \bar{w}} = \frac{(w_0 - \bar{w})}{\gamma^2} h(w_0),$$

so

$$EF_{ij} = Fh(w_0)\left\{ G_{ij} + \frac{(w_0 - \bar{w})}{\gamma^2} G_i G_j \right\}. \tag{xi}$$

Equations (x) and (xi) then yield Eqs. (12)–(14) of the text.

APPENDIX 2: COMPARATIVE STATICS ANALYSIS

The firm in its choice of x_1, x_2 takes w^*, n (the Authority's control variables) and F, β (the courts control variables) as parametric. Let

$$D = E\pi_{11} E\pi_{22} - E\pi_{12}^2 > 0$$

and denote a general parameter as $\eta \, (= w^*, n, F, \beta)$. The general comparative static equation is then

$$D\frac{dx_i^*}{d\eta} = -E\pi_{i\eta} E\pi_{jj} + E_{\pi_{j\eta}} E\pi_{ij}, \quad i \neq j, \, i, j = 1, 2, \tag{i}$$

where $dx_i^*/d\eta$ denotes the comparative statics derivative. The terms $E\pi_{ij}$, $E\pi_{i\eta}$, can be easily expressed in terms of EF_{ij}, $EF_{i\eta}$ and these are straightforward to calculate. We summarise the results as follows: Let

$$Y \equiv F^2 h(w_0)(G_2 G_{12} - G_1 G_{22})/D \tag{ii}$$

$$Z \equiv \left[pq'' FG_2 + F^2 h(w_0)(G_1 G_{12} - G_2 G_{11}) \right]/D. \tag{iii}$$

The comparative static equations may then be written as

$$\frac{dx_1^*}{dF} = Y\frac{h(w_0)}{F}; \quad \frac{dx_2^*}{dF} = Z\frac{h(w_0)}{F} \tag{iv}$$

$$\frac{dx_1^*}{d\eta} = Y\frac{\partial h(w_0)}{\partial \eta}; \quad \frac{\partial x_2^*}{\partial \eta} = Z\frac{\partial h(w_0)}{\partial \eta} \tag{v}$$

$$\eta = w^*, n, \beta.$$

To obtain qualitative information on (iv) and (v) requires qualitative information on

126

BEAVIS AND DOBBS

TABLE II

η	$\dfrac{\partial w_0}{\partial \eta}$	$\dfrac{\partial \gamma}{\partial \eta}$	$\dfrac{\partial \bar{w}}{\partial \eta}$
w^*	1	0	0
n	$-\phi_\beta \dfrac{\gamma}{2n}$	$\dfrac{-\gamma}{2n}$	0
β	$\gamma\phi_\beta'$	0	0

Y, Z, and $\partial h(w_0)/\partial \eta$. The signs of Y and Z are discussed in Section 5 of the main text. However, we also require the sign of $\partial h(w_0)/\partial \eta$ for each case, where

$$\frac{\partial h(w_0)}{\partial \eta} = \frac{\partial h(w_0)}{\partial w_0}\frac{\partial w_0}{\partial \eta} + \frac{\partial h(w_0)}{\partial \gamma}\frac{\partial \gamma}{\partial \eta} + \frac{\partial h(w_0)}{\partial \bar{w}}\frac{\partial \bar{w}}{\partial \eta}, \tag{vi}$$

(for $\eta = w^*, n, \beta$; see Table II) and

$$\frac{\partial h(w_0)}{\partial w_0} = -h(w_0)\left(\frac{w_0 - \bar{w}}{\gamma^2}\right)$$

$$\frac{\partial h(w_0)}{\partial \gamma} = h(w_0)\left[(w_0 - \bar{w})^2 - \gamma^2\right]/\gamma^2. \tag{vii}$$

Hence

$$\frac{\partial h(w_0)}{\partial w^*} = -h(w_0)\left(\frac{w_0 - \bar{w}}{\gamma^2}\right) \tag{viii}$$

$$\frac{\partial h(w_0)}{\partial n} = h(w_0)\left[(w_0 - \bar{w})(\bar{w} - w^*) + \gamma^2\right]/\gamma^2 2n \tag{ix}$$

$$\frac{\partial h(w_0)}{\partial \beta} = -h(w_0)\frac{(w_0 - \bar{w})}{\gamma}\phi_\beta'. \tag{x}$$

Equations (viii)–(x) in conjunction with (ii) and (iii) give the full expressions for the comparative statics derivatives in (iv) and (v). The signs of Y and Z and hence (iv) and (v) are discussed in Section 5.

REFERENCES

1. Z. Adar and J. M. Griffin, Uncertainty and the choice of pollution control instruments, *J. Environ. Econom. Management* **3**, 178–188 (1976).
2. W. J. Baumol and W. E. Oates, "The Theory of Environmental Policy," Prentice–Hall, Englewood Cliffs, N.J. (1975).
3. B. Beavis and M. Walker, Achieving environmental standards with stochastic discharges, *J. Environ. Econom. Management*, **10**, 103–111 (1983).
4. H. Cramer, "Mathematical Methods of Statistics," Princeton Univ. Press, Princeton, N.J. (1946).
5. D. Dewees, Instrument choice in environmental policy, *Econom. Inquiry* **21**, 53–71 (1983).

6. P. B. Downing and W. D. Watson, Jr. The economics of enforcing air pollution controls, *J. Environ. Econom. Management* **1**, 219–236 (1974).

7. J. D. Harford, Firm behaviour under imperfectly enforceable pollution standards and taxes, *J. Environ. Econom. Management* **5**, 26–43 (1978).

8. N. J. Ireland, Ideal prices vs. prices vs. quantities, *Rev. Econom. Stud.* **44**, 183–186 (1977).

9. J. J. Laffont, More prices vs. quantities, *Rev. Econom. Stud.* **44** 177–182 (1977).

10. K. E. Maler, "Environmental Economics: A Theoretical Inquiry," Johns Hopkins, Baltimore, MD, (1974).

11. National Water Council, "River Water Quality: The Next Stage" (1977).

12. M. Roberts and M. Spence, Effluent charges and licenses under uncertainty, *J. Pub. Econom.* **5**, 193–208 (1976).

13. J. Weitzman, Prices vs. quantities, *Rev. Econom. Stud.* **41**, 477–491 (1974).

14. G. Yohe, "A Comparison of Price Controls and Quantity Controls Under Uncertainty," Garland, New York, (1979).

15. G. Yohe, Should sliding controls be the next generation of pollution controls? *J. Pub. Econom.* **15** 251–267 (1981).

[17]

Polluters' Profits and Political Response: Direct Controls Versus Taxes

By James M. Buchanan and Gordon Tullock*

Economists of divergent political persuasions agree on the superior efficacy of penalty taxes as instruments for controlling significant external diseconomies which involve the interaction of many parties. However, political leaders and bureaucratic administrators, charged with doing something about these problems, appear to favor direct controls. Our purpose in this paper is to present a positive theory of externality control that explains the observed frequency of direct regulation as opposed to penalty taxes or charges. In the public-choice theory of policy,[1] the interests of those who are subjected to the control instruments must be taken into account as well as the interests of those affected by the external diseconomies. As we develop this theory of policy, we shall also emphasize an elementary efficiency basis for preferring taxes and charges which heretofore has been neglected by economists.

I

Consider a competitive industry in long-run equilibrium, one that is composed of a large number of n identical producing firms. There are no productive inputs specific to this industry, which itself is sufficiently small relative to the economy to insure that the long-run supply curve is horizontal. Expansions and contractions in demand for the product invoke changes in the number of firms, each one of which returns to the same least-cost position after adjustment. Assume that, from this initial position, knowledge is discovered which indicates that the industry's product creates an undesirable environmental side effect. This external diseconomy is directly related to output, and we assume there is no technology available that will allow alternative means of producing the private good without the accompanying public bad. We further assume that the external damage function is linear with respect to industry output; the same quantity of public bad per unit of private good is generated regardless of quantity.[2] We assume that this damage can be measured and monitored with accuracy.

This setting has been deliberately idealized for the application of a penalty tax or surcharge. By assessing a tax (which can be computed with accuracy) per unit of output on all firms in the industry, the government can insure that profit-maximizing decisions lead to a new and lower industry output that is Pareto optimal. In the short run, firms will undergo losses. In the long run, firms will leave the industry and a new equilibrium will be reached when remaining firms are again making normal returns on investment. The price of the product to consumers will have gone up by the full amount of the penalty tax.

* Center for Study of Public Choice, Virginia Polytechnic Institute and State University. We wish to thank the National Science Foundation for research support. Needless to say, the opinions expressed are our own.

[1] Charles Goetz imposes a public-choice framework on externality control, but his analysis is limited to the determination of quantity under the penalty-tax alternative.

[2] This assumption simplifies the means of imposing a corrective tax. For some of the complexities, see Otto Davis and Andrew Whinston and Stanislaw Wellisz.

No one could dispute the efficacy of the tax in attaining the efficient solution, but we should note that in this setting, the same result would seem to be equally well insured by direct regulation. Policy makers with knowledge of individual demand functions, the production functions for firms and for the industry, and external damage functions, could readily compute and specify the Pareto-efficient quantity of industry output.[3] Since all firms are identical in the extreme model considered here, the policy makers could simply assign to each firm a determinate share in the targeted industry output. This would then require that each firm reduce its own rate of output by X percent, that indicated by the difference between its initial equilibrium output and that output which is allocated under the socially efficient industry regulation.[4]

Few of the standard arguments for the penalty tax apply in this setting. These arguments have been concentrated on the difficulties in defining an efficient industry output in addition to measuring external damages and on the difficulty in securing data about firm and industry production and cost functions. With accurately measured damage, an appropriate tax will insure an efficient solution without requiring that this solution itself be independently computed. Or, under a target or standards approach, a total quantity may be computed, and a tax may be chosen as the device to achieve this in the absence of knowledge about the production functions of firms.[5]

[3] See Allen Kneese and Blair Bower, p. 135.

[4] No problems are created by dropping the assumption that firms are identical so long as we retain the assumption that production functions are known to the regulator.

[5] This is the approach taken by William Baumol, who proposes that a target level of output be selected and a tax used to insure the attainment of this target in an efficient manner.

In the full information model, none of these arguments is applicable. There is, however, an important economic basis for favoring the penalty tax over the direct control instrument, one that has been neglected by economists. The penalty tax remains the preferred instrument on strict efficiency grounds, but, perhaps more significantly, it will also facilitate the enforcement of results once they are computed.[6] Under the appropriately chosen penalty tax, firms attain equilibrium only at the efficient quantity of industry output. Each firm that remains in the industry after the imposition of the tax attains long-run adjustment at the lowest point on its average cost curve only after a sufficient number of firms have left the industry. At this equilibrium, there is no incentive for any firm to modify its rate of output in the short run by varying the rate of use of plant or to vary output in the long run by changing firm size. There is no incentive for resources to enter or to exit from the industry. So long as the tax is collected, there is relatively little policing required.

This orthodox price theory paradigm enables the differences between the penalty-tax instrument and direct regulation to be seen clearly. Suppose that, instead of levying the ideal penalty tax, the fully informed policy makers choose to direct all firms in the initial competitive equilibrium to reduce output to the assigned levels required to attain the targeted efficiency goal for the industry. No tax is levied. Consider Figure 1, which depicts the situation for the individual firm. The initial competitive equilibrium is attained when each firm produces an output, q_i. Under regulation it is directed to produce only q_0, but no tax is levied. At output q_0, with an unchanged number of firms, price is above

[6] See George Hay. His discussion of the comparison of import quotas and tariffs on oil raises several issues that are closely related to those treated in this paper.

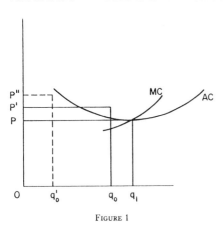

FIGURE 1

marginal cost (for example price is at P'). Therefore, the firm is not in short-run equilibrium, and would if it could expand output within the confines of its existing plant. More importantly, although each firm will be producing the output quota assigned to it at a somewhat higher cost than required for efficiency reasons, there may still be an incentive for resources to enter the industry. The administrator faces a policing task that is dimensionally different from that under the tax. He must insure that individual firms do not violate the quotas assigned, and he must somehow prevent new entrants. To the extent that the administrator fails in either of these tasks, the results aimed for will not be obtained. Output quotas will be exceeded, and the targeted level of industry production overreached.

If the administrator assigns enforceable quotas to existing firms and successfully prevents entrants, the targeted industry results may be attained, but there may remain efficiency loss since the industry output will be produced at higher average cost than necessary if firms face U-shaped long-run average cost curves. Ideally, regulation may have to be accompanied by the

assignment of full production quotas to a selected number of the initial firms in the industry. This policy will keep these favored firms in marginal adjustment with no incentives for in-firm adjustments that might defeat the purpose of the regulation. But even more than under general quota assignment there will be strong incentives for firms to enter the industry and to secure at least some share of the rents that the restriction of industry output generates. If the response to this pressure should be that of reassigning quota shares within the unchanging and targeted industry output so as to allow all potential entrants some share, while keeping all firms, actual and potential, on an equal quota basis, the final result may be equivalent to the familiar cartel equilibrium. No firm will be earning more than normal returns, but the industry will be characterized by too many firms, each of which produces its assigned output inefficiently.

II

When we examine the behavioral adjustments to the policy instruments in the manner sketched out above, a theory of policy emerges. Regulation is less desirable on efficiency grounds even in the presence of full information, but this instrument will be preferred by those whose behavior is to be subjected to either one or the other of the two policy instruments. Consider the position of the single firm in the fully competitive industry, depicted in Figure 1. Under the imposition of the tax, short-run losses are necessarily incurred, and the firm reattains normal returns only after a sufficient number of its competitors have shifted resources to other industries. The tax reduces the present value of the firm's potential earnings stream, whether the particular firm remains in the industry after adjustment or withdraws its investment and shifts to alternative employment. In terms of their own private interests, owners of firms in the industry along with employees will oppose the tax. By contrast, under regulation firms may well secure pecuniary gains from the imposition of direct controls that reduce total industry output. To the extent that the restriction is achieved by the assignment of production quotas to existing firms, net profits may be present even for the short term and are more likely to arise after adjustments in plant. In effect, regulation in this sense is the directional equivalent of cartel formation provided that the individual firm's assigned quota falls within the limited range over which average cost falls below price. Such a range must, of course, exist, but regulatory constraints may possibly be severe enough to shift firms into positions where short-term, and even possibly long-term, losses are present, despite increased output price. Such a result is depicted by a restriction to q_0' in Figure 1, with price at P''.

Despite the motivation which each firm has to violate assigned quotas under regulation, it remains in the interest of firms to seek regulatory policy that will enforce the quotas. If existing firms foresee the difficulty of restricting entry, and if they predict that governmental policy makers will be required to accommodate all entrants, the incentive to support restriction by regulation remains even if its force is somewhat lower. In final cartel equilibrium, all the firms will be making no more than normal returns. But during the adjustment to this equilibrium, above-normal returns may well be available to all firms that hold production quotas. Even if severe restriction forces short-term losses on firms, these losses will be less than those under the tax. Rents over this period may well be positive, and even if negative, they will be less negative than those suffered under the tax alternative. Therefore, producing firms will always oppose any impo-

sition of a penalty tax. However, they may well favor direct regulation restricting industry output, even if no consideration at all is given to the imposition of a tax. And, when faced with an either/or choice, they will always prefer regulation to the tax.

III

There is a difference between the two idealized solutions that has not yet been discussed, and when this is recognized, the basis of a positive hypothesis about policy choice may appear to vanish. Allocationally, direct regulation can produce results equivalent to the penalty tax, providing that we neglect enforcement cost differentials. *Distributionally*, however, the results differ. The imposition of tax means that government collects revenues (save in the case where tax rates are prohibitive) and these must be spent. Those who anticipate benefits from the utilization of tax revenues, whether from the provision of publicly supplied goods or from the reduction in other tax levies, should prefer the tax alternative and they should make this preference known in the political process. To the extent that the beneficiaries include all or substantially all members of the community, the penalty tax should carry the day. Politicians, in responding to citizenry pressures, should heed the larger number of beneficiaries and not the disgruntled members of one particular industry. This political choice setting is, however, the familiar one in which a small, concentrated, identifiable, and intensely interested pressure group may exert more influence on political choice making than the much larger majority of persons, each of whom might expect to secure benefits in the second order of smalls.

There is an additional reason for predicting this result with respect to an innovatory policy of externality control. The penalty tax amounts to a legislated change in property rights, and as such it will be viewed as confiscatory by owners and employees in the affected industry. Legislative bodies, even if they operate formally on majoritarian principles, may be reluctant to impose what seems to be punitive taxation. When, therefore, the regulation alternative to the penalty tax is known to exist, and when representatives of the affected industry are observed strongly to prefer this alternative, the temptation placed on the legislator to choose the direct control policy may be overwhelming, even if he is an economic theorist and a good one. Widely accepted ethical norms may support this stance; imposed destruction of property values may suggest the justice of compensation.[7]

If policy alternatives should be conceived in a genuine Wicksellian framework, the political economist might still expect that the superior penalty tax should command support. If the economist ties his recommendation for the penalty tax to an accompanying return of tax revenues to those in the industry who suffer potential capital losses, he might be more successful than he has been in proposing unilateral or one-sided application of policy norms. If revenues are used to subsidize those in the industry subjected to capital losses from the tax, and if these subsidies are unrelated to rates of output, a two-sided tax subsidy arrangement can remove the industry source of opposition while still insuring efficient results. In this respect, however, economists themselves have failed to pass muster. Relatively few modern economists who have engaged in policy advocacy have been willing to accept the Wicksellian methodological framework which does, of course, require that some putative legitimacy be assigned to rights existent in the status quo.[8]

[7] For a comprehensive discussion of just compensation, see Frank Michelman.

[8] For a specific discussion of the Wicksellian approach, see Buchanan (1959).

IV

To this point we have developed a theory of policy for product-generated external diseconomies, the setting which potentially counterposes the interest of members of a single producing industry against substantially all persons in the community. External diseconomies may, however, arise in consumption rather than in production, and these may be general. For purposes of analysis, we may assume that all persons find themselves in a situation of reciprocal external diseconomies. Traffic congestion may be a familiar case in point.

The question is one of determining whether or not persons in this sort of interaction, acting through the political processes of the community, will impose on *themselves* either a penalty tax or direct regulation. We retain the full information assumption introduced in the production externality model. For simplicity here, consider a two-person model in which each person consumes the same quantity of good or carries out the same quantity of activity in the precontrol equilibrium, but in which demand elasticities differ. Figure 2 depicts the initial equilibrium at E with each person consuming quantity Q. The

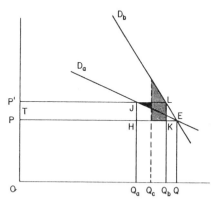

Figure 2

existence of the reciprocal external dis-economy is discovered. The community may impose an accurately measured pen-alty tax in the amount T, in which case A will reduce consumption to Q_a and B will reduce consumption to Q_b. Total consumption is reduced from $2Q$ to (Q_a+Q_b), but both A and B remain in equilibrium. At the new price P', which includes tax, neither person desires to consume more or less than the indicated quantities. The government collects tax revenues in the amount $[2(PP'JH)+HJLK]$. Alternatively, the community may simply assign a restricted quantity quota to each person. If the government possesses full information about demand functions it can reduce A's quota to Q_a, and B's quota to Q_b, securing results that are allocatively identical to those secured by the tax. However, under the quota, both A and B will find themselves out of equilibrium; both will, if allowed quantity adjustment, prefer to expand their rate of consumption.

It will be useful to examine the ideal tax against the quota scheme outlined above, which we may call the idealized quota scheme. If individuals expect no returns at all from tax revenues in the form of cash subsidies, public goods benefits, or reductions in other taxes, both A and B will clearly prefer the direct regulation. The loss in consumers' surplus under this alternative is small relative to that which would be lost under the penalty tax. Each person willingly trades off marginal quantity adjustment for the more favorable inframarginal terms offered under direct regulation, given our assumptions that both instruments achieve the same overall externality control objective.

Under extreme fiscal illusion, individuals may ignore benefits from tax revenues, but consistent methodological precept requires that we allow persons to recognize the benefit side of the fiscal account, at least to some degree. Let us allow all revenues

under the penalty tax to be returned in equal shares to all taxpayers. Under this arrangement, each person expects to get back one-half of the amount measured as indicated above for Figure 2. Simplifying, each expects to get back the amount $PP'JH$, which he personally pays in, plus one-half of the amount measured by the rectangle $JHKL$, all of which is paid in by B. From an examination of Figure 2, it is clear that individual A will favor the penalty tax under these assumptions. The situation for individual B is different; he will prefer direct regulation. He will secure a differential gain measured by the horizontally shaded area in Figure 2, which is equal to the differential loss that individual A will suffer under this alternative. The policy result, insofar as it is influenced by the two parties, is a standoff under this idealized tax and idealized quota system comparison.

For constitutional and other reasons, control institutions operating within a democratic order could scarcely embody disproportionate quota assignments. A more plausible regulation alternative would assign quotas proportionate to initial rates of consumption, designed to reduce overall consumption to the level indicated by target criteria. The comparison of this alternative with the ideal tax arrangement is facilitated by the construction of Figure 2 where the initial rates of consumption are equal. In this new scheme, each person is assigned a quota Q_c, which he is allowed to purchase at the initial price P. We want to compare this arrangement with the ideal tax, again under the assumption that revenues are fully returned in equal per head subsidies. As in the first scheme, both persons are in disequilibrium at quantity Q_c and price P. The difference between this model and the idealized quota scheme lies in the fact that at Q_c, the marginal evaluations differ as between the two persons. There are unexploited gains from trade, even under the determined overall quantity restriction.

It will be mutually advantageous for the two persons to exchange quotas and money, but, at this point, we assume that such exchanges do not take place, either because they are prohibited or because transactions costs are too high. Individual A will continue to favor the tax alternative but his differential gains will be smaller than under the idealized quota scheme. In the model now considered, A's differential gains under the ideal tax are measured by the blacked-in triangle in Figure 2. Individual B may or may not favor the quota, as in the earlier model. His choice as between the two alternatives, the ideal tax on the one hand and the restriction to Q_c at price P on the other, will depend on the comparative sizes of the two areas shown as horizontally and vertically shaded in Figure 2. As drawn, he will tend to favor the quota scheme, but it is clearly possible that the triangular area could exceed the rectangular one if B's demand curve is sufficiently steep in slope. In any case, the choice alternatives for both persons are less different in the net than those represented by the ideal tax and the idealized quota.

While holding all of the remaining assumptions of the model, we now drop the assumption that no exchange of quotas takes place between A and B. To facilitate the geometrical illustration, Figure 3 essentially blows up the relevant part of Figure 2. With each party initially assigned a consumption quota of Q_c, individual A will be willing to sell units to individual B for any price above his marginal evaluation. Hence, the lowest possible supply price schedule that individual B confronts is that shown by the line RL in Figure 3. The maximum price that individual B is willing to pay for additional units of quota is his marginal evaluation, shown by SL. The gains-from-trade are measured by the triangular area RLS. The distribution of

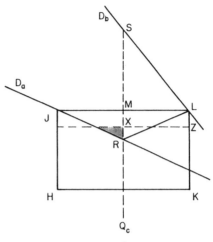

FIGURE 3

these gains will, of course, be settled in the strict two-man setting by relative bargaining skills, but let us assume that individual B, the buyer, wants to purchase consumption quota units from A, but also to do so in such a way that individual A will come to prefer this system over the tax. To accomplish this, he must insure that A gets a share of the net gains at least equal to the area RML on Figure 3. Individual B, the buyer, retains gains of MSL under this division of the spoils. But in this arrangement, both persons are indifferent as between the policy alternatives. The system is on the Pareto frontier, and the quota scheme plus the exchange process produces allocative and distributive results identical to those generated under the ideal tax. This becomes the analogue of the Coase theorem in the context that we are examining.[9]

V

These somewhat inconclusive results may seem to provide anything but a posi-

[9] See Ronald Coase. For a related extension of the Coase theorem, see Buchanan (1973).

146   **THE AMERICAN ECONOMIC REVIEW**   **MARCH 1975**

tive theory of policy akin to that presented with respect to production externalities. The comparisons are, however, a necessary stage in developing such a theory. Recall that we have made these comparisons under the most favorable possible assumption concerning anticipated return of revenues under the penalty tax. In the real world, individuals will not anticipate that these will be returned dollar-for-dollar, and they will tend to place at least some discount on the value of benefits that they expect.

Let us say that each person expects an aggregate benefit value of only 80 cents on the dollar from tax revenues collected under the penalty tax. Consider what this single change does to the results of the last comparison made, that which involves proportionate quota assignments along with a free market in quotas. In this case, individual B, the buyer, can offer individual A, the seller, more than the amount required to make him prefer the quota alternative, while himself continuing to secure differential benefit under this alternative. Individual A's differential gains from the ideal penalty tax are reduced to the shaded area in Figure 3. By paying individual A the amount measured by RML, he has improved A's position relative to the penalty tax. And, in the process, he has retained for himself a differential gain measured by the area $MXZL$. Both persons in full knowledge of the alternatives will prefer the quota system, and political leaders will presumably respond by opting for regulation.

The same reasoning can readily be extended to apply to any quota system. In the idealized quota assignment first considered, we demonstrated that one person would favor the penalty tax and the other the quota. Individual A, who favors the penalty tax, loses no consumer's surplus, and he does expect to secure an income

transfer through the return of tax revenues. When we modify the assumptions concerning expectations of the value of returned revenues or benefits, however, this conclusion need not hold. Individual A will, of course, expect to get back in benefits some part of the tax revenues paid in by B that is in excess of that contributed by A himself. If, however, individual A applies the same discount factor to all revenues collected, the deadweight loss may more than offset the income transfer effect. Examination of Figure 2 indicates that under the 80 percent assumption, one-fifth of the area measured by $PP'JH$ will represent deadweight loss to A from the revenues that he pays in. This deadweight loss may well be larger than the measure of the income transfer that he expects, which amounts to 80 percent of the horizontally shaded area in Figure 2. Once we introduce any plausible discount factor into the expectation of individuals concerning the return of tax revenues, it is relatively easy to demonstrate situations under which both persons may be led by private self-interest to favor the direct regulation alternative.

VI

We have developed a positive theory of externality control policy for both the production and consumption interactions under highly abstract and simplified models which allow us to isolate influences on policy formation which have been neglected. Decisions on the alternative policy instruments in democratic governments are surely influenced by the preferences of those who are subjected to them. The public-choice approach, which concentrates attention on the individual's choice as between policy instruments, allows us to construct hypotheses that explain the prevalence of direct regulation.[10] For economists who continue to support the penalty tax alternative, the analysis suggests that they had best become good Wicksellians and begin to search out and invent institutional arrangements that will make the penalty tax acceptable to those who are primarily affected.

[10] Much of the analysis developed in this paper can be applied more or less directly to policy alternatives proposed in the energy crisis of late 1973 and early 1974. For such application, see Buchanan and Nicolaus Tideman.

REFERENCES

W. J. Baumol, "On Taxation and the Control of Externalities," *Amer. Econ. Rev.*, June 1972, *62*, 307–22.

J. M. Buchanan, "Positive Economics, Welfare Economics, and Political Economy," *J. Law Econ.*, Oct. 1959, *2*, 124–38.

———, "The Coase Theorem and the Theory of the State," *Natur. Resources J.*, Oct. 1973, *13*, 579–94.

——— and N. Tideman, "Gasoline Rationing and Market Pricing: Public Choice in Political Democracy," research pap. no. 808231-1-12, Center for Study of Public Choice, Virginia Polytechnic Inst. and State Univ., Jan. 1974.

R. H. Coase, "The Problem of Social Cost," *J. Law Econ.*, Oct. 1960, *3*, 1–44.

O. A. Davis and A. Whinston, "Externalities, Welfare, and the Theory of Games," *J. Polit. Econ.*, June 1962, *70*, 241–62.

C. J. Goetz, "Political Equilibrium vs. Economic Efficiency in Effluent Pricing," in J. R. Conner and E. Loehman, eds., *Economic Decisionmaking for Environmental Control*, Gainesville 1973.

G. A. Hay, "Import Controls on Foreign Oil: Tariff or Quota?," *Amer. Econ. Rev.*, Sept. 1971, *61*, 688–91.

A. V. Kneese and B. T. Bower, *Managing Water Quality: Economics, Technology, Institutions*, Baltimore 1968.

F. J. Michelman, "Property Utility, and Fairness: Comments on the Ethical Foundations of 'Just Compensation' Law," *Harvard Law Rev.*, Apr. 1967, *80*, 1165–1257.

S. Wellisz, "On External Diseconomies and the Government-Assisted Invisible Hand," *Economica*, Nov. 1964, *31*, 345–62.

[18]

Differentiated Regulation: The Case of Auto Emissions Standards

Howard K. Gruenspecht*

Differentiated regulation occurs when a regulatory authority applies different regulatory criteria across subgroups of a regulated sector. An extreme form of differentiation, grandfathering, fixes the regulatory standard for each regulated unit as of its entry date. Grandfathering plays an important role in economic and social regulatory programs such as occupational licensure, building codes, and most product safety standards. Other programs, notably the environmental and occupational safety programs of EPA and OSHA, incorporate some retroactive regulation but generally apply tougher standards for newer facilities. Despite the widespread use of differentiated regulatory tools, little attention has been paid to their effect on investment and scrapping decisions in the regulated sector. Neoclassical capital theory suggests that bias against new sources in regulation will reduce investment in new facilities and lengthen the economic lifetime of old ones. Once the effect of differentiation on the composition of the capital stock is taken into account, tighter new source standards may actually be counterproductive, increasing the aggregate level of the regulated externality rather than reducing it.

This paper presents an analysis of automotive emissions regulation that explicitly considers the link between regulation and the composition of the vehicle stock. The adoption of more stringent emissions standards for new vehicles is shown to prolong the retention of old, high-emission-rate vehicles in the fleet. For this reason, tighter new source standards actually increase aggregate emissions in the short run. This result is of general as well as specific interest given the widespread use of differentiated regulatory

tools incorporating a bias against new sources in many regulatory programs.

I. An Economic Model of Scrapping

Regulation affects fleet composition through its impact on purchase and scrapping decisions. Consider the scrapping decisions made by an optimizing individual who consumes both nondurables and the service of autos and other durables. Suppose all vehicles of the same vintage and model (i.e., 1977 Ford Pintos) provide an identical service stream. Then, a necessary condition for optimality is that an individual car is scrapped if and only if its market value in operable condition, P_i, minus its scrap value, SV_i, is less than the cost of the repairs needed to restore it to operating condition, $PR*R$, where R and PR are, respectively, the repair quantity and the repair price. This "rational scrapping" condition is necessary for optimal consumption plans because any given capital service stream obtained without it can be duplicated at lower cost through a combination of rational scrapping and used car purchase. If all vehicles of each model and vintage draw their repair realizations at each date from a common repair incidence distribution, $f_i(R)$, and scrapping behavior is rational, the scrappage rate, d_i, for each vehicle type equals one minus the integral of the repair incidence distribution over the range where repair is rational:

$$(1) \qquad d_i = 1 - \int_0^{(P_i - SV_i)/PR} f_i(R)\,dR.$$

Given (1), d_i is inversely related to P_i holding the other prices fixed. This relationship, but not (1) itself, would survive the transition to a model that included transaction costs, information asymmetries and other relevant "real world" features.

*Graduate School of Industrial Administration, Carnegie-Mellon University.

How does this relate to regulation? New and used cars are substitutes. Any increase in the price or operating cost of new cars, including one attributable to regulation, causes substitution towards old ones. Starting from an initial equilibrium in all vehicle markets, such an increase would result in excess demand for used cars. As the supply of used cars is perfectly inelastic at any point in time, prices must rise to clear used car markets. If PR, SV_i and $f_i(R)$ are unaffected, scrapping rates for used cars fall.

The above discussion and the fuller development in my earlier study suggest a structural model in which used car prices, used car scrapping rates, and new car sales are all endogenous. In such a structural model P_i, SV_i, PR, and proxies for $f_i(R)$ alone would explain scrapping rates. New car price developments would affect scrappage rates solely through their influence on used car prices. However, the poor quality and limited coverage of the available used car price data led me to estimate a reduced-form model in which new car prices and other structural determinants of used car prices and new car sales enter the scrapping rate specification directly. The regression results, which are reported in my earlier study, were largely in accord with expectations. Scrappage rates for all age groups were inversely related to real new car prices, with the elasticity of scrappage rates with respect to real new car prices highest at the tails of the used car age distribution. One explanation of the high elasticity for the oldest cars is that their low market value causes the critical repair value, $(P_i - SV_i)/PR$, to fall in the region of minor repairs, where $f_i(R)$ is likely to be densest. Even a small change in P_i induces a significant scrapping response if $f_i(R)$ is locally dense.

How has regulation affected new car prices and user costs? In 1981 new car emissions standards for CO and NO_x were made more stringent. The CO standard was changed from 7.0 to 3.4 grams/mile, while the NO_x standard was changed from 2.0 to 1.0 grams/mile. The capital cost of the 1981 increment of regulation is approximately $350 per vehicle, a figure in the middle of the range defined by recent BLS, GM, and EPA

estimates. I assume, conservatively, that the only operating cost penalty associated with the 1981 regulatory increment stems from a loss in fuel economy, estimated to be 5 percent on the basis of comparisons between U.S. and Canadian cars identical in all respects except the level of emissions control. The present discounted value of the fuel economy penalty over the first four years of a vehicle's lifetime is $125. Thus, if new car buyers have a four-year horizon the total cost penalty associated with the 1981 regulatory increment is $475. Since the costs associated with previous increments of regulation have fallen over time, this cost penalty is assumed to fall by 50 percent beginning with the 1986 model year as fixed costs are amortized and technical progress occurs.

II. Emissions Results

The total cost penalty estimate may be used in conjunction with the scrappage elasticity estimates derived from the regression analysis to calculate the effect of the adoption of more stringent new car standards on scrapping rates. These changes are not large. However, given the wide disparity in emissions rates across vintages, they exert a significant effect on short run emissions outcomes.

Table 1 reports annual emissions and sales paths through 1990 for the current policy (CURRENT) and two hypothetical alternatives. The first, NONSPS, is identical to CURRENT except the switch to more stringent new car emissions standards beginning in the 1981 model year is omitted. The second, BOUNTY, omits the 1981 shift in standards and adds a $250 payment to owners of fifteen-year old cars who scrap their vehicles. This offer raises the vintage 15 scrappage rate from 27.5 to 75 percent. Emission paths for all three options are calculated using the EPA Mobile Source Emissions Model modified to take account of the link between regulation and fleet composition. Aggregate vehicle miles travelled (VMT) and the total number of vehicles are held constant across the model runs. The implied price elasticity of demand for new cars under this assumption and the scrap-

TABLE 1—AGGREGATE EMISSIONS AND SALES PERFORMANCE UNDER ALTERNATIVE REGULATORY POLICIES[a]
(Percentage differences from *CURRENT* policy)

	HC		CO		NOX		SALES	
Year	*NONSPS*	*BOUNTY*	*NONSPS*	*BOUNTY*	*NONSPS*	*BOUNTY*	*NONSPS*	*BOUNTY*
1981	0.83	1.32	0.70	−1.01	0.86	1.03	4.09	4.32
1982	−1.86	−2.42	−1.01	−1.51	5.47	5.47	3.35	3.59
1983	−2.29	−2.92	−1.60	−1.88	9.76	9.76	2.72	2.97
1984	−1.64	−2.80	−1.15	−1.71	14.35	14.13	2.20	2.44
1985	−0.52	−1.57	0.81	−1.27	18.39	17.93	1.84	2.09
1986	2.63	1.17	0.79	−0.14	23.00	22.52	−0.66	−0.40
1987	5.81	4.19	2.27	0.99	26.95	26.20	−0.32	−0.09
1988	9.54	8.13	3.57	2.68	29.87	29.35	0.09	0.33
1989	12.93	11.79	4.46	3.72	32.10	31.56	0.56	0.81
1990	16.19	14.98	5.31	4.51	33.33	32.80	0.93	1.19

[a] In all cases, the impact of fluctuations in sales not attributable to regulation has been suppressed. This procedure, which isolates the role of differentiated regulation, does not affect differences across alternatives, which are of primary interest here, but may cause discrepancies between simulated and actual levels.

page elasticity estimates derived from the scrappage regression is approximately 1, a value consistent with results obtained in the extensive literature on new car demand. This outcome suggests that the scrappage elasticity estimates used in this analysis are empirically reasonable.

The Table 1 emissions paths are interesting in several respects. First HC and CO emissions through 1984 are lower for *NONSPS* than for *CURRENT*, despite the fact that *CURRENT* entails a more stringent CO standard and the same HC standard as *NONSPS*. Although the advantage of *NONSPS* over *CURRENT* is not large in percentage terms, the fact that it exists at all is noteworthy given the large incremental investment in emissions control capital associated with the *CURRENT* policy over the 1981–84 period. Moreover, *BOUNTY*, which is designed to accelerate the retirement of high-emission-rate vehicles, yields lower HC and CO emissions than both *CURRENT* and *NONSPS* through 1985. In contrast to the $475 per new vehicle cost penalty associated with *CURRENT*, *BOUNTY* could be fully financed by a tax of approximately $20 on each new vehicle sold. Another advantage of "old source" approaches, such as *BOUNTY*, is that their parameters can be adjusted to take account of local conditions: regions with dirty air can offer higher retirement incentives or extend the offer to younger cars.

III. Conclusion

One lesson for regulatory analysts is that both technical factors, such as the emission rates for each type of capital, and economic factors, such as scrapping decisions, are important determinants of regulatory outcomes. Models that include only technical factors can yield seriously flawed projections. The present model also highlights the economic interaction among multiple externality regulation programs operating in the same sector. For example, if newer cars are more fuel efficient than older ones, the adoption of more stringent emissions standards, which reduces the replacement rate, works against the goal of reducing fuel consumption. This economic interaction is over and above the technical interaction between multiple objectives that has long been recognized in the engineering and policy literature.

Although tighter new source standards are ultimately reflected in lower aggregate emissions, their impact on investment plans can result in undesirable short-run outcomes. Clearly, other regulatory tools are needed as substitutes for or supplements to the differentially stringent regulation of new sources. Policies designed to promote the retirement of old capital are particularly attractive. Given the degree of differentiation in many current regulatory programs, such policies are likely to be the cheapest source of short-run externality reductions. In addition, they can serve as a substitute for a premature committment to a major new regulatory initiative, buying time for a close examination of its scientific basis and hopefully allowing for a reduction in the misallocation of scarce regulatory resources.

Where new source performance standards are the only tool available to regulators, regulatory decisions involve a tradeoff between short- and long-run externality levels as well as the usual tradeoff between costs and externality levels. Indeed, the short run can be quite long when the externality sources subject to regulation have very low rates of physical deterioration and technical obsolescence. New source regulation is particularly inappropriate in such cases; nevertheless it has been used extensively in both building codes and emissions standards for power plants, which obviously fit the above description. Moreover, some present regulations are completely unjustifiable once their impact on the composition of the capital stock is taken into account. For example, the legislatively mandated requirement that all new cars beginning with the 1984 model year be capable of meeting all emissions standards at high altitudes will raise the cost, and therefore the price, of new vehicles without improving their low altitude emissions performance. Since higher new vehicle prices increase the average age of the fleet through the mechanism outlined above, this standard will raise emissions in low altitude areas above the level that would prevail under the current and less costly standard that allows manufacturers to produce specially calibrated and equipped vehicles for high altitude areas with air pollution problems. As the emissions performance of any vehicle deteriorates with age, the aggregate emissions level depends on

the average age of the vehicle stock even if all vehicles in the fleet are subject to an identical standard when new. Therefore, the adverse effect of the 1984 requirement on the low altitude emissions level is a permanent one.

Finally, higher new source user costs resulting from differentiated standards are only one among many factors affecting investment and replacement decisions. Other factors, which are not under the control of the regulatory authority, can have important implications for emissions outcomes in view of the wide disparity in emissions performance across vintages. For example, if the new vehicle sales performance of 1978–79 had been duplicated in 1980–81, HC, CO, and NO_x emissions in 1981 would have been, respectively, 7.4, 7.2, and 2.4 percent lower than the levels attained under the actual sales history. When regulation incorporates a bias against new sources, an economic climate conducive to the rapid turnover of the capital stock furthers the achievement of our short-run regulatory objectives. A fuller discussion of the theory of differentiated regulation and the empirical application to automotive emissions controls is available in my earlier study.

REFERENCES

Gruenspecht, Howard K., "Differentiated Social Regulation in Theory and Practice," doctoral dissertation draft, Yale University, 1981.

Lave, Lester, "Conflicting Objectives in Regulating the Automobile," *Science*, May 22, 1981, 893–99.

Users Guide to Mobile 2, U.S. Environmental Protection Agency, EPA 460/3-81-006.

[19]

Standards versus Standards: The Effects of Different Pollution Restrictions

By Gloria E. Helfand[*]

When economists refer to pollution standards, they almost universally mean uniform restrictions on pollution emissions (e.g., William J. Baumol and Wallace E. Oates, 1975 Ch. 13; Susan Rose-Ackerman, 1973; David Besanko, 1978). However, in practice, standards take many forms: not only emissions restrictions, but restrictions on pollution per unit of output or per unit of an input, restrictions on the use of a polluting input, or mandated use of a particular pollution-control technology. These specifications of regulations will have a variety of effects on a firm's resource-allocation decisions. For instance, allowing pollution per unit of output as the standard instead of directly restricting the level of pollution gives the firm some choice: in addition to reducing pollution, it can increase its output to "dilute" the pollution.

This paper examines the effects of five different forms of pollution standards on input decisions, the level of production, and firm profits. The results show that the different standards, by providing firms with different incentives, change the firm's allocation decisions and affect the relative profitability of the standards. Section I reviews the existing literature on the incentives of different instruments. The following section describes the model to be used. Five different forms of pollution constraints are individually examined in Section III using a graphical approach. Their relative effects,

*Assistant Professor, Department of Agricultural Economics, University of California, Davis, Davis, CA 95616. I thank Peter Berck, Larry Karp, Jonathan Rubin, David Zilberman, Susanne Scotchner, seminar participants at the University of Illinois, the University of California at Berkeley, the University of California at Davis, and Duke University, and two anonymous referees for their helpful comments. This is Giannini Foundation Paper No. 950.

as well as some special cases, are assessed in Section IV. Finally, Section V raises the case of different standards when firms are not identical.

I. Different Standards and What Is Known of Them

The forms of standards used in actual regulations vary tremendously. A review of various state and federal regulations would show that standards are set in terms of such units as the total level of emissions per unit time, the amount of emissions per unit of output or input, requirements for certain pollution-control technologies, and restrictions on polluting inputs (Clifford S. Russell et al., 1986 pp. 17–19; Helfand, 1988 Ch. 2).

While the pollution-control literature includes analysis involving many of these forms, the correspondence between frequency of use in the literature and frequency of use in the standards is not perfect. For instance, while the literature includes examples of standards expressed as restrictions on output (e.g., James M. Buchanan and Gordon Tullock, 1975), no example of such a standard was found in any regulations reviewed. Many national regulations are set in terms of emissions per unit of output or input, but relatively few articles use that form of standard (Eithan Hochman and David Zilberman [1978] is an exception). Standards with such varied forms must inevitably provide firms with different incentives. Studies that examine the effects of a particular standard may not generalize to other forms of standards.

A few studies have examined pieces of this problem, but no one study has looked in a general way at the range of standards in use. Jon D. Harford and Gordon Karp (1983) compared the efficiency of standards expressed as pollution per unit of output

VOL. 81 NO. 3 *HELFAND: POLLUTION RESTRICTIONS* 623

and pollution per unit of input. Their model, which held output and total emissions fixed, found that the pollution-per-output standard was more efficient because it least distorted the input mix.

Vinod Thomas (1980) compared the welfare costs of an emissions standard to several other forms of regulation for the steel industry near Chicago: a set level of pollution-control expenditures, restricting the polluting input (fuel), and restricting emissions as a function of fuel. The welfare costs of these other policies, except the input restriction, were estimated to be 30–40 percent higher than those of the emissions standard, though only 1–1.4 percent of the value of steel output; direct restriction of the fuel input had a welfare cost of about 30 percent of the value of output. Thomas concluded that the choice of a standard could have a significant effect on the efficiency costs of environmental regulation.

Besanko (1987) analyzed the effects of restricting pollution (a "performance" standard) versus mandating a specific level of pollution-control technology (a "design" standard). For N symmetric Cournot firms, the performance standard resulted in higher profits for the individual firms, but output was higher under the design standard. If social welfare is defined as the sum of producer and consumer surplus, neither standard unambiguously produced higher welfare than the other: the greater output under the design standard enhanced consumer surplus, even as it reduced producer surplus.

The articles by Harford and Karp and by Besanko look only at subsets of possible standards, while the article by Thomas examines the effect of a range of standards only for one industry. Because the models used in these studies have different assumptions, the results are not directly related. The present study reviews a range of standards in the context of one model to compare their effects unambiguously.

II. The One-Firm Model

The model used here involves one firm, facing a horizontal output demand curve and using two inputs, x_1 and x_2, with horizontal supply curves. While the assumption that there are only two inputs is a simplification, made to keep the comparative statics and the graphical presentation simple, the problem is likely to generalize to N inputs with few differences. The assumption of a horizontal output demand curve is more limiting. Made to keep the problem tractable and to permit a graphical presentation, it is realistic only for a good whose world price is unaffected by production in this country. It deserves to be lifted in future work.

The firm is assumed to produce an output f using the production function $f(x_1, x_2)$. It is initially assumed that $f_i > 0$ for $i = 1,2;$[1] that is, both inputs contribute to production. Additionally, $f_{ii} < 0$ for both i, and the Hessian of the production function is negative definite; that is, both inputs and the production function are subject to diminishing marginal returns.

The firm also produces pollution $A(x_1, x_2)$. It is initially assumed that input 1 increases pollution (i.e., $A_1 > 0$), while input 2 decreases pollution ($A_2 < 0$). For instance, input 1 can be thought of as coal used in electricity generation, and input 2 as water used both to turn the turbines for electricity and to cool thermal discharges.

The firm is assumed to maximize profits while facing an output price p and input prices w_1 and w_2, corresponding to inputs 1 and 2, respectively. They are assumed to be fixed for this analysis.

Before a pollution-control constraint is imposed, the firm is assumed to maximize profits, π. The unconstrained problem becomes

$$(1) \quad \max_{x_1, x_2} \pi = pf(x_1, x_2) - w_1 x_1 - w_2 x_2$$

which results in a solution $x^0 = (x_1^0, x_2^0)$; it is unique, since the production function is strictly concave.

---

[1] Subscripts in this and future cases denote derivatives with respect to i.

624   THE AMERICAN ECONOMIC REVIEW   JUNE 1991

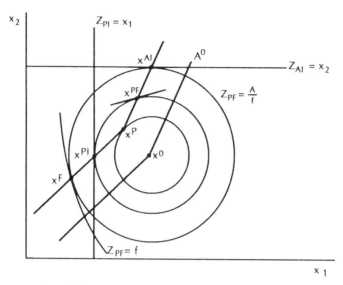

FIGURE 1. EFFECTS OF DIFFERENT POLLUTION-CONTROL STANDARDS

This result can be displayed on a diagram like one used for isoquant analysis, such as Figure 1. The polluting input, x_1, is measured along the horizontal axis, while the pollution-abating input, x_2, is measured along the vertical axis. The profit-maximizing combination of these inputs is found at x^0, which implicitly lies at the tangency of an isoquant $f^0 = f(x_1^0, x_2^0)$ with an isocost line whose slope is $-w_1/w_2$.

Because of the uniqueness of this maximum, isoprofit contours (i.e., contours representing input combinations yielding the same level of profits) can be drawn on the diagram as well. That these contours are convex rings can be derived from the assumed strict concavity of the production function. The maximum-profits point x^0 can be considered the top of a profits hill. Isoprofit contours away from this point are decreasing levels of profits with movement down the profits hill.

An isopollution line (i.e., a line representing input combinations giving a fixed level of pollution) can be added to this diagram. Totally differentiating the function $A^0 = A(x_1, x_2)$ gives the slope of this line as $dx_2/dx_1 = -A_1/A_2$, which is positive since $A_1 > 0$ and $A_2 < 0$. In other words, as use of the polluting input increases, use of the abating input must also increase in order to maintain a constant level of pollution. Isopollution contours above and to the left of another contour represent lower levels of pollution: for a given amount of the pollution-abating input, pollution decreases as use of the polluting input decreases. In Figure 1, the initial level of pollution is represented by the line $A^0 = A(x_1^0, x_2^0)$. The line A^P, above and to the left of the original contour, represents a lower level of pollution.

III. The Constraints

Five different kinds of pollution-control standards will be examined here: a fixed level of emissions, a fixed level of emissions per unit of output, a fixed level of emissions per unit of an input, a fixed level of output, and a fixed level of an input.[2] In two of these cases (pollution per input and fixing the level of an input), two subcases will be analyzed, reflecting the different inputs used here. These formulations include most of the standards analyzed by Thomas (1980), Harford and Karp (1983), and Besanko (1987), as well as others used in either the pollution-control literature or environmental regulation.

One way to analyze the effects of each constraint relative to the baseline of no regulation is to set up the profit-maximization problem, differentiate totally the first-order conditions, and use comparative statics to analyze the effects of the constraint on the level of input use, output, profits, and pollution. Identical results can be derived from a graphical analysis. Since a graphical approach is easier to interpret, it will be used to present the results; the comparative-statics analysis can be obtained from the author.

A. *Standard as a Set Level of Emissions*

Let Z_P be the numerical standard set when emissions are regulated by the amount of total pollution permissible in a certain period of time. It can be represented as a constraint on the profit function with the form $A \leq Z_P$.

Figure 1 presents the graphical interpretation of this problem. The point $x^0 = (x_1^0, x_2^0)$ is the previously described unconstrained-maximum-profits locus. The addition of the constraint requires that the firm

choose inputs that will place it on a leftward isopollution line, such as A^P. Since the firm will nevertheless seek to maximize profits, it will choose the point of the new isopollution line that achieves the highest level of profits: the point where the isopollution line is tangent to the highest attainable isoprofit line. This point is represented by $x^P = (x_1^P, x_2^P)$.

In general, as Figure 1 shows, use of the polluting input should decrease, and use of the pollution-abating input should increase. These results are expected, since the purpose of the constraint is to reduce pollution. However, use of the polluting input may actually increase, or use of the abating input may decrease. These seemingly perverse results depend on the actual shape of the isoprofit contours: the tangency between the isoprofit and isopollution line may lie above and to the right, or below and to the left, of x^0. Mathematically, these results depend on the sign and magnitude of the f_{12} term, the effect on the marginal product of input 1 as use of input 2 changes. All that is known theoretically about this term is that $f_{11}f_{22} > (f_{12})^2$ for a strictly concave production function. This information is not sufficient to determine more about the effects of this constraint on input use.

Because use of one input is increasing and use of the other is decreasing, the effects on production are ambiguous. Profits are unambiguously lower under this standard, since the firm can always do better in an unconstrained situation than in a constrained one. Pollution, by definition of the constraint, is reduced.

B. *Standard as Emissions per Unit of Output*

Let Z_{PF} be the standard expressed as a set level of pollution per unit of output, or $A/f \leq Z_{PF}$. Totally differentiating the constraint equation gives the slope of the constraint line, $dx_2/dx_1 = -(fA_1 - f_1A)/(fA_2 - f_2A)$. The denominator is negative according to the assumptions on signs; $fA_1 - f_1A$ is positive if $A_1/(A/x_1) > f_1/(f/x_1)$. If A is convex in input 1 (indicating that

---

[2]These standards are not always as different as they first appear. For instance, a regulatory agency may develop a standard based on a particular cleanup technology (fixing an abating input) but express that standard in other units, such as pollution per unit of output. Indeed, the technology standard may de facto be mandated by the pollution-per-output standard. This analysis assumes that expression of the standard as pollution per unit gives the firm the choice of using alternative approaches.

pollution increases more rapidly as use of the polluting input increases), the marginal amount of pollution is greater than the average amount at any given point, and $A_1/(A/x_1)$ is greater than one. On the other hand, f is assumed to be concave in input 1; by analogous reasoning, the marginal product is less than the average product, and the ratio is less than 1. Therefore, $fA_1 - f_1A$ is positive, and the slope of this standard is positive.

The slope of this constraint line can be compared to the slope of the isopollution contour by subtraction:

$$\frac{dx_2}{dx_1}(Z_{PF}) - \frac{dx_2}{dx_1}(Z_P) = \frac{(f_1 A_2 - f_2 A_1)}{(fA_2 - f_2 A)A_2} < 0$$

by the assumed signs of these terms. Though this line slopes upward, it slopes upward at a lesser angle than does the isopollution contour.

This new constraint is also illustrated in Figure 1. If the standard is normalized to achieve the same level of pollution as standard P, its constraint line must intersect the isopollution line at the constraint's tangency with the highest possible isoprofit contour. This fact, plus the fact that its slope is shallower than that of the isopollution line, can establish the location of this constraint. If it were along the same isoprofit contour as is standard P, then the tangency would be to the right of the tangency for standard P, x^P; however, if it were along that isoprofit contour, then it could not be on the same isopollution line, since it would lie above that isoprofit contour except at x^P. Therefore, the combination of inputs that maximizes profits while achieving standard PF, $x^{PF} = (x_1^{PF}, x_2^{PF})$, is above and to the right of the combination for standard P along the isopollution line A^P.

In the "normal" case, use of the polluting input should drop, and use of the abating input should increase. As with the pollution standard, though, it is possible that use of

the polluting input may increase or that use of the abating input may decrease. These results depend, as with the pollution constraint, on the sign and the magnitude of the f_{12} term. The effects on production remain ambiguous; profits unambiguously decrease. Finally, the effects of this standard on pollution levels are somewhat ambiguous, depending on the sign and magnitude of the f_{12} term: if production increases more rapidly than pollution, then this constraint could lead to the perverse result that pollution increases with its imposition.

C. *Standard as Emissions per Unit of a Specified Input*

The mathematical representation of this standard is $A/x_j \leq Z_{PJ}$, $J = 1$ or 2, where J is the subscript for the input in terms of which pollution is measured. Two cases are possible here: regulating pollution per unit of a pollution-causing input, such as restricting the amount of sulfur dioxide emissions per ton of coal used for electricity; or regulating pollution per unit of a pollution-reducing input, such as regulating biological oxygen demand in water pollution per unit of water in a production process. Wesley A. Magat et al. (1986 p. 35) note that the Environmental Protection Agency (EPA) preferred a pollution-per-output standard to the latter form, out of concern that plants would meet the standard solely by dilution and not by cleanup.

Let PPI represent the situation in which pollution is regulated per unit of the polluting input, and let PAI represent pollution regulated in terms of the pollution-abating input. The slope for $Z_{PPI} = A/x_1$ is $-(A_1 x_1 - A)/A_2 x_1$ which is positive if $A_1 > A/x_1$. By the same reasoning used for $f_1 A - fA_1$ in the pollution-per-output case above, the marginal pollution at any given level of inputs is greater than the average level if A is convex in the polluting inputs. Thus, $A_1 x_1 - A$ is positive, and this slope is positive. For $Z_{PAI} = A/x_2$, the slope is $-A_1 x_2/(A_2 x_2 - A) > 0$. As with pollution per output, the slopes of these constraint lines can be compared with that of the isopollution line:

$$(2) \quad \frac{dx_2}{dx_1}(Z_{PPI}) - \frac{dx_2}{dx_1}(Z_P) = \frac{A}{A_2 x_1} < 0$$

$$\frac{dx_2}{dx_1}(Z_{PAI}) - \frac{dx_2}{dx_1}(Z_P)$$

$$= \frac{AA_1}{A_2(A - A_2 x_2)} < 0.$$

Clearly the slopes of the constraint lines are less than the latter slope. However, no clear comparison of slopes is possible between these two standards or between either of these standards and the pollution-per-output standard: none of the following comparisons can be signed:

$$(3) \quad \frac{dx_2}{dx_1}(Z_{PPI}) - \frac{dx_2}{dx_1}(Z_{PAI})$$

$$= \frac{A(A_1 x_1 + A_2 x_2 - A)}{A_2 x_1(A_2 x_2 - A)}$$

$$\frac{dx_2}{dx_1}(Z_{PPI}) - \frac{dx_2}{dx_1}(Z_{PF})$$

$$= \frac{A[A_2(f - f_1 x_1) + f_2(A_1 x_1 - A)]}{A_2 x_1(fA_2 - f_2 A)}$$

$$\frac{dx_2}{dx_1}(Z_{PAI}) - \frac{dx_2}{dx_1}(Z_{PF})$$

$$= \frac{A[A_1(f_2 x_2 - f) + f_1(A - A_2 x_2)]}{(A_2 x_2 - A)(fA_2 - f_2 A)}.$$

The numerator of the slope difference between standards PPI and PAI would be zero if the pollution function had constant returns to scale (since, by Euler's law, $A = A_1 x_1 + A_2 x_2$ if the function has constant returns to scale); however, without assumptions about the returns to scale of this function, it cannot be signed. It nevertheless indicates that these returns determine any

differences between these functions: increasing returns makes the PAI standard line steeper, and decreasing returns make PPI have a steeper slope. Somewhat similarly, if both the production and the pollution functions have constant returns to scale, then the numerators of the comparisons between either pollution-per-input standard and the pollution-per-output standard would be zero as well.[3] Thus, comparison of either standard with the constraint line for standard PF requires further knowledge of the relative returns to scale for both the pollution and the production functions.

Because the constraint lines of these three standards (pollution per unit of output and pollution per unit of either input) cannot be readily distinguished from each other, the remainder of this analysis will not distinguish among them. These three formulations of the standard will be referred to as the "dilution" standards, since they all involve measuring pollution diluted by either output or input. The point x^{PF} in Figure 1 thus reflects the effects of standards PPI and PAI as well as it reflects the effects of standard PF.

D. *Standard as a Set Level of Total Output*

This standard forces the firm to reduce output; input substitution alone is inadequate. The constraint is now $f \le Z_F$. This formulation of the standard mandates that the firm operate on an isoquant closer to the origin than the profit-maximizing isoquant. The firm will operate with the input mix given by the point of tangency of that downward-sloping isoquant with the highest possible isoprofit curve, represented by the point $x^F = (x_1^F, x_2^F)$. Because of the negative slope of the isoquant, this tangency must be below and to the left of x^P.

---

[3] With constant returns, $f - f_1 x_1 = f_2 x_2$, and $A_1 x_1 - A = -A_2 x_2$; the numerator of $(dx_2/dx_1)(Z_{PPI}) - (dx_2/dx_1)(Z_{PF})$ becomes zero. A similar analysis can be done for the terms in the comparison of standard PAI and standard PF.

Use of both inputs will generally decrease, though as before, exceptions are possible, depending on f_{12}. By definition, output decreases, and profits decline as usual, because the firm is constrained away from the optimum.

It should be noted that pollution need not decrease with this standard: the new optimal input mix could end up on an isopollution curve either to the left or to the right of the initial curve, depending on whether the firm uses the polluting input or the abating input more intensively as production is reduced. It is assumed in the drawing of Figure 1 that the lower level of pollution will be achieved by a reduction in output.

E. *Standard as a Set Amount of a Specified Input*

This standard takes two forms. A maximum can be set on the use of a polluting input (standard PI); alternatively, imposing a minimum level on the use of a pollution-abating input (standard AI) captures the effect of imposing a particular pollution-control technology on a firm. Standard PI will be represented as $x_1 \le Z_{PI}$, and standard AI will be represented as $x_2 \ge Z_{AI}$.

Graphically, the limitation on use of the polluting input is a vertical line at the chosen level of input 1; the requirement for a minimum use of the pollution-abating input results in a horizontal line at the chosen level of input 2. The vertical slope of standard PI places its optimum, x^{PI}, along A^P between x^P and x^F. The zero slope of standard AI places x^{AI} above and to the right of x^{PF} along A^P. These are shown in Figure 1.

For case PI, the limit on use of the polluting input obviously causes a reduction in the use of input 1. The effect on input 2 is ambiguous, depending entirely on the sign of f_{12}. Production will be lower if use of input 2 is not increased much, is unaffected, or decreases. Profits are obviously reduced, as in all cases. Finally, as with the output standard, pollution could either increase or decrease with the formulation of the standard, depending on the slope of the isopol-

lution line and the shape of the isoprofit contours.

For case AI, use of input 1 may either increase or decrease, depending on the sign of f_{12}. If f_{12} is nonnegative or not very negative, then production will increase through use of this standard, in contrast to the effects of standard PI. Profits decrease as usual. Finally, pollution could either increase or decrease with the use of this standard.

In sum, these different forms of pollution constraints have some similar patterns, but each has its own peculiarities. While standard AI, the minimum requirement on the abating input, may actually increase output, standard PI, the maximum requirement on the polluting input, and standard F, the restriction on output, should decrease production. The other standards have ambiguous effects on output, since they all tend to decrease use of the polluting input and to increase use of the pollution-abating input increase. Pollution unambiguously decreases only for standard P and for the dilution and input standards if f_{12} is "small" (i.e., small enough that terms involving it do not affect the sign of the function as a whole) or zero. For standard F, the effects of the restrictions on pollution are completely ambiguous. For all standards except AI, use of the polluting input is expected to drop as the pollution-control constraint is tightened; for all standards except F and PI, use of the pollution-abating input should increase.

IV. Comparisons of the Different Standards

The previous section describes the results of the different standards as they are imposed. As noted, most of the results have the same sign and are therefore initially indistinguishable. However, if the standards are normalized to achieve the same level of pollution, as they have been in Figure 1, comparisons among the standards can easily be made.

This normalization is necessary because the standards differ in form—indeed, even in units of measurement. A change of a given magnitude in one standard is not equivalent to a change of the same magnitude in another standard. For instance, changing Z_P by one unit will change the level of pollution by one unit; however, changing Z_{PF} by one unit will result in a one-unit change in pollution per unit of output, and the results are not obviously comparable. The standards have to be normalized to reflect a common level of change. The normalization used here is to make each standard cause an identical change in the amount of pollution. The standards have been ordered along the isopollution line A^P by comparing the slopes implied by the different standards, as already discussed.

This procedure reveals a strict ordering among the standards for levels of input use and levels of output as well as some information on relative profits. The standard that most reduces input use and output levels is standard F, the restriction on output, followed by standard PI, the restriction on the polluting input. Standard PI gives the firm higher profits than does standard F. Although standard P, the restriction on pollution itself, gives the highest level of profits among any of these standards, it has lower levels of input use and output than do the dilution standards, which in turn have lower levels of input use and output (though higher profits) than result from mandating a minimum amount of the pollution-abating input.[4]

In sum, even if the individual comparative-statics results for almost all the standards are ambiguous, the orderings among the standards are clear. Though a direct restriction on pollution offers the highest profit to a firm, mandating a minimum amount of the pollution-abating input leads to the highest production, while restricting output or the polluting input causes lower levels of production.

---

[4]The relative levels of profits between standards F and PF, standards F and AI standards PI and PF, and standards PI and AI, cannot be determined from this analysis. On the diagram, standards F and AI and standards PI and PF are drawn on the same isoprofit contours only to keep the diagram simple.

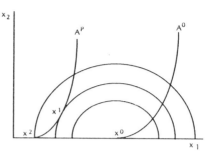

FIGURE 2. EFFECTS OF DIFFERENT STANDARDS WHEN THE ABATING INPUT DOES NOT CONTRIBUTE TO PRODUCTION

Special Cases

These findings depend on some specific assumptions underlying the model. In particular, the assumption that the second input contributes both to production and to pollution abatement affects the differences among the standards studied here. Analysis of two special cases of the model provides some interesting contrasts with these findings.

(i) $f_2 = 0$.—In this case, it is assumed that the pollution-abating input does not contribute to production; rather, it only affects pollution. For instance, x_2 may be a pollution device attached to a smokestack that does not otherwise intervene in the production process. The isoquants are now vertical lines (since production only depends on input 1), with the unconstrained profit-maximizing point on the x_1 axis (i.e., where no use of x_2 is required, since x_2 does not contribute to production and is costly). As seen in Figure 2, assuming $f_2 = 0$ reduces the standards to two forms: x^1, which represents the effects of a restriction of pollution (Z_P), of pollution per output (Z_{PF}), of pollution per input (Z_{PPI} or Z_{PAI}), or a required amount of the abating input (Z_{AI}); and x^2, which results from either a restriction on the polluting input (Z_{PI}) or a restriction on output (Z_F) (since output results only from this input). Similarly to the

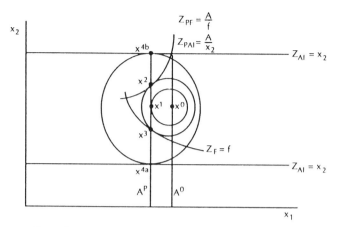

FIGURE 3. EFFECTS OF DIFFERENT STANDARDS WHEN THE SECOND INPUT
DOES NOT ABATE POLLUTION

basic model, the standards yielding x^1 lead to higher levels of input use, output, and profits than the standards yielding x^2.

(ii) $A_2 = 0$.—It is now assumed that the second input in the process does not abate pollution but has a neutral effect; the only way to reduce pollution is to reduce the level of the polluting input. As shown in Figure 3, the isoprofit contours are the same as in the basic model; however, the isopollution lines are vertical, since they are affected only by use of x_1. Now, with all standards normalized to achieve A^P, restricting pollution per unit of output (Z_{PF}) and restricting pollution per unit of the second input (Z_{PAI}) have the same effect, a high level of production (x^2); restricting pollution (Z_P), restricting pollution per unit of the polluting input (Z_{PPI}), and restricting the polluting input (Z_{PI}) lead to the highest profits (x^1); and restricting output (Z_F) leads to lower levels of production (x^3). In interesting contrast to the basic case, a positive f_{12} (at point x^{4a}) causes a mandate on the second input (Z_{AI}) to have the most severe output reduction: since x_2 now does not affect pollution directly, it can only achieve the standard by inducing a reduction in use of x_1. If f_{12} is negative (at x^{4b}),

this standard leads to the greatest increase in production, again by inducing a change in the level of x_1.

These special cases demonstrate the dependence of these results on the assumptions of the model. If the abating input does not affect production (as would, for instance, a device attached only to a smokestack or an outfall pipe), then the standards collapse into two different levels of effects. If the only way of abating pollution is to reduce the level of the polluting input, then a range of standards is maintained, though it is not the same range as in the basic model. Any application of this model should consider the specific assumptions of the case to be analyzed, in order to account for these differences.

V. The Case of Heterogeneous Firms

The original analysis assumed that an individual firm was being regulated and that the different standards would all result in the same total level of pollution. While this analysis provides insights into the incentives of these different standards, it is not realistic. In fact, the EPA and state regulatory bodies usually set standards that apply uniformly to firms of a specific industry. Within

VOL. 81 NO. 3   *HELFAND: POLLUTION RESTRICTIONS*   631

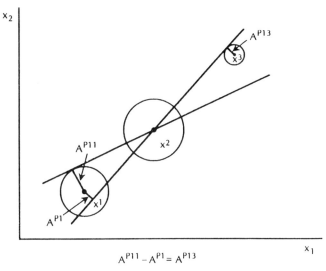

$$A^{P11} - A^{P1} = A^{P13}$$

FIGURE 4. UNIFORM STANDARDS WITH HETEROGENEOUS FIRMS

an industry, individual firms are likely to vary by technology, size, or other factors. Given that only one form of standard is used to regulate an industry, it is unlikely that the results for the individual-firm case will carry over to the case of heterogeneous firms. This analysis will sketch some of the issues involved in choosing a standard for a number of different firms.

Assume that an industry contains three firms, all of which have the characteristics of the individual firm in the previous analysis, but which differ in optimal scale. In other words, firm 1's profit-maximizing level of output (x^1) uses less x_1 and less x_2, and produces less output, than firm 2's profit-maximizing level of output (x^2), which in turn uses less x_1 and x_2 and produces less output than firm 3's profit-maximizing level of output, (x^3) (see Fig. 4).

Further assume that the pollution function $A(x_1, x_2)$ previously described is the same for all three firms. That is, only the inputs cause or abate pollution; the size of the firm itself does not have a direct effect on how much a firm pollutes. Thus, if firm 1

chose to operate with the same input mix as firm 2, it would pollute the same amount as firm 2. Under normal operating conditions, though, the level of pollution will change with the firm, since the firms use different levels of x_1 and x_2.

This assumption simplifies a graphical analysis. Because most of the standards addressed in this study (in particular, the pollution standard, the pollution-per-input standards, and the input standards) rely only on the levels of the inputs and the pollution function; they do not depend on the level of output and, therefore, do not depend on the technology of the firm. As a result, one constraint can be drawn onto an isoquant diagram and will apply equally to all firms.[5]

---

[5]The pollution-per-output standard and the output standard will clearly have effects that depend on what technology a firm uses. For instance, these firms may have isoquants that cross each other. With these two standards, a single constraint line cannot be used, complicating a graphical analysis. For present purposes, these constraints will not be considered. While ignoring the output constraint does not have significant

Because the pollution-per-input standard[6] is commonly used, while the pollution standard is heavily used in the economics literature, this analysis will focus on these two standards.

Firms are in compliance with the standards if they lie above or to the left of the constraint lines appropriate for the form of standard chosen. They initially violate the standards and must readjust their operations if they lie below or to the right of the constraint lines.

Figure 4 shows the constraint lines for these two standards. As in the single-firm analysis, the pollution constraint slopes upward, while the pollution-per-input standard slopes upward but more shallowly. (These relationships still hold, since both of these constraints depend only on the levels of input use.)

Where a firm lies on the diagram determines whether it has to adjust its production to comply with the constraints. In this diagram, firm 2 is in exact compliance with both standards; firm 1 satisfies the pollution standard but not the pollution-per-input standard; and firm 3 meets the pollution-per-input standard but not the pollution standard.

In the one-firm analysis, all the standards were oriented to produce the same total emissions from the one firm. Here, instead of making emissions per firm constant, the total level of pollution from all the firms will be held constant. This analysis will describe how to make such a normalization.

Let Z^P be the level of pollution specified by the pollution standard, and let A^{ij} represent the absolute value of the deviation from Z^P for firm j ($j = 1, 2, 3$) under standard i ($i = P, PI$; where PI is pollution per unit of input for this analysis). Under the

pollution standard, firm 3 will have to reduce its pollution to Z^P to meet the standard, while firm 2 already pollutes exactly Z^P. Firm 1 already has lower emissions than this standard: it underpollutes by A^{P1}. Thus, total pollution under the pollution standard (TP_P) is $TP_P = 3Z^P - A^{P1}$.

Under the pollution-per-input standard, firm 1 has to clean up to meet the standard, even with its relatively low emissions. In contrast, firm 3 satisfies the standard and does not have to abate its emissions, though it pollutes more than Z^P. Total emissions under this standard (TP_{PI}) are

$$(4) \qquad TP_{PI} = 3Z_P - A^{PI1} + A^{PI3}.$$

For total emissions to be the same under both standards, then $TP_P = TP_{PI}$, or

$$(5) \qquad A^{PI1} - A^{P1} = A^{PI3}.$$

This normalization is approximated in Figure 4. In general, an exact normalization would require more complete knowledge of how changing the level of pollution changes the distance between the isopollution lines.

The effects of the different standards on the levels of input use, output, and profits for an individual firm relative to the nonpollution baseline follow the patterns of the one-firm analysis, with a notable exception: the small firm is completely unaffected by the pollution standard, while the large firm must adjust its behavior; in contrast, the pollution-per-input standard forces the small firm to adjust its behavior and leaves the large firm unaffected.

Comparisons between the standards are more interesting, but they are also more difficult to make, since they depend heavily on the shapes of the isoprofit contours. In the one-firm analysis, the pollution-per-input standard uses more of both inputs than the pollution standard, resulting in more output, though it gives the firm lower profits. With heterogeneous firms, however, similar conclusions cannot be drawn. Indeed, virtually any change in aggregate input use is possible. For instance, the shallower slope of the pollution-per-input standard may appear to induce less reduction in the polluting input and more increase in the

---

policy results, since this standard is not commonly used, further analysis of the common pollution-per-output standard is desirable.

[6]As seen in the one-firm analysis, the effects of measuring pollution per unit of the abating input are difficult to distinguish from the effects of measuring pollution per unit of the polluting input. For the purposes of this analysis, no distinction is drawn between the two.

abating input for the small firm than the pollution standard induces for the large firm. However, the "perverse" cases of the one-firm analysis can reverse this appearance. If the large firm increases its use of the polluting input or the small firm reduces its use of the abating input, then the relative change in total use of the inputs under the two standards cannot be determined. The shapes of the isoprofit contours for the firms, as well as the exact slopes of the two constraints, will determine the relative effects.

Because input comparisons cannot be made definitively, the effects of the different standards on output cannot be assessed. Finally, the relative effects on aggregate profits cannot be determined. If one firm has a shallower "profit hill" than the other (i.e., if profits for one firm decrease less rapidly than for the other when the profit-maximizing input mix is no longer available), then that firm is less likely to be severely hurt by either pollution standard. There is no obvious reason for smaller firms to have steeper profit hills than larger firms; as a result, no clear comparison can be made.

Perhaps the major conclusion from Figure 4 is that the effects of the different standards depend on the size of the firm in question. If pollution increases with output, even if not in direct proportion, then the large firm is more likely to have to adjust to a pollution standard than will the small firm.[7] In contrast, a pollution-per-input standard "dilutes" the effect of pollution increasing with scale, because use of the input also increases with scale; the large firm is thus less likely to be penalized under this standard than under the pollution standard. The small firm may violate the pollution-per-input standard, even though it pollutes relatively small amounts.

[7]As drawn, pollution does increase with output, since the expansion path for firm scale has a flatter slope than the isopollution line. For output to increase while pollution declines, the slope of this expansion path would have to be steeper than the slope of the isopollution line. While such a scenario is possible if the expansion path is relatively intensive in input 2, it is generally more likely that larger firms pollute more than smaller firms.

These two standards provide different incentives for firms entering and exiting the industry. If pollution does increase with the size of the firm, then the pollution standard will put more pressure on large firms; they are more likely to exit, and small firms are more likely to enter. In contrast, the pollution-per-input standard will lead to larger firms entering the industry and smaller firms exiting. Ironically, this standard could lead to increased pollution, since large firms can comply with the pollution-per-input standard and still pollute more than small firms. The PI standard would have to become increasingly strict to maintain a lower level of emissions, which would in turn induce new firms to be even larger. The uniform pollution standard would have to be adjusted as the number of firms in the industry changed, but it would obviously not lead to more pollution per firm.

VI. Summary and Conclusions

This study has reviewed the effects of five different forms of pollution-control standards. It has shown in the basic model that a direct restriction on pollution leads to the highest level of profits and efficiency when all pollution standards are set to achieve the same level of emissions; a mandate for use of a pollution-abating input leads to highest output, followed by pollution per unit of output or input; and a restriction on production most reduces output, followed by a restriction on the polluting input. Besanko (1987) similarly found that firms profit most from a pollution restriction, though a mandate for an abating input might increase social welfare by increasing output. Thomas's (1980) simulation found the welfare costs of a pollution restriction to be lower than those of other standards, especially those of a mandate on the polluting input. If the results of this study apply to his model as well, then the combination of reduced profits for a firm and reduced output lead to more severe effects by reducing both consumer and producer surplus. Finally, Harford and Karp (1983) focused on comparisons of a pollution-per-output standard versus pollution per input and found the former to be more efficient than the latter, since it leads to less distortion in input use. The present study, in contrast, found no easily distinguishable differences between these standards (without assumptions on the relative returns to scale of the production and pollution functions): both produce more output than the pollution standard, though both also produce lower profits. The disparity between studies is probably caused by Harford and Karp's assumption of fixed output: while only input ratios could be inefficient in their study, the output level can be inefficient here as well. Given that the results of the other studies are all derived from different models, the similarities in the results outweigh the differences.

The present analysis has also shown that altering the underlying assumptions of the model affects the comparisons of the standards. Any application of this model must consider the specific characteristics of the case to be studied in order to account for the differences caused by different assumptions.

Finally, when a single regulation is imposed on an industry composed of nonidentical firms, the chosen standard will not affect all firms identically. The scale, input intensity, and pollution intensity of the firm, among other factors, will determine whether it will satisfy various standards. The chosen standard can thus have an effect on the optimal structure of the firm and may play a role in the entry and exit decisions of firms.

REFERENCES

Baumol, William J. and Oates, Wallace E., *The Theory of Environmental Policy*, Englewood Cliffs, NJ: Prentice-Hall, 1975.

Besanko, David, "Performance versus Design Standards in the Regulation of Pollution," *Journal of Public Economics*, October 1987, *34*, 19–44.

Buchanan, James M. and Tullock, Gordon, "Polluters' Profits and Political Response: Direct Controls versus Taxes," *American Economic Review*, March 1975, *65*, 139–47.

Harford, Jon D. and Karp, Gordon, "The Effects and Efficiencies of Different Pollution Standards," *Eastern Economics Journal*, April-June 1983, *9*, 79–89.

Helfand, Gloria E., "Standards versus Standards: The Incentive and Efficiency Effects of Pollution Control Restrictions," Ph.D. Dissertation, Department of Agricultural and Resource Economics, University of California at Berkeley, 1988.

Hochman, Eithan and Zilberman, David, "Examination of Environmental Policies Using Production and Pollution Microparameter Distributions," *Econometrica*, July 1978, *46*, 739–60.

Magat, Wesley A., Krupnick, Alan J., and Harrington, Winston, *Rules in the Making: A Statistical Analysis of Regulatory Agency Behavior*, Washington, DC: Resources for the Future, 1986.

Rose-Ackerman, Susan, "Effluent Charges: A Critique," *Canadian Journal of Economics*, November 1973, *6*, 512–28.

Russell, Clifford S., Harrington, Winston, and Vaughan, William J., *Enforcing Pollution Control Laws*, Washington, DC: Resources for the Future, 1986.

Thomas, Vinod, "Welfare Cost of Pollution Control," *Journal of Environmental Economics and Management*, June 1980, *7*, 90–102.

Part III
Implementation Issues

[20]

COMPARATIVE ANALYSIS OF ALTERNATIVE POLICY INSTRUMENTS

PETER BOHM

University of Stockholm

and

CLIFFORD S. RUSSELL*

Resources for the Future, Washington

1. Introduction

The choice of instruments for environmental policy implementation has had a special place in applied economics since the 1920s, when Pigou suggested the use of taxes on negative external effects and subsidies on positive external effects to correct allocative distortions. This is understandable, for at least at first glance it is a problem that appears to offer a nearly perfect target for our skills. Because of the importance here of external effects and public goods, and because the policies and thus the associated implementation strategies have had to be devised *de novo*, it has seemed an area to which economic insights, independent of other disciplines and unfettered by tedious historical baggage, could make very great contributions.

To some extent, of course, this has been true. Sophisticated theoretical work has contributed to the understanding of the characteristics of particular instruments. Empirical studies have produced estimates of the actual cost advantages to be expected from the adoption of instruments favored by economists instead of those being put in place by policy-makers. But therein lies the rub; those policy-makers have for the most part stubbornly refused to accept and act on the basis of the theory offered and the supporting empirical work. Overall, economists seem to have been perceived as gadflies, ignoring or misunderstanding the real situation and thus producing largely irrelevant criticisms of the instruments actually chosen, along with impractical, even politically dangerous, prescriptions for change. (Although see the comments on European experience below.)

* The second author is grateful for support from the Alfred P. Sloane and Andrew W. Mellon Foundations for his work on alternatives to direct regulation in environmental policy. Both authors wish to thank Alan Kneese, John Mullahy, and especially T.H. Tietenberg for helpful critiques of earlier drafts.

Handbook of Natural Resource and Energy Economics, vol. I, edited by A.V. Kneese and J.L. Sweeney
© *Elsevier Science Publishers B.V., 1985*

As with all such standoffs between the research and policy worlds, some truth resides with both sides. As this review will seek to show, economists have achieved some of their fundamental results by ignoring crucial features of the physical world and by abstracting from the full complexity of the economic world. These concessions to simplicty have made most of the arguments that are easiest to explain, and hence potentially easiest to sell, if not wrong at least seriously misleading. Further, and whether rightly or wrongly, economists seem to have refused to take seriously the political implications of some of their favorite prescriptions. These implications include both straightforward matters of cost distribution and more subtle problems of ethical content. At the same time, the developers and supporters of the regulatory systems currently in place in the United States and many other industrial countries have themselves been guilty of misleading arguments, and some of these will be pointed out in what follows. The core of good sense in economic criticisms of command and control regulation and in economic prescriptions for more flexible incentive systems should not permanently be obscured by the rhetorical flourishes of those who favor systems with strong and explicit moral overtones or who have narrower interests in the evolving status quo.

The structure of this chapter is designed to accomplish four specific goals as part of our broader aim of clarifying the contribution of economic analysis to the debate over the instruments of environmental policy. First, we shall describe the general situation in which environmental policy goals must be achieved. An appreciation of the complexity of this situation will provide a base from which to consider both past error and actual special cases. Second, we shall define a set of dimensions along which policy instruments may usefully be judged. These include: static efficiency, centralized information and computation requirements, enforceability, dynamic incentive effects, flexibility in the face of exogenous change, and implications for goals other than efficiency. In the process, we intend to make explicit the irreducible political content of choices among policy instruments and thus the reasons that technical arguments on the other dimensions will not be decisive in the political arena. Third, we shall briefly review both some major non-economic attempts to evade the complexity of the general case and the record of adoptions of explicitly economic prescriptions.

Finally, following this background tour, we shall return to examine more carefully some of the economic complexities associated with a variety of instruments and problems.

Section 2 will concentrate on administratively (or legislatively) set prices and taxes designed to influence behavior.

Section 3 will be devoted to instruments that complete the set of markets – that is, where commodities or rights are administratively defined and prices are set by decentralized bargains among the actors subject to the policy (owning or wishing to own the rights).

Section 4 will deal with various forms of deposit–refund systems and performance bonds, as well as liability rules.

Section 5 will take up specifications of behavior and other instruments involving direct intervention in the behavior of the actors subject to the policy.

Section 6, finally, will discuss moral suasion as a policy instrument, primarily in contexts where there are significant constraints on the set of policy alternatives.

1.1. Some definitions and background assumptions

The position adopted in this chapter is that choice of policy goal and choice of instrument or implementation system are essentially separable problems. And, for the most part, the discussion here will take as given goals or standards for ambient environmental quality (air, surface or groundwater, landscape or whatever). The conceptually preferable position, that both goal and instrument must be chosen simultaneously in a grand meta-benefit/cost analysis, is for now operationally quite hopeless. Further, we shall usually assume the existence of an agency of government charged with meeting the standards.

The essence of the agency's problem of attaining chosen ambient quality standards is that the actions of many individual and independent actors (firms, households, other government units) affect actual environmental quality. The actors will differ among themselves in production technologies and product mixes (where these words are interpreted broadly enough to include such activities as home space heating and sewage treatment plant operation). In the most general case, the environmental effect of each actor is different from that of every other actor and more than one combination of actions by all the actors will result in meeting the standard.[1] The combinations will differ both in total cost to society and in the way any particular cost is distributed across the set of actors.

These actors are all assumed to be "rational" and self-interested.[2] For the agency to succeed in attaining the ambient environmental standard it must in

---

[1] The most common environmental policy problems involve as "actions" discharges of pollution into part of the natural environment. The effects of each source's actions depend on the characteristics of the environment (stream flow, water temperature, and so forth for water pollutants; wind speed and direction, terrain, hours of sunlight and so forth, for air pollutants). More generally, "actions" can include such diverse matters as the construction of ugly buildings, the use of farming methods that disrupt natural terrestrial ecosystems, or the placing of radioactive wastes into trenches or caverns. We usually will take "effects" to be measured relative to the ambient environmental quality standards at specified monitoring points. A more elegant treatment would involve measuring effects as damages (to human health and welfare, the ecological support systems, and so forth, but such an approach to policy implementation is not yet practically significant.

[2] Some criticisms of policy instruments that allow the actors flexibility, such as charges per unit of emissions of pollutant, appear to arise from the opposite assumption – that dischargers of pollution will act in an economically irrational way and pay a higher charge bill than would be optimal just for the pleasure of polluting. It seems possible that these critics have confused the position of a wealthy person confronted by a consumption tax with that of a firm.

some way induce at least some of the actors to take actions contrary to their narrow self-interests as defined in the pre-policy world of relative prices and constraints. The costs of these actions may involve both real resource use and transfers that are costs only to the payors, not to the larger society. The assumptions imply that when an actor, more particularly a firm, is faced with orders or charges ordained by the agency, it will respond in a way to maximize its present value in the long run. In the short run, we can usually capture all that is important by assuming the minimization of costs for given output, location, technology and factor prices, but subject to the new constraint or taking account of the new price, as on pollution discharge.[3] The difference between short and long run is the range of adjustments available. In the long run, the discharger can seek a new location, production process and pollution control technology, even entirely new products to make.

A full description of the background setting for the discussion of environmental policy instruments must include the fact that the agency cannot costlessly know

[3] Again it will be useful to tie this notion down by reference to the most common problem, pollution discharge. The cost to an actor of adjusting to orders or prices imposed by the agency will be captured in a cost-of-discharge-reduction function. This function shows the marginal (or total) cost of reducing discharge by an amount, R, below its level in the absence of any agency initiative. For many short run purposes it will be convenient to assume that this function is defined for fixed output, though more generally, output and all factor input decisions are made simultaneously with the decision about R. Thus, more generally, the two problems, one for an emission charge and one for an emission limit, may be written as follows:

Charge

$$\max_{\text{s.t.}} p'Z - q'Y - e(X - R)$$
$$0 = F(Z, Y, X - R),$$
$$Z, Y, X - R, R \geq 0.$$

Limit

$$\max_{\text{s.t.}} p'Z - q'Y$$
$$0 = F(Z, Y, X - R),$$
$$X - R \leq L,$$
$$Z, Y, X - R, R, \geq 0,$$

where Z is a vector of outputs, with prices p'; Y is a vector of inputs other than pollution discharge services, with prices q'; X, uncontrolled discharges; R, discharge reduction; e the emission charge; and L the emission limit.

In the short run, X is implicitly defined by the problem:

$$\max_{\text{s.t.}} p'Z - q'Y$$
$$0 = F(Z, Y, X),$$
$$Z, Y, X \geq 0.$$

This notion of "discharge reduction" can be expanded just as we expanded the notion of "discharge".

what each of the actors is actually doing at any particular time or on average over any period. Checking the behavior of the actors against applicable regulatory orders, or determining what is owed by way of emission charges is another resource-using problem, one we shall refer to as the monitoring problem. The subsequent matter of punishing violators of orders, or those in some way abusing the charge system, we call the enforcement problem. The monitoring problem is made especially difficult, again in the most general case, by the character of "pollution discharges". These are for the most part invisible to the unaided human senses. Furthermore, once a unit of discharge has left the discharge point it is in general not attributable to any particular discharger. Thus, measurement must occur at the point of exit (or before) if it is to occur at all.[4]

Finally, the entire problem of environmental policy implementation is embedded in changing natural (atmospheric land surface, and aquatic) and economic worlds. These changes occur on the short run, stochastic scale of wind and weather shifts as well as the secular scale of changing tastes and technology. As the world changes, with ambient quality standard held constant, the implementation problem changes. If a particular set of actions by dischargers results in meeting the ambient quality constraint under conditions A, those actions may fail to produce an acceptable result under conditions B, perhaps because sources have moved or production levels have changed in response to changing tastes or resource prices, or simply because the natural systems involved do not dilute and disperse the discharges in the same way under B as under A.

1.2. Dimensions for judging environmental policy instruments

The above description of the general situation in which environmental policies must be achieved suggests several dimensions along which potential instruments for achievement may be judged.

(1) Static efficiency. The efficient implementation system achieves the chosen goal at least resource cost. This dimension is almost always interpreted in a static sense and that will be the approach here. "Static" means, as a practical matter, that we assume an unchanging environmental goal and allow only for the first round of reaction to the implementation orders or incentives; that is, discharge reductions with fixed technology and location for each discharger.

(2) Information intensity. This dimension involves the attempt to measure, at least qualitatively, how much data and what level of predictive modeling skills must be available to the pollution control agency to use the implementation system in question. Its importance lies in the desire for efficiency coupled with our assumption of many different actors affecting the environment differently. As a

---

[4] This, we repeat, is the general case and may be qualified in a number of ways. See below, footnote 47.

general matter, efficiency will require that each actor be given an individually tailored order (such as a discharge reduction order) or be faced with an individually tailored price for discharges or subsidy for discharge reductions. Finding the full set of such tailor-made instruments requires either an information- and computation-intensive "model" of the situation to be regulated or a very difficult trial and error process.

(3) Ease of monitoring and enforcement. This refers to the relative difficulty of making and interpreting the measurements of discharges necessary to judge compliance, prepare bills, or audit self-reporting. These measurements are complicated not only by the features of invisibility and inherent "fugitiveness" already mentioned, but by the variability of discharges with production levels, equipment malfunctions and operator actions; by the imprecision of measurement devices and the discrete sampling techniques used in many such devices; and by the awkwardness involved in obtaining entry to a discharger's premises and setting up elaborate equipment in order to take the samples. The overall effect is to make it very expensive for the agency to use common measurement methods frequently enough to produce any reasonable probability of detecting a true violation of a time-averaged discharge standard (or to check the payment for an emission charge over a similar period).

Enforcement actions to prod violators back into line may include administrative fines, civil or criminal court proceedings and penalties, or more indirect actions, such as blacklisting. The relation between the enforcement penalties and methods and the monitoring activities (and hence the probability of detection) is important in defining the incentives for compliance with the chosen instruments, but the choice of enforcement methods may reasonably be seen as a second-order version of the choice of the instruments themselves and is therefore treated only cursorily in what follows.

(4) Flexibility in the face of economic change. Here the interest is in the ease with which the implementation system adjusts to maintain the given ambient goal when exogenous changes occur in tastes, technologies, resource use, or other features of economic activity. The fundamental distinction is between a system that adjusts through decentralized actions of the regulated dischargers – firms, households, and other government units – and one that must be adjusted through recalculation and imposition by the agency of the new discharge standards or required emission charges. The advantages of flexibility in this sense include the avoidance not only of information gathering and computations, but also of the inevitable political interference with changes in the system.[5]

(5) Dynamic incentives. This involves the actions encouraged by the instrument in the longer run. One important distinction here is between instruments that

---

[5] This judgement assumes that the ambient quality standard is a legitimately chosen policy goal. Tinkering with the implementation system, while aimed at changing only cost shares, may affect the society's ability to achieve the goal itself.

encourage the search for and adoption of new, environment-saving technology and those that encourage retention and operation of existing plants. Another is between instruments that distort relative factor prices, as by making capital-intensive methods artificially cheap, and those that do not. A third distinction of some interest is between instruments that provide incentives for dischargers to move and those that do not.[6]

(6) Political considerations. Several political considerations affect society's choice of policy instrument at least as much as cogent arguments about their relative merits on the above dimensions. Three are especially important. The first is distributional, the second ethical, and the third relates to broader economic stabilization concerns.

At the simplest level, it is clear that the matter of cost distribution is intimately linked to the political viability of alternative ways of meeting collective environmental goals. Because we choose a distribution of benefits when we choose the goals, and because we have no mechanisms (other than the very creakiest mechanisms) for redistributing incomes (and thus benefits), a choice of cost distribution implies a fixed pattern, of net benefits for that broad area of environmental policy.[7] If an analysis of the distribution of costs and benefits shows that a majority of voters or the members of some powerful or vocal voting block will probably incur net costs from the policy, one would certainly be tempted to predict at least a rocky road for it.[8] Note further that, from the point of view of the payor, "mere transfers", such as emission charge payments, are part of the cost of an instrument.

The ethical features of environmental policy instruments include, most prominently, the message conveyed and the extent to which the actors in the system are allowed to choose among alternative actions. One widely held view is that environmental policy should involve stigmatizing pollution, whether as a crime against nature or against other persons. [See, for example, the arguments in Kelman (1981) and those of Railton (1984).] In this view, regulatory orders backed up by criminal sanctions have the proper flavor, while charge systems that make "buying pollution" just like buying labor services, are immoral. A related matter is that of choice. While freeing up discharger choices is usually at the heart

---

[6] This distinction may be illustrated by the difference between an efficient emission charge system, which must be based on individual charges, tailored to location, and a uniform charge system. Under the former there necessarily are possibilities for movement to lower charge locations. (Though, as a practical matter, the anticipated savings might only rarely be large enough to justify the cost of moving.)

[7] As already discussed, in the longer run, by changing residence, job, asset portfolio, or habits, individuals can change their own net benefits from a particular policy (goal plus method of accomplishment).

[8] Because neither costs nor benefits of environmental quality improvements are easy for individuals to determine, most may find it hard to judge where their self interest stands once any option is operating. Thus, predictions may create opposition that would not otherwise exist. Indeed, the idea that opposition *ought* to exist may be itself enough to do in a plan.

of economists' arguments for the efficiency properties of economic incentive instruments, the very provision of such choice appears ethically undesirable to others. If pollution discharge is wrong *per se* (as opposed to being wrong only when done in excess of a discharge standard or as part of a fraud in the context of an emission charge system) then there should at least be no choice about how much of the wrong each actor is allowed to produce. And, indeed, there should be no confusion about "allowable" being equivalent to "acceptable". For those who hold discharge to be wrong, there is no acceptable discharge goal this side of zero, and the only acceptable dynamic incentive is one aimed at that goal. (On ethical questions in environmental policy more broadly, see Chapter 5 of this Handbook.)

Aspects of stabilization policy may also play an important role. At least, this was the case in certain European countries during the 1970s, where municipal waste water treatment installations were improved partly for reasons of environmental policy, partly to counteract recession in the building industry. [See, for example; OECD (1978).]

1.3. Avoiding the complications: Shortcuts from goals to behavior

The practical difficulties of the general case, which imply that advancing along one of the above dimensions usually means giving up something on another (or on several others), may be seen as the inspirations for attempts to construct shortcuts for society to follow. Some of the features of these shortcuts will reappear in later discussions, but for now they may be viewed simply as special cases, in which goal and instrument collapse into a single entity.

One such case involves pure technology standards. The actors in the situation are required to adopt particular treatment (or production) technologies. Whatever discharges result from the adoption are accepted, and the ambient goal implicitly becomes whatever is achieved when all sources comply. This approach has the advantage of appearing easy to monitor (though operation is different from installation and the "easy" monitoring only applies to installation).

The technology standard may be extended to the long run and in the process appear to capture some of the ethical high ground while at the same time seeming to provide desirable incentives. This shortcut amounts to the injunction to "do your best" at all times – in particular to adopt better technology as it becomes available. This seems to force each discharger inexorably toward zero discharge. But, of course, since much technical change is endogenous to the system of incentives, and since this policy implies fresh costs for new technologies with no rewards, "do your best" seems very likely to have the effect of slowing progress toward lower discharges.

A third shortcut is to use an emission charge as a revenue-raising device for a program of environmental quality improvement based on government projects or subsidies. Here, the charge is related to some characteristic of the discharger that will be relatively insensitive to it – for example, output – with the actual unit charge usually based on a rule of thumb relation between output and discharge. In this case, the expectation is that output, and hence charge payments will remain unchanged. In such plans the revenue collected is usually intended to be used for projects such as regional sewers and treatment plants, or treatment plant subsidies to individual dischargers, designed to improve environmental quality. The facilities become, *de facto*, the policy goal, and the environmental quality they produce is accepted willy nilly. Another possibility is to treat the revenue as part of the state's general revenue. For a discussion of the reduction in excess burden from a tax system achievable under this second alternative, see Terkla (1984).

1.4. Historical perspective: Notes on chosen approaches

The residuals from human production and consumption activities have always found their way to the natural environment. And even in long-vanished ages of sparse populations and small scale production units the disposal of these residuals could create local pollution problems in the sense of significant negative externalities. There is no lack of anecdotal evidence of the seriousness of these problems, especially in large cities [Baumol and Oates (1979)]. What does appear to be lacking is evidence that prices (charges) or markets were invited to play any role in dealing with these problems. Regulatory orders backed by fines, imprisonment, or physical punishment, seem to have predominated as policy instruments, though certainly those orders could be quite sophisticated.[9]

What changed over time was the source and geographic scope of pollution problems; not the method of trying to correct them. In the nineteenth century, when rapid industrialization was producing very large air and water pollution problems all over Europe, and in the northeastern United States, it seems that slightly more modern versions of the ancient prohibitions were the medicine first prescribed [e.g. the historical sections in Johnson and Brown (1976) dealing with France, Germany, Hungary, Great Britain and Sweden]. When these manifestly failed, an effort was made to require licenses (permissions, consents, contracts) by the terms of which some limits could be placed on private and municipal dischargers [Richardson et al. (1983.)].

The first significant move away from simple prescription of particular activities in pollution control policy seems to have come very early in this century in Germany, when the first water management cooperatives or Genossenschaften

---

[9] For example, Parker (1976) reports that the record of the manorial court for the Chatteris Manor, including the village of Foxton (in England) contains a number of rules constraining pollution of the brook that ran through the village. Householders were prohibited from allowing ducks or geese to "frequent" the brook, from washing linen "clothes" in the brook, and from draining household wastes into the brook except after 8 at night. All rules were backed by specified fines per offense.

were authorized for river basins in the North Rhine-Westphalia state [Johnson and Brown (1976), Kneese (1964)]. Instead of attempting to forbid the inevitable waste disposal, these organizations set out to deal with it collectively, through sewer and treatment-plant construction, assessing the costs of these efforts to their members. Of even more interest to latter day economists, the method of cost assessment was (and still is) based on the waste load each member generated. Because of the units (money per unit waste) this charge-back method looks very like an effluent charge and has been described as such by many commentators. But, as we shall see below, the common arguments for the social desirability of an effluent or emission charge are based on quite different goals and system design. Therefore, however much we may admire the audacity of the Germans who broke with at least 1000 years of traditional approaches, we really must wait even longer to see a charging scheme designed with incentive rather than revenue-raising effects in mind.[10]

Implementation programs closely related to the pioneering work of the German Genossenschaften exist now in several European countries, including the Netherlands, France and Hungary [Johnson and Brown (1976)]. Sewer services charges, which are a narrower version of the same approach are widely used in the United States and the United Kingdom [Elliott (1973), Urban Systems (1979), Webb and Woodfield (1981)]. But it appears that only in the Federal Republic of Germany has a national system of charges, designed explicitly to have an incentive effect, been put in place. This system was created by the national law passed in 1976 which will take full effect in 1986 [Bower et al. (1981), Brown (1982)]. This charge is linked to permit terms and compliance therewith, but is *not* based on the costs of collective treatment works.

These European countries are exceptions, however. The United States, for example, has not adopted an emission charge system for dealing with any pollution problem (a sketch of the approaches actually adopted is found in the previous chapter). While any number of proposals for charge applications have been made, both at federal and state levels, none has survived to the stage of implementation. Examples of these failed initiatives include [Baumol and Oates (1979) and Zeckhauser (1981)]:

(1) A 1970 proposal for a national tax on lead in gasoline.

(2) A 1970 citizen's initiative in Maine that put a BOD effluent charge on the ballot as a referendum item [Freeman (1970a), (1970b)].

(3) A 1971 proposal for a national effluent charge based on biochemical oxygen demand (BOD).

---

[10] Arguments along this line are made by Johnson and Brown (1976) and by Bower et al. (1981). The collective decision-making process of the Genossenschaften is of some considerable interest in its own right, with the dischargers themselves dominating the boards that decide on quality levels, treatment efforts, and hence necessary charges [Klevorick and Kramer (1973)].

(4) A 1972 SO_2 Emissions Tax Proposal. [This proposal was resurrected by Senator Durenberger of Minnesota, as another alternative for dealing with the problem of acid rain. *Inside EPA* (1983).]

(5) The 1972 Vermont law establishing effluent charges for organic discharges to natural waters. (This law was never put into effect, though neither was it, to our knowledge, repealed.)

Rather, the U.S. system of pollution control developed since the Second World War, and very largely since the late 1960s, contains modern versions of the consent or permit approach.[11] At the present time, however, administrative initiatives are creating many of the features of a *marketable* permit system out of the raw material of the original legislation. These new features will be mentioned below when we discuss marketable permits generally.

It is perhaps too extreme to say that the new German national effluent charge law is the only economic incentive system for pollution control ever successfully legislated. A major exception is the so-called "bottle bill" or deposit–refund system aimed at litter pollution by drink containers. Such laws (and similar ones concerning waste lubrication oil, junked cars, etc.) have been successfully put in place in several states of the United States and many European countries and do constitute explicit attempts to influence polluting behavior through economic incentives [Bohm (1981)]. The fact remains, however, that over the long sweep of history direct regulations (prohibitions, specifications of behavior, nonmarketable permits to discharge) have been the instruments of actual choice for dealing with pollution, whether from geese in village brooks or petroleum refineries on major rivers. Unlike commodity prices and markets, which existed before economists began analyzing them, administratively set prices or legislatively created markets do not appear to have sprung up as intuitive responses to externality problems. Quite the reverse; even after sustained intellectual development of these concepts during the period from 1960, we can find few examples of their application.

Let us turn now to more careful consideration of what that development has been about and to the ongoing debate over whether these newer instruments are or are not to be preferred to one or another version of the traditional approaches.

2. Instruments in the form of prices

The use of prices as instruments of environmental policy began to receive serious, and for the most part favorable, attention from economists in the mid 1960s. The most important early work is generally acknowledged to be that of Kneese

---

[11] More will be said about this system below, but for a reasonably full description, see for water pollution control, Freeman (1978) and for air pollution control, Lave and Omenn (1981).

[especially Kneese (1964)].[12] The theme of this section will be to show how the extremely attractive and compelling case made by Kneese has had to be modified as inconvenient elements of reality were explicitly recognized and dealt with. Because the literature on charges is enormous, matching the broad range of specific questions that has captured the interests of economists, we shall be forced to choose only a few of many lines of argument we might trace. Our choices are based on judgements about practical importance, not necessarily on number of pages devoted to the issue in the literature. We do, however, in the footnotes refer the reader to other disputes.

After we have discussed effluent or emission charges in principle, we shall turn to the design of models for the calculation of optimal charges in realistically complex situations. We shall then be able to report some of the results obtained from those models when they are directed to questions of the relative costs of alternative implementation systems.

To this point, the section will have been couched in static terms, and even our complications of the Kneese model will have assumed away a number of further important considerations relevant to instrument choice. The remainder of the discussion will be devoted to these other matters and will parallel the list of dimensions of judgement offered in the introduction. That is, we consider enforceability, flexibility in the face of exogenous change, dynamic incentive effects, and political implications of alternative instruments.

2.1. Arguments in the static case

For expository convenience in this and certain other sections let us construct a very simple model of an environmental policy problem involving two dischargers of a single residual, a natural environment receiving their discharges, receptors (unspecified in number) suffering damage from the resulting environmental degradation, and two potential monitoring stations at which that degradation can be measured.[13] We shall call the dischargers 1 and 2, and the potential monitoring

---

[12] As a matter of intellectual history, it would be interesting to trace the development of the emission-charge idea from Pigou's statements to Kneese's influential book, with its very practical air. This is not attempted here, but we do note in passing that an even earlier RFF book contained a paper by Gulick (1958) in which the use of prices to "determine interrelationships, priorities, and comparative needs and desires" was advocated in the context of resource problems, including pollution, in the modern city.

[13] A first judgement is implicit in our choice of model structure. It is that a partial equilibrium model can be a useful tool in examining questions of instrument performance. It is not a judgement that will receive universal assent, for general equilibrium treatment allows consideration of the consumption effects of output reductions due to a tax and can thus provide important perspectives about the appropriate instrument for controlling a monopolist and about the symmetry or asymmetry of charges and subsidies. Thus, Mäler's (1974a) comprehensive and insightful treatment of issues related to instrument choice is couched in terms of a general equilibrium model. So is Fisher's (1981)

stations A and B. The following quantities and functions will be central to our purpose:

$X_1, X_2 \equiv$ raw waste loads generated per unit time at the two sources,

$R_1, R_2 \equiv$ reductions in the raw waste loads achieved at the sources, as for example by recycling,

$D_1, D_2 \equiv$ discharges from the two sources per unit time,

so that $D_i \equiv X_i - R_i$,   (1)

$C_1(R_1), C_2(R_2) \equiv$ costs of pollution reduction at the two sources
$$\left(\text{assume } dC_i/dR_i > 0;\ d^2C_i/dR_i^2 \geq 0\right),$$

$f(D_1, D_2) \equiv$ damages suffered by receptors of the pollution
$$\left(\text{assume } \partial f/\partial D_i > 0 \text{ and } \partial^2 f/\partial D_i^2 \geq 0\right).$$

Sometimes we shall wish to consider ambient quality standards rather than assuming a damage function is known. For this purpose we define:

$S_A, S_B \equiv$ ambient environmental standards at the monitoring points.

The pollution control agency's problem for our simple region may be written in general terms as:

$$\min f(D_1, D_2) + C_1(R_1) + C_2(R_2) \tag{2}$$

or, by (1):

$$\min f(D_1, D_2) + C_1(X_1 - D_1) + C_2(X_2 - D_2). \tag{2a}$$

With this apparatus in hand it is possible easily to explore the "Kneese case" for charges and several of the most important qualifications to it. The classic case for emission charges depends on two assumptions: that f is linear and that the locations of the sources does not matter to their relative roles in damage production. Then the problem in (2a) becomes

$$\min a(D_1 + D_2) + C_1(X_1 - D_1) + C_2(X_2 - D_2). \tag{3}$$

The first-order conditions for an optimum are:

$$a - dC_1/dR_1 = 0 \quad \text{and} \quad a - dC_2/dR_2 = 0.$$

Thus, if the authority knows the linear damage function it can announce the optimal charge, a, without knowledge of the sources' cost functions. It is easy to see that if each firm minimizes cost, its response to this charge a will be the "proper" one, and $dC_i/dR_i = a$ will be true for $i = 1, 2$. The emission charge

---

more recent and much simpler discussion. Examples of papers addressing specific issues in a general equilibrium framework include: Burrows (1981) on controlling the monopolistic polluter; Sims (1981) on the asymmetry of subsidies and charges in the short run; Meselman (1982) also on subsidies and charges; and Harford and Ogura (1983) on charges and standards.

approach therefore boasts a powerful combination of static efficiency and information economy.

The first part of this case that we shall examine is the assumption that $D_1 + D_2$ is the appropriate argument for the damage function. Consider, for example, the possibility that damages are measured at a particular point (say a riverside park) and that one source is farther upstream from the park than the other. If the residual involved is not entirely conservative, the appropriate (still linear) damage function form should be $a(\alpha_1 D_1 + \alpha_2 D_2)$ with $0 \le \alpha_1$, $\alpha_2 < 1$ and $\alpha_1 \ne \alpha_2$.[14]

Then the first order conditions are:

$$a\alpha_1 - dC_1/dR_1 = 0 \quad \text{and} \quad a\alpha_2 - dC_2/dR_2 = 0. \tag{4}$$

They tell us that the optimal charges must be tailored to the location of each source (location matters) but that the authority can still announce optimal charges without knowledge of the cost functions *if* it knows both the damage function and the action of the environment on the discharges (captured in the α_i, often referred to as "transfer coefficients"). This result holds even if damages are measured at more than one point and added to get total regional damages, and if the sources affect the damage function arguments differently at each such location. Thus, if total damages are given by

$$a_A(\alpha_{1A} D_1 + \alpha_{2A} D_2) + a_B(\alpha_{1B} D_1 + \alpha_{2B} D_2) + \cdots a_N(\alpha_{1N} D_1 + \alpha_{2N} D_2),$$

then the optimal charge for source 1 is

$$a_A \alpha_{1A} + a_B \alpha_{1B} + \cdots + a_N \alpha_{1N}$$

The generalization to M sources is also straightforward.

The classical case for charges begins to unravel as soon as we drop the assumption of linear damage functions. Then the optimal charge is not independent of the optimal discharge levels and, in general, cost functions must be known to the agency. In the simplest such case, the damage function is non-linear in $D_1 + D_2$, and the sources have identical cost functions so that if $D_1 = D_2$, then $C_1(R_1) = C_2(R_2)$. Then it can be shown that at the optimum $D_1 = D_2$, and $dC_1/dR_1 = dC_2/dR_2$, and a single emission charge is optimal. But the optimal

---

[14] Notice that the form $\alpha_1 D_1 + \alpha_2 D_2$ arises whenever *either* the residual in question is nonconservative in the environment (e.g, is chemically changed or physically settles out between source and receptors) *or* where we are not in a position to measure the total contribution of a source to ambient quality by looking at a finite number of monitoring points. This latter condition differentiates the general air pollution problem from the general water pollution problem, because diffusion in the atmosphere results in the "loss" of discharged residuals.

Notice also that we are assuming a particularly simple form of the environmental model implicitly embedded in our problem. In general the effect of D_1 on the ambient quality at the damage measurement point may depend both on the level of D_1 and on the levels of all the other discharges. In this general case, things are even more difficult than we shall see them to be below for linear transfer functions.

D_1 and D_2, and hence the optimal charges depend on the cost function parameters. The agency's information requirement is immediately vastly greater.[15]

If either the cost functions are not identical or the locations of the sources matter, then the optimal charges must be source specific and depend on knowledge of the cost functions.

Any number of commentators in the early charges literature pointed out that not only was the assumption of *linear* damage functions unrealistic, but the very idea of any known and accepted damage function was more than economic knowledge could (perhaps ever) support. The point was certainly valid when made, and the reader is free to reach a conclusion on its current validity on the basis of the relevant chapters in this Handbook; our interest is in the line of analysis inspired by it (see especially Chapters 7 and 16). This is the line that takes ambient quality standards, chosen by some exogenous (probably political) process as representing the goals of environmental policy and sees charges as instruments for realizing those goals.

In this context, the agency's problem becomes:

$$\min_{\text{s.t.}} \left[C_1(X_1 - D_1) + C_2(X_2 - D_2) \right] \tag{5}$$

$$g(D_1, D_2) \le S_A$$

for two dischargers and a standard defined at a single point. For M dischargers and N standards, the problem becomes:

$$\min_{\text{s.t.}} \sum_i C_i(X_i - D_i)$$

$$g_A(D_1, D_2, \ldots, D_M) \le S_A$$

$$\vdots \tag{6}$$

$$g_N(D_1, D_2, \ldots, D_M) \le S_N.$$

In the simplest case, location is assumed not to matter, and only one standard is specified. Then on the basis of our other assumptions we can assume that the standard will be exactly satisfied, and a simple Lagrangian formulation suffices. The agency's problem is:

$$\min L = C_1(X_1 - D_1) + C_2(X_2 - D_2) - \lambda(D_1 + D_2 - S_A). \tag{7}$$

---

[15] The fact that, in the simpler case, a single emission charge applies might suggest that a trial and error process for seeking the optimum would work. The problem is that only by being able to measure costs and benefits at each trial would the agency be able to decide whether its last trial produced an improvement. Certainly measuring costs and benefits at a point does not require knowledge of the functions over their ranges, but the distinction in terms of required centralized data seems minor.

The first-order conditions are:

$$\frac{\partial L}{\partial D_1} = -\frac{dC_1}{dR_1} - \lambda = 0,$$

$$\frac{\partial L}{\partial D_2} = -\frac{dC_2}{dR_2} - \lambda = 0,$$

$$D_1 + D_2 = S_A,$$

from which we see that $dC_1/dR_1 = dC_2/dR_2 = -\lambda$. Thus, a single charge is optimal, but it can be found only on the basis of knowledge of costs or through a trial and error process. The latter is possible because after each trial the result can be observed at the monitoring point and there is no necessity for the agency actually to measure costs at all. (Of course, even though the proper charge could in principle be found via trial and error, the "errors" imply higher overall costs because of lumpy and at least partially irreversible investments. Thus, the results of proper charges set on the first try are not the same as those achieved without the knowledge necessary to that accomplishment.)

The introduction either of location differences or of a non-conservative residual complicates matters, but not fatally in principle, so long as a single standard (one monitoring point) is still all we have to worry about. The constraint in the agency's problem becomes $\alpha_1 D_1 + \alpha_2 D_2 \leq S_A$, and the first-order conditions from the Lagrangian problem are:

$$\frac{\partial L}{\partial D_1} = -\frac{dC_1}{dR_1} - \lambda\alpha_1 = 0,$$

$$\frac{\partial L}{\partial D_2} = -\frac{dC_2}{dR_2} - \lambda\alpha_2 = 0, \tag{8}$$

$$\alpha_1 D_1 + \alpha_2 D_2 = S_A,$$

so that, for example,

$$\frac{dC_2}{dR_2} = \frac{\alpha_2}{\alpha_1} \cdot \frac{dC_1}{dR_1}.$$

This result leaves us with some hope for trial and error, because even though charges must be individually tailored, the ratio of any two optimal source-specific charges is the ratio of the sources' transfer coefficients. Thus, trial and error could proceed on the basis of a single "numeraire" charge.

Similarly, if there is more than one standard to be met, but every source affects every monitoring point exactly the same, a single charge for all sources is still optimal. The agency's monitoring problem is more difficult because it must check each point at which a standard is defined, but it can still, in principle at least, perform a simple trial-and-error exercise based on iterations on one charge.

As soon as we both introduce multiple monitoring points and allow location to matter, however, any practical possibility of trial and error disappears. Thus, in this most realistic case, an optimal effluent charge system depends on the agency having knowledge of source cost functions and calculating a set of individually tailored charges. To see why this is so, consider our simple example with a second standard (monitoring point). The agency's problem is:

$$\min L = C_1(X_1 - D_1) + C_2(X_2 - D_2) - \lambda_A(\alpha_{1A}D_1 + \alpha_{2A}D_2 - S_A)$$
$$- \lambda_B(\alpha_{1B}D_1 + \alpha_{2B}D_2 - S_B), \tag{9}$$

and from the first-order conditions:

$$\frac{dC_1/dR_1}{dC_2/dR_2} = \frac{\lambda_A\alpha_{1A} + \lambda_B\alpha_{1B}}{\lambda_A\alpha_{2A} + \lambda_B\alpha_{2B}}. \tag{10}$$

Thus, even if both constraints could be exactly satisfied so that the shadow prices, λ_j, were non-zero, those shadow prices would not be known without a full solution. And without the shadow prices as weights, the ratio of the marginal costs cannot be calculated, even when the agency knows the transfer coefficients. Thus, trial and error cannot proceed on the basis of a single numeraire related in a known and constant way to each other optimal marginal cost (charge). This difficulty is compounded when there are many sources and monitoring points, because quality at some of the latter will inevitably be better than specified by the standard when the standard is not violated at any monitoring point. The corresponding λ's are zero, but which are zero is not known in advance.[16] Thus, while an actual trial-and-error process could lead to a feasible charge set (one that produced the desired ambient quality), it will not in general produce the cost effective outcome.

2.2. Modeling of the realistic static case

These last observations carry us to the end of our discussion of the simple static case and its complication. Overall we have seen just how restrictive are the assumptions that support the classical case for charges – that static regional efficiency can be attained with no knowledge by the agency of the cost functions of individual sources. Two natural enough questions are: If calculation of individually tailored charges is usually going to be necessary, just how hard is it likely to be? And how much difference will various charging systems make? For example, if individually tailored charges are optimal, but a single region-wide charge were actually to be applied to all sources, how much additional cost would be incurred?

The answers to these questions turn out to be specific to particular regions (because specific locations and the nature of the local environments matter); and

---

[16] Trial and error is difficult because of the large number of "knobs" available to twist in a multi-source region. If each of only 10 sources could control to each of only three levels of discharge, there would be over 59 000 possibilities for an initial trial. That first trial might eliminate some fraction of the options as either infeasible or unnecessarily strict, but finding a feasible and even modestly efficient option might easily involve many very expensive trials.

to particular pollution problems (because the cost-of-control functions differ among residuals as does the behavior of the discharges in the environment).[17] We shall confine ourselves, however, to two examples of modeling efforts designed to mimic realistic regional environmental quality management problems, attempting thereby at least to give an indication of the variations likely to be encountered.[18] These models were chosen because they can be compared both here, where effluent charges are at issue, and in Section 3, where our attention turns to marketable permits of various kinds. After the very briefest of descriptions of the models, we shall summarize some of the lessons learned from them.

The two models we shall use for comparison were both constructed in the early 1970s when enthusiasm for such exercises, and the regional efficiency solutions to which they might lead, was sufficiently great to sustain the costs of development and computation. One, the Atkinson and Lewis model, is of major point sources of particulates in the St. Louis region [Atkinson and Lewis (1974a, 1974b)]. The other is a multimedia, multiresidual model of the Lower Delaware Valley region (referred to here for brevity as Philadelphia) [Spofford et al. (1976)]. The differences and similarities of the models are highlighted in Table 10.1; and there we can see that the biggest differences are in size and complexity. The RFF model contains many more point sources, other residuals discharged both to water and air, and interactions among residuals in treatment processes.[19] In structure, however, and in the important matter of atmospheric dispersion modeling, the two models are similar.

In Table 10.2 some key comparisons are summarized. A policy of uniform percentage reduction orders for all sources sufficient to achieve the desired standard at the worst polluted monitoring station is taken to be the benchmark for compliance costs. (This policy is close enough to that embodied in most U.S. State Implementation Plans (SIP) for air pollution control that we shall follow the studies and refer to it by this acronym.) The other two policy instruments are a regionally (or zonally) uniform emission charge and an optimal effluent charge set. The latter, of course, involves different charges at each source reflecting their different locations relative to the binding ambient quality constraints. Atkinson and Lewis look at primary and secondary particulate standards, while Spofford examines only primary standards, but has results for both particulates and SO_2.

---

[17] In actual cases, removal processes often display such complications as economies of scale and joint removal of two or more residuals, so that the seeking of optimal regional management solutions, including optimally tailored charges, is much more difficult than our simple example hints at. See, for example, Russell (1973) and Russell and Vaughan (1976) on industrial pollution reduction costs.

[18] See, for other examples: on organic water pollution control in the Delaware estuary, Kneese and Bower (1972) and Johnson (1967); On water pollution control in Wisconsin's Fox River, O'Neil (1980); On SO_2 control in Nashville, Tennessee, Teller (1967); On nitrogen oxide emissions control in Chicago, Seskin, Anderson and Reid (1983); On chlorofluorocarbon control, Palmer et al. (1980); On phosphorus runoff control, Jacobs and Casler (1979).

[19] The sources of cost function data also differ for the major sources. The RFF model uses specially constructed industrial LP models to derive the regional model control vectors for steel mills, petroleum refineries and power plants.

Table 10.1
Summary of model structures and data bases

Model	Basic structure	No. of sources	Residuals included	Sources of cost data	Date of cost data	Basis for air pollution dispersion model	Number of monitoring points
Atkinson & Lewis (St. Louis)	Separable LP	27 point sources (All industrial: 9 power plants, 2 pet. refineries, 4 feed & grain mills)	Particulates	IPP Model[b]	Unclear (probably 1970)	Steady-state Gaussian diffusion (Martin & Tikvart)	9
Spofford (Philadelphia)	LP[a]	183 point sources (124 industrial: 17 powerplants, 7 pet. refineries, 5 steel mills; 57 Area sources home & commercial heating)	Air Particulates SO$_2$ Water Biochemical Oxygen Demand (nitrogeneous and carbonaceous)	Specially constructed plant LPs[c] IPP model[b]	Roughly 1970	Steady-state Gaussian diffusion (Martin & Tikvart)	57

[a] The original version was non-linear but the results reported in Table 10.2 come from a new, linear version.

[b] The Implementation Planning Program was designed to operate on air quality control region inventories and to allow the user to specify different control options, producing an estimate of control costs for the region and predicting resulting levels of ambient quality.

[c] See Russell (1973) and Russell and Vaughan (1976) for published examples.

Sources: Scott E. Atkinson and Donald H. Lewis (1974) *A Cost Evaluation of Alternative Air Quality Control Strategies*, Report No. EPA 600/5-74-003 (USEPA, Washington Environmental Research Center, Washington). Walter O. Spofford, Jr., Clifford S. Russell, and Robert A. Kelly (1976) *Environmental Quality Management* (Washington: Resources for the Future, Washington). Walter O. Spofford, Jr. (forthcoming) "Properties of Alternative Source Control Policies: Case Study of the Lower Delaware Valley", unpublished manuscript in progress, Resources for the Future.

When particulate matter is the residual of concern, both models produce similar results. Compliance costs are highest for the uniform roll-back approach and lowest (of course) for the optimal charge set. A regionally uniform charge produces intermediate compliance costs in each model. Notice also that as the number of zones increases in Spofford's model, the costs for a zonally uniform charge fall toward the optimal charge result. For the primary standard (75 μg/m^3) there is even surprising agreement between the models on the relative costs under each of these instruments, though the absolute size of Spofford's costs are very much higher, reflecting a larger number of sources and worse initial quality level. The same pattern holds when Atkinson and Lewis examine the costs of meeting the secondary standard (60 μg/m^3). The cost savings achievable by

Table 10.2
Compliance costs and emission charges under different emission charge systems in two regional models

Policy instruments	Atkinson and Lewis (St. Louis)				Spofford Lower Delaware Valley[d]			
	Particulates				Particulates		SO$_2$	
	Secondary Std.		Primary Std.		Primary Std.		Primary Std.	
	10^6/yr	Relative to SIP	10^6/yr	Relative to SIP	10^6/yr	Relative to SIP[e]	10^6/yr	Relative to SIP[e]
Compliance costs								
SIP/uniform percent reductions[a]								
Single regional zone	$8.3		$2.0		$158.0		210.5	
Three zones[b]					115.9	0.73	202.6	0.96
Eleven zones[b]					63.9	0.40	167.5	0.80
Uniform emission charge								
Single regional zone	3.8	0.46	0.3	0.15	14.2	0.09	252.0	1.20
Three zones					12.8	0.08	193.5	0.91
Eleven zones					10.4	0.06	138.5	0.66
Optimal charges[c]	1.9	0.23	0.07	0.03	7.2	0.05	118.5	0.56
Total out-of-pocket costs (including effluent charges)								
SIP/uniform percent reduction								
Single regional zone	$8.3		$2.0		158.0		210.5	
Three zones					115.9	0.73	202.6	0.96
Eleven zones					63.9	0.40	167.5	0.80
Uniform emission charge								
Single regional zone	6.7	0.81	1.3	0.64	54.7	0.34	504.1	2.39
Three zones					49.7	0.31	431.7	2.05
Eleven zones					32.7	0.21	295.9	1.40
Optimal charges	3.5	0.43	0.3	0.15	[Optimal charge payments not available.]			

[a] The two models have slightly different versions of this option. Atkinson and Lewis begin with industry-specific, technology-based emission standards and then reduce all discharges uniformly to reach the ambient standards. Spofford begins from the actual 1970 emission levels and reduces all discharges by uniform percentages (regionally or by zone) until the standard is met.
[b] Spofford's zones correspond to political jurisdictions: 3 states and 11 counties.
[c] The optimal charge solution corresponds to the least-cost solution when applied to both the point and area sources in the region.
[d] The Lower Delaware Valley region, which stretches from Wilmington, Del. to Trenton, N.J., includes eleven counties in Delaware, Pennsylvania, and New Jersey.
[e] The numbers in this column are calculated relative to the region-wide SIP alternatives.

going to a more efficient policy instrument are sufficiently great in both models (and for both standards in the Atkinson and Lewis work) that, even allowing for the out-of-pocket emission charge payments, it would be possible to make every discharger in the region better off through a suitable transfer arrangement.

When, however, we look at Spofford's results for SO_2 primary standards (80 $\mu g/m^3$), a very different pattern emerges. The regionally uniform emission charge produces a less efficient outcome than the uniform roll-back. Under zonally uniform charges, the more familiar pattern reasserts itself, but in all cases the call is a close one. In no case is the cost saving enough that the sum of compliance costs and emission charges is less than the compliance cost under the uniform roll-back instrument. This pattern of results happens to depend on Spofford's inclusion of home and commercial sources (the area sources) of SO_2 emissions among those subject to the charge. It can be shown, however, using a simple model like the one used in our earlier discussion of effluent charge properties, that the uniform emission charge is more likely to produce a costlier regional solution whenever sources with high marginal costs of discharge reduction have large impacts on ambient concentrations at the monitoring (standard) point. This is true in the Delaware model, where petroleum refineries with very steep marginal costs of SO_2 reduction (at the high reduction levels required) are also sited very close to the critical monitoring point. The addition of the relatively low marginal cost home and commercial heating sources far from the critical monitoring point accentuates this tendency and produces the result observed by Spofford.

Thus, how seriously one takes the Spofford results depends to some extent, though by no means entirely, on how seriously one takes the idea of applying an emission charge to small dispersed sources. (Note that such application could be via a fuel sulphur-content charge, so need not depend on unrealistic assumptions about monitoring and enforcement capabilities.) Nonetheless, the fact that in particular circumstances such results *can* be observed should make us cautious about general rule ranking policy instruments. While it is true that a tailored charge set can produce large savings, it is not always true that a uniform charge can improve on a simple regulator approach – even when we confine our attention to compliance costs. When we add in potentially massive transfer payments produced by the charge we can understand why sources might be extremely reluctant to see this instrument adopted. The only general rule would seem to be that if we want to explore alternatives in real settings we ought to do so with models first and only after we have an idea of the range of useful options, propose policy changes.[20]

2.3. Other dimensions of judgement

As important as static efficiency and economy of centralized information may be in the economic literature on environmental policy instruments, we must consider

[20] However, on the legal issues surrounding actual use of models for computing optimal or other charge sets, see Case (1982).

other dimensions of judgement as well. And these dimensions can well be more important to the adoption and long-term success of an instrument than the more familiar arguments.

2.3.1. *Ease of monitoring and enforcement*

Monitoring pollution sources to ascertain that they are paying the proper emission charge is a difficult problem. But a central point, as we see it, is that the monitoring problem is no harder if an emission charge is involved than if compliance with emission standards or permit terms is the concern. Thus, criticisms of emission charges based on the claim that compliance is harder to monitor are incorrect when the alternatives are also concerned with limiting the discharge of residuals per unit time. However, in a richer model including not only the statistical nature of the monitoring problem but also the decentralization of monitoring and enforcement activities and the possibility of polluter actions to conceal true discharge levels, Linder and McBride (1984) have identified certain drawbacks to a charge system not shared by a discharge standard. These include possible encouragement for less aggressive monitoring.

2.3.2. *Flexibility in the face of exogenous change*

It is first necessary to be clear about what counts as "flexibility". We shall use that word to mean the ease with which the system maintains the desired ambient standards as the economy changes. The most important measures of "ease" are first the amount of information the agency has to have and the amount of calculation it has to do to produce the appropriate set of incentives for a new situation and, second, the extent to which adjustments involve a return to a politically sensitive decision-making process.

In the restricted situation in which charges are both efficient and independent of costs (known, linear damage functions) the case for charges remains impressive. In fact, the same charge remains optimal after the addition of a new source or the expansion or shutdown of an existing source so long as change does not shift the marginal damage function. This automatic adjustment is thus based on allowing changes in discharges and ambient quality levels while maintaining marginal damage equal to marginal cost at each source.

If the policy goal is to maintain an ambient standard at a single monitoring point, after a change the charge must be adjusted, but the convenient relationship among optimal charges based on transfer coefficients is still there to take advantage of.

In the general case, where location matters and ambient standards are the goal of environmental policy (or where damage functions are either unknown or non-linear) emission charges do not protect ambient quality unless they are adjusted by the agency as change occurs. Such adjustment requires new calculations if the charges are to be efficient. (And then, because the charges are individually tailored, each charge is a fresh chance for political action.)[21] If the actual charges used are uniform and set by trial and error, adjustment will involve the expense of error, and, in addition, static efficiency will not be achieved.

2.3.3. Dynamic incentives

In the matter of incentive to technical change, the simple general rule may be summarized as follows. If compliance with an order is costly and if there is some choice of how to comply (what equipment or technique to use) then there will be an incentive for the source faced with the order to seek cheaper ways of complying in the long run. It is also true that for any particular source, an incentive system that puts a value on the discharge remaining after control will create a greater incentive to change than will a regulation specifying that same level of discharge.[22] We shall return to this matter when discussing the regulatory approach in Section 5.

2.3.4. Political considerations

Two broad questions should be dealt with here: distributional problems and ethical arguments. As for the first, it is obvious that emission charges in their pure form are bound to cost any particular source more than would a simple emission standard designed to achieve the same discharge at that source. Such evidence as that from cost models, both simple and complex [e.g. Vaughan and Russell

---

[21] For a discussion of the inevitability of political bargaining over emission charges, see the fine discussion by Majone (1976).

[22] This is easy to show. In the figure below, the firm's initial marginal cost-of-discharge-reduction cure is MC_0. Assume it is complying with an order to discharge no more than D_0. This could also be achieved by the agency charging a *fee* of e_0 per unit of discharge. The order costs the firm area A, the cost of control to D_0. The charge would cost it area $A + B$, the control cost *plus* the total fee paid on remaining discharges. If, as shown in the second panel, the firm can find a way to reduce its costs to MC_1, it saves C under the order system and $C + G$ under the charge.

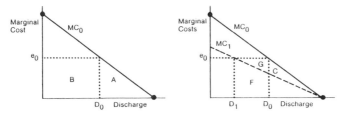

The new discharge, D_1, under the charge system is lower as well. This result also applies to marketable permits, for the permit price corresponds to the charge even though it may not be paid out of pocket by the originally permitted source. This argument is set out more fully and formally by Wenders (1975). For a slightly different view, see Magat (1978). And for another analytical approach, see Reppetto (1979).

(1976)], suggest that out-of-pocket costs of achieving a particular emission level at a source may easily be doubled by charge payments. On the other hand, the appropriate comparison should be the regional setting for a given ambient standard and real policy instrument alternative. Thus, if the efficient set of charges is contrasted with an inefficient set of emission standards, such as that obtained by imposing uniform percent reduction requirements on all dischargers, it is, as we have seen, an open and region-specific empirical question whether or not the savings from better distribution of pollution control effort will leave none, some, or all of the sources better off under the efficient charge, even after allowing for the charge payment itself. The real political problem here may be that dischargers doubt that the efficient charge set would ever be found or applied and see that an inefficient charge has a much increased chance of just costing them more money for the same results in the short run.[23]

One response to this political problem has been the proposal to use the revenue from charges to subsidize other acts of environmental protection. Another response has been concern over whether or not those revenues should be used to compensate the sufferers of damages from the remaining pollution. Certainly the idea has political appeal and seems to provide a symmetry otherwise lacking in the charges approach. But economists appear to have agreed after some debate that this symmetry would in fact be undesirable from an efficiency point of view; that while polluters should in principle pay charges equal to the marginal social damages they cause, damaged parties should absorb those damages without compensation and not be subjected to the incentive to increase exposure to pollution to collect (additional) compensation. [See, for example, Baumol and Oates (1975), Fisher (1981) and Olsen and Zeckhauser (1970). For a discussion of some ethical aspects of the compensation issue see Chapter 5 of this Handbook.] Only slightly more palatable to economists, but politically attractive, is the alternative already mentioned of using charge receipts to finance pollution control investments, especially those of an inherently collective nature such as in-stream aeration facilities or low flow augmentation dams.

The second political question, that of ethical stance, will be mentioned only briefly. The question arises because to many people pollution is wrong, not morally neutral.[24] These people do not want to see decisions about pollution placed on a footing symmetric with the firm's decisions about purchasing "normal" inputs such as labor services or packing cases. They want pollution stigmatized by strongly worded laws with strictly defined discharge limitations and criminal penalties for violations. The polluter's ability to choose how to react to a charge, the heart of the economist's efficiency case, is also the heart of the

---

[23] Distributional impacts on competitive industries are analyzed under a variety of assumptions by Dewees (1983).

[24] There is no evidence about what part of the general population feels this way, but Kelman's interviews with congressional staff members and active Washington environmentalists reveal a preponderance of this view among Democratic staffers and the environmentalists [Kelman (1981)].

environmentalist's opposition (for further discussion see Chapter 5 of this Handbook).[25]

A summary of this discussion of emission charges as a policy instrument for pollution control reveals a distinctly mixed bag of features. Certainly the classical position, in which static efficiency, information economy, and automatic adjustment to exogenous change can all be obtained at once, rests on very restrictive assumptions. In the more general case, static efficiency must be purchased at the cost of both information economy and flexibility in the face of change. Beyond that, emission charges suffer in the political arena from their distributional disadvantages (potentially large transfers imposed on polluters) and their ethical "flavor", which is apparently entirely too neutral to suit those who judge pollution to be a moral rather than a technical problem of market failure. In later sections we shall see how other instruments look under the same light.

3. Incentives to complete the set of markets: Tradeable rights

In practice the commonest form of policy instrument for environmental policy is the pollution permit, the terms of which usually embody either technological specifications or discharge limitations. We have explored some of the advantages and disadvantages of replacing such specifications with administratively set prices on discharges. Another possibility is to create a situation in which prices are attached to discharges by a decentralized, market-like process. To achieve this permits must be tradable among the interested parties, and the supply of permits must be less than the potential demand at zero price.

The idea of a marketable permit system appears to have occurred first to Crocker (1966) and to have been set out more completely by Dales (1968a, 1968b). It amounts to the dual of the emission charge idea – quantities instead of prices are set administratively; prices instead of discharge totals result from the free choices of those subject to the system. Its development in the literature has roughly paralleled that of charges. Early formulations were simple and compelling but later analysis showed that introducing complications reflecting features of reality reduced that apparent attractiveness. [For an excellent recent review, see Tietenberg (1980).] Just as with the charges, alternative versions of marketable

[25] It is worth noting in passing that the early writers may have unwittingly encouraged the views that those who favor economic instruments are basically insensitive to the health of the environment. For example, Kneese in his classic 1964 work, gives as examples of policies leading, potentially at least, to more efficient regional policies, the dedication of an entire river to waste disposal (the open sewer idea) and the storage of residuals for discharge in times of high assimilative capacity. In an illustrative example he also appears to sanction pollution-caused fish kills if the costs of cleanup are not exceeded by the damages to downstream commercial fishing interests. None of these are intrinsically wrapped up with emission charges and any or all might or might not be justifiable on the basis of efficiency analysis in a particular situation. But the political realities in the United States at least have made it clear that these are unacceptable alternatives. Their appearance in a fundamentally important statement of the value of emission charges probably tainted the latter.

permits have been defined, having different static properties and different implications for information and calculation loads. In discussing these cases and their properties the idea of duality will provide a useful benchmark, though in some cases this must be interpreted broadly.

Moving beyond the simple static context to complex (but still static) regional models, we shall observe as expected that employing marketable permit systems can lead to substantial costs savings compared to regulatory methods.[26] When we expand the horizon to include other dimensions of judgement, such as flexibility in the face of exogenous change, we shall continue to find a broad notion of duality useful for putting our findings in perspective. We shall, however, find that in some respects charges and marketable permits have the same properties while in others they are different without being dual. As before, at each stage we shall pick and choose among the many issues that have interested economists but shall endeavor to provide citations where we avoid discussion.

3.1. Simple static cases: Efficiency and information

Strictly speaking the benefit-based arguments for charges do not have duals in the set of marketable permit systems. It is when we introduce constraints on quality that we find price- and quantity-setting systems to be dual. But it is worthwhile nonetheless to observe that some early (and even not so early) statements of the case for marketable permits introduced an assumption that was conceptually equivalent to assuming knowledge of the benefit functions. This was the assumption that environmentalists (those with tastes for a clean natural environment) could and would combine to buy and retire pollution rights, thus carrying the system toward a socially optimum level of pollution analogous to that reached by the optimal benefit-based emission charge set.[27] But this assumption is fully as unrealistic as that involving benefit functions. The problems of public goods and free riders that imply no markets in environmental services, hence no demand or benefit functions available from directly observable behavior, imply that such combinations would be very difficult to establish. Even the analogy of the environmental groups, which combine thousands of individuals into potent forces for pollution control, cannot help us here. These groups succeed through highly

---

[26] In modeling studies of permit systems the model is almost always asked to produce the optimal (post trade) allocation of a fixed supply of permits and not to mimic the set of trades that leads there. As we shall see, it is not necessary for the control agency to have a complete model to introduce a statically optimal permit system even in the general case when such modeling *is* necessary to find an optimal emission charge set.

[27] Emphasis should be placed on "analogous". Because the social choice process contemplated by this argument for marketable permits is completely different from that involved in voting for standards or even calculating an optimal result using costs and benefits "to whomsoever they may accrue", there is no reason to expect the same quality levels to be thrown up by the three processes – assuming for the moment that environmentally minded citizens *could* combine to purchase and retire rights.

leveraged lobbying and litigation, not competition in the market. Rough calculations strongly suggest that all the national environmental groups in combination could make only a small dent in the pollution problem of any single large urbanized region if they had to do it by purchase of rights [Oppenheimer and Russell (1983)].[28]

More realistically, marketable permit systems are seen as potential instruments for achieving chosen ambient quality goals.[29] Corresponding to the single regional emission charge, which we saw was optimal only in very special circumstances, the simplest marketable permits system involves a single regional total emission limitation and a market for emission rights equally valid at any location in the region. These permits trade at a single regional price. Such a system can produce the desired ambient quality at least cost when location of discharge does not matter. The optimum level of total discharge for the given ambient standard could in principle be found by trial and error – the largest total just allowing the standard to be met.

In other situations, specifying a regional emission total and permitting trades among all dischargers on a pound-for-pound basis is not optimal, just as the single charge is not. While market transactions would result in an allocation of the permits such that resource costs were minimized for that total, one of the following would be unavoidable.

(a) the ambient quality goal would not be met; or

(b) if the initial total were chosen so that no conceivable set of trades could result in violation of the ambient standard, then the cost of meeting the standard would certainly be higher than necessary; or

(c) even if the total were greater, so that the standard were met only after some particular predicted trades, there would in general be a cheaper way of meeting it.

Trial and error could, however, be used to find a total allocation such that after trading the ambient standard was nowhere violated.[30] The trials would involve specification of the total permits to be allocated and observation of resulting quality. The same problems of extra cost arising from irreversible investment decisions arise here as in the use of trial and error with a charge system.

At the other end of the scale is the ambient rights system where the rights specified and traded are rights to cause pollution by particular amounts (usually assumed to be steady state concentrations) at the specified monitoring points. In

---

[28] Another reason that rights markets are unlikely to achieve a socially efficient outcome is that interfirm pollution effects may produce nonconvexities in production sets of nonpolluters who are allowed to buy and sell permits. Multiple optima then exist and the final result will be sensitive to the amount of rights issued initially. See Crone (1983) and Tietenberg (1983).

Rose (1973) analyzed systems of permit allocation using iterative bids and responses keyed to a known non-linear damage function. This provides another link to the optimal charge literature.

[29] We postpone for now the matter of how the permits might be initially distributed. This is discussed briefly under distributional matters in Section 3.3 below.

[30] How carefully the standard is protected depends on how many monitoring points are specified. The fewer of these the higher the chance that an after-trade allocation will result in an undetected violation (a "hot spot").

the ideal ambient rights system the agency simply defines rights totals at each ambient monitoring point equal to the difference between the desired standard and the contributions of all sources not subject to the system.[31] It has been shown by Montgomery (1972) that from any original allocation of these ambient rights the least cost regional solution can be reached by decentralized trading.

This system, then, sounds very appealing. Virtually nothing need be known by the agency except what amount of ambient quality "capacity" is available to be allocated. The market does the rest, without central calculation. The problem is that the *decentralized* information problem is formidable. Each source must simultaneously decide its optimal moves in each of the several markets, because any changes in its discharge rate simultaneously affects its need for ambient rights in every market.[32] If each source can be assumed to be a price taker in every market, the system looks like a set of competitive factor markets and we can invoke familiar market stability theorems to reassure ourselves that the optimal trading *could* go on. With only a few large buyers and sellers in each market, however, the practical chances for optimal decentralized results fall substantially. Thus, the centralized information intensity of the optimal charge system has its "dual" in the decentralized information problem of multiple markets in ambient quality permits.

Compromises between the extremes have been proposed. In zoned emission permit systems [e.g. Tietenberg (1974), Atkinson and Tietenberg (1982)] the region is divided into subregions, emission permit subtotals are allocated to subregions, and within subregions to sources. A source can trade pound-for-pound within its subregion and not at all outside it. If the initial allocation does not violate the standard, the zoned system raises the chances that no allowable set of subsequent trades will do so.

The zoned system raises in a more insistent way a problem we have so far ignored: market thinness. Tradable permit systems depend for their desirable properties on trades taking place and on these trades being sufficiently frequent to

---

[31] Such sources are usually termed "background", meaning such contributions as those blowing (or flowing) in from other regions or those from natural, uncontrollable sources in the region. More completely, however, the allocated permit totals can only equal whatever is left at each station when all sources not required to hold permits are operating in accordance with assumed regulatory requirements. Thus, the contribution of home heating discharge to regional SO_2 and particulates could be estimated using assumptions about fuel quality requirements.

[32] For a given initial allocation of ambient quality rights, q_{ij}^0 to each source i at each point j, each source must solve the problem:

$$\min_{\text{s.t.}} C_i\left(X_i - D_i^1\right) + \sum p_{q_j}\left(\Delta q_{ij}\right)$$

$$\alpha_{ij} D_i^1 \le q_{ij}^0 + \Delta q_{ij}, \quad \text{for all } j,$$

where $C(\cdot)$, D, X, and α are as defined in Section 2 and superscripts denote before and after trading situations, q_{ij} is the initial allocation of ambient quality permits at point j to source i, Δq_{ij} is the change through trade in the number of permits held at j by i, where purchases are plus and sales are minus and p_{q_j} is the price of permits at point j.

establish a market clearing price (regional emission permits) or a number of market clearing prices (ambient permits, zoned emission permits). If there are only a few sources in each market there may be no transactions for many periods because of capital commitments in particular production or discharge control process. Or transactions may be distorted by monopolistic or monopsonistic (duopolistic or duopsonistic) behavior by a dominant source or sources. These problems are major concerns of Hahn and of Cass et al., who have worked on designing an SO_2 discharge permit system for the South Coast Air Basin in California [Hahn (1980), Cass et al. (1980) and Hahn and Noll (1981)]. See also Russell (1981) for some preliminary evidence on numbers of sources and the supply and demand for permits. Several workers in this field, for example Tietenberg (1974) and David et al. (1980) have advocated periodic expiration of rights, making them like leases rather than freehold properties, with the idea that when some or all of a source's permits expired it would be forced into the market to obtain replacements.[33]

Another compromise emission permit system depends on "trading ratios" related to the source-specific transfer coefficients. If it is possible to identify a hot spot in advance, the coefficients relating all source discharges to that point can be used. Then, if source i sells to source j e_i units of emission permits, source j can use (discharge) $\alpha_{ik}/\alpha_{jk}\ (e_i)$ units where k is the designator of the potential hot spot.[34]

3.2. Evidence from regional models

It will be useful here, as it was in our discussion of emission charges, to introduce some evidence from realistic regional models. In order to maintain comparability

---

If the p_{qj} were exogenously given, this calculation would be straightforward for any source. But for the decentralized system the p_{q_j} are only implicitly defined by the market-clearing relations:

$$\sum_i \Delta q_{ij} = 0, \quad \text{for all } j,$$

and

$$\sum_i \sum_j p_{q_j} \Delta q_{ij} = 0.$$

[33] This strategy is also liked by some writers to the maintenance of agency "flexibility" – the ability to retire permits without the cost or fuss of litigation over the taking of property. See the discussion below under flexibility.

[34] A complete system of implicit trading ratios constraining trades has been suggested by Krupnick and Oates (1981) and Krupnick, Oates and Van de Verg (1983) who refer to it as an "offset system". This scheme protects ambient quality at all monitoring points (points for which transfer coefficients are available). In fact, however, the constraints faced by each source in deciding how to trade seem to be equivalent to those involved in the ambient permit system when source-specific constraints are combined with the zero net creation of permits at each monitoring point. For a system aimed at maintaining the status quo quality if that is better than the standard, see McGartland and Oates (1983).

Table 10.3
Compliance costs under different marketable permit systems in two regional models

| | Atkinson and Tietenberg (St. Louis) Particulates | | | | Spofford (Philadelphia) | | | |
| | | | | | Particulates Primary Std | | SO$_2$ Primary Std | |
	2 μg/m^{3d} 10^6/yr	Rel to SIP	10 μg/m^{3d} 10^6/yr	Rel to SIP	10^6/yr	Rel to SIP	10^6/yr	Rel to SIP
SIP/uniform percent reduction[a]	$9.8		$6.2		$158.0		$210.5	
Emission permits[b]								
Single zone	8.0	0.82	1.5	0.24	16.0	0.10	199.4	0.95
Three zones[a]	6.9	0.70	1.5	0.24	16.1	0.10	204.6	0.97
Six zones	8.6	0.88	1.8	0.29				
Eleven zones					23.3	0.15	215.2	1.02
Ambient permits								
Single Market	3.5	0.36	0.6	0.10	–		–	
Multiple Markets	3.1	0.32	0.5	0.08	9.7	0.06	177.1	0.84

[a] The Atkinson and Tietenberg SIP strategy involved first assigning to each of 27 sources an emission level based on application of control strategies represented in SIP guideline documents. To produce the level of ambient pollution at the worst receptor point shown in the table, further necessary reductions were made on an equal-percentage-reduction basis. Spofford's version of this policy involves equal percentage reductions at all sources from a base of 1970 inventory emissions.

[b] For the emission permit and ambient permit systems, Spofford imposes fuel quality regulations on home and commercial heating activities. These activities do *not* participate in the permit markets.

[c] Atkinson and Tietenberg report on two slightly different versions of a three-zone permit system. The costs reported here are a rough average of those reported in their article (Figure 4) for the two versions.

[d] Contribution to annual average concentration of suspended particulates at receptor location with worst quality of the 27 point sources modeled. Nothing is said about what value of this indicator might correspond to the primary air quality standard of 75 μg/m^3. Results are given for levels of this indicator from roughly 2 to 12 μg/m^3.

Sources: Scott E. Atkinson and T.H. Tietenberg (1982) "The Empirical Properties of Two Classes of Designs for Transferable Discharge Permit Markets", *Journal of Environmental Economics and Management* 9, no. 2, 101–121.

Walter O. Spofford Jr. (1983) "Properties of Alternative Source Control Policies: Case Study of the Lower Delaware Valley", Resources for the Future, unpublished report.

across instruments, we shall again concentrate on two such modeling projects: Atkinson and Tietenberg's work on particulate control in the St. Louis region;[35] and Spofford's analysis of particulate and SO$_2$ control in the Lower Delaware Valley Region (Philadelphia).[36]

Some control-cost results from these two modeling exercises are summarized in Table 10.3. These must be interpreted with caution, because the ambient stan-

[35] The Atkinson/Tietenberg work is based on the same model as that of Atkinson and Lewis (1974), used in the emission charge section.

[36] Again, this by no means exhausts the possibilities. Other studies providing modeling evidence include: deLucia (1974) on BOD discharge permits for the Mohawk River in New York; Cass et al. (1980) and Hahn and Noll (1981) on SO$_2$ discharge permits in the South Coast Air Basin in California; Eheart (1980) on BOD discharge permits for the Willamette River in Oregon; David et al. (1980) on phosphorus discharge permits for Lake Michigan; O'Neil et al. (1981) on BOD discharge permits for the Fox River in Wisconsin.

dards imposed on the models were not the same. Atkinson and Tietenberg (A&T) report the contribution of 27 major point sources to quality degradation at the receptor location with the worst air quality. Spofford imposes the primary air quality standards at each of 57 monitoring points in the region.

Subject to this caveat, however, the pattern of results is of some interest. Most obviously, the A&T results for the less strict "standard" (10 $\mu g/m^3$) look very like Spofford's results for the primary particulate standard. Either type of permit represents a very large improvement over the SIP/uniform percentage reduction policy, with the ambient permit system costing 10 percent or less of the strictly regulatory approach (in terms of real compliance costs only).

For both the stricter particulate standard in A&T's work and in the SO_2 example from Spofford, however, the relative cost differences change. In the former these drop off less dramatically. In the latter, the emission permit systems represent either no cost improvement or only the tiniest of improvements, and even the ambient permits are well over 80 percent as expensive as the SIP policy.

Thus, again it appears that the rankings of policy instruments, even in static efficiency terms, will in general depend on the residual in question, the strictness of the ambient standard being contemplated, and the characteristics of the regional economy and environment. We cannot even be certain that the theoretically best ambient permit system will be the lowest cost alternative because of the important of such small sources as home heating, for which permit requirements and trading seem completely out of the question.

3.3. Other dimensions of judgement

Marketable permit systems display both similarities to and differences from emission charges when judged on such dimensions as ease of monitoring and enforcement, flexibility in the face of exogenous change, dynamic incentives, and political attributes. We consider these in turn in this section.

3.3.1. Monitoring and enforcement

Monitoring an emission permit, marketable or not, defined in terms of allowable emissions per unit time, is the same problem as monitoring for emission charge billing. When permits are marketable, the problem may be compounded by the necessity of being current with completed trades. And this extra difficulty might be exploited by dischargers trading in the short run to stay one jump ahead of agency monitoring teams.[37] Problems are compounded if trades are allowed between conventional sources such as stacks and hard-to-monitor sources such as dirt roads and refuse piles.

---

[37] This strategy could be foiled by requiring long minimum holding periods, but this would have to be backed up by a complete, real-time inventory of all permits. David et al. (1980) propose that all trades take place only at quarterly auctions as another strategy to assist in monitoring for compliance.

One extra fillip accompanies an ambient permit system, however. The current state of technology does not in general allow us to differentiate the contributions of specific dischargers to concentrations of pollutants observed at an ambient monitoring point.[38] This means that monitoring for compliance in this case must also involve monitoring of discharges. That is, a portfolio of ambient permits must be translated into an effective discharge permit by use of an agreed-on regional environmental model.

3.3.2. Flexibility in the face of change

This is a dimension on which a marketable permit system seems to have a distinct advantage. Once established, and assuming necessary monitoring and enforcement effort, a permit system maintains either discharge totals (regional or zonal) or ambient quality standards without constant intervention and recalculation by the agency. If the demand for permits shifts because of regional growth or decay, this will be reflected in the market prices of permits. Permit reallocation takes place as firms find it in their interest to reduce discharges and sell permits to new entrants or expanding resident firms.

With reallocation through trading of emission permits goes the continued danger of new hot spots.[39] This danger, plus the thought that the initial allocation might be regretted for other reasons, has inspired several analysts to push for a different kind of flexibility – bureaucratic ability to reduce the total of emission permits outstanding without compensation [e.g. Tietenberg (1974), deLucia (1974)]. This flexibility would be obtained by automatic and periodic expiration of rights (e.g. one-fifth might expire every year). There would be no obligation to reissue the same number that expired, and in some systems, all new rights would be auctioned. This particular form of flexibility seems to threaten the real long run advantages of permit systems, however, for decisions to buy and sell permits would become shorter run matters if expropriation after only 5 years were a real possibility.

3.3.3. Dynamic incentives

In principle, the incentives for technical change provided by permits correspond to those produced by charges. In either case, reducing discharges produces a

---

[38] But see footnote 47 on inferring discharges from the multiple sources affecting multiple monitoring points on the basis of knowledge of the elements in each discharge stream.

[39] Notice that by a suitably conservative choice of initial allocation the agency could avoid all possibility of hot spots no matter what the pattern of trading. This would in general imply a very severe restriction on total permits and thus a high control cost. One place for modeling, then, as in our empirical section, is to identify the efficient post-trade pattern of discharges so that the initial total allocation can be such as to produce the desired ambient standard under that spatial pattern of discharge. Thus, costly information again can substitute for costly discharge control.

monetary gain to the source. However, it may be difficult to sell any substantial number of permits, especially if the market is thin; hence, a (full) monetary gain may not be captured as easily under the permit system as under the charge system. Moreover, for very strict initial allocations of emission permits designed to avoid hot spots under any possible set of trades, the permit price will be higher than that implied, for example, by an ambient quality permit system. Thus, static inefficiency can produce larger long-run incentives to reduce discharges.

3.3.4. Political considerations

The distribution of costs under a marketable permit system depends on both the number of permits originally allocated and on how the allocation is done. Auction systems are conceivable [e.g. Rose (1974), Repetto (1979) and, for a "Vickrey Auction", Collinge and Bailey (1983)] and produce a result similar to emission charges, with all sources being out of pocket for both control costs and permit costs. More likely would seem to be free initial allocation, either in proportion to original, uncontrolled emissions or to a projected equilibrium allocation. In either case, the value of the issued permits is a windfall to the existing sources. This may purchase their acceptance of such a system, where they seemed likely (though not certain) to oppose an emission charge. The other side to this coin is the opposition such a windfall is likely to create among environmentalists – and, indeed, others.

The other political consideration we have mentioned is the extent to which the instrument stigmatizes polluting activity and appears to give the polluter no choice but to clean up. On this scale, the marketable permit looks modestly preferable to the charge. The chance to pay and pollute without committing a "violation" is limited by the total number of available permits, not merely by the arguments of economists who assume rational cost-minimizing behavior. While permit violations are possible the very use of the word "violations" indicates that such behavior is considered wrong and presumably subject to punishment.

3.4. A real-world approximation

More significant than intricate efficiency arguments, modeling exercises, and speculation about political considerations is the fact that an approximation to a marketable emission permit system is now in place for air pollution control in the United States. This system has been developed out of a combination of necessity and imagination by the USEPA and certain of its contractors. It has three major components:

(1) *Offsets* – arrangements that allow a new or expanding source to buy into an area by paying for the reduction of emissions at other sources. The reductions must more than balance the new source's emissions, and the new source must

meet applicable regulatory requirements such as new source performance standards. [See, for example, Liroff (1980).]

(2) *Bubbles* – originally designed to let a single complex plant balance its
pollution control effort among its several stacks in such a way as to reduce its
costs while simultaneously reducing its emissions. The idea is basically to relax
specific regulatory requirements at one or another high cost process in return for
extra effort at a place where extra removal comes cheap. The idea has subsequently been expanded to allow multiplant bubbles which amount to permit
trades among existing sources (e.g. Brady and Morrison, (1982)].

(3) *Emission Reduction Credit Banking* – This feature allows sources that have
opportunities to reduce emissions but no current markets in which to sell the
freed up "permits" to bank them for later use or sale [Brady and Morrison
(1982)]. The system represented by these related features is complicated and
constrained by the apparatus of direct regulation that has been retained. This
apparatus limits the extent of control effort relaxations a source can buy, limits
the circumstances in which trades can take place (both in terms of source
compliance with regulatory requirements and of regional compliance with ambient quality standards) and introduces separate and to some extent inevitably *ad
hoc* approval procedures for each desired trade. On the other hand, the regulatory
apparatus introduces possibilities for unwanted outcomes. For example, existing
permits under State Implementation Plans may allow sources far more discharges
than they are using or indeed have ever used. These excess emissions are
apparently available for trade and the results have been damned as "paper
offsets" when used [Liroff (1980)].

An analysis of the actual cost and discharge results of operation of this system
must wait on more experience. What seems likely at this point is that many
proposed and approved trades have involved notional or paper offsets or their
equivalent in bubbles – as when dirt roads are to be oiled to cut ground level dust
in exchange for relaxation in high-level particulate emission requirements. On the
other hand, the mere existence of the system and experience with its operation
can give us confidence to go on into better structures.

4. Other incentive systems

4.1. Deposit–refund systems and performance bonds

As we have seen, remedies such as charges or marketable permits require that
discharges be monitored. This may not be feasible in practice, i.e. when the
sources of environmental degradation are numerous and/or mobile. Moreover, a
system of charges or marketable permits provides incentives for concealing the
volume of discharges, which may jeopardize reliable monitoring. For these

reasons, such systems are not likely to work in many cases, such as releases of freons from automobile air conditioners; improper disposal of mercury batteries or waste lubrication oil and other hazardous material by individuals; or littering, be it beer cans or abandoned cars.

Similarly, the establishment of property rights through appropriate liability assignments (discussed below in this section) runs into many problems that limit its use. For example, proof of guilt is required and often difficult to establish even in cases where proof of innocence would be easy, had it been required or had there been incentives to provide such evidence. Furthermore, the exact implication of liability may be unclear, in particular concerning the size of indemnities for a given type of violation, which makes the deterrent role of this instrument unclear. In addition, if the probable size of indemnities exceeds the net worth of the violator, the incentive effect on behavior as well as compensation to the injured party (when relevant) may be insufficient.

General deposit–refund systems may be a better instrument in such situations. These systems imply that the potential injurer is subjected to a tax (deposit) in the amount of the potential damage and receives a subsidy (refund), equally large in terms of present value, if certain conditions are met, e.g. proof that a product is returned to a specified place or that a specified type of damage has not occurred. Thus, such systems introduce a price for the right to inflict detrimental effects on the environment and a (negative) price if this right is not used. As a special case, the government may not be involved at all, instead the separate tax payment is set to zero and the subsidy payment is required to be made by a non-government party engaged in damage-related transactions with the potential injurer (for example, sellers of beverages in certain types of containers). This party would typically respond by increasing prices for such transactions and by introducing a "deposit" as part of the new price. The resulting arrangement amounts to a so-called "mandatory deposit" where the sole requirement is that a refund be made (e.g. mandatory deposits on beverage containers). As another special case, the potential injurer may be allowed to transfer the liability to pay the *net* tax/deposit, i.e. when the conditions for a subsidy/refund are not met, to a trusted third party such as a bank or an insurance company. This amounts to a "performance bond" for which the potential injurer will have to pay some interest or a premium [Bohm (1981)].

Deposit–refund systems may perform better than alternative instruments in that (a) they also work when the act of environmental degradation is not directly observable or when the potential injurers are numerous and/or mobile, (b) they simplify the proof of compliance in some cases, (c) they specify the (maximum) economic consequences of noncompliance, (d) actual or expected damages are covered by actual payments, at least in principle, and (e) in certain applications they may stimulate people other than those directly involved to reduce the effects on the environment (such as scavengers in the case of refunds on littered items).

In addition, as compared to alternative economic incentive systems such as pure charges or pure subsidies, deposit–refund systems have properties that would make them more attractive from the politician's point of view. Charges have sometimes been avoided because of fears that low-income people would be found to be hit relatively hard by such measures. In contrast, taxes/deposits are balanced by the right to subsidies/refunds which would leave nominal income unaffected. Indeed, the refund incentives may be particularly strong for low-income people and allow them to make income gains on balance. Subsidies have to be financed by government revenue and are disliked for that reason. In contrast, the specific taxes/deposits cover the subsidies/refund in a deposit–refund system.

Thus, deposit–refund systems – when applicable – can be said to provide the advantages of an economic incentive system, while avoiding some of the political disadvantages of the "traditional" forms of such incentive schemes. The applicability of such systems requires that it is technically feasible and not prohibitively expensive to establish proof of absence of pollution from the potential polluter.

4.1.1. Forms of deposit–refund systems

4.1.1.1. Adjusting market-generated systems.. The fact that deposit–refund systems (or refund offers) are found in the market economy indicates that there exists empirical experience with such systems. The reasons for the emergence of market-generated systems are diverse, ranging from a reuse value (e.g. old tires, containers) or a recycling value (e.g. lead batteries) to price differentiation or the speeding up of replacement purchases by refund offers to old customers. Thus, the rationale for voluntary refund offers may be that the reuse or recycling value (V) is positive or simply that a refund prospect (R) stimulates demand; in the latter case, V may be negative.

As some monopoly element is likely to be present when an individual firm makes a refund offer, we may write the profit function as

$$\Pi = p(Q, R)Q - c(Q) + r(R)(V - R)Q,$$

where Q is output;[40] $p(Q, R)$ the inverted demand function; $c(Q)$ the cost function; and $r(R)$ the return rate. The return rate will be determined by the individual consumer's (i) disposal options, where c_d^i is the total unit disposal cost without a return alternative and c_R^i the corresponding cost of returning the product. Consumers whose $c_R^i - c_d^i$ falls short of R will be assumed to choose the return alternative.

---

[40] To fix ideas, the output may be considered as a quantity of bottled beverages or mercury batteries. Later when we deal with government-initiated deposit–refund systems, a better illustration may be provided by the production of freons; here, freons (chlorofluorocarbons) could be returned for a refund (instead of being released into the atmosphere) when cooling equipment is being serviced or scrapped.

The introduction of a refund offer can normally be expected to raise prices. However, for a given demand effect $\partial p / \partial R > 0$, and a given effect on the return rate, $dr/dR > 0$, a sufficiently high V value will cause equilibrium prices to fall [see Bohm (1981, ch. 2)]. Regardless of this outcome it is up to the firm to announce that part of the price now is a "deposit" $D = R$.

If non-returns typically create negative external effects in the amount of E (expected environmental hazards or extra waste treatment costs), the firm's optimal $R(R_F)$ may not give rise to a socially optimal return rate. Assuming for simplicity that V also equals the social reuse value and that second best complications from the monopolistic behavior of the firm can be disregarded, the socially optimal $R(R_S)$ would equal either $E + V$ (where $V \gtreqless 0$), if the return rate is less than 100 percent, or the lowest level $R < E + V$ at which a 100 percent return rate is attained. Thus, if $R_F < R_S$, an adjustment of the market refund rate may be called for. A "mandatory deposit" in the amount of R_S may, however, create problems, as the firm would lose when refunds are set at a level other than R_F. Hence, the firm might want to obstruct the system by making returns from consumers more cumbersome (increasing c_R^i). If so, either measures specifying the obligations of the firm would have to be taken or the government would have to become financially involved in the administration of the system. The latter alternative could be designed as a full-fledged deposit–refund system with a tax imposed on output in the amount of $D = R_G = R_S - V = E$ and a subsidy per unit returned in the amount of R_G.

4.1.1.2. Government-initiated systems. If no market-generated return system is in operation, but the disposal of used products gives rise to negative marginal external effects (E), which would be avoided if the used products were returned, a deposit–refund system of the type just mentioned could be introduced by the government. Assuming that the industry is competitive, $V \gtreqless 0$ would be the market price, equal to the firms' value, of a returned product, whereas the socially optimal payment (R_S) for a returned unit equals (at most) $E + V$. If so, a tax/subsidy in the amount of $D = R_G = R$ would be appropriate.

As consumers whose disposal cost difference $c_R^i - c_d^i$ exceeds the total payment, $R_S = R_G + V$ will continue to use the traditional disposal option, then, at the margin, social costs of traditional disposal equal social costs incurred by the return alternative, i.e. $E + V = c_R^i - c_d^i = R_S$. In the general case, the shift to this optimum will give rise to various distribution effects, where the losers will be (a) the producers, as producer price net of D is likely to go down (by $\Delta p < 0$) and (b) those consumers whose $c_R^i - c_d^i > R_S + \Delta p$. The winners include the remaining consumers, scavengers (who now may pick up discarded units for a refund), and taxpayers, to the extent that total deposits exceed total refunds.

So far we have discussed deposit–refund systems for *consumers* of products which may create environmental effects when disposed of (mercury and cadmium batteries, beverage containers, tires, junked cars, used "white goods", lubrication oil, freons in air conditioners and refrigerators, etc.). Similarly, deposit–refund systems may be designed for *producers* to check hazardous emissions of chemicals into the air and waterways or dumping of toxic wastes, in particular when proper treatment of such releases or wastes is expensive and improper disposal is easy to conceal. If the potential emissions or wastes are related to certain inputs in a straightforward fashion (such as potential sulphur emissions to the input of high-sulphur fuel oil), a tax/deposit could be levied on these inputs and a subsidy/refund paid for the quantity of chemicals (e.g. sulphur) or toxic material transferred to a specified type of processing firms. Here, as well as for other deposit–refund systems, a precondition for a well-functioning application is that there are sufficient safeguards against abuses, e.g. that ordinary sulphur cannot be bought and passed off as sulphur extracted from stack gases.

4.1.1.3. Performance bonds. Producer-oriented deposit–refund systems can be used to control other kinds of detrimental effects on the environment than those explicitly discussed so far [Solow (1971)]. First, restoration of production sites after shut-down may be required to avoid unwarranted permanent eyesores or accident risks (strip mining sites, junk yards, etc.) Second, we have the vast problem of safeguarding against *a priori* unknown environmental effects of new products, in particular new chemicals, or new production processes. When applying the principle of deposit refund to such cases, the producer could be required to pay a deposit, determined by a court estimate of the likely maximum restoration costs or the maximum damages (in general or specific respects), to be refunded if certain conditions are met. In this way, society is protected against incomplete restoration because of intentional or unintentional bankruptcies. Moreover, the firm's operation will now be planned with respect to future restoration costs as well. And in the case of potential risks of innovation, this creates an alternative to awaiting the results of a test administered or supervised by the government. In this way, the introduction of the new products or processes would not be delayed. This alternative may be attractive to the innovating firm because the firm may have gathered information – and now definitely has an incentive to gather such information from the beginning of its R & D activities – implying that no harm is likely to result. Therefore, the firm may be willing to market the product or start using the new process with the specified financial responsibility, and both the firm and its customers may be better off [Bohm (1981, ch. 4)].

Although we take it for granted that the government will not trust a firm to meet its obligations without a financial commitment, in either of our two cases, it is conceivable that other parties which *are* trusted by the government would like to assume the financial responsibility involved. Thus, for example, the firm may convince a bank or an insurance company that the new product is safe. Or the firm may reveal its product secrets only to such a party but not to a public authority. If so, banks or insurance companies may assume the liability at a price. Thus, by using the risk-shifting mechanisms of the credit or insurance markets, the deposit–refund system can be transformed from a cash deposit version to a performance bond version, or firms could be allowed to choose either of these two versions. In other words, whenever the environmental effects potentially attributable to an individual decision-maker, and hence that individual's deposit, become sufficiently large and the transaction costs of the credit or insurance markets are no longer prohibitive, deposit–refund systems are likely to take the form of a performance bond system.

4.2. Liability

Another possibility for providing incentives to polluters or potential polluters is to make them liable for the actual damage they cause but without demanding a deposit or performance bond. To some extent, of course, they have been liable right along, at least in common-law countries; and remain liable even after the enactment of regulatory legislation aimed at pollution control. This liability arises under the common law of private and public nuisance and is enforceable through the courts; by damaged parties in the former case and, for the most part, by governments in the latter case [Boger (1975)]. The "natural" occurrence of this instrument and thus its apparent independence of government regulatory activity have made it attractive to those who favor minimal government interferences with the functioning of the market system.

   The theoretical literature dealing with liability as a policy instrument for the most part descends from the important and challenging theorem of Coase on the irrelevance of property rights to efficiency outcomes in environmental conflicts [Coase (1960)]. This line of descent is hardly surprising, since the right to enjoy property free from external interference and the entitlement to liability for interferences that do occur are closely related though distinct possibilities for dealing with conflicts over the use of property generally [Calabresi and Melamed (1972), Bromley (1978)].

   This literature is interested in the efficiency properties of these alternative principles and in their comparison with explicit government intervention of the classical Pigouvian sort (e.g. Brown (1973), Polinsky (1979)]. For the most part it confines itself to the case of small numbers of both polluters and damaged parties, though alternative assumptions about the availability of information and the behavior of those parties in bargaining (cooperative or not) are explored. In addition, the costs of enforcement through the courts and the problems of proof of damage for liability purposes are generally ignored. In these circumstances property and liability approaches have been shown to be roughly equivalent in efficiency properties and in terms of protection of the entitlement at issue and both have been shown superior to Pigouvian taxes when behavior is not cooperative [Polinsky (1979)].

Unfortunately, the restriction to small numbers, which frees one from the internal information and decision problems that would be faced by, say, a river basin's population if it tried to act collectively to stop a polluter of the river, and the ignoring of real costs of enforcement make these results of limited interest. Moreover, writers on liability seem, rather surprisingly, to have ignored the problem of incentives for damage-seeking behavior created by the liability payments to damaged parties. As we noted above in Section 2, in discussing the possibility of using emission-charge revenues to indemify pollutees, the conclusion of writers in that literature has been that such a policy would be incorrect. The largest difference between this conventional charge-payment idea and a legislated liability system would be that arising from uncertainty about whether damage would occur or not and whether if occurring they would be compensated. It would still seem to be the case that the more successful a liability system were in guaranteeing compensation, the stronger the incentives it would provide to potential damaged parties. Finally, the problems raised by the uncertainty itself have been disregarded in this literature's comparisons of liability rules and Pigouvian taxes. If discovery and ultimate proof of responsibility are uncertain, the polluter must face a potential payment adjusted to provide the socially correct signal, given that no payment at all may be required even if an incident occurs and damages result. [For a discussion of a related situation, see Shavell (1982).]

Somewhat more to the point is work such as that of Wittman (1977), focusing on the role of monitoring costs in choosing between prior regulation and expost liability for attacking public problems. This points the way to some interesting considerations relevant to choosing liability as a policy instrument. It also emphasizes the close relation between a system of expost liability and some of the deposit–refund arrangements just discussed.

Thus, a liability system, despite its drawbacks, may be a desirable way to approach problems for which information, in any of several senses, is scarce and expensive. For example, take a case in which the prospective damages of some contemplated action (introduction of a new drug or construction and use of a hazardous waste dump site) cannot be estimated even in a meaningful probabilistic sense. This might be true if the experts' prior subjective probability density function were uniform over a very wide range from zero to some catastrophic loss. This provides a weak basis for choosing a particular set of regulations (deciding on a drug ban or on a landfill design requirement) or for setting a fee (for wastes dumped at the site). In such circumstances a designation of strict liability could be appealing. The liability payment might be guaranteed by a performance bond or insurance policy as described in the preceding subsection. It would provide incentives for the active party to engage in information gathering and to take some actions aimed at prevention, at least those where the costs are small and the information or prevention effects are likely to be substantial.

If monitoring of actions to avoid causing damage (e.g. discharge reduction or spill prevention activities) is expensive or technically very difficult, but the sources of actual discharges or spills could be identified expost, a liability rule might usefully substitute for a regulatory rule. If monitoring the quantity of discharge, as opposed to the mere existence of some discharge, is expensive, or if the problem is with spills seen as stochastic discharges, so that fixed fees per unit discharge are difficult to apply, liability may again hold promise.

Notice that even in these rather special situations, the choice of a liability approach is not without serious disadvantages, however. Unless some special process of enforcement were set up, damaged parties would still: suffer real and possibly very serious damage; have to hire lawyers and go to court to claim their entitlements; and have to prove the connection between their damages and the act of the active party. The first of these three requirements must be seen as a political strike against the liability instrument. Policies that appear to prevent damages are surely easier to sell to an electorate than policies that depend for their working on a more or less ironclad guarantee that damage will be compensated by the polluter.

The third requirement, that a connection between action and damage be proved, also looms large as a potential difficulty. If the drug we originally hypothesized could only have one type of ill-effect or if the landfill were completely isolated from other sources of ground-water pollution, the position of the damaged party would be most clear cut. But, if the drug might produce long-delayed symptoms that could also be attributable to naturally occurring disease, or if the landfill site is surrounded by other industrial and commercial establishments (and perhaps even old landfills) proof of the cause–effect relation may be very difficult or even impossible. Special standards of proof (one or another version of a "rebuttable presumption" of causality) may be established to get around this obstacle in particular circumstances, but this course is circumscribed. If every case of X arising within T years among residents of area Y is by fiat to be attributed to our landfill, we must be quite sure that X arises only rarely without *any* obvious cause. Furthermore, we cannot thereafter similarly attribute X to another competing cause should we wish to use the liability instrument in other contexts near Y. This limitation has its most obvious meaning where some ubiguitous cause, such as sulfate air pollution of a metropolitan area is to be attacked by a compensation scheme amounting to the imposition at joint liability on the polluters of the region.

Where these difficulties of proof can be overcome, and where the political objection to a damage-accepting policy can also be overcome, liability as an instrument of policy does offer some dynamic advantages. It is self-adjusting in the face of exogenous change. For example, as technology changes, the polluters can adjust their actions to reflect the new balance between avoidance cost and expected damages. And it provides a continuing incentive to seek new technologies reducing expected damages.

An actual strict liability system, where damages are hard to estimate and preventive action hard to monitor, has been established in the United States Outer Continental shelf oil tract leasing program. Liability for damage from spills attaches to lease owners, and some information is available on the likelihood that a spill in a particular block would affect either fishing grounds or beaches. For a brief description and some preliminary evidence that the value of leases reflects an estimate of potential liability costs (or the costs of their avoidance) see Opaluch and Grigalunas (1983).

For a brief discussion of the problems of liability law in the context of damages from toxic substances, and for suggestions on moving away from that law toward "no-fault" victim compensation funds, see Trauberman (1983). It would appear that the desire to make compensation easy to obtain conflicts with the desire to impose incentives for improving disposal systems on individual generators of hazards. An attempt to make the two goals more compatible is the proposal to fund the U.S. Superfund (for the restoration of hazardous waste disposal sites and other compensation-type activities) from a tax on hazardous waste disposal rather than chemical feedstock use [AWPR (1983)].

5. Regulation

By regulation we mean essentially "a directive to *individual* decision-makers requiring them to set one or more output or input quantities at some specified levels or prohibiting them from exceeding (or falling short of) some specified levels" [Baumol and Oates (1975)]. As pointed out earlier, regulation has been the form of environmental policy preferred by politicians throughout the industrial world. We begin by presenting what appears to be the main arguments for this choice (Section 5.1). The different forms of regulation and their efficiency effects are then discussed (Sections 5.2 and 5.3). Finally, we analyze how the drawbacks of regulations in some applications can be mitigated or eliminated by certain modifications and, in particular, by introducing some complementary element of economic incentive systems (Section 5.4).

5.1. Why politicians prefer regulation

As we shall see in the next section, in some situations regulation emerges as an efficient instrument of environmental policy. However, efficiency aspects alone do not explain why governments in most countries have relied mainly on regulatory instruments in this field. It is hardly an easy undertaking to pin down what these other considerations have been. Different reasons seem to have been invoked in different nations and in different policy situations as well as at different points in time. In addition, important reasons may not have been explicitly invoked at all, implying that their identification becomes guesswork and possibly subject to tendentious interpretations. An attempt to identify the factors which influence the choice of policy instruments is nevertheless central to a discussion of environmental policy alternatives in the real world. This statement is partly due to the fact that not all – perhaps not even one – of these factors are irrelevant from the point of view of the complete set of policy goals and the policy constraints existing in a democratic political environment.

In passing, it should be pointed out that the dominance of direct command and control instruments can be observed not only in environmental but also in other policy areas such as occupational health and safety, consumer protection and transportation. It appears that taxes and charges have rarely been introduced as instruments to control specific activities; and even more rarely have they been designed to achieve a specified goal with respect to such activities. The principal long-term function of these "economic incentive" systems has been to withdraw purchasing power from consumers and firms in order to finance the activities of the public sector. (The economic incentive system of subsidies, on the other hand, seems to have been used as an intentional control instrument, although the transfer of purchasing power may have been an important complementary reason for such a policy.) But this principal function of taxes and charges only increases economists' doubts about why governments "avoid" the use of charges in environmental policy when these charges, in contrast to those instruments now in force would provide revenue to the government without, in principle, any deadweight loss or excess burden.

We now try to identify some of the main reasons why politicians have taken the regulatory approach to environmental policy; most of the reasons that originate in the technical characteristics of environmental problems are left to the next subsection.

(1) In many countries, economists play a minor part, if any, in the administrative groundwork of environmental policy. If the administrators have a background in science, technology or law, the economic aspects will not always be taken into account for obvious reasons. And especially when members of the legal profession dominate the higher echelons of the executive agencies, instruments of the law and of traditional law enforcement are more likely to have the upper hand.

(2) Still, economists have made their voices heard and have confronted politicians with efficiency arguments in favor of economic incentive instruments. One reason why the impact has been small seems to be that these arguments are "sophisticated" and rely on an understanding of the market mechanism and of

the "indirect" effects of prices. In contrast, bans and other forms of regulation are often geared precisely to the activity which is to be controlled. Even when politicians grasp the implications of the alternative policy solutions, they may feel that their constituents would not and that they have to settle for the policy which will receive broad support from the general public.

(3) Financial considerations can prompt the government to prefer regulation to economic incentive instruments. This is obvious for the case of subsidies, but it may concern the case of charges as well. The fact that the effects of effluent charges on ambient quality are uncertain means that government revenue from such charges is also uncertain. This is a drawback from the point of view of budget administration.

(4) Taking other specific policy goals into account can favor the regulatory alternative. Thus, charges will add to inflation, whereas regulation may not do so to the same extent.

(5) Charges can have clearcut distribution effects, which the government may be hesitant to accept. This is so, for example, when low-income groups are in a position where reductions in real income are judged to be unacceptable and when a charge system would hit the consumption of this group. The fact that the distribution effects of regulation are less conspicuous of course does not mean that they are unimportant or even that they are less objectionable than those of a corresponding system of charges.[41] As the time profile of the price effects of charges and regulation, respectively, may be quite different – say, higher prices in the short run with charges than with direct control measures, and vice versa in the long run – an adequate consideration of the distribution aspects becomes quite difficult. But from a political point of view, the short-term distribution effects may be judged to be the most important ones, and here regulation is likely to perform better.

(6) Moving into the sphere of environmental policy proper, it is important to note that, if successful, regulation of discharges or the production processes of polluters will, in general, result in a more certain effect on ambient quality than charges levied on pollutants. As we saw in Section 2, unless the cost function for the reduction of discharges is known, directly or after a trial-and-error process, the effect of a given effluent charge is uncertain. We return to this important aspect in what follows.

(7) Regulation, even when it is less direct that we just suggested, has the aura of being a "no-nonsense" instrument, adequate for the control of serious environmental problems. In contrast, charges have often been viewed as an imperfect obstacle to continued environmental degradation and even as a "license to pollute".

---

[41] See, for example, White (1982, pp. 88, 89), where he estimates that both costs and benefits of the regulation of automobile exhaust emissions in the United States have been regressive. See also Pearce (1983).

(8) It may have come as a surprise to those who hold the view that charges provide a "license to pollute" that the polluting firms and their trade associations seem to prefer regulation to charges. This, in itself, may be enough to make the government choose a regulatory approach. There are at least three reasons for polluters to take a position in favor of regulation:

(a) If charges were to be set at levels which would produce the same reduction of discharges as a regulation in the long run as well as in the short run, it is of course worse for the polluters if they, in addition, would have to pay fees.[42]

(b) As regulation in general can be said to be more uncompromising for the polluters than charges, government is more inclined to listen to the views of the polluters or their representatives before any action is taken. In this process, the polluters may expect to have some influence on the design and stringency of the regulation.[43]

(c) In certain countries, the legal process of introducing new regulations implies drawnout negotiations and provides ample opportunity for appeals. In this way, government intervention may be delayed for a considerable period of time to the benefit of the polluters.

5.2. Forms of regulation: Static efficiency and information

In what follows we take the polluter to be a producer. (This terminology is formally adequate even for a polluting household which obviously not only consumes but also produces effects on others.) The main reason for this choice is a practical one; more often than charges or subsidies, regulation has been and, on administrative grounds, must be aimed at firms.

If a set of effluent charges can be determined so that given ambient standards are met, it is obvious that the same result can be achieved by regulating individual sources of pollution, provided the necessary information is available.[44] Thus, if such charges would make producer A reduce his discharges by 90 percent next year and producer B by 1 percent (due to higher removal costs), effluent standards for the two sources could be so specified.[45] If it were known that the charges would lead to the introduction of a new abatement technology in firm A five years from now and in firm B two years from now, design standards for the two firms

---

[42] On this point, and the possibility that cost savings would more than make up for the added out of the pocket cost of charge payments, see the discussion of Spofford's study above.

[43] For a discussion of the influence of business on regulation in the United States see Quirk (1981). For a different view, see Linder and McBride (1984).

[44] For a simple presentation see, for example, Tresch (1981, pp. 164–168).

[45] However, the *optimal* volume of pollution will, in general, vary with the policy instrument used. See Harford and Ogura (1983).

could be so specified. What may differ between the two alternative policies are the costs of administration, monitoring and enforcement. Once we observe that the necessary information is not *freely* available, however, an even more important difference between the two policies is seen to be the information cost or the availability of the necessary information at any cost. If the necessary information is not attainable, the two alternatives are no longer comparable on a cost-effectiveness basis; policy benefits as well as compliance costs may differ as well. To complicate matters further, given the information constraint, these differences cannot be known in complete detail.

This sets the stage for evaluating the static efficiency of regulation. What are its benefits, compliance costs, information costs, and administrative, monitoring and enforcement costs? Space does not allow us to cover all these aspects, nor does the literature or at least our knowledge of it. Instead we observe the different principal forms of environmental regulation, essentially in the order of decreasing degrees of freedom for the regulated parties, commenting on that appears to be the characteristic differences in the dimensions just referred to.

5.2.1. Forcing the polluter and the pollutee to negotiate

This regulatory approach obviously requires that the parties involved be either few in number or organized in such a way that they emerge as only a few negotiating parties. The two-party case is probably the only one pertinent for this kind of mild regulation. At the one extreme, negotiations would develop similar to those within a merger and lead to an efficient solution. Such an outcome would imply that both parties have free access to relevant information about one another. This outcome is likely only for parties engaged in activities about which there is common knowledge. At the other extreme, information and bargaining strength are unevenly distributed between the parties so that the outcome may be far from a first-best optimum, say close to status quo but with significant negotiating costs being incurred.

Thus, legislation that forces a polluter and a pollutee to negotiate a settlement can be an efficient policy under certain conditions. These conditions would include, in addition to complete information about relevant costs on both sides, sufficiently small monitoring costs, small compliance costs for the polluter, and the threat of alternative measures if a settlement satisfactory to the pollutee is not reached. One important case where this kind of regulation is not likely to be an efficient policy should be mentioned. If the information as specified is far from complete, while the authorities can extract the necessary information at low costs, other solutions, such as a more interventionist form of regulation, may be preferable.

5.2.2. Performance standards

A form of regulation that provides the polluter with maximum freedom of compliance is the establishment of effluent standards for pollutants. Assuming that monitoring does not cause any significant problems and that information about compliance costs is available to the regulator at low costs, this kind of performance standard is likely to qualify as an efficient instrument. It should be noted, however, that the determination of *optimal* effluent standards requires at least as much information as the determination of optimal effluent charges [Mäler (1974b)].

Even when little is known about compliance costs, effluent standards may be more efficient than alternative instruments such as effluent charges.[46] One reason is that the costs of the trial-and-error process of adjusting charges to meet the given standard may be high (see Section 2). Another reason arises when, in a given air- or watershed, there are several polluters, whose discharges have different "transfer coefficients" (the α_i in Section 2.1). As the optimal charges must be source specific in this case, effluent standards would perform at least equally well. Temporary fluctuations in the assimilative capacity of the environment, giving rise to occasional environmental crises, would call for either "unrealistically" frequent changes in charge levels or more constant and occasionally too high charge levels. In such cases, a flexible effluent standard has been suggested as a feasible and more efficient solution [Baumol and Oates (1975, ch. 11), Baumol and Oates (1979, ch. 20) and Howe and Lee (1983)].

Another instance when performance standards can be an efficient instrument has been discussed above in the context of marketable permits (Section 3). In the simplest version of such a system, pollutants released from all members of a given set of sources are taken to have the same environmental impact. Although the initial distribution of pollution rights is specified according to source, the transferability of these rights makes the regulation area-specific instead of source-specific.

Turning to applications of performance standards where inefficiency is likely to result even in the short run, we should note at least the following three cases. First, we have the traditional showcase of inefficient standards, where different polluters with the same environmental effect per unit of pollutant discharged have different marginal removal costs at the individual standards assigned to them (e.g. a 50 percent reduction of discharges for all polluters). Here, a given reduction of pollution is achieved at a higher total costs than would be the case for a uniform charge per pollutant which would equalize marginal removal costs for all polluters.

---

[46] For a pathbreaking analysis of charges vs. standards in the presence of uncertainty about, *inter alia*, compliance costs, see Weitzman (1974). See also the survey article by Yohe (1977). And for a recent extension to more general functional forms and error structures, see Watson and Ridker (1984).

Second, effluent standards are often differentiated between old and new sources of pollution. For example, a producer who operates an existing plant is exempted from pollution control, whereas new or remodeled plants are subjected to emission limits. This application of performance standards provides an incentive to keep old plants for a longer period of time than would be the case, for example, under a system of effluent charges. Obviously, this is a special case of the problem discussed in the preceding paragraph, with effluent standards allocated to polluters regardless of marginal removal costs.

Finally, when monitoring the discharges of pollutants is costly, neither effluent charges nor effluent standards may be the optimal policy choice. The special difficulty of the monitoring problem should be elaborated at this point. This difficulty may be ascribed to five features and applies, as already observed, to charges and marketable permits as well as performance standards.

(1) All emissions are fugitive in the sense that once outside the source's stack or wastewater pipe they are lost to measurement. They leave no trail unless some human agency intervenes.[47] Thus, we cannot monitor at our leisure if we really wish to know what is and has been going on.

(2) Discharges very randomly because of random equipment breakdowns, shifts in product mix or input quality, and changes in production levels at the source. These variations, it must be stressed, are separate from any intention the discharger might have to cheat; even the best corporate citizen can suffer a breakdown of a precipitator in vastly increased emissions. This randomness has itself two implications. First, we cannot usefully think of emission standards as simple fixed numbers. The appropriate orders for a region must take into account source variations and the probability of ambient standard violations. In addition, the orders must recognize in one way or another that in adjusting to the order (or to an economic incentive) the source must balance probability of violation against cost of controlling or narrowing its range of variation.[48] Second, the rules for

---

[47] This statement must be qualified in two ways. Remote monitoring equipment makes it possible to measure concentrations of certain residuals in a stack plume, though these methods are neither simple nor precise. [See Williamson (1981)]. Somewhat more tenuous is the technique of using ambient quality levels and discharge composition to infer discharges, though it might in some cases provide a defensible check on self monitored data. See Courtney, Frank and Powell (1981) and Gordon, (1980). More generally, some residuals are disposed of in "packages" – for example, drums of hazardous pollutants.

[48] At its simplest this means that if the agency orders a source to hold its dischargers below D at all times, the source must actually aim at a target or mean discharge value far enough below D that random occurrences of excess emissions will be so infrequent as to be ignored. How far below D the target emission must be depends on the width of possible swings in discharge, the costs of control, and the penalties for detected violations. If the regulatory agency wants to see the source emit D on average, it must redefine a violation. For example, if it knew the distribution of actual discharges around the source's target, it might define a violation as any discharge greater than $D + K$. K would reflect how closely the source could control its emissions and would be matched to an appropriate penalty reflecting the costs of this control and the acceptable probability of really high emissions (greater than $D + K$).

identifying violations must be consistent with the statement of the discharge limitation orders [e.g. Beavis and Walker (1983)].

(3) Some pollutants are measured using "batch" or discrete sampling techniques.[49] This means that the choice of discharge limitation order and the source's optimal reaction to it should both be complicated by the choice of sampling regime (how often to sample and how many individual samples to draw at a time).[50]

(4) Monitoring instruments are inevitably imprecise – that is, they measure with some error. This further complicates the task of defining and finding real violations.

(5) All the above features of the monitoring problem take on a different cast when we drop the implicit assumption that sources *will* try to obey their discharge limitation orders. Cheating will be worthwhile if the probability of detection and the penalty for a detected violation do not together provide a strong enough incentive. Where intermittent agency monitoring visits are involved, we further have to reckon with legal problems of access to sample, whether (and how much) advance notice is required, and how hard it is for the source to adjust discharges up and down – to avoid being caught cheating. Given these monitoring problems, regulatory orders other than simple discharge limits may be preferable.

5.2.3. Regulating decision variables correlated with emissions

If certain inputs or outputs are perfectly correlated with the volume of pollutants discharged and less costly for the government to monitor, indirect control is more efficient than direct control. This may be true even when correlation is less than perfect, but the advantages of indirect regulation may be limited to the short term and may not even hold for the period during which the firm's basic production process remains unchanged. The correlation between emissions of pollutants and the variable monitored may be based on the inspection of a plant or a piece of equipment when new (see, for example, standards for noise and exhaust emissions from new vehicles) or when carefully maintained with respect to releases of pollutants. This performance may not be representative at later stages of operation or when it is no longer worthwhile for the firm to undertake maintenance. Thus, if the government is forced to rely on information provided by the polluters, the reduction in monitoring costs from making control indirect may be outweighed by the imperfections of such information.

---

[49] It appears that continuous sampling methods with automatic recording are being developed for more and more pollution types, so this difficulty may tend to disappear as time goes on [APCA (1981)].

[50] Sampling size and frequency, given the source's distribution of discharges and the characteristics of the tests performed, define the probabilities of missing the violations and of finding false violations [Vaughan and Russell (1983)].

5.2.4. Design standards

When direct as well as indirect monitoring of releases of pollutants is unreliable, expensive or technically infeasible, requirements that producers use a specific technology become an obvious candidate for optimal policy. Such a policy has been used in practice in a large number of cases. For example, it is often difficult to monitor the source of air or noise pollution. Measuring emissions of BOD in waste water has proved expensive. In such cases, producers can be required to use particular production processes or input qualities (e.g. low-sulphur fuel). Alternatively, they can be required to install a specific kind of abatement or purification process or be forced to reprocess certain kinds of wastes. Or they can be required to transfer certain wastes to publicly owned purification plants, without (as it often happens) being charged the full costs of waste treatment.[51] As a less specific kind of design standard (at the time of the regulatory decision), dischargers may be required to apply the "best practicable technology" (BPT) or "best available technology" (BAT) at some given future date.

Design standards can be efficient policy not only for reasons of low monitoring costs. They also provide a way to save information costs among polluters. When there is no doubt about the most efficient solution to meeting a certain performance standard, a design standard is the obvious policy choice [Crandall (1979)].

But, when there are doubts about the most efficient approach to meeting a performance standard, the requirement that a specific technology be used is likely to cause misallocation of resources. For all firms in an industry, a series of small adjustments of the existing production processes or simply reduced output may turn out to be less costly alternatives to the required production process or abatement technology. More often perhaps, different firms in an industry have different least-cost solutions to the reduction of discharges accomplished by a certain design standard [see, for example, OECD (1982a, 1982b)].

Many of the political aspects discussed in the preceding subsection may explain why politicians often prefer the design standard solution. Installation of purification equipment is the "natural" policy if you want wastes to contain a smaller volume of pollutants; moreover, it may appear as an effective instrument if you want to satisfy the environmentally conscious general public, etc. Above all, perhaps, design standards are believed to contribute to protection of the environment with a high degree of certainty. However, there is evidence that the security provided by design standards in environmental policy is false or exaggerated in a number of cases. Thus, as touched upon earlier, the amount of actual discharges for which the required process was designed may be exceeded dramatically [see,

---

[51] This may be seen as a combination of a design standard and a subsidy. It is a subsidy in the sense that all costs of the regulation are not borne by the regulated party. Combinations of this kind have been quite popular with policy-makers, involving either lump-sum subsidies or subsidizing a part (or percentage) of the costs incurred, e.g. a percentage of the installation costs for the equipment required.

for example, Mäler (1974b)]. And equipment which meets certain standards when leaving the producer may be tampered with by the user; although peripheral to the case of design standards, the difference between emission levels for new cars and actual in-use emissions is a good illustration [White (1982)]. There are also indications that stricter standards for new equipment are circumvented by increasingly frequent modification of the equipment when in use.[52]

In many cases where design standards have not proved effective in practice, the problem has not so much been the standards themselves as the way they are enforced or checked. Thus, inspection of plants or equipment when in use can improve the results of design standards. However, the advantages in terms of low administrative costs that this kind of regulation was credited with may be lost in the process.

5.2.5. Bans on products or processes

Outright bans may appear to be the strictest form of regulation. Banning the production (or use) of a product which has no close substitutes is a case that supports this view. But close substitutes are often available at low extra costs (as is illustrated, for example, by the appearance of other propellants for aerosol sprays when chlorofluorocarbons were banned in certain countries). And this may be true when bans are imposed on certain inputs, such as high-sulphur fuel in certain areas. Moreover, when bans take the form of zoning or curfews, compliance costs may be small, because alternatives remain open to the regulated party. This is so in particular when bans are announced well in advance. In this perspective, design standards rather than bans represent the most severe type of regulatory constraint.

It follows from what we just said that bans on products or processes may be an efficient policy instrument when there are close substitutes at low additional costs. Moreover, bans – and even more, design standards – may make economies of scale in the production of the substitutes (the required or nonbanned equipment) materialize faster than through the market mechanism by itself. In fact, nonconvexities in production may prevent the market mechanism from ever reaching a point which is less harmful for the environment, and, at the same time, less costly; in such a case, regulation may be the obvious way to eliminate, as it were, the two market failures.

A similar case of non-convexities appears when the pollution problem is only latent, but still the source of inefficient resource allocation. This is the case, for example, where an existing plant pollutes the environment so that certain other activities sensitive to the pollution have never been established in the vicinity, although the social surplus would be higher if they were than if the existing firm

---

[52] See Broder (1982, ch. 5) for the case of noise emissions from motorcycles.

were kept there.[53] Charges are not likely to work in this situation, especially not if they should reflect the value of the latent externalities; an arrangement along such lines might incite blackmail or at least create insurmountable information problems. A ban on pollution, e.g. in the form of zoning, is perhaps the obvious choice of policy in this "no-pollutee case", given that the optimum form of land use has been identified.

The traditional case for bans is, of course, when environmental standards call for the elimination of a certain kind of discharges, such as highly toxic substances. In addition, even though zero pollution from a particular type of activity is not called for, a ban may be chosen for administrative reasons, e.g. because it is immediately apparent when the ban has been broken.

5.2.6. *Collective facilities – a digression*

Government investments in facilities for environmental protection (sewers, waste treatment plants, walls for protection against motorway noise, etc) or government restoration activities (cleaning up, reforestation, reaeration of lakes, etc.) bear some resemblance to the regulatory solution and may be discussed at this point.

The analytical background for government protection and restoration activities can be briefly outlined as follows. If costs of protection/restoration fall short of the value of the corresponding environmental damages, there is a case for protection/restoration. Furthermore, if collective protection/restoration activities are less costly than environmental protection administered by the polluters individually, the collective alternative is favored. To implement this kind of policy, it may be sufficient for the government to ban certain kinds of discharges into the environment, provided that this ban actually institutes voluntary actions leading to the emergence of the optimal, collective arrangement. An illustrative example here could be the emergence of privately owned refuse collection activities as a consequence of such a ban.

Privately owned facilities of this kind may not materialize for reasons of administrative complexity or when the protection involved is a pure public good, instigating free-rider behavior among individual members of the common-interest group. Or organization costs may simply be believed to be too high, e.g. due to fears that several competing units may be established (at least temporarily) for a private-good kind of activity subjected to large economies of scale. Or a privately owned natural monopoly, once established, may charge monopoly prices. For

---

[53] For example, the existing firm A runs at a profit of $1 million per year. The "other activities", if firm A were absent, would run at an aggregate profit of $2 million per year. However, when A is present, they would not be able to make a profit due to pollution from A. Moreover, costs of organizing these other activities or lack of available funds bar the formation of an interest group which could buy firm A and shut it down. Or, there may be space for a new firm A' to locate in the area once firm A is shut down. Hence, for several reasons, the market cannot make the optimum allocation materialize.

reasons such as these, the government may prefer to give the protection activities a public-utility status with the government control accompanying such a status; or the activities may be operated directly by the government. To implement such a choice, the government may want to complement the ban with, or have it replaced by, a design standard requiring the polluters to be connected to a central waste treatment plant.[54]

In other instances of government provision of collective facilities, no act of regulation may be involved. This is the case, for example, with most forms of restoration campaigns as well as with all improvements of existing waste treatment plants. To evaluate whether such activities are worthwhile it is only required that they meet the relevant cost–benefit criterion.

5.3. Regulation and dynamic efficiency

In the preceding section, our primary objective was to describe the principal forms of regulation and their static or short-term efficiency characteristics. In this section, we discuss regulatory instruments with respect to efficiency over time. Economists' evaluations of environmental regulation have to a large extent concentrated on this aspect. Here we discuss the following three issues: adaption to changes in exogenous variables, incentives to develop new forms of pollution-abatement technology, and effects on market structure and competition.

5.3.1. Environmental regulation in the presence of exogenous changes

Efficiency over time requires, in principle, that policy be adapted to exogenous changes in environmental costs as well as compliance or removal costs, subject to administrative and other specific costs associated with policy change. As mentioned above these costs of policy change may be lower for regulatory instruments than for economic incentive systems in the context of short-term fluctuations in the assimilative capacity of the environment. This might extend to the case of exogenous long-term fluctuations as well. In practice, however, regulation may not be administered with sufficient flexibility to take advantage of this potential. This is likely to be true at least for certain forms of regulation such as design standards, for which the regulatory process may be very slow.

If it turns out that regulation and economic incentive systems in fact tend to be equally inflexible over time, we may investigate the relative merits of the two policy approaches when confronted by exogenous changes. Assume a situation where a system of effluent charges and a system of effluent standards would be

---

[54] For a discussion of the choice between pollution charges leading to individually administered protection and forcing or simply allowing polluters to connect to centralized waste treatment activities, see Bohm (1972).

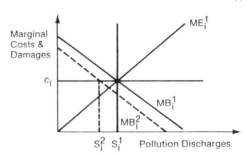

Figure 10.1. Charges, standards, and change in the marginal costs of reducing pollution discharges.

equally efficient and successful in guiding the economy to a short-term optimum position. See the standard S_i^1 for firm i and the uniform charge per unit of pollutant c_i in Figure 10.1. Emissions are brought to the point where the initial marginal benefit curve of the polluter (MB_i^1), also the reverse of the marginal compliance cost curve, intersects the curve for the marginal environmental effects (ME_i^1). Now, we should note first of all that neither the inflexible charge system (with unit charges independent of emission volume) nor the inflexible standard would allow any adjustments when the estimated marginal environmental effects (ME_i) change. On the other hand, when external factors influence the marginal compliance costs (MB_i), some adjustment will automatically take place in the charges case. (MB_i^2 produces discharges equal to S_i^2 when a charge is applied.) But as long as the charges do not change to perfectly reflect the marginal environmental effects, these adjustments may not be preferable to the absence of adjustments in the standard case. It is clearly seen that the outcome will depend on the extent to which both the marginal environmental effects and the marginal compliance costs change in the relevant interval. The charge system will tend to perform better than the standard if the marginal environmental effects are sufficiently close to being constant around the initial optimum point. Conversely, if these effects rise at a sufficiently high rate at this point, the inflexible standard will be the least imperfect instrument of the two [see Weitzman (1974) and Yohe (1977)].

5.3.2. *Endogenous adjustments of compliance costs*

When subjected to a given policy, the polluter has, in principle, a larger number of adjustment options open to him, the longer the adjustment period. Furthermore, if there are incentives for the polluter to develop new forms of adjustments – something which may be influenced by policy design – additional options may emerge over time. For these two reasons, compliance costs of a given policy will in general be lower in the long run than in the short run. This also means that compliance costs at the time the policy takes effect will be lower if advance notice of the policy is given [see Kneese and Schultze (1975, pp. 79–80)]. And this may be increasingly important the fewer the options allowed by the policy. For example, compliance with a design standard introduced on short notice may be very costly to the firm; say if the plant has just been remodeled. In contrast, an effluent standard – and even more an effluent charge – may allow the firm to make a much less costly temporary adjustment and introduce the technology implied by the design standard at a later stage, assuming this standard is the most efficient form of long-term adjustment.

Incentives to develop new options diminish the smaller the scope of adjustment allowed by the policy, *ceteris paribus*. Thus, with effluent charges, a maximum number of compliance alternatives are acceptable and hence, technological R & D may be pursued in any direction. At the other extreme, a design standard leaves no room for innovation. Or this is so at least if policy cannot easily be redesigned should new and superior ways to meet a given ambient standard happen to be developed. The important aspect from the incentive point of view is, of course, to what extent the firm believes it to be possible to influence policy by developing new and more efficient technology.

Moreover, once the polluter has adjusted to the new piece of regulation, there is no longer any incentive for him to attempt reaching a lower level of pollution than that implied by the regulation (be it a performance or a design standard), even when such a reduction would be valuable to society. Charges, on the other hand, provide such an incentive although its size may be nonoptimal (e.g. too large in the situation portrayed in footnote 22). Certain forms of regulation may even actually discourage the development or introduction of innovations. Thus, establishing shifting BAT standards for an industry creates perverse incentives for innovation.[55]

Although no real-world policy instrument can be expected to send correct signals to guide the long-term adjustment of pollution abatement and the development of new abatement technology, regulation and especially design standards are likely to perform much worse than economic incentive systems in these

---

[55] For example, in the U.S. Clean Water Act explicitly, and at least in the rhetoric surrounding the Clean Air Act, improvements in technology are supposed to trigger tightening of the standards [Clean Water Act, Section 302d in Government Institute (1980)]. This reduces the incentive to seek cost-reducing technical improvements in production process or treatment equipment, and under some circumstances may eliminate the incentive altogether. A very simple way of looking at this process uses the figure in footnote 22. When technology is improved, and marginal cost falls to MC_1, the ratchetting-down requirement implies a new lower discharge standard. Let us say that the rule for choosing this level is to maintain equal marginal costs (e_0) before and after. Then, after technical change, the standard would be D_1, and the net savings to the firm would be C-F. In this figure, area F will always be greater than area C, so there is a *disincentive* to innovate. More generally, the existence of the additional cost, F, will at least reduce the positive incentive to innovate.

respects. Thus, a policy that relies on regulatory intervention tends to make the long-term costs of attaining a given ambient quality unnecessarily high. This does not mean, however, that environmental regulation must lead to a reduction in productivity as commonly measured. In fact, there are some indications that increasingly stringent effluent standards have operated as a challenge to industry and spurred an innovation response whereby both pollution has been diminished and productivity has increased [OECD (1982a, 1982b)]. This is not to say, of course, that policy instruments, which allow a still larger freedom of adjustment and provide stronger incentives for developing new ways of reducing pollution, would not have performed even better.

5.3.3. *The effects of regulation on market structure*

If industry has an influence on the design of environmental regulation and the larger firms play a prominent role in this process, the result may be unfavorable for the smaller competitors in the industry. Moreover, the use of design standards requiring new production processes or the installation of expensive pollution-abatement technology may hit small firms particularly hard.[56]

If regulation tends to disfavor certain types of firms in an industry, the effect may be that competition is reduced [see Buchanan and Tullock (1975) and Dewees (1983)]. This effect may be particularly serious if mainly innovative firms (e.g. small growing firms) are hit hard by regulation. Moreover, if control is tighter for new firms, competition and innovation in the industry may be reduced still further [OECD (1982b)]. All this would contribute to maintaining a high level of direct as well as indirect compliance costs of regulation in the long run.

5.4. *Modifying the performance of regulatory instruments*

Some ways to improve the efficiency of the regulatory approach follow from our discussion in the preceding section. First of all, we saw that adding dynamic efficiency aspects to the static ones presented in Section 5.2 suggests that regulatory design be shifted towards forms which allow more freedom of adjustment. Second, advance notice of a given piece of regulation tends to reduce compliance costs. Third, design standards and other inflexible forms of regulation may be less costly to society if government shows a willingness to redesign its rulemaking when new solutions for protecting the environment emerge. In this way, the regulated party may be given an incentive to undertake R&D of new pollution-abatement technologies. In contrast, the use of BAT standards and a tendency to introduce stricter standards for industries that have developed less

---

[56] See Grabowski and Vernon (1978) for examples from the field of consumer product safety regulation.

harmful production processes are likely to impede innovation. Hence, compliance costs for a given ambient quality are increased or political ambitions with respect to ambient quality may have to be lowered.

In addition to modifications of the type suggested above, the regulatory system may be improved by introducing elements from economic incentive schemes. In this way, the high degree of certainty as to the effects of regulation, which – right or wrong – seems to be decisive for policy choice in the real world, can be obtained along with a stimulus towards efficiency that otherwise may be absent.

First, it should be noted that an economic incentive element is in fact already incorporated into most forms of regulation. If a polluter fails to comply with the directive given to him, he may be fined for doing so. A disadvantage of this regulatory design is, however, that the exact penalty level often is not known beforehand.

The problem of uncertain penalties would be eliminated if regulated parties were confronted with explicit, punitive *non-compliance fees* [see, for example, Viscusi (1979)]. That is, the polluter is formally allowed to exceed the standard given to him and will do so if his compliance costs are high.[57] Although regulation might seem less stringent as a consequence of such a system, it should be noted that this kind of legalized non-compliance allows standards to be set at a more demanding level than otherwise.

In practice, the application of noncompliance fees is often subject to severe imperfections. Thus, the fee is frequently calculated to equal the regulated party's gain from non-compliance; in other words, the fee is not punitive. Given that non-compliance is not always detected and that the regulated party's gain is likely to be underestimated by an outside party such as the government, this kind of policy can hardly be conceived of as rational. For example, it is difficult to see why the polluter would pay any attention to the standard imposed under these circumstances, unless, of course, there were additional and diffuse costs of stigmatization embedded in non-compliance.

As another form of incentive element, effluent charges could be levied on the polluter along with an effluent standard.[58] Assuming that the standard is binding when initially introduced, the effect of the charge would be to promote a future reduction in pollution below the level of the standard. This would increase long-run efficiency, provided, of course, that the value of further reductions in pollution were sufficiently high. Alternatively, reduction in pollution below the

---

[57] This idea, which in the United States originated as a practical policy in Connecticut and came to Washington with Douglas Costle, former Administrator of EPA, is now part of the Clean Air Act. (Section 120 of the Clean Air Act is devoted to a noncompliance penalty system.) See also, Drayton (1980). It allows EPA administratively to assess, on a source not complying with discharge regulations, a penalty equal to what the agency calculates the source would save through its noncompliance.

[58] For a version of an optimal mixed program of this kind see Baumol and Oates (1975, pp. 162–171).

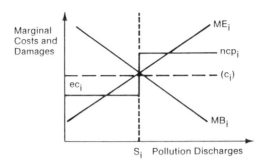

Figure 10.2. Combination of charge, standard, and non-compliance penalty.

level of the standard could be subsidized. The overall effects of a subsidy would differ from those of an equally large effluent charge, unless the income effect of the charge/subsidy could be disregarded and the shadow price of a dollar of government revenue equalled a dollar.

In principle, such a combination of standards and charges (subsidies) would provide the advantages of both systems, i.e. the relative certainty of a maximum limit to pollution and an incentive for the polluter to seek ways to reduce pollution still further in the short as well as in the long run. Furthermore, the standard could be combined with both (a subsidy or) an effluent charge (ec) and a noncompliance penalty (ncp), exceeding the level of the effluent charge.[59] Given sufficient information about the (nonconstant) marginal environmental effects of discharges and about the frequency distribution of the curve displaying the polluter's marginal benefits of discharges, such a system could be more efficient than a pure system of uniform effluent charges (see Figure 10.2). However, a prerequisite for such an outcome is that discharges from one polluter do not significantly alter the marginal environmental effects of discharges from another.

To sum up: although the actual use of environmental regulation appears to be based largely on factors unrelated to efficiency, there are as we have seen a number of instances in which efficiency aspects call for regulation instead of economic incentive schemes. But when emphasis is placed on long term efficiency and on the strength of the profit motive for seeking innovations in pollution abatement, economic incentives become more important. How much more im-

[59] See Roberts and Spence (1976). A fee-subsidy system was developed by James Smith and his colleagues at the City of Philadelphia Air Management Services and is reported in *Feasibility Study: A Fee/Subsidy System for Controlling Sulfur Dioxide Emissions*; a multiple-volume set of working papers by various authors [Philadelphia Air Management Services (1981)]. The aim here is zero net revenue raising (zero net transfer cost to polluters) and the mechanism is a combination of a specified emission level for each source, a fee for emissions over that level, and a subsidy for reductions in emissions below the chosen level.

portant is a matter of belief in the innovating capacity of the polluters and special firms developing pollution abatement equipment. However, combining standards and economic incentive schemes opens up a possibility to extract some of the best from both approaches. Such a combination can be obtained with a system of marketable permits, as discussed in Section 3. Or, a set of charges or subsidies and/or noncompliance penalties can be added to the set of standards. But for this to be meaningful, the standards must be of a type that allows some freedom of adjustment.

6. Moral suasion

As we saw in the preceding section, the choice of environmental policy instruments may be influenced by a number of "non-economic" factors. As a special but probably not unusual case, the policy-maker is confronted with a situation where there are definite constraints on the set of policy instruments. The origin of such constraints may be found in the political interpretations of public opinion. Thus, for example, it may become clear or interpreted as clear that charges on polluters are out-of-bounds, politically speaking, whereas subsidies are not or vice versa.

Estimates of compliance costs, employment effects, etc. made by interest groups often play a prominent role in the formation of such constraints. Typically, these estimates are based on insights that outsiders, and among them the government, cannot check. In particular, the effect of pollution charges on employment and the volume of exports of an industry may be greatly exaggerated by industry representatives without anyone else being able to prove that these estimates are biased and even less, of course, to ascertain the extent of which they are biased. [See Sonstelie and Portney (1983) for possible solutions to some of these problems.]

Thus, political constraints on environmental policy (to be distinguished from observing other goals of economic policy such as distribution goals) may be in force and turn the choice of optimal policy into a second best problem. In the limiting case, all stringent political actions to meet certain government policy goals may be blocked by such constraints. In that case, only actions that are voluntary on the part of the polluters are open to government influence. We now turn to a brief discussion of this "policy of moral suasion", which has occasionally been used and, in some cases, has proved to be effective.

Government initiatives to influence behavior on a voluntary basis can hardly be expected to be effective in all instances of environmental protection. If the environmental hazards are not conspicuous or dramatic enough, moral pressure may not materialize among any significant number of people. Similarly, when it is generally felt that the formal or moral responsibility rests with an identifiable party, others may not be easily convinced to take action. But many cases of environmental degradation are characterized by a lack of well-defined property rights and hence by unclear responsibility. In such cases, protection of the environment may be seen as a moral concern for people in general.

Attempts to influence the behavior of *individuals and households* can gain support from existing attitudes and social valuations related to the environmental issue involved. Thus, an attempt to make people abstain from buying fur coats to protect endangered species may receive firm backing from people who cannot afford to buy them. In other cases, where voluntary actions to protect the environment are conspicuous, such actions may be supported by feelings of cooperation and shared interests. This is probably more true for non-government initiatives than for the government-initiated attempts to influence behavior, which we are concerned with here. But this distinction may be less relevant for certain countries or local areas with a tradition of consensus on a large number of social issues. Thus, for example, attempts to make people voluntarily return used mercury and nickel-cadmium batteries to sellers have been fairly successful in some countries [see, for example, OECD (1981)]. A more general problem is that, unless new habits have had time to be formed, moral suasion may be effective only for a short period – as long as the arguments seem new and compelling.[60]

The likelihood of persuading *firms* to take voluntary action of reduce pollution (without the backing of a threat of harsher measures) is even smaller. Firms under the pressure of competition can be assumed to pay attention to arguments without a legal or economic content only when their costs of reducing pollution are negligible. Exceptions will be found when a conspicuous attempt to take moral arguments into account would serve the purpose of sales promotion, as when consumers have been building up a demand for new products with less negative effects on the environment (such as low phosphate detergents). But in such cases, unless the new product happens to be as effective, attractive and inexpensive as the original one, it is the consumers who pay the costs.

It should be noted in this connection that the relation between voluntary actions and constraints on policy may be the opposite of the one assumed here. Thus, firms may support voluntary programs among consumers or take voluntary actions on their own as an offensive measure to block the government from using more stringent and more effective policy instruments in the future.

So far we have discussed whether it is worthwhile for the government to undertake moral suasion when other instruments are blocked. But as pointed out by Baumol and Oates (1979), there are instances when such a policy is in fact more efficient than other instruments. First, this may occur when the monitoring required for economic incentive schemes or regulation is ruled out as being

---

[60] See, however, Baumol and Oates (1979, ch. 19) for examples where voluntary actions have remained in effect for longer periods of time.

technically infeasible or prohibitively expensive. For example, improper disposal of hazardous material into the sewage system is difficult to control by such methods, as is littering or careless use of open fire in wilderness areas. Here, moral suasion may be more efficient than realistic versions of other instruments. That this approach can also be quite effective is supported to some extent by experience from campaigns against littering in Scandinavia and "Smokey the Bear" forest-fire prevention campaigns in the United States.

Second, in certain cases of environmental catastrophies or immediate risks of such catastrophies (e.g. extremely hazardous smog levels), ordinary policy instruments may be too cumbersome or simply too slow. Again, there are examples which show that government appeals for voluntary actions can work fast and have an important impact in such voluntary situations.

To sum up, there are indications that, in certain cases, it may be worthwhile for the government to rely on moral suasion when alternative measures are blocked for political reasons. In addition, even when more sophisticated policy alternatives are available, there are cases when moral suasion emerges as an efficient policy instrument.

7. Concluding remarks

The message of this chapter may be seen as either negative or positive, depending on the perspective of the reader. The negative version is that no general statements can be made about the relative desirability of alternative policy instruments once we consider such practical complications as that location matters, that monitoring is costly, and that exogenous change occurs in technology, regional economies, and natural environmental systems. The positive way of stating this result is to stress that all the alternatives are promising in some situations. Even design standards have a place in the armamentarium of the environmental policy-maker. If the classic case for the absolute superiority of effluent charges is flawed by the simplicity of the necessary assumptions, the arguments for the superiority of rigid forms of regulations suffer equally from unstated assumptions and static views of the world. There is no substitute for careful analysis of the available alternatives in the specific policy context at issue.

That said, however, we are still tempted to stress the advantages of economic incentive systems in the long-run context, at least as a complement to a regulatory approach. The extra push toward the development of new production and discharge reduction technology provided by these instruments seems likely to dwarf in importance the short-run, and to some extent illusory, advantages to be gained by specifying actions or stigmatizing pollution at any non-zero level. Furthermore, we believe it worthwhile expanding the fields of application contemplated for such relatively unexplored instruments as deposit–refund systems.

Some exploration and experimentation can be done in real policy problems, but in many instances realistically complicated models will, we anticipate, provide insights currently lacking because of the simplicity of available theoretical models and the narrowness of actual experience.

References

Adar, Z. and J.M. Griffin (1976) "Uncertainty and Choice of Pollution Control Instruments", *Journal of Environmental Economics and Management* 3, no. 3, 178–188.

APCA (1981) *Continuous Emission Monitoring: Design, Operation, and Experience* (Air Pollution Control Association, Pittsburgh).

Atkinson, Scott E. and Donald H. Lewis (1974a) *A Cost Evaluation of Alternative Air Quality Control Strategies*, Report No. EPA 600/5-74-003, (USEPA, Washington Environmental Research Center, Washington).

Atkinson, Scott E. and Donald H. Lewis (1974b) "A Cost-Effective Analysis of Alternative Air Quality Control Strategies", *Journal of Environmental Economics and Management* 1, 237–250.

Atkinson, Scott E. and T.H. Tietenberg (1982) "The Empirical Properties of Two Classes of Designs for Transferable Discharge Permit Markets", *Journal of Environmental Economics and Management* 9, no. 2.

AWPR (Air/Water Pollution Report) (1983) *Chemical Industry, Environmentalists Support Tax on Waste Disposers to Finance Superfund* (May) p. 202.

Baumol, William J. and Wallace E. Oates (1971) "The Use of Standards and Prices for Protection of the Environment", *Swedish Journal of Economics*, 73, no. 1.

Baumol, William J. and Wallace E. Oates (1975) *The Theory of Environmental Policy* (Prentice-Hall, Englewood Cliffs).

Baumol, William J. and Wallace E. Oates (1979) *Economics, Environmental Policy, and the Quality of Life* (Prentice-Hall, Englewood Cliffs).

Beavis, Brian and Martin Walker (1983) "Achieving Environmental Standards with Stochastic Discharges", *Journal of Environmental Economics and Management*, 10, no. 2, 103–111.

Boger, Kenneth S. (1975) "The Common Law of Public Nuisance in State Environmental Litigation", *Environmental Affairs* 4, no. 2, 367–392.

Bohm, Peter (1972) "Pollution: Taxation or Purification?", *Kyklos*, Vol. XXV, no. 3, pp. 501–517.

Bohm, Peter (1981) *Deposit–Refund Systems: Theory and Applications to Environmental, Conservation, and Consumer Policy* (Baltimore: Johns Hopkins University Press for Resources for the Future).

Bower, B.T., Remi Barre, Jochen Kuhner and Clifford S. Russell (1981) *Incentives in Water Quality Management: France and the Ruhr Area* (Resources for the Future, Washington).

Brady, Gordon L. and Richard E. Morrison (1982) *Emissions Trading: An Overview of the EPA Policy Statement.* (Washington, D.C., National Science Foundation) Policy and Research Analysis Rep. 82-2 (May).

Broder, Ivy E. (1982) "Regulation of Transportation Noise", Working Paper no. 18, American Enterprise Institute, Washington, mimeo.

Bromley, Daniel W. (1978) "Property Rules, Liability Rules and Environmental Economics", *Journal of Economic Issues* 12, 43–60.

Brown, Gardner (1982) "The Effluent Charge System in the Federal Republic of Germany", unpublished report, University of Washington, Department of Economics, Seattle.

Brown, John Prather (1973) "Toward an Economic Theory of Liability", *Journal of Legal Studies* 2 no. 2, 323–349.

Buchanan, James M. and G. Tullock (1975) "Polluters' Profits and Political Response: Direct Control versus Taxes", *American Economic Review* 65, no. 1, 139–147.

Burrows, Paul (1981) "Controlling the Monopolistic Polluter: Nihilism or Eclecticism", *Journal of Environmental Economics and Management* 8, no. 4, 372–380.

Calabresi, Giudo and A. Douglas Melamed (1972) "Property Rules, Liability Rules, and Inalienability: One View of the Cathedral", *Harvard Law Review* 85, 1089–1128.

Case, Charles D. (1982) "Problems in Judicial Review Arising from the Use of Computer Models and Other Quantitative Methodologies in Environmental Decisionmaking", *Boston College Environmental Affairs Law Review* 10, no. 2, 251–363.

Cass, Glen et al. (1980) "Implementing Tradeable Emissions Licenses: Sulfur in the Los Angeles Air Shed", Environmental Quality Laboratory, California Institute of Technology, for the National Commission on Air Quality.

Coase, Ronald H. (1960) "The Problem of Social Cost", *Journal of Law and Economics* 3, no. 1, 1–44.

Collinge, Robert A. and Martin J. Bailey (1983) "Optimal Quasi-Market Choice in the Presence of Pollution Externalities", *Journal of Environmental Economics and Management* 10, no. 3, 221–232.

Courant, Carl (1980) "Emission Reductions From Shutdowns: Their Role in Banking and Trading Systems", U.S.E.P.A., unpublished.

Courtney, F.E., C.W. Frank and J.M. Powell (1981) "Integration of Modelling, Monitoring, and Laboratory Observation to Determine Reasons for Air Quality Violations", *Environmental Monitoring and Assessment* 1, no. 2, 107–118.

Crandall, Robert W. (1979) "Reforming Social Regulation," The Brookings Institution, Washington, mimeo.

Crocker, Thomas D. (1966) "The Structuring of Atmospheric Pollution Control Systems", in: H. Wolozin (ed.), *The Economics of Air Pollution* (W.W. Norton, New York).

Crone, Theodore M. (1983) "Transferable Discharge Permits and the Control of Stationary Air Pollution: Comments", *Land Economics* 59, no. 1, 123–125.

Dales, J.H. (1968a) "Land, Water and Ownership", *Canadian Journal of Economics* 1, 797–804.

Dales, J.H. (1968b) *Pollution, Property and Prices* (Toronto: University Press).

David, Martin, W. Eheart, E. Joeres and E. David (1980) "Marketable Permits for the Control of Phosphorous Effluent into Lake Michigan", *Water Resources Research* 16, no. 2, 263–270.

deLucia, R.J. (1974) *Evaluation of Marketable Effluent Permit Systems*, Office of Research and Development, U.S. Environmental Protection Agency (Government Printing Office, Washington).

Dewees, Donald N. (1983) "Instrument Choice in Environmental Policy", *Economic Inquiry* 21, no. 1, 53–71.

Drayton, William (1980) "Economic Law Enforcement", *The Harvard Environmental Law Review* 4, no. 1, 1–40.

Eheart, J. Wayland (1980) "Cost-Efficiency of Transferable Discharge Permits for the Control of BOD Discharges", *Water Resources Research* 16, 980–986.

Elliott, Ralph D. (1973) "Economic Study of the Effect of Municipal Sewer Surcharges on Industrial Wastes and Water Usage", *Water Resources Research* 9, no. 5, 1121–1131.

Fisher, Anthony C. (1981) *Resource and Environmental Economics* (Cambridge University Press, Cambridge).

Førsund, Finn R. (1972) "Allocation in Space and Environmental Pollution", *Swedish Journal of Economics* 74, 19–34.

Førsund, Finn R. (1973) "Externalities, Environmental Pollution, and Allocation in Space: A General Equilibrium Approach", *Regional and Urban Economics* 3, no. 1, 3–32.

Freeman, A. Myrick III (1970a) "Should Maine Polluters Pay "By the Pound?", Maine *Times*, 27 (February).

Freeman, A. Myrick III (1970b) "An Amendment of Title 38 of the Maine Revised Statutes to Institute a System of Effluent Charges to Abate Water Pollution", unpublished.

Freeman, A. Myrick III (1978) "Air and Water Pollution Policy", in: Paul Portney (ed.), *Current Issues in U.S. Environmental Policy* (Johns Hopkins University Press for Resources for the Future, Baltimore).

Ginberg, P. and Grant W. Schaumberg, Jr. (1979) *Economic Incentive Systems for the Control of Hydrocarbon Emissions from Stationary Sources*, Report by Meta Systems, Cambridge, Massachusetts, to U.S. Council on Environmental Quality (September).

Gordon, Glen E. (1980) "Receptor Models", *Environmental Science and Technology* 14, no. 1, 792–800.

Government Institute (1980) *Environmental Statutes: 1980 Edition*, (Government Institutes, Washington).

Grabowski, Henry G. and John M. Vernon (1978) "Consumer Product Safety Regulation", *American Economic Review* 68, no. 2, 284–289.

Gulick, Luther (1958) "The City's Challenge in Resource Use", in: H Jarrett (ed.), *Perspectives on Conservation*, (Johns Hopkins University Press for Resources for the Future, Baltimore).

Hahn, Robert W. (1980) "On the Applicability of Market Solutions to Environmental Problems", unpublished. EQL paper 314-40 (California Institute of Technology, Pasadena) (February).

Hahn, Robert W. and Roger Noll (1981) "Implementing Tradable Emissions Permits", a paper prepared for the Conference on Reforming Government Regulation: Alternative Strategies to Social Regulatory Policy. (California Institute of Technology, EQL, Pasadena) (February).

Harford, J. and S. Ogura (1983) "Pollution Taxes and Standards: A Continuum of Quasi-optimal Solutions", *Journal of Environmental Economics and Management* 10, no. 1, 1–17.

Howe, Charles W. and Dwight R. Lee (1983) "Priority Pollution Rights: Adapting Pollution Control to a Variable Environment", *Land Economics* 59, no. 2, 141–149.

Inside EPA (1983) "Durenberger May Draft Acid Rain Bill to Tax SO_2 Emissions Across Nation", 20, 4.

Jacobs, J.J. and G.L. Casler (1979) "Internalizing Externalities of Phosphorus Discharges from Crop Production to Surface Water: Effluent Taxes versus Uniform Restrictions", *American Journal of Agricultural Economics* 61, 309–312.

Johnson, Edwin (1967) "A Study in the Economics of Water Quality Management", *Water Resources Research* 3, no. 2.

Johnson, Ralph W. and Gardner M. Brown (1976) *Cleaning Up Europe's Waters*, (Praeger, N.Y.).

Kelman, Steven (1981) *What Price Incentives?* (Auburn House Publishing Company, Boston).

Klevorick, Alvin K. and Gerald H. Kramer (1973) "Social Choice on Pollution Management: The Genossenschaften", *Journal of Public Economics* 2, 101–146.

Kneese, Allen V. (1964) *The Economics of Regional Water Quality Management* (Johns Hopkins University Press for Resources for the Future, Baltimore).

Kneese, Allen V. and Blair T. Bower (1972) *Managing Water Quality: Economics, Technology, Institutions* (Johns Hopkins University Press for Resources for the Future, Baltimore).

Kneese, Allen V. and Charles E. Schultze (1975) *Pollution, Prices, and Public Policy* (The Brookings Institution, Washington).

Krupnick, Alan J. and Wallace E. Oates (1981) "On the Design of a Market for Air-Pollution Rights: The Spatial Problem", unpublished paper, University of Maryland, Department of Economics.

Krupnick, Alan J., Wallace E. Oates and Eric Van De Verg (1983) "On Marketable Air Pollution Permits: The Case for a System of Pollution Offsets", *Journal of Environmental Economics and Management* 10, no. 3, 233–247.

Kurtzweg, Jerry A. and Cristina J. Nelson (1980) "Clean Air and Economic Development: An Urban Initiative", *Journal of the Air Pollution Control Association* no. 11, 1187–1193.

Lave, Lester and Gilbert Omenn (1981) *Clearing the Air: Reforming the Clean Air Act* (The Brookings Institution, Washington).

Lee, Dwight (1982) "Environmental Versus Political Pollution", International Institute for Economic Research, Original Paper, no. 39, Los Angeles.

Lewis, Tracy (1981) "Markets and Environmental Management with a Storable Pollutant", *Journal of Environmental Economics and Management* 8, no. 1, 11–18.

Linder, Stephen H. and Mark E. McBride (1984) "Enforcement Costs and Regulatory Reform: The Agency and Firm Response", 11, no. 4, 327–346.

Liroff, Richard A. (1980) *Air Pollution Offsets: Trading, Selling, and Banking* (Conservation Foundation, Washington).

Magat, Wesley A. (1978) "Pollution Control and Technological Advance: A Dynamic Model of the Firm", *Journal of Environmental Economics and Management* 5, no. 1, 1–25.

Majone, Giandomenico (1976) "Choice Among Policy Instruments for Pollution Control", 2, 589–613, June.

Mäler, Karl-Goran (1974a) *Environmental Economics: A Theoretical Inquiry* (Johns Hopkins University Press for Resources for the Future, Baltimore).

Mäler, Karl-Goran (1974b) "Effluent Charges versus Effluent Standards", in: J. Rothenberg and G. Heggie (eds.), *The Management of Water Quality and the Environment* (Macmillan, London).

McGartland, Albert H. and Wallace E. Oates (1983) "Marketable Permits for the Prevention of Environmental Deterioration", University of Maryland (Department of Economics) Working Paper 83-11.

Meselman, Stuart (1982) "Production Externalities and Corrective Subsidies: A General Equilibrium Analysis", *Journal of Environmental Economics and Management* 9, no. 2, 186–193.

Mills, Edwin S. (1966) "Economic Incentives in Air-Pollution Control", in: Harold Wolozin (ed.), *The Economics of Air Pollution* (W.W. Norton, New York).

Montgomery, David (1972) "Markets in Licenses and Efficient Pollution Control Programs", *Journal of Economic Theory*, 5, 395–418.

Moore, I. Christina (1980) "Implementation of Transferable Discharge Permits When Permit Levels Vary According to Flow and Temperature: A Study of the Fox River, Wisconsin", Department of Civil and Environmental Engineering (University of Wisconsin, Madison) Masters Thesis.

OECD (1978) *Employment and Environment Policy - A State of the Art Review*, ENV/ECO77.5 (OECD, Paris).

OECD (1981) *Economic Instruments in Solid Waste Management* (OECD, Paris).

OECD (1982a) *Environmental Policy and Technical Change*, report by The Secretariat of the Environment Committee (OECD, Paris) mimeo (November).

OECD (1982b) *Impact of Environmental Policies on Technological Change: A Synthesis of the Industrial Case Studies*, ENV/ECO82.5 (OECD, Paris) mimeo.

Olsen, Mancur, Jr. and Richard Zeckhauser (1970) "The Efficient Production of External Economies", *American Economic Review* 60 512–517.

O'Neil, William (1980) "Pollution Permits and Markets for Water Quality", Unpublished Ph.d dissertation, (University of Wisconsin, Madison).

O'Neil, William, Martin David, Christina Moore and Erhard Joeres (1981) "Transferable Discharge Permits and Economic Efficiency", Workshop Sheries 8107, Madison, Wisconsin, Social Science Research Institute. (May).

Opaluch, James J. and Thomas A. Grigalunas (1983) "Controlling Stochastic Pollution Events Through Liability Rules: Some Evidence from OCS Leasing", Staff Paper 83-13, Department of Resource Economics (University of Rhode Island, Kingston).

Oppenheimer, Joe. A. and Clifford S. Russell (1983) "A Tempest in a Teapot: The Analysis and Evaluation of Environmental Groups Trading in Markets for Pollution Permits", in E. Joeres and M. David (eds.), *Buying a Better Environment*. Land Economics, Monograph #6, (University of Wisconsin Press, Madison).

Palmer, Adele R., William E. Mooz, Timothy H. Quinn and Kathleen A. Wolf (1980) *Economic Implications of Regulating Chlorofluorocarbon Emissions from Nonaerosol Applications*, Rand Corporation Report R-2524-EPA (June).

Parker, Roland (1976) *The Common Stream* (Granada Publishing, London).

Pearce, David W. (1983) *The Distribution of the Costs and Benefits of Environmental Policy*, Report from the Environmental Committee of the OECD (OECD, Paris), mimeo.

Philadelphia Air Management Services (1981) *Feasibility Study: A Fee/Subsidy System for Controlling Sulfur Dioxide Emissions in Philadelphia Air*, esp. vol. II, *Economic Theory* (Air Management Services, Philadelphia).

Polinsky, A. Mitchell (1979) "Controlling Externalities and Protecting Entitlements: Property Right, Liability Rule, and Tax-Subsidy Approaches", *Journal of Legal Studies* 8, no. 1, 1–48.

Quirk, Paul G. (1981) *Industry Influence on Federal Regulatory Agencies* (Princeton, N.J.: Princeton University Press).

Railton, Peter (1984) "Locke, Stock, and Peril: Natural Property Rights, Pollution, and Risk", *To Breathe Freely: Risk, Consent and Air*, to be published by Rowman and Allanheld (eds.), Forthcoming Fall of 1984.

Reppeto, Robert (1979) "Economic Systems for the Allocation of Entitlements to Degrade Air Quality in Clean Air Areas Under Prevention of Significant Deterioration Rules", unpublished, Harvard School of Public Health, (October).

Richardson, Genevra with Anthony Ogus and Paul Burrows (1983) *Policing Pollution*, (Oxford University Press).

Roberts, Marc. J. and Michael Spence (1976) "Effluent Charges and Licenses under Uncertainty", *Journal of Public Economics*, 5, no. 3, 4, 193–208.

Rose, Marshall (1973) "Market Problems in the Distribution of Emission Rights", *Water Resources Research* 9, no. 5, 1132–1144.

Ruff, Larry E. (1970) "The Economic Common Sense of Pollution", *The Public Interest* no. 19, 69–85.

Russell, Clifford S. (1973) *Residuals Management in Industry: A Case Study from Petroleum Refining* (Johns Hopkins University Press for Resources for the Future, Baltimore).

Russell, Clifford S. (1979) "What Can We Get from Effluent Charges?" *Policy Analysis* 5, no. 2 155–180, Spring.

Russell, Clifford S. (1981) "Controlled Trading of Pollution Permits," *Environmental Science and Technology* 15, no. 1, 24–28.

Russell, Clifford S. (1984) "Achieving Air Pollution in Three Different Settings", in: *To Breathe Freely: Risk, Consent and Air* (Rowman and Allanheld).

Russell, Clifford S. and William J. Vaughan (1976) *Steel Production: Processes, Products and Residuals* (Johns Hopkins University Press, Baltimore).

Seskin, Eugene P., Robert J. Anderson, Jr. and Robert O. Reid (1983) "An Empirical Analysis of Economic Strategies for Controlling Air Pollution", *Journal of Environmental Economics and Management* 10, no. 2, 112–124.

Shavell, Steven (1982) "The Social Versus the Private Incentive to Bring Suit in a Costly Legal System", *Journal of Legal Studies* XI 333–339.

Sims, W.A. (1981) "Note: The Short-Run Asymmetry of Pollution Subsidies and Charges", *Journal of Environmental Economics and Management* 8, no. 4, 395–399.

Solow, Robert M. (1971) "The Economist's Approach to Pollution and Its Control," *Science*, 173 (August 6).

Sonstelie, Jon C. and Paul R. Portney (1983) "Truth or Consequences: Cost Revelation and Regulation", *Journal of Policy Analysis and Management* 2, no. 2, 280–284.

Spofford, Walter O. Jr., Clifford S. Russell and Robert A. Kelly (1976) *Environmental Quality Management: An Application to the Lower Delaware Valley* (Resources for the Future, Washington).

Spofford, Walter O. Jr. (1983) "Properties of Alternative Source Control Policies: Case Study of the Lower Delaware Valley", Resources for the Future, unpublished report.

Teller, Azriel (1967) "Air Pollution Abatement: Economic Rationality and Reality", *Daedalus* 96, no. 4, 1082–1098.

Terkla, David (1984) "The Efficiency Value of Effluent Tax Revenues", *Journal of Environmental Economics and Management* 11, no. 2, 107–123.

Tietenberg, T.H. (1974) "The Design of Property Rights for Air Pollution Control", *Public Policy* 22, 275–292.

Tietenberg, T.H. (1980) "Transferable Discharge Permits and the Control of Stationary Source Air Pollution: A Survey and Synthesis", *Land Economics* 56, no. 4, 391–416.

Tietenberg, T.H. (1983) "Transferable Discharge Permits and the Control of Stationary Source Air Pollution: Reply," *Land Economics* 59, no. 1, 128–130.

Trauberman, Jeffrey (1983) "Toxic Substances and the Chemical Victim", *Environmental Law* (Summer) p. 1–3, and 8, 9.

Tresch, Richard W. (1981) *Public Finance: A Normative Theory* (Business Publications, Inc., Plano).

Urban Systems Research and Engineering, Inc. (1979) *Responses to Local Sewer Charges and Surcharges*, Report to the U.S. Council on Environmental Quality, Contract No. EQ8AC029.

Vaughan, William J. and Clifford S. Russell (1983) "Monitoring Point Sources of Pollution: Answers and More Questions from Statistical Quality Control", *The American Statistician* 37, no. 4, pt. 2, 476–487.

Viscusi, W. Kip (1979) "Impact of Occupational Safety and Health Regulation", *Bell Journal of Economics* 10, no. 1, 117–140.

Watson, William D. and Ronald G. Ridker (1984) "Losses from Effluent Taxes and Quotas Under Uncertainty", forthcoming in the *Journal of Environmental Economics and Management*.

Webb, Michael G. and Robert Woodfield (1981) "Standards and Charges in the Control of Trade Effluent Discharges to Public Sewers in England and Wales", *Journal of Environmental Economics and Management* 8, no. 3, 272–286.

Weitzman, Martin L. (1974) "Prices vs Quantities", *Review of Economic Studies* XLI (4), no. 128, 477–491.

Wenders, John T. (1975) "Methods of Pollution Control and the Rate of Change in Pollution Abatement Technology", *Water Resource Research*, 11, no. 3, 393–396.

White, Lawrence J. (1982) *The Regulation of Air Pollutant Emissions from Motor Vehicles* (American Enterprise Institute, Washington).

Williamson, M.R. (1981) "SO_2 and NO_2 Mass Emission Surveys: An Application of Remote Sensing", in: Air Pollution Control Association, *Continuous Emission Monitoring: Design, Operation, and Experience* (APCA, Pittsburgh).

Wittman, Donald (1977) "Prior Regulation Versus Post Liability: The Choice Between Input and Monitoring Cost", *Journal of Legal Studies* 6, 193–211.

Yohe, Gary W. (1976) Substitution and the Control of Pollution: A Comparison of Effluent Charges and Quantity Standards Under Uncertainty", *Journal of Environmental Economics and Management* 3, no. 4, 312–324.

Yohe, Gary W. (1977) "Comparisons of Price and Quantity Controls: A Survey", *Journal of Comparative Economics* 1, no. 3, 213–233.

Zeckhauser, Richard (1981) "Preferred Policies When There is a Concern for Probability of Adoption", *Journal of Environmental Economics and Management* 8, no. 3, 215–237.

[21]

Policy Studies Review, Spring 1988, Vol. 7, No. 3, 500-518

Hans Th.A. Bressers

A COMPARISON OF THE EFFECTIVENESS OF INCENTIVES AND DIRECTIVES: THE CASE OF DUTCH WATER QUALITY POLICY

The Netherlands is a small country, approximately the size of Maryland and Delaware combined. Yet, it has much industry and, with a population of some 14 million people, it has one of the highest population densities in the world. Criss-crossed with thousands of miles of waterways of all sizes, it is also a country rich in surface waters. By 1970, the year the Pollution of Surface Waters Act went into effect, industry and private households were producing roughly 45 millions of population equivalents[1] of oxygen-consuming organic pollution. As a result, overall water quality in the country had, in plain terms, become rotten--both figuratively and literally! But by 1980, organic pollution caused by industrial production had declined by two-thirds. Almost half of the remaining organic pollution, from both industry and households, was removed in sewage treatment plants, many of which had been newly built. Everywhere biologically "dead" waters revived. In light of the results achieved, water quality policy is regarded in The Netherlands as one of the few examples of effective governmental intervention (Hoogerwerf, 1985).

Government intervention can take many forms and in the case of Dutch water quality policy many different instruments are involved. Production of collective goods by the state (the treatment of sewage in plants owned by local water boards) has been a major component of this policy. In this article, however, I want to focus on those aspects of the policy that deal with the regulation of industrial water pollution. This regulation works with a mix of different policy instruments, including not only instruments of direct regulation but also the much-disputed instrument of indirect regulation, the effluent charge. As far as the level of charges levied, the Dutch system of effluent charges is the most substantial in the world (Johnson & Brown, 1976). It is also one of the oldest systems in operation.

The use of effluent charges as an instrument of regulatory policy has been the object of much dispute, as much in The Netherlands (Bressers, 1985) as in the United States (some interesting examples from the high point of this discussion in the USA are: White, 1976; Majone, 1976; Rose-Ackerman, 1977; McSpadden-Wenner, 1978). The controversy between advocates and opponents of replacing directives by incentive strategies in various fields of public intervention has always been rather heated, though carried on more in terms of theory than empirical evidence drawn from experience with charges in actual operation (cf. Mitnick, 1980, pp. 373-383). Much like permit trading, an incentive approach actually used in the United States, regulatory effluent charges in The Netherlands more or less "sneaked in through the back door" (Oates, 1983). The Dutch system of

---

The author is grateful to Dr. Kenneth Hanf for his valuable help with the translation of this article.

water quality charges had originally been designed to fulfill a revenue raising function. Indeed, until the publication (Bressers, 1983) of the results of the research on the actual operation of this system of charges reported on in this article, many environmental organizations, political parties and students of environmental policy (especially lawyers), doubted whether these effluent charges had any regulatory effect at all.[2]

A BRIEF SURVEY OF DUTCH WATER QUALITY POLICY

In an effort to combat pollution at the source, the 1970 Pollution of Surface Waters Act prohibits all nonlicensed discharges into surface waters. Indirect polluters, those discharging into a sewage collector, must meet the conditions imposed by municipalities for those hooked up to the collection system. In turn, the municipalities are often subject to conditions from the water authorities regarding the quality of the sewage that they discharge into the treatment plants of the water boards. In the Netherlands, the water boards own and operate the treatment plants. In the past decade, an enormous effort was made to expand effluent treatment capacity, with the result that the amount of organic pollution removed from sewages discharges increased to 25 percent in 1975 and to 46 percent in 1980. This removal rate put The Netherlands more or less on par with its neighboring countries.

The "polluter pays" principle is the basis for the effluent charges of the Pollution of Surface Waters Act. Officially, the sole purpose of these charges is to raise the money needed to finance sewage treatment measures. In view of the ambitious program for expanding treatment capacity, these charges involve large amounts of money. In 1971, they generated revenues of about $30 million; in 1975 approximately $200 million; in 1980 almost $450 million; and in 1985, $550 million.

In The Netherlands each firm has to pay a fee roughly according to the amount of pollution it produces. The amount of pollution is calculated in pollution units called "population equivalents." A population equivalent is defined as being equivalent to the amount of organic pollution in wastewater normally produced by one person. The fee per pollution unit rose sharply during the period the charge system has been in effect (moving from about $6/unit to about $30/unit). The fee charged varies according to the region within which the firm is located. These regional differences are not, however, based on different environmental conditions or quality objectives but rather reflect regionally different costs of building and operating treatment plants.

The unique features of The Netherlands system make it an interesting example of the use of charges. The Dutch system of effluent charges has been in operation since 1970 and, in terms of the level of the charges, is more than twice as large as the comparable German program (Brown & Johnson, 1983). But the most distinguishing feature of the Dutch system is that its use as a regulatory instrument has been "accidental." It was not intended to work in this way. Originally, the charges were to be used to finance the construction and operation of sewage treatment plants. In this sense, they did not replace the official intervention strategy of direct regulation. Given this situation, the Dutch case provides a unique opportunity to examine the effects of these two approaches as they were applied to the same case. It also was possible to collect for the period 1975–1980 data on industrial pollution by type of pollution, branch of industry and water district.[3] In addition, it was possible to collect information on the degree to which different water authorities employed various types of policy instruments. On the basis of these data, it was possible to draw inferences regarding the relative effectiveness of effluent charges and direct regulation based on a number of different statistical analyses.

In the following section, the degree of effectiveness, our dependent variable, will be defined. In section three, we will then look at three statistical analyses of the impact of the policy instruments used. These analyses are supplemented in section four by two expert assessments of these impacts. The final section presents some general conclusions regarding the Dutch experience with effluent charges.

Goal Attainment and Effectiveness: What has been accomplished?

The first question to be answered empirically regards trends in Dutch industrial water pollution during the period under investigation. These trends are examined in relation to the two most important types of industrial water pollution: oxygen-consuming organic pollution measured in PEs (so-called population equivalents) and heavy metal pollution.

In 1975, the year in which the first Multi-Year Indicative Program on water pollution control was issued, Dutch industry still produced a staggering 19,670,000 population equivalents (PE) of oxygen-consuming organic pollution. This was substantially more than the pollution produced by all Dutch households together. By 1980, the end of the period covered by the program, the total amount of industrially produced organic pollution had dropped to 14,331,000 PE. This represented a reduction of more than five million PE, or 27 percent. A decrease of roughly this magnitude had been projected by the 1975 policy program. There were, however, enormous regional differences. In some areas (water board districts) organic industrial pollution had actually increased, whereas in other areas this type of pollution had been reduced by more than one-half.

The 1975 program had set no explicit targets regarding the abatement of industrial water pollution with heavy metals. As was the case with organic pollution, there have also been huge regional differences in the amount of reduction attained in heavy metal pollution. In general, however, it can be concluded that industrial pollution with heavy metals has, on the whole, been reduced by approximately one-half in the period involved.

MEASURING EFFECTIVENESS I: MULTIPLE TIME SERIES

Without further analysis it is not clear whether these decreases in industrial wastewater pollution are in fact the results of the water quality policy pursued by the government. As is often the case in evaluating policy, it also was not possible in this case to employ a truly experimental research design in examining this question. For this reason, a number of other methods have been combined to determine the extent to which levels of pollution reduction attained can be attributed to the policy measures taken.

We first carried out a time-series analysis of the trends in industrial wastewater containing biodegradable organic substances. Until 1970, the year in which the water quality law went into effect, organic industrial pollution increased. Thereafter there was a sharp, continuous decline. (see Figure 1.) However, even such evidence of a striking decline does not, in itself, provide ironclad proof of a causal relation between the policy pursued since 1970 and the pollution reduction achieved, the more so as the policy was not abruptly introduced, but was implemented gradually. Before such a relationship can be established, it is necessary to consider alternative explanations for the decline observed.

Figure 1

Index Figures on the Amount of Industrial Production (solid line) and

Oxygen-Consuming Industrial Pollution in Industrial Wastewater (dotted line)

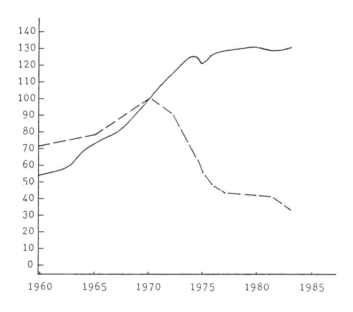

The extent of pollution is obviously influenced by decisions taken by the industrial firms. In this connection, the relevant decisions are regarding:
- the overall level of production (to increase or decrease production)
- the amount of pollution produced per unit of production (to produce with more or less pollution).

The firm may decide to reduce production because the costs of water quality control are so high that maintaining the present level of production is not profitable. A 1972 Dutch Central Planning Agency study concluded that production level in 1983 would be 4.5 percent less due to the direct and indirect costs of water quality policy. It is, however, better to be cautious in this regard and to attribute any change in the level of production not to water quality policy but instead to such factors as economic recession, making production decline an alternative explanation for the pollution decline observed.

Figure 1 shows that, in fact, the volume of production did not decline, not even in the recessionary period of 1975-1980. At the same time,

504   Policy Studies Review, Spring 1988, 7:3

however, there was a dramatic decrease in pollution between 1970 and 1975, a time when the economy was booming. If we look at the separate branches of industry, in order to make sure that this overall decrease in pollution is not the result of a decline in production in some of the heavily polluting industries, we find that there are only a few industries that have experienced a decrease in production. Even if we assume that any decrease in pollution leads directly to a proportionate reduction in pollution (an overly-optimistic assumption!), the production decreases in these industries would result in a decrease in total oxygen-consuming industrial pollution of only 2 to 3 percent as compared with the actually recorded overall decline of 27 percent. A similar situation can be found with regard to pollution with heavy metals. Given the fact that production has, on the whole, tended to increase instead of decline, we can conclude that pollution per product unit has decreased more sharply than total pollution and that the level of pollution per product unit has decreased more sharply than total pollution and that the level of pollution reduction attained tends to underestimate the effectiveness of water quality policy.

MEASURING EFFECTIVENESS 2:
FURTHER ANALYSIS OF ALTERNATIVE EXPLANATIONS

Theoretically, changes in pollution per product unit might very well be accounted for by other factors apart from policy outputs. Before we can say anything about their relative influence, we must first list the different kinds of factors that may have an impact on per unit pollution. We have constructed such an inventory using five dimensions of decision making in the mode of the "subjectively rational actor" (Hennipman, 1945; cf. Simon, 1955). Figure 2 provides an overview of these dimensions and the alternative explanations available to account for reductions in per unit pollution. Below we present a summary of the conclusions reached by applying Michael Scriven's "modus operandi" method (Scriven, 1976) to the examination of these different factors.

The conclusions drawn from this analysis of alternative explanations all point consistently in the same direction: the factors considered cannot in themselves account for the amount of pollution reduction attained. They will, however, tend to further or impede the impact of water quality policy on the amount of pollution produced. "Environmentally friendly" technological developments (see "a," Figure 2) do not just appear out of nowhere. Since other factors have always given the impulse for such developments, they cannot themselves be viewed as an independent cause of pollution abatement levels. However, technological development can sometimes be indispensable for the realization of the abatement that companies, for other reasons (such as in response to policy measures), deemed necessary. In the case of heavy metal pollution, developments with respect to the price of these materials cannot account for the amount of abatement achieved since, in the period under investigation, these prices declined instead of rising (see "b," Figure 2). It is unlikely that the environmental awareness of industrial leaders and the population in general (see "c and d," Figure 2) has been an independent cause of pollution abatement, among other reasons because of the "logic of collective action" (Olson, 1971). But it is conceivable that the extent to which and the rate at which policy leads to results will be influenced by the level of environmental awareness. In addition, the impact of information provided by nongovernmental bodies regarding abatement measure (see "e," Figure 2) should be considered marginal (supplementing other factors). Information is necessary but in itself hardly ever sufficient for companies to decide to reduce pollution. This is so because these measures usually lead to higher production costs.

Figure 2
Inventory of Alternative Explanations of Reductions in Pollution per Produce Unit

Decisionmaking dimension	Alternative explanatory factor
1. new behavioral alternatives	a. independent technical developments
2. fewer alternatives	
3. change in properties of alternatives	b. increase value waste matter as raw material
	c. increase environmental awareness with population
4. change in valuation of properties of alternatives	d. increase environmental awareness with companies
5. information on presence and properties of alternatives	e. information given by nongovernment institutions

---

CONCLUSIONS

After the Water Pollution Act had gone into effect, pollution of Dutch industrial wastewater sharply decreased, not only in the initial period, when simple and cheap abatement methods could be applied, but also later, when the costs of abatement measures were higher. Time series analyses, supplemented by analyses using the "modus operandi" method, provide support for the conclusion that the level of pollution reduction achieved can be attributed to the policy pursued. The answer to the main question of this paper, i.e. which policy instruments have had the greatest effect on the abatement of pollution in industrial wastewater, obviously depends on determining, first of all, how important policy as such has been in bringing about a reduction in industrial water pollution. If the empirically observed decrease in pollution cannot be accounted for by other factors, then it would seem, logically speaking, that policy measures taken must account for it. As the next sections will demonstrate, this has indeed been the case in The Netherlands. The substantial reduction in industrial wastewater pollution has been the result of Dutch water quality policy. Since that policy contains a "mix" of different measures, the questions remains as to which of the various instruments applied has contributed most to the policy's apparent effectiveness.

THE CONTRIBUTION OF DIRECTIVES AND INCENTIVES 1:
A STATISTICAL ASSESSMENT

In the previous section we noted that there have been substantial regional differences in the extent to which the policy objectives have been achieved. However, these differences cannot, without further investigation, be explained by the extent to which the individual water boards applied the various policy instruments at their disposal. First of all, the water districts themselves vary greatly in size and thus in pollution loads. Along with 17 rather large districts, there were, during the research period, 23 of rather small area. In 1975, 97 percent of the total volume of organic pollution was produced in these 17 large districts. Since the percentage decreases in pollution often reached extreme values in the 23 smaller areas, based, however, on only a few observations (firms), these areas have been excluded from the analysis reported below. The impact of developments in these districts on the results of the analysis would have been out of proportion to their actual importance.

Differences in goal attainment among the remaining areas can be accounted for by two sets of factors. On the one hand, they can be partly explained in terms of differences in the economic structure of different water districts with regard to the relative weight of particular branches of industry. In the case of heavy metal pollution, differences in abatement levels may be attributed to the relative share of various metals in the total volume of pollution. Using data on the share of different branches of industry in the regional economy and, for heavy metals, the share of various substances in the regional pollution in 1975, and on the average decrease in pollution per industry and substance in The Netherlands between 1975 and 1980, the amount of pollution decrease that could be expected on the basis of the structure of regional economy was calculated for each water district.[4] This expected value was closely correlated to actual decrease of pollution. In the case of organic pollution the Pearson's r was +.79 (r^2 = .62) and in the case of heavy metals pollution it was +.74 (r^2 = .55).

Effluent Charges and Branch Differences in Pollution Decrease

In 1969, the first time organic pollution of industrial wastewater was measured by industrial branch, it was determined that fourteen industries accounted for 90 percent of this pollution. These fourteen industries form the research units in the analysis described below. Since they accounted for such a large part of the pollution, this analysis is not so much a sample study but rather a semipopulation study. Nevertheless, in view of the small number of units involved, a word of caution is in order when drawing conclusions based on this sample of industries.[5]

There are a number of factors that could account for the differences found between these industries with regard to the percent of decrease in their organic pollution. Most obviously, decreases or increases in the level of production of the different industries could affect the volume of pollution produced in the period under investigation. A second cause of differences in abatement rates can be found in the fact that in a particular industry it is much more difficult, and thus more expensive, to produce more "cleanly" than it is in another industry. Therefore, in addition to the increase or decrease in production, relative abatement costs can play an important role in determining the level of pollution reduction achieved. Finally and (in connection with the focus of this study) most importantly, it is also conceivable that effluent charges could lead to differences in abatement between industries. The more pollution units per unit of production value that are caused by an industry, the greater the impact that the charges will have on production costs. This can mean that some industries will bear charges that are significantly greater than other industries. Efforts to reduce the number of pollution units produced will be greater to the extent that a given industry is liable for higher charges compared to its production value. In the discussion that follows this factor will be referred to, for convenience, as the "charge factor."

The extent to which organic pollution of industrial wastewater decreased between 1969 and 1980 appears to be related to the three above-mentioned factors as follows: production increase, $r = -.21$; abatement costs, $r = .50$; and charge factor, $r = .73$. The signs of these relations are all in the anticipated direction. The strong correlation with the charge factor is very striking. When the three factors are related simultaneously to pollution decrease, they account statistically for 63 percent of the decrease. The regression equation across the 14 major branches was estimated as: $P = .33$ $C - .32$ AC $- .01$ Prod, $R^2 = .63$ where

> P = percent decrease in discharge 1969-1980
> C = charge factor calculated as pollution (in 1000 PE) liable to charges divided by total production value in millions of Dutch guilders in 1969
> AC = estimated average cost of treatment per branch
> Prod = percent change in output during 1969-1980

The standardized contribution (beta) of the three factors is: production increase, -0.30; abatement costs, -0.25; and charge factor, 0.61.

In the case of two of the fourteen industries (potato-starch industry and the stock-raising industry), charges had less influence than was to be expected on the basis of their charge factor. In anticipation of ultimately being hooked up to a "smeerpipe" (literally, a dirt pipe), the potato starch industry had been allowed to continue with untreated discharges. During this time, the industry was paying much lower charge-rates than were other industries. The stock-raising industry is composed of thousands of relatively small farms. Here the charges are seldom calculated exactly on the basis of the amount of pollution actually produced. Hence it is to be expected that the impact of charges will probably be less in this case than in other industries with a similar charge factor. When these two industries are left out of the analysis, the correlation of the charge factor with the decrease in pollution is even stronger: $r = .84$. The correlation with the other two factors remains about the same, with production increase showing a r of $-.24$ and abatement costs an r of $-.48$. These three factors, taken together, account for 76 percent of the differences in the decrease of organic pollution in industrial wastewater between these twelve industries.

Policy Instruments and Regional Differences
in the Decrease of Organic Pollution

An important part of the regional variation in pollution abatement cannot be explained by differences in the industrial structure of the regions. If the conclusion of the above section is correct, this leftover variation can be explained entirely in terms of the degree to which different policy instruments are being used. The degree to which these different instruments are actually used will also vary greatly from region to region. In this section, we will examine the extent to which this explanation is accurate.

As in the discussion of the alternative explanations of pollution decline, policy instruments can also be grouped in terms of factors that influence

508   Policy Studies Review, Spring 1988, 7:3

decision making by subjectively rational actors.[6] All instruments mentioned are used simultaneously in Dutch water quality policy.

As can be seen in Figure 3, prohibitions have not been linked with reducing the number of behavioral alternatives available to the decision maker. Although this would appear to be the aim of such directives, it would be unrealistic to assume that they are always successful in this regard. What effect prohibitions in fact have upon behavior will have to be established empirically for a given situation.

Figure 3
Inventory of Types of Policy Instruments

Factor in decision model	Type of policy instrument
1. new behavioral alternatives	a. promotion of technical know-how development
2. fewer alternatives	
3. change in properties of alternatives	b. directives 1 (permits and prohibitions) c. directives 2 (inspection and the courts) d. incentives (effluent charges)
4. change in valuation of properties of alternatives	e. deliberation, persuasion and negotiations without specific legal grounds
5. information on presence and properties of alternatives	f. advice to companies about properties of existing abatement techniques

Since we are particularly interested in effluent charges and since, on the basis of the branch analysis above, we expect to find a clear impact of charges on decisions affecting the amount of pollution, we will examine the impact of this instrument first and then look at the other instruments.[7]

The dependent variable for this part of the analysis is the "relative success of abatement." This variable has been calculated as the difference between the actual percent of abatement and the percent of abatement expected in view of the industrial structure of the region. Regional variation in the charge level cannot, however, straightforwardly be related to this variable. The reason for this is as follows.

The relative success of abatement has been calculated for the period 1975-1980. Prior to this time, however, pollution of industrial wastewater had already declined substantially. Our analysis of interindustry variations (see above) makes it plausible to assume that abatement in this period was also largely the result of charges. The 1975 amount of pollution was already more or less' in equilibrium with the charge level of that time. For this reason, the charge factor influencing the abatement of pollution of industrial wastewater is indicated much better by the difference between the rate before 1975 and the rate at the end of the period 1975-1980 than by the level of the charge in any one of the preceding years.

However, it cannot be determined on theoretical grounds for which period the charge increase should be used as independent variable. On the one hand, there is a time-lag between the stimulus to abatement efforts and the installation and putting into operation of the relevant control measures. On the other hand, it is not implausible to expect a certain degree of anticipation by the companies to charges and rate increases (Bressers, 1983-2; Ewringmann, Kibat, Schafhausen, Fedderson, Krickel, Perdelwitz, & Strauss, 1980). It is better, therefore, to decide on empirical grounds which period should be considered in determining the rate increase. This period starts in the year in which the rate of the charge has the highest negative correlation with the relative success of abatement between 1975 and 1980. The rate in that year can be taken as the best indicator for the extent to which abatement efforts before 1975 negatively affected abatement activities in the subsequent period by the resulting increase of marginal costs of further abatement. The rate charged in that year should then be subtracted from the charge at the end of the period with the highest positive correlation in order to get the best possible indicator of the charge factor that stimulated abatement of pollution in the 1975-1980 period.

As expected, the level of the charge prior around 1975 is negatively correlated with the relative success of abatement of organic pollution of industrial wastewater in the years 1975-1980. This negative relationship is strongest with the 1974 charge ($r = -.53$). The correlations gradually become more positive up to and including 1980. The 1980 rate of charge shows a clear positive correlation with relative abatement success ($r = +.53$). As an indicator for the influence of effluent charge on relative abatement success, we will, therefore, take the increase in the rate charged in the period 1974-1980. For two of the 17 water districts examined this indicator could not be calculated.

The increase in the rate charged in the period 1974-1980 has a very strong positive correlation with the relative abatement success for the years between 1975 and 1980 ($r = +.86$, $n = 15$, $p = .000$). This relation is not the result of one or two extreme values for the units analyzed, although the two water quality districts with respectively the most and the least relative success weaken the strength of the relationship somewhat. Without these two observations the correlation is even stronger: $r = +.92$ ($n = 13$, $p = .000$). Further analysis has showed that this correlation cannot be attributed to water authorities reacting on abatement efforts by industry by raising charge rates to keep the same amount of revenues.[8]

With such high correlations between the charge factor and the pollution decrease it is natural that the relationships with the indicators of the great variety of other policy instruments used are absent or only weak. Among all these indicators, only water board permit-giving to municipalities and abatement schemes show up with relatively substantial but still modest positive correlations with the relative abatement success.

The relative abatement success, the dependent variable in all these analyses, has been defined as the percentage decrease of oxygen-consuming pollution of industrial wastewater, corrected to take the regional structure of industry into account. It should, therefore, be possible to explain the variation in the percentage decrease of organic pollution by: (a) the decrease to be expected as a result of differences in the regional structure of industry; (b) the increase in the rate charged; (c) the number of municipalities with a permit issued by water boards (thus giving these authorities indirect influence on sewage permits issued to individual plants); (d) the weighted number of abatement plans.

As it turns out, 96 percent of the percentage decrease of organic water pollution can be explained, statistically, by the first two factors (R^2 after correction is .95). Ninety-eight per cent can be explained by all four factors taken together, but this is due mainly to the fact that one research unit has been dropped from the calculation. If we take this smaller group of 14 units, the two first-named factors explain 97 percent of the differences in pollution decrease.

The regression formula used to explain 96 percent of the difference in pollution decrease is:

$$y = 0.96x_1 + 1.55x_2 - 35.6 \text{ (constant)} \qquad R^2 = .96, \ n = 15.$$
$$(0.08) \ (0.20)$$

Where

x̄ y = percent decrease of the pollution of industrial wastewater
x̄ x_1 = percent decrease to be expected on the basis of the regional structure of industry
x̄ x_2 = increase in the rate charged 1974-1980 (in Dutch guilders)

In brackets the standard errors. This means that one percent of actual abatement corresponds to approximately one percent abatement to be expected on the basis of the regional structure of industry (which is an indication that the calculation is not distorted) and with an increase in rate of 1.5.

Policy Instruments and Regional Differences
in Decrease in Heavy Metal Pollution

Two water boards do not impose effluent charges on heavy metal polluters. Their results are not much different from those of the other districts. However, as a matter of fact, in their negotiations with industry, when demanding quite substantial reductions in pollution level, these boards did use the possibility of introducing effluent charges as threat. In this sense, their effectiveness in achieving decreases in heavy metal pollution is not totally independent of the effluent charge system. We will not include these districts in the following analysis.

The indicators for policy outputs used here are, for the most part, the same as those used in looking at organic pollution or they have been calculated in a similar manner. The indicators for permits issued and for technology, advice and informal negotiations are different. The indicator for permits has been calculated as the percent of companies that discharge a taxable amount of heavy metals having a permit that stipulates conditions for these discharges in 1975, the percent with such permits in 1980, and the increase in these numbers between 1975 and 1980. Technology development, advice and information negotiations are also indicated by the proportion of major dischargers contacted instead of by the weighted number.

The increase of the charge rate in the period 1975-1980 is clearly related to the relative success of the abatement of heavy metal pollution of industrial wastewater, although the relationship is not as evident as with the abatement of organic pollution. The $r = .65$, for an $n = 13$ and a $p = .008$.

Among the other indicators of the instruments used weighted number of formal reports by pollution inspectors and the proportion of companies with which an abatement plan was drawn up were positively correlated with the relative abatement success with heavy metal pollution.

The weighted number of inspectors turns out not to be related to abatement success. The same goes for both the number of infringements detected by them (as an indicator of their activity) and the number of times they threatened official enforcement actions. However, the situation is different with regard to the number of times these inspectors in fact officially report such infringements and use the police to enforce compliance. The weighted number of formal reports is positively related to the relative abatement success ($r = .30$, $n = 14$, n.s.). If we take the rate of increase for effluent charges into account, the correlation is even stronger ($pr = .43$, $n = 13$, $p = .082$).

The proportion of companies with which an abatement plan was drawn up is clearly related to the relative abatement success. Although this relation is at first relatively weak ($r = .30$, $n = 14$, n.s.) if we take into account the influence of the increase of rate charges, the correlation is stronger ($pr = .55$, $n = 13$, $p = .031$).

The proportion of companies with which an abatement plan was drawn up is clearly related to the relative abatement success. Although this relation is at first relatively weak ($r = .30$, $n = 14$, n.s.) if we take into account the influence of the increase of rate charges, the correlation is stronger ($pr = .55$, $n = 13$, $p = .031$).

The combination of policy outputs that accounts for the greatest part of the variation in relative success in abating pollution by heavy metals includes (after a correction for the number of independent variables):

- the increase in the charge rate 1975-1980;
- the proportion of heavy metal dischargers with an abatement plan;
- the weighted number of official reports on noncompliance.

Together these policy outputs explain statistically 82 percent of the difference in relative abatement success. These three policy outputs together with the abatement that could have been expected in view of the regional structure of industry and the shares of the various kinds of metal statistically account for 91 percent of the variation in decrease in heavy metal pollution (R^2 after correction is .87).

The standardized contribution (beta) of the different factors to this are:

x_1 – abatement expected (%)	.33
x_2 – increase in rate charged (in guilders)	.56
x_3 – proportion of companies with abatement plan (5)	.47
x_4 – weighted number of official reports on noncompliance	.36

The regression equation itself runs as follows:

$$y = \underset{(0.30)}{0.73x_1} + \underset{(0.80)}{3.35x_2} + \underset{(0.11)}{0.42x_3} + \underset{(1.93)}{6.24x_4} = 78.1$$

$R^2 = 0.91$, $n = 14$ (In brackets, the standard errors)

Given that B of x_2 (charges) is more than twice as high for heavy metals as it was for organic pollution, it can be concluded that the relative level of charges compared with abatement costs is much lower for heavy metals than for organic pollution. For most heavy metals, one kilogram is equal to one population equivalent of organic pollution. In general, the relatively low level of charges on heavy metals is acknowledged by that staff of the water boards. In spite of this, the charge appears to provide the primary impetus for pollution abatement also in the case of heavy metals.

512   Policy Studies Review, Spring 1988, 7:3

THE CONTRIBUTION OF DIRECTIVES AND INCENTIVES 2:
AN EXPERT ASSESSMENT

Thus far three independent statistical analyses have shown that effluent charges have been a quite effective instrument of Dutch water quality policy. The contribution of each of the policy instruments applied to the substantial reductions of industrial wastewater pollution in The Netherlands can also be determined with the help of the assessments by insiders. In this section we will compare the results of the statistical analysis with the assessment of experts from both the water boards and industry regarding the relative effectiveness of these instruments. Both approaches have their strong and weak points (Reichardt & Cook, 1979). However, should both methods lead to the same conclusion, our confidence would be increased in the reliability of this conclusion. As was the case with the statistical analysis, independent analyses based on different sets of data were used: on the one hand, the opinions of negotiators from the water boards and, on the other hand, representatives of the companies who were responsible for water quality.

Assessment of Policy Instrument by Policy Implementors

In a questionnaire sent to the regional water boards, administrators were asked to indicate how much influence they thought the various policy instruments had on the abatement of industrial wastewater pollution. For both organic pollution and pollution with heavy metals the respondents could choose from five answers. The category "Not applied" could be checked to indicate that a particular instrument had not been used with regard to that type of pollution. In practice, "No answer" also nearly always indicated the instrument in question had not been used at all or very little. This was the case for the seven smaller water authorities that did not fill in the questionnaire completely, adding such comments as "no experience" and the like. Regarding heavy metals, the question was irrelevant for seventeen smaller water boards since there were no companies in their areas known to discharge heavy metals. Table 1 shows the answers given to this question.

In general these results correspond with those of the statistical analyses, at least as far as the main points are concerned. Charges emerge as the most influential policy instrument for dealing with organic pollution, whereas in the case of heavy metals we see that the respondents attributed equal amounts of influence to a broader range of instruments. There are, however, some small but interesting differences between the results of the questionnaire and those of the statistical analyses.

According to the statistical analysis, charges also played the most important role in decreasing heavy metals pollution, although to a lesser degree than was the case with oxygen-consuming pollution. Why was it that the regional water quality administrators recognized the importance of charges for the abatement of organic pollution, but not for decreasing pollution with heavy metals? The explanation seems to be that these officials can clearly see that there are companies that are going to abate organic pollution because of charges, but hardly any companies are going to abate their heavy metal pollution solely because of the effluent charges. Discharging heavy metals is simply too cheap. Abatement plans and inspections are almost always necessary to persuade heavy-metal polluters to take appropriate action. What these water quality administrators do not see (indeed, what they individually cannot see) is that the success of these abatement plans is much greater in districts where charges were increased than in other areas. In both cases, companies have to be persuaded by the water authorities to get started; in some cases, however, officials must continue to exert pressure to ensure that the agreed-upon measures are in fact carried out. Even when they cannot in themselves motivate companies to take abatement action, charges apparently play an important role in facili- tating the task of the water quality administrators in getting abatement measures implemented. This conclusion opens interesting perspectives on the possibility of applying charges not only as an alternative for, but also as a complement to direct regulation.

Table 1

Assessment by Regional Water Quality Administrators of
the Effectiveness of Policy Instruments

Policy Instrument	Type of Pollution	Extent of Influence					#	Total
		++ ++	++ +	+ +	+	0		
a. Permits	Organic	2	12	14	3	2	7	40
	Heavy metals	5	8	8	0	2	17	40
b. Inspection and the courts	Organic	1	16	13	2	1	7	40
	Heavy metals	2	10	10	1	0	17	40
c. Effluent charges	Organic	16	14	3	0	0	7	40
	Heavy metals	1	9	9	2	2	17	40
d. Promotion of technological developments	Organic	0	3	14	12	4	7	40
	Heavy metals	1	2	12	5	3	17	40
e. Advice	Organic	3	9	14	6	1	7	40
	Heavy metals	2	7	10	2	2	17	40
f. Informal negotiations	Organic	5	15	11	2	0	7	40
	Heavy metals	4	5	12	2	0	17	40

Another point calling for attention is the role played by informal negoti- ations. It is clear from the questionnaire than water board officials at- tached greater importance to this instrument that we would have anticipated on the basis of the statistical analysis. It appears that this apparent discrepancy can be accounted for by the fact that negotiations do not constitute a separate instrument in addition to other policy instruments, but rather represent a manner in which the total "policy mix" is applied. In this regard there are two interesting aspects of informal negotiations. In the first place, they serve as a sort of "lubricant" to oil the machinery put into operation by the other policy instruments discussed above (compare for Germany: Hucke, 1978; and for Great Britain: Vogel, 1983, pp. 1-2). The following statement by a water board official during one of the oral inter- views that followed the written questionnaire is illustrative of this function: "When I'm going to have a talk with a company about the abatement of their discharges, I always take my pocket calculator along. I calculate their potential savings on charges and invariably get an interesting conversation started."

This, primarily informative, function of negotiations is especially important when other circumstances and the policy instruments applied have an
influence strong enough to make abatement seem worthwhile to the companies to begin with. This situation will tend to occur more often in the case
of organic pollution than with heavy metals. The implementation process
with regard to organic waste matter is characterized by relatively small
differences of opinion between administrators and companies regarding the
amount of abatement aimed at. This is due to the pressure of the effluent
charges, which make reductions of discharges economically attractive.

Heavy metals negotiations not only fulfill an informative function; here
they also represents the way in which power is wielded. This strategy
reflects the fact that in such cases there are often quite substantial differences of opinion between water quality administrators and the companies
regarding the desirable amount of further discharge reduction. In these
situations the balance of power is such that neither party is in a position to
impose its will upon the other. It appears then that unless the administrator
is in a position to exert enough power and to apply a policy instrument
vis-a-vis the company so that he can achieve his abatement goals without
delay, it is quite rational for him to apply the available policy instruments
selectively within a framework of deliberation and negotiation with industry.

The Assessment of Policy Instruments by the Regulated Industries

In addition to the analyses mentioned above, we also made use of a
study by two Dutch fiscal lawyers, Schuurman and Tegelaar (1983). This
study, based on the subjective assessments of company officials directly
involved with pollution abatement, was based on 150 firms constituting
one-fifth of the category "large polluters." The data were collected in
interviews with one or more functionaries at the staff or executive level who
were responsible for company policy with regard to water pollution or who
were closely involved in the implementation of these policies. In general,
these officials can be considered the conversation partners of the respondents from the preceding analysis of the water board administrators.

The results of this study show clearly that in the period 1975-1980
effluent charges had a substantial effect on the abatement of pollution by
the companies investigated. For nearly 60 percent of the companies interviewed, the charge levied was the decisive factor in the decision to take
abatement measures. The pollution decrease realized by these measures
amounted to more than 3.2 million population equivalents out of a total
reduction of pollution of 4 million PEs. This means that 80 percent of
pollution abatement realized in the period 1975-1980 can, according to the
companies investigated, be explained in terms of the firms' response to the
effluent charge, with the remaining 20 per cent attributable to other factors. The most important noncharge instrument mentioned are the discharge conditions contained in the permits. For more than 20 percent of
the firms studied this factor was given as the immediate reason for taking
abatement measures. The pollution decrease realized by these measures,
however, amounted at most to 800,000 PEs. According to the company
officials interviewed, it is expected that charges will also be an important
reason for taking further abatement measures.

In this analysis, charges once again emerge as by far the most influential
instrument, but somewhat less dominating with regard to organic wastewater
pollution than in the statistical analysis. Two factors should, however, be

kept in mind when comparing these two analyses. First of all, there seems to be a certain tendency in oral interviews to give the socially desirable answer that something is being done for environmental protection, not because it is cheaper but because one has a social responsibility to do it, or because the democratically elected authorities have asked for it. This will lead to a certain underestimation of the influence of charges on abatement actions. Secondly, charges also can play an important role in cases where discharge regulations are the decisive motive for measures taken. Without charge, the difference between demands made by water boards and what industry deems economically feasible would have been much greater and, consequently, the resistance of the firms to these demands much stronger.

CONCLUSION

Taken together these analyses lead to the remarkable conclusion that the substantial reduction of pollution of Dutch industrial wastewater between 1975 and 1980 has been much more the result of a policy instrument, effluent charges, that was not officially designed for this purpose, than a result of the use of the policy instrument, direct regulation, specifically intended to achieve this objective.

The Dutch case, with the most substantial system of effluent charges in the world, shows the enormous potential of this policy instrument. In the research reported here, I did not come across any evidence that would suggest that this result can be explained in terms of specific national characteristics of The Netherlands. Two points, however, deserve attention.

The implementation of effluent charges can be hampered by the need for a devoted executive body that is determined to actually collect the charges. Central government policies are often frustrated by lack of cooperation by local authorities appointed to implement them (Bressers & Honigh, 1986). This problem did not arise in Holland, thanks to the fact that their effluent charges function as so-called revenue raising charges. Through the charge system, the water authorities are collecting their own money, needed very much for the building and exploitation of treatment plants, a task they were eager to undertake.

Another major problem with effluent charges is the massive amount of information that is required to assess the fee each company has to pay. Some authors even see this as the most important reason to discard this policy instrument altogether. In Holland, this problem was reduced by not charging the millions of households and small industrial polluters (less than 10 PE) in proportion to the actual pollution they caused. Having relatively few opportunities of limiting pollution, this category of polluters is of minor importance to the instrument's regulating power. The amount of information required is again substantially reduced by basing the assessment of medium-sized polluters (usually between 10 and 100 PE) not on samples of their effluent, but on a coefficient table. On the basis of easily obtainable data such as the amount of water used by the firm or the amount of certain raw materials it processes with that expertly calculated coefficient table the probable amount of pollution can be established accurately. However, the incentive to reduce pollution remains intact. Companies that feel they are overrated on the coefficient table can request their effluent to be sampled and to be charged on the basis of the results. Not always this leads to permanent sampling. Sometimes only an adjustment of the relevant

516   Policy Studies Review, Spring 1988, 7:3

coefficient is made. All these pragmatic adjustments make it possible to implement the charges at the cost of only a few percent of the total revenue, without diminishing substantially the instrument's regulating power (as could be shown).

Implementation problems being solvable and the way in which effluent charges influence the negotiation process being consistent with what we would have been led to expect by the economics literature, the greatest problem seems to be the feasibility of introducing such charges. In the words of Senator Domenici: "Charges have only one problem: they lack political support" (Domenici, 1982). Perhaps the key question is: Who's afraid of effective government?

NOTES

[1] A population equivalent is the amount of organic pollution equivalent to the average organic pollution caused by one person in a normal household, which number of PEs is therefore equal to the population. For industrial pollution the number of PEs is predominantly calculated on basis of BOD and COD figures.

[2] The publication of the study in 1983 had a substantial impact on the stance taken by the Dutch environmentalists and led to the proposal in parliament to extend the coverage of the charges. Time, however, has been not on their side due to the most severe postwar economic crisis Holland has experienced in recent years.

[3] Data of low aggregation level were obtained from the task force for the preparation of the "Indicative Multi-Year Program 1980-1984: The fight against water pollution," Ministry of Transport, Public Works and Water-affairs, 1981 (in particular data on pollution emissions) and from the Dutch Central Statistical Office (in particular parts of the branch data and the data on prices of heavy metals).

[4] The following formula was used for this calculation in the case of organic pollution; in the case of heavy metal pollution, a similar formula was used, expanded to take the relative shares of different metals into account.

$$P_i \text{ exp.} = x1_i b1 + x_{2i} b_2 + \dots + x_{ni} b_n$$
$$(i = 1 \dots\dots m)$$

Where
  P exp. = expected percent of decrease in pollution of industrial
            wastewater 1975-1980
  1 ... n = 7 branches: 6 highly polluting branches and one including
            all other branches of industry
  x1 ... n = the proportion of pollution contributed by each industrial
            branch in the regional economy in 1975.
  b1 ... n = per cent decrease in pollution by each branch
  1 ... m = the 17 major water districts.

[5] A more extended version of this analysis can be found in Bressers (1983, chap. 2). Methodological problems, especially those in connection with the operationalization of the relevant theoretical notions to a model of

analysis, are dealt with there in greater detail. Here we have limited ourselves to a survey of findings and a reflection on conclusions drawn from the analysis.

[6] For the purpose of effectiveness research, policy instruments should preferably be classified on the basis of the way in which they influence and not on the more usual basis of ex-ante estimates of the degree of coercion that goes along with them (cf. Nagel, 1975). In the latter type of classification, dependent and independent variables get mixed up. See for an extended classification of policy instruments on the basis of their "modus operandi," Bressers and Klok (in press).

[7] The seventeen areas analyzed here accounted for nearly all water pollution in 1975. Nevertheless, there are certain problems involved in drawing conclusions based on the statistical analysis of seventeen cases. This is especially so since a few scores are missing for a number of independent variables so that the number of units available for analysis is reduced even further. Since we are dealing with a semipopulation study, the danger is particularly great that one or more extreme values will determine the direction and strength of the correlations. In order to control for this, a plot was made of each bivariate relation so that the possible influence of extreme values could be determined. As a matter of fact the scores of the cases on the dependent and the independent variables spread out quite normally, with no real outlayers and some concentration around the center of the range.

[8] We have calculated that part of the rate increase between 1974 and 1980 that was necessary to prevent the pollution abatement attained from resulting in a reduction in revenue from effluent charges. This turned out to range from 3 percent to 20 percent of the total increase (5% on the average). Next, this part of the rate increase was subtracted from the total increase in the period 1974-80 so that what remained was only the increase that cannot be attributed to the decrease in pollution. The remaining increase in rates charged was almost as strongly related to the relative abatement success as was the total increase: $4 = .83$ ($n = 15$, $p = .000$). This seems to indicate that the increase in charges is a cause rather than an effect of pollution abatement of industrial wastewater.

REFERENCES

Bressers, H.Th.A. (1983). The role of effluent charges in Dutch water quality policy. In Downing & Hanf (Eds.), *International comparisons in implementing pollution laws* (pp. 143-168). Boston: Kluwer-Nijhoff.

Bressers, H.Th.A. & Honigh, M. (1986, June). A comparative approach to the explanation of policy effects. *International Social Science Journal.*

Bressers, H.Th.A. & Klok, P-J. (in press). *Fundamentals for a theory of policy instruments*, paper International Congress on environmental policy in a market economy, Agricultural University Wageningen (forthcoming in *International Journal of Social Economics*).

Bressers, J.Th.A. (1983). *Beleidseffectiviteit en waterkwaliteitsbeleid* (Policy effectiveness and water quality policy), dissertation. Enschede: University of Twente.

518   Policy Studies Review, Spring 1988, 7:3

Bressers, J.Th.A. (1985). *Milieu op de markt* (Environment on the market), Amsterdam: Kobra.

Brown, G.M., Jr. & Johnson, R.W. (1983). *The effluent charge system in the federal republic of Germany and its potential application in the US.* Seattle: University of Washington.

Domenici, P.V. (1982, December). Emissions trading: The subtle heresy. *The Environmental Forum.*

Ewringmann, D.K.K., Schafhausen, F.J., Fedderson, F., Krickel, W., Perdelwitz, D., & Strauss, W. (1980). *Die Abwasserabgabe als Investitionsanreiz: Auswirkungen des 7a WHG und des Abwasserab gabegezetzes auf Investions planung und -abwicklung.* Berlin: Erich Schmidt Verlag.

Hennipman, P. (1945). *Economisch motief en economisch principe.* Amsterdam.

Hoogerwerf, A. (1985). The anatomy of collective failure in The Netherlands. In Patton (Ed.), *Culture and evaluation* (pp. 47-60). San Francisco-Washington-London: Jossey-Bass, 47-60.

Hucke, J. (1978). Bargaining in regulative policy implementation: The case of air and water pollution control. *Environmental Policy and Law* (pp. 109-115).

Johnson, R.W., & Brown, G.M. (1976). *Cleaning up Europe's waters: Economics, management, policies.* Preager publ.

Majone, G. (1976, Fall). Choice among policy instruments for pollution control. *Policy Analysis,* 589-613.

McSpadden-Wenner, L. (1978, Winter). Pollution control: Implementation alternatives. *Policy Analysis,* 47-65.

Mitnick, B.M. (1980). *The political economy of regulation: Creating, designing and removing regulatory forms.* New York: Columbia University Press.

Oates, W.E. (1983). *Economic incentives for environmental management: The recent US experience.* University of Maryland.

Nagel, S.S. (1975). Incentives for compliance with environmental law. In Milbrath & Inscho (Eds.), *The politics of environmental policy* (pp. 74-94). Beverly Hills: Sage.

Olson, M. (1971, 1965). *The logic of collective action: Public goods and the theory of groups.* Cambridge, MA: MIT Press.

Rose-Ackerman, S. (1977). Market models for water pollution control: Their strength and weaknesses. *Public Policy,* 383-406.

Reichardt, C.S., & Cook, T.D. (1979). Beyond qualitative versus quantitative methods. In Cook & Reichardt (Eds.), *Qualitative and quantitative methods in evaluation research* (pp. 7-32). Beverly Hills: Sage.

Schuurman, J., & Tegelaar, J. (1983). De regulerende werking van de verontreinigingsheffing oppervlaktewateren: Een kwantificering. *Weekblad voor fiscaal recht (Fiscal Law Weekly),* 529-550.

Scriven, M. (1976). Maximizing the power of causal investigations: The modus operandi method. In Glass (Ed.), *Evaluation Studies Review Annual* (pp. 101-119). Beverly Hills: Sage.

Simon, H.A. (1955, February). A behavioral model of rational choice. *Quarterly Journal of Economics,* 88-106.

Vogel, D. (1983). Cooperative regulation: Environmental protection in the United States and Great Britain. *Public Administration Bulletin,* 65-78.

White, L.J. (1976). Effluent charges as a faster means of achieving pollution abatement. *Public Policy,* 111-125.

[22]

The Importance of Exposure in Evaluating and Designing Environmental Regulations: A Case Study

By Albert L. Nichols*

Two themes dominate most economists' critiques of current environmental policy: regulations are promulgated with little attention to benefits and costs, and hence are likely to be set at the "wrong" level; and regulations are not cost effective because they employ inefficient instruments, standards, that fail to take account of variations in control costs. The typical prescriptions are that regulation be based on the principles of benefit-cost or cost-effectiveness analysis, and that standards be replaced by economic incentive schemes, such as emission charges or marketable emission permits.

In this paper I apply the general prescriptions to a specific case, benzene emissions from maleic anhydride plants, the subject of a standard proposed by the Environmental Protection Agency (EPA) in April 1980. The results illustrate how benefit-cost techniques can provide useful guidance to decision makers, even with limited information, and how incentive-based approaches can provide workable and efficient alternatives to standards. The case study also suggests, however, that regulators and economists both need to shift their foci from emissions to damages, that efficient regulations must be sensitive to variations across sources in marginal damages, as well as to differences in the costs of controlling emissions.

Section I provides some background and presents estimates of the costs and benefits of control. Section II analyzes the proposed standard and suggests two more cost-effective strategies based on standards. Section

III considers two types of charges, one levied on emissions and the other on exposure, and shows that the latter will be more efficient. Section IV summarizes the results and examines the impact of information obtained after the standard was proposed.

I. Background, Costs, and Benefits

Benzene is a major industrial chemical, and also a constituent of gasoline. It has been recognized as a hazardous chemical for many decades, but until recently concern focused on noncarcinogenic effects that only occur at levels far above those found in the ambient air. In 1977, however, new epidemiological evidence implicating benzene as a leukemogen led the Occupational Safety and Health Administration (OSHA) to propose a tenfold reduction in the occupational exposure limit and EPA to list it as a "hazardous air pollutant" under Section 112 of the Clean Air Act.

The major sources of benzene emissions are automobiles, service stations, and chemical manufacturing plants. Preliminary EPA estimates indicated that over half of the benzene emitted in chemical manufacturing comes from a handful of maleic anhydride plants that use benzene as a feedstock. Moreover, within each plant virtually all of the benzene is emitted from a single process vent than can be controlled using any one of several well-known technologies. Based on these facts, maleic anhydride plants appeared to be ideal candidates for regulation.

The proposed standard prohibits any benzene emissions from new maleic anhydride plants; due to an increase in the relative price of benzene, however, new plants are expected to use an alternative feedstock regardless of the regulation. The standard would limit existing plants to 0.3 kg benzene emissions per 100 kg benzene input, requir-

*Assistant professor, School of Government, Harvard University. Due to space limitations, most references are omitted. For full documentation and a more extended discussion of the issues in this paper, see my report to the Environmental Protection Agency, which I will provide on request to interested readers. Most of the cost and exposure estimates are based on the standard support document (U.S. EPA).

TABLE 1—PLANT CHARACTERISTICS AND CONTROL COSTS

Plant	Capacity (10^6 kg/year)	Current Control (%)	Control Cost (10^3/year) 97%	Control Cost (10^3/year) 99%
Ashland	27.2	0	410	425
Denka	22.7	97	0	369
Koppers	15.4	99	0	0
Monsanto	38.1	0	542	560
Reichold (IL)	20.0	90	320	333
Reichold (NJ)	13.6	97	0	245
Tenneco	11.8	0	213	224
U.S. Steel	38.5	90	547	565
Total			2,032	2,721

Source: U.S. EPA.

ing approximately a 97 percent reduction from uncontrolled levels. Continuous monitoring of emission levels would be required. When the standard was proposed, EPA identified ten existing maleic anhydride plants. Two do not use benzene, and three already meet the standard due to state regulation of hydrocarbons and hazardous air pollutants. Two of the other five plants, however, already achieve 90 percent control, so the reductions in emissions at those plants would be minimal.

Section 112 specifies that standards be set to assure an "adequate margin of safety," which for carcinogens such as benzene, EPA interprets as requiring Best Available Technology (*BAT*) at a minimum. EPA considered two control levels for defining *BAT*: 97 and 99 percent. Table 1 reports capacities, current control levels, and estimated annualized control costs at both levels. The cost estimates are for carbon absorption, the cheaper of the two control techniques investigated by EPA, and assume full-capacity operation. Note that if a plant already meets a control level (for example, Denka at 97 percent or Koppers at 99 percent), estimated costs are zero. Where existing controls do not achieve the desired level (for example, U.S. Steel at 97 percent or Denka at 99 percent), however, EPA believes that all-new equipment will be required. The plant-specific cost estimates are of dubious accuracy; the results of an engineering analysis of a single, hypothetical model plant were scaled

to each plant only on the basis of production capacity.

Estimating the benefits of control is far more problematical. The primary uncertainty concerns the risk of leukemia from low-dose exposure. The epidemiological studies indicating leukemogenicity are all of workers exposed to concentrations several orders of magnitude higher than those found in the ambient air. Some scientists argue that the threshold for benzene-induced leukemia is well in excess of the pre-1977 OSHA limit of ten parts-per-*million* (*ppm*), so that ambient concentrations of several parts-per-*billion* (*ppb*) pose no threat. But others argue that thresholds do not exist for carcinogens, so that any exposure to benzene poses a risk of leukemia. Even among those who hold the no-threshold position, however, there is considerable disagreement about the appropriate dose-response function.

The EPA's Carcinogen Assessment Group (*CAG*) estimates a risk factor of $.339 \times 10^{-6}$ leukemia deaths per *ppb*-person-year of exposure, based on three epidemiological studies and the assumption that lifetime excess risk is proportional to exposure. That risk estimate implies that exposing ten million people to 1 *ppb* (or one million to 10 *ppb*) for one year would result eventually in an average of 3.4 extra leukemia deaths. The linear dose-response model, however, is the most conservative of the extrapolative models generally considered; estimated from the same data, the others typically generate low-dose risk estimates several orders of magnitude lower. Moreover, critics suggest that the *CAG* made several questionable assumptions about the epidemiological studies that bias upward the risk estimate. Using alternative plausible assumptions, but still the linear model and the same three studies, two EPA analysts derived a risk factor four times lower than the *CAG*'s, and a physician-consultant for industry argues that it should be ten times lower. Although a firm risk estimate cannot be made, these criticisms, together with the conservatism of the linear model, suggest that the *CAG* estimate is too high, and should not be viewed as an expected value.

The ideal approach would be to estimate the risk under each dose-response model, and then compute an expected value based on expert assessment of the probability that each model is correct. Unfortunately, we do not have the information to undertake this task. Because the nonlinear models predict minute risks, however, we know that such an expected-risk function would be very close to linear so long as we assigned a nonnegligible probability to the linear model being correct. Thus, I assume that *expected* benefits will be proportional to the reduction in total exposure, though the coefficient will be smaller than that derived solely from the linear model.

Estimating the benefits also requires assigning a value to "lives saved." A wide range of values has been suggested, from less than $100,000 based on discounted earnings, to several million dollars in some studies of occupational risk premiums. Most fall well below $1 million. Martin J. Bailey has combined data from several willingness-to-pay studies, for example, to estimate an "intermediate" value of $360,000 per life saved in 1978 dollars. We are left with great uncertainty about the benefits of reducing benzene exposure. Even if we combine the *CAG* risk factor of $.339 \times 10^{-6}$ with a value per life saved of $3 million, however, we obtain a "high" benefit estimate of only $1 per *ppb*-person-year. Strong cases can be made for using values one or more orders of magnitude lower.

To predict the effects of emission controls on exposure, EPA conducted dispersion

TABLE 2—REDUCTIONS IN EXPOSURE

Plant	Exposure Factor (*ppb*-years/kg)	Reduction in Exposure (10^3 *ppb*-years/year)	
		97%	99%
Ashland	.017	28.9	30.0
Denka	.248	0.0	4.5
Koppers	.162	0.0	0.0
Monsanto	.391	943.	979.
Reichold (IL)	.008	0.5	0.9
Reichold (NJ)	.384	0.0	4.1
Tenneco	.195	146.	151.
U.S. Steel	.179	21.0	37.8
Total		1,139	1,208

Sources: U.S. EPA; and my calculations.

modeling to estimate concentrations around a model plant, and then used plant-specific population data to estimate the exposure factors shown in Table 2. Multiplying those exposure factors by the predicted reductions in emissions (assuming full-capacity utilization) gives the reductions in total exposures shown in Table 2. Note that the exposure factor, a measure of the marginal benefit of controlling emissions, varies by a factor of almost 50, and that the Monsanto plant accounts for over 80 percent of the reduction in total exposure at either level of control.

II. The Proposed Standard and Alternatives

Based on the estimates in Tables 1 and 2, the proposed standard (97 percent) has a cost-effectiveness ratio of $2,032/1,139 =$ $1.78 per *ppb*-person-year of exposure reduction. (With the *CAG* risk estimate that translates to $5.3 million per life saved.) The other alternative considered by EPA, 99 percent control, has only a slightly higher average cost-effectiveness ratio, but the more relevant marginal ratio indicates that to justify tightening the standard from 97 to 99 percent requires a valuation in excess of $10 per *ppb*-person-year.

If we compare these cost-effectiveness ratios to the high benefit estimate of $1 per *ppb*-person-year discussed earlier, even the 97 percent standard cannot be justified on benefit-cost grounds. Plant-specific cost-effectiveness ratios at 97 percent control cover several orders of magnitude, however, from a low of $.57 for the Monsanto plant to a high of $640 for the Reichold (IL) plant. This wide range reflects differences in both the marginal costs and the marginal benefits of controlling emissions. Monsanto is the largest plant currently without controls, and it also has the highest exposure factor. Reichold (IL), in contrast, already achieves 90 percent control and has the lowest exposure factor.

Lowering the standard to 90 percent would permit the two plants that already achieve that level to avoid the expense of installing new control equipment. Unfortunately, EPA has not estimated the cost of 90 percent controls for the three plants that would need them. We can, however, make a conservative estimate of the benefits of shifting to a 90 percent standard by assuming that it would not reduce control costs for those three plants, but that it would reduce benefits. Under these assumptions, a 90 percent standard would cost $1.2 million and reduce exposure by 1.1 million *ppb*-person-years annually, for a cost-effectiveness ratio of $1.09. Relative to the 97 percent standard, a 6 percent reduction in benefits yields a 43 percent reduction in costs; to justify EPA's 97 percent proposal requires a value of over $11.50 per *ppb*-person-year.

The 90 percent standard improves cost effectiveness by screening out plants with unusually high costs per unit of emissions controlled. Another approach would be to screen out those plants where the *benefits* per unit of emissions controlled are low. Suppose the standard depended on the marginal damage caused by emissions. Carried to the limit, that strategy would result in plant-specific standards. Consider, however, a simpler and more workable version. If we divided the eight plants into two equal-size groups based on the exposure factors in Table 2, and imposed a 97 percent standard only on the four "high-exposure" plants, Monsanto and Tenneco would require new controls, reducing exposure by 1.1 million *ppb*-person-years at a cost of $755,000, for a cost-effectiveness ratio of $.69. Extending the 97 percent requirement to the four low-exposure plants (i.e., reverting to the uniform

97 percent standard) would more than double costs, while increasing benefits by less than 5 percent, for a marginal cost-effectiveness ratio in excess of $25. Thus, modifying the standard to take account of differences in marginal damages could yield substantial gains in efficiency.

III. Incentive-Based Alternatives

The obvious incentive-based alternative to the uniform standard proposed by EPA is a uniform emission charge. Such a charge automatically allocates control efforts so that costs are minimized for any given total reduction in emissions. Table 3 ranks plant-control combinations in estimated order of cost per unit of emissions controlled. (Although 97 percent is less costly per unit of emissions reduction when applied uniformly to all plants, 99 percent is more cost effective for each individual plant, with the exception of Tenneco.) Table 3 also reports estimated costs per unit of exposure reduction.

In theory, the emission charge should be set equal to marginal damage, which in this case is λE, where λ is the damage per *ppb*-person-year of exposure and E is the exposure factor. If, for simplicity, we use the average exposure factor of .2 *ppb*-person-years/kg, the emission charge would be .2 λ. This approach, however, leads to some anomalies. In particular, if $\lambda < .224/.2 = \$1.13$, the charge will not induce Monsanto to control, though controlling Monsanto

TABLE 3—PLANT-SPECIFIC COST-EFFECTIVENESS RATIOS

Plant	Control Level (%)	Cost-effectiveness	
		Emissions ($/kg)	Exposure ($/*ppb*-year)
Monsanto	99	.224	.57
Ashland	99	.238	14.2
Tenneco	97	.285	1.46
Tenneco	99	.383	1.96
U.S. Steel	99	2.68	15.0
Reichold (IL)	99	3.04	366.
Denka	99	20.5	82.8
Reichold (NJ)	99	22.7	59.2

Source: My calculations.

yields net benefits if $\lambda > .57$. Conversely, if $\lambda > .238/.2 = \$1.19$, the charge will induce Ashland to control, though that is not optimal unless $\lambda > \$14.2$. Moreover, even with a more sophisticated approach, it would be difficult for a uniform emission charge to induce Monsanto to control without also affecting Ashland, because their costs per kilogram controlled differ by only 6 percent. An emission charge that led both to control at 99 percent would reduce exposure by 1.0 million *ppb*-person-years at a cost of just under $1 million, for a cost-effectiveness ratio of $.98 per *ppb*-person-year. While more efficient than the standard proposed by EPA, such a charge is dominated by the differential, high-exposure standard discussed earlier.

Efficiency can be achieved with a uniform charge, however, if it is levied on exposure rather than emissions. A charge of $1 per *ppb*-person-year, for example, corresponding to the high benefit estimate, is predicted to induce 99 percent control at Monsanto alone, for a cost-effectiveness ratio of $.57. Under an exposure charge, the next candidate for control would be Tenneco, as is appropriate given that plant's costs and exposure factor.

In practice, of course, exposure cannot be measured directly. Emissions can be monitored on an ongoing basis, however, and combined with plant-specific exposure factors to yield exposure measures on which charges could be levied. Analytically, a uniform exposure charge is equivalent to plant-specific emission charges that are proportional to exposure factors. The exposure charge may be more acceptable, however, because it is uniform and it makes clearer that firm's payments are proportional to the expected damages they cause.

IV. Conclusions

The proposed standard does not appear to be justified by benefit-cost criteria, even if we use relatively high estimates of the benefits of control. Several alternatives provide roughly the same benefits, but at substantially lower costs. Even the most efficient approach, however, is unlikely to yield positive net benefits.

The case study illustrates the importance of focusing analysis (and regulations) on damages rather than emissions. Given wide variations in marginal damages, reductions in emissions are a poor proxy for benefits. Modifying standards to differentiate between high- and low-exposure sites can improve efficiency substantially. Incentive schemes also need to take account of differences in exposure. An exposure charge offers cost-effective control and robustness in the face of uncertainty about control costs. Moreover, an exposure charge can be extended easily to additional source categories, without the need for detailed control cost estimates. Where expected damages are proportional to exposure, as in this case, the optimal exposure charge is independent of control costs.

New information received by EPA after it proposed the standard illustrates the importance of adopting strategies that are robust in the face of uncertainty. Monsanto and Tenneco, the only uncontrolled plants not located in lightly populated areas, will control at the 97 percent level (due to state requirements), even if EPA does not regulate. An additional maleic anhydride plant has been identified, a relatively small plant operated by Pfizer in a lightly populated area. Thus the standard would affect only four plants, two that already control at 90 percent and two with very low exposure factors. As a result, estimated costs fall 27 percent (to $1.5 million) but the benefits fall 95 percent (to 62,000 *ppb*-person-years), raising the cost-effectiveness ratio by a factor of 13, from $1.78 to almost $24 per *ppb*-person-year. An analysis that focused on emissions would not capture the significance of these changes; the cost per kilogram of benzene controlled rises by less than 50 percent.

In addition to the changes described above, the Chemical Manufacturers' Association (CMA) has disputed EPA's cost estimates. For the four plants now likely to be affected if the standard is imposed, CMA estimates a cost of $5.5 million annually, which raises the cost per *ppb*-person-year to $88. Although cost estimates supplied by industry must be viewed with some skepticism, CMA's supporting arguments suggest that EPA's figures are indeed too low. Moreover, even if we discount the CMA estimates entirely, the other information is sufficient to confirm the inefficiency of the proposed standard.

REFERENCES

Bailey, Martin J., *Reducing Risks to Life*, Washington: American Enterprise Institute, 1980.

Nichols, Albert L., "Alternative Strategies for Regulating Airborne Benzene," in T.C. Schelling et al., *Incentive Arrangements for Environmental Protection*, report submitted to U.S. EPA under grant no R8054464010, January 1981.

U.S. Environmental Protection Agency, *Benzene Emissions from Maleic Anhydride Industry — Background Information for Proposed Standards*, Research Triangle Park: Emission Standards and Engineering Division, EPA-450/3-80-001, February 1980.

[23]

The *Net* Benefits of Incentive-Based Regulation:
A Case Study of Environmental Standard Setting

By WALLACE E. OATES, PAUL R. PORTNEY, AND ALBERT M. MCGARTLAND*

Economists interested in environmental, safety, and health regulation have long argued that decentralized, incentive-based (or IB) policies are more efficient than centralized, command-and-control (or CAC) approaches (see Charles Schultze, 1977, for instance). These arguments generally have been based on the assumption that IB policies will accomplish the same goals as their CAC counterparts, but at less cost to society.

However, some of those touting IB policies have overlooked an important point: environmental, workplace, or even product-safety standards typically take the form of maximum permissible concentrations of harmful substances, so that compliance only requires that all monitoring points or samples register readings below these critical levels. For this reason, IB policies typically assign a shadow price of zero to improvements that exceed the standard(s), while more crude CAC policies generally result in "overcontrol" beyond the standards. If there is no value to this overcontrol, CAC policies will not improve at all on IB approaches and will indeed be more expensive. If, however, reduced concentrations below the level of the standards bring with them further improvements in health or the environment, CAC approaches will produce greater benefits than IB approaches. Thus, a fair comparison between the two necessitates that any additional benefits associated with CAC policies be offset against the cost advantages enjoyed by their IB counterparts.

Although this possibility has been recognized by others (Scott Atkinson and T. H. Tietenberg, 1982, for instance), it is little appreciated and its empirical significance has never been ascertained. That is our purpose here. We do so by developing data on the costs and benefits of controlling a common air pollutant, total suspended particulates (or TSP), in Baltimore. By comparing the TSP levels likely under both IB and CAC approaches, we are able to estimate the marginal costs and benefits associated with a variety of alternative air quality standards which take the form of maximum permissible concentrations. This in turn allows us to determine the *net* benefits arising from the two kinds of regimes.

In the next section we present a simple conceptual framework for our analysis. Section II describes the estimation of the costs and benefits of our hypothesized air pollution controls in Baltimore. Section III presents our somewhat surprising findings, and Section IV discusses those findings and their potential significance for regulation in the "real world."

*The authors' affiliations are, respectively, Department of Economics and Bureau of Business and Economic Research, University of Maryland, and University Fellow, Resources for the Future; vice president, Resources for the Future; and Economist, Abt Associates. The views in this paper are those of the authors and do not necessarily reflect those of the organizations with which they are affiliated. We are very grateful to Karen Clay, Julie Kurland, and especially Stephen McGonegal for their invaluable assistance with the empirical work. In addition, we wish to thank Ann Fisher, Kerry Smith, our colleagues at Resources for the Future, and two referees for their most helpful comments on earlier drafts of this paper. Finally, we are indebted to the National Science Foundation and the Andrew W. Mellon Foundation for their support of this research.

I. The Conceptual Framework

Before turning to our Baltimore data, it will be helpful to set the problem in a more general framework. Suppose that we have a specific "region"—it could be an air shed, a system of waterways, or even the ambient environment in a large factory—in which there are m sources of pollution, each of which is fixed in location. Environmental quality is defined in terms of pollutant concentrations at each of n "receptor points" in the region. We can thus measure environmental quality by a vector $Q = (q_1, q_2, \ldots, q_n)$ whose elements indicate the concentration of the pollutant at each of the receptors. This, incidentally, makes one important, if obvious, point: the "level" of environmental quality is actually a set of pollutant concentrations at different points in the region—it is not (for most pollutants) simply a single level of pollution.

The dispersion of emissions from the m sources in the region is described by an $m \times n$ matrix of unit diffusion (or transfer) coefficients:

$$D = \ldots d_{ij} \ldots,$$

where d_{ij} indicates the increase in pollutant concentration at receptor j from an additional unit of emissions of the pollutant by source i. If we denote by e_i the level of emissions by source i, we can then describe the pattern of waste emissions in the region by the vector $E = (e_1, e_2, \ldots, e_m)$. The levels of pollution at the various receptor points can then be determined by mapping the vector of emissions through the diffusion matrix:

$$ED = Q.$$

Finally, we introduce the abatement cost function: $C_i(e_i)$ is the cost to source i of holding its emissions to e_i.

Let us suppose that some standard for environmental (or workplace) quality has been set—we take it for now as predetermined. The standard takes the form of a maximum permissible level of pollutant concentration at any receptor point in the region. There are various regulatory strategies that an environmental agency might pursue to comply with the standard. Following a "command-and-control" (CAC) approach, the agency might specify abatement technologies for the sources. Suppose, as is common practice, that it required all similar sources to adopt the same control procedures and tightened up these procedures until the standard was everywhere satisfied. Such a control program would result in a specific vector of emissions from sources —call it E_c. And this vector would map through the diffusion matrix into a vector Q_c of pollutant concentrations.

Note that in virtually all cases the standard will be binding at only one or a few receptors. Most receptors will have pollutant concentrations below that required by the standard so that environmental quality at most points in the air shed, waterway, or workplace will exceed that prescribed by the standards. *It is also clear that the resulting vector of environmental quality (and the associated levels of damages and control costs) depends on the specific regulatory program adopted by the agency.*

Suppose instead that the environmental agency pursues an IB strategy. By this we mean that it seeks that vector of emissions (E_1) that can attain the standard at the minimum aggregate abatement cost:

$$\text{Min } \Sigma C(e_i)$$

$$\text{s.t.} \quad ED \leq Q^*$$

$$E \geq 0,$$

where Q^* is the upper bound on allowable pollutant concentrations. There are various ways this might be done, including the use of effluent fees or transferable discharge permits. Such a program will, by definition, achieve the standard at a cost less than (or equal to) our CAC program. And this will involve a different vector of emissions. In general (as existing studies show), the IB vector will entail higher levels of emissions and higher levels of pollutant concentrations at nonbinding receptor points than will the CAC solution. This is not surprising, since the cost-minimization procedure assigns a zero shadow price to any additions to pollutant concentrations so long as the standard is not exceeded. In its search to reduce abatement costs, the IB approach effectively makes use of any "excess" environmental capacity to allow increased emissions. *Thus, for any given Q^*, we expect in general to find* *levels of emissions, concentrations, and damages that are higher under the IB solution than under the CAC outcome.*

The levels of both benefits and control costs associated with a particular standard will, in consequence, tend to be higher under a CAC than under an IB regime. Just how the levels of *net* benefits and the optimal standard will compare under these approaches is not something we can determine a priori; it is an empirical matter. To get a sense of the magnitudes involved, we investigate in the succeeding sections the benefit and cost functions for a specific air pollutant in the Baltimore region.

II. Estimating Benefits and Costs

To estimate the marginal costs of TSP control under the two regimes, we used a model developed by McGartland (1983, 1984) which reflects the technological control possibilities, associated particulate reduction efficiencies, and costs for about 400 actual sources in Baltimore. The marginal abatement cost function under the IB approach reflects, for each possible standard considered, the least-cost combination of control options across all particulate sources that ensures attainment at all receptors. To estimate marginal costs for the CAC regime, we adopted the basic spirit of the regulatory strategy used in Baltimore. First, all sources were categorized and similar sources grouped together—for instance, industrial coal-fired boilers, grain shipping facilities, etc. Then, marginal costs for additional control were estimated for each source *category*. Finally, when additional controls were required to reduce particulate levels, the source *category* with the lowest cost-per-ton was targeted for further regulation; all sources within that category were required to adopt the same technology regardless of their individual costs or location.

To estimate the marginal benefits associated with alternative standards, we first assigned the 1980 population of the Baltimore metropolitan area to one of the 23 receptors in the area using the geographic coordinates of each census tract and each receptor. This gave us an "exposed population" by receptor

ranging from as few as 3,800 people assigned to one receptor to more than 180,000 at another.

Given these exposures, we calculated marginal benefits from successively tighter TSP standards for four different categories: reduced premature mortality, reduced morbidity, reduced soiling damages to households, and improved visibility. For each category, the changes in TSP levels that would accompany successively tighter standards were first translated into physical improvements (fewer sick days, fewer "statistical" lives lost, reduced soiling, and increased visibility). To do so, we relied primarily on the peer-reviewed dose-response studies actually used by the EPA in setting the national air quality standard for particulates. We then monetized these physical improvements using recent studies on the valuation of premature mortality, morbidity, soiling, and visibility.[1] While we have made what we feel are the best estimates possible for marginal benefits and costs, the real value of the analysis lies in the comparisons between the IB and CAC approaches. Such comparisons are more important and more legitimate than inferences about the actual levels of benefits and costs.

To summarize, given a hypothetical change in the TSP standard for Baltimore, the cost model is used to determine how that change will be accomplished technologically under both the CAC and IB approaches. The model not only determines the pattern of controls, emissions reductions, and associated costs under each regime, but also produces a vector of ambient TSP levels at each of the 23 receptors. By comparing this vector with the preexisting one, we can determine the change in air quality at each receptor. Using the mortality, morbidity, soiling, and visibility

---

[1] We will provide upon request a detailed description of the methods used to calculate the benefit functions. Briefly, the mortality and morbidity benefit estimates are based on cross-section and time-series epidemiological studies of the effects of particulate matter on health coupled with valuations of $2 million per life "saved," $100 per lost work day, and $25 for each restricted activity day. Soiling and visibility benefits are estimated in an analogous way.

"dose-response" functions, we translate the physical changes in air quality into welfare improvements and, simultaneously, value them to arrive at estimates of marginal benefits. That is the process behind the empirical results presented below.

III. The Findings

We report in Table 1 and depict in Figures 1 and 2 our basic results. As we move from left to right along the horizontal axes in the figures, we encounter successively more stringent *standards* as indicated by lower permissible maximum concentrations of TSP. Consider, for example, a TSP standard of 100 $\mu g/m^3$ under the IB case. We see from Table 1 that the marginal control costs of moving from a standard of 105 $\mu g/m^3$ to the more stringent standard of 100 $\mu g/m^3$ are $1.82 million, while the associated marginal benefits are $8.53 million. These values appear in Figure 1 as points on the MC and MB curves at a TSP standard of 100.[2]

A cursory examination of the table and the accompanying figures suggests, first, that were we to select an "optimum" standard

---

[2] The marginal benefit and abatement cost curves are reasonably well behaved for the least-cost case in Figure 1. Marginal benefits remain roughly constant over most of the relevant range with some tendency to tail off after a standard of 85 $\mu g/m^3$ is achieved. The relative constancy of marginal benefits results primarily from the fact that the dose-response functions that we use (based on EPA documents) are linear over the range of air quality standards that we consider. The occasional "ups and downs" in the MB curve reflect the differing degree to which individual receptors are controlled as we move to successively more stringent standards.

Marginal abatement costs remain low and well below marginal benefits for less stringent standards, but begin to rise rapidly after a standard of 90 $\mu g/m^3$ is reached. The functions are not so well behaved for the CAC case in Figure 2. In particular, the marginal cost curve exhibits a large "hump" around a standard of 90 $\mu g/m^3$. This hump has its source partly in our rule for regulatory behavior. It turns out that to go from a standard of 95 $\mu g/m^3$ to 90 $\mu g/m^3$ requires the adoption of some additional and rather costly control measures by a large number of sources; our CAC rule necessitates that these measures be applied to a whole class of polluting sources (irrespective of location), resulting in a sharp increase in control costs.

under each system by equating marginal benefits and costs, the IB approach would give us a more stringent standard than the CAC regime. From Table 1, we see that the "optimum" standard under the IB case is 90 $\mu g/m^3$, while for the CAC case this standard is only 100 $\mu g/m^3$.[3] This result appears to confirm a point that environmental economists have long argued: the adoption of less costly control techniques should make it possible to attain higher levels of environmental quality.

This inference, however, is misleading. The source of the confusion is the natural inclination to associate air quality *standards* with air quality *levels*. But as we have seen, these are not the same thing. We cannot emphasize this distinction enough: an air quality standard maps into a vector of pollutant concentrations and the mapping itself depends, as we have seen, upon the regulatory regime. While it is true for our Baltimore case that the IB "optimum" would lead us to select the more stringent *standard* for air quality, it does not necessarily follow that this would actually result in better air quality throughout the area. Standards do not provide an unambiguous measure of air quality; they are ceilings on permissible levels of pollutant concentrations—most receptors will have concentrations well below the standard. Thus, *the same standard on the horizontal axis in our figures will produce a different vector of air quality under our two regulatory systems.*

To provide a better sense of these differences, we present in Table 2 estimated TSP concentrations for the various standards for a representative sample of our 23 Baltimore

---

[3] It is interesting that the "optimum" standard for Baltimore for the IB case is quite close to the EPA primary standard for TSP concentrations of about 85 $\mu g/m^3$. (With a standard deviation of roughly 1.5, the EPA primary standard of 75 $\mu g/m^3$ expressed as a geometric mean translates roughly into a standard of 85 $\mu g/m^3$ as an arithmetic mean). As we indicated earlier, we should not make too much of this, for there is considerable uncertainty surrounding the estimates we have used for the benefit and cost functions. What is of more interest is the comparison between the IB and CAC cases.

TABLE 1—MARGINAL CONTROL COSTS (MC) AND MARGINAL BENEFITS
(MB) UNDER THE INCENTIVE-BASED AND CAC SYSTEMS
(IN MILLIONS OF 1980 DOLLARS)

| Standard | Incentive-Based Case | | |
	MC	MB	(MB-MC)
115	1.36	7.25	5.89
110	1.90	12.94	11.04
105	2.63	9.09	6.46
100	1.82	8.53	6.71
95	4.60	13.22	8.62
90	8.66	15.14	6.48
85	20.98	16.37	−4.61
83	35.23	3.88	−31.35

| Standard | Command & Control Case | | |
	MC	MB	(MB-MC)
115	0.50	2.18	1.68
110	2.45	10.52	8.07
105	3.32	9.69	6.37
100	9.14	11.48	2.34
95	15.06	7.51	−7.55
90	54.67	10.00	−44.67
85	16.00	6.49	−9.51
83	9.95	1.19	−8.76

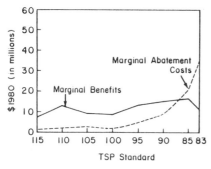

FIGURE 1. LEAST-COST CASE

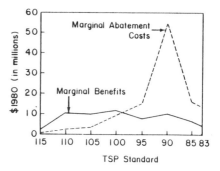

FIGURE 2. COMMAND AND CONTROL CASE

receptors under the IB and CAC regimes. Consider, for example, the TSP levels at receptor 1 under each system. We find that for receptor 1 the TSP level under CAC for a standard of 100 $\mu g/m^3$ is 61.4 $\mu g/m^3$; under the IB "optimum" of 90 $\mu g/m^3$, the TSP level at receptor 1 is 61.6 $\mu g/m^3$. For this particular receptor, then, the less stringent standard under CAC actually results in a higher level of air quality than does the more stringent standard under the least-cost outcome. This is not the case for all the receptors. For example, the binding receptors will obviously have higher TSP concentrations where the standard is less stringent (compare, for example, the TSP levels at receptor 5). Table 2 also makes clear the wide variation in air quality among the various recep-

TABLE 2—TSP CONCENTRATION BY RECEPTOR

Receptor	120	115	110	105	100	95	90	85	83
				Incentive-Based Case					
1.	67.8	67.4	66.2	66.0	65.3	63.7	61.6	59.3	58.6
2.	64.6	63.7	62.2	61.8	60.9	58.7	55.5	51.7	50.9
3.	56.2	56.0	55.5	55.5	55.3	54.6	53.7	52.5	52.2
4.	116.3	113.8	107.8	104.3	100.0	95.5	90.0	85.0	84.0[a]
5.	119.7	115.3	110.4	105.5	100.0	95.2	89.5	84.7	83.5
6.	52.4	51.6	49.1	47.5	46.0	43.4	40.9	38.2	37.6
7.	120.0	114.9	110.4	101.0	99.6	93.0	79.5	53.3	45.4
8.	105.3	102.8	98.9	97.7	95.1	90.4	83.8	74.1	70.1
				Command & Control Case					
1.	65.1	65.0	64.0	62.9	61.4	60.6	59.3	58.4	58.3
2.	60.7	60.4	58.9	57.2	54.9	53.6	52.0	50.7	50.5
3.	54.5	54.4	54.1	53.8	53.2	52.9	52.5	52.1	52.1
4.	109.9	108.7	103.1	99.5	95.2	92.0	87.8	85.0	84.0[a]
5.	120.8	115.5	109.8	104.7	99.6	95.0	89.3	84.2	83.4
6.	45.8	45.5	44.0	42.7	41.0	39.9	38.5	37.6	37.4
7.	106.0	105.8	102.9	94.7	71.9	64.9	58.1	43.9	42.7
8.	97.0	96.5	92.8	88.8	82.5	78.1	73.9	69.8	69.2
				Population-Weighted Averages of Receptor TSP Levels					
IB	77.4	75.7	72.9	70.9	69.0	66.2	62.9	59.3	58.5
CAC	71.1	60.6	68.3	66.2	63.7	62.0	59.9	58.5	58.2

[a]Although the standard is 83 $\mu g/m^3$, there are no controls in the model capable of reducing air pollution at this receptor.

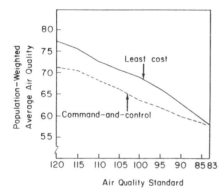

FIGURE 3. POPULATION-WEIGHTED AVERAGE AIR QUALITY UNDER THE LEAST COST AND COMMAND AND CONTROL SYSTEMS

tors; the TSP concentrations at receptors 1, 2, 3, and 6, for example, are far below the standard—in some instances the concentrations are less than one-half of the standard.

In the bottom rows of the table and in Figure 3, we present a summary measure of air quality: a population-weighted average of TSP levels at all 23 receptor points. A comparison of these weighted averages for the "optima" under the two systems reveals that the IB outcome yields a weighted average TSP level of 62.9 $\mu g/m^3$ as compared to the weighted average of 63.7 $\mu g/m^3$ under CAC. Thus, average air quality is only very slightly (probably negligibly) higher under the IB "optimum" than under the CAC "optimum." This is in sharp contrast to the large difference under the two systems in the "optimal" *standard*.

Our first result then is that although the IB regime results in a more stringent "optimal" standard, there is really little difference in overall air quality under the "optima" of our two systems.

The second issue concerns the *total net benefits* of pollution control under the two systems. This calculation is more problematic. The data in Table 1 only allow us to compute the *marginal* net benefits for successively tighter air quality standards beginning with a standard of 120 $\mu g/m^3$. Ideally, we should compare the net benefits of going

TABLE 3—A COMPARISON OF THE CUMULATIVE NET BENEFITS
UNDER THE TWO SYSTEMS (MILLIONS OF 1980 DOLLARS)

1. *Incentive-Based Case: Net Benefits from Moving from a Standard of 120 $\mu g/m^3$ to the "Optimal" Standard of 90 $\mu g/m^3$*

Cumulative MB	$66.17
Cumulative MC	20.97
Cumulative Net Benefits	$45.20

2. *Command & Control Case: Net Benefits from Moving from a Standard of 120 $\mu g/m^3$ to the "Optimal" Standard of 100 $\mu g/m^3$*

Cumulative MB	$33.86
Cumulative MC	15.41
Cumulative Net Benefits	$18.45

3. *Adjustment of Net Benefits Under the CAC System*

Cumulative Net Benefits Under CAC	$18.45
Less: Baseline Control Costs in Excess of IB Case	7.81
Plus: Baseline Benefits in Excess of IB Case	28.67
Adjusted Cumulative Net Benefits	$39.31

from a baseline of the totally uncontrolled level of emissions to the optimal standard under each system. (The uncontrolled outcome involves very high levels of TSP concentrations of around 500 $\mu g/m^3$).

We feel that the benefit functions cannot legitimately be extended to value changes in air quality over such extreme levels of TSP concentrations. Consequently, we chose as a baseline the vector of air quality that results from a standard of 120 $\mu g/m^3$. For the IB system, when we sum the differences between the MB and MC curves from this baseline to the "optimal" level of 90 $\mu g/m^3$, we find that the net benefits (from our arbitrary baseline) are roughly $45 million.

Two adjustments must be made to calculate a comparable net benefit estimate for the CAC regime. Recall that, for any given standard (including our baseline), the resulting vector of air quality under the CAC outcome indicates cleaner air than under the IB result. Therefore, we cannot sum the differences between the MB and MC curves from 120 $\mu g/m^3$ to the "optimal" standard of 100 $\mu g/m^3$ and compare this estimate to the IB net benefit calculation. The two calculations have different starting points.

To make the CAC "starting point" comparable to that under the IB system, we must

make both a cost and a benefit adjustment. Turning first to the cost adjustment, we find that to go from the uncontrolled level of about 500 $\mu g/m^3$ to our baseline of 120 $\mu g/m^3$, it costs an estimated $7.81 million more under the CAC approach than under the IB system. So for purposes of comparison, we must add to the cumulative costs of the CAC system this additional sum of $7.81 million.

Second, we must make a benefit adjustment. Although we do not attempt to measure the benefits from moving from the uncontrolled state to the baseline for reasons discussed earlier, we must account for the cleaner air (and correspondingly higher benefits) that the CAC outcome provides at the baseline standard. Our benefit functions yield an estimate of $28.67 million for the value of the differentially higher level of air quality produced by the CAC relative to the IB outcome at our baseline standard of 120 $\mu g/m^3$. After making these two adjustments so that the CAC starting point is equivalent to that under the IB regime, we find that the *net* benefits of the CAC scheme are roughly $39 million.

Table 3 summarizes these net benefit calculations including the adjustments needed to permit comparisons between the IB and

CAC outcomes. It is important to interpret these numbers properly. We emphasize that they do *not* provide estimates of the cumulative net benefits under each system; in fact, they greatly underestimate these net benefits because they omit any valuation of the benefits provided by the improvement in air quality from the uncontrolled state to the baseline. And these benefits are no doubt very large. We omit them because (as mentioned earlier) we are not comfortable using our benefit functions to value changes over such extreme levels of pollution. But these benefit figures have been omitted from the estimates for both systems so that cumulative benefits are understated by the same sum for the IB and CAC cases. The figures can thus be used legitimately to compare the net benefits under the two systems.

When we do this, we find that the difference between the cumulative net benefits under the two systems is quite small. As Table 3 shows, the cumulative net benefits under the IB outcome exceed those under the CAC case by only about $6 million when evaluated at their respective "optima."

It is interesting to contrast this comparison with one in which no consideration is given to the differentials in benefits under the two systems. Suppose, for example, that we were to choose a standard of 100 $\mu g/m^3$ and were simply to compare the costs under the two systems of achieving that standard (under the implicit and mistaken assumption that air quality is the same in both cases). Our computations indicate that the attainment of this standard would cost $32.7 million under the IB system as compared to $48.1 million under the CAC regime. We would thus conclude that the CAC approach costs about half again as much to attain the same outcome. But, as we have seen, the outcomes are far from the same.[4]

Finally, we stress once again that while we believe that our findings provide a legitimate basis for comparison between our two prototypical systems, the absolute levels of benefits and costs associated with the various standards must not be taken very seriously. Some sensitivity analysis using upper and lower bounds for our benefits estimates suggests that the "optimal" standard under both systems is quite sensitive to our choice of benefits measures.

IV. Concluding Remarks

The theme of this paper is that IB policies designed to achieve prescribed regulatory standards at least cost may not be so obviously superior to CAC approaches as has been supposed. This will be the case when CAC policies are designed with at least one eye on cost savings—as they sometimes are —and when reductions below the level of the relevant environmental, workplace, or product standards result in beneficial effects. In these cases, the "overcontrol" that makes CAC policies more expensive also makes them more efficacious.

---

[4]Although there are considerable differences in the outcomes under our two systems over most of the range of alternative standards for air quality, the outcomes converge as air quality approaches relatively high levels. When we reach a TSP concentration of 83 $\mu g/m^3$, the highest of the standards indicated in our figures and tables, the outcomes under the IB and CAC systems are

virtually the same (as is evident in the population-weighted air quality in Figure 3). This occurs because we have now reached the point at which virtually all sources are controlling their emissions to the maximum degree possible under existing control technologies. This manifests itself in the IB case by very rapidly increasing marginal control costs, which are the result of having to introduce control measures in suburban areas in an attempt to reduce pollutant concentrations at the binding receptors which are located in the center city.

These rapidly increasing marginal control costs under the least-cost system have an interesting implication for the design of a regulatory system for pollution control. Following Martin Weitzman's seminal paper (1974) on the choice between quantity and price instruments, our results suggest a strong preference for an effluent fee system over a system of marketable emission permits. In a setting of uncertainty regarding the true benefit and cost functions, a mistake in setting the environmental standard is likely to result in a more costly error when the marginal cost curve is steep relative to the marginal benefit curve in the relevant region. A look at Figure 1 indicates that this is indeed the case for our pollutant at the least-cost outcome. This suggests that the environmental authority should employ the fee approach where errors are likely to be less costly.

One problem with this conclusion, of course, is that neither approach results in the economically optimal outcome in the full sense. We have been careful to describe the standard for which marginal benefits equal marginal costs under either system as the "optimal" outcome, using quotation marks to emphasize that the standard is optimal only with respect to that system. But it is clear that there is an economically superior outcome, namely the Pareto-efficient solution.

One may legitimately ask why we do not reject both of the suboptimal regimes we examine in favor of the first-best outcome. After all, we would seem to have all the information needed to determine the first-best solution. Our response to this question is twofold. First, while it is admittedly an easy matter, in principle, to characterize the first-best outcome, calculating it is a more complicated matter. The characterization is straightforward: the first-order conditions for the economic optimum would have us determine an emissions vector such that *for each source* the marginal benefits from an additional unit of abatement equal marginal cost. This would involve a very complicated general-equilibrium calculation, one that could easily entail a multiplicity of local optima. Nevertheless, with sufficient ingenuity and patience, one might determine this outcome. This solution, incidentally, would typically involve assigning abatement techniques on a polluter-by-polluter basis. Alternatively, one might design some differentiated set of effluent fees to induce the requisite pattern of abatement behavior.

While this may be conceivable in principle, we find it very difficult to see how an environmental agency could implement such procedures. And this is our second, and more basic, point. As the spirit of this paper suggests, we have sought to consider those alternatives that appear feasible in an actual policy setting. The achievement of some selected standards for environmental quality either through a command-and-control approach or through a general incentive-based approach represent alternatives *with precedent* in the policy arena. While the former is certainly the more common, there are now

programs (such as EPA's Controlled Trading Program) that make some use of economic incentives to attain the specified environmental standards.

In the choice between the two approaches, we have argued that the case in the literature has been biased in favor of IB measures. However, it is important to put this contention in the proper perspective. One reason that the CAC outcome fares so well in our analysis is that Baltimore air quality authorities employed a somewhat sophisticated and relatively cost-effective procedure for TSP control. Under a less enlightened regulatory regime, control costs could be much higher for equivalent air quality levels.[5]

This brings us to what we see as the basic implication of our findings. They suggest that a carefully designed and implemented CAC system may stack up reasonably well relative to a feasible IB counterpart. However, where CAC standards and implementation procedures are motivated primarily by political considerations (for example, the avoidance of plant closings or of unpopular increases in the cost of local power), CAC policies will get bad marks in comparison to the IB alternatives. Badly designed CAC measures, in short, will yield bad outcomes.

Where, in contrast, economic analysis plays a larger role in CAC standard setting and program design—as it did in Baltimore and does for certain programs at EPA—one may have to take a harder look at such

---

[5]As one referee pointed out, our results are also sensitive to the choice of pollutant. To take the extreme case, if there were "perfect mixing" such that an emission of the pollutant at any place contributed equally to pollutant concentrations at all receptor points, then air quality would everywhere be the same. The standard would be binding at all receptors, and there would be no overcontrol under the CAC system to provide differential benefits relative to the IB counterpart. However, where the effects of the pollutant become more highly localized, overcontrol at some receptors will tend to be more pronounced. Our sense is that the pollutant we have used in this study is somewhere to the middle of the "localization spectrum." While volatile organic compounds, for example, exhibit somewhat greater mixing propensities than TSP, many other pollutants including carbon monoxide, "air toxics" (like lead and various chemicals), and certain workplace agents have far more localized effects than TSP.

approaches. Efforts by economists to make CAC measures more effective may, for particular programs, produce outcomes that compare quite well with IB alternatives. Particularly when we take into account real-world regulatory institutions that require uniformity of fees (or in other ways reduce the flexibility needed to achieve the full advantages of the IB approach), incentive-based programs may not clearly dominate well-designed CAC measures.[6]

---

[6]We should also note that the argument in this paper relates solely to the *static* efficiency properties of the alternative approaches. Over the longer haul, it is of great importance that we have a system that embodies the appropriate incentives for research and development of new abatement technologies. IB approaches, as economists have long argued, have compelling advantages over typical CAC regimes on this count. Even here, however, there is some scope for designing CAC programs in a way that encourages, rather than impedes, R&D efforts.

REFERENCES

Atkinson, Scott and Tietenberg, T. H., "The Empirical Properties of Two Classes of Designs for Transferable Discharge Permit Markets," *Journal of Environmental Economics and Management*, June 1982, *9*, 101–21.

McGartland, Albert M., "The Cost Structure of the Total Suspended Particulate Emission Reduction Credit Market," *Baltimore Region Emission Report*, Vol. *V.*, Regional Planning Council, Baltimore, MD, 1983.

_____ , *Marketable Permit Systems for Air Pollution Control: An Empirical Study*, unpublished doctoral dissertation, University of Maryland, College Park, 1984.

Schultze, Charles L., *The Public Use of Private Interest*, Washington: The Brookings Institution, 1977.

Weitzman, Martin L., "Prices vs. Quantities," *Review of Economic Studies*, October 1974, *41*, 477–91.

OXFORD REVIEW OF ECONOMIC POLICY, VOL. 6, NO. 1

ECONOMIC INSTRUMENTS FOR ENVIRONMENTAL REGULATION

T. H. TIETENBERG
Colby College

I. INTRODUCTION

As recently as a decade ago environmental regulators and lobbying groups with a special interest in environmental protection looked upon the market system as a powerful adversary. That the market unleashed powerful forces was widely recognized and that those forces clearly acted to degrade the environment was widely lamented. Conflict and confrontation became the battle cry for those groups seeking to protect the environment as they set out to block market forces whenever possible.

Among the more enlightened participants in the environmental policy process the air of confrontation and conflict has now begun to recede in many parts of the world. Leading environmental groups and regulators have come to realize that the power of the market can be harnessed and channelled toward the achievement of environmental goals, through an economic incentives approach to regulation. Forward-looking business people have come to appreciate the fact that cost-effective regulation can make them more competitive in the global market-place than regulations which impose higher-than-necessary control costs.

The change in attitude has been triggered by a recognition that this former adversary, the market, can be turned into a powerful ally. In contrast to the traditional regulatory approach, which makes mandatory particular forms of behaviour or specific technological choices, the economic incentive approach allows more flexibility in how the environmental goal is reached. By changing the incentives an individual agent faces, the best private choice can be made to coincide with the best social choice. Rather than relying on the regulatory authority to identify the best course of action, the individual agent can use his or her typically superior information to select the best means of meeting an assigned emission reduction responsibility. This flexibility achieves environmental goals at lower cost, which, in turn, makes the goals easier to achieve and easier to establish.

One indicator of the growing support for the use of economic incentive approaches for environmental control in the United States is the favourable treatment it has recently received both in the popular business[1] and environmental[2] press. Some public interest environmental organizations have now even adopted economic incentive approaches as a core part of their strategy for protecting the environment.[3]

In response to this support the emissions trading concept has recently been applied to reducing the lead content in gasoline, to controlling both ozone depletion and non-point sources of water pollution, and was also prominently featured in the Bush administration proposals for reducing acid rain and smog unveiled in June 1989.

Our knowledge about economic incentive approaches has grown rapidly in the two decades in which they have received serious analytical attention. Not only have the theoretical models become more focused and the empirical work more detailed, but we have now had over a decade of experience with emissions trading in the US and emission charges in Europe.

As the world community becomes increasingly conscious of both the need to tighten environmental controls and the local economic perils associated with tighter controls in a highly competitive global market-place, it seems a propitious time to stand back and to organize what we have learned about this practical and promising approach to pollution control that may be especially relevant to current circumstances. In this paper I will draw upon economic theory, empirical studies, and actual experience with implementation to provide a brief overview of some of the major lessons we have learned about two economic incentive approaches—emissions trading and emission charges—as well as their relationships to the more traditional regulatory policy.[4]

II. THE POLICY CONTEXT

(i) Emissions Trading

Stripped to its bare essentials, the US Clean Air Act[5] relies upon a *command-and-control* approach to controlling pollution. Ambient standards establish the highest allowable concentration of the pollutant in the ambient air for each conventional pollutant. To reach these prescribed ambient standards, emission standards (legal emission ceilings) are imposed on a large number of specific emission points such as stacks, vents, or storage tanks. Following a survey of the technological options of control, the control authority selects a favoured control technology and calculates the amount of emission reduction achievable by that technology as the basis for setting the emission standard. Technologies yielding larger amounts of control (and, hence, supporting more stringent emission standards) are selected for new emitters and for existing emitters in areas where is it very difficult to meet the ambient standard. The responsibility for defining and enforcing these standards is shared in legislatively specified ways between the national government and the various state governments.

The emissions trading programme attempts to inject more flexibility into the manner in which the objectives of the Clean Air Act are met by allowing sources a much wider range of choice in how they satisfy their legal pollution control responsibilities than possible in the command-and-control approach. Any source choosing to reduce emissions at any discharge point more than required by its emission standard can apply to the control authority for certification of the excess control as an 'emission reduction credit' (ERC). Defined in terms of a specific amount of a particular pollutant, the certified emissions reduction credit can be used to satisfy emission standards at other (presumably more expensive to control) discharge points controlled by the creating source or it can be sold to

---

[1] See, for example, Main (1988).

[2] See, for example, Stavins (1989).

[3] See the various issues in Volume XX of the EDF Letter, a report to members of the Environmental Defense Fund.

[4] In the limited space permitted by this paper only a few highlights can be illustrated. All of the details of the proofs and the empirical work can be found in the references listed at the end of the paper. For a comprehensive summary of this work see Tietenberg (1980), Liroff (1980), Bohm and Russell (1985), Tietenberg (1985), Liroff (1986), Dudek and Palmisano (1988), Hahn (1989), Hahn and Hester (1989a and 1989b), and Tietenberg (1989b).

[5] The US Clean Air Act (42 U.S.C. 7401–642) was first passed in 1955. The central thrust of the approach described in this paragraph was inititated by the Clean Air Act Amendments of 1970 with mid-course corrections provided by the Clean Air Act Amendments of 1977.

other sources. By making these credits transferable, the US Environmental Protection Agency (EPA) has allowed sources to find the cheapest means of satisfying their requirements, even if the cheapest means are under the control of another firm. The ERC is the currency used in emissions trading, while the offset, bubble, emissions banking, and netting policies govern how this currency can be stored and spent.[6]

The *offset policy* requires major new or expanding sources in 'non-attainment' areas (those areas with air quality worse than the ambient standards) to secure sufficient offsetting emission reductions (by acquiring ERCs) from existing firms so that the air is cleaner after their entry or expansion than before.[7] Prior to this policy no new firms were allowed to enter non-attainment areas on the grounds they would interfere with attaining the ambient standards. By introducing the offset policy EPA allowed economic growth to continue while assuring progress toward attainment.

The *bubble policy* receives its unusual name from the fact that it treats multiple emission points controlled by existing emitters (as opposed to those expanding or entering an area for the first time) as if they were enclosed in a bubble. Under this policy only the total emissions of each pollutant leaving the bubble are regulated. While the total leaving the bubble must be not larger than the total permitted by adding up all the corresponding emission standards within the bubble (and in some cases the total must be 20 per cent lower), emitters are free to control some discharge points less than dictated by the corresponding emission standard as long as sufficient compensating ERCs are obtained from other discharge points within the bubble. In essence sources are free to choose the mix of control among the discharge points as long as the overall emission reduction requirements are satisfied. Multi-plant bubbles are allowed, opening the possibility for trading ERCs among very different kinds of emitters.

Netting allows modifying or expanding sources (but not new sources) to escape from the need to

meet the requirements of the rather stringent new source review process (including the need to acquire offsets) so long as any net increase in emissions (counting any ERCs earned elsewhere in the plant) is below an established threshold. In so far as it allows firms to escape particular regulatory requirements by using ERCs to remain under the threshold which triggers applicability, netting is more properly considered regulatory relief than regulatory reform.

Emissions banking allows firms to store certified ERCs for subsequent use in the offset, bubble, or netting programmes or for sale to others.

Although comprehensive data on the effects of the programme do not exist because substantial proportions of it are administered by local areas and no one collects information in a systematic way, some of the major aspects of the experience are clear.[8]

• The programme has unquestionably and substantially reduced the costs of complying with the requirements of the Clean Air Act. Most estimates place the accumulated capital savings for all components of the programme at over $10 billion. This does not include the recurring savings in operating cost. On the other hand the programme has not produced the magnitude of cost savings that was anticipated by its strongest proponents at its inception.

• The level of compliance with the basic provisions of the Clean Air Act has increased. The emissions trading programme increased the possible means for compliance and sources have responded.

• Somewhere between 7,000 and 12,000 trading transactions have been consummated. Each of these transactions was voluntary and for the participants represented an improvement over the traditional regulatory approach. Several of these transactions involved the introduction of innovative control technologies.

• The vast majority of emissions trading transactions have involved large pollution sources trading emissions reduction credits either created by excess

---

[6] The details of this policy can be found in 'Emissions Trading Policy Statement' 51 *Federal Register* 43829 (4 December 1986).

[7] Offsets are also required for major modifications in areas which have attained the standards if the modifications jeopardize attainment.

[8] See, for example, Tietenberg (1985), Hahn and Hester (1989a and 1989b), and Dudek and Palmisano (1988).

control of uniformly mixed pollutants (those for which the location of emission is not an important policy concern) or involving facilities in close proximity to one another.

• Though air quality has certainly improved for most of the covered pollutants, it is virtually impossible to say how much of the improvement can be attributed to the emissions trading programme. The emissions trading programme complements the traditional regulatory approach, rather than replaces it. Therefore, while it can claim to have hastened compliance with the basic provisions of the act and in some cases to have encouraged improvements beyond the act, improved air quality resulted from the package taken together, rather than from any specific component.

(ii) Emissions Charges

Emission charges are used in both Europe and Japan, though more commonly to control water pollution than air pollution.[9] Currently effluent charges are being used to control water pollution in France, Italy, Germany, and the Netherlands. In both France and the Netherlands the charges are designed to raise revenue for the purpose of funding activities specifically designed to improve water quality.

In Germany dischargers are required to meet minimum standards of waste water treatment for a number of defined pollutants. Simultaneously a fee is levied on every unit of discharge depending on the quantity and noxiousness of the effluent. Dischargers meeting or exceeding state-of-the-art effluent standards have to pay only half the normal rate.

The Italian effluent charge system was mainly designed to encourage polluters to achieve provisional effluent standards as soon as possible. The charge is nine times higher for firms that do not meet the prescribed standards than for firms that do meet them. This charge system was designed only to facilitate the transition to the prescribed standards so it is scheduled to expire once full compliance has been achieved.[10]

Air pollution emission charges have been implemented by France and Japan. The French air pollution charge was designed to encourage the early adoption of pollution control equipment with the revenues returned to those paying the charge as a subsidy for installing the equipment. In Japan the emission charge is designed to raise revenue to compensate victims of air pollution. The charge rate is determined primarily by the cost of the compensation programme in the previous year and the amount of remaining emissions over which this cost can be applied *pro rata*.

Charges have also been used in Sweden to increase the rate at which consumers would purchase cars equipped with a catalytic converter. Cars not equipped with a catalytic converter were taxed, while new cars equipped with a catalytic converter were subsidized.

While data are limited a few highlights seem clear:

• Economists typically envisage two types of effluent or emissions charges. The first, an efficiency charge, is designed to produce an efficient outcome by forcing the polluter to compensate completely for all damage caused. The second, a cost-effective charge, is designed to achieve a predefined ambient standard at the lowest possible control cost. In practice, few, if any, implemented programmes fit either of these designs.

• Despite being designed mainly to raise revenue, effluent charges have typically improved water quality. Though the improvements in most cases have been small, apparently due to the low level at which the effluent charge rate is set, the Netherlands, with its higher effective rates, reports rather large improvements. Air pollution charges typically have not had much effect on air quality because the rates are too low and, in the case of France, most of the revenue is returned to the polluting sources.

---

[9] See Anderson (1977), Brown and Johnson (1984), Bressers (1988), Vos (1989), Opschoor and Vos (1989), and Sprenger (1989).

[10] The initial deadline for expiration was 1986, but it has since been postponed.

T. Tietenberg

• The revenue from charges is typically earmarked for specific environmental purposes rather than contributed to the general revenue as a means of reducing the reliance on taxes that produce more distortions in resource allocation.

• The Swedish tax on heavily polluting vehicles and subsidy for new low polluting vehicles was very successful in introducing low polluting vehicles into the automobile population at a much faster than normal rate. The policy was not revenue neutral, however; owing to the success of the programme in altering vehicle choices, the subsidy payments greatly exceeded the tax revenue.

III. FIRST PRINCIPLES

Theory can help us understand the characteristics of these economic approaches in the most favourable circumstances for their use and assist in the process of designing the instruments for maximum effectiveness. Because of the dualistic nature of emission charges and emission reduction credits,[11] implications about emission charges and emissions trading flow from the same body of theory.

Drawing conclusions about either of these approaches from this type of analysis, however, must be done with care because operational versions typically differ considerably from the idealized versions modelled by the theory. For example, not all trades that would be allowed in an ideal emissions trading programme are allowed in the current US emissions trading programme. Similarly the types of emissions charges actually imposed differ considerably from their ideal versions, particularly in the design of the rate structure and the process for adjusting rates over time.

Assuming all participants are cost-minimizers, a 'well-defined' emissions trading or emission charge

system could cost-effectively allocate the control responsibility for meeting a predefined pollution target among the various pollution sources despite incomplete information on the control possibilities by the regulatory authorities.[12]

The intuition behind this powerful proposition is not difficult to grasp. Cost-minimizing firms seek to minimize the sum of (*a*) either ERC acquisition costs or payments of emission charges and (*b*) control costs. Minimization will occur when the marginal cost of control is set equal to the emission reduction credit price or the emission charge. Since all cost-minimizing sources would choose to control until their marginal control costs were equal to the same price or charge, marginal control costs would be equalized across all discharge points, precisely the condition required for cost-effectiveness.[13]

Emission charges could also sustain a cost effective allocation of the control responsibility for meeting a predefined pollution target, but only if the control authority knew the correct level of the charge to impose or was willing to engage in an iterative trial-and-error process over time to find the correct level. Emissions trading does not face this problem because the price level is established by the market, not the control authority.[14]

Though derived in the rarified world of theory, the practical importance of this theorem should not be underestimated. Economic incentive approaches offer a unique opportunity for regulators to solve a fundamental dilemma. The control authorities' desire to allocate the responsibility for control cost-effectively is inevitably frustrated by a lack of information sufficient to achieve this objective. Economic incentive approaches create a system of incentives in which those who have the best knowledge about control opportunities, the environmental managers for the industries, are encouraged to use

---

[11] Under fairly general conditions any allocation of control responsibility achieved by an emissions trading programme could also be achieved by a suitably designed system of emission charges and vice versa.

[12] For the formal demonstration of this proposition see Baumol and Oates (1975), Montgomery (1972), and Tietenberg (1985).

[13] It should be noted that while the allocation is cost-effective, it is not necessarily efficient (the amount of pollution indicated by a benefit–cost comparison). It would only be efficient if the predetermined target happened to coincide with the efficient amount of pollution. Nothing guarantees this outcome.

[14] See Tietenberg (1988) for a more detailed explanation of this point.

OXFORD REVIEW OF ECONOMIC POLICY, VOL. 6, NO. 1

that knowledge to achieve environmental objectives at minimum cost. Information barriers do not preclude effective regulation.

What constitutes a 'well-defined' emissions trading or emission charge system depends crucially on the attributes of the pollutant being controlled.[15]

To be consistent with a cost-effective allocation of the control responsibility, the policy instruments would have to be defined in different ways for different types of pollutants. Two differentiating characteristics are of particular relevance. Approaches designed to control pollutants which are uniformly mixed in the atmosphere (such as volatile organic compounds, one type of precursor for ozone formation) can be defined simply in terms of a rate of emissions flow per unit time. Economic incentive approaches sharing this design characteristic are called *emission trades* or *emission charges*.

Instrument design is somewhat more difficult when the pollution target being pursued is defined in terms of concentrations measured at a number of specific receptor locations (such as particulates). In this case the cost-effective trade or charge design must take into account the *location* of the emissions (including injection height) as well as the *magnitude* of emissions. As long as the control authorities can define for each emitter a vector of transfer coefficients, which translate the effect of a unit increase of emissions by that emitter into an increase in concentration at each of the affected receptors, receptor-specific trades or charges can be defined which will allocate the responsibility cost-effectively. The design which is consistent with cost-effectiveness in this context is called an *ambient trade* or an *ambient charge*.

Unfortunately, while the design of the ambient ERC is not very complicated,[16] implementing the markets within which these ERCs would be traded is rather complicated. In particular for each unit of planned emissions an emitter would have to acquire separate ERCs for each affected receptor. When the number of receptors is large, the result is a rather complicated set of transactions. Similarly, establishing the correct rate structure for the charges in

this context is particularly difficult because the set of charges which will satisfy the ambient air quality constraints is not unique; even a trial-and-error system would not necessarily result in the correct matrix of ambient charges being put into effect.

As long as markets are competitive and transactions costs are low, the trading benchmark in an emissions trading approach does not affect the ultimate cost-effective allocation of control responsibility. When markets are non-competitive or transactions costs are high, however, the final allocation of control responsibility is affected.[17] Emission charge approaches do not face this problem.

Once the control authority has decided how much pollution of each type will be allowed, it must then decide how to allocate the operating permits among the sources. In theory emission reduction credits could either be auctioned off, with the sources purchasing them from the control authority at the market-clearing price, or (as in the US programme) created by the sources as surplus reductions over and above a predetermined set of emission standards. (Because this latter approach favours older sources over newer sources, it is known as 'grandfathering'.) The proposition suggests that either approach will ultimately result in a cost-effective allocation of the control responsibility among the various polluters as long as they are all price-takers, transactions costs are low, and ERCs are fully transferable. Any allocation of emission standards in a grandfathered approach is compatible with cost-effectiveness because the after-market in which firms can buy or sell ERCs corrects any problems with the initial allocation. This is a significant finding because it implies that under the right conditions the control authority can use this initial allocation of emissions standards to pursue distributional goals without interfering with cost-effectiveness.

When firms are price-setters rather than price-takers, however, cost-effectiveness will only be achieved if the control authority initially allocates the emission standards so a cost-effective allocation would be achieved even in the absence of any trading. (Implementing this particular allocation would, of course, require regulators to have com-

---

[15] For the technical details supporting this proposition see Montgomery (1972), and Tietenberg (1985).

[16] Each permit allows the holder to degrade the concentration level at the corresponding receptor by one unit.

[17] See Hahn (1984) for the mathematical treatment of this point. Further discussions can be found in Tietenberg (1985) and Misiolek and Elder (1989).

22

plete information on control costs for all sources, an unlikely prospect.) In this special case cost-effectiveness would be achieved even in the presence of one or more price-setting firms because no trading would take place, eliminating the possibility of exploiting any market power.

For all other emission standard assignments an active market would exist, offering the opportunity for price-setting behaviour. The larger is the deviation of the price setting source's emission standard from its cost-effective allocation, the larger is the deviation of ultimate control costs from the least-cost allocation. When the price-setting source is initially allocated an insufficiently stringent emission standard, it can inflict higher control costs on others by withholding some ERCs from the market. When an excessively stringent emission standard is imposed on a price-setting source, however, it necessarily bears a higher control cost as the means of reducing demand (and, hence, prices) for the ERCs.

Similar problems exist when transactions costs are high. High transactions costs preclude or reduce trading activity by diminishing the gains from trade. When the costs of consummating a transaction exceed its potential gains, the incentive to participate in emissions trading is lost.

IV. LESSONS FROM EMPIRICAL RESEARCH

A vast majority, though not all, of the relevant empirical studies have found the control costs to be substantially higher with the regulatory command-and-control system than the least cost means of allocating the control responsibility.

While theory tells us unambiguously that the command-and-control system will not be cost-effective except by coincidence, it cannot tell us the magnitude of the excess costs. The empirical work

cited in Table 1 adds the important information that the excess costs are typically very large.[18] This is an important finding because it provides the motivation for introducing a reform programme; the potential social gains (in terms of reduced control cost) from breaking away from the status quo are sufficient to justify the trouble. Although the estimates of the excess costs attributable to a command and control presented in Table 1 overstate the cost savings that would be achieved by even an ideal economic incentive approach (a point discussed in more detail below), the general conclusion that the potential cost savings from adopting economic incentive approaches are large seems accurate even after correcting for overstatement.

Economic incentive approaches which raise revenue (charges or auction ERC markets) offer an additional benefit—they allow the revenue raised from these policies to substitute for revenue raised in more traditional ways. Whereas it is well known that traditional revenue-raising approaches distort resource allocation, producing inefficiency, economic incentive approaches enhance efficiency. Some empirical work based on the US economy suggests that substituting economic incentive means of raising revenue for more traditional means could produce significant efficiency gains.[19]

When high degrees of control are necessary, ERC prices or charge levels would be correspondingly high. The financial outlays associated with acquiring ERCs in an auction market or paying charges on uncontrolled emissions would be sufficiently large that sources would typically have lower financial burdens with the traditional command-and-control approach than with these particular economic incentive approaches. Only a 'grandfathered' trading system would guarantee that sources would be no worse off than under the command-and-control system.[20]

Financial burden is a significant concern in a highly competitive global market-place. Firms bearing

---

[18] A value of 1.0 in the last column of Table 1 would indicate that the traditional regulatory approach was cost-effective. A value of 4.0 would indicate that the traditional regulatory approach results in an allocation of the control responsibility which is four times as expensive as necessary to reach the stipulated pollution target.

[19] See Terkla (1984).

[20] See Atkinson and Tietenberg (1982, 1984), Hahn (1984), Harrison (1983), Krupnick (1986), Lyon (1982), Palmer *et al.*(1980), Roach *et al.* (1981), Seskin *et al.* (1983), and Shapiro and Warhit (1983) for the individual studies, and Tietenberg (1985) for a summary of the evidence.

OXFORD REVIEW OF ECONOMIC POLICY, VOL. 6, NO. 1

Table 1
Empirical Studies of Air Pollution Control

Study	Pollutants Covered	Geographic Area	CAC Benchmark	Ratio of CAC Cost to Least Cost
Atkinson and Lewis	Particulates	St Louis	SIP regulations	6.00[a]
Roach *et al.*	Sulphur dioxide	Four corners in Utah	SIP regulations Colorado, Arizona, and New Mexico	4.25
Hahn and Noll	Sulphates	Los Angeles	California emission standards	1.07
Krupnick	Nitrogen dioxide regulations	Baltimore	Proposed RACT	5.96[b]
Seskin *et al.*	Nitrogen dioxide regulations	Chicago	Proposed RACT	14.40[b]
McGartland	Particulates	Baltimore	SIP regulations	4.18
Spofford	Sulphur Dioxide	Lower Delaware Valley	Uniform percentage regulations	1.78
	Particulates	Lower Delaware Valley	Uniform percentage regulations	22.00
Harrison	Airport noise	United States	Mandatory retrofit	1.72[c]
Maloney and Yandle	Hydrocarbons	All domestic DuPont plants	Uniform percentage reduction	4.15[d]
Palmer *et al.*	CFC emissions from non-aerosol applications	United States	Proposed emission standards	1.96

Notes:

CAC = command and control, the traditional regulatory approach.

SIP = state implementation plan.

RACT = reasonably available control technologies, a set of standards imposed on existing sources in non-attainment areas.

[a] Based on a 40 $\mu g/m^3$ at worst receptor.

[b] Based on a short-term, one-hour average of 250 $\mu g/m^3$.

[c] Because it is a benefit–cost study instead of a cost-effectiveness study, the Harrison comparison of the command-and-control approach with the least-cost allocation involves different benefit levels. Specifically, the benefit levels associated with the least-cost allocation are only 82 per cent of those associated with the command-and-control allocation. To produce cost estimates based on more comparable benefits, as a first approximation the least-cost allocation was divided by 0.82 and the resulting number was compared with the command-and-control cost.

[d] Based on 85 per cent reduction of emissions from all sources.

T. Tietenberg

large financial burdens would be placed at a competitive disadvantage when forced to compete with firms not bearing those burdens. Their costs would be higher.

From the point of view of the source required to control its emissions, two components of financial burden are significant: (a) control costs and (b) expenditures on permits or emission charges. While only the former represent real resource costs to society as a whole (the latter are merely transferred from one group in society to another), both represent a financial burden to the source. The empirical evidence suggests that when an auction market is used to distribute ERCs (or, equivalently, when all uncontrolled emissions are subject to an emissions charge), the ERC expenditures (charge outlays) would frequently be larger in magnitude than the control costs; the sources would spend more on ERCs (or pay more in charges) than they would on the control equipment. Under the traditional command-and-control system firms make no financial outlays to the government. Although control costs are necessarily higher with the command-and-control system than with an economic incentive approach, they are not so high as to outweigh the additional financial outlays required in an auction market permit system (or an emissions tax system). For this reason existing sources could be expected vehemently to oppose an auction market or emission charges despite their social appeal, unless the revenue derived is used in a manner which is approved by the sources, and the sources with which it competes are required to absorb similar expenses. When environmental policies are not coordinated across national boundaries, this latter condition would be particularly difficult to meet.

In the absence of either a politically popular way to use the revenue or assurances that competitors will face similar financial burdens, this political opposition could be substantially reduced by grandfathering. Under grandfathering, sources have only to purchase any additional ERCs they may need to meet their assigned emission standard (as opposed to purchasing sufficient ERCs or paying charges to cover all uncontrolled emissions in an auction market). Grandfathering is *de facto* the approach taken in the US emissions trading programme.

Grandfathering has its disadvantages. Because ERCs become very valuable, especially in the face of stringent air quality regulations, sources selling emission reduction credits would be able to command very high prices. By placing heavy restrictions on the amount of emissions, the control authority is creating wealth for existing firms *vis-à-vis* new firms.

Although reserving some ERCs for new firms is possible (by assigning more stringent emission standards than needed to reach attainment and using the 'surplus' air quality to create government-held ERCs), this option is rarely exercised in practice. In the United States under the offset policy firms typically have to purchase sufficient ERCs to more than cover all uncontrolled emissions, while existing firms only have to purchase enough to comply with their assigned emission standard. Thus grandfathering imposes a bias against new sources in the sense that their financial burden is greater than that of an otherwise identical existing source, even if the two sources install exactly the same emission control devices. This new source bias could retard the introduction of new facilities and new technologies by reducing the cost advantage of building new facilities which embody the latest innovations.

While it is clear from theory that larger trading areas offer the opportunities for larger potential cost savings in an emissions trading programme, some empirical work suggests that substantial savings can be achieved in emissions trading even when the trading areas are rather small.

The point of this finding is *not* that small trading areas are fine; they do retard progress toward the standard. Rather, when political considerations allow only small trading areas or nothing, emissions trading still can play a significant role.

Sometimes political considerations demand a trading area which is smaller than the ideal design. Whether large trading areas are essential for the effective use of this policy is therefore of some relevance. In general, the larger the trading area, the larger would be the potential cost savings due to a wider set of cost reduction opportunities that would

OXFORD REVIEW OF ECONOMIC POLICY, VOL. 6, NO. 1

become available. The empirical question is how sensitive the cost estimates are to the size of the trading areas.

One study of utilities found that even allowing a plant to trade among discharge points within that plant could save from 30 to 60 per cent of the costs of complying with new sulphur oxide reduction regulations, compared to a situation where no trading whatsoever was permitted.[21] Expanding the trading possibilities to other utilities within the same state permitted a further reduction of 20 per cent, while allowing interstate trading permitted another 15 per cent reduction in costs. If this study is replicated in other circumstances, it would appear that even small trading areas offer the opportunity for significant cost reduction.[22]

Although only a few studies of the empirical impact of market power on emissions trading have been accomplished, their results are consistent with a finding that market power does not seem to have a large effect on regional control costs in most realistic situations.[23]

Even in areas having especially stringent controls, the available evidence suggests that price manipulation is not a serious problem. In an auction market the price-setting source reduces its financial burden by purchasing fewer ERCs in order to drive the price down. To compensate for the smaller number of ERCs purchased, the price-setting source must spend more on controlling its own pollution, limiting the gains from price manipulation. Although these actions could have a rather large impact on *regional financial burden*, they would under normal circumstances have a rather small effect on *regional control costs*. Estimates typically suggest that control costs would rise by less than 1 per cent if market power were exercised by one or more firms.

It should not be surprising that price manipulation could have rather dramatic effects on regional financial burden in an auction market, since the cost of *all* ERCs is affected, not merely those purchased by the price-setting source. The perhaps more surprising result is that control costs are quite insensitive to price-setting behaviour. This is due to the fact that the only control cost change is the net difference between the new larger control burden borne by the price searcher and the correspondingly smaller burden borne by the sources having larger-than-normal allocations of permits. Only the costs of the marginal units are affected.

Within the class of grandfathered distribution rules, some emission standard allocations create a larger potential for strategic price behaviour than others. In general the larger the divergence between the control responsibility assigned to the price-searching source by the emission standards and the cost-effective allocation of control responsibility, the larger the potential for market power. When allocated too little responsibility by the control authority, price-searching firms can exercise power on the selling side of the market, and when allocated too much, they can exercise power on the buying side of the market.

According to the existing studies it takes a rather considerable divergence from the cost-effective allocation of control responsibility to produce much difference in regional control costs. In practice the deviations from the least cost allocation caused by market power pale in comparison to the much larger potential cost reductions achievable by implementing emissions trading.[24]

V. LESSONS FROM IMPLEMENTATION

Though the number of transactions consummated under the Emissions Trading Program has been large, it has been smaller than expected. Part of this failure to fulfil expectations can be explained as the result of unrealistically inflated expectations. More restrictive regulatory decisions than expected and

---

[21] ICF, Inc. (1989).

[22] As indicated below, the fact that so many emissions trades have actually taken place within the same plant or among contiguous plants provides some confirmation for this result.

[23] For individual studies see de Lucia (1974), Hahn (1984), Stahl, Bergman and Mäler (1988), and Maloney and Yandle (1984). For a survey of the evidence see Tietenberg (1985).

[24] Strategic price behaviour is not the only potential source of market power problems. Firms could conceivably use permit markets to drive competitors out of business. See Misiolek and Elder (1989). For an analysis which concludes that this problem is relatively rare and can be dealt with on a case-by-case basis should it arise, see Tietenberg (1985).

higher than expected transaction costs also bear some responsibility.

The models used to calculate the potential cost savings were not (and are not) completely adequate guides to reality. The cost functions in these models are invariably *ex ante* cost functions. They implicitly assume that the modelled plant can be built from scratch and can incorporate the best technology. In practice, of course, many existing sources cannot retrofit these technologies and therefore their *ex post* control options are much more limited than implied by the models.

The models also assume all trades are multilateral and are simultaneously consummated, whereas actual trades are usually bilateral and sequential. The distinction is important for non-uniformly mixed pollutants;[25] bilateral trades frequently are constrained by regulatory concerns about decreasing air quality at the site of the acquiring source. Because multilateral trades would typically incorporate compensating reductions coming from other nearby sources, these concerns normally do not arise when trades are multilateral and simultaneous. In essence the models implicitly assume an idealized market process, which is only remotely approximated by actual transactions.

In addition some non-negligible proportion of the expected cost savings recorded by the models for non-uniformly mixed pollutants is attributable to the substantially larger amounts of emissions allowed by the modelled permit equilibrium.[26] For example, the cost estimates imply that the control authority is allowed to arrange the control responsibility in *any* fashion that satisfies the ambient air quality standards. In practice the models allocate more uncontrolled emissions to sources with tall stacks because those emissions can be exported. Exported emissions avoid control costs without affecting the readings at the local monitors. That portion of the cost savings estimated by the models in Table 1 which is due to allowing increased emissions is not acceptable to regulators. Some

recent work has suggested that the benefits received from the additional emission control required by the command and control approach may be justified by the net benefits received.[27] The regulatory refusal to allow emission increases was apparently consistent with efficiency,[28] but it was not consistent with the magnitiude of cost savings anticipated by the models.

Certain types of trades assumed permissible by the models are prohibited by actual trading rules. New sources, for example, are not allowed to satisfy the New Source Performance Standards (which imply a particular control technology) by choosing some less stringent control option and making up the difference with acquired emission reduction credits; they must install the degree of technological control necessary to meet the standard. Typically this is the same technology used by EPA to define the standard in the first place.

A lot of uncertainty is associated with emission reduction credit transactions since they depend so heavily on administrative action. All trades must be approved by the control authorities. If the authorities are not co-operative or at least consistent, the value of the created emission reduction credits could be diminished or even destroyed.

For non-uniformly mixed pollutants, trades between geographically separated sources will only be approved after dispersion modelling has been accomplished by the applicants. Not only is this modelling expensive, it frequently ends up raising questions which ultimately lead to the transaction being denied. Few trades requiring this modelling have been consummated.

Trading activity has also been inhibited by the paucity of emission banks. The US system allows states to establish emission banks, but does not require them to do so. As of 1986 only seven of the fifty states had established these banks. For sources in the rest of the states the act of creating emission credits is undervalued because the credits cannot be

---

[25] See Tietenberg and Atkinson (1989) for a demonstration that this is an empirically significant point.

[26] This is demonstrated in Atkinson and Tietenberg (1987).

[27] See Oates, Portney, and McGartland (1988).

[28] Not all of the cost savings, of course, is due to the capability to increase emissions. The remaining portion of the savings, which is due to taking advantage of opportunities to control a given level of emissions at a lower cost, is still substantial and can be captured by a well-designed permit system which does not allow emissions to increase beyond the command-and-control benchmark. See the calculations in Atkinson and Tietenberg (1987).

27

OXFORD REVIEW OF ECONOMIC POLICY, VOL. 6, NO. 1

legally held for future use. The supply of emission reduction credits is hence less than would be estimated by the models.

The Emissions Trading Program seems to have worked particularly well for trades involving uniformly mixed pollutants and for trades of non-uniformly mixed pollutants involving contiguous discharge points.

It is not surprising that most consummated trades have been internal (where the buyer and seller share a common corporate parent) rather than external. Not only are the uncertainties associated with inter-firm transfers avoided, but most internal trades involve contiguous facilities. Trades between contiguous facilities do not trigger a requirement for dispersion modelling.[29]

It is also not surprising that the plurality of consummated trades involve volatile organic compounds, which are uniformly mixed pollutants. Since dispersion modelling is not required for uniformly mixed pollutants even when the trading sources are somewhat distant from one another, trades involving these pollutants are cheaper to consummate. Additionally emissions trades involving uniformly mixed pollutants do not jeopardize local air quality since the location of the emissions is not a matter of policy consequence.

The establishment of the Emissions Trading Program has encouraged technological progress in pollution control. Although generally the degree of progress has been modest, it has been more dramatic in areas where emission reductions have been sufficiently stringent as to restrict the availability of emission reduction credits created by more traditional means.[30]

Theory would lead us to expect more technological progress with emissions trading than with a command-and-control policy because it changes the incentives so drastically. Under a command-and-control approach technological changes discovered by the control authority typically lead to more

stringent standards (and higher costs) for the sources. Sources have little incentive to innovate and a good deal of incentive to hide potential innovations from the control authority. With emissions trading, on the other hand, innovations allowing excess reductions create saleable emission reduction credits.

The evidence suggests that the expectations based on this theory have been borne out to a limited degree in the operating programme. The most prominent example of technological change has been the substitution of water-based solvents for solvents containing volatile organic compounds. Though somewhat more expensive, this substitution made economic sense once the programme was introduced.

It should probably not be surprising that the number of new innovations stimulated by the programme is rather small. As long as cheaper ways of creating credits within existing processes (fuel substitution, for example) are available, it would be unreasonable to expect large investments in new technologies with unproven reliabilities. On the other hand as the degree of control rises and the supply of readily available credits dries up, the demand for new technologies would be expected to rise as well. This expectation seems to have been borne out in those areas where unusually low air quality or stringent regulatory rules have served to limit the available credits.[31]

This is an important point. Those who fail to consider the dynamic advantages of an economic incentive approach sometimes suggest that if few credits would be traded, implementing a system of this type has no purpose. In fact it has a substantial purpose—the encouragement of new technologies to meet the increasingly stringent standards.

Introducing the Emissions Trading Program has provided an opportunity to control sources which can reduce emissions relatively cheaply, but which under the traditional policy were under-regulated due either to their financially precarious position or the fact that they were not subject to regulation.[32]

[29] The fact that so many trades have taken place between contiguous discharge points serves as confirmation that substantial savings can be achieved even if the geographic boundaries of the trading area are quite restricted.

[30] For more details see Tietenberg (1985), Maleug (1989), and Dudek and Palmisano (1988).

[31] For the experience in California see Dudek and Palmisano (1988).

[32] See Tietenberg (1985).

T. Tietenberg

Due to the social distress caused by any resulting unemployment, the control authorities and the courts are understandably reluctant to enforce stringent emission standards against firms which would not be able to pass higher costs on to customers without considerable loss of production. Since many of these sources could control emissions at a lower marginal cost than other sources, their political immunity from control makes regional control costs higher than necessary; other sources have to control their own emissions to a higher degree (at a higher marginal cost) to compensate.

Due to its ability to separate the issue of who pays for the reduction from the issue of which discharge points are to be controlled, the emissions trading programme provides a way to secure those low cost reductions. The command-and-control policy would assign, as normal, a very low (perhaps zero) emission reduction to any previously unregulated firm. Once emissions trading had been established, however, it would be in the interest of this firm to control emissions further, selling the resulting emission reduction credits. As long as the revenues from the sale at least covered the cost, this transaction could profit, or at least not hurt, the seller. Because these reductions could be achieved at a lower cost than ratcheting up the degree of control on already heavily controlled sources, non-immune sources would find purchasing the credits cheaper than controlling their own emissions to a higher degree. Everyone benefits from controlling these previously under-regulated sources.

Another unique attribute of an emissions trading approach is the capability it offers sources for leasing credits.[33]

Leasing offers an enormously useful degree of flexibility which is not available with other policy approaches to pollution control. The usefulness of leasing derives from the fact that some sources, utilities in particular, have patterns of emission that vary over time while allowable emissions remain constant. In a typical situation, for example, suppose an older utility would, in the absence of control, be emitting heavily. In the normal course of a utility expansion cycle the older plant would

subsequently experience substantially reduced emissions when the utility constructed a new plant and shifted a major part of the load away from the older plant to the new plant. Ultimately growth in demand on the system would increase the emissions again for the older plant as its capacity would once again be needed. The implication of this temporal pattern is that during the middle period, as its own emissions fell well below allowable emissions, this utility could lease excess emission credits to another facility, recalling them as its own need rose with demand growth. Indeed one empirical study of the pattern of the utility demand for and supply of acid rain reduction credits over time suggests that leasing is a critical component of any cost-effective control strategy, a component that neither the traditional approach nor emission charges can offer.[34]

Leasing also provides a way for about-to-be-retired sources to participate in the reduction programme. Under the traditional approach once the deadline for compliance had been reached the utility would either have to retire the unit early or to install expensive control equipment which would be rendered useless once the unit was retired. By leasing credits for the short period to retirement, the unit could remain in compliance without taking either of those drastic steps; it would, however, be sharing in the cost of installing the extra equipment in the leasing utility. Leased credits facilitate an efficient transition into the new regime of more stringent controls.

Unless the process to determine the level of an effluent or emissions charge includes some automatic means of temporal adjustment, the tendency is for the real rate (adjusted for inflation) to decline over time.[35] This problem is particularly serious in areas with economic growth where increasing real rates would be the desired outcome.

In contrast to emissions trading where ERC prices respond automatically to changing market conditions, emission charges have to be determined by an administrative process. When the function of the charge is to raise revenue for a particular purpose, charge rates will be determined by the costs of

[33] See Feldman and Raufer (1987) and Tietenberg (1989a).
[34] Feldman and Raufer (1987).
[35] For further information see Vos (1989) and Sprenger (1989).

29

achieving that purpose; when the costs of achieving the purpose rise, the level of the charge must rise to secure the additional revenue.[36]

Sometimes that process produces an unintended dynamic. In Japan, for example, the charge is calculated on the basis of the amount of compensation paid to victims of air pollution in the previous year. While the amount of compensation has been increasing, the amount of emissions (the base to which the charge is applied) has been decreasing. As a result unexpectedly high charge rates are necessary in order to raise sufficient revenue for the compensation system.

In countries where the tax revenue feeds into the general budget, increases in the level of the charge require a specific administrative act. Evidently it is difficult to raise these rates in practice, since charges have commonly even failed to keep pace with inflation, much less growth in the number of sources. The unintended result is eventual environmental deterioration.

VI. CONCLUDING COMMENTS

Our experience with economic incentive programmes has demonstrated that they have had, and can continue to have, a positive role in environmental policy in the future. I would submit the issue is no longer *whether* they have a role to play, but rather *what kind* of role they should play. The available experience with operating versions of these programmes allows us to draw some specific conclusions which facilitate defining the boundaries for the optimal use of economic incentive approaches in general and for distinguishing the emissions trading and emission charges approaches in particular.

Emissions trading integrates particularly smoothly into any policy structure which is based either directly (through emission standards) or indirectly (through mandated technology or input limitations) on regulating emissions. In this case emission limitations embedded in the operating licences can serve as the trading benchmark if grandfathering is adopted.

Emissions charges work particularly well when transactions costs associated with bargaining are high. It appears that much of the trading activity in the United States has involved large corporations. Emissions trading is probably not equally applicable to large and small pollution sources. The transaction costs are sufficiently high that only large trades can absorb them without jeopardizing the gains from trade. For this reason charges seem a more appropriate instrument when sources are individually small, but numerous (such as residences or automobiles). Charges also work well as a device for increasing the rate of adoption of new technologies and for raising revenue to subsidize environmentally benign projects.

Emissions trading seems to work especially well for uniformly mixed pollutants. No diffusion modelling is necessary and regulators do not have to worry about trades creating 'hot spots' or localized areas of high pollution concentration. Trades can be on a one-to-one basis.

Because emissions trading allows the issue of who will pay for the control to be separated from who will install the control, it introduces an additional degree of flexibility. This flexibility is particularly important in non-attainment areas since marginal control costs are so high. Sources which would not normally be controlled because they could not afford to implement the controls without going out of business, can be controlled with emissions trading. The revenue derived from the sale of emission reduction credits can be used to finance the controls, effectively preventing bankruptcy.

Because it is quantity based, emissions trading also offers a unique possibility for leasing. Leasing is particularly valuable when the temporal pattern of emissions varies across sources. As discussed above this appears generally to be the case with utilities. When a firm plans to shut down one plant in the near future and to build a new one, leasing credits is a vastly superior alternative to the temporary installation of equipment in the old plant which would be useless when the plant was retired. The useful life of this temporary control equipment would be wastefully short.

---

[36] While it is theoretically possible (depending on the elasticity of demand for pollution abatement) for a rise in the tax to produce less revenue, this has typically not been the case.

We have also learned that ERC transactions have higher transactions costs than we previously understood. Regulators must validate every trade. When non-uniformly mixed pollutants are involved, the transactions costs associated with estimating the air quality effects are particularly high. Delegating responsibility for trade approval to lower levels of government may in principle speed up the approval process, but unless the bureaucrats in the lower level of government support the programme the gain may be negligible.

Emissions trading places more importance on the operating permits and emissions inventories than other approaches. To the extent those are deficient the potential for trades that protect air quality may be lost. Firms which have actual levels of emissions substantially below allowable emissions find themselves with a trading opportunity which, if exploited, could degrade air quality. The trading benchmark has to be defined carefully.

There can be little doubt that the emissions trading programme in the US has improved upon the command-and-control programme that preceded it. The documented cost savings are large and the flexibility provided has been important. Similarly emissions charges have achieved their own measure of success in Europe. To be sure the programmes are far from perfect, but the flaws should be kept in perspective. In no way should they overshadow the impressive accomplishments. Although economic incentive approaches lose their Utopian lustre upon closer inspection, they have none the less made a lasting contribution to environmental policy.

The role for economic incentive approaches should grow in the future if for no other reason than the fact that the international pollution problems which are currently commanding centre-stage fall within the domains where economic incentive policies have been most successful. Significantly many of the problems of the future, such as reducing tropospheric ozone, preventing stratospheric ozone depletion, moderating global warming, and increasing acid rain control, involve pollutants that can be treated as uniformly mixed, facilitating the use of economic incentives. In addition larger trading areas facilitate greater cost reductions than smaller trading areas. This also augers well for the use of emissions trading as part of the strategy to control many future pollution problems because the natural trading areas are all very large indeed. Acid rain, stratospheric ozone depletion, and greenhouse gases could (indeed should!) involve trading areas that transcend national boundaries. For greenhouse and ozone depletion gases, the trading areas should be global in scope. Finally, it seems clear that the pivotal role of carbon dioxide in global warming may require some fairly drastic changes in energy use, including changes in personal transportation, and ultimately land use patterns. Some form of charges could play an important role in facilitating this transformation.

We live in an age when the call for tighter environmental controls intensifies with each new discovery of yet another injury modern society is inflicting on the planet. But resistance to additional controls is also growing with the recognition that compliance with each new set of controls is more expensive than the last. While economic incentive approaches to environmental control offer no panacea, they frequently do offer a practical way to achieve environmental goals more flexibly and at lower cost than more traditional regulatory approaches. That is a compelling virtue.

31

OXFORD REVIEW OF ECONOMIC POLICY, VOL. 6, NO. 1

REFERENCES

Anderson, F. R. *et al.*(1977), *Environmental Improvement Through Economic Incentives*, Baltimore, The Johns Hopkins University Press for Resources for the Future, Inc.

Atkinson, S. E. and Lewis, D. H. (1974), 'A Cost-Effectiveness Analysis of Alternative Air Quality Control Strategies', *Journal of Environmental Economics and Management*, 1, 237–50.

— and Tietenberg, T. H. (1982), 'The Empirical Properties of Two Classes of Designs for Transferable Discharge Permit Markets', *Journal of Environmental Economics and Management*, 9, 101–21.

— — (1984), 'Approaches for Reaching Ambient Standards in Non-Attainment Areas: Financial Burden and Efficiency Considerations', *Land Economics*, 60, 148–59.

— — (1987), 'Economic Implications of Emission Trading Rules for Local and Regional Pollutants', *Canadian Journal of Economics*, 20, 370–86.

Baumol, W. J., and Oates, W. E. (1975), *The Theory of Environmental Policy*, Englewood Clifs, N.J., Prentice Hall.

Bohm, P. and Russell, C. (1985), 'Comparative Analysis of Alternative Policy Instruments', in A. V. Kneese and J. L. Sweeney (eds.), *Handbook of Natural Resource and Energy Economics*, Vol. 1, 395–460, Amsterdam, North-Holland.

Bressers, H. T. A. (1988), 'A Comparison of the Effectiveness of Incentives and Directives: The Case of Dutch Water Quality Policy', *Policy Studies Review*, 7, 500–18.

Brown, G. M. Jr. and Johnson, R. W. (1984), 'Pollution Control by Effluent Charges: It Works in the Federal Republic of Germany, Why Not in the United States?', *Natural Resources Journal*, 24, 929–66.

de Lucia, R. J. (1974), *An Evaluation of Marketable Effluent Permit Systems*, Report No. EPA–600/5–74–030 to the US Environmental Protection Agency (September).

Dudek, D. J. and Palmisano, J. (1988), 'Emissions Trading: Why is this Throughbred Hobbled?', *Columbia Journal of Environmental Law*, 13, 217–56.

Feldman, S. L. and Raufer, R. K. (1987), *Emissions Trading and Acid Rain Implementing a Market Approach to Pollution Control*, Totowa, N.J., Rowman & Littlefield.

Hahn, R. W. (1984), 'Market Power and Transferable Property Rights', *Quarterly Journal of Economics*, 99, 753–65.

— (1989), 'Economic Prescriptions for Environmental Problems: How the Patient Followed the Doctor's Orders', *The Journal of Economic Perspectives*, 3, 95–114.

— and Noll, R. G. (1982), 'Designing a Market for Tradeable Emission Permits', in W. A. Magat (ed.), *Reform of Environmental Regulation*, Cambridge, Mass., Ballinger.

— and Hester, G. L. (1989a), 'Where Did All the Markets Go? An Analysis of EPA's Emission Trading Program', *Yale Journal of Regulation*, 6, 109–53.

— — (1989b), 'Marketable Permits: Lessons from Theory and Practice', *Ecology Law Quarterly*, 16, 361–406.

Harrison, D., Jr. (1983), 'Case Study 1: The Regulation of Aircraft Noise', in Thomas C. Schelling (ed.), *Incentives for Environmental Protection*, Cambridge, Mass, MIT Press.

ICF Resources, Inc. (1989), 'Economic, Environmental, and Coal Market Impacts of SO2 Emissions Trading Under Alternative Acid Rain Control Proposals', a report prepared for the Regulatory Innovations Staff, USEPA (March).

Krupnick, A. J. (1986), 'Costs of Alternative Policies for the Control of Nitrogen Dioxide in Baltimore', *Journal of Environmental Economics and Management*, 13, 189–97.

Liroff, R. A. (1980), *Air Pollution Offsets: Trading. Selling and Banking*, Washington, D.C., Conservation Foundation.

— (1986), *Reforming Air Pollution Regulation: The Toil and Trouble of EPA's Bubble*, Washington D.C., Conservation Foundation.

Lyon, R. M. (1982), 'Auctions and Alternative Procedures for Allocating Pollution Rights', *Land Economics*, 58, 16–32.

McGartland, A. M. (1984), 'Marketable Permit Systems for Air Pollution Control: an Empirical Study', Ph.D. dissertation, University of Maryland.

Main, J. (1988), 'Here Comes the Big Cleanup', *Fortune*, 21 November, 102.

Maleug, David A. (1989), 'Emission Trading and the Incentive to Adopt New Pollution Abatement Technology', *Journal of Environmental Economics and Management*, 16, 52–7.

Maloney, M. T. and Yandle, B. (1984), 'Estimation of the Cost of Air Pollution Control Regulation', *Journal of Environmental Economics and Management*, 11, 244–63.

T. Tietenberg

Misiolek, W. S. and Elder, H. W. (1989), 'Exclusionary Manipulation of Markets for Pollution Rights', *Journal of Environmental Economics and Management*, 16, 156–66.

Montgomery, W. D. (1972), 'Markets in Licences and Efficient Pollution Control Programs', *Journal of Economic Theory*, 5, 395–418.

Oates, W. E., Portney, P. R., and McGartland, A. M. (1988), 'The Net Benefits of Incentive-Based Regulation: The Case of Environmental Standard Setting in the Real World', Resources for the Future Working Paper December.

Opschoor, J. B. and Vos, H. B. (1989), *The Application of Economic Instruments for Environmental Protection in OECD Countries*, Paris, OECD.

Palmer, A. R., Mooz, W. E., Quinn, T. H., and Wolf, K. A. (1980), *Economic Implications of Regulating Chlorofluorocarbon Emissions from Nonaerosol Applications*, Report No. R–2524–EPA prepared for the US Environmental Protection Agency by the Rand Corporation, June.

Roach F., Kolstad, C., Kneese, A. V., Tobin, R., and Williams, M. (1981), 'Alternative Air Quality Policy Options in the Four Corners Region', *Southwestern Review*, 1, 29–58.

Seskin, E. P., Anderson, R. J., Jr., and Reid, R. O. (1983), 'An Empirical Analysis of Economic Strategies for Controlling Air Pollution', *Journal of Environmental Economics and Management*, 10, 112–24.

Shapiro, M. and Warhit, E. (1983) 'Marketable Permits: The Case of Chlorofluorocarbons', *Natural Resource Journal*, 23, 577–91.

Spofford, W. O., Jr. (1984), 'Efficiency Properties of Alternative Source Control Policies for Meeting Ambient Air Quality Standards: An Empirical Application to the Lower Deleware Valley', Discussion paper D–118, Washington D.C., Resources for the Future.

Sprenger, R. U. (1989), 'Economic Incentives in Environmental Policies: The Case of West Germany', a paper presented at the Symposium on Economic Instruments in Environmental Protection Policies, Stockholm, Sweden (June).

Stahl, I., Bergman, L., and Mäler, K. G. (1988), 'An Experimental Game on Marketable Emission Permits for Hydro-carbons in the Gothenburg Area', Research Paper No. 6359, Stockholm School of Economics (December).

Stavins, R. N. (1989), 'Harnessing Market Forces to Protect the Environment', *Environment*, 31, 4–7, 28–35.

Terkla, D. (1984), 'The Efficiency Value of Effluent Tax Revenues', *Journal of Environmental Economics and Management*, 11, 107—23.

Tietenberg, T. H. (1980), 'Transferable Discharge Permits and the Control of Stationary Source Air Pollution: A Survey and Synthesis', *Land Economics*, 56, 391–416.

—   (1985), *Emissions Trading: An Excercise in Reforming Pollution Policy*, Washington, D.C., Resources for the Future.

—   (1988), *Environmental and Natural Resource Economics*, 2nd edn., Glenview, Illinois, Scott, Foresman and Company.

—   (1989a), 'Acid Rain Reduction Credits', *Challenge*, 32, 25–9.

—   (1989b), 'Marketable Permits in the U.S.: A Decade of Experience', in Karl W. Roskamp (ed.), *Public Finance and the Performance of Enterprises*, Detroit, MI, Wayne State University Press.

—   and Atkinson, S. E. (1989), 'Bilateral, Sequential Trading and the Cost-Effectiveness of the Bubble Policy', Colby College Working Paper (August).

Vos, H. B. (1989), 'The Application and Efficiency of Economic Instruments: Experiences in OECD Member Countries', a paper presented at the Symposium on Economic Instruments in Environmental Protection Policies, Stockholm, Sweden (June).

[25]

Second-best Regulation
of Road Transport Externalities

By Erik Verhoef, Peter Nijkamp and Piet Rietveld*

1. Introduction

In the 1920s, economists like Pigou (1920) and Knight (1924) recognised that "road pricing" offers the first-best solution for optimising congested road-traffic flows. Since then, the severity of traffic congestion has dramatically increased, turning it from a matter of academic interest into one of the most serious problems affecting urbanised areas and transport arteries. In addition, along with the growing levels of road transport, other external costs of road transport (notably environmental effects, noise annoyance and accidents) have become matters of increasing relevance. After seventy years, economists' answers to such market failures in road transport typically still rely heavily on the concept of Pigouvian taxes.

In principle, various types of road charging do exist. From a welfare economic point of view, a system of Electronic Road Pricing (ERP) is generally seen as the first-best technical mechanism for charging regulatory road charges. The reason for its superiority is that the regulatory tax can be differentiated according to the various dimensions affecting the actual marginal external costs of each trip, such as the length of the trip, the time of driving, the route followed and the vehicle used. Hence, individual drivers can be confronted with first-best incentives, inducing optimal (efficient) behavioural responses (assuming full information). However, although the well-known Hong Kong experiment has demonstrated that it is technically possible to operate an ERP scheme successfully nowadays (see Dawson and Catling, 1986; and Hau, 1992), various social and political impediments appear to prevent ERP from being widely introduced and accepted, certainly in the short run. Consequently, recent publications on road pricing have focused on the question of its political feasibility (Borins, 1988; Evans, 1992; Giuliano, 1992; Lave, 1994; and Verhoef, 1994); on methods of introducing road pricing schemes (Goodwin, 1989; Jones, 1991; May, 1992; Poole, 1992; Small, 1992a); and on various alternatives to ERP in traffic demand regulation,[1] such as fuel taxes (Mohring, 1989) and parking policies (Arnott, de Palma and Lindsey, 1991; Glazer and Niskanen, 1992; and Verhoef, Nijkamp and Rietveld, 1995).

* Free University, Amsterdam. Erik Verhoef is also affiliated to the Tinbergen Institute and participates in the VSB-fonds sponsored research project "Transport and Environment". The authors thank an anonymous referee for very useful comments.

[1] See Button (1992) for an overview.

May 1995   Journal of Transport Economics and Policy

The present paper belongs to this last category. It is concerned with the welfare economic characteristics of second-best alternatives to ERP in managing road transport demand. In particular, attention is focused on the consequences of the regulator's associated incapability of optimal tax differentiation between different types of road users. Throughout the paper, the analyses are restricted to *economic instruments* only (being second-best regulatory user *fees*). The additional shortcomings of physical restrictions in traffic regulations are discussed in Verhoef, Nijkamp and Rietveld (1995). Discussions of related second-best topics in transport can be found in Wilson (1983), and d'Ouville and McDonald (1990) on optimal road capacity supply with suboptimal congestion pricing; Braid (1989) and Arnott, de Palma and Lindsey (1990) on uniform or step-wise pricing of a bottleneck; and Arnott (1979), Sullivan (1983) and Fujita (1989, chapter 7.4) on congestion policies through urban land use policies. Two classic examples on second-best regulation in transport are Lévy-Lambert (1968) and Marchand (1968), studying optimal congestion pricing with an untolled alternative. Finally, it may be noteworthy that the issues considered in this paper bear close resemblance to those discussed in the literature on optimal taxation (see, for instance, Diamond, 1973; Sandmo, 1976; and Atkinson and Stiglitz, 1980).

The structure of the paper is as follows. In Section 2, the optimal undifferentiated fee is derived, and in Section 3 its welfare effects are compared with first-best regulation. Section 4 discusses some aspects related to demand interdependencies, and Section 5 considers a simple form of cost interdependencies, where the regulator is not capable of taxing all road users. Section 6 offers some concluding remarks.

2. The Optimal Common Regulatory Fee

The basic issue considered in this and the next section concerns a road network which is used by different groups of road users, each having their own group-specific level of external costs. We may, for instance, consider *vehicle-specific* externalities. Many types of external costs of road transport depend on the technical characteristics of the vehicle used (for instance, emissions of pollutants, or noise). Efficiency requires regulatory fees to be dependent on such characteristics (that is, on the actual emissions). However, it will often be practically impossible to operate such a delicate tax structure when applying second-best regulatory instruments. For instance, regulatory parking fees (meant to regulate traffic flows rather than parking as such) are in general not suitable for adaptation according to the type of vehicle parked (except for a rough distinction into private cars, vans, buses and trucks); nor are policies such as peak-hour permits. Likewise, many of the suggested alternatives to electronic road pricing are often not capable of differentiation according to individual *trip lengths*. This certainly holds for regulatory parking policies, which are implemented at the end of each trip, and therefore do not enable the regulator to differentiate according to the distance driven. It also holds for instruments such as peak-hour permits. Many (if not all) external costs of road transport, however, are dependent on trip length. Comparable problems may arise with externalities which are *route-* or *time-specific*.

In this section, the optimal common regulatory fee (the optimal second-best undifferentiated or coarse fee, as opposed to the optimal first-best differentiated or fine fee) for such cases is derived. First, we consider an *environmental externality*; then an *intra-sectoral externality* (congestion) is discussed.

2.1 The optimal common regulatory fee for an environmental externality

Suppose we have M different groups of car drivers (denoted $m = 1,2,...,M$), jointly using a certain road network. Each group has its own specific marginal private cost of driving c_m and its own specific marginal external (environmental) cost $\partial E(N_1, N_2, ..., N_M)/\partial N_m$ (conveniently denoted $\varepsilon_m(\bullet)$ hereafter); N_m gives the number of trips made by group m. The groups are ordered according to increasing marginal external costs: $\varepsilon_j > \varepsilon_i$ if $j > i$. Now, suppose that the regulatory body wishes to reduce emissions by means of some second-best undifferentiated regulatory tax policy. More ambitiously, it wishes to find the *optimal common regulatory fee*. This fee maximises social welfare under the inherent limitation of the policy, being the impossibility of first-best tax differentiation. Such an optimal common fee can be found by solving the following Lagrangian:

$$L = \sum_{m=1}^{M} \int_0^{N_m} D_m(n_m)\,dn_m - \sum_{m=1}^{M} c_m {\bullet} N_m - E(N_1, N_2, ..., N_m) + \sum_{m=1}^{M} \lambda_m {\bullet} [D_m(N_m) - c_m - f] \quad (1)$$

The first term defines total benefits in the system as the relevant areas under the inverse demand curves D_m. The next term gives the total private cost of driving, for all groups defined as the average cost of a trip c_m, multiplied by the number of trips. Note that we assume an uncongested road network; average private costs are constant, and are therefore equal to marginal private costs. The third term indicates the total environmental external cost E. Finally, the restrictions λ_m indicate that in any equilibrium, for all groups, the inverse demand equals the sum of the marginal private cost c_m and the common regulatory fee (f) (that is, for all groups, the marginal driver has a zero net surplus). The first-order conditions are as follows (with an apostrophe denoting the first derivative):

$$\frac{\partial L}{\partial N_m} = D_m(N_m) - c_m - \varepsilon_m(\bullet) + \lambda_m {\bullet} D'_m(N_m) = 0 \qquad m = 1,2,...,M \qquad (1a)$$

$$\frac{\partial L}{\partial f} = -\sum_{m=1}^{M} \lambda_m = 0 \qquad (1b)$$

$$\frac{\partial L}{\partial \lambda_m} = D_m(N_m) - c_m - f = 0 \qquad m = 1,2,...,M \qquad (1c)$$

The parameters λ_m cause these first-order conditions to differ from the first-best optimal ones. In particular, when in the restrictions in the Lagrangian (1) the common fee f is replaced by differentiated fees r_m, the first-order conditions include $\lambda_m = 0$ for all m. Hence, the first-best optimum is given by $r_m = \varepsilon_m$, which are the standard Pigouvian fees equal to marginal external costs. In the second-best case under consideration here, however, equations (1a) imply the following expressions for λ_m:

$$\lambda_m = -\frac{D_m(N_m) - c_m - \varepsilon_m(\bullet)}{D'_m(N_m)}$$

Therefore, using (1b) and (1c):

May 1995   Journal of Transport Economics and Policy

$$\sum_{m=1}^{M} \frac{f - \varepsilon_m(\bullet)}{D'_m(N_m)} = 0$$

which gives the following expression for the optimal common fee:

$$f = \sum_{m=1}^{M} \frac{[\varepsilon_m(\bullet)]/[D'_m(N_m)]}{\sum_{i=1}^{M} 1/[D'_i(N_i)]} \qquad (2)$$

Equation (2) shows that the optimal common fee is a weighted average of the marginal environmental costs for all groups. Hence, with constant marginal external costs, the optimal common fee is a weighted average of the individual optimal Pigouvian fees (r_m) that would be charged under first-best regulation. This is an appealing result, which in fact bears close resemblance to other results obtained in the literature on optimal taxation (see, for instance, Sandmo, 1976; and Diamond, 1973). The weight attached to the individual optimal fee for a certain group is inversely related to the slope of the demand curve of that group *in the optimum*. This makes sense: the flatter the demand curve, the more distortive deviations from the individual optimal fee are, and hence the larger the weight should be. The slopes of the demand curves in fact capture both the *relative importance* of the groups and their *demand elasticities* (the larger the group and the more elastic the demand, the flatter the demand curve). In order to disentangle these two effects, we use the definition of demand elasticity η (dropping the argument N_m):

$$\eta_m = \frac{dN_m}{dD_m} \cdot \frac{D_m}{N_m}$$

where dN_m/dD_m is simply defined as $1/(dD_m/dN_m)$, and find that the weight w_m attached to group m amounts to:

$$w_m = \frac{1/D'_m}{\sum_{i=1}^{M}(1/D'_i)} = \frac{(\eta_m \bullet N_m)/D_m}{\sum_{i=1}^{M}[(\eta_i \bullet N_i)/D_i]} \qquad (3)$$

It may be illustrative to note that, when all demand functions have the following form:

$$D_m(N_m) = \alpha \bullet \ln N_m + \beta_m$$

(where α has the same negative value for all groups), $dD_m/dN_m = \alpha/N_m$ and therefore $\eta_m/D_m = 1/\alpha$ for all values of N_m. This results in weights which are proportional to the group sizes in the (second-best) optimum.

Finally, the fact that the private costs of driving may differ among the groups does not affect the outcome. Such relative cost differences are internal, and therefore do not enter the expression for the optimal common fee.

2.2 The optimal common congestion fee

The foregoing analysis gives the general results for second-best regulation concerning an *environmental externality* (that is, an external cost posed upon actors outside the population of car drivers). We now turn to a special *intra-sectoral externality* associated with road transport: *congestion*. As a matter of fact, this is often seen as one of the most important negative externalities associated with road transport, and traffic regulation is often mainly concerned with the reduction (or optimisation) of congestion. In its most basic form, congestion can be introduced by making the private cost of driving within each group dependent on the number of drivers in that group. Hence, for S groups, we may formulate the following Lagrangian:

$$L = \sum_{s=1}^{S} \int_{0}^{N_s} D_s(n_s) dn_s - \sum_{s=1}^{S} N_s \cdot c_s(N_s) + \sum_{s=1}^{S} \lambda_s \cdot [D_s(N_s) - c_s(N_s) - f] \qquad (4)$$

The particular form of the cost functions (the second term) can be interpreted as follows: total costs within a group s (say, C_s) is defined as the group size (N_s) times the average cost (c_s), which in turn depends on the group size. Furthermore, the restrictions show that the individual road user bases his behaviour on the average cost of driving — average social costs are perceived as marginal private costs. Note that the cost functions are assumed to be independent (that is, groups are defined so as to rule out cost interdependencies). Groups may be thought of as users of different radial access roads to a city centre, served by the same parking space where regulatory parking levies are charged. Alternatively, groups may be different cohorts, sufficiently separated in time so as to avoid interdependencies. Clearly, this is a rather restrictive assumption. However, when allowing for inter-group congestion effects, the expression for the optimal common fee can become quite complicated and difficult to interpret (see the Appendix). A relatively simple form of cost-interdependencies is discussed in Section 5 below.

It is easy to show that first-best regulation involves fees r_s, which are equal to the marginal external congestion costs (being the difference between the average and marginal cost of driving):

$$r_s = N_s \cdot c'_s(N_s)$$

However, the regulatory body has to find the optimal common congestion fee by solving (2). The first-order conditions are:

$$\frac{\partial L}{\partial N_s} = D_s(N_s) - c_s(N_s) - N_s \cdot c'_s(N_s) + \lambda_s \cdot [D'_s(N_s) - c'_s(N_s)] = 0 \qquad s = 1,2,\ldots,S$$

$$\frac{\partial L}{\partial f} = - \sum_{s=1}^{S} \lambda_s = 0$$

$$\frac{\partial L}{\partial \lambda_s} = D_s(N_s) - c_s(N_s) - f = 0 \qquad s = 1,2,\ldots,S$$

which may be solved to yield the following optimal common fee:

$$f = \sum_{s=1}^{S} \frac{[N_s \cdot c'_s(N_s)]/[D'_s(N_s) - c'_s(N_s)]}{\sum_{i=1}^{S} 1/[D'_i(N_i) - c'_i(N_i)]} \qquad (5)$$

Like (2), (5) gives a weighted average of the marginal external congestion costs in the second-best optimum. However, the weights are somewhat different, having turned into:

$$w_s = \frac{[1/(D'_s - c'_s)]}{\sum_{i=1}^{S} [1/(D'_i - c'_i)]} = \frac{[1/(d'_s + c'_s)]}{\sum_{i=1}^{S} [1/(d'_i + c'_i)]} \qquad (6)$$

where d' gives the absolute value of the slope of the demand curve ($d' = -D'$). A group

May 1995   Journal of Transport Economics and Policy

now receives a larger weight, not only the flatter the demand curve (the smaller d_s'), but also the flatter the average cost function (the smaller c_s') in the second-best optimum. The reason is that the weights should be such that the (private) welfare losses due to deviations of the common fee from the optimal individual fees are minimised. With congestion, such welfare losses do not only depend on the slopes of the demand curves, but on those of the average cost functions as well. The steeper both curves, the less responsive the group size to deviations of the optimal common fee from the optimal individual fee, and therefore the smaller the group's weight (relative to the other weights).

3. Evaluating Second-best Policies

3.1 The relative efficiency of second-best policies
A crucial question is, of course, how second-best policies as described by equations (2) and (5) relate to the first-best solution of charging all groups taxes equal to their marginal external costs. In order to shed some light on this issue, we will use the following index of relative welfare improvement, ω:

$$\omega = \frac{W_P - W_0}{W_R - W_0}$$

where W_P is social welfare (social benefits minus social costs) under second-best policies of charging the optimal common fee, W_R is welfare under the first-best policy of differentiated regulation, and W_0 is welfare under non-intervention.[2]

By definition, $W_0 \leq W_P \leq W_R$. Welfare under second-best regulation is at least as high as under non-intervention, which can always be realised by setting $f=0$. Likewise, welfare under first-best policies is at least as high as under (optimal) second-best policies, since this latter can always be realised by setting equal fees for all groups. However, $W_0 < W_P$ when $f>0$; and $W_P < W_R$ if for any $i \neq j$, $r_i \neq r_j$ (where r_m gives the first-best optimal fee for group m) and $D'_i \neq -\infty$ and $D'_j \neq -\infty$. Finally, in order to avoid $\omega = 0/0$, we have to assume that $W_0 < W_R$ (that is, we assume an externality to exist and not all demands to be perfectly inelastic). Under these conditions ($W_0 \leq W_P \leq W_R$ and $W_0 < W_R$), ω may range between zero and one. When $\omega=0$, it is impossible to increase welfare by using second-best policies. On the other hand, $\omega=1$ means that second-best regulation yields the same welfare improvement as does first-best regulation, implying that the instruments are perfect substitutes.

In order to develop an expression for ω, we first consider a simple two-group model with group-specific marginal environmental external costs $\varepsilon_i(N_i)$ ($i=1,2$). Consider the non-trivial case where $0 < \varepsilon_1 < \varepsilon_2$ and neither demand curve is perfectly inelastic. Hence, $0 < r_1 < f < r_2$ (f is a weighted average of r_1 and r_2). Let $N_i(c_i)$ give the number of trips made by group i under non-intervention. $N_i(c_i+f)$ and $N_i(c_i+r_i)$ give the number of trips under second-best and first-best policies, respectively (all these values satisfy the inverse demand relations $D_i(N_i)$). The following two inequalities hold: $N_1(c_1+f) < N_1(c_1+r_1) < N_1(c_1)$; and $N_2(c_2+r_2) < N_2(c_2+f) < N_2(c_2)$. Both first-best and second-best policies lead to reduc-

---

[2]   Note that ω is concerned with efficiency considerations only. The redistributive effects of the respective policies are not considered. Furthermore, differences in implementation and intervention costs are ignored.

Second-best Regulation of Road Transport Externalities   E. Verhoef *et al.*

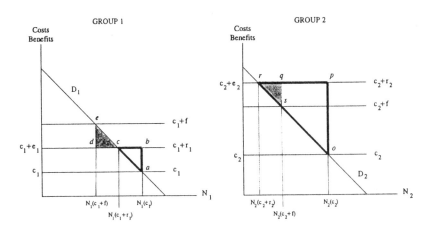

Figure 1
The Welfare Effects of Optimal First-best Regulation (bold) versus
Optimal Second-best Regulation (bold minus shaded) for two Groups

tions in participation by both groups, albeit that group 1 is more restricted under second-best policies (since $r_1 < f$), while group 2 is more restricted under first-best policies (since $f < r_2$). In this case, ω has the following general form:

$$\omega = \frac{I_1[N_1(c_1 + r_1)N_1(c_1)] + I_1[N_1(c_1 + f)N_1(c_1 + r_1)] + I_2[N_2(c_2 + f)N_2(c_2)]}{I_1[N_1(c_1 + r_1)N_1(c_1)] + I_2[N_2(c_2 + f)N_2(c_2)] + I_2[N_2(c_2 + r_2)N_2(c_2 + f)]} \quad (7)$$

where $I_i(N_i)$ is defined as the integral of the function resulting from summing the marginal private cost plus the marginal external cost, minus the marginal benefit. The first terms in the numerator and denominator are identical, giving the increase in social welfare that can be realised by levying group 1 their optimal fee r_1. The second term in the numerator, however, gives the welfare loss due to charging the higher second-best fee f, rather than r_1. The third term in the numerator, identical to the second term in the denominator, gives the welfare gain resulting from charging group 2 the fee f. The third term in the denominator, however, gives the additional welfare gain that may be realised by charging group 2 the higher fee r_2, rather than f. All terms are positive, except the second term in the numerator.

Figure 1 provides an illustration for the case where both groups have linear demand curves (D_i), constant marginal private cost (c_i) and constant marginal external cost (e_i). The maximum possible increase in social welfare in comparison with non-intervention is given by the sum of the surfaces of the two bold triangles *abc* and *opr*. This gain can be realised by charging the first-best fees $r_1 = e_1$ and $r_2 = e_2$. Under second-best regulation, equation (2) may yield an optimal common fee $f(r_1 < f < r_2)$. In comparison with the first-

May 1995   Journal of Transport Economics and Policy

best solution, this policy involves welfare losses as given by the shaded areas *ced* and *sqr*. Hence, in this case, we find the following expression for ω:

$$\omega = \frac{abc - ced + opqs}{abc + opqs + sq\ r}$$

Clearly, the smaller the surface of the shaded areas in comparison with the surface of the bold triangles, the closer to one ω is. In terms of the more general expression (7), ω is closer to one, the smaller the second term in the numerator and the third in the denominator are in comparison to the other terms.

Before turning to the factors determining the value of ω, however, we generalise (7) for the m group case. For that purpose, a distinction should be made between those groups which are *undercharged* and those which are *overcharged* under second-best policies. The critical (possibly fictional) group μ, which draws the distinction between these two, is determined by that particular value of m for which the optimal first-best fee is equal to the optimal second-best fee. For instance, we may assume the external cost function to be linear in some physical factor Γ: $E = \varepsilon \cdot \Gamma$, and Γ to be additively separable in group-specific factors γ_m:

$$\Gamma = \sum_{m=1}^{M} \gamma_m \cdot N_m$$

Groups may, for example, be classified according to trip length γ_m. The above equations then mean that the total distance driven is equal to the summation over all possible trip lengths γ_m times the number of trips of that length, while total external cost depends linearly on total mileage driven. Alternatively, Γ may give total noise emissions over a day, where the different weights γ_m indicate that the severity of noise annoyance varies over the day. In such cases, the marginal external costs can be written as $\varepsilon \cdot \gamma_m$, and μ is that particular group for which:

$$\varepsilon \cdot \gamma_\mu = \varepsilon \cdot \sum_{m=1}^{M} \frac{\gamma_m / D'_m}{\sum_{i=1}^{M} (1/D'_i)} \Rightarrow \gamma_\mu = \sum_{m=1}^{M} (\gamma_m \cdot w_m)$$

At any rate, we may generalise (7) as follows:

$$\omega = \frac{\sum_{m=1}^{\mu} I_m[N_m(c_m+r_m), N_m(c_m)] + \sum_{m=1}^{\mu} I_m[N_m(c_m+f), N_m(c_m+r_m)] + \sum_{m=\mu}^{M} I_m[N_m(c_m+f), N_m(c_m)]}{\sum_{m=1}^{\mu} I_m[N_m(c_m+r_m), N_m(c_m)] + \sum_{m=\mu}^{M} I_m[N_m(c_m+f), N_m(c_m)] + \sum_{m=\mu}^{M} I_m[N_m(c_m+r_m), N_m(c_m+f)]}$$

(8)

Equation (8) confirms the intuitive expectation that second-best policies are a better substitute than first-best policies, when the total population of car drivers is more homogeneous in terms of external cost generation. Obviously, given the demand structure, ω is closer to 1 the smaller the spread in the terms r_m (that is, the nearer the optimal common fee approaches both the minimum and maximum optimal individual fees). In the extreme where $f = r_1 = r_2 = \ldots = r_M$, $\omega = 1$.

The effect of the demand structure on ω, given the values r_m, is a bit less clear-cut. Speaking in absolute terms, deviations of the common fee from the first-best fees are less distortive (and hence absolute welfare losses are smaller) the more inelastic all demands and the smaller all groups are. However, since ω is a relative measure, this observation does not get us much further. It may be noted, though, that the more concave D_m for groups $m<\mu$, and the more convex D_m for groups $m>\mu$, the smaller the relative welfare losses will be (represented by the two "distortion terms" in (8)), and hence the larger ω will be (see also Figure 1). Furthermore, ω will be closer to one, the steeper the demand curves of the more extreme groups (in the optimum) in comparison to those near to μ. As indicated in the previous section, these slopes jointly capture the effects of demand elasticities and group sizes. For instance, suppose that in the optimum the distributions of the factors r_m and of the slopes of demand curves are perfectly symmetric, with μ in the centre of both distributions. The optimal second-best fee will then be r_m. However, the value of ω will be larger, the more inelastic the demands and the smaller the relative importance of the groups in the tails are in comparison to those in the centre of the distribution. The extreme $\omega=1$ is only attainable if all but one of the demand curves are perfectly inelastic and f is therefore set equal to the optimal individual fee for that remaining group. On the other hand, $\omega=0$ only occurs if one of the groups (say group 1) has zero external marginal cost and a perfectly elastic demand curve (the group is infinitely large and completely homogeneous). Second-best regulation then becomes a very unattractive option as it is completely frustrated by $f=r_1=0$. For all other cases, $0<\omega<1$.

When considering the optimal common congestion fee (5), the indices m in (8) may simply be replaced by s, which also have to be assigned according to increasing optimal individual congestion fees, with σ (like μ) being the group for which the second-best fee equals the first-best congestion fee. Again, ω is closer to 1, the smaller the spread in optimal individual first-best fees and the steeper the demand curves in both tails of the distribution are (for groups with either very low or very high congestion levels) as compared with the slopes in the centre. The same holds for the slopes of the average cost functions, which also determine the welfare losses due to uniform taxation in this case. However, one should not ignore the interplay between these slopes and the optimal individual congestion fees, which determine the assignment of the indices s in the first place. In particular, with "comparable" group sizes, a group with a steep average cost function would never be in the left tail of the distribution.

3.2 The effectiveness of second-best environmental policies

A question which is especially relevant from an environmental point of view, is how the absolute reduction in emissions under second-best policies compares to the reduction under a scheme of first-best regulation. Intuition tells us that the former is likely to be smaller, since second-best policies involve certain welfare losses. Hence, the cost of reducing the externality is larger, and we expect a lower optimal reduction in total emissions for any $0\leq\omega<1$. In general, a policy change from second-best to first-best regulation leads to a change in total environmental cost which is equal to:

$$\Delta E = E[N_1(c_1+r_1),N_2(c_2+r_2),\ldots,N_M(c_M+r_M)] - E[N_1(c_1+f),N_2(c_2+f),\ldots,N_M(c_M+f)]$$

In order to keep the analysis manageable, consider a model with linear demand curves and

May 1995   Journal of Transport Economics and Policy

linear, additively separable external cost (as discussed above). The change in total environmental cost, resulting from a policy change from second-best to first-best policies, amounts to:

$$\Delta E = \sum_{m=1}^{M} \Delta N_m \cdot \varepsilon \cdot \gamma_m = \sum_{m=1}^{M} (r_m - f) \cdot \frac{1}{D'_m} \cdot \varepsilon \cdot \gamma_m \qquad (9)$$

For the relatively low-externality groups $(m<\mu)$, $r_m<f$, and the relevant term in (9) is positive, indicating the effect of the *increase* in emissions of these groups due to a higher participation. Conversely, for $m>\mu$, $r_m>f$, and the relevant term in (9) is negative, indicating the effect of the *decrease* in emissions of these groups due to lower participation. The overall effect on total emissions can be found by substitution of the optimal second-best fee f given in (2) into (9), and using $r_m = \varepsilon \cdot \gamma_m$:

$$\Delta E = \sum_{m=1}^{M} \left[\left(\varepsilon \cdot \gamma_m - \varepsilon \sum_{i=1}^{M} \frac{\gamma_i / D'_i}{\sum_{j=1}^{M} (1/D'_j)} \right) \cdot \frac{1}{D'_m} \cdot \varepsilon \cdot \gamma_m \right]$$

$$= \varepsilon^2 \cdot \sum_{m=1}^{M} \left[\frac{\gamma_m^2}{D'_m} - \frac{\gamma_m}{D'_m} \left(\sum_{i=1}^{M} \gamma_i \cdot w_i \right) \right] \qquad (9')$$

It is not easy to determine the sign of (9') at first sight. After some manipulation, however, we find that this expression can be rewritten as:

$$\Delta E = 1/2 \cdot \varepsilon^2 \cdot \left[\sum_{i=1}^{M} \sum_{j=1}^{M} \left(\frac{(\gamma_i - \gamma_j)^2}{\sum_{k=1}^{M} (D'_i \cdot D'_j) / D'_k} \right) \right] \qquad (9'')$$

For instance, for three groups, equation (9'') turns into:

$$\Delta E = \varepsilon^2 \cdot \left(\frac{(\gamma_1 - \gamma_2)^2}{D'_1 + D'_2 + \frac{D'_1 \cdot D'_2}{D'_3}} + \frac{(\gamma_1 - \gamma_3)^2}{D'_1 + D'_3 + \frac{D'_1 \cdot D'_3}{D'_2}} + \frac{(\gamma_2 - \gamma_3)^2}{D'_2 + D'_3 + \frac{D'_2 \cdot D'_3}{D'_1}} \right)$$

The expression in (9'') is always negative (or, in some extreme cases, equal to zero). Therefore, first-best differentiated regulation usually involves lower total environmental cost (and lower total emissions Γ) than do second-best policies — at least with linear demand and linear, additively separable, external cost. (However, it is not difficult to construct corner solutions in which second-best policies actually do lead to a lower level of total emissions than does first-best regulation.)

Since congestion is an intra-sectoral external cost, the benefits of optimising the externality are also to be reaped within the sector. Therefore, it is less useful to consider the consequences of both types of policy on the optimal level of the total external cost. From a welfare point of view, it is sufficient to recognise that, when ω is smaller than 1, the potential welfare gains from second-best instruments are smaller than those resulting from first-best regulation.[3]

---

[3]   However, one would intuitively expect the total level of congestion generally to be smaller under first-best policies for the same reason that the cost of reducing congestion is larger under second-best regulation due to the associated welfare losses.

4. Group Choice and Modal Choice:
Considering Demand Interdependencies

Because of the nature of the Lagrangians (1) and (4), some possibly important behavioural responses to the kind of regulation applied (that is, first-best versus second-best) are ignored. In particular, the postulated independent, stable demand curves for all groups, suggest that car drivers do not consider the possibility of switching group in response to the form of regulation applied. However, this possibility may additionally affect the performance of second-best policy instruments.[4] Analytically, the issues of *group choice* and *modal choice* (in terms of whether to leave the road system) are closely related: whereas the former involves exiting one group and entering another one, the latter merely involves exiting the initial group without entering another.

In transport markets such as those considered in this paper, it seems implausible to specify continuous individual demand functions, including continuous "cross-relations" between the different groups. Rather, individual behaviour is more adequately described by considering binary choices of whether to make a certain trip, based on the benefits and costs of doing so. Furthermore, the different groups are likely to be mutually exclusive for each individual driver; per trip, he chooses only one route (and therewith one trip length), one departure time, and one vehicle. Consequently, when considering the issues of modal and group choice in relation to the kind of regulation applied, the only assumption concerning individual behaviour that seems generally valid is that the (rational) individual car driver i chooses that group m where his (expected) net surplus (benefits $B_m{}^i$ minus (group-specific) private cost c_m minus any regulatory fee t_m) is maximised, provided it exceeds some threshold (money metric) indirect utility $Q_0{}^i$ associated with leaving the road system. This latter determinant depends on the utility associated with the best alternative available, which may be either an alternative mode such as public transport, or refraining from making the trip at all.[5] Therefore, individual behaviour may be characterised as:

$$MAX\{MAX_m\{B_m{}^i - c_m - t_m\};Q_0{}^i\} \tag{10}$$

Via the regulatory fee t_m (which may be either the first-best fee r_m or the optimal common fee f), the prevailing form of regulation to some extent determines the individual net surpluses enjoyed. By definition, under first-best policies, actors face the optimal incentives for group and modal choice. Actors choose a group by maximising the

---

[4]   The potential relevance of such behavioural responses increases, the easier they can be made. For instance, changes in departure times or routes may involve lower costs than changes in trip length (requiring changes in spatial activity patterns), change of mode or change of vehicle. However, we will ignore such transitional costs.

[5]   After the drivers have chosen a group, the discontinuous individual demand functions can still be aggregated to more or less continuous demand functions on the group level, simply by ordering the group members according to decreasing willingness to pay (as was done in Section 2). However, within each group, the individual willingnesses to pay to make a trip in one of the other groups may of course vary unsystematically among the group members. We therefore refrain from setting up general analyses such as those in Section 2 (which require continuous functions), and restrict the discussion to a comparison of incentives faced by the actors under first-best and second-best regulation.

May 1995   Journal of Transport Economics and Policy

difference between benefits $B_m{}^i$ and the generalised costs $c^g{}_m$, being the sum of private costs and the regulatory fee. Since under first-best policies these generalised costs are by definition equal to the social costs, individual utility-maximising behaviour maximises social welfare, and the optimal distribution of road users over the groups and the optimal number of road users within each group will result. Under second-best regulation, however, the incentives given for switching from high-externality groups to low-externality groups are far from optimal.

For the environmental externality, the optimal common fee provides no incentives at all to switch from one group to another, since the factors $t_m = f$ are equal for all groups. This fee reduces the net surpluses achievable in each possible group by equal amounts. Therefore, actors belonging to one group under non-intervention will either remain in that group or leave the system. Hence, the incentive to switch from a "high externality group" j to a "low externality group" i misses out a factor:

$$\varepsilon_j(\bullet) - \varepsilon_i(\bullet)$$

Therefore, second-best uniform taxation not only leads to a larger (smaller) than optimal reduction in the sizes of low (high) externality groups via the direct effect discussed in the foregoing sections. This is the "modal choice effect" that (in terms of equation (8)), for $m < \mu$, the excess reduction in the number of trips is $N_m(c_m + r_m) - N_m(c_m + f)$; and for $m > \mu$, the reduction in car use falls short with $N_m(c_m + f) - N_m(c_m + r_m)$. In addition, there is the more dynamic indirect effect (the "group choice effect") of actors facing insufficient (that is, *no*) incentives to change their behaviour in terms of switching to low externality groups.

For congestion, the modal choice effect is analogous to the effect mentioned above. That is, for $s < \sigma$, the excess reduction in the number of trips is $N_s(c_s + r_s) - N_s(c_s + f)$; and for $s > \sigma$, the reduction in car use falls short with $N_s(c_s + f) - N_s(c_s + r_s)$. However, the group choice effect may actually even be counterproductive for common congestion fees. Again, the uniform tax in the first instance leads to increases in generalised costs which are equal for all groups, suggesting a lack of incentive to switch from a "high congestion group" j to a "low congestion group" i equal to:

$$N_j \bullet c'_j(N_j) - N_i \bullet c'_i(N_i)$$

However, the common fee may very well lead to larger reductions in the average cost c_m for high congestion groups than for low congestion groups. If this is the case, the generalised costs for less congested groups increase more strongly than for more congested groups. This may lead to the adverse indirect effect of drivers who initially belonged to low congestion groups switching to more congested groups (see (10)). Whether this adverse "group choice effect" can occur depends on the slopes of the demand and average cost curves of the initial groups. Suppose there are only two congestion groups: a low congestion group i and a high congestion group j. The common fee f, by definition, leads to reductions in the number of trips made by both groups such that, in terms of the initial demand and average cost curves:

$$\int_{N_i(c_i)}^{N_i(c_i+f)} \left(\frac{\mathrm{d}D_i}{\mathrm{d}N_i} - \frac{\mathrm{d}c_i}{\mathrm{d}N_i} \right) \mathrm{d}N_i = f = \int_{N_j(c_j)}^{N_j(c_j+f)} \left(\frac{\mathrm{d}D_j}{\mathrm{d}N_j} - \frac{\mathrm{d}c_j}{\mathrm{d}N_j} \right) \mathrm{d}N_j$$

Second-best Regulation of Road Transport Externalities   E. Verhoef *et al.*

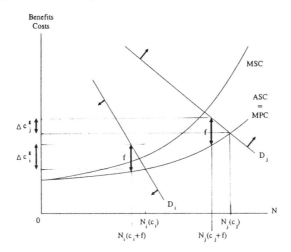

Figure 2
Adverse Group Choice Effects Resulting from a Common Congestion Fee

Since j is assumed to be the high congestion group, $N_j \cdot dc_j/dN_j > N_i \cdot dc_i/dN_i$, and $dc_j/dN_j >$ dc_i/dN_i if congestion rises progressively with group size. Then, the decrease in c_j exceeds the decrease in c_i unless the (absolute value of) the average slope (over the relevant range) of the demand curve D_j exceeds that of D_i, at least to the same extent as the slope of the average cost curve c_j exceeds that of c_i. This, of course, need certainly not be the case; especially since high congestion is often associated with large groups, other things being equal, leading to a flattening of the demand curve.

For instance, Figure 2 shows a road network used during a peak hour by group j and in the off-peak by group i. The curves D give demand, MSC is the marginal social cost and $ASC=MPC$ gives average social cost = marginal private cost. With a common congestion fee f (such as a time-independent regulatory parking fee), which is drawn as some weighted average of what the optimal differentiated fees would be, the generalised cost for group i increases more than for group j: $\Delta c^g_i > \Delta c^g_j$. Therefore, some drivers, initially using the network during the off-peak, may actually be attracted to the peak hour, causing an inward shift of D_i and an outward shift of D_j (as illustrated by the arrows). As a consequence, the reduction in congestion during the peak (the off-peak) may be even further below (above) the optimal reduction than indicated by the modal choice effect alone. Of course, these shifts may in turn affect the optimal common fee, an effect from which we abstain.

May 1995   Journal of Transport Economics and Policy

5. Second-best Charging Mechanisms: A Simple Form of Cost Interdependencies

A last, somewhat different reason why second-best instruments may be an imperfect substitute for optimally-differentiated regulation occurs when a certain number of actors is not confronted with the second-best fee at all. Apart from sheer fraud (which may of course also occur under first-best policies), this may happen when second-best fees are levied via second-best charging mechanisms. For instance, road users who use private parking spaces will not be confronted with a regulatory parking fee.[6] Especially when the aim of the policy is to optimise congestion, the implications are not immediately clear. At first sight, one might argue that those who do park on public space should be taxed more heavily, in order to try and realise the optimal reduction in traffic flows as closely as possible. However, the taxation of the public parkers leads to a reduction in congestion, and therefore to a reduction in the marginal private cost of making trips, thus inducing additional traffic from private parkers. There is clearly a welfare loss associated with this process, as some road users (public parkers) with a certain willingness to pay who are "taxed off the road" will to some extent be replaced by other road users (private parkers) with a (necessarily) lower willingness to pay. This type of problem introduces a simple form of cost interdependency into our model; a more complex case is discussed in the Appendix.

When tolled road users (group 1) and untolled road users (group 2) jointly use a congested road network, where the average cost of driving depends on the total number of road users $N=N_1+N_2$ (implying that the marginal cost of adding an extra road user is equal for both groups), the regulator faces the following Lagrangian:

$$L = \int_0^{M_1} D_1(n_1)dn_1 + \int_0^{M_2} D_2(n_2)dn_2 - N{\cdot}c(N)$$
$$+ \lambda_1{\cdot}[D_1(N_1) - c(N) - f] + \lambda_2{\cdot}[D_2(N_2) - c(N)] \tag{11}$$

with: $N=N_1+N_2$.

The first-order conditions are:

$$\frac{\partial L}{\partial N_1} = D_1(N_1) - c(N) - N{\cdot}c'(N) + \lambda_1{\cdot}[D'_1(N_1) - c'(N)] - \lambda_2{\cdot}c'(N) = 0$$

$$\frac{\partial L}{\partial N_2} = D_2(N_2) - c(N) - N{\cdot}c'(N) - \lambda_1{\cdot}c'(N) + \lambda_2{\cdot}[D'_2(N_2) - c'(N)] = 0$$

$$\frac{\partial L}{\partial f} = -\lambda_1 = 0$$

$$\frac{\partial L}{\partial \lambda_1} = D_1(N_1) - c(N) - f = 0$$

$$\frac{\partial L}{\partial \lambda_2} = D_2(N_2) - c(N) = 0$$

---

[6] The type of problem considered in this section is not specific to parking. Another example would be congestion pricing with the constraint that commercial vehicles be exempt, as in fact happened originally in Singapore (we owe this example to the anonymous referee).

Using $\lambda_1 = 0$, we find:

$$f = N \cdot c'(N) + \lambda_2 \cdot c'(N).$$

Solving for λ_2 yields:

$$\lambda_2 = -\frac{N \cdot c'(N)}{c'(N) - D'_2(N_2)}$$

Substitution of λ_2 then yields the following optimal second-best fee:[7]

$$f = N \cdot c'(N) \cdot \left[\frac{-D'_2(N_2)}{c'(N) - D'_2(N_2)} \right] \qquad (12)$$

A comparison of (12) with the standard congestion fee (see Section 2) shows how the presence of untolled road users affects the result. The optimal second-best fee now depends on the slope of the demand curve of the untolled users, which in turn depends on their group size and demand elasticities, as outlined in Section 2.

In the one extreme, where the untolled users have a perfectly elastic demand in the optimum (group 2 is infinitely large and completely homogeneous), the term between the large brackets in (12) is zero, and therefore the optimal second-best fee is zero. Any positive fee will only lead to welfare losses, since any tolled driver who is "taxed off the road" will be replaced by an untolled driver with a lower willingness to pay for using the road network. Congestion therefore remains unaffected, while the total benefit decreases. Therefore, the best thing the regulator can do is avoid levying any fee at all. At the other extreme, where the untolled road users have a perfectly inelastic demand in the optimum,[8] the term between the large brackets in (12) is equal to 1, and (12) reduces to:

$$f = N \cdot c'(N) \qquad (12')$$

This optimal fee is similar to the standard congestion fee derived in Section 2. In this case, however, the optimal fee for the tolled users depends on the (fixed) number of untolled users. In a sense, the presence of untolled drivers can be thought of as being a mere restriction on the road network capacity, leading, other things being equal, to a lower optimal number of tolled road users and therefore to a higher second-best regulatory fee.

In any situation between the two extreme cases mentioned above, the optimal fee simply depends on the joint effects of the slope of the untolled drivers' demand curve (the flatter their demand curve, the lower the optimal fee) and, of course, on the overall second-best demand in relation to the capacity of the network.

Clearly, the presence of users who are not subject to regulatory policies seriously harms the performance of such policies. In the first place, not every road user responsible for (and subject to) congestion can be reached with such policies. Moreover, and perhaps less obviously, regulatory policies will be frustrated by the fact that the users who are "taxed off the road" may be replaced by untolled users, representing a lower willingness to pay for road use and consequently representing lower benefits. This process involves welfare losses which have to be weighed against the welfare gains of second-best regulatory policies in terms of reducing congestion.

---

[7]   This result is equivalent to the one obtained by Glazer and Niskanen (1992); see their equation (18).

[8]   An important reason why, in the case of regulatory parking policies, the untolled private parkers' demand could be perfectly inelastic in the second-best optimum is a binding restriction on private parking space.

May 1995   Journal of Transport Economics and Policy

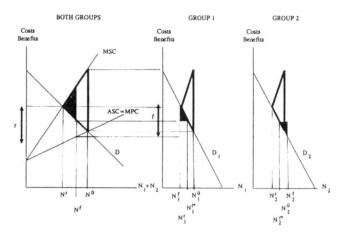

Figure 3

*The Welfare Effects of First-best Congestion Fees
versus the Optimal Second-best Fee in Presence of
Both Tolled Users (Group 1) and Untolled Users (Group 2)*

Figure 3 gives a diagrammatic representation of the situation considered above. In the left panel, the demand (D), average social cost (ASC), which equals marginal private cost (MPC), and the marginal social cost (MSC) curves are drawn for the entire population of road users. The middle and right panel give the demand curves for group 1 (the tolled users) and group 2 (the untolled users), respectively. Optimal first-best regulation implies a levy r for both groups, leading to a reduction in the total number of trips from N^0 to N^r and an increase in social welfare equal to the bold triangle in the left panel (or the sum of the two bold triangles in the middle and right panel). However, the use of the optimal second-best fee as given by (12) leads to a reduction in the number of trips made by group 1 from N_1^0 to N_1^f (rather than N_1^r) and an increase in the number of trips made by group 2 from N_2^0 to N_2^f (rather than the optimal reduction to N_2^r). The total reduction in number of trips is therefore only from N^0 to N^f, and the use of this policy misses out on the potential welfare gains as given by the shaded triangle in the left panel. Moreover, the reduction from N^0 to N^f is not accomplished in the most efficient way, which would have been reductions to N_1^{f*} and N_2^{f*}, respectively, where the marginal benefits for both groups are equalised. Therefore, the two black triangles give additional welfare losses of optimal second-best regulation in comparison with optimal first-best differentiated regulation. The index of relative welfare improvement ω will in this case be as follows:

162

$$\omega = \frac{I(N^f, N^0) - [B_1(N_1^f, N_1^{\{*\}}) - D_1(N_1^{f*})\cdot(N_1^{f*}-N_1^f)] - [D_2(N_2^{f*})\cdot(N_2^f-N_2^{f*}) - B_2(N_2^{f*}, N_2^f)]}{I(N^r, N^f) + I(N^f, N^0)}$$

(13)

The function I is as defined before, but it is now specified for the entire population of road users; and $B_i(\bullet)$ gives the integral of the inverse demand function i over the specified range. From (13), it follows that ω is closer to 1, the steeper the demand curve of group 2, either due to a lower demand elasticity or a smaller group size. This reduces the value of the first term in the denominator, since f in (12) approaches r. Furthermore, the second and third term in the numerator then approach zero.

Finally, the total reduction in the number of trips under second-best policies is in this case obviously always smaller than under first-best regulation whenever the demand of group 2 is not perfectly inelastic. First, group 1 is confronted with a lower fee; and secondly, group 2 even enjoys a fall in the cost of driving.

6. Conclusions

In this paper, the welfare economic characteristics of second-best alternatives to first-best differentiated electronic road pricing were investigated. We considered economic instruments only. Second-best regulation will often suffer from the impossibility of optimal tax differentiation between different types of road users in cases where first-best policies do call for such differentiation — for instance, when external costs are trip-length-, time-, route- or vehicle-specific.

The optimal common fee turns out to be a weighted average of the optimal first-best differentiated fees. For an environmental externality, the optimal weight assigned to a group depends on the inverse of the slope of its demand curve in the optimum, and is therefore positively related to its (second-best) optimal group size and to its demand elasticity in the second-best optimum. For the optimal common congestion fee, the weight is also (inversely) related to the slope of the average cost function in the second-best optimum.

By application of an index of relative welfare improvement, we were able to compare the welfare effects of second-best policies to those resulting from first-best policies. It turned out that second-best instruments become a less favourable substitute, the larger the differences in marginal external costs, and the larger the relative importance and the demand elasticities of the groups in the extremes. It was also shown that the optimal reduction in the externality is usually smaller when second-best instruments are used, than under first-best regulation.

A second-best common fee results in non-optimal incentives in terms of modal choice, in that low externality groups are overcharged and high externality groups are undercharged. In terms of group choice, a common fee completely fails to provide (optimal) incentives to switch from high externality to low externality groups for an environmental externality. A common congestion fee may even create adverse incentives for group choice.

May 1995   Journal of Transport Economics and Policy

In the last section, it was shown that the performance of a second-best congestion fee, suffering from the incapability of charging all road users on a network, critically depends on the relative importance and the demand elasticity of the group which remains unaffected by the policy.

Armed with the results obtained in this paper, it is easier to form a judgement on the relative desirability of second-best alternatives to road pricing, whenever suggested.

References

Arnott, R. J. (1979): "Unpriced Transport Congestion". *Journal of Economic Theory*, 21, pp.294-316.

Arnott, R., A. de Palma and R. Lindsey (1990): "Economics of a Bottleneck". *Journal of Urban Economics*, 27, pp.11-30.

Arnott, R., A. de Palma and R. Lindsey (1991): "A Temporal and Spatial Equilibrium Analysis of Commuter Parking". *Journal of Public Economics*, 45, pp.301-35.

Atkinson, A. B. and J. E. Stiglitz (1980): *Lectures on Public Economics*. London: McGraw-Hill.

Braid, R. M. (1989): "Uniform versus Peak-Load Pricing of a Bottleneck with Elastic Demand". *Journal of Urban Economics*, 26, pp.320-27.

Borins, S. F. (1988): "Electronic Road Pricing: An Idea Whose Time May Never Come". *Transportation Research*, 22A, pp.37-44.

Button, K. J. (1992): "Alternatives to Road Pricing". Paper presented to the OECD/ECMT/NFP/GVF *Conference on The Use of Economic Instruments in Urban Travel Management*, Basel.

Dawson, J. A. L. and I. Catling (1986): "Electronic Road Pricing in Hong Kong". *Transportation Research*, 20A, pp.129-34.

Diamond, P. J. (1973): "Consumption Externalities and Imperfectly Corrective Pricing". *Bell Journal of Economics and Management Science*, 4, pp.526-38.

d'Ouville, E. L. and J. F. McDonald (1990): "Optimal Road Capacity with a Suboptimal Congestion Toll". *Journal of Urban Economics*, 28, pp.34-49.

Evans, A. W. (1992): "Road Congestion Pricing: When is it a Good Policy?" *Journal of Transport Economics and Policy*, 26, pp.213-43.

Fujita, M. (1989): *Urban Economic Theory: Land Use and City Size*. Cambridge: Cambridge University Press.

Giuliano, G. (1992): "An Assessment of the Political Acceptability of Congestion Pricing". *Transportation*, 19, 4, pp.335-58.

Glazer, A. and E. Niskanen (1992): "Parking Fees and Congestion". *Regional Science and Urban Economics*, 22, pp.123-32.

Goodwin, P. B. (1989): "The Rule of Three: A Possible Solution to the Political Problem of Competing Objectives for Road Pricing". *Traffic Engineering and Control*, 30, pp.495-97.

Hau, T. D. (1992): *Congestion Charging Mechanisms: An Evaluation of Current Practice*. Preliminary Draft, Transport Division, the World Bank, Washington (monograph).

Jones, P. (1991): "Gaining Public Support for Road Pricing through a Package Approach". *Traffic Engineering and Control*, 32, pp.194-96.

Knight, F. H. (1924): "Some Fallacies in the Interpretation of Social Cost". *Quarterly Journal of Economics*, 38, pp.564-74.

Lave, C. (1994): "The Demand Curve under Road Pricing and the Problem of Political Feasibility". *Transportation Research*, vol.28A, no.2, pp.83-91.

Lévy-Lambert, H. (1968): "Tarification des Services à Qualité Variable: Application aux Péages de Circulation". *Econometrica*, 36, pp.564-74.

Marchand, M. (1968): "A Note on Optimal Tolls in an Imperfect Environment". *Econometrica*, 36, pp.575-81.

May, A. D. (1992): "Road Pricing: An International Perspective". *Transportation*, vol.19, p.313-33.

Second-best Regulation of Road Transport Externalities   E. Verhoef *et al.*

Mohring, H. (1989): "The Role of Fuel Taxes in Controlling Congestion". In: *Transport Policy, Management and Technology Towards 2001*. Proceedings of the Fifth World Conference on Transport Research, vol. 1: "The Role of Public Sector in Transport", pp.243-57. Yokohama: WTCR.

Pigou, A. C. (1920): *Wealth and Welfare*. London, Macmillan.

Poole, R. W. (Jr) (1992): "Introducing Congestion Pricing on a New Toll Road". *Transportation*, 19, pp.383-96.

Sandmo, A. (1976): "Optimal Taxation: An Introduction to the Literature". *Journal of Public Economics*, 6, pp.37-54.

Small, K. A. (1992a): "Using the Revenues from Congestion Pricing". *Transportation*, 19, pp.359-81.

Small, K. A. (1992b): "Urban Transportation Economics". *Fundamentals of Pure and Applied Economics*, 51, Harwood, Chur etc.

Sullivan, A. M. (1983): "Second-Best Policies for Congestion Externalities". *Journal of Urban Economics*, 14, pp.105-23.

Verhoef, E. T. (1994): "The Demand Curve under Road Pricing and the Problem of Political Feasibility: A Comment". Forthcoming in *Transportation Research*.

Verhoef, E. T., P. Nijkamp and P. Rietveld (1995): "The Economics of Parking Management Systems: The (Im)Possibilities of Parking Policies in Traffic Regulation". TRACE discussion paper TI93-254, Tinbergen Institute, Amsterdam-Rotterdam. Forthcoming in *Transportation Research*, 29A, 2, pp.141-56.

Wilson, J. D. (1983): "Optimal Road Capacity in the Presence of Unpriced Congestion". *Journal of Urban Economics*, 13, pp.337-57.

Appendix
The Optimal Common Fee with Cost Interdependencies

In equations (4) and (5) (Section 2), the optimal common congestion fee is derived for S different groups having their own level of intra-group congestion, ignoring the possibility of inter-group congestion effects. When allowing for congestion effects not only *within* groups but also *between* groups (such as congestion at road crossings when considering route-specific congestion, or groups with different trip lengths using one joint network), the Lagrangian in (4) may be rewritten as follows:

$$L = \sum_{s=1}^{S} \int_{0}^{N_s} D_s(n_s) dn_s - \sum_{s=1}^{S} N_s \cdot c_s(N_1, N_2, \ldots, N_S) + \sum_{s=1}^{S} \lambda_s \cdot [D_s(N_s) - c_s(N_1, N_2, \ldots, N_S) - f] \quad \text{(A1)}$$

The cost functions (in the second term) have changed so as to allow the average cost of group $s(c_s)$ to depend not only on the size of the group itself, but (possibly) on all group sizes. Consequently, optimal discriminatory road pricing involves road prices r_s equal to:

$$r_s = \sum_{i=1}^{S} N_i \cdot \frac{\partial c_i(\bullet)}{\partial N_s}$$

The optimal common fee can be found by solving the set of first-order conditions to (A1):

$$\frac{\partial L}{\partial N_s} = D_s(N_s) - c_s(\bullet) - \sum_{i=1}^{S} N_i \cdot \frac{\partial c_i(\bullet)}{\partial N_s} + \lambda_s \cdot \frac{d D_s(N_s)}{d N_s} - \sum_{i=1}^{S} \lambda_i \cdot \frac{\partial c_i(\bullet)}{\partial N_s} = 0 \quad s = 1,2,\ldots,S \quad \text{(A2)}$$

$$\frac{\partial L}{\partial f} = -\sum_{s=1}^{S} \lambda_s = 0 \quad \text{(A3)}$$

165

May 1995   Journal of Transport Economics and Policy

$$\frac{\partial \mathcal{L}}{\partial \lambda_s} = D_s(N_s) - c_s(\bullet) - f = 0 \quad s = 1,2,\dots,S \tag{A4}$$

Contrary to the problem in (4), it is not possible to solve for each λ_s in terms of $\partial D_s/\partial N_s$ and $\partial c_s/\partial N_s$ only. In matrix notation, the set of first-order conditions can be represented as follows:

$$
\begin{bmatrix}
\frac{dD_1}{dN_1} - \frac{\partial c_2}{\partial N_1} - \frac{\partial c_1}{\partial N_1} & \cdots & \frac{\partial c_s}{\partial N_1} \\
-\frac{\partial c_1}{\partial N_2} & \frac{dD_2}{dN_2} - \frac{\partial c_2}{\partial N_2} & \cdots & \frac{\partial c_3}{\partial N_2} \\
\vdots & \vdots & \ddots & \vdots \\
-\frac{\partial c_1}{\partial N_s} & -\frac{\partial c_2}{\partial N_s} & \cdots & \frac{dD_s}{dN_s} - \frac{\partial c_s}{\partial N_s} \\
-1 & -1 & \cdots & -1
\end{bmatrix}
\begin{bmatrix}
\lambda_1 \\ \lambda_2 \\ \vdots \\ \lambda_S
\end{bmatrix}
-
\begin{bmatrix}
f - \sum_{i=1}^{S} N_i \bullet \frac{\partial c_i}{\partial N_1} \\
f - \sum_{i=1}^{S} N_i \bullet \frac{\partial c_i}{\partial N_2} \\
\vdots \\
f - \sum_{i=1}^{S} N_i \bullet \frac{\partial c_i}{\partial N_S} \\
0
\end{bmatrix}
\tag{A5}
$$

By leaving out the bottom row of the matrix and of the vector to the right of the equal sign, we can use Cramer's rule and solve for each λ_s in terms of the remaining elements of the vector on the right and elements of the matrix. Next, using (A3), we may rewrite one of the conditions (A2) as:

$$f - \sum_{i=1}^{S} N_i \bullet \frac{\partial c_i}{\partial N_s} - \sum_{t \neq s} \lambda_t \left(\frac{dD_s}{dN_s} - \frac{\partial c_s}{\partial N_s} + \frac{\partial c_t}{\partial N_s} \right) = 0 \tag{A6}$$

It is then a matter of substitution of the expressions found for the terms λ_s and rewriting (A6) to find the expression for the optimal common fee. For two groups, the expression for the optimal common fee is still quite straightforward:

$$
f = \left(N_1 \bullet \frac{\partial c_1}{\partial N_1} + N_2 \bullet \frac{\partial c_2}{\partial N_1} \right) \bullet \frac{\dfrac{1}{\dfrac{dD_1}{dN_1} - \dfrac{\partial c_1}{\partial N_1} + \dfrac{\partial c_2}{\partial N_1}}}{\dfrac{1}{\dfrac{dD_1}{dN_1} - \dfrac{\partial c_1}{\partial N_1} + \dfrac{\partial c_2}{\partial N_1}} + \dfrac{1}{\dfrac{dD_2}{dN_2} - \dfrac{\partial c_2}{\partial N_2} + \dfrac{\partial c_1}{\partial N_2}}}
$$
$$
+ \left(N_1 \bullet \frac{\partial c_1}{\partial N_2} + N_2 \bullet \frac{\partial c_2}{\partial N_2} \right) \bullet \frac{\dfrac{1}{\dfrac{dD_2}{dN_2} - \dfrac{\partial c_2}{\partial N_2} + \dfrac{\partial c_1}{\partial N_2}}}{\dfrac{1}{\dfrac{dD_1}{dN_1} - \dfrac{\partial c_1}{\partial N_1} + \dfrac{\partial c_2}{\partial N_1}} + \dfrac{1}{\dfrac{dD_2}{dN_2} - \dfrac{\partial c_2}{\partial N_2} + \dfrac{\partial c_1}{\partial N_2}}}
\tag{A7}
$$

For more than two cohorts, the expression for the optimal common fee becomes increasingly complicated, due to the appearance of an increasing number of cross-effects. For instance, for three cohorts, it can be shown that:

$$
f = \frac{\displaystyle\sum_{s=1}^{3} \frac{\displaystyle\sum_{t=1}^{3} N_t \cdot \frac{\partial c_t}{\partial N_s}}{\displaystyle\prod_{u,v \neq s, u \neq v} \left[\left(\frac{dD_s}{dN_s} - \frac{\partial c_s}{\partial N_s} \right) \cdot \left(\frac{dD_u}{dN_u} - \frac{\partial c_u}{\partial N_u} + \frac{\partial c_v}{\partial N_u} \right) + \frac{\partial c_v}{\partial N_s} \cdot \left(\frac{dD_u}{dN_u} - \frac{\partial c_u}{\partial N_u} + \frac{\partial c_s}{\partial N_u} \right) + \frac{\partial c_u}{\partial N_s} \cdot \left(\frac{\partial c_v}{\partial N_u} - \frac{\partial c_s}{\partial N_u} \right) \right]}}{\displaystyle\sum_{s=1}^{3} \frac{1}{\displaystyle\prod_{u,v \neq s, u \neq v} \left[\left(\frac{dD_s}{dN_s} - \frac{\partial c_s}{\partial N_s} \right) \cdot \left(\frac{dD_u}{dN_u} - \frac{\partial c_u}{\partial N_u} + \frac{\partial c_v}{\partial N_u} \right) + \frac{\partial c_v}{\partial N_s} \cdot \left(\frac{dD_u}{dN_u} - \frac{\partial c_u}{\partial N_u} + \frac{\partial c_s}{\partial N_u} \right) + \frac{\partial c_u}{\partial N_s} \cdot \left(\frac{\partial c_v}{\partial N_u} - \frac{\partial c_s}{\partial N_u} \right) \right]}}
$$

(A8)

Comparison of (A7) and (A8) with (5) shows how the cross-effects associated with inter-group congestion effects affect the weights in the optimal common fee. In Section 2, it was concluded that a group receives a larger weight, both the flatter its demand curve and its average cost curve. The reason was that the optimal weights should be such that the (private) welfare losses due to deviations of the common fee from the optimal individual fees are minimised. In the presence of inter-group congestion effects, the welfare losses due to such deviations obviously also depend on the extent to which a group influences congestion (and therefore welfare) in other groups. For two groups, a group's weight decreases, the stronger the cross-effect it has on the other group; the terms reflecting the cross-effects have a sign opposite to those reflecting own-welfare effects. This is easy to understand once it is realised that the decrease of a group's weight simply implies a relative increase of the other group's weight. In fact, (A7) says that the regulator's concern with welfare losses in a certain group A due to second-best intervention should be reduced, and his concern with losses in group B should be increased, (a) the smaller the welfare losses within group A itself due to such second-best intervention, and (b) the more strongly group A influences welfare in group B compared to the opposite effect. Equation (A8) reflects a comparable message, albeit that it is even more difficult to put the content of (A8) into words, due to the appearance of all the possible cross-effects in the expression.

Date of receipt of final manuscript: July 1994

Part IV
Enforcement

[26]

Enforcement of Pollution Regulations in a Declining Industry[1]

MARY E. DEILY

Department of Economics, Rauch Business Center 37, Lehigh University,
Bethlehem, Pennsylvania 18015

AND

WAYNE B. GRAY

Department of Economics, Clark University, 950 Main St., Worcester, Massachusetts 01610

Received October 30, 1990; revised April 1, 1991

A regulatory agency enforcing compliance in a declining industry might recognize that certain plants would close rather than comply, imposing large costs on the local community. Data on EPA enforcement activity in the U.S. steel industry are examined for evidence of this with a two-equation model linking EPA enforcement decisions and firms' plant-closing decisions. The results indicate that the EPA directed fewer enforcement actions toward plants with a high predicted probability of closing and plants that were major employers in their community; also, plants predicted to face relatively heavy enforcement were more likely to close. © 1991 Academic Press, Inc.

I. INTRODUCTION

Previous research has established that pollution-control legislation and regulations favor struggling industries and slower-growing regions. The government's sensitivity to politically awkward trade-offs between pollution control and jobs reveals itself in special deals stretching out the compliance schedules of individual industries, in regulations biased against new sources, and in congressional voting patterns on policies like PSD (prevention of significant deterioration) (Crandall [4]; Pashigian [25]).[2] Given this history, it seems likely that the potential political costs of pollution control have also influenced regulators' enforcement decisions.

This paper presents the first empirical study of the EPA's enforcement activity at the plant level. We examine the EPA's enforcement actions at U.S. steel plants during the years 1977–1986 for evidence that enforcement was responsive to the possible economic disruption from plant closings.[3] During this period, the EPA faced the problem of enforcing new, higher air quality standards on an industry that was a major polluter, but that was also undergoing severe contraction. We

---

[1] Part of this research was completed while Mary E. Deily was a Visiting Scholar at the Federal Reserve Bank of Cleveland. Financial support was also provided by National Science Foundation grant SES-8921277.

[2] Also, see Eberts and Fogarty [9] for an estimate of the differences in the productivity cost of pollution regulation across regions.

[3] We use "EPA" in this paper to refer to both the individual state pollution agencies and the federal agency. Much of the enforcement is actually done by the state agencies, under EPA supervision.

estimate a simultaneous system of EPA enforcement decisions and firms' plant-closing decisions to test the hypothesis that the EPA directed less enforcement activity toward plants that were likely to close, or that were located where a closing would generate higher-than-average adjustment costs.

We model the regulator as wishing to reduce steel industry air pollution while at the same time avoiding the political costs of even appearing to be the deciding factor in a firm's plant-closing decision. Thus the expected political cost of enforcement depends on the probability that a plant will close and on the amount of local disruption that will occur if the plant does close. Firms, in turn, must decide which plants to close during the contraction, and they include a prediction of future enforcement activity by the EPA in their assessments of future plant profits.

We estimate our two-equation model using data on 49 plants, which together represented virtually all the capacity of the U.S. integrated steel industry in 1976.[4] While restricting our study to plants from a single industry reduces our sample size, it allows us to eliminate the possible effects of inter-industry variation, arising, for example, from technology or pollution control costs, that would be difficult to control.

The steel industry has several characteristics that make it an appropriate test case. First, the industry produces a great deal of pollution, forcing the EPA to take some action toward it.[5] Second, steel plants employ large amounts of workers, increasing the potential for large local adjustment costs. Third, the industry contracted sharply during the test period: of the 49 plants in our sample, 21 closed by 1987. Studying an industry undergoing such very sharp contraction increases the likelihood of observing EPA sensitivity to potential adjustment costs.

We discuss the model of EPA enforcement and firms' plant-closing decisions in Section II. In Section III we review the estimation procedures used and describe the specification. The results are discussed in Section IV, and Section V is the conclusion.

II. MODEL

Assume that new anti-pollution legislation tightening air quality standards is enacted. Assume further that one of the major polluters is an industry that is contracting because of declining demand. We first discuss the EPA enforcement decision in this situation, and then the firms' plant-closing decisions.

The Enforcement Decision

We use a model of regulatory behavior in which the regulator allocates enforcement resources so as to maximize net political support (Stigler [30]; Peltzman

[4]The steel industry can be roughly split into integrated firms and minimill firms. Unless otherwise stated, all references to the steel industry are references to the integrated firms.

[5]"The iron and steel industry ... may be responsible for as much as 10 percent of all particulate air emissions ... ." "How Federal Policies Affect the Steel Industry: A Special Study," p. 43, Congressional Budget Office, U.S. Govt. Printing Office, Washington, DC (February 1987).

[26]).[6] We single out two main sources of political reaction to EPA activity. The first is the general public, which we assume would prefer reduced levels of pollution. Thus, other things equal, the regulator will attempt to reduce as much pollution as possible per unit of regulatory resource expended by directing more enforcement toward plants emitting higher levels of pollution.

The public's perceived benefit from pollution reduction may vary from place to place, however, depending on local conditions: we expect that support for reducing pollution will be greater in more heavily polluted areas. We thus predict more enforcement activity toward plants emitting more pollution, and toward plants in more polluted areas, ceteris paribus.[7]

As the benefits of pollution reduction are diffused across a largely unorganized general public, we expect that opposition to regulatory enforcement arising from smaller groups facing the possibility of sharp losses may have a stronger impact on enforcement decisions. One such potential loser is the firm: bringing a plant into compliance with regulations may impose costs in extra investment, increased operating costs, and reduced productivity.[8]

Given a set schedule of fines, a firm has more incentive to resist agency enforcement efforts the greater the cost is of bringing a plant into compliance. In the limit in which either the fines in the case of non-compliance or the costs of bringing the plant into full compliance reduce the present value of the plant below zero, the firm would shut down the plant rather than comply. The agency might thus find itself using scarce resources to enforce a compliance that could be very costly in terms of political support. We therefore predict that less enforcement activity will be directed toward plants requiring greater firm expenditures to bring into compliance in a declining industry, since inducing compliance has a higher expected cost in agency resources and in possible loss of political support.[9]

The second source of opposition to agency activity is employees and other local citizens that are threatened by a plant-closing. When plants close workers become unemployed and the local community loses income. In some cases local govern-

---

[6]Regulators could in principle have set less stringent standards for marginal plants (driven by the same goal of maximizing net political support). Fewer enforcement actions toward such plants might then reflect differences in initial standards, rather than in enforcement behavior. However, conversations with regulatory officials indicate that the State Implementation Plans developed in the 1970s did not include such accommodations for marginal plants. Since then, plant-specific exceptions have generally involved delays in required compliance dates rather than relaxation of the standards themselves.

[7]The benefits from enforcement would also be affected by the plant's compliance status, and firms' compliance decisions may well be affected by expected enforcement. In an earlier version of this paper (Deily and Gray [7]) we attempted to study steel firm compliance separately by including a third equation for this decision, in the manner of Bartel and Thomas [2]. We were unsuccessful, however, due to poor data on compliance status.

[8]See Crandall [4] and Gollup and Roberts [13]. See also Gray ([15, 16]) for evidence that industries facing heavy enforcement tended to have lower productivity growth.

[9]Such an outcome is not necessarily inefficient: as previous authors have noted, the net social benefit of pollution control is maximized if the EPA enforces regulations so as to equalize marginal abatement costs. See Gollup and Roberts [14], and the citations listed therein. However, an alternative possibility is that the regulatory agency is using a more sophisticated dynamic strategy that induces greater compliance by all firms by specifically concentrating on firms that do not comply, casting them into "purgatory," where they would receive more inspections and higher fines, and from whence they could escape only with continual compliance (Scholz [28]; Harrington [18]).

ments must reduce provision of such services as police and fire.[10] The political costs of having regulatory enforcement behavior cause, or even appear to cause, a plant to shut down may induce regulators to direct their enforcement toward plants for which the probability of closing, and the adjustment cost if closing occurs, are lower. Further, aside from the purely political considerations, it may be more efficient to avoid spending resources to enforce compliance at a plant that is likely to close, since time is likely to take care of the problem.

In summary, the regulatory agency enforces compliance across plants in a declining industry so as to reduce pollution at minimal political cost. The agency receives general political support for reducing pollution levels, but will enforce regulations less rigorously when the cost of bringing a plant into compliance is very high, or when a plant is in danger of closing anyway, particularly if adjustment costs for local communities associated with a plant-closing would be high.

The Plant-Closing Decision

Several recent articles have examined the factors that determine which plants exit from a contracting industry. Theoretical work has focused on examining strategic behavior among oligopolists in a declining industry (Ghemawat and Nalebuff [11, 12]; Reynolds [27]; Whinston [33]). Industry studies have analyzed the effect of plant characteristics and of firm size or diversification on exit from contracting industries (Franklin [10]; Harrigan [17]; Deily [6]; Lieberman [23]; Baden-Fuller [1]).

Firms in a declining industry minimize their losses by closing plants with the lowest expected long-run net revenues. Net revenues depend on a plant's production costs and the competition that it faces. A plant with an older capital stock is more likely to be closed than a plant with a newer capital stock, ceteris paribus, simply because major re-investment decisions should arise in the older plant first (Stigler [29]). (This effect will be exaggerated, of course, if newer capital is more efficient.) Thus, plants with the lowest expected revenue, the highest expected production costs, and the oldest capital should be the most likely to close.

Government efforts to control pollution may affect the pattern of exit from the industry in two ways. First, the costs of complying with pollution regulations may vary across plants. These costs include capital expenditures for pollution control equipment or for retrofitting current capital, operating costs of pollution equipment, and any lost productivity of the original capital due to pollution control efforts. The higher the cost, the lower the plant's expected net revenues, and the more likely it is to close.

Second, the expected compliance cost depends on the level of enforcement that a firm expects to encounter at each plant. Other things equal, plants expected to face more enforcement pressure will need to spend more for pollution control. Plants facing more expected enforcement activity will have lower expected net revenues because of higher expected compliance costs; such plants are more likely to close. Thus, variation in enforcement levels also may affect the pattern of exit.

---

[10]"When Bethlehem Steel decided to shut its Lackawanna, N.Y., plant, idling 7300, Lackawanna authorities began planning layoffs of fire and police and other government workers: half the municipal budget came from Bethlehem's $6 million in taxes." David Nyhan, "Crisis in Steel and for a Way of Life," *Boston Globe*, page 1 and ff., 1/30/83.

In summary, firms minimize their losses by closing plants with the lowest expected net revenues, the oldest capital, and the highest expected compliance costs. Firms are more likely to close plants where more enforcement is expected to occur, since enforcement is likely to raise compliance costs.

III. ECONOMETRICS AND SPECIFICATION

Our model has two endogenous variables: EPA enforcement at a plant and the plant-closing decision of the firm. These decisions are linked in our model because enforcement at a plant depends on the expected probability that the owning-firm will close the plant, and because the probability of a plant-closing depends on the expected cost of compliance at the plant, determined in part by the expected level of enforcement.

We estimate the system using a two-stage, instrumental variables method, instead of a more efficient full-information maximum likelihood estimator, because the enforcement equation has a continuous dependent variable and is estimated with panel data, while the plant-closing equation has a dichotomous dependent variable and is estimated with cross-section data. We use the first-stage equations to generate predicted values for each decision, and then use the predicted values as instruments in the second-stage (structural) equations.

In the first-stage estimations, the plant-closing probabilities are estimated using a logit model, while an ordinary regression model is used for the enforcement decision. All equations include a number of variables that are fixed for each plant: its location, product mix and size, the age of its capital in 1976, the amount of emissions it produces, and the cost of bringing the plant into full compliance. All equations also include the plant's employment relative to the size of its local labor market, which is measured in 1976 for the cross-sectional equations, and annually for the panel. The enforcement equation also includes year dummies and local unemployment rates.[11]

One final adjustment is required to generate a full set of instruments for the second-stage equations. The information about enforcement over time must be compressed into a single number for the cross-section (plant-closing) estimation. But when a firm decides to close a plant, no more enforcement is directed toward that plant, potentially skewing the enforcement measure in the plant-closing equation. Therefore, we use the estimated enforcement equation to predict enforcement in all years for every plant. We then sum the 11 years of predicted enforcement for each plant, including those that close during the sample period.

The first-stage estimations, and the additional adjustments, yield the instruments: PCLOSE, the predicted probability of closing, and PLENFSUM, the sum of predicted enforcement activity over the whole sample period. We now discuss the structural equations in which these variables are used.

---

[11] The area unemployment rate is not included in the other equation because it varies substantially during the sample (so the 1976 value would not pick up cross sectional differences) and may be affected if the plant is closed during the period (so the average unemployment during the sample might not be exogenous).

Enforcement

We use the specification for the enforcement equation,

$$\text{LENF}_{i,t} = f(\text{LEMIT}_i, \quad \text{ATTAIN}_i, \quad \text{PCLOSE}_i, \quad \text{LRELEMP}_{i,t}, \quad \text{CNTYU}_{i,t},$$

$$\text{COMPCAP}_i, \quad 10 \text{ YEAR dummies,} \quad 14 \text{ STATE dummies}), \quad (1)$$

where i indexes plants and t indexes time. The dependent variable, LENF, is the log of the number of enforcement actions directed toward the plant each year by agencies regulating air pollution. Agency actions range from letters and phone calls to inspections and enforcement orders.[12]

The first two variables, LEMIT and ATTAIN, are measures of the potential for pollution damage represented by a plant. LEMIT, the log of the annual tons of pollutants emitted by a plant (an average of 1981 and 1985 values), is included to control for variation across plants in their pollution production. For each plant, emissions of three major pollutants, particulates, sulfur dioxide, and nitrogen oxides, were summed to get a single measure of the pollution potential of the plant.[13]

The variable ATTAIN is a dummy variable indicating that the plant is located in an area that is meeting its air-quality standards. Assuming that the EPA wishes to maintain general public support by reducing pollution enough to meet air-quality standards, it will do so most efficiently by allocating enforcement resources toward plants producing greater amounts of pollution in areas that have not attained minimal air-quality standards. Thus, we expect the coefficient on LEMIT to be positive, and that of ATTAIN to be negative.

We use four variables to measure the cost of the compliance that enforcement is designed to produce. The first, the variable COMPCAP, is an estimate of the total expenditure (per ton of plant capacity) needed to bring a plant into full compliance.[14] Since higher compliance costs will be associated with a lower payoff in reduced pollution per enforcement resource expended (due to greater firm resistance), and may also involve political costs, the coefficient of this variable is expected to be negative.

The other three variables model the potential local adjustment costs that are the focus of this paper: PCLOSE is the predicted probability that the plant will close

---

[12]Since we lack systematic information about the relative importance of these actions, all are included and each receives equal weight. We obtain similar results if only "serious actions" (inspections, penalties, and enforcement orders, which represent 49% of total actions) are counted.

[13]We summed the pollutants because we had no measure of the relative damage caused by each pollutant, and because the pollutants were of similar magnitudes. Further, emissions of the three pollutants are highly correlated. Including the three pollutants together in the equation results in three positive coefficients (as expected) with only one coefficient significant; estimating the equation using only one pollutant at a time results in a significant positive coefficient in all three cases.

[14]We do not know what pollution control expenditures had already been undertaken at the plant, so COMPCAP does not measure incremental compliance costs. Rather, we use estimates in Temple, Barker and Sloane, Inc. [31] of the total national expenditures necessary to bring each piece of steel equipment into compliance with air pollution regulations. Dividing these totals by national capacities of each type of equipment provides a national average cost for bringing each type of equipment into compliance. Total costs for individual plants were then calculated by aggregating the appropriate compliance costs according to the equipment present in the plant, weighted by the capacity of that equipment.

sometime during the sample period, indicating whether the plant is "near the borderline" of being closed; LRELEMP is the log of the ratio of employment at the plant to employment in the local labor market; and CNTYU, the local area unemployment rate, captures the difficulty that laid-off workers might have in finding their next job. All three variables are expected to have negative coefficients.

Finally, the enforcement effort directed toward a particular plant depends on the total amount of enforcement being carried out (or at least, being recorded) during each year in each state.[15] Equation (1) controls for variation in regulation over time and across states with YEAR and STATE dummies. But with only 45 plants in 15 states, the state dummies greatly reduce the explanatory power of the other plant-specific variables. As an alternative, we re-estimate Eq. (1), replacing the STATE dummies with a single variable, LSTATEAV, which measures the total enforcement done in a state during a year. In addition to preserving degrees of freedom, this formulation has the advantage of picking up changes in enforcement within each state over time as well as controlling for variation across states.

Plant Closing

We base our model of steel firms' plant-closing decisions on the estimations reported in Deily [6]. In that work, the probability of exit is determined by a set of plant characteristics only; owning-firm characteristics appear to have been relatively unimportant in this industry. Further, since the individual characteristics of steel plants were quite stable over time, they are measured just once (at the start of the contraction), and the plant-closing probabilities are then estimated from the resulting cross-section data set. The model correctly predicts whether or not a plant survived the decline in over 80% of the cases.

We use an augmented version of this model in the current paper that includes the effect of regulatory enforcement. The following specification is employed,

$$\text{CLOSE}_i = f(\text{COAST}_i, \quad \text{SHAPES}_i, \quad \text{LCAPNEW}_i, \quad \text{LCAP}_i,$$

$$\text{COMPCAP}_i, \quad \text{PLENFSUM}_i), \qquad (2)$$

where i indexes plants. The dependent variable, CLOSE, is dichotomous, equal to one for those plants that closed by the end of 1987, and zero for those that remained open.

The first two variables are proxies for a plant's expected long-run revenues, based on the competition faced by the plant. COAST is a dummy variable indicating plants that are located on the coast, facing more import competition. SHAPES measures the percentage of a plant's product mix composed of plates, structural shapes and pilings, and bars and bar shapes. Plants producing these products face a relatively larger decrease in demand because minimills (a competing source of supply) produce some of the products, while others are made for industries that were themselves undergoing contraction (e.g., railroads). The coefficients of both variables should be positive.

---

[15]The amount of enforcement carried out varies from year to year as enforcement budgets change, from state to state due to differing state policies, and within states over time as state policies evolve.

The variable LCAPNEW, the percentage of a plant's capacity that is new, controls for variation among plants in age of capital stock. LCAP, the annual steel producing capacity of the plant, controls for scale economies in steel production. The coefficients of both variables should be negative.

The final two variables are related to the costs the plant will face in complying with EPA regulations. COMPCAP, the cost of fully complying with pollution regulations, controls for variation in potential expenditures for pollution control. PLENFSUM, the sum of predicted logs of enforcement for the plant, measures the pressure to comply that the plant is expected to face. The coefficients on COMP-CAP and PLENFSUM should be positive.

IV. RESULTS

Table I presents the means and variances of the data used for each independent variable, for the two dependent variables of the first-stage estimations, and for the two instruments PCLOSE and PLENFSUM.

TABLE I
Means and Standard Deviations[a]

Variable	Enforcement ($N = 412$)	Plant-closing ($N = 49$)
ATTAIN	0.17	—
	(0.38)	
CNTYU	8.92	—
	(3.50)	
COAST	0.11	0.14
	(0.31)	(0.35)
COMPCAP	26.04	25.55
	(8.55)	(9.01)
LCAP	1.05	0.93
	(0.59)	(0.69)
LCAPNEW	3.49	3.25
	(0.98)	(1.26)
LEMIT	9.00	8.89
	(1.19)	(1.21)
LRELEMP	−3.55	−3.62
	(1.67)	(1.74)
LSTATEAV	0.59	—
	(0.31)	
PCLOSE ∗ CNTYU	2.87	—
	(2.68)	
SHAPES	0.32	0.36
	(0.32)	(0.34)
CLOSE	—	0.43
		(0.50)
LENF	2.05	—
	(1.34)	
PCLOSE	0.34	—
	(0.29)	
PLENFSUM	—	22.84
		(5.92)

[a]Variables beginning with the letter L are in logs.

268   DEILY AND GRAY

TABLE II
First-Stage Estimations[a]

	Dependent variable	
	LENF[b]	CLOSE
INTERCEPT	− 1.57**	− 6.06
	(0.75)	(5.49)
COAST	− 0.12	0.94
	(0.19)	(1.20)
SHAPES	0.27	2.68*
	(0.18)	(1.37)
LCAP	0.53**	− 2.54**
	(0.12)	(1.12)
LCAPNEW	0.15**	− 0.49
	(0.06)	(0.32)
CNTYU	0.17**	—
	(0.02)	
ATTAIN	− 0.14	—
	(0.15)	
LEMIT	0.13*	1.11*
	(0.07)	(0.65)
LRELEMP	− 0.14**	− 0.27
	(0.04)	(0.28)
COMPCAP	0.004	− 0.09
	(0.007)	(0.05)
Adj R-squ	0.43	—
F statistic	17.07	—
LL	—	− 21.70
Correct predictions	—	82%
N	412	49

[a] Standard errors are in parentheses.
[b] Estimated equation included 10 year dummies.
* Significant at the 10% level, 2-tail test.
** Significant at the 5% level, 2-tail test.

The results of the first-stage estimations are shown in Table II. The enforcement equation does well, with several significant variables explaining over 43% of the variance in (the log of) enforcement actions. The plant-closing equation also does well, with several variables contributing to predict correctly 82% of the plant closing decisions.

The second-stage equations are presented in Tables III and IV. In general, the interactions between the decisions are as expected, although some of the exogenous variables offer significant surprises. We first discuss the enforcement equation estimations reported in Table III. The estimation in column 1 includes 14 state dummies to control for variation in state-level regulatory behavior, while the estimation in column 2 replaces the dummies with the variable LSTATEAV. The estimations are quite similar, but we prefer the more parsimonious Model 2, and refer to its results in the following discussion.

Turning to the main thesis of the paper, we find evidence that enforcement behavior is indeed influenced by potential adjustment costs to local communities. The coefficient of PCLOSE is negative and significant; enforcement activity drops

TABLE III
Second-Stage Estimations: Enforcement[a]

	Dependent variable: LENF		
	(1)[b]	(2)[c]	(3)[c]
INTERCEPT	−2.06**	−2.40**	−2.56**
	(0.81)	(0.65)	(0.63)
PCLOSE	−0.74**	−0.65**	1.58**
	(0.29)	(0.20)	(0.54)
LRELEMP	−0.14**	−0.19**	−0.15**
	(0.05)	(0.04)	(0.04)
CNTYU	0.13**	0.17**	0.24**
	(0.03)	(0.02)	(0.03)
PCLOSE∗CNTYU	—	—	−0.25**
			(0.06)
LEMIT	0.33**	0.28**	0.25**
	(0.07)	(0.05)	(0.05)
ATTAIN	−0.05	−0.25*	−0.25*
	(0.18)	(0.14)	(0.14)
COMPCAP	−0.01	−0.006	−0.001
	(0.01)	(0.007)	(0.007)
LSTATEAV	—	1.00**	0.77**
		(0.25)	(0.25)
Adj R-squ	0.46	0.42	0.45
F statistic	12.61	18.51	19.38
N	412	412	412

[a]Standard errors are in parentheses.
[b]Estimated equation included 10 year dummies and 14 state dummies.
[c]Estimated equation included 10 year dummies.
*Significant at the 10% level, 2-tail test.
**Significant at the 5% level, 2-tail test.

by 6.5% for each 10 percentage-point increase in the probability of closing. The coefficient of the variable LRELEMP is also negative and significant, indicating that a 10% increase in employment size relative to the community work force decreases enforcement actions by 1.9%. However, the estimated coefficient of the third adjustment-cost variable, CNTYU, is positive and significant. This indicates that plants in high-unemployment counties receive *more* enforcement actions rather than fewer.

We investigate this unexpected result by adding an interaction term, PCLOSE∗CNTYU, to the enforcement equation, and re-estimating (column 3). The coefficient on the interacted term is negative and significant, while the coefficient of PCLOSE becomes positive and significant. These results indicate that the tendency for marginal plants to face less enforcement (seen in column 2) is concentrated in counties with high unemployment. Regulators seem to be "skewing" their enforcement more in these counties than in others, with greater enforcement on average, but much less for the plants that are in danger of closing.

Note that this is *not* a reflection of inter-state differences, since variables are included to control for that variation. Rather, enforcement levels are varying for different counties in the same state. This could be due to unmeasured variation in

270   DEILY AND GRAY

TABLE IV
Second-Stage Estimation: Plant-Closing[a]

Dependent variable: CLOSE		Marginal effects[b]	
		COAST = 0	COAST = 1
INTERCEPT	−0.87	—	—
	(2.69)		
PLENFSUM	0.40**	0.086	0.019
	(0.16)		
COAST	3.72**	—	—
	(1.74)		
SHAPES	3.09**	0.669	0.145
	(1.48)		
LCAP	−4.23**	−0.917	−0.199
	(1.50)		
LCAPNEW	−1.07**	−0.232	−0.050
	(0.43)		
COMPCAP	−0.10*	−0.022	−0.005
	(0.06)		
LL	−19.27		
Correct predictions	88%		
N	49		

[a]Standard errors are in parentheses.
[b]Calculated at sample means.
*Significant at the 10% level, 2-tail test.
**Significant at the 5% level, 2-tail test.

the benefits from reducing emissions. To the extent that high-unemployment areas tend to be more populous or more polluted, the benefits from reducing emissions in such areas may be greater.[16]

The other variables in the enforcement equation hold few surprises. The coefficient on LEMIT is positive and always significant, indicating that a plant producing 10% more emissions will receive 2.8% more enforcement activity. The coefficient of ATTAIN is negative as expected, though only significant at the 10% level. It indicates that plants located in areas that attained mandated levels of air quality experienced, on average, 22% fewer enforcement actions each year.

The coefficient of COMPCAP is negative as expected but insignificant, perhaps indicating that regulators do not pay much attention to variation in abatement costs across plants. LSTATEAV's coefficient is significant, and close to one, indicating it measures overall shifts in enforcement affecting all plants proportionally.

Estimation results for the plant-closing equation are presented in column 1 of Table IV; columns 2 and 3 present the corresponding marginal effects of each variable on the probability of closing, column 2 for inland plants and column 3 for

[16]One might think that the concentration of steel plants in a few high-unemployment states is also part of the answer, but note that including state dummies or state unemployment rates does not affect the result.

coastal plants. Since 86% of the plants were located inland, we concentrate on the results in columns 1 and 2.

The estimated coefficient on PLENFSUM, which is positive and significant, indicates that firms are more likely to close plants that are expected to face more enforcement in the future. A 12% increase in expected enforcement increases the probability of closing by 1 percentage point. As expected, the estimated coefficients of the plant-characteristics variables indicate that small, old, coastal plants producing bars and structural shapes are significantly more likely to close. The one unexpected result is the negative and marginally significant coefficient on compliance costs, which suggests that firms were less likely to close plants that cost more to bring into compliance.[17]

V. CONCLUSIONS AND FUTURE WORK

The evidence presented here indicates that air-pollution regulators generally allocate their enforcement activity as if they want to avoid causing local adjustment costs. Plants that were "large" in their local labor market encountered less enforcement pressure. On average, plants with a higher probability of closing also experienced less enforcement pressure. However, plants in high unemployment counties are found, unaccountably, to face greater enforcement.

This pattern of enforcement, with stronger plants bearing more of the costs, reinforces previous work indicating that the regulatory burden has been heavier in faster growing, high-employment regions. However, since we include controls for state-level enforcement, our results are not based on regional differences, but on plant-specific differences in the probability of closing. Thus, for two plants in the same state, the plant with the greater risk of closing faces less enforcement, ceteris paribus.

Our results also show that firms' plant-closing decisions during this period of drastic industry decline were influenced by the enforcement activity of regulators. Plants predicted to face more enforcement were more likely to close. This provides further support for the sensitivity of regulators to the probability of a plant's closing.

We plan to extend this research along a number of lines. First, we can test whether our results hold in another regulatory area by looking at OSHA's enforcement activity in these same steel plants. Second, we plan to look at similar data for other declining industries, to see if enforcement is sensitive to their economic conditions. Finally, we will examine firms' responses to enforcement on a company-wide basis, looking at compliance decisions as well as plant-closing decisions, to test whether these decisions are interrelated across plants within a company and how the decisions are related to the economic health of the firm. Insight into these matters will add to our understanding of the complex relationship between regulatory and firm decision-making.

---

[17]This may come from the use of total rather than incremental compliance costs, or from problems with data quality. Alternatively, it may reflect a technological coincidence. Large plants producing flat-rolled products were most likely to survive, but these plants also generally operated their own coke ovens, and coke ovens represented the largest single compliance expenditure for steel plants (over 40% of the total compliance costs faced by the industry (Temple, Barker and Sloane, Inc. [31])).

APPENDIX: DATA

The sample is the 48 steelmaking plants owned by the integrated producers listed by the Institute for Iron and Steel Studies (IISS) [20], plus the Portsmouth, Ohio, plant of Cyclops. (Small electric-arc based plants and plants producing mainly specialty steels are excluded.) The EPA data sets we use have the plant as the unit of observation and list each plant's name, street address, city, county, state, and industry. We used this information to find the records belonging to the plants in our sample.

The enforcement data are from the EPA's Compliance Data System (CDS). We use the CDS from early 1983 and the CDS from early 1987; the 1983 data are needed because plants are eventually removed from the CDS after they close. Only 3 of the 49 plants are not found on either data set (2 closed in 1977 and the third in 1978). The plants in our sample faced a total of 9316 enforcement actions during the period 1977–1986.

Data on plant closings are from Hogan [19], corporate reports, and phone calls to companies. A plant closed if its steel-making furnaces were shut down; three plants that experienced capacity reductions of over 65% were also counted as closed plants. In all, 21 of 49 plants (43%) closed.

Emissions data are from the EPA's National Emissions Data System (NEDS). As with the CDS, we use two versions of the NEDS, the end-of-year tapes from 1981 and 1985. Data from NEDS include the amount of a plant's emissions of five major pollutants; the three we use (particulates, sulfur dioxide, and nitrogen oxides) were regularly present for steel plants.

The two NEDS tapes, plus emissions data on the pollutants contained in the two CDS data sets, gave us three or four possible measures of plant-level emissions for each pollutant for most plants. (No emissions data were available for two plants; they were assigned the predicted value from a regression of log(total emissions) on log(capacity), log(employment), log(new capital) and a dummy for electric-arc furnaces, which produce much less pollution.) We used the median of the available values for each pollutant (the figures varied substantially across the four data sets) which gave us three measures of emissions at each plant.

The 1986 CDS indicates whether a plant is in an Air Quality Control Region that is attaining its standards for each of the major air pollutants. We say that a plant is located in an "attainment" area if the area is attaining the standards for any of the three pollutants. In our sample, only seven plants were located in an attainment area.

The cost of full compliance is calculated for each plant using estimates from Temple, Barker and Sloane, Inc. [31] of the total cost to the industry of bringing each major piece of equipment into full compliance by 1984. For each type of equipment, we took the total expected capital cost through 1985 and divided it by the gross capacity expected to exist in 1985 to get a cost per unit annual capacity.

Data on the actual equipment in each plant and on the equipment's capacity (both as near to 1976 as was possible) are from Deily [5], IISS [22], annual reports, and various issues of "Directory of Iron and Steel Works of the U.S. and Canada" [8] and of Cordero and Serjeantson [3]. The cost of full compliance for a plant was calculated as the cost (per ton of capacity) of bringing each particular type of equipment into compliance, times the equipment's capacity, summed over the different types of equipment in each plant. This sum, divided by plant capacity in

1976, is our estimate of the cost of bringing the plant into compliance, in 1980 dollars per ton of plant capacity. (The costs of operating the equipment are not included, but are highly correlated with the capital costs.)

Average state-level enforcement rates were calculated from the CDS data as the total number of enforcement actions in the state, minus the number of enforcement actions directed toward the particular plant, divided by the number of all other plants on the CDS for that state.

Information on the local labor market is from the Bureau of Labor Statistic publication "Employment and Unemployment in States and Local Areas" [32], in annual editions from 1976–1986. Labor market tightness is measured by the unemployment rate in the county where the plant is located. The size of the local labor market is measured by the number of people employed in the county.

Data on employment at the plant is from the 1975–1976, 1979–1980, and 1981–1982 issues of "Marketing Economics, Key Plants" [24]. We assigned the numbers in an issue to the first year of the issue (i.e., 1975, 1979, and 1981), and interpolated the remaining years of the sample, with the 1981 value used for 1981–1986.

Coastal plants were those located on or near the East, West, or Gulf coasts. The SHAPES variable is the percentage of a plant's hot-rolled capacity used to produce plates, structural shapes and pilings, and bars and bar shapes. The data are from the 1962 issue of Cordero and Serjeantson [3]. A plant's size is its annual raw-steel capacity in 1976 (IISS [20, 21]).

The percentage of new (post-1959) capacity in each of four major departments (coke-making, blast furnace, steel furnace, and primary rolling/continuous casting) was calculated, and the sum was divided by the number of these departments a plant contained. Thus, the figure is the percentage of each plant that is "new," adjusted for the number of departments located at the plant and for the amount of replacement within a department. The necessary data on investments made at each plant and on the capacity of each department in a plant are from Deily [5], IISS [22], annual reports, and various issues of "Directory of Iron and Steel Works of the U.S. and Canada" [8] and of Cordero and Serjeantson [3].

REFERENCES

1. C. W. F. Baden-Fuller, Exit from declining industries and the case of steel castings, *Econom. J.* **99**, 949–961 (1989).
2. A. P. Bartel and L. G. Thomas, Direct and indirect effects of regulations: A new look at OSHA's impact, *J. Law Econom.* **28**, 1–25 (1985).
3. R. Cordero and R. Serjeantson (Eds.), "Iron and Steel Works of the World," Metal Bulletin Books, Ltd., New York (various years).
4. R. W. Crandall, "Controlling Industrial Pollution," Brookings Institution, Washington, DC (1983).
5. M. E. Deily, Investment activity and the exit decision, *Rev. Econom. Statist.* **70**, 595–602 (1988).
6. M. E. Deily, "The Impact of Firm Characteristics on Plant-Closing Decisions," Federal Reserve Bank of Cleveland Working Paper No. 8803 (1988).
7. M. E. Deily and W. B. Gray, "Enforcement of Pollution Regulations in a Declining Industry," Federal Reserve Bank of Cleveland Working Paper No. 8912 (1989).
8. "Directory of Iron and Steel Works of the U.S. and Canada," American Iron and Steel Institute, Washington, DC (various years).
9. R. W. Eberts and M. S. Fogarty, The differential effects of federal regulations on regional productivity, Federal Reserve Bank of Cleveland Mimeo (1987).

10. P. J. Franklin, Some observations on exit from the motor insurance industry, 1966–1972, *J. Ind. Econom.* **22**, 299–313 (1974).
11. P. Ghemawat and B. Nalebuff, Exit, *Rand J. Econom.* **16**, 184–194 (1985).
12. P. Ghemawat and B. Nalebuff, The devolution of declining industries, *Quart. J. Econom.* **105**, 167–186 (1990).
13. F. M. Gollup and M. J. Roberts, Environmental regulations and productivity growth: The case of fossil-fueled electric power generation, *J. Polit. Econom.* **91**, 654–674 (1983).
14. F. M. Gollup and M. J. Roberts, Cost-minimizing regulation of sulfur emissions: Regional gains in electric power, *Rev. Econom. Statist.* **67**, 81–90 (1985).
15. W. B. Gray, "Productivity versus OSHA and EPA Regulations," UMI Research Press, Ann Arbor, MI (1986).
16. W. B. Gray, The cost of regulation: OSHA, EPA, and the productivity slowdown, *Amer. Econom. Rev.* **77**, 998–1006 (1987).
17. K. R. Harrigan, "Strategies for Declining Businesses," Lexington Books, Lexington, MA (1980).
18. W. Harrington, Enforcement leverage when penalties are restricted, *J. Public Econom.* **37**, 29–53 (1988).
19. W. T. Hogan, SJ, "Minimills and Integrated Mills: A Comparison of Steelmaking in the United States," Lexington Books, Lexington, MA (1987).
20. Institute for Iron and Steel Studies, "IISS Commentary: Special Reports I," Institute for Iron and Steel Studies, Greenbrook, NJ (1977).
21. Institute for Iron and Steel Studies, "Steel Plants U.S.A.: Raw Steelmaking Capacities, 1960 and 1973–1980," Institute for Iron and Steel Studies, Greenbrook, NJ (1979).
22. Institute for Iron and Steel Studies, "Steel Industry in Brief: Databook, U.S.A. 1983," Institute for Iron and Steel Studies, Greenbrook, NJ (1983).
23. M. B. Lieberman, Exit from declining industries: "Shakeout" or "stakeout"?, *Rand J. Econom.* **21**, 538–554 (1990).
24. Marketing Economics Institute, "Marketing Economics, Key Plants," Marketing Economics Institute, New York (various issues).
25. B. P. Pashigian, Environmental regulation: Whose self-interests are being protected?, *Econom. Inquiry* **23**, 551–584 (1985).
26. S. Peltzman, Toward a more general theory of regulation, *J. Law Econom.* **19**, 211–240 (1976).
27. S. S. Reynolds, Plant closings and exit behavior in declining industries, *Economica* **55**, 493–503 (1988).
28. J. T. Scholz, Cooperation, deterrence, and the ecology of regulatory enforcement, *Law & Society Rev.* **18**, 179–224 (1984).
29. G. J. Stigler, "The Theory of Price," 3rd ed., MacMillan Co., New York (1966).
30. G. J. Stigler, The theory of economic regulation, *Bell J. Econom. Management Sci.* **2**, 3–21 (1971).
31. Temple, Barker and Sloane, Inc., "An Economic Analysis of Final Effluent Limitations Guidelines, New Source Performance Standards, and Pretreatment Standards for the Iron and Steel Manufacturing Point Source Category," NTIS PB82231291, Environmental Protection Agency, Washington, DC (May 1982).
32. United States Bureau of Labor Statistics, "Employment and Unemployment in States and Local Areas," U.S. Govt. Printing Office, Washington, DC (various issues).
33. M. D. Whinston, Exit with multiplant firms, *Rand J. Econom.* **9**, 74–94 (1988).

[27]

JOURNAL OF ENVIRONMENTAL ECONOMICS AND MANAGEMENT **24**, 229–240 (1993)

Managerial Incentives and Environmental Compliance[1]

H. LANDIS GABEL AND BERNARD SINCLAIR-DESGAGNÉ

INSEAD, Boulevard de Constance, 77305 Fontainebleau Cedex, France

Received October 10, 1991; revised March 18, 1992

This paper uses a principal-agent model to analyze the role of monetary incentives to implement corporate environmental policy. The model assumes that the agent has to split a limited amount of effort between two tasks: profit enhancement and environmental risk reduction. We find that when the agent's effort constraint is not binding, wages should increase with performance in each task. Moreover, the sharpness of the optimal monetary incentives for a given task should depend positively on the principal's eagerness to influence performance on this task, *and* on the accuracy of the monitoring technology. When the agent's effort constraint is binding, however, the necessity for the principal to provide insurance to the agent may make it inefficient to link managerial effort expended on environmental risk reduction to the corporate compensation system. © 1993 Academic Press, Inc.

1. INTRODUCTION

Misallocation of environmental resources can often be imputed to market failure. This is surely an uncontroversial statement among environmental economists. Indeed, much of the policy-oriented research in the field over the last decades aims mainly at finding ways to improve the functioning of markets (see, for example, Baumol and Oates [1], Dales [2], Kneese and Shultze [7]). There is no doubt that this work has had a significant influence on public policy (Tietenberg [14]).

Many environmental resources, however, are actually allocated *within* firms. Organizational failure, i.e., systematic deviation from the common assumption that firms behave as unitary and rational *personae fictae*, can then also be a major source of environmental problems. In fact, as we argue below, behind the corporate veil lie inescapable sources of misallocation of environmental resources, which are analogous to externalities which are well-understood in the context of market-mediated transactions. Paradoxically, little attention has been paid in the environmental economics literature to the organizational shortcomings that affect the allocation of environmental resources. Yet, just as public policy tools like Pigouvian taxes, rules of civil liability, and marketable property rights might remedy market failures, there are corporate policy tools that might alleviate organizational failures. In addition to considering the correct *economic incentives* in a decentralized

[1]The first version of this paper was presented at the annual meeting of the European Association of Environmental and Resource Economists, Stockholm School of Economics, June 10–14, 1991. We thank the participants of our session, particularly David Miltz and Tom Tietenberg, for helpful comments. We are also grateful to our colleagues Anil Gaba, Xavier de Groote, and Albert Angehrn for instructive conversations, and to two referees for thought-provoking suggestions. This research was supported by INSEAD's Management of Environmental Resources programme, Grant 2097R.

229

230   GABEL AND SINCLAIR-DESGAGNÉ

economy, environmental economists should examine as well the appropriate *contractual incentives* faced by managers and employees in a firm.

In a recent article, Xepapadeas [15] analyzed environmental policy-making concerning pollution in a setting in which the regulatory agency could not perfectly monitor the compliance of regulated firms. Such situations of hidden action or *moral hazard* abound within firms. Chief executives can only approximately measure how much effort a manager puts into reducing the risks of an environmental accident.[2] How should contracts be designed in order to alleviate an environmental moral hazard?

The issue of corporate versus individual liability has been analyzed in some recent works (see, for instance, Segerson and Tietenberg [12]) using a principal-agent model. In this paper, we consider corporate monetary incentives, and we also use the principal-agent framework. Our model, however, differs from those traditionally studied in the literature (for a survey, see Hart and Holmström [4] or Sappington [11]) in two important aspects. First, the agent (i.e., the manager) does not have one but *two tasks* to perform; they relate to the enhancement of expected profit and to the reduction of risk, respectively. Second, the agent has a *limited amount of effort* to split between tasks. Such "multi-task principal-agent" models were recently introduced by Holmström and Milgrom [5]. Contrary to the class of models they study, however, we allow for the realistic possibility that information signals are non-additive and non-normally distributed, and that the agent does not have constant absolute risk aversion.

Although the focus of this paper is strictly within the corporation, we believe that research behind the corporate veil has value not only for corporate environmental policy but for public policy as well. As Segerson and Tietenberg pointed out, and as it is acknowledged by actual enforcement policy, environmental regulatory agencies do not limit themselves to correcting market prices. They have at their disposal a variety of enforcement tools that can reach inside the corporation including individual liability, command and control policies, and organizational remedies.[3] We believe that efficient and effective use of these policy instruments must be founded on the understanding of the decision-making processes that underly corporate action.

The paper is structured as follows. The next section discusses the problem of environmental compliance from the viewpoint of a corporation's top management. We assume that top management wants to reduce environmental emissions or the possibility of an environmental accident, but that the delegation of tasks and discretion to subordinates makes actual compliance uncertain.[4] This motivates the principal-agent model that is presented in the third section. Based on this model, qualitative propositions concerning salary-based incentive schemes to promote

---

[2] Environmental audits, about which more will be said in this paper, are one, albeit imperfect, means of measurement.

[3] To the best of our knowledge, organizational remedies have not been used under the U.S. environmental laws, but they have been in other countries. In Germany, for example, most large firms must by law designate a top manager who is responsible for compliance with environmental regulations and who must present the company's compliance plans to the government. Similar organizational remedies have been used in the U.S., but to ensure compliance in other areas such as antitrust laws, equal employment opportunity laws, etc.

[4] This is not to deny that significant agency problems exist between the public and the firm, but to some extent they have already been analyzed whereas those within the firm have not.

environmental compliance objectives are derived in Section 4. In the plausible case where the principal assigns a positive value to the possibility of influencing the agent's effort on each task, it is found that wages should increase with assessed performance. The rate of increase should actually be higher for the task that the principal is more eager to influence *and* that can be more accurately monitored. On the other hand, in a situation where total managerial effort is at its maximal level and managers are willing to work more than the principal wishes, managers' salaries should not vary with environmental risk-reducing performance. In this case, the appropriate way to fashion the incentives necessary to make managers focus on risk reduction should be based on other means. The section concludes with a comment on the use of environmental audits as a management control mechanism. Section 5 sketches how variants of our model could be employed to examine how job design, individual liability, and corporate culture might be used to promote "greener" management. Section 6 concludes the paper.

2. THE CHALLENGE OF ENVIRONMENTAL COMPLIANCE

Most of the literature in environmental economics adopts a "black box" view of the firm. The firm maximizes profit given the prices of its inputs and outputs. Hence, the environmental policy maker's sole challenge is to ensure that correct prices are set.

Public policy has not yet accomplished this objective, however, nor is it likely to do so in the near future. Prices of environmental resources are still badly distorted, so that alternative channels of enforcement are widely used. The mix of progressively more threatening enforcement measures, both civil and criminal, and the growing pressure from customers, communities, employees, and non-governmental organizations has now made greener management *de rigueur* in corporate board rooms.

The fact that corporate environmental compliance constitutes a serious practical challenge can be inferred from the rising number of environmental management consulting firms (more than 200 currently exist in the UK, for example), from the proliferation of business school educational programs on the subject,[5] from the concerns expressed by business organizations like the International Chamber of Commerce, the Conference Board of Europe, and the U.S. Business Roundtable, and from the growing attention of business policy academics.[6]

The growing concern for greener management, however, indicates that neither academic research nor managerial experience are yet able to provide guidance on how "green" objectives can best be operationalized within a large and complex

---

[5] As an indication of this proliferation, a conference held at the authors' institution in October 1990 for teachers in such programs drew 60 participants. The authors are unaware that any similar conference was previously held. Had one been organized several years earlier, it is unlikely that it would have had more than a handful of attendees. In fact, many institutions dedicated to executive education on environmental matters have been created in Europe and in the United States over the last 2 years.

[6] For instance, the 1991 annual meeting of the Strategic Management Association was devoted to "The Greening of Strategy."

232   GABEL AND SINCLAIR-DESGAGNÉ

corporation's systems of planning, management, and control. There is a need for formal studies on the various ways of implementing corporate environmental policies.

Management surveys [see Refs. 3, 16–20] describe several common practices to fulfill green management objectives. Environmental audits are the most frequent. Typically, their goal is to make an assessment of the company's performance relative to legal requirements. A second objective is to locate the risks. These objectives alone, however, are generally recognized as insufficient. As noted by *Business International* [20, p. 107],

> Without an overreaching environmental management system, auditing will achieve little. It is not a panacea, only a tool, offering at best a critical examination of performance. It is then up to management to provide the commitment and resources to act on the analysis to safeguard the environment.

In a similar vein, the International Chamber of Commerce [19, p. 3] states that

> Environmental auditing should not be seen in isolation but as just one, albeit very important, element in a comprehensive approach to environmental management.

A more comprehensive approach might link the results of environmental audits to the firm's incentive compensation system. In addition to profit objectives, line managers could be given non-profit goals like limiting effluents or the risk of environmental accidents. Audits could be made of performance toward these goals, and the managers' salaries could depend in part on assessed performance. Although none of the surveys of management practice gives an example of a firm that directly and formally links employee compensation to audit results, many firms do so informally. As expressed in a survey by *The Economist* ([16], p. 23),

> [Audits] allow chief executives to set goals for subsidiaries: get your reported emissions down to such-and-such level, or lose a bit of your bonus.

It is this policy of making managers' compensation depend in part on environmental performance that we analyze in the upcoming sections.

Another approach to compliance is via the firm's accounting and control system. Accounting measures may be modified in such a way as to mimic the market. For example, a salesman's commission may be reduced for environmentally hazardous products to spur substitution towards less dangerous products. Overhead costs for insurance premia or for the legal department may be allocated to specific products to change the financial incentives to produce them. Fictitious accounting costs (from the firm's viewpoint) may be charged against certain products to reduce their internal economic attractiveness.

Finally, task reorganization is often observed. Staff experts may be employed with responsibility, authority, and financial resources to handle the investment in and operations of environmental controls. Also, decisions with environmental implications may be centralized and shifted up in the corporate hierarchy.

All these practices implicitly recognize that in large complex organizations, senior management necessarily delegates decision-making to subordinate managers and other employees whose behavior cannot be perfectly controlled. Hence, incomplete information, hidden action, incentives, and moral hazard are pervasive constraints of contractual arrangements.

3. A PRINCIPAL-AGENT MODEL

Imagine that a corporation's principal—its chief executive officer (CEO) or chairman of the board—is worried that risks of environmental accident lurk throughout the company. These risks could be reduced if subordinate line managers were to spend more effort tackling them, but effort so spent would be at the expense of activities that generate the profit figures on which the managers' compensation is currently based.[7] Furthermore, the CEO cannot perfectly observe the amount of effort managers would allocate to the various tasks. Effort can only be inferred indirectly and imperfectly through some measure of performance. How should the CEO link managerial compensation to performance with respect to environmental risk reduction?

The CEO's problem can be modeled as follows. Let a line manager, whom we shall call the agent, select actions in the compact polyhedron $\mathbf{A} = \{\mathbf{a} = (a_1, a_2) | a_1 + a_2 \leq 1, \mathbf{a} \geq \mathbf{0}\}$ of $\mathbf{R}^2$. That is, the agent/manager splits his effort between profit-seeking and environmental risk-reducing activities. The manager's "effort" can be defined as the integral of his hourly levels of attention over a week.[8] There is an objective upper bound on the agent's effort, which is determined by the fixed, contractual working week and the current, well-documented limitations on human information processing (Newell and Simon [10]). The first and second components of an $\mathbf{a}$ in $\mathbf{A}$ are then the relative (i.e., normalized by the upper bound on effort) amounts of effort devoted to profit enhancement and environmental risk reduction, respectively.

The action chosen, i.e., the allocation of effort, is known only by the agent. The agent's actions are nonetheless costlessly (but imperfectly) monitored by the principal and generate a compound discrete score

$$\mathbf{s} \in \mathbf{S} = \{\mathbf{s} = (s_1, s_2) | s_1 = 0, 1, \ldots, S_1; s_2 = 0, 1, \ldots, S_2\}.$$

Let the conditional likelihood of a score $\mathbf{s}$ with respect to an action $\mathbf{a}$ be given by $p(\mathbf{s}|\mathbf{a})$.

A standard assumption about the probability distributions $p(\cdot|\mathbf{a})$ is that a higher score $\mathbf{s}$ makes a greater effort $\mathbf{a}$ more likely. Formally, this means that the likelihoods $p(\mathbf{s}|\mathbf{a})$ display the monotone likelihood ratio property (Milgrom [9]), a multivariate version of which is the following.

ASSUMPTION: *The ratios* $p_t(\mathbf{s}|\mathbf{a})/p(\mathbf{s}|\mathbf{a})$ *are non-decreasing in* $\mathbf{s}$, *for every* t *and* $\mathbf{a}$.

The index t refers to the action—profit-making or risk-reducing. The subscript t in $p_t(\mathbf{s}|\mathbf{a})$ denotes the first-order partial derivative with respect to a_t ($t = 1, 2$) at $\mathbf{a}$.

Observing the performance score $\mathbf{s}$, the CEO may estimate the profit (before wage costs) to be $\pi(\mathbf{s})$, where $\pi(\cdot)$ increases with $\mathbf{s}$. In this model the

---

[7]The literature on executive compensation systems is generally critical of compensation based on short-run accounting profits. Yet, such compensation is very common (see Kaplan and Atkinson [6]) and we will assume it here.

[8]This notion of effort is then similar to the notion of "work" in physics, which is the integral of "power" (measured in watts) with respect to time.

principal/CEO is risk neutral.[9] The agent/manager, on the other hand, is attributed a separable Von Neumann–Morgenstern utility function over income and action, given by $v(y) - c(\mathbf{a})$. The function $v(\cdot)$ is defined on the real line; it is strictly increasing and displays strict risk aversion in income. The cost function $c(\cdot)$ is convex.

The principal's objective is to set a contingent wage schedule $w = w(\mathbf{s})$ that maximizes her expected utility, given that the agent will optimize over his action space and that he can always get a (status quo) utility level V^* in some other employment. This yields the following *multi-task principal-agent problem*.

$$\max_{a,w} \sum_{s \in S} p(s|a)(\pi(s) - w(s))$$

subject to

$$a \in \arg \max_{\hat{a} \in A} \sum_{s \in S} p(s|\hat{a})v(w(s)) - c(\hat{a})$$

$$\sum_{s \in S} p(s|a)v(w(s)) - c(a) \geq V^*. \tag{1}$$

The first constraint above implies incentive compatibility and the second implies individual rationality. As a technical matter, the incentive compatibility constraint is hardly tractable, for it involves a continuum of inequalities. Under some conditions, however, this constraint can be replaced by the first-order necessary conditions for $\mathbf{a}$ to be the agent's optimal action.[10] This "first-order approach" yields the following equivalent formulation of the multi-task principal-agent problem, which is easier to solve.

$$\max_{a,w,\gamma} \sum_{s \in S} p(s|a)(\pi(s) - w(s))$$

subject to

$$\sum_{s \in S} p_t(s|a)v(w(s)) - c_t(a) - \gamma \geq 0$$

$$\text{for } t = 1, 2, \tag{2}$$

$$\left(1 - \sum_{t=1}^{2} a_t\right)\gamma \leq 0, \qquad \gamma \geq 0, \qquad a \in A;$$

$$\sum_{s \in S} p(s|a)v(w(s)) - c(a) \geq V^*$$

γ denotes the agent's shadow price of effort.

Our next task is to explore the qualitative properties of an optimal compensation scheme.

---

[9]The results of Section 4 would still be valid if we assumed instead that $\pi(\cdot)$ is a *realized* profit level, increasing in $\mathbf{s}$, and that the principal is risk averse.

[10]These conditions are essentially that all functions $v(\cdot)$, $p(s| \cdot)$, and $c(\cdot)$ be twice continuously differentiable, and that the probability of a signal being less than or equal to s decrease, at a decreasing rate, as $\mathbf{a}$ grows. See Sinclair-Desgagné [13].

4. ENVIRONMENTAL RISK-REDUCING PERFORMANCE AND SALARIES

Under the usual constraints qualification assumptions, a solution to problem (2) must satisfy the following Kuhn–Tucker conditions.

$$\frac{1}{v'(w(s))} = \mu + \frac{\sum\limits_{t=1}^{2} \delta_t p_t(s|a)}{p(s|a)} \tag{3}$$

for every $w(s)$, and

$$\sum_{s \in S} p_t(s|a)(\pi(s) - w(s)) + \sum_{r=1}^{2} \delta_r \left\{ \sum_{s \in S} p_{rt}(s|a) v(w(s)) - c_{rt}(a) \right\}$$

$$+ \mu \left\{ \sum_{s \in S} p_t(s|a) v(w(s)) - c_t(a) - \gamma \right\} - \lambda - \rho\gamma + \mu\gamma \le 0 \qquad \text{if } a_t = 0$$

$$= 0 \qquad \text{if } a_t > 0$$

for $t = 1, 2;$ \hfill (4)

$$- \sum_{r=1}^{2} \delta_r + \rho \left(1 - \sum_{t=1}^{2} a_t \right) \le 0 \qquad \text{if } \gamma = 0;$$

$$= 0 \qquad \text{if } \gamma > 0$$

$$\rho \le 0, \qquad \lambda \ge 0, \qquad \delta_r \ge 0$$

In these expressions, $v'(w(s))$ denotes the first-order derivative of v at $w(s)$, and $p_{rt}(s|a)$, $c_{rt}(a)$ stand for second-order partial derivatives with respect to a_r and a_t at a. λ is the principal's shadow price for the agent's effort; ρ is the opportunity cost she would incur if the agent were to waste an amount of effort that is worth one utile to him; δ_r denotes the shadow price of the incentive compatibility constraint for task r, i.e., the increase in the principal's utility of a marginal deviation from the agent's utility-maximizing level of effort on task r; and μ is the principal's shadow price for the individual rationality constraint, i.e., the increase in the principal's utility from a marginal decrease in the agent's reservation utility V^*.

It is intuitively appealing that the optimal wage increases with an improvement in the profit-enhancing or the risk-reducing scores. As the next proposition shows, this holds provided all the incentive compatibility constraints are binding.

PROPOSITION 1. *If all the δ_r's are positive, the optimal wage schedule $w(s)$ is non-decreasing in s.*

Proof. First note that positive δ_1 and δ_2 ensure that a solution to problem (2) is also a solution to the multi-task principal-agent problem (see Sinclair-Desgagné [13]).

The statement then follows from equation (3) and the previous assumption concerning $p(s|a)$. \hfill Q.E.D.

What can we say now about the sharpness of the optimal monetary incentives, i.e., the slope of the optimal wage schedule? Consider two distinct vectors of measurements (s_1^+, s_2^+) and (s_1, s_2). Subtracting equation (3) at (s_1^+, s_2^+) from equation (3) at (s_1, s_2) gives

$$
\frac{1}{v'(w(s_1^+, s_2^+))} - \frac{1}{v'(w(s_1, s_2))}
$$
$$
= \delta_1 \left[\frac{p_1(s_1^+, s_2^+ | a_1, a_2)}{p(s_1^+, s_2^+ | a_1, a_2)} - \frac{p_1(s_1, s_2 | a_1, a_2)}{p(s_1, s_2 | a_1, a_2)} \right] \qquad (5)
$$
$$
+ \delta_2 \left[\frac{p_2(s_1^+, s_2^+ | a_1, a_2)}{p(s_1^+, s_2^+) | a_1, a_2)} - \frac{p_2(s_1, s_2 | a_1, a_2)}{p(s_1, s_2 | a_1, a_2)} \right].
$$

This formula is equivalent to

$$
\frac{1}{v'(w(s_1^+, s_2^+))} - \frac{1}{v'(w(s_1, s_2))}
$$
$$
= \delta_1 \left[\frac{\partial}{\partial a_1} \log \frac{p(s_1^+, s_2^+ | a_1, a_2)/p(s_1^+, s_2^+ | \bar{a}_1^+, \bar{a}_2^+)}{p(s_1, s_2 | a_1, a_2)/p(s_1, s_2 | \bar{a}_1, \bar{a}_2)} \right] \qquad (6)
$$
$$
+ \delta_2 \left[\frac{\partial}{\partial a_2} \log \frac{p(s_1^+, s_2^+ | a_1, a_2)/p(s_1^+, s_2^+ | \bar{a}_1^+, \bar{a}_2^+)}{p(s_1, s_2 | a_1, a_2)/p(s_1, s_2 | \bar{a}_1, \bar{a}_2)} \right],
$$

where the symbol "$-$" refers to the maximand over $\mathbf{A}$ of the corresponding likelihood function. In this context, one can interpret the shadow prices δ_1 and δ_2 as the magnitudes of the principal's demand for monitoring performance on the profit-enhancing and the risk-reducing activity, respectively. Note that the partial derivatives of the log-likelihood ratios measure precisely the capability of the monitoring technology to detect changes in effort. *Equation (6) then links the salary-performance gradient to the relative intensities of the principal's demand for monitoring performance on each activity, and to the accuracy of the monitoring technology.*

Put $s_1^+ = s_1 + 1$, $s_2^+ = s_2 + 1$, and denote the wage schedule's partial finite differences as

$$
\Delta_1 = w(s_1^+, s_2) - w(s_1, s_2),
$$
$$
\Delta_2 = w(s_1, s_2^+) - w(s_1, s_2).
$$

The next statement follows directly from equation (6).

PROPOSITION 2. *Let δ_1, δ_2 be positive. Suppose that, for all s_1, s_2*

$$
\frac{\partial}{\partial a_1} \log \frac{p(s_1, s_2^+ | a_1, a_2)/p(s_1, s_2^+ | \bar{a}_1, \bar{a}_2^+)}{p(s_1, s_2 | a_1, a_2)/p(s_1, s_2 | \bar{a}_1, \bar{a}_2)} \approx 0, \qquad (7)
$$

$$
\frac{\partial}{\partial a_2} \log \frac{p(s_1^+, s_2 | a_1, a_2)/p(s_1^+, s_2 | \bar{a}_1^+, \bar{a}_2)}{p(s_1, s_2 | a_1, a_2)/p(s_1, s_2 | \bar{a}_1, \bar{a}_2)} \approx 0. \qquad (8)
$$

Then $\Delta_1 \geq \Delta_2$ if and only if

$$
\delta_1 \left[\frac{\partial}{\partial a_1} \log \frac{p(s_1^+, s_2 | a_1, a_2) / p(s_1^+, s_2 | \bar{a}_1^+, \bar{a}_2)}{p(s_1, s_2 | a_1, a_2) / p(s_1, s_2 | \bar{a}_1, \bar{a}_2)} \right]
$$
$$
\geq \delta_2 \left[\frac{\partial}{\partial a_2} \log \frac{p(s_1, s_2^+ | a_1, a_2) / p(s_1, s_2^+ | \bar{a}_1, \bar{a}_2^+)}{p(s_1, s_2 | a_1, a_2) / p(s_1, s_2 | \bar{a}_1, \bar{a}_2)} \right].
$$

(9)

Condition (7) asserts that an increase in the score s_2 cannot be used to infer anything about a_1—the effort expended on the profit-seeking activity. Similarly, condition (8) means that an improvement in s_1 says nothing about a_2—the effort expended on reducing the probability of an environmental accident. Hence, the measurements s_1 and s_2 refer exclusively to the profit-enhancing and to the risk-reducing activities, respectively. Proposition 2 then says that the slope of the optimal wage schedule with respect to the environmental risk reduction score, relative to the slope with respect to the profit enhancement score, depends on the relative intensity of the demand for monitoring environmental risk reduction *and* on the relative capability of the monitoring technology to detect changes in the risk reducing effort.

Our finding that relatively flat salary schedules are appropriate for activities that are relatively hard to monitor is consistent with Holmström and Milgrom's [5] theoretical result. Note that this is inherently the case for activities that affect risk compared to activities that yield routinely reported current profit. Our framework and result are more general, however. Furthermore, we shall now see that the appropriateness of flat wage schedules might not be due to any explicit difference in monitoring accuracy. Incentive pay for environmental risk reduction is actually irrelevant when managers are "workaholics."

Positive shadow prices on the incentive constraints mean that the manager would spontaneously resist spending extra effort on any task (i.e., deviating from his current optimum); he must then assign a value of 0 to incremental potential effort. Suppose, on the contrary, that the manager would like to spend extra effort but that his constraint has been reached. Then γ is positive. In this case the following proposition shows that, under some circumstances, wages should not vary when only the performance score on the risk reduction task improves.

PROPOSITION 3. *Let the manager's effort be fully extended, so that $\gamma > 0$. If also*

$$
\lambda \leq (\mu - \rho)\gamma, \tag{10}
$$

then

$$
\text{for all } s_1, s_2, s_2': \quad w(s_1, s_2) = w(s_1, s_2'). \tag{11}
$$

Proof. Condition (10) ensures again that a solution to problem (2) also solves the multi-task principal-agent problem (see Sinclair-Desgagné [13]).

Now, a positive shadow price γ implies that $\delta_1 = \delta_2 = 0$ by the next to last line of (4). Thus, equation (3) can be satisfied only if (11) is true.   Q.E.D.

Condition (10) has an interesting interpretation. Suppose that the agent is given 1% more potential effort. This may happen, for instance, through a longer working week or an improvement of the current information technology. The agent's total amount of available effort, relative to the previous amount, is now 101%. The principal's and the agent's utilities increase then by λ and γ, respectively. Expressed in terms of the principal's utiles, however, the agent's additional satisfaction is $(\mu - \rho)\gamma$. Inequality (10) thus says that the agent is more eager than the principal to get this extra unit of effort. Such a situation could be inferred from the agent's eagerness to lengthen his contractual working week or to invest in better information technology (setting aside, of course, the problems of adverse selection that this entails).

Proposition 3 then has an intuitive rationale. A positive shadow price for managerial effort implies that the agent is using 100% of his effort, and would like to use more but cannot. The incentive constraints are then irrelevant, so $\delta_1 = \delta_2 = 0$. Interaction between the principal and the agent is now limited to risk sharing, but efficient risk sharing requires that their marginal rates of substitution between the various income levels be equal. Given that the principal is risk neutral, this can only be achieved if the wage schedule is invariant in the risk reduction score.

Another way to look at the result of proposition 3 is as follows. Under the stated conditions ($\gamma > 0$ and inequality (10)), suppose that the wage schedule were instead tied to the score on environmental risk reduction. This would increase the amount of environmental risk borne by the agent. In order to avoid low s_2 scores the risk averse agent would then devote more effort to risk reduction and less to profit enhancement than the risk neutral principal would wish.

One specific implication of proposition 3 is that linking salaries and environmental audits may be unsuited as a means of managerial control. If managerial effort is at an upper bound, the CEO should simply appraise profit performance and pay contingent wages $w^*(s_1)$ that mimic the solution $w(s_1, s_2)$ to problem (2). That is,

$$\text{for all } s_1: \qquad w^*(s_1) = w(s_1, 0). \qquad (12)$$

5. OTHER INCENTIVE SCHEMES AND DIRECTIONS FOR FUTURE RESEARCH

If incentive compensation and monitoring are insufficient for controlling environmental risks in large decentralized firms, what other means are there at the disposal of corporate principals? In this section we will consider how several of the alternatives discussed in section 2 could be explored using the multi-task principal-agent framework we have laid out.

An alternative to assigning both profit and environmental risk reduction responsibilities to a single agent is to assign the different tasks to different agents. This is common in the business arena as was noted in section 2. Holmström and Milgrom [5] argue that tasks in which performance is relatively easy to assess should be assigned to one manager, while tasks that are inherently difficult to monitor should be assigned to another manager. The former's compensation should be incentive-based while the latter's should not. In our context, this would imply that profit should be the responsibility of line managers with incentive contracts, while environmental risk reduction should be assigned to personnel under a fixed-salary contract. Is this splitting of tasks and responsibilities according to a "monitoring criterion" reasonable in the environmental context? In our model, the problem of job design could be tackled by introducing a second agent and expanding the set of actions and performance scores accordingly. This would greatly increase the complexity of the problem, and the Kuhn–Tucker conditions, in particular, would become intricate. Yet qualitative insights yielding propositions like those above should still be possible.

Another frequently observed alternative is to shift some corporate liability to the individual agent. This is a corporate policy analogous to the public policy of individual criminal penalties. Companies can and do, for instance, make it clear to employees that negligent conduct causing environmental liability is ground for disciplinary action, and that legal aid and indemnification of fines would be denied to any employee personally liable for an environmental accident. To analyze such a case, liability costs functions $c_P(D)$ and $c_A(D)$, increasing with the extent of environmental damage D, would have to be subtracted from the principal's and the agent's respective utility functions; and a conditional likelihood $q(D|\mathbf{a})$ would have to be specified. Qualitative insights should then again be possible.

Finally, companies can and do place great emphasis on corporate "culture." In a recent article, Kreps [8] defined corporate culture as a focal point in a noisy reputation game. Corporate reputation might influence the agent's reservation utility V^*; it might figure as an argument of the agent's utility function; or it might protect the company from harsh penalties in the event of an environmental accident. Also, in the presence of ambiguity, a focal point would indicate a specific likelihood $p(\mathbf{s}|\mathbf{a})$, i.e., it would indicate the types of inference on $\mathbf{a}$ the principal may draw from observing $\mathbf{s}$.

6. CONCLUSIONS

We have argued that environmental economists and policy makers should pay more attention to organizational failures, just as they have long been concerned with market failures. As a beginning to a formal study of important organizational failures and their remedies, we have examined how incentive compensation systems can and should be devised to deal with the trade-off that managers often face between improving current profits and reducing the risk of environmental accidents.

We have presented a multi-task principal-agent model and have used it to assess the relevance of incentive pay linked to performance on environmental risk reduction. One main result was that monetary incentives should become stronger, as the principal becomes more eager to promote environmental risk-reducing activities relative to activities that enhance profit *and* as the monitoring technology concerning environmental risk reduction becomes relatively more accurate. When total managerial effort reaches its peak, however, it might no longer be appropriate to make wages vary with the observed reduction of environmental risk. Under some conditions, an incentive wage should only apply to profit performance.

Although we formally modeled only incentive compensation in this paper, our principal-agent model can be used to address other issues such as investment in improved systems for appraising environmental risks, task division between different agents, corporate imposition of personal liability on agents, and the role of corporate culture as a means of governing agents' behavior.

240   GABEL AND SINCLAIR-DESGAGNÉ

REFERENCES

1. W. Baumol and W. Oates, "The Theory of Environmental Policy," Cambridge Univ. Press, Cambridge, UK (1988).
2. J. Dales, "Pollution, Property, and Prices," University Press, Toronto (1968).
3. M. Flaherty and A. Rappaport, "Multinational corporations and the environment: A survey of global practices," The Center for Environmental Management, Tufts University (1991).
4. O. Hart and B. Holmström, The theory of contracts, *in* "Advances in Economic Theory, Fifth World Congress," (T. Bewley, Ed.), Cambridge Univ. Press, MA (1987).
5. B. Holmström and P. Milgrom, "Multi-task principal-agent analysis: Incentive contract, asset ownership and job design," *J. Law, Econom., Organiz.* **7**, 24–52 (1991).
6. R. S. Kaplan and A. A. Atkinson, "Advanced Management Accounting," 2nd ed., Prentice–Hall, Englewood Cliffs, N.J. (1989).
7. A. Kneese and C. Schultze, "Pollution, Prices, and Public Policy," The Brookings Institution, Washington, D.C. (1975).
8. D. M. Kreps, Corporate culture and economic theory, *in* "Perspectives on Positive Political Economy," (J. E. Alt and K. A. Shepsle, Eds.), Cambridge Univ. Press (1990).
9. P. Milgrom, Good news and bad news: representation theorems and applications, *Bell J. Econom.* **12**, 380–391 (1981).
10. A. Newell and H. A. Simon, "Human Problem Solving," Prentice–Hall, Englewood Cliffs, N.J. (1972).
11. D. Sappington, Incentives in principal–agent relationships, *J. Econom. Perspectives* **5**, 45–66 (1991).
12. K. Segerson and T. Tietenberg, "The structure of penalties in environmental enforcement: an economic analysis," paper presented at the Environmental Law Workshop, Woods Hole, MA (1991).
13. B. Sinclair-Desgagné, "The first-order approach to multi-task principal–agent problems," INSEAD (1991).
14. T. Tietenberg, Economic instruments for environmental regulation, *Oxford Rev. Econom. Pol.* **6**, 17–33 (1990).
15. A. P. Xepapadeas, Environmental policy under imperfect information: Incentives and moral hazard, *J. Environ. Econom. Management* **20**, 113–126 (1991).
16. A survey of industry and the environment, *The Economist*, 23 (1990).
17. "Companies' Organization and Public Information to Deal with Environmental Issues," United Nations Environment Program (1990).
18. "Corporate response to the environmental challenge," McKinsey and Company (1991).
19. "Environmental auditing," Technical Report Series No. 2, International Chamber of Commerce (1991).
20. "Managing the Environment. The Greening of European Business." Business International Ltd, London, UK (1990).

[28]

JOURNAL OF ENVIRONMENTAL ECONOMICS AND MANAGEMENT 5, 26–43 (1978)

Firm Behavior Under Imperfectly Enforceable Pollution Standards and Taxes

Jon D. Harford [1]

Department of Economics, State University of New York at Albany, New York 12222

Received November 2, 1976

Assuming expected profit maximization, the behavior of the firm under imperfectly enforceable pollution standards is examined. Among other results, it is found that cost subsidies can reduce the size of violation and amount of wastes, and that the shape of the expected penalty function determines the direction of the firm response to tighter standards. Under imperfectly enforceable pollution taxes, it is found, among other results, that the firm's actual level of wastes is independent of proportional changes in the expected penalty for pollution tax evasion, and that the marginal cost of actual waste reduction equals the unit tax on reported wastes. Some normative aspects of the results are discussed.

I. INTRODUCTION

In this paper we present simple theoretical models of firm behavior under imperfectly enforceable pollution standards and pollution taxes, respectively. Such models are motivated by the nature of the current and past pollution control efforts in this country and encouraged intellectually by coexisting literatures on approaches to externalities and the economics of crime. Although the model presented here has some strong simplifying assumptions, it presents a point of view distinctly different from most treatments of externalities and allows derivation of some interesting results.

The approach to pollution control adopted by the federal government has by and large been based upon the use of standards, both ambient and emission (or effluent). In the case of air pollution, the Clean Air Act of 1970 gives the Environmental Protection Agency the power to set emission standards on a wide variety of sources with the goal of meeting ambient air quality standards. In the case of water pollution, The Federal Water Pollution Control Act Amendments of 1972 give broad power to the EPA to set effluent standards on practically all industrial and municipal sources with the goal of meeting ambient water quality standards, which, according to legislative intent, are to become increasingly stringent.

In the case of both air and water pollution standards it has been the case that these standards have not always been complied with. For support of this understatement with respect to water pollution controls one should consult Kneese's

[1] The author gratefully acknowledges support from the Research Foundation of the State of New York for the researching and writing of this paper.

26

paper in Bain [2] and Zwick's [15] even less charitable view. In fact, major aspects of the recent water pollution legislation have been aimed at reducing the costs of enforcing pollution standards by making the EPA's power greater and more sharply defined. Both pieces of legislation mentioned have clauses setting penalties for violations of standards.

It may be that once full monitoring of pollution sources is effective, the probability of successfully violating standards without being caught and punished will be close to zero. However, past experience does not support the view that perfect enforcement of pollution standards would be of trivial costs in comparison with the benefits.

Because of the above considerations it is of interest to examine the behavior of a profit maximizing firm under imperfectly enforced pollution standards. However, it is also of interest to examine the behavior of the firm under imperfectly enforceable pollution taxes. One reason would be the substantial literature that has been devoted to proving that under perfect competition the tax per unit of pollution equal to marginal damages with no compensation of victims is the Pareto optimal policy (and that in general virtually all other solutions cannot attain Pareto optimality). Baumol and Oates [4] present a general proof of this proposition and discuss some of the substantial literature that led up to this conclusion. A second reason for studying the effect on the firm of an imperfectly enforceable pollution tax is that it provides us with a comparison set of results for the standards case. We find, in fact, that the pollution tax case does have interesting properties, the main one being a type of separability between the amount of pollution released and the amount of unreported pollution, or tax evasion, that occurs.

If there is a second major federal approach to pollution control it would be that of providing cost subsidies. For firms this is done mostly through the special form of accelerated depreciation allowed on pollution control capital as provided for in the Tax Reform Act of 1969. Such subsidies, by themselves, would seem to provide no positive incentive for pollution control. They merely reduce the cost of doing something which adds nothing to revenues. However, in a world of imperfectly enforceable controls, we can show that such subsidies reduce the tendency to violate standards.

Lastly, we will point out some implications of the analysis for the characterization of optimal levels of pollution control. Since firms that violate standards and evade taxes change the trade-offs in the economy one would expect a change in the set of efficient resource allocations in comparison with a world of perfectly costless enforcement or perfectly compliant firms.

In Section II we will examine the expected profit maximizing firm's response to imperfectly enforceable pollution standards. In Section III we examine the firm's behavior under evadeable pollution taxes. In Section IV we discuss some of the normative implications of the analysis, and Section V presents some conclusions.

II. THE FIRM RESPONSE TO STANDARDS

Let us consider a firm which produces a good x from which it receives revenue $R(x)$. The firm may be either a perfect competitor of some form of imperfect competitor. The costs C of producing the good x depend, not only on the quantity of x, but also on the amount of emitted wastes w. So by definition $C \equiv$

$C(x, w)$. This cost function is taken to reflect all of the process changes which would reduce created wastes and any efforts which would "neutralize" wastes already created. Emitted wastes are those which are created and not neutralized, and all references to firm wastes (or actual wastes) will mean emitted wastes as here defined.

We will indicate partial derivatives by subscripts. Thus, marginal output costs and marginal revenue are C_x and R_x, respectively. Under usual circumstances we would expect $|R_{xx} - C_{xx}| < 0$, where the subscripts indicate second partial derivatives, which we will assume exist for all of the relevant functions unless noted otherwise. In order for the firm to find it profitable to emit pollution in the absence of any external constraints it must be true that C_w is negative over a range from zero pollution up to some level w_o. "w_o" is the point where $C_w = 0$, and would be the amount of pollution released in the absence of any controls. We may assume that C_w becomes positive after this point; that is, it would increase costs to release any more pollution. Of course, the firm would never operate on this portion of its cost curve. Consistent with this description, we will assume that $C_{ww} > 0$. If costs were separable in pollution and output we would have $C_{xw} = 0$, but it is more reasonable to assume that $C_{xw} = C_{wx} < 0$, over the relevant range of the cost function.

Now assume that a particular pollution control board (PCB) or the EPA imposes a particular allowable level of released wastes, labeled w_s, which is less than what the firm would have otherwise chosen. The PCB wishes to see this standard complied with. Therefore it sets penalties in the form of fines (and perhaps prison terms) and expends resources on activities which create a probability of discovering and convicting standards violators greater than zero but less than one. For present purposes, the discovery of a violation, any adjudication process, and the receipt of a penalty for the violation will be treated as one event. For any size of violation, $v = (w - w_s)$, the firm sees the probability of discovery and punishment and the level of fine as parameters.

Generally we would expect that any fine f will be a positive function of the size of violation. It is also plausible that the probability p of detecting a violation will be a positive function of the size of violation. Thus we would suppose that f_v and p_v are ordinarily greater than zero. We will be assuming risk neutrality for the firm, so that the expected fine, $g = pf$, is all that matters to the firm. To facilitate the analysis and reduce the number of terms let us write $g = F_s PG(v)$, where F_s and P are shift parameters for the general size of fines under standards and the general level of the probability of detection and punishment of violators, respectively. Because of their nature, the terms F_s and P can be taken to be equal to one without loss of generality. Thus, $G(v) = f(v)p(v)$, and it is assumed to be continuous and twice differentiable; and

$$G' = f'p + p'f \tag{1a}$$
$$G'' = f''p + 2p'f' + p''f, \tag{1b}$$

where the primes indicate the order of differentiation.

It is conceivable that there could be a discontinuity at $v = 0$. This would require that there be a sudden jump from zero to some positive value in the probability of punishment as one goes from no violation to one of arbitrarily small size *and* that there be some minimum level of fine. We mention this as a possibility since such a discontinuity would tend to produce extreme solutions.

As a practical matter, however, laws are seldom enforced in such a way as to produce a significant discontinuity of this sort. Before going on it should be noted that both f'' and p'' may be negative and still have G'' positive.

The structure of the fines is something to be determined by the PCB and presents an optimization problem. It may be suggested that characterizing the optimal structure of fines is an interesting problem and quite possibly a very difficult one in general. The probability of detection as a function of the size of violation might be either something given in the nature of the production function of detection, or a result of conscious allocation of resources. For a situation involving different types of crimes, the latter interpretation is more plausible (and the concept of crime size less clear) while for crimes of the same type this writer would lean toward the former interpretation. In either case, we assume the firm views these functions as given, and, given risk neutrality, is concerned only with the product of these functions and its derivatives.

In order to complete our picture, let us assume that the firm receives a subsidy at a rate b, where b is between zero and one, for the costs of eliminating wastes. In reality subsidies usually apply only to explicit capital expenditures on pollution control and seldom or never to the costs of all potential ways of reducing pollution, which would include process changes and non-capital inputs into waste neutralization. Indeed, it would be difficult to divide costs between output and pollution control if costs are not additively separable in the technical sense. We will ignore this practical difficulty and assume the total subsidy is $b(C(x, w) - C(x, w_0))$ where w_0 is taken to be the level of wastes that the firm would produce in the absence of pollution controls and cost subsidies. Since w_0 is a constant, we may define $C(x, w_0) = C^o(x)$.

Given the previous definitions, expected profits, Π_s, is of the following form:

$$\Pi_s = R(x) - C(x, w) + b[C(x, w) - C^o(x)] - PF_sG(v). \qquad (2)$$

The firm is assumed to choose output x and the level of waste w so as to maximize expected profits. The assumption of risk neutrality is not perfectly general, of course, but even for nonrisk neutral firms risk neutrality is a good approximation for relatively small risks, which is probably a fair description of most risks associated with violating pollution standards.[2]

The first order necessary conditions for an interior maximum are:

$$R_x - C_x + b(C_x - C_x^o) = 0 \qquad (3a)$$

$$-(1 - b)C_w - PF_sG' = 0. \qquad (3b)$$

Condition (3a) says that marginal revenue should equal marginal costs less the subsidized fraction of the difference between the marginal costs of output at the base level of waste and the marginal costs of output at the chosen level of

---

[2] Given the basic tenet of portfolio theory that one can evaluate the risk of any asset only in the context of its relationship to the array of all assets, it is difficult to see how one would make rigorous sense out of the idea of either risk averseness or risk preference for a firm that may be owned by thousands of stockholders with differing portfolios and preferences for risk. The present assumption can at least be defended on the grounds that in the long run (assuming survival of the firm and some version of the law of large numbers) the risk neutral firm will have had greater average profits than either the risk averse or risk preferring firm. To put it another way, if all profits were reinvested for all firms at some constant rate of return, then the risk neutral firms would eventually be the largest.

wastes. If the cost function is separable in x and w, the last term is zero and we get the conventional answer that marginal revenue should equal marginal cost. If $C_{xw} < 0$, we would expect $(C_x - C_{x^o}) > 0$, and therefore, that marginal revenue would be less than marginal costs. With such a cost function and subsidy one ends up subsidizing output as well as pollution control. This is no doubt part of the reason why most actual cost subsidies are for types of pollution control equipment that have no obvious connection to the production of output. This, of course, means that pollution control is not achieved in a least cost manner from society's viewpoint.

Condition (3b) tells us that the unsubsidized fraction of marginal costs of reducing wastes should equal the increase in the expected monetary value of the fine with a one unit increase in actual waste. Clearly it may often be the case that equality in (3b) does not hold. For example, if both the fine and the probability of receiving a penalty are fixed and independent of the violation size, then $G' = 0$, and $C_w = 0$ will appear to be the maximizing condition, which is what it would be in the absence of standards. However, if the expected fine is high enough one will get the other "corner" solution where $w = w_s$.

The two possibilities for the case just mentioned are illustrated in Fig. 1. We have drawn costs as a function of wastes while holding output constant. For simplicity we have assumed a zero subsidy in this and following figures. Two different fixed sizes of expected fine are indicated by G_1 and G_2. The level of costs under complete compliance is indicated by C_2, and under no effort at compliance the costs are C_1. Since $C_1 + G_1 < C_2$, the firm makes no effort at pollution control under expected fine G_1. Under expected fine G_2, the firm chooses zero violation since $C_1 + G_2 > C_2$.

In general, the possibilities for corner solutions depend upon the relative heights and shapes of the cost and expected fine functions, as well as the levels of w_s and b. Figure 2 shows a case where an interior solution, characterized by $w_o > w^o > w_s$, exists. The full compliance level of costs OM_2 is greater than the costs at the minimum point of the $C + G_3$ curve where wastes are w^o and costs are at height OM_1. It is due to the increasing nature of the expected fine and the decreasing marginal costs of waste prevention that w^o is to the left of w_o.

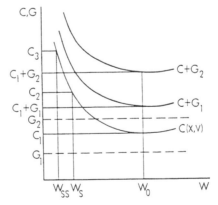

Fig. 1. Choice of violation size with constant expected fine.

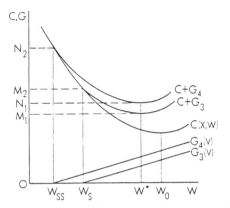

Fɪɢ. 2. Choice of violation size with a constant slope expected fine function.

Careful perusal of variations on Fig. 2 indicates that as long as $G'' \geqq 0$, a (further) requirement for a positive violation level is that

$$-C_w \big|_{w=w_s} > G' \big|_{v=0},$$

given a zero subsidy and the other assumptions on the cost function. Since the marginal costs of reducing wastes decrease as wastes increase, while the slope of the expected fine function is here assumed to be non-decreasing, we must have marginal costs of waste reduction at the allowed level of wastes greater than the rate of increase in the expected penalty with respect to the first unit of illegal wastes in order to have a positive violation.

To gain further insight into the possibilities of corner solutions we shall now consider the second order conditions for a maximum. These conditions are that the determinant of the Hessian matrix,

$$H_s = \begin{vmatrix} (R_{xx} - (1-b)C_{xx} - bC_{xx}{}^{o}) & -(1-b)C_{xw} \\ -(1-b)C_{xw} & -(1-b)C_{ww} - PF_sG'' \end{vmatrix} > 0, \quad (4)$$

and that the terms on the diagonal be negative. This expression has the common implication that the cross partials of the cost function should not be too large in comparison with the repeated partials in w and x. A further implication is that G'' cannot be too strongly negative or the second order conditions will not hold. One may note that the linear expected fine function in Fig. 2 implies that its second derivative is zero, and thus it is consistent with satisfaction of the second order conditions. The satisfaction of the second order conditions ensures us that there is at most one interior maximum, which will be the global maximum if there is no discontinuity in the expected fine function at the zero violation level.

We may now list and discuss a number of comparative static results showing rates of change of controlled variables with respect to parameter changes. In the following results we have taken advantage of the normalization $P = F_s = 1$.

$$\frac{\partial w}{\partial P} = \frac{\partial w}{\partial F_s} = \frac{[R_{xx} - (1-b)C_{xx} - bC_{xx}{}^{o}]G'}{H_s} < 0 \qquad (5a)$$

32   JON D. HARFORD

$$\frac{\partial w}{\partial w_s} = \frac{-[R_{zz} - (1-b)C_{zz} - bC_{zz}{}^\circ]G''}{H_s} \gtrless 0, \qquad \text{as} \qquad G'' \gtrless 0. \qquad (5b)$$

$$\frac{\partial w}{\partial b} = \frac{-[R_{zz} + (1-b)C_{zz} - bC_{zz}{}^\circ]C_w + (1-b)C_{zw}(C_z - C_z{}^\circ)}{H_s} < 0 \qquad (5c)$$

$$\frac{\partial x}{\partial P} = \frac{\partial x}{\partial F_s} = \frac{(1-b)C_{zw}G'}{H_s} < 0 \qquad (5d)$$

$$\frac{\partial x}{\partial w_s} = \frac{-(1-b)C_{zw}G''}{H_s} \gtrless 0 \qquad (5e)$$

as $G'' \gtrless 0$, assuming $C_{zw} < 0$.

$$\frac{\partial x}{\partial b} = \frac{(C_z - C_z{}^\circ)[(1-b)C_{ww} + G''] - (1-b)C_w C_{zw}}{H_s} \gtrless 0. \qquad (5f)$$

These results are based on the assumption that both the first and second order conditions for a maximum are satisfied. On this basis the relationships in (5a) tell us that wastes created and released will decrease with increases in the level of fines or increases in the probability of detection at all sizes of violation. In other words, a general increase in expected fines decreases the actual level of wastes and the size of violation which the firm chooses. We also find from (5b) that an increase in the cost subsidy will reduce the size of any standards violation, a result due basically to the fact that the relative cost of complying with the standard has been reduced. This provides a rationale for the commonly used cost subsidies, which economists have tended to ignore or presume to be useless in situations where none of the benefits of reducing pollution are internal to the firm.

On the output side we find that increased fines and probability of detection tend to reduce the expected profit maximizing level of output to the extent that marginal costs of output are increased by reductions in the emitted wastes. Thus, without separability we would tend to find that limitations on the amount of wastes would tend to reduce the profit maximizing level of output for the firm.

Interestingly enough, the direction of change in waste and output levels with respect to a change in the allowable wastes under the standard is dependent upon G'', the second derivative of the expected fine function with respect to the size of the violation. If the slope of G is increasing, then waste and output will move in the same direction as allowable wastes under the standard. This is the more intuitive result. One may conjecture that it is likely though not certain that this would be the most relevant case. If $G'' > 0$, it can be plausibly argued from the relevant expressions that the increase in wastes will be less than the allowable wastes under the standard, i.e., that the size of violation will decrease with a loosening of the standard.

However, there is nothing in the second order conditions which definitely rules out the case where the slope of the expected fine function is decreasing. In this case we find that making the standard marginally stricter will increase the actual amount of waste released by the firm. To give the reader a better feel for the situation we refer to Fig. 2, which illustrates what occurs in the border line case where the slope of the expected penalty function is constant. The shift in the wastes standard from w_s to w_{ss} has the effect of shifting upward the curve representing the sum of costs plus the expected penalty to $C + G_4$. Thus costs at w° increase from a height of OM_1 to ON_1. However, due to the parallel nature of the shift, ON_1 now represents the new minimum of the sum of costs plus expected penalty. In other words, there is no change in actual wastes emitted. The size of the violation increases in the same magnitude as the reduction in the allowed level of wastes.

Finally, we note that the change in output with respect to a change in the cost subsidy rate is in general ambiguous in sign. This is due to the existence of two counteracting effects. The subsidy reduces the cost of pollution control which allows more output for the same pollution control efforts and total costs. However, because the subsidy encourages less pollution it indirectly increases the marginal costs of output. Which effect dominates depends upon the size of practically all of our parameters and the first and second partial derivatives of the cost function. In the special case of separability between output and wastes in the cost function, we can assert that the change in output with respect to a change in the subsidy rate will be zero.

All of these comparative static results are under the assumption that the first order conditions are satisfied with equality and the second order conditions hold. If, on the other hand, we do not have an interior solution, we would still expect that small changes in the parameter values would have either no effect, or effect waste and output levels in the direction indicated by the comparative statics results.

This expectation tends to hold up even with respect to variations in w_s in the case where complying with the standards maximizes expected profits. One might be tempted to expect that permitted and actual wastes would move in the same direction regardless of the shape of the expected fine function. That this may not be the case is illustrated within Fig. 1. When the allowable level of wastes is decreased from w_s to w_{ss} under the assumption that the expected fine is a constant G_2 we find that overall expected costs are now lowest at the level of w_o wastes, i.e., costs at full compliance with the stricter standard C_3 are greater than $(C_1 + G_2)$. Thus, in this case, increasing the strictness of the standard increases the actual level of wastes from w_s to w_o. To put it another way, the firm may go from full compliance to totally ignoring the standard as it increases in strictness if the expected fine does not increase rapidly enough with the size of violation. For the special case illustrated by Fig. 1, Anderson [1] preceded us in discovering this result.

All of the previous results are about individual firms and, therefore, one needs to know something about industry structure in order to discuss effects of pollution standards on aggregate of firms. For firms operating in an environment where expected long run economic profits should be zero, average costs become important for determining the amount of wastes which the industry will generate since average costs determine the size of the industry in terms of output. Thus, even in the seemingly perverse case where tightening standards increases the firm's waste creation, we would expect (and Figs. 1 and 2 confirm) that the firm's average costs would increase. This would tend to reduce the size of a competitive industry and work in the opposite direction from the "counterproductive" effect on the individual firm.

In the same vein, an increase in the cost subsidy rate for pollution control would be expected to reduce the average costs of output for the firm even though the marginal effect is uncertain. Thus the size of the industry would be larger with the subsidy, ceterus paribus, even though each individual firm may produce less pollution. It is possible that the larger industry size would increase pollution relatively more than any reduction in the pollution per unit of output that the cost subsidy might induce. The overall effect will depend upon a number of factors including the elasticity of the market demand for the output.

III. EVADEABLE POLLUTION TAXES

It is of interest to compare the kind of results we obtain with pollution standards with those we would get in the case where a tax per unit of waste emitted is imposed. To make the comparison reasonable, it is assumed that the pollution tax is evadeable. This may be interpreted as meaning that actual wastes released, w, are greater than the reported wastes, w_r, where the tax is directly on reported wastes only. The firm's costs C are a function of ouput x and the actual wastes released w in the same manner assumed in the case of standards. (We no longer assume any cost subsidy.) All of the remarks about the signs of the derivatives of the cost function still apply.

The unit tax t is applied to w_r so that total pollution tax paid equals t times w_r. In order to prevent the firm from evading all taxes by reporting zero released wastes, the pollution taxing agency imposes penalties for evasion of pollution taxes. The size of the evasion will be measured by $v = (w - w_r)$, where we re-use the letter v to reflect the size of violation in this different case. There is some justification for this procedure. In the cases of standards, the reported wastes would always be w_s since there is no incentive to have actual or reported wastes less than w_s, and reported wastes above w_s would be asking for a penalty. So even in the standards case the violation size v could be interpreted as the difference between actual and reported wastes.

Again we will use the function $G(v) = pf$ to reflect the shape of the expected fine as a function of the violation size where p and f have the same meaning as before. The shift parameter of the probability of detection and punishment will again be labeled P, which will be taken to be equal to one. We shall distinguish between two components of what we shall call the expected penalty function. The first component consists of a fine shift parameter F, also taken to be equal to one, multiplied by $PG(v)$ to reflect the expected fine size. The second component will be the tax that will be collected on unreported but discovered wastes. Thus, in addition to any expected fine and the tax on reported wastes, there is an expected tax on unreported wastes of $Ppvt$. To reduce the number of terms in later expressions we will define $B = pv$. We note that

$$B' = p + p'v \qquad (6a)$$

$$B'' = 2p' + p''v. \qquad (6b)$$

If we again use $R(x)$ to denote total revenue as a function of output, we may write expected profits as

$$\Pi_t = R(x) - C(x, w) - tw_r - FPG(v) - PB(v)t. \qquad (7)$$

The firm now has three control variables, x, w, and w_r, with which it attempts to maximize expected profits. Accordingly, we will have three first order necessary conditions for a maximum, which are:

$$R_z - C_z = 0, \tag{8a}$$

$$-C_w - FPG' - PB't = -C_w - G' - B't = 0, \tag{8b}$$

$$-t + FPG' + PB't = -t + G' + B't = 0. \tag{8c}$$

Condition (8a) informs us that marginal revenue should equal marginal cost, the same type of result one would get if there were no pollution tax (although the level of marginal costs will be affected by the actual level of pollution control). Condition (8b) indicates that the marginal cost of reducing actual waste should be equal to the marginal increase in the expected penalty (expected fine plus additional tax) with a unit increase in violation size. Condition (8c) tells us that the increase in expected penalty from a unit decrease in reported wastes should equal the unit tax on reported wastes.

If these conditions hold, it is obvious but interesting the $-C_w = t$. In other words, the marginal cost of actual pollution reduction by the firm will equal the unit tax on *reported* pollution. This implies that the actual waste level of the firm does not directly depend upon the size of our shift parameters for the fine or the probability of discovery of the violation. Furthermore, if the expected punishment levels are generally so high that the firm maximizes expected profits by having actual and reported wastes the same $[(G' + B't) > t$ at all positive v would cause this situation], then it would still clearly be true that the marginal costs of waste reduction would equal the unit tax on reported wastes, since reported and actual wastes are the same.

At the other extreme of corner solutions, however, this result would not hold. If expected punishment costs are sufficiently low and increase so slowly that the unit tax on reported wastes is everywhere greater than the increase in expected penalty with respect to violation size, one would get a case where reported wastes are equal to zero, and the marginal costs of actual waste reduction would be equal to the smaller magnitude of the increase in expected penalty costs.

Practically speaking, corner solutions where reported wastes are zero are unlikely. It would be irrational to set penalties so low that no pollution tax at all was collected. Moreover, if it is obvious that every firm generates some wastes, reporting zero wastes would be a clear signal of a violation.

Deriving the Hessian matrix of second order derivatives we can characterize the second order conditions for a maximum in terms of restrictions on the signs of its determinant and principal minors. We have the restrictions that the determinant

$$H_l = \begin{vmatrix} (R_{zz} - C_{zz}) & -C_{zw} & 0 \\ -C_{zw} & (-C_{ww} - A) & A \\ 0 & A & -A \end{vmatrix} < 0, \tag{9}$$

where $A = G'' + B''t$, and that all the diagonal elements must be negative, while all second order minors should be positive. These restrictions imply that the expression represented by A must be positive for an interior maximum to be unique, and a global maximum if there is no discontinuity in the expected penalty function at the zero violation level. Thus, the expected penalty function must have an increasing slope to satisfy the second order conditions and provide the possibility of an interior solution in the pollution tax case, whereas the expected fine function did not have to have a positive curvature in order for an interior solution to exist. It is still true, however, that the expected fine function, by itself, does not have to have a positive curvature if the function B has a sufficiently strong positive curvature.

Although we have a larger matrix, the comparative statics analysis turns out to be simpler in some ways than in the standards case. Equating the shift parameters to one we have the following results:

$$\frac{\partial w}{\partial P} = \frac{\partial w}{\partial F} = 0 \tag{10a}$$

$$\frac{\partial x}{\partial P} = \frac{\partial x}{\partial F} = 0 \tag{10b}$$

$$\frac{\partial w}{\partial t} = \frac{-(R_{zz} - C_{zz})A}{H_t} < 0 \tag{10c}$$

$$\frac{\partial x}{\partial t} = \frac{-C_{zw}A}{H_t} < 0 \tag{10d}$$

$$\frac{\partial w_r}{\partial F} = \frac{G'[(R_{zz} - C_{zz})C_{ww} + C_{zw}^2]}{H_t} > 0 \tag{10e}$$

$$\frac{\partial w_r}{\partial P} = \frac{(G' + B't)}{G'} \left(\frac{\partial w_r}{\partial F} \right) > 0 \tag{10f}$$

$$\frac{\partial w_r}{\partial t} = \frac{-[(R_{zz} - C_{zz})C_{ww} + C_{zw}^2](1 - B')}{H_t} + \frac{\partial w}{\partial t} < 0 \tag{10g}$$

$$\frac{\partial v}{\partial t} = \frac{\partial w}{\partial t} - \frac{\partial w_r}{\partial t} > 0 \tag{10h}$$

First of all, quite consistently with previous statements, we find that the actual level of wastes released is independent of the level of fines or probability of detection and punishment. Small changes in P and F have no effect on w. Furthermore, small changes in P and F have no effect on the firm's level of output. This follows quite intuitively; if the optimal level of actual waste does not change, then costs as a function of output would not change, and, therefore, marginal revenue would equal marginal costs at the same output level both before and after any shift in fines or the probability of punishment.

We find that actual wastes decline as the tax on reported wastes is increased. This occurs because of the connection of both actual and reported wastes to the rate of change in the expected penalty with respect to violation size. If costs are not separable in output and wastes, we find that output is a negative function of the tax on wastes. If costs are separable, then the firm's output will not be directly affected by the tax on reported wastes.

Result (10f) indicates that increases in the probability of capture at all sizes of violation will increase reported wastes. Since actual wastes are not affected by such a parameter shift, this implies a reduction in violation size at the same absolute rate. Using this result and explicitly writing out and simplifying (10f), one can derive the following relationship between the elasticity of violation size with respect to the general probability of detection, and the elasticity of the slope of the expected penalty function.

$$\left(\frac{\partial v}{\partial P}\right)\frac{P}{v} \equiv e_{P_v} = \frac{-1}{e_{vz}} \equiv -\left(\frac{Z}{Z'}\right)\frac{P}{v} \tag{11}$$

where $Z = G' + B't$, and use is made of the normalization $P = 1$.

This relationship says that the elasticity of violation size (in absolute value) with respect to a proportional shift in the probability of punishment is inversely related to the relative positive curvature of expected punishment as a function of violation size. To say it another way, if the rate of increase in the increase in expected punishment with violation size is relatively small, then a proportional increase in the probability of being fined at all levels of punishment will elicit a relatively large decrease in violation size.

The rate of change in reported wastes with respect to a change in the tax on reported wastes may be broken up into two parts: A term exactly equal to the effect on actual wastes, and a term which reflects the direct tax and penalty effects of a tax shift. It has already been determined that an increase in the pollution tax reduces actual wastes, and accordingly this component reduces the reported wastes to the same extent. The other term in (10g) will be negative if $(1-B')$ is positive. An examination of the first order condition (8c) indicates that this must be true if the slope of the expected fine function is positive, which we assume to be the case. Since both terms are negative, reported wastes will decline to a greater extent than actual wastes when the pollution tax is increased. This, as (10h) indicates, means that an increase in the pollution tax increases the size of the violation that the firm chooses. As one might expect, raising the tax rate causes more evasion and presumably a more difficult enforcement problem.

Given the insensitivity of actual wastes to the general level of fines and the probability of punishment, one might be tempted to conclude that their exact magnitude is of little or no consequence over a considerable range. This idea has a rough validity for all firms in the short run, and firms in noncompetitive industries in the long run. It is definitely not true in the long run for those firms in industries where expected long run economic profits tend toward zero. In these industries it is the long run average costs which determine the industry size in terms of output. A lower level of enforcement of payment of pollution taxes may not change the actual level of wastes emitted by any firm, but the implied lower level of some combination of pollution taxes, and penalty payments will imply lower average costs and more firms in the industry. In this regard it should be noted that, ceteris paribus, firms with the same expected fine function will have lower average costs under the imposition of standards than under the imposition of pollution taxes. It is roughly on this point, as Buchanan and Tullock [7] explain, that standards fail to produce a Pareto optimal (or pollution tax equivalent) resolution to a pollution externality under perfect competition.

This model may not be irrelevant to the current situation in the area of water pollution control. The recent federal legislation states that municipalities must charge fees to firms for accepting their wastes, fees which must be designed to cover the costs of treating those wastes. If the firm were not legally allowed to channel any wastes directly into public waters, then, with proper interpretation, this model might almost directly apply. The fee for municipal acceptance of firms' effluent would correspond to the tax on reported, in this case delivered, wastes. Violations might take the form of illegal dumping of wastes in the river, or channeling more wastes to the municipal treatment plant than the firm pays for in effluent fees. (More complicated and subtle versions of the same problem might involve the firm reneging on pretreatment requirements or negotiating reductions in other types of taxes (such as the local property tax) to compensate for the newly instituted effluent fees.) Even if the firm is allowed to release a fixed level of wastes, w_s, directly into public waters, we may simply define the size of violation as $v_o = [w - (w_s + w_r)]$, and the previous results are not qualitatively altered, although chosen values of the control variables would differ.

If one modifies the definition of the violation size as just suggested, one gets the additional results that $\partial w_r/\partial w_s = -1$, and $\partial w/\partial w_s = 0$. The former result says that there would be a one to one reduction in reported waste (subject to effluent fees by the municipality) with any increase in the allowable standard of wastes disposable directly into public waters. Put another way this result is quite striking. It tells us that making the standard stricter would not increase the violation size, but simply cause an increase in the amount of wastes going to the treatment plant and *paying* the effluent fee. The latter result indicates that changing the standard would have no effect on the actual amount of wastes created, which is in contrast to the pure standards case, but quite in spirit with our results in the pollution tax case.

Inevitably one runs into difficulty in trying to fit reality to a model. Relevant here is the fact that legislatively described fines are often in terms of days of violation rather than being geared to the quantity of pollutant. Even so, or perhaps because of this, the actual level of fines is left to the discretion of the courts within the constraint of maximum and minimum fines.[3] This methodology could produce its own peculiarities. Another complicating factor is the actual variety and interactability of real world wastes. And, with at least two, and perhaps three, levels of government (with varying views of benefits and costs) participating in pollution control with regard to the same set of potentially geographically mobile firms, the actual situation becomes extremely complicated.

IV. NORMATIVE IMPLICATIONS

The analysis heretofore has been predictive in the sense of simply drawing out the implications for a firm's behavior under expected profit maximization. We will now attempt a discussion of the normative implications of an analysis that includes the possibility of evading the pollution taxes, which are imposed pre-

[3] For example, within Title III, Section 309 of the Federal Water Pollution Control Act Amendments of 1972 it is stated that any firm or municipality (or the persons responsible) violating an applicable pollution standard ". . . shall be punished by a fine of not less than $2250 nor more than $25,000 per day of violation, or by imprisonment for not more than one year, or by both." Except for the expectation that the courts will use their discretion wisely in gearing fines to the actual damages caused by the pollution, this would appear to be an unsatisfactory way to relate fines to the severity of the violation.

sumably to reduce an externality to an efficient level. The discussion will speak in terms of a pollution tax rather than a standard, since it is the former which has been most favored by economists for its potential to improve efficiency. We will use the intuitive, but not completely rigorous, concepts of damages and costs measured in dollar values. Hopefully this problem may be treated within a rigorous general equilibrium framework at some future point. At this point, however, the technical problems with more precise approach appear too complex to explore.

The results derived indicate that the larger the pollution tax, the greater the evasion of that tax. This points to increasing enforcement costs as attempts are made to reduce actual pollution. We therefore have increasing marginal costs of reducing pollution which include additional costs of enforcing pollution taxes (or standards). This would argue for a reinterpretation of the rule that pollution should be eliminated to the point where marginal damages are equal to the marginal costs of treatment or prevention. Under present assumptions we should have marginal damages equal to the marginal costs of treatment or prevention plus the marginal costs of enforcing the treatment or prevention.

This point can be argued somewhat more rigorously by a modification of Becker's [5] approach to the optimal level of crime and law enforcement, which is to minimize the sum of the damage caused by crime plus the costs of capturing and punishing criminals. In this context we wish to minimize the sum of the damages caused by pollution plus the costs of treating or preventing pollution plus the costs of enforcing the taxes or other instruments by which pollution control is encouraged.

Let us define the value of the damage done to society by an aggregate level of waste W as $D(W)$. On the basis of previous analysis we would expect aggregate wastes to be a decreasing function of the unit pollution tax t. It is also consistent with previous results to assert that the aggregate level of waste should be a decreasing function of the general level of fines F and the general probability of detection and punishment for an evasion of pollution taxes, P. (P and F are now viewed as control variables of the government rather than as parameters.) Previous results indicate that the later two factors may ordinarily have no effect on the actual level of the firm, but, as we have argued before, expected punishment levels will effect the number of firms (and thus aggregate wastes) in a competitive industry via the effects on expected average costs. The costs of removing or preventing wastes are an increasing function of how much lower wastes are than than they would be if no efforts were made to incur or prevent wastes. This base level of wastes will be labeled W_o, while the aggregate costs of neutralizing wastes will be $S = S(Y)$, where $Y = (W_o - W)$.

Costs of enforcement of the pollution tax, E, will be an increasing function of the probability of capturing violators, P, and an index of the number of violators, V. No analysis of the shape of the function relating the probability of detection and the size of the violation will be attempted. However, the variable V is taken to reflect some index of the number and size of violations, which would in general depend upon the ease of detecting violations of different sizes. The level of the index of violations will itself be a negative function of the probability of detection and punishment, and the general level of fines F, but it should be a positive function of the height of the unit tax t on reported wastes. Thus, we have $E = E[P, V(P, F, t)]$.

Lastly, consistent with Becker, it will be assumed that there are social costs of punishment. These social costs are assumed to be a linear function of the general probability of punishment times the general magnitude of punishment. Specifically, total punishment costs are hPF, where h is a parameter which reflects general population size and other factors. Again, for simplicity, we avoid consideration of the structure of fines over the violation size. Given the fact that punishment in this context is likely to be a fine, the assumption that there are non-negligible social costs to punishment may not be very appealing to some. Harris [9] suggests another rationale for social costs of punishment: that of wrongful punishment. Occasionally, the innocent may be fined and this can be looked upon as causing social costs. On a more general note Stigler [12] points to the idea of marginal deterrence rather than social costs of punishment as the critical factor in the determination the appropriate level of punishment for any crime. This may be interpreted as saying that the higher the punishment for a given type of crime the greater the occurrence of other types of crimes with their attendant social costs. Further discussion of how, and how well, these rationales serve in arguing for social costs of punishment is not warranted here, except to say that Stigler's line of thought quickly leads us back to the tricky problem of determining an optimal structure of punishment.[4] For the moment we will simply work with our simple assumption.

We may now write the sum of pollution damage costs plus the costs of treatment/prevention plus the costs of enforcement of pollution taxes as

$$L = D(W) + S(Y) + E(P, V) + hPF. \tag{12}$$

The social goal is to minimize total social costs through the choice of F, P, and t. The three first order necessary conditions for minimum social costs are: where the subscripts indicate partial derivatives.

$$(D_W - S_Y)W_t + E_V V_t = 0 \tag{13a}$$

$$(D_W - S_Y)W_P + E_P + E_V V_P + hF = 0 \tag{13b}$$

$$(D_W - S_Y)W_F + E_V V_F + hP = 0, \tag{13c}$$

For present purposes it is condition (13a) that is of main interest, and so we will forego any comment on conditions (13b) and (13c), except to say that they indicate the usual balancing of marginal damages and marginal costs as they relate to the variables P and F, respectively. Condition (13a) can be re-arranged to read

$$D_W = S_Y - (E_V V_t)/W_t. \tag{14}$$

---

[4] If one examines the concept of marginal deterrence closely it becomes evident that, unless there is some maximum feasible costless punishment, the need for marginal deterrence alone will not always lead one to a finite punishment for any crime. The basic reason for this is that one can mathematically conceive of infinite *rates* of punishment. In the context of this paper, one might have an infinite amount of fine per unit of violation and thus preserve marginal deterrence without having finite punishments. Clearly, in any practical sense, there is a maximum feasible costless punishment. Assuming for the moment that fines have zero social costs, it is clear that any fine cannot exceed the economic worth of the firm or individual fined. Given that there is a maximum feasible costless punishment, the concept of marginal deterrence should enable one to reasonably consider what an optimal structure of punishments should be.

Given our assumptions about the various functional relationships involved, the expression to the right of the minus sign is itself negative. This implies that the damages caused by one more unit of waste should be greater than the costs of physically eliminating the unit by the amount of the extra expenditure on enforcement induced by the increase in tax evasion caused by the additional pollution tax required to reduce pollution by the last unit.

Of course, a full assessment of the normative implications of pollution controls must recognize that the imperfections may be just as great in enforcing non-pollution types of taxes and controls. Imperfections in controls are simply another set of factors to consider in forming policy toward the existence of pollution types of externalities, along with market structure, various uncertainties, and dynamic considerations.

V. CONCLUSION

This paper has presented models of an expected profit maximizing firm under imperfectly enforceable pollution standards and under imperfectly enforceable pollution taxes. We have found that under standards increasing the expected level of the penalty will reduce the level of wastes released by the firm, but that increasing the strictness of the standard will only reduce the firm's wastes if the slope of the expected fine function with respect to the size of the standards violation is increasing. We have also found that the use of cost subsidies for pollution control expenses can serve the useful function of reducing the level of the firm's violation of the standard and thus its actual level of wastes.

Under imperfectly enforceable pollution taxes the analysis has established the neat result that the marginal costs of pollution reduction by the firm will be equated to the constant rate of the pollution tax as long as the slope of the expected penalty function is increasing. This implies that on the firm level the amount of pollution tax evasion is independent of the actual wastes, which are determined by the pollution tax rate. The level of tax evasion is determined by equating the increase in expected penalty with respect to a unit reduction in reported wastes to the decrease in tax paid. Therefore, the general level of fines and the probability of punishment affect reported wastes but not actual wastes. Further results indicate that increasing the tax rate on reported wastes will reduce actual wastes released by the firm, but it will reduce reported wastes even more, implying an increase in pollution tax evasion.

We have also suggested that, with proper reinterpretation of the violation size as the difference between actual wastes and an allowable standard of wastes plus an amount of (reported) wastes going to a fee charging treatment plant, one can apply the pollution tax model to current situation in the area of water pollution. Under this reinterpretation, changes in the allowable standard of wastes has no effect on wastes leaving the firm, or violation size, but are offset exactly by the amount of wastes going to the treatment plant.

In considering the implications of these results for policy it has been mentioned that effects of various policies on the expected average costs of firms are important and may not always work in the same direction as the effect on the individual firm's marginal decision. Furthermore, in Section IV, we suggested a reinterpretation of the intuitive rule of efficiency in pollution control that the marginal damages of pollution should equal the marginal costs of eliminating pollution. The version which this paper suggests is that the marginal damages should be equal to the marginal costs of physically eliminating the pollution plus the additional costs of enforcing any pollution control instrument to the extra degree required to induce the unit reduction in pollution.

Throughout this paper we have assumed that the pollution control agency faces firms which believe that they cannot effect the penalty structure or subsidy rate of the agency. If the number of different polluters is small, this assumption may not be valid. Firms may adopt various kinds of strategic behavior and threats in order to affect the penalty structure. They may threaten to go out of business, or to move to a region where the agency does not have authority. Firms might collude to violate pollution control laws simultaneously, thereby overloading the agency's ability to enforce its laws with any effectiveness. The agency may be able to adopt various counterstrategies in this regulatory duopoly game. The outcome of such a situation is not easily determined, but there are counterparts to it in many types of regulation. Assuming that it is desirable to minimize the possibilities of such strategic behavior on the part of firms, it appears preferable to formulate and administer pollution controls at the most inclusive level of government possible.

This paper raises a number of (other) unsolved problems. One of the more interesting ones is the development of the concept of an optimal structure of penalties. This will likely involve an examination of both the structure of fines and penalties per se, and the technical and resource allocational nature of the probability of detecting and punishing violations of different sizes. Another problem would be to develop a more explicit and rigorous model to analyze the optimal level of pollution, pollution taxes, and enforcement of pollution taxes than that developed in Section IV of this paper. That analysis might be interpreted as suggesting that one should have a lower pollution tax rate in a case where there are enforcement problems than when there are not, but such a conclusion is not explicit, and we suspect that it may not always be true. A more explicit analysis would clarify this and other significant ambiguities.

REFERENCES

1. Robert J. Anderson, Jr., Environmental management economics, *in* "Regional Environmental Management: Selected Proceedings of the National Conference," (Coate and Bonner, Eds.), pp. 201–208, John Wiley and Sons, New York (1975).
2. J. S. Bain (Ed.), "Environmental Decay," Little Brown and Company, New York, (1973).
3. W. J. Baumol, On taxation and the control of externalities, *Amer. Econ. Rev.* 62, 307–322 (1972).
4. W. J. Baumol and W. E. Oates, "The Theory of Environmental Policy," Prentice-Hall, Englewood Cliffs (1975).
5. G. Becker, Crime and punishment: An economic approach, *J. Polit. Econ.* 6, 169–217 (1968).
6. J. M. Buchanan, External diseconomies, corrective taxes, and market structure, *Amer. Econ. Rev.* 59, 174–177 (1969).
7. J. H. Buchanan and G. Tullock, Polluters profits and political response: Direct controls versus taxes, *Amer. Econ. Rev.* 65, 139–147 (1975).
8. M. E. Darby and E. Karni, Free competition and the optimal amount of fraud, *J. Law Econ.* 67–88 (1973).
9. J. R. Harris, On the economics of law and order, *J. Polit. Econ.* 78, 165–174 (1970).

IMPERFECTLY ENFORCEABLE STANDARDS 43

10. S. Rose-Ackerman, The economics of corruption, *J. Pub. Econ.* 4, 187–203 (1975).
11. W. Schulze and R. C. d'Arge, The Coase proposition, information constraints, and long-run equilibrium, *Amer. Econ. Rev.* 64, 763–772 (1974).
12. G. J. Stigler, The optimum enforcement of the laws, *J. Polit. Econ.* 78, 526–536 (1970).
13. U. S. Congress, "Clean Air Amendments of 1970," (PL91-604) USGPO, Washington, D. C. 1970.
14. U. S. Congress, "The Federal Water Pollution Control Act Amendments of 1972," (PL92-500) USGPO, Washington, D. C. (1972).
15. D. Zwick, "Water Wasteland," Bantam, New York (1972).

[29]

Journal of Public Economics 37 (1988) 29–53. North-Holland

ENFORCEMENT LEVERAGE WHEN PENALTIES ARE RESTRICTED

Winston HARRINGTON*

Resources for the Future, Inc., 1616 P Street, N.W., Washington, DC 20036, USA

Received January 1987, revised version received April 1988

1. Introduction

In the United States, empirical studies of the enforcement of continuous compliance with environmental regulations, especially air and water pollution regulations, have repeatedly demonstrated the following:

(i) For most sources the frequency of surveillance is quite low.

(ii) Even when violations are discovered, fines or other penalties are rarely assessed in most states.

(iii) Sources are, nonetheless, thought to be in compliance a large part of the time.[1]

The evidence for the apparently low level of enforcement activity is found in a number of case studies of state and local enforcement of air and water quality regulations conducted jointly by the Environmental Protection Agency (EPA) and Council on Environmental Quality (CEG), together with periodic surveys of implementation of air and water quality legislation conducted by the General Accounting Office (GAO).[2] These surveys show

*This research was partially supported by the National Science Foundation, Grant No. PRA 8413311. I would also like to thank, without implicating, Carol A. Jones, Arun Malik, Wallace E. Oates, Paul Portney, Clifford S. Russell, Robert Schwab and Michael Toman for discussion and comments on earlier drafts.

[1] Because several different environmental regulations or requirements are enforced, 'compliance' is a somewhat ambiguous term, and depending on context may mean (1) initial compliance, in which the objective is to force the regulated source to install the abatement equipment that enables the regulation to be met, (2) compliance with reporting requirements, which aim to force the firm to meet regulatory requirements on the reporting of data to the authorities, or (3) continuous compliance, which attempts to force the source to keep emission discharges within regulatory limits. There have been well-publicized examples of heavy penalties for violation of environmental regulations, but these are largely confined to the first two categories. Continuous compliance with discharge limitations, which is of course what determines environmental quality, is hardly enforced via penalties at all.

[2] The results of these case studies are summarized in U.S. EPA (1981a, 1981b) and in Russell, Harrington and Vaughan (1986) and U.S. GAO (1982). Also see U.S. GAO (1979), Downing and Kimball (1982), Willick and Windle (1973), Environmental Law Institute (1975), Harrington (1981) and Russell (1982), and Abbey and Harrington (1981) for reports on some other enforcement surveys.

30   W. Harrington, Enforcement leverage

Table 1

State enforcement activity (annual averages, 1978–1983).

| State | NOVs issued | Civil actions brought | Penalties | | |
			Number of penalties assessed	Average of penalties assessed	Average penalty per NOV
Colorado	124	3.6	0.5	$120	$0.48
Connecticut	800	2.3	21.5	363[a]	9.75
Indiana	59[b]	NA	21	4,050[c]	1,442.37
Kentucky	194	5.2	5.2	2,520[d]	67.52
Massachusetts	NA	0	0	0	0
Minnesota	41	NA	10	10,900	2,658.53
Nebraska	59[e]	NA	0.2	200	0.67
Nevada	31.5	0.3	2.3	45	3.33
New Jersey	1,167	NA	350	1,430	428.45
Oregon	197	NA	30.7	705	110.00
Pennsylvania	NA	NA	176[f]	1,480	–
Rhode Island	5	7.2	0	0	0
South Carolina	68[g]	5.5	2.2	24,250[h]	NA
South Dakota	17	1.2	0.3	1,000	20.00
Tennessee	193[i]	8.4	0	0	0
Virginia	161	7.8	3	200	3.79
Wisconsin	80.5[j]	13.5	7.7	7,951	760.00

Note: NA = not available.

[a]Refers to amounts assessed; over the same period actual collections were 62 percent of assessments.

[b]Includes both NOVs and compliance orders.

[c]Excludes one penalty of $415,000, which was cancelled when company bought equivalent amount of air pollution equipment.

[d]Excludes performance bonds (two required for total of $45,000 in last five years).

[e]Excludes Lincoln and Omaha.

[f]Only data from the last quarter of 1983 was readily available. This figure is an extrapolation to an annual average.

[g]No NOVs were issued before 18 April 1983. From 1 July 1983 to 31 March 1984, 51 NOVs were issued.

[h]Excludes one fine of $1,700 per month until compliance was restored and one fine of $250,000 dropped in lieu of a donation of a like sum to a technical college.

[i]Does not include NOVs from Continuous Monitoring Data, which were extremely numerous.

[j]1980–1983 only.

that the typical source can expect to be inspected on the order of once or twice each year. When a violation is discovered, by far the most common response is for the agency to send the firm a Notice of Violation (NOV), ordering it to return to compliance but taking no further action. The reticence to use penalties is exhibited in table 1, which reports the results of an RFF survey of state-level enforcement activity conducted in 1984. As shown, most states levied penalties for less than 5 percent of the notices of violation (NOVs) issued each year, although some states, such as Pennsylvania and

New Jersey, made frequent use of penalties. Also, the size of the penalties is generally very small.

Many of these same case studies also provide evidence for the third assertion. States are required to report annually on the compliance status of major stationary sources of air pollution, and these reports routinely indicate that well over 90 percent of all sources are in compliance. Spot-checks by the EPA (which used to have a program to reinspect a random 10 percent of sources reported by the states to be in compliance) have shown that most of such sources actually were in compliance. During 1977, for example, 22 percent of such sources were found to be in violation by the EPA [U.S. GAO (1982)]. In water quality enforcement, surveys by GAO in 1978 and 1980 revealed that 55 percent of industrial waste dischargers and 34 percent of municipal waste dischargers had committed 'serious' violations of permit requirements at some time during the past year [U.S. GAO (1982)]. Note, however, that while violations are frequent, they are far from universal.

Other surveys have disclosed similar results. The EPA–CEQ studies mentioned above collectively estimated that the sources examined were in violation of the standards about 9 percent of the time [U.S. EPA (1981b)]. However, the authors of these studies made the assumption that those sources were in compliance except when available plant data indicated otherwise, so that they probably underestimated the time in violation. In New Mexico, about 30–40 percent of all surveillance visits by agency personnel uncovered emission violations [Harrington (1981)].

But even though empirical support for all three statements can be found, together they seem mutually contradictory. Indeed, the truth of the three statements seems to violate both common sense and most existing models of environmental enforcement. For if enforcement activity is carried on at such a low level, and if violations are rarely punished even if discovered, why would any firm bother to comply? Reconciliation is indeed difficult in static economic models of environmental enforcement, in which the penalty facing the firm depends on the firm's rate of emissions, not on its previous compliance record.

All three observations, however, can be made consistent with a dynamic repeated-game model in which the regulated firm and the enforcement agency can react to previous actions by the other.

In this paper such a model will be described and its properties examined. The enforcement agency sets the parameters of the game, which in turn determine the average rate of compliance, the firm's expected compliance cost and the agency's expected enforcement cost. A key feature of this game is the assumption that the size of the penalty that the agency can impose in any period is restricted.

The interest in such a model is justified in part by the very low level of fines currently imposed for violations of continuous compliance. The rarity of penalties might be attributable to a number of causes. In most states there is a restriction on size of penalties that can be levied each day (usually $1000 or $5000). In addition, agencies can often cite violations occurring only on the day of surveillance events, even though the inspector may believe the violation has been occurring for some time.[3] Even when a maximum fine is not imposed by statute, there may be a practical or political limit to the size of penalties. Severe but rarely-imposed penalties might seem capricious and unfair. Also, there is an upper limit to the fine that can be imposed on any given firm such that the firm is not driven into bankruptcy [Braithwaite (1982)]. In any event, levying penalties can be a costly activity for the agency, just as surveillance is.

In section 2 the enforcement game is described and analyzed from the firm's point of view. The firm's optimum strategy is a two-state Markov decision problem, one for which parameters selected beforehand by the agency determine transition probabilities and payoffs, and hence the firm's compliance rate. The main result of this section is that the firm may have an incentive to comply with regulation even though its cost of compliance each period exceeds the expected penalty for violation, or even the maximum penalty that can be levied in any period. The property will be called 'leverage'. Section 3 discusses the optimum choice of parameters for the agency. It will be shown that when these optimum parameters are selected, no penalties are ever collected for violations. In addition, compared to state-independent enforcement schemes this model allows a target compliance level for less enforcement effort, although the advantage is not very great if the target compliance rate is high. In section 4 it is shown that introducing a third state into the model increases the cost-effectiveness off enforcement by the agency by turning the game into a 'threat game'. Finally, in section 5 these results will be discussed in light of the empirical findings that motivated the paper in the first place.

2. A dynamic model of firm behavior

The economic approach to enforcement began with the seminal article by Becker (1968). In contrast to the then-prevailing views in criminology, Becker assumed rational economizing behavior among criminals, whose expectations of gains from illicit activity had to be countered by an expectation that some violators would be caught and punished. The structure of penalties is thus an important determinant of the crime rate.

Unlike criminal activity, violation of environmental regulations is usually inadvertent rather than willful, especially after initial compliance has been

---

[3]In contrast, large fines have been assessed for violations of initial compliance or reporting violations, which are relatively easy to detect and also easy to infer more than one day's violation.

achieved. Nonetheless, plant operators have choices in maintenance and operation of abatement equipment, choices that can strongly affect the frequency, duration and magnitude of violations. Environmental regulations must therefore be continuously enforced, and the application of Becker's insights to this problem is straightforward.

Most of the early enforcement literature is static [e.g. Downing and Watson (1974), Harford (1978), Storey and McCabe (1980), Viscusi and Zeckhauser (1979), and Linder and McBride (1984)]. In a static analysis there is no way for the agency and the firm to react to each other's actions; with the expected penalty a function of the rate of violation, the firm makes a single choice of the rate of violation. Penalties for violating emission regulations are treated like any other cost, whereupon the optimal rate of violation of environmental regulations by a firm occurs when the marginal gain from violation equals to marginal expected penalty.

It is certainly consistent with this approach to consider enforcement a simple two-person game, as, for example, Brams and Davis (1983) have done in the context of arms control. The players of the game are the enforcement agency and a single regulated, risk-neutral firm. In a given play of the game, the agency chooses between two actions: to inspect or not to inspect. The firm also chooses between two actions: to comply or to violate the regulation. Compliance has the same cost for the firm regardless of what the agency does. This compliance cost can be avoided by violating the regulation, but if the agency inspects, thus discovering the violation, the firm must pay the cost to return to compliance plus a penalty. The agency, in contrast, is rewarded for every period that the source is in compliance (or penalized whenever the source is in violation). Thus, the agency's objective is to minimize the frequency of violation, subject to a fixed enforcement budget.

In a single play of this game the agency inspects the firm with a certain probability. If the agency announces beforehand what the inspection probability will be, the best strategy for the firm is nonrandom: the firm is better off complying with probability one if $pF > c$, where p is the inspection frequency, F the fine for violation, and c the cost of compliance. Otherwise it violates. Unfortunately, the expected penalty required to achieve this result seems to be very high, implying much higher inspection frequencies and penalties than actually experienced in environmental regulation.

But if the enforcement game is played repeatedly, the agency can alter the expected penalty and the inspection frequency based on the firm's past performance. Landsberger and Meilijson (1982) have shown, in a model of income tax compliance, that an enforcement regime in which the probability of an audit depends on the outcome of the most recent audit is more cost-effective than a system in which the audit frequency is independent of past audit outcomes. That is, greater tax revenues can be collected for the same level of enforcement resources. In Landsberger and Meilijson's model all taxpayers are placed in one of two states, with movement from one state to the other depending on the occurrence and outcome of an audit. Greenberg (1984) added to this model a third state, from which no escape is allowed, so that once in this state one faces a sure audit each period. He demonstrated a result that seems, as the author himself admitted, almost too good to be true: regardless of the agency's budget, the rate of violations can be made arbitrarily small (though greater than zero).

The free-lunch aspect of this model was attributed primarily to a hidden assumption, namely that the agency could determine with perfect accuracy whether the taxpayer was cheating. If false positives are possible, every taxpayer is eventually in the third state, forcing the agency to audit everyone every period. As Greenberg observed, to prevent this outcome some escape from the third state would have to be allowed, and this would impose some limits on the performance of the model. This false positive problem in Greenberg's model is examined in the context of environmental regulation in Russell (1984) and Russell, Harrington and Vaughan (1986).

There is another 'hidden' assumption that is also important in the performance of this model: Greenberg assumes that there is no maximum panalty, or at least that the penalty that the agency can impose exceeds the firm's per-period compliance cost. That is to say, if the firm is certain that an inspection will occur, the penalty for a violation is assumed large enough so that it will choose to comply.

In other words, it is assumed that the penalty can be made as large as is necessary. This recalls the treatment of the static (i.e. state-independent) model examined by earlier authors. In the static model, the penalty can always be made large enough so that the firm will always comply. (In the notation used earlier, the firm will always comply if $F > c/p$.) To be sure, the penalty must meet a much less stringent requirement, for Greenberg's model only requires $F > c$.

Here that assumption will be relaxed. The interest will be in the relationship between the firm's compliance cost and the average level of compliance that can be achieved when both enforcement budgets and the maximum feasible penalty are limited. It will also be evident just what the addition of a third state adds to the cost-effectiveness of enforcement.

Suppose, then, that the agency classifies firms into two groups, one of which faces more severe enforcement than the other. Each firm can then move from one group to the other based on its performance. More precisely, let G_1 and G_2 denote the two groups of firms, and suppose the inspection probability in G_i is p_i and the penalty for violation is F_i, with $p_1 < p_2$ and $F_1 < F_2$. Violations discovered in G_1 are punished by exile into G_2 and compliance discovered in G_2 is rewarded with the chance of a return to G_1. Let u denote the probability that a firm found in compliance in G_2 is returned to G_1. Thus, the agency and firm are players in a pair of linked

Table 2
Payoff matrices for the enforcement games.

	Group 1		Group 2	
	Comply	Violate	Comply	Violate
No inspection	c	0	c	0
Inspection	c	F_1, $\to G_2$	c $P(\to G_1) = u$	F_2

games with payoff matrices as shown in table 2 (only payoffs to the firm are shown).

Before examining the properties of this model several simplifying assumptions are to be noted. First, in many real-world applications the rate of emissions is continuous, and hence the degree of noncompliance is important. In these models, however, the firm has only two discrete choices: compliance or noncompliance. Second, the probability of an inspection is assumed not to be affected by the incidence of a violation. But often the occurrence of a violation can be detected, or at least suspected, offsite, raising the probability of an inspection. Third, it is assumed that an inspection will determine without error whether a violation has occurred. This assumption is discussed above.

From the firm's standpoint this monitoring and enforcement scheme poses a Markov decision problem, as discussed in Kohlas (1982). The firm moves from group to group according to transition probabilities that depend not only on the current state of the system but on the action taken during that period by the firm (i.e. to comply or not). In addition, each period the firm receives a payoff (or incurs a cost) that likewise depends on the state of the system and the action taken. A *decision* at any time is a map from states to actions, and a *policy* is a sequence of decisions over time.

Let r_{ti}^a denote the payoff if action a is chosen when in state i at time t. If these payoffs are constant over time, i.e.

$$r_{ti}^a = r_i^a, \quad t = 0, 1, 2, \ldots,$$

and if future payoffs are discounted by a discount factor β, $0 \leq \beta < 1$, the following ergodic theorem can be proved [see Kohlas (1982)]: (a) the expected gains of any policy over an infinite time horizon are finite, (b) the optimum policy is a *stationary* policy [i.e. if $\{f_0, f_1, \ldots\}$ is optimum then $f_n(i) = g(i)$ for some decision rule g], and (c) the optimum policy g is independent of the initial state of the system (although the expected gains of that policy are not).

Applied to the monitoring problem, the states are the two groups G_1 and

G_2. A policy for the firm is a map $f:\{1,2\}\rightarrow\{0,1\}$ of states into decisions to comply with or violate the regulations, and the firm's problem is to choose the policy that minimizes the expected cost over future periods. The stationary property means that there are four policies possible.

The transition probabilities p_{ij}^a are given as follows:

	$a=0$ (comply)			$a=1$ (cheat)	
	G_1	G_2		G_1	G_2
G_1	1	0	G_1	$1-p_1$	p_1
G_2	$p_2 u$	$1-p_2 u$	G_2	0	1

Denote the four available policies by $f_{00}, f_{01}, f_{10}, f_{11}$, where, for example, f_{00} is the policy of complying when in both states and f_{01} is the policy of complying if in G_1 and cheating if in G_2. Let $E^{ij}(m)$ denote the expected cost, in present value, of policy f_{ij} when initially in state m. By the stationary property, the expected present value must be the cost this period plus the expected present value discounted one period. For example, consider the present value of f_{00}, the policy of compliance in both groups. Thus,

$$E^{00}(1)=c+\beta E^{00}(1),$$

$$E^{00}(2)=c+\beta p_2 u E^{00}(1)+\beta(1-p_2 u)E^{00}(2).$$

Solution of these simultaneous equations gives the present value of the policy f_{00}. The four sets of simultaneous equations giving the present values of the four policies can be conveniently displayed as follows:

Group 1	Group 2
Comply(0): $E_1=c+\beta E_1$	$E_2=c+\beta p_2 u E_1+\beta(1-p_2)E_2$
Violate(1): $E_1=(1-p_1)\beta E_1+p_1(F_1+\beta E_2)$	$E_2=p_2 F_2+\beta E_2$

To evaluate policy f_{ij}, solve the system of equations formed by taking the ith equation column 1 and the jth equation of column 2.

The optimum policy g takes on values $g(k)=v_k$, $k=1,2$, that satisfy:

$$v_k=\min_{i,j} E^{ij}(k).$$

The ergodic theorem assures that such a policy exists.

To find v_1 and v_2, first find the solution of the four simultaneous equation systems for the $E^{ij}(m)$, which are given in table 3. Examination of these solutions reveals several characteristics this game, as follows.

Environmental Instruments and Institutions

Table 3

Expected cost of alternative policies.

Policy	$E^{ij}(1)$	$E^{ij}(2)$
f_{00}	$\dfrac{c}{1-\beta}$	$\dfrac{c}{1-\beta}$
f_{10}	$\dfrac{cp_1\beta + p_1 F_1(1-\beta+p_2 u\beta)}{(1-\beta)(1-\beta+p_1\beta+p_2 u\beta)}$	$\dfrac{c(1-(1-p_1)\beta)+\beta p_2 u p_1 F_1}{(1-\beta)(1-\beta+p_1\beta+p_2 u\beta)}$
f_{11}	$\dfrac{p_1 F_1(1-\beta)+\beta p_1 p_2 F_2}{(1-\beta)(1-(1-p_1)\beta)}$	$\dfrac{p_2 F_2}{1-\beta}$

Lemma 1. f_{01} *is never optimal.*

This follows because f_{01} is dominated by f_{00} when $c < p_2 F_2$ and by f_{11} when $c \geq p_2 F_2$.

Lemma 2. For each initial state k, the remaining $E^{ij}(k)$, considered as functions of the compliance cost c, are of the form $A + Bc$, where

for E^{00}; $A = 0, B > 0$,

for E^{10}; $A > 0, B > 0$,

for E^{11}; $A > 0, B = 0$,

Lemma 3. When $c = L_0 = p_1 F_1$, $E^{00} = E^{10} < E^{11}$, and the firm is indifferent between f_{00} and f_{10}.

L_0 is found by setting $E^{00}(2)$ equal to $E^{10}(2)$ and solving for c (the expected values for group 1 could also be used).

Lemma 4. Similarly, $E^{10} = E^{11} < E^{00}$ when $c = L_1$, where

$$L_1 = p_2 F_2 + \frac{p_2 \beta u(p_2 F_2 - p_1 F_1)}{[1-(1-p_1)\beta]}. \tag{1}$$

These results can be summarized in the following proposition.

Proposition 1. If g denotes the optimal policy, then

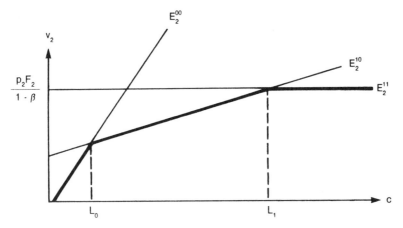

Fig. 1. Expected cost of the firm's optimum policy as a function of compliance cost.

$$g = \begin{cases} f_{00}, & \text{if } c \le L_0, \\ f_{10}, & \text{if } L_0 \le c \le L_1, \\ f_{11}, & \text{if } L_1 \le c. \end{cases}$$

Inasmuch as each f_{ij} is a linear function of c, the expected costs of the optimum policy, $v_1(c)$ and $v_2(c)$, are piecewise linear functions of c, as shown in fig. 1 for $k=2$ (for $k=1$ the threshold values L_0 and L_1 are the same but the expected cost is lower).

Since $p_2 F_2 \ge p_1 F_1$, the threshold L_1 must be at least as great as $p_2 F_2$, the expected penalty in Group 2, and will equal $p_2 F_2$ only in degenerate cases:

(i)   no monitoring in Group 2 $(p_2 = 0)$;
(ii) perfect myopia $(\beta = 0)$;
(iii) no escape from Group 2 $(u = 0)$;
(iv) no difference in expected penalty between the groups.

In all other cases we must have $L_1 > p_2 F_2$.

Proposition 1 classifies firms as f_{00}, f_{10} or f_{11} firms depending on its optimum strategy, which in turn depends on its compliance costs and the enforcement parameters chosen by the enforcement agency.

Let C_{ij} be the probability that an f_{ij} firm is in compliance. Then $C_{00} = 1$ and $C_{11} = 0$. An f_{10} firm is more interesting, because it complies only part of the time. When f_{10} is optimum, the system has the following transition matrix:

	G_1	G_2
G_1	$1 - p_1$	p_1
G_2	$p_2 u$	$1 - p_2 u$

In this case, the firm cheats when it is in G_1 and complies in G_2. The irony of this model is that the firms with the status of 'good guys' are the ones that can afford to cheat, while those with a 'bad guy' reputation comply until moved back into the G_1.

For an f_{10} firm the frequency of compliance in steady state is the stationary probability of being in G_2, or

$$C_{10} = \frac{p_1}{p_1 + p_2 u}. \tag{2}$$

Likewise, let I_{ij} be the probability that an f_{ij} firm is inspected. Clearly, $I_{00} = p_1$, and $I_{11} = p_2$, because f_{00} firms are eventually in Group 1 with probability one, and f_{11} firms are eventually in Group 2. Also,

$$I_{10} = \frac{p_1 p_2 (1 + u)}{p_1 + p_2 u}. \tag{3}$$

Note that $C_{00} > C_{10} > C_{11}$, but, since $p_1 < p_2$, $I_{00} < I_{10} < I_{11}$. Thus, firms with better compliance records are those with less frequent surveillance.

The expected fine per period paid by an f_{10} firm is $p_1 F_1 (1 - C_{10})$. If $F_1 = 0$, no fines will be collected, yet the firm will be in compliance part of the time. In a static model, by way of contrast, absence of fines requires either perfect compliance or perfect noncompliance. Furthermore, the firm's compliance cost c can easily exceed the expected fine in G_2 (which is $p_2 F_2$), because $L_1 > p_2 F_2$. Thus, in this dynamic model compliance can be achieved even though the expected penalty would not be large enough to ensure compliance in a static model. The reason, of course, is that compliance in G_2 allows the firm eventually to return to G_1, where the cost of cheating is much lower.

3. Formulating agency enforcement policy

In this section I describe the policy selection of an enforcement agency playing the two-group game described above, and consider the consequences of adding a third group as described by Greenberg and Russell. The enforcement agency wishes to minimize the resources devoted to monitoring and enforcement, consistent with achieving a target compliance rate by the regulated firm, assumed to have a known compliance cost c^*. To carry out this task it can modify five variables: the two monitoring frequencies, p_1 and p_2, the two penalties, F_1 and F_2, and the probability u of being allowed back into Group 1 after being found in compliance. $T = 1/p_2 u$ is the expected duration in Group 2, assuming compliance. Suppose the target long-run (i.e.

steady state) compliance rate is Z, and suppose that the maximum fine the agency is allowed to impose is F^*.

The agency has at least three objectives in setting up its enforcement policy. (Of course, it will in general be impossible to optimize on all three objectives simultaneously, and the agency's problem is one of constrained optimization, optimizing on one objective while meeting target levels on the other two.) It wishes to minimize the average inspection rate I^*, it wishes to make the average rate of compliance as high as possible, and it wishes to give firms with the highest compliance cost possible the incentive to comply. In the terminology of the two-group model discussed in the previous section, this last objective boils down to making L_2 as large as possible. This property will be called leverage, and is defined as follows. Suppose a target compliance rate Z is given, and let W be the set of all compliance costs such that any firm with a compliance cost $c \in W$ complies with probability exceeding Z. Now put $\Delta = \text{lub}(W) - F$ and call Δ the leverage of the enforcement policy for the compliance rate Z.

If the agency's target is perfect compliance ($Z = 1$), then the inspection resources must be large enough so that $I^* F^* > c^*$, just as in the static model. Also necessary for perfect compliance is for the maximum allowable fine to exceed c^*. At the other extreme, the agency cannot elicit any compliance at all if c^* exceeds $F^*/(1 - \beta)$, because the cost of complying even once exceeds the present value of all future fines. If $c^* < F^*/(1 - \beta)$, then some degree of compliance can be achieved.[4]

Suppose $c^* < F^*/(1 - \beta)$, so that some level of compliance is feasible and suppose the target $Z < 1$ is given. The agency's problem is to find parameters (p_1, p_2, F_1, F_2, u) to minimize:

$$I_{10} = \frac{p_1 p_2 (1 + u)}{p_1 + p_2 u} \tag{4}$$

subject to the following constraints:

$$L_1 = p_2 F_2 + \frac{p_2 \beta u (p_2 F_2 - p_1 F_1)}{[1 - (1 - p_1)\beta]} \geq c^*, \tag{5}$$

$$C_{10} = \frac{p_1}{p_1 + p_2 u} \geq Z, \tag{6}$$

$$0 \leq F_1 \leq F^*; \quad 0 \leq F_2 \leq F^*.$$

[4] Let $F_1 = 0$, $F_2 = F^*$, $p_2 = u = 1$, and $p_1 = \varepsilon$. By choosing ε small enough, L_1 can be made as close as desired to $F^*/(1 - \beta)$ and we can assure that $I_{10} < I^*$. Of course, that makes C_{10} pretty close to zero, though still positive.

Although a full solution to this optimization problem is complicated and would probably not provide much insight, several interesting observations can be made.

Observation 1. The optimum penalties F_1 and F_2 must be 0 and F^*, respectively. This follows because the penalties do not affect the values of I_{10} or C_{10}, and the choice of fines maximizes L_1 for a given choice of the other parameters.

Observation 2. At the optimum no fines are ever collected, because the firm violates only in Group 1, where $F_1 = 0$.

Observation 3. It also follows that an increase in the penalties does not affect the rate of compliance, unless it changes the compliance class to which the firm belongs (e.g. a f_{10} firm becomes a f_{00} firm).

Observation 4. If the compliance cost exceeds the maximum allowable fine $(c^* > F^*)$, then the maximum feasible target compliance rate Z is F^*/c^*. To see this, consider the unexpected role of p_2, the inspection frequency in Group 2. Increasing p_2 might seem to decrease the average rate of compliance, ceteris paribus, because a firm in G_2 is already in compliance, and an increase in the inspection rate in that group is just an enhanced opportunity to be returned to G_1 to do some cheating. Although this may suggest making p_2 as small as possible, there is a minimum value for p_2 consistent with Z. This minimum value may be found by considering the numerator and denominator in the expression (1) defining L_1:

Numerator: $p_2 \beta u (p_2 F_2 - p_1 F_1) \leqq p_2 u (p_2 F^*)$, since $\beta \leqq 1$, $F_1 = 0$ and $F_2 = F^*$.
Denominator: $1 - (1 - p_1)\beta = 1 - \beta + p_1 \beta \geqq (1 - \beta)p_1 + p_1 \beta = p_1$. Thus,

$$c^* \leqq L_1 \leqq p_2 F^* \left[1 + \frac{p_2 u}{p_1} \right] = \frac{p_2 F^*}{C_{10}} \leqq \frac{p_2 F^*}{Z}.$$

This means that p_2 must be at least $Z c^*/F^*$. In particular, if $c^* > F^*$, the maximum feasible target compliance rate Z is F^*/c^*, in which case $p_2 = 1$.

Observation 5. Let t be the compliance rate that can be achieved in a state-independent enforcement strategy with the same resources I^* that achieve a compliance rate Z in the two group model. Then $Z \geqq t \geqq Z^2/(1 + u)$.

This observation concerns the comparison of this state-dependent enforcement strategy with one that is independent of prior behavior by the firm. As noted above, this repeated game can achieve some degree of compliance for firms that would have incentives to be always in violation in the standard state-independent model. Even for those firms for which the standard model achieves some compliance, the optimum choice of parameters for the repeated game achieves a greater compliance rate for a given inspection budget. However, the greater the target compliance rate Z, the less the advantage of the repeated game over a state-independent model, because Z and Z^2 become increasingly close.

Consider first a *random* enforcement strategy, in which the agency inspects in each period with probability I^*, regardless of past outcomes. If the firm's compliance cost c^* is less than the expected penalty I^*F^*, then the firm complies at all times; if $c^* > I^*F^*$, then the firm never complies. With a nonrandom strategy the agency can achieve partial compliance with a state-independent model even when $I^*F^* < c^* < F^*$, in the following way. The agency announces that for some fraction t of the year, inspections will occur with a frequency of c^*/F^*, and the rest of the year no inspections will take place. Assuming the firm knows which is which, its incentive will be to comply during the first interval and violate during the second. If the average inspection rate for the year is I^*, the average compliance rate is:

$$t = I^*F^*/c^*.$$

To compare this level of compliance to that achieved with equal inspection resources in the two-group model, note that from (4) and (6):

$$I^* = Zp_2(1 + u),$$

and from Observation 4 above:

$$c^* \leq p_2 F^*/Z.$$

Substitution into the expression for t yields:

$$t \geq Z^2(1 + u). \tag{7}$$

This expression says that the closer Z gets to 1, the less difference there is between Z and t. The bound (7) is fairly tight when β is close to unity (i.e. a low discount rate) and optimal values for the parameters p_1, p_2 and u are chosen. For example, suppose target enforcement rate $Z = 0.8$, compliance cost $c^* = 50$ per month, the maximum penalty $F^* = 100$, and the discount rate is 10 percent per year (which makes $\beta = 0.992$ per month). Then the optimum values of p_1, p_2 and u (per month) are approximately:

$$p_1 = 0.05, \quad p_2 = 0.41, \quad u = 0.03,$$

and the minimum inspection frequency is $I^* = 0.341$. Applying these resources in a state-independent model we would have $t = 0.68$ and $Z^2(1 + u) = 0.66$.

To summarize the distinction between the state-independent and state-dependent models discussed above, it has been shown that the state-independent model can offer an incentive for partial compliance as long as compliance cost c^* does not exceed the maximum per-period penalty F^*. The compliance rate that can be achieved by the agency depends on the rate of inspection, but it can be as high as 1 if inspections are sufficiently frequent. The state-dependent model can achieve partial compliance even if $c^* > F^*$; in fact, c^* may be as large as $F^*/(1 - \beta)$ before the firm can be given no incentive to comply at all. When $c^* > F^*$, however, Z must be less than F^*/c^*.

Besides having greater leverage, the state-dependent model is more cost-effective than the state-independent model when $c^* < F^*$. Unfortunately, as the desired compliance rate approaches unity, this advantage begins to disappear. This raises the question of what the addition of a third group might accomplish in this same context of restricted penalties.

4. The advantages of a three-group model

As noted above, Greenberg proposed a third, absorbing state for his tax compliance model. The strength of this model was that it allows the agency to threaten a noncomplying firm with a dire result (eternal compliance) without having to commit the vast resources (constant surveillance) that would ordinarily be necessary to achieve that result. By making such a threat, arbitrarily good compliance could be achieved with arbitrarily small budgets. However, that outcome depends on the assumption that penalties are large enough to induce compliance when an inspection is certain (i.e. $F^* > c$). In this section the cost-effectiveness and leverage of this model are examined when that assumption no longer holds.

It will be assumed that, as in Greenberg's model, the third state is an absorbing state, from which no escape is possible and which faces inspections with certainty. The fine in the third group F_3 cannot exceed the maximum allowable fine F^*; in fact, it will be assumed that the penalty in Groups 2

Table 4

Expected cost of alternative policies: Three-group model.

Policy	$E^{ijk}(1)$	$E^{ijk}(2)$	$E^{ijk}(3)$
f_{000}	$\dfrac{c}{1-\beta}$	$\dfrac{c}{1-\beta}$	$\dfrac{c}{1-\beta}$
f_{100}	$\dfrac{cp_1\beta+p_1F_1(1-\beta+p_2u\beta)}{(1-\beta)(1-\beta+p_1\beta+p_2u\beta)}$	$\dfrac{c(1-(1-p_1)\beta)+\beta p_2up_1F_1}{(1-\beta)(1-\beta+p_1\beta+p_2u\beta)}$	$\dfrac{c}{1-\beta}$
f_{110}	$\dfrac{(1-(1-p_2)\beta)(1-\beta)p_1F_1+(1-\beta)p_1p_2\beta F_2+p_1p_2\beta^2c}{(1-(1-p_1)\beta)(1-(1-p_2)\beta)(1-\beta)}$	$\dfrac{p_2F_2(1-\beta)+p_2\beta c}{(1-(1-p_2)\beta)(1-\beta)}$	$\dfrac{c}{1-\beta}$
f_{111}	$\dfrac{(1-(1-p_2)\beta)(1-\beta)p_1F_1+(1-\beta)p_1p_2\beta F_2+p_1p_2\beta^2F_2}{(1-(1-p_1)\beta)(1-(1-p_2)\beta)(1-\beta)}$	$\dfrac{p_2F_2(1-\beta)+p_2\beta F_2}{(1-(1-p_2)\beta)(1-\beta)}$	$\dfrac{c}{1-\beta}$

and 3 will both be the same, or $F_2=F_3$. Analogous to the two-group case, let $f_{ijk}(i,j,k=0,1)$ denote the eight possible policies for the firm, and $E^{ijk}(m)$ the expected cost of f_{ijk} when m is the initial state ($m=1,2,3$). As before, the $E^{ijk}(m)$ are found by solving the system of simultaneous equations formed by taking the appropriate equation from the table below (the ith equation of the first column, the jth equation of the second, and the kth of the third):

Group 1	Group 2	Group 3
Comply: $E_1=c+\beta E_1$	$E_2=c+\beta p_2uE_1+\beta(1-p_2u)E_2$	$E_3=c+\beta E_3$
Violate: $E_1=(1-p_1)\beta E_1$		

$$+p_1(F_1+\beta E_2) \quad E_2=p_2(F_2+\beta E_3)+(1-p_2)E_2 \quad E_3=F_2+\beta E_3$$

Four of these policies are clearly nonoptimal regardless of the enforcement parameters or the firm's compliance cost: f_{001}, f_{010}, f_{101} and f_{011}. For f_{001} and f_{101}, in fact, the absorbing group is never entered because the firm complies in Group 2. The policies f_{010} and f_{011} are extensions of the dominated policy in the two-group model and are also dominated in this model. For the remaining four policies, the expected costs in each initial state are given as functions of c and the agency enforcement parameters in table 4.

Unlike the two-group case, the steady-state compliance and inspection frequencies for the other four policies depend on the initial state. If a firm starts in Group 3, it can never get out regardless of the policy adopted. In addition, a firm using policy f_{110} or f_{111} is eventually in Group 3 with certainty; the inspection rate is 1 and the compliance rate is either 1 (f_{110}) or 0 (f_{111}). The stationary probabilities for the other two policies are similar in form to those encountered in the two-group model. For f_{000} the inspection rate is p_1 and the compliance rate is 1, while for f_{100} the inspection rate is $I_{10}=p_1p_2(1+u)/(p_1+p_2u)$ and the compliance rate is $C_{10}=p_1/(p_1+p_2u)$.

To find the firm's best policy as a function of its compliance cost c, proceed as in the two group case, finding the point of indifference for each pair of policies. The results can again be summarized in a series of lemmas.

Lemma 5. The $E^{ijk}(m)$ are of the form $A + Bc$, where

for f_{000}: $A = 0, B > 0$,

for f_{100}: $A > 0, B > 0$,

for f_{110}: $A > 0, B > 0$,

for f_{111}: $A > 0, B = 0$,

Lemma 6. When the E^{ijk} are evaluated for particular values of c, we have the following:

for $c = p_1 F_1$: $E^{000} = E^{100} < E^{110}$,

for $c = F_2$: $E^{110} = E^{111}$,

for $c = L_2$: $E^{100} = E^{111}$,

where

$$L_2 = \frac{p_2 F_2}{1 - (1 - p_2)\beta} + \frac{\beta p_2 u(p_2 F_2 - p_1 F_1)}{(1 - (1 - p_1)\beta)(1 - (1 - p_2)\beta)} = \frac{L_1}{1 - (1 - p_2)\beta}. \qquad (10)$$

Again, the expected cost is a piecewise linear function. If $L_2 \geq F_2$ the expected cost functions for the optimal policy, v_1 and v_2,[5] consist of three linear segments, as shown in fig. 2(a). The policy f_{110} is dominated and is never optimal. On the other hand, if $L_2 < F_2$, as shown in fig. 2(b), the optimal policy consists of four segments. f_{110} is no longer dominated. To summarize:

Proposition 2. (a) If $L_2 \geq F_2$, then the optimum policy g is:

$$g = \begin{cases} f_{000}, & \text{if } c \leq L_0, \\ f_{100}, & \text{if } L_0 \leq c \leq L_2, \\ f_{111}, & \text{if } L_2 \leq c. \end{cases}$$

[5]The expected cost function v_3 for when Group 3 is the initial state is ignored; it is not very interesting since escape from Group 3 is impossible.

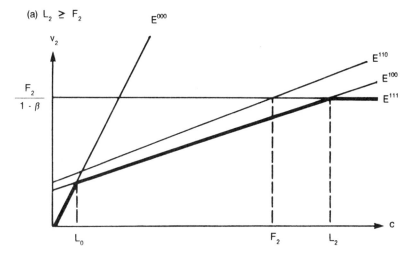

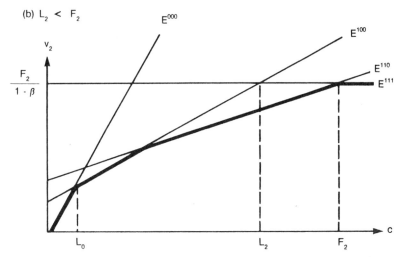

Fig. 2. Expected cost of the firm's optimum policy as a function of compliance cost: three-group model.

(b) If $L_2 < F_2$, then the optimum policy is:

$$g = \begin{cases} f_{000}, & \text{if } c \leq L_0, \\ f_{100}, & \text{if } L_0 \leq c < w, \\ f_{110}, & \text{if } w \leq c < L_2, \\ f_{111}, & \text{if } L_2 \leq c, \end{cases}$$

Table 5

Inspection resources required to achieve target compliance
rates: Two-group vs. three-group models.

I.	Maximum fine, F^*	100
	Compliance cost, c^*	150
	Target compliance rate, Z	0.6
	Discount factor, β	0.992

Two groups	Three groups
$p_1 = 0.062$	$p_1 = 0.078$
$p_2 = 0.943$	$p_2 = 0.103$
$u = 0.044$	$u = 0.503$
$I = 0.591$	$I = 0.093$

II.	Maximum fine, F^*	100
	Compliance cost, c^*	100
	Target compliance rate, Z	0.9
	Discount factor, β	0.992

Two groups	Three groups
$p_1 = 0.082$	$p_1 = 0.069$
$p_2 = 0.908$	$p_2 = 0.074$
$u = 0.010$	$u = 0.103$
$I = 0.826$	$I = 0.074$

where w is the value of c where $E^{100}(1) = E^{110}(1)$ *and* $E^{100}(2) = E^{110}(2)$.

Given a firm with compliance cost c^*, the agency's problem, as in the two-group case, is to find the optimum parameters so that the inspection budget I is minimized subject to meeting the steady-state target compliance frequency Z. Naturally, the agency wishes to avoid setting parameters that might give the firm an incentive to adopt the policy f_{111}. This is the worst of all worlds for the agency, because it must inspect every period and yet the firm continues to violate. The policy f_{110} is almost as bad, for constant inspections are again required. The implication is that the agency must choose parameters to ensure that L_2 exceeds both F_2 and the compliance cost c^*.

The addition of a third group can result in a spectacular reduction in the minimum resources required to achieve a given level of compliance. This advantage in cost-effectiveness is illustrated by two numerical examples, shown in table 5. In the first, $F^* = 100$, $c^* = 150$, $\beta = 0.992$ and the target compliance rate Z is 0.6. The group fines are set so that, for the two-group model, $F_1 = 0$ and $F_2 = F^* = 100$, and for the three-group model, $F_1 = 0$ and $F_2 = F_3 = F^* = 100$. As shown, the steady-state inspection rate I falls from 0.591 to 0.093 when the third group is added, primarily due to the sharp decline in p_2, the inspection rate in the second group. This represents a reduction of nearly 85 percent. This advantage remains even when the target compliance rate is high. In the second example, $Z=0.9$, $c^*=100$ and the inspection rate required with three groups is 91 percent less than that required with two groups.

Yet even though the three-group model is much more cost-effective than the two-group model, the leverage is the same. Just as in the two-group model, the maximum feasible compliance frequency Z that can be achieved when the firm's compliance cost is c^* and the maximum penalty is F^* is F^*/c^*. To see this, consider expression (10) defining L_2. Since $1-(1-p_2)\beta>p_2$,

$$\frac{p_2 F_2}{1-(1-p_2)\beta}<F_2,$$

and

$$\frac{\beta p_2 u(p_2 F_2 - p_1 F_1)}{(1-(1-p_1)\beta)(1-(1-p_2)\beta)}<\frac{\beta u p_2 F_2}{1-(1-p_1)\beta}.$$

In addition, $1-(1-p_1)\beta>p_1$ and $p_2 u/p_1=(1-Z)/Z$ therefore,

$$\frac{\beta u p_2 F_2}{1-(1-p_1)\beta}<\left[\frac{1-Z}{Z}\right]F_2.$$

Hence, $L_2<F_2+F_2(1-Z)/Z=F_2/Z$.

This is why, in example II of table 5, the compliance cost was assumed to be 100 and not 150 as in example I. With $F^*=100$ and $c^*=150$, a compliance rate of 0.9 is not feasible in either the two-group or the three-group model.

5. Conclusion

In this paper the behavior of enforcement agencies and regulated firms has been analyzed in a repeated game setting in which each can react to the previous actions of the other. It has been argued that the two-group state-dependent model described in sections 2 and 3 offers an explanation for a number of characteristics of the current enforcement of air and water quality regulations, some of which are rather difficult to explain using the standard, static model. In addition, these adaptive models, especially the three-state model described in section 4, are more cost effective than the nonadaptive or state-independent models that comprise the bulk of the economic literature on enforcement of environmental regulations. In effect, these models work by offering a bribe and threatening a penalty at the same time.

The salient behavioral characteristics of the state-dependent enforcement model are, first, that firms are divided into classes depending on compliance cost. Low-cost firms (f_{00}) are always in compliance, and high-cost firms (f_{11}) are always in violation. In the middle is the class f_{10}, consisting of firms that move in and out of compliance, depending on the results of the most recent inspection. In fact, studies of compliance do show that many firms are apparently always in compliance, and there are a few firms that are found to be in violation at nearly every inspection and that are subject to incessant complaints of nearby residents. However, it is not clear that compliance cost is the most important determinant of a firm's behavior.[6] Among other determinants is the firm's financial health: firms on the brink of bankruptcy, or plants about to be closed by parent firms, appear much more likely to commit violations.

Second, in these models at least partial compliance is achieved without ever having penalties imposed on violators. As shown above, this requires the penalty in G_1 (i.e. F_1) to be set to zero. As noted in the Introduction, compliance without penalties seems to be a common characteristic of air and water quality enforcement by state governments in the United States. Note, however, that if $F_1 = 0$, then the group of perfectly-complying firms consists only of those with zero compliance costs. It would not be unheard of for the relevant compliance costs to be zero, since to achieve compliance the firm may have installed abatement equipment with very low operating cost or modified the production process to eliminate residuals generation.

It should be noted also that there are other explanations of compliance without penalties, some of which would apply to particular firms, but hardly to all firms. An elaborate model is not required to explain the compliance of firms with zero compliance cost or firms that for whatever reason may be inherently law-abiding, for example. A competing explanation with a wider application is the existence of other penalties for noncompliance besides money fines. As suggested in Downing (1984) and Harrington (1985), a noncomplying firm may face bad publicity; it may be ordered to conduct costly stack tests; it may be forced to attend time-consuming meetings or conduct maintenance operations at inconvenient times. Such informal incentives may often be important, but are difficult to identify. No one has demonstrated empirically that an unofficial sanction such as fear of adverse publicity actually has improved continuous compliance with environmental regulations.

Third, these models 'work' better if the desired rate of compliance is not too close to unity. The two-group model has an advantage in cost-effectiveness over a state-independent model that vanishes as the compliance rate approaches unity. In both the state-dependent models examined, the leverage depended on the target compliance rate: the lower the compliance

---

[6]The relevant cost, of course, is the variable operating cost. Fixed costs are excluded because they are unaffected by compliance.

frequency, the higher the compliance cost that could be accommodated. If, as this suggests, the marginal cost of enforcing compliance is increasing, then agencies might be willing to tolerate a high rate of noncompliance to minimize costs. And in fact, results of case studies of enforcement do find low compliance rates, although again there may be a number of alternative explanations for this observation.

On the other hand, there are some apparent truths from these case studies that do not match up well with comparable aspects of the state-dependent model, although often only modest extensions of the model are required. For example, in the model fines are assessed upon the second consecutive violation of the regulation, whereas in reality, sources are often caught violating standards over and over again before a penalty is exacted. This phenomenon can be imitated by introducing a parameter u', analogous to u, governing the transition from Group 1 to Group 2. Alternatively, one could visualize a model with firms divided into n groups, with no penalties being assessed for violations if in the first $m < n$ groups. But this alteration would probably introduce computational unwieldiness without a corresponding gain in insight.

In short, these models are consistent with several broad observations about enforcement and compliance of air and water quality regulations, or can be made so with minor alterations. But although such consistencies are suggestive, they do not constitute a proper test for a theory. Instead, one should ask: Do these models generate any testable hypotheses that distinguish them from other theories? Are the behavioral assumptions that underlie the model reasonable?

It turns out that testable hypotheses are hard to come by, though hardly nonexistent. Let me propose the following two. First, according to (6) the rate of compliance for any f_{10} firm depends not on the cost of compliance, but on the transition probabilities in the two groups, p_1 and p_2, plus the parameter u. Assuming that the parameters for all firms are the same, the rate of compliance for such firms will be the same also.

Second, and of more interest, is the effect of finding a violation on subsequent behavior by the firm. If the firm is inspected and found to be in violation, then the probability of a violation at the next inspection is reduced. If found in compliance, on the other hand, the probability of a violation is *increased*. That is, suppose we have a record of inspections of a firm over time, $(I_1 I_2 I_3 ...)$, where $I_n = 1$ if the plant was found in violation of the nth inspection and 0 otherwise. Then

$$P\{I_n = 0 \,|\, I_{n-1} = 0\} < P\{I_{n-1} = 0\} < P\{I_n = 0 \,|\, I_{n-1} = 1\}.$$

In static models, by way of contrast, the outcome of the current inspection is not affected by the outcome of the previous inspection. This effect, if it exists, may be difficult to find in practice, because the inspection may cause the firm to revise upward its estimate of the probability of an inspection. An inspection could in this case lead to improved compliance even if the firm was found in compliance, just as borderline speeders often slow down upon seeing a highway patrol car.

As for the reasonableness of the assumptions of the model, enforcement authorities admittedly do not formally and consciously pursue a game-theoretic policy of any sort. But they do tend to adjust the inspection frequency and penalties to the past performance of sources. In addition, enforcement officials typically have a very good idea who the 'bad actors' are. These sources are burdened with more frequent surveillance and greater likelihood of being fined for violations. Also, in the assessment of penalties the recalcitrance of the source is probably more decisive than the likely environmental consequences of a violation. The two-group model may capture the essence of agency behavior, even it it is not the formal procedure used.

It may also be questioned whether firms are so cynical as to commit wilful violations just when they think they can get away with it. This objection is another way of saying that firms are not pure cost minimizers. Perhaps not, but cost minimization remains a useful simplifying assumption. As noted earlier, violations result from a conjunction of numerous stochastic events, such as variation in input quality, process upsets, and breakdowns in abatement equipment, rather than deliberate acts. A firm has considerable discretion in the care with which abatement equipment is operated and maintained. It stands to reason that its diligence would be the greater during those times when violations were likely to be costly.

The assumption that there is a maximum penalty also deserves comment. On paper, most environmental agencies probably do possess sufficient power to force firms to comply with regulations. Even when fines are limited, agencies may seek an injunction shutting the plant down. To take such an extreme action, however, is costly for the agency and the outcome is uncertain. Melnick (1982) has shown how reluctant U.S. courts have been to impose draconian measures on recalcitrant firms when such measures would result in unemployment. Thus, enforcement agencies often have the authority to take drastic action, but it is not a step that would be taken lightly. Given such limitations it makes sense to consider what can be achieved with limited penalties.

I think there is yet another assumption being made in these models that is perhaps less obvious but probably more questionable: both the firm and the agency are assumed to know a lot about the other. The firm is assumed to know what enforcement parameters the agency is using, and to know, at any instant in time, to which group it is assigned. For its part, the agency is assumed to know the firm's cost of compliance. In the real world these assumption may not hold. This means that both the agency and the firm can make mistakes, the agency in the original choice of the parameters of the game and the firm in its choice of strategy, as well as in its choice of action in any given period. Furthermore, both will know this and will tend to act cautiously. For example, suppose a firm had been in compliance for some time after a violation. How is it to tell when it is 'safe' to relax and begin to commit violations again? Presumably inferences could be formed from the pattern of inspections, but necessarily there will be a delay, for some time will elapse between the time the firm's status is changed and the firm can infer the fact. During this period of delay the firm would be in compliance, thus raising the firm's overall compliance rate, but the arractiveness of returning to Group 1 would be reduced.

Incorporation of this reality into the model would allow the investigation of an interesting theoretical question: Who would benefit if both parties were to conceal from the other the requisite information? One can visualize a second game, to be played before continuous enforcement begins, in which the agency announces the enforcement parameters to be used, the firm announces its compliance cost, and each must decide whether to believe the other.

In contrast to the two-group model, the three-group model discussed in section 4 is not represented as a description of current enforcement practice. There probably are not any enforcement authorities that threaten repeat violators with perpetual surveillance. Nonetheless, this model has much to recommend it, for it promises to enforce high compliance rates with relatively small enforcement budgets. When penalties are limited the performance of the model is not arbitrarily high, as in Greenberg's original model. The limitation on penalty size appears to affect the maximum compliance rate that can be achieved, but the inspection resources required remain quite small, though not arbitrarily so. Practical implementation of this model would require some way of allowing firms to escape from the third group, for otherwise most firms would eventually end up there. Leakage into the absorbing state would occur either because of inevitable mistakes by the agency in determining violations, as Greenberg and Russell have pointed out, or inadvertent violations by the firm. Allowing an escape from Group 3 would, no doubt, compromise to some degree the cost-effectiveness of the model. Determination of just how much is a subject of future research.

References

Abbey, David and W. Harrington, 1981, Air quality regulation and management in the four corners states, Southwestern Journal of Economics and Business 1, no. 2.
Becker, Gary S., 1968, Crime and punishment: An economic analysis, Journal of Political Economy 76, no. 2.

Braithwaite, John, 1982, The limits of economism in controlling harmful corporate conduct, Law and Society Review 16, no. 3.

Brams, Stephen J. and M.D. Davis, 1983, The verification problem in arms control: A theoretical analysis, Economic Research Report 83-12 (New York University Department of Economics, New York, NY).

Downing, Paul, 1984, Environmental economics and policy (Little Brown & Co., Boston, MA).

Downing, Paul and James Kimball, 1982, Enforcing pollution control laws in the United States, Policy Studies Journal 11, no. 1, Sept.

Downing, Paul and W.D. Watson, 1974, The economics of enforcing air pollution controls, Journal of Environmental Economics and Management 1, 219–236.

Environmental Law Institute, 1975, Enforcement of federal and state water pollution controls (Report prepared for the National Commission on Water Quality, Washington, DC).

Greenberg, Joseph, 1984, Avoiding tax avoidance: A (repeated) game-theoretic approach, Journal of Economic Theory 32, no. 1, 1–13.

Harford, John, 1978, Firm behavior under imperfectly enforceable standards and taxes, Journal of Environmental Economics and Management 5, no. 1.

Harrington, Winston, 1981, The regulatory approach to air quality management: A case study of New Mexico (Resources for the Future, Washington, DC).

Harrington, Winston, 1985, Enforcement of continuous compliance with air quality regulations (unpublished Ph.D. dissertation, University of North Carolina, Chapel Hill, NC).

Kohlas, J., 1982, Stochastic methods of operations research (Cambridge University Press, New York).

Landsberger, Michael and Isaac Meilijson, 1982, Incentive generating state dependent penalty system, Journal of Public Economics 19, 333–352.

Linder, Stephen H. and Mark E. McBride, 1984, Enforcement costs and regulatory reform: The agency and firm response, Journal of Environmental Economics and Management 11, no. 4, 327–346.

Melnick, Shep, 1982, Regulation and the Courts: The case of the Clean Air Act (The Brookings Institution, Washington, DC).

Russell, Clifford S., 1982, Pollution monitoring survey summary report (Resources for the Future, Washington, DC).

Russell, Clifford S., 1984, Imperfect monitoring of sources of externalities: Lessons from single and multiple play games, Discussion paper QE84-01 (Resources for the Future, Washington, DC).

Russell, Clifford S., W. Harrington and W.J. Vaughan, 1986, Enforcing pollution control laws (Johns Hopkins University Press, Baltimore, MD).

Storey, D.J. and P.J. McCabe, 1980, The criminal waste discharger, Scottish Journal of Political Economy 27, no. 1.

U.S. Environmental Protection Agency, 1981a, Profile of nine state and local air pollution agencies (Office of Planning and Evaluation, The Agency, Washington, DC).

U.S. Environmental Protection Agency, 1981b, Characterization of air pollution control equipment operation and maintenance problems (Office of Planning and Evaluation, The Agency, DC).

U.S. General Accounting Office, 1979, Improvements needed in controlling major air pollution sources, Discussion paper CED-78-165 (The Agency, Washington, DC).

U.S. General Accounting Office, 1982, Cleaning up the environment: Progress achieved but major unresolved issues remain, Discussion paper CED-82-72 (The Agency, Washington, DC).

Viscusi, W. Kip and R.J. Zeckhauser, 1979, Optimal standards with incomplete enforcement, Public Policy 27, no. 4.

Willick, W. and T. Windle, 1973, Rule enforcement by the Los Angeles air pollution control district, Ecology Law Quarterly 3, 507–534.

30

EFFECTIVENESS OF THE EPA'S REGULATORY ENFORCEMENT: THE CASE OF INDUSTRIAL EFFLUENT STANDARDS*

WESLEY A. MAGAT and W. KIP VISCUSI
Duke University

I. INTRODUCTION

IN the almost two decades since the initial wave of social regulation, the academic literature documented very few, if any, instances of a health, safety, or environmental regulation being an unqualified success. Indeed, in most cases, the problem is even more fundamental. The typical analysis of government regulation found that the regulation did not even fulfill its primary mission, much less pass a more demanding benefit-cost test.

This absence of a well-documented case study of effective social regulation may be due, in part, to the particular set of regulations selected for analysis. There is certainly no inherent economic reason why such regulations cannot play a productive role in our economy. In the case of environmental quality, for example, the externality problems being addressed are not handled well by markets, implying that government regulation has at least the potential for playing a beneficial role. However, this potential will not be realized if the regulations are ill conceived or not effectively enforced, or if the environmental problem has no feasible solution.

A brief review of past regulatory experiences may be instructive to put in better perspective the Environmental Protection Agency's (EPA) water pollution control effort—the focus of this article. Most of these detailed evaluations have been done with respect to agencies other than the EPA.

* This work was completed under two Cooperative Agreements from the U.S. Environmental Protection Agency, one to Duke University (CR811902-02) and one to Northwestern University (CR81302-01). We thank Alan Carlin, who was invaluable in facilitating the research process, and the many people at the EPA who so kindly assisted us with obtaining the data and understanding the agency's enforcement process. Anil Gaba provided superb computer programming support, Mark Dreyfus assisted in the data collection, and an anonymous referee offered several useful suggestions.

[*Journal of Law & Economics,* vol. XXXIII (October 1990)]

Environmental Instruments and Institutions

Although there have been some treatments of EPA regulations in the academic literature,[1] as well as some assessments within the government,[2] none of these evaluations have been undertaken with the same degree of statistical rigor and detailed empirical analysis that characterize analyses of health and safety regulations.

In large part, this lack of attention stems from the greater difficulty in constructing an environmental data base.[3] The decentralized nature of polluting activity, some of which is clandestine, makes pollution levels more difficult to monitor than compliance with, for example, safety cap requirements. These difficulties posed for external evaluation may also generate monitoring problems for the agency's enforcement staff. An important issue to be addressed here is whether the prolonged process required for us to amass a sound environmental data base for the purpose of external analysis is a reflection of underlying intrinsic difficulties in the monitoring and enforcement of EPA regulations.

The past assessments of health and safety regulations indicated that regulations were ineffective in promoting their objectives for two general reasons. The first is ineffectively designed regulatory policies. Thus, even though there is compliance with the regulatory requirements, little or no beneficial effect has been observed.

The seat belt requirements of the National Highway Traffic Safety Administration are one exhaustively studied instance. Since many drivers do not use seat belts, and those that do may alter their driving habits, the regulation has not produced the dramatic reduction in injuries and fatalities that the proponents of the regulation envisioned. Although some studies suggest no significant effect,[4] while others suggest a modest beneficial effect,[5] the overall implication is that seat belts have not pro-

---

[1] Robert W. Crandall, Controlling Industrial Pollution: The Economics and Politics of Clean Air (1983); Paul MacAvoy, The Regulation of Air Pollutant Emissions from Plants and Factories (1981); and B. Peter Pashigian, Environmental Regulation: Whose Self-Interests Are Being Protected? 23 Econ. Inquiry 551 (1985), are excellent examples of such contributions.

[2] See, for example, U.S. General Accounting Office, Wastewater Dischargers Are Not Complying with EPA Pollution Control Permits (1983); and U.S. General Accounting Office, Water Pollution: Application of National Cleanup Standards to the Pulp and Paper Industry (March 1987).

[3] See Crandall, *supra* note 1, for discussion of many of the problems confronted with respect to air pollution data.

[4] For data supporting this conclusion, see Sam Peltzman, The Effects of Auto Safety Regulation, 83 J. Pol. Econ. 677 (1975).

[5] Among the best of the optimistic assessments of seat belt regulations is that of Robert W. Crandall and John D. Graham, Automobile Safety Regulation and Offsetting Behavior: Some Empirical Estimates, 74 Am. Econ. Rev. 328 (1984).

duced large reductions in injury and fatality rates because those designing the policy did not consider the crucial behavioral link involving drivers.

A similar effect has been observed with respect to the Consumer Product Safety Commission's safety requirements,[6] and, more generally, there is evidence that consumer product safety regulations are not sufficiently effective or extensive to substantially affect product safety. Manufacturers have complied with the regulatory standards, but consumer safety has not been enhanced.

Much the same story is true in the pharmaceutical area. Pharmacists and doctors have complied with the U.S. prescription requirements for drugs, with only occasional notable violations. Nevertheless, in terms of the effect of prescriptions on health, no significant health effects of these requirements have been observed either for the United States or elsewhere in the world.[7]

The second reason for regulatory failure is the lack of enforcement. For example, the Occupational Safety and Health Administration (OSHA) has extensive regulatory requirements but traditionally enforced them quite laxly. Indeed, the inspection rates are so low (less than one inspection per century per firm) and the penalties are so small (only $6 million annually) that there are few incentives for compliance. The result is, at best, a very modest effect on safety outcomes.[8]

The EPA water pollution regulations—the focus of this study—represent an interesting departure from past patterns of regulatory failure. First, the nature of the regulations—discharge limits—relates directly to the policy objective of controlling pollution, and there is no potential for offsetting behavioral responses. If the pollution standards are binding and enforced, they should improve water quality. Second, the enforcement effort is so extensive that enforcement should affect firms' compliance. In the pulp and paper industry, which we will analyze, the EPA averages roughly one inspection annually per major pollution source. In addition, firms are required to file monthly discharge monitoring reports, providing one of the most thorough monitoring capabilities of any health, safety, or

[6] See W. Kip Viscusi, Consumer Behavior and the Safety Effects of Product Safety Regulation, 18 J. Law & Econ. 527 (1985).

[7] For supporting data, see Sam Peltzman, The Health Effects of Mandatory Prescriptions, 30 J. Law & Econ. 2 (1987).

[8] The most extensive analysis is that in W. Kip Viscusi, The Impact of Occupational Safety and Health Regulation, 1973–1983, 17 Rand J. Econ. 567 (1986). Analyses of earlier periods of OSHA enforcement are provided in Ann P. Bartel & Lacy Glen Thomas, Direct and Indirect Effects of OSHA Regulations, 28 J. Law & Econ. 1 (1985), and in Robert S. Smith, The Impact of OSHA Inspections on Manufacturing Injury Rates, 14 J. Human Resources 145 (1979).

environmental agency. Prior to the 1987 revisions of the Clean Water Act,[9] one potential weak link was that EPA officials could not directly assess penalties for noncompliance. They could, however, seek the imposition of substantial penalties through court action.

In the subsequent sections, we describe the nature of the EPA enforcement of water pollution regulations in the pulp and paper industry and the original data base we created for this study. Using information from EPA and industry sources, we constructed a longitudinal data base by firm, permitting a detailed evaluation of the effects of EPA inspections and their associated enforcement actions on the behavior of pulp and paper plants. As the empirical results will indicate, we find diverse evidence of significant EPA effects on the polluting and reporting activities of firms in the pulp and paper industry.

II.   Enforcement of Water Pollution Regulations in the Pulp and Paper Industry

In choosing to study the enforcement of environmental regulations by the U.S. Environmental Protection Agency and by state environmental agencies, we could have chosen several different media. Only for water pollution was it possible to find a relatively complete data base of pollution discharge measurements by source and a data base on enforcement actions at these same plants. The same informational base that permits us to provide a sound empirical analysis also assists the EPA in its effort to monitor and enforce compliance. Overall, it is believed that more than 90 percent of all major water discharges are in compliance with EPA standards, as contrasted with estimated compliance rates as low as 20 percent for toxic and hazardous substance regulation.[10] Thus, one should be cautious in generalizing from the EPA's record in water pollution to other pollution problems. The investigation reported here should be regarded as an examination of an important and representative component of one of the EPA's most effective regulatory programs.

Since the data on inspections were much more complete than on other enforcement actions, such as administrative orders, notices of violations,

---

[9] Section 314 of the Federal Water Quality Act of 1987 authorizes the use of administrative penalties that can be assessed directly by the EPA.

[10] For supporting data, see Cheryl Wasserman, Improving the Efficiency and Effectiveness of Compliance Monitoring and Enforcement of Environmental Policies, United States: A National Review, Organization for Economic and Cooperative Development (1984).

warning letters, and telephone calls,[11] we focus on the relationship between plant inspections and water pollution discharge levels. This emphasis on inspections also accords with our a priori views regarding the role of different enforcement instruments since inspections are one of the most important components of any enforcement program and thus merit special attention.

To measure the relationship between inspections and subsequent compliance, we examine one industry, pulp and paper. This industry is the country's largest discharger of conventional pollutants, such as organic waste and sediment,[12] and has a long history of water pollution enforcement efforts by various governmental agencies. There is no reason to believe that the effectiveness of inspections in the pulp and paper industry differs markedly from that in other industries regulated by the EPA. Also, by concentrating on one industry, we avoid the problem of controlling for interindustry differences in the stringency of regulations, differences in the nature of the pollution, and differences in the technologies for compliance.

The EPA traditionally focuses on the control of Biological Oxygen Demand (BOD) because it is the most damaging conventional pollutant discharged by the pulp and paper industry.[13] Most inspections examine BOD levels in addition to other pollutants of interest for a given plant. Also, the technologies that control BOD discharges tend to reduce the levels of other pollutants, which means that the relationship between inspections and BOD discharge reductions ought to be similar to the relationship between inspections and discharge reductions for other pollutants.

The pulp and paper industry consists of hundreds of companies operating plants in thirty states within seven of the ten EPA regions in the country. The EPA Permit Compliance System (PCS) data base described below lists 418 separate sources of pollutant discharge in the industry. Biological Oxygen Demand, Total Suspended Solids (TSS), and the pH levels of discharges are the three main conventional pollutants controlled, although in recent years Congress has initiated new regulatory efforts to also control toxic pollutants.

---

[11] One reason for the completeness of the data on inspections is that the EPA regional offices are not credited with conducting an inspection until it is coded into the central data base. See U.S. Environmental Protection Agency, Office of Water Enforcement, NPDES Inspection Manual, at iii (June 1984).

[12] U.S. General Accounting Office, Water Pollution, *supra* note 2, at 8.

[13] BOD is the standard measure of the organic pollutant content of water.

If the EPA set water pollution standards in the same manner as seat belt regulations or OSHA standards, a description of the regulatory constraints would be straightforward. In the seat belt and OSHA cases, firms face well-defined requirements on the technology or work environment. All firms must comply with the same set of regulations, such as ensuring that punch presses have the specified guards. There has been little change over time in the nature of the standards, except that some new regulations have been added. In contrast, EPA water pollution standards involve permissible pollution amounts that vary across firms and over time.

The 1972 Federal Water Pollution Control Act amendments set the framework for regulation on industrial water pollution. The act required that all sources discharging into the navigable waters of the country meet discharge standards based on the application of the "best practicable control technology" (BPT) by July 1, 1977, while complying with standards based on the "best available technology economically achievable" (BAT) by July 1, 1983.

In 1977 the act was amended again, pushing back the 1983 deadline to July 1, 1984, and substituting a more complicated requirement. Conventional pollutants such as BOD and TSS were to meet standards based on the adoption of the best conventional technology (BCT), while toxic pollutants were to meet standards based on the best available technology (BAT).

The final BPT and BAT standards for various subcategories of the pulp and paper industry were promulgated on three separate dates: May 9, 1974; May 29, 1974; and January 6, 1977. The final BCT standards were issued on December 17, 1986, and left the BPT standards for BOD control unchanged. The BPT standards generally set limitations on the quantities of BOD that a plant could discharge per pound of pulp or paper produced.[14] However, the allowable discharges of BOD from each source were derived by multiplying this effluent limitation by the number of pounds of pulp or paper produced per day at the plant. This latter number formed the basis of the National Pollutant Discharge Elimination System (NPDES) permit required of each discharger. Since our empirical study covers the period from the first quarter of 1982 through the first quarter of 1985, the NPDES permits restricting BOD discharge were based on the 1977 BPT standards.

The EPA possesses the authority to issue the NPDES permits, but the authority has been delegated to thirty-seven states meeting specified fed-

---

[14] For a formal description and analysis of the BPT rule-making process, see Wesley A. Magat *et al.*, Rules in the Making: A Statistical Analysis of Regulatory Agency Behavior (1986).

eral criteria. States approved to issue NPDES permits also assume responsibility for their enforcement, which means inspecting the plants and taking action against sources found to be out of compliance. For states not approved to run their own permit systems, the EPA issues and enforces the permits.

An important aspect of the permit process should be emphasized. The EPA and the states do not set uniform permit levels irrespective of the industry characteristics associated with the pollution source. Each standard is industry specific and represents pollution levels that are potentially achievable with available technologies.

Each source must regularly measure its pollution discharge levels and report its actual discharges of each pollutant in its permit on a monthly basis through a Discharge Monitoring Report (DMR). If a source is out of compliance with the effluent standards in its permit, it is also required to file a noncompliance report. The states and EPA regional offices send the DMRs to the EPA, which enters them into the PCS data base to serve as a basis for tracking compliance. In addition, the EPA requires that Quarterly Non-compliance Reports (QNCR) be filed by each state and region to identify sources out of compliance. In the empirical study that follows, we use the reported BOD discharge levels in the DMRs to measure the effects of inspections on BOD discharge levels.

Because the sources are required to report their pollutant discharge levels on a monthly basis, the on-site inspections play a somewhat different role than inspections carried out by other regulatory agencies, such as an OSHA inspection of an industrial site. The latter inspections constitute the primary basis for the agency to check compliance with its regulations and to have a visible presence in the workplace. In contrast, EPA or state-run inspections of industrial water pollution sources create a similar visible presence, but they provide only a secondary source of information about compliance because the monthly DMRs address the compliance question directly. Some NPDES permit inspections do test whether the DMR discharge levels are reported accurately and honestly, and they provide an incentive for firms to submit DMRs more frequently.

The difference between EPA inspections and OSHA inspections has also been narrowing over the years. Although the Bureau of Labor Statistics does not release the mandated injury reports to OSHA for compliance purposes, OSHA now gathers this information through on-site records checks to target its inspections. This procedure represents a partial and more time-consuming variant of the DMR process. Firms with good injury records are exempt from OSHA inspections.

The EPA inspections directly address one or more of the following items: the existence of an up-to-date permit, the installation of the abatement equipment necessary for compliance with the permit, management plans and practices, the preparation and maintenance of records, the correct operation of the abatement equipment, and the conduct of sampling and sample analysis. As a recent EPA report to the Organization for Economic and Cooperative Development (OECD) explains, "Despite widespread self-monitoring, inspections remain the backbone of agency compliance monitoring programs. . . . inspections are the government's main tool for officially assessing compliance, and for assuring quality control and lending credibility to self-monitoring programs. The independent evaluation provided by a government inspection is the key."[15]

The EPA carries out three main types of inspections—compliance sampling inspections, compliance evaluation inspections, and performance audit inspections. Compliance sampling inspections require approximately thirty workdays of time to complete and involve actual sampling of the effluent at the plant, as well as an examination of the company's record-keeping system, its testing procedures, and its treatment system. In contrast, the compliance evaluation inspections take only about three workdays to complete. They involve no sampling, but the inspectors do examine the company's treatment facilities, monitoring methods, and records. The performance audit inspections require about twelve days to complete and consist of the same practices used in the compliance evaluation inspection, plus observation of the permittee going through the steps in the self-monitoring process from sample collection and flow measurement through laboratory analyses, data workup, and reporting. In addition, the performance audit inspector may leave a check sample for the permittee to analyze.

Based on the discharge reports in the DMRs and in the QNCRs, as well as on the findings of inspections, the EPA or the approved state agencies take enforcement actions against violators. Informal actions include telephone calls, warning letters, and notices of violation, as well as inspections. If these measures do not achieve the intended results, the control agencies can proceed with formal actions such as administrative orders, permit revision, formal listing of companies as ineligible for government contracts, grants, and loans; and, finally, civil and criminal judicial responses.

Court action is a lengthy process involving the Justice Department that is started only as a last resort. Under Section 309(e) of the 1977 Clean Water Act, civil penalties could have been awarded up to a level of $10,000 per day, while criminal penalties could have ranged from $2,500

---

[15] Wasserman, *supra* note 10, at III-7.

to $25,000 for the first violation and up to $50,000 for the second violation.[16] In addition, first violations could have led to imprisonment of up to one year, with up to two years of imprisonment for the second violation. During the period from January 1, 1975, to July 1, 1985, the EPA commenced 64 judicial actions in the pulp and paper industry. Of these, 42 cases resulted in fines, and four were still pending at the end of the period. The fines varied from $1,500 to $750,000, with an average of $89,437. Because the regions lacked the incentives to report regularly enforcement actions other than inspections into the PCS data base, we concentrate our study on the effectiveness of the inspections on bringing firms into compliance with their permits.

The inspections variable is intended to be a proxy for the overall enforcement effort associated with an inspection and all subsequent enforcement actions. The financial penalties associated with noncompliance may be much greater than is indicated by the fines actually assessed since these fines do not reflect the potential losses due to noncompliance. Indeed, to take the extreme case, if the sanctions were so great that enforcement effort was fully effective, no instances of noncompliance or penalty assessments would be observed. The enforcement measure to be used in the regression equations should reflect the expected sanctions, not the level of sanctions observed after the deterrence effect has operated. The EPA inspections variable will serve both as a measure of the formal contact of the EPA with each firm as well as the institutional trigger that will generate additional enforcement sanctions for noncomplying firms.

III. The Sample and the Variables

The Data Base

The PCS data base, which we utilize in our analysis, lists 418 separate sources in the pulp and paper industry in its Inspections file. Under half of these sources—194—had BOD discharges. Of this group, seventy-seven submitted DMR measurements for BOD discharge into the Measurement file. As a result, the data set that we used includes seventy-seven of the 194 major sources of BOD in the pulp and paper industry.

The information that is missing was not governed by an entirely random process. The sources that were not included either did not enter their DMRs into the PCS data base, or they did not submit DMRs including

---

[16] Under the Federal Water Quality Act of 1987, the maximum civil penalty rose to $25,000 per day and the maximum criminal penalty increased to $50,000 for the first violation and $100,000 for the second violation.

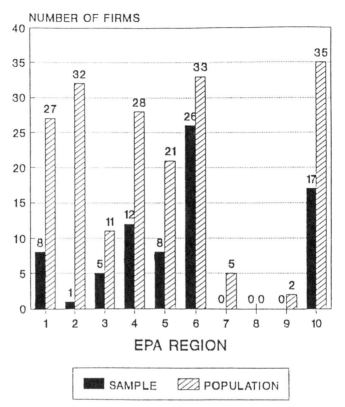

FIGURE 1.—Sample and population firms by EPA region; $N = 194$

BOD measurements during the period under study, or they discharged pollutants other than BOD.

The principal factor affecting the missing sources is the time required before the records could be computerized. Officials at the EPA believe that most of the other sources not in the PCS data base did submit DMR data to the EPA or the states, but they were not entered into the PCS data base because the system was not yet operational and the states and regions were not required to enter the data.[17] This view is consistent with the regional patterns of sources included in our data base. Seven of the nine EPA regions with pulp and paper mills are represented in our data base (see Figure 1). In addition, nineteen of the twenty-seven states with

---

[17] This view is based on discussions with EPA analysts most familiar with the PCS data base and the pulp and paper industry. Even though some of the states did not have the capability to enter the DMR data into the PCS data base, they regularly screened the data and summarized them in the QNCRs.

TABLE 1

NUMBER OF FIRMS IN SAMPLE AND POPULATION BY STATE

State	Firms in Sample	Firms in Population
Alaska	1	2
Alabama	0	10
Arkansas	8	8
California	0	2
Florida	8	9
Idaho	1	1
Iowa	0	2
Illinois	2	3
Indiana	1	4
Kentucky	3	3
Louisiana	9	13
Maryland	0	1
Massachusetts	3	16
Minnesota	1	5
Missouri	0	3
New Hampshire	5	11
New Jersey	1	10
New York	0	21
North Carolina	0	4
Ohio	4	9
Oklahoma	3	3
Oregon	7	14
Pennsylvania	5	10
Puerto Rico	0	1
Tennessee	1	2
Texas	6	9
Washington	8	18

pulp and paper mills are captured in the sample (see Table 1). The principal selectivity process is that some states and two EPA regions did not computerize their DMR reports by 1982:1. If computerization is positively correlated with more effective enforcement (which we have no way of knowing and no a priori reason to suspect), our empirical results will tend to overstate the effectiveness of water pollution enforcement overall. However, it should be noted that, even if there is a selectivity bias, it should create much less bias than a scenario in which the firms missing from the data base were determined on a firm-by-firm basis based on the quality of the DMR data supplied by each firm.

The firms in our sample tend to be larger than those without data. Available data on 170 out of the 194 pulp and paper plants show that the mean output level of firms in the sample is 62 percent higher than those outside the sample and that 54 percent of the industry output is produced by the firms in our sample. We do not have the data to measure the share of industry pollution discharges from our sample, but to the extent that output levels are correlated with pollution discharge levels, about half of the industry BOD discharges would have been produced by the firms in our sample.

Thus, based on the available evidence, we conclude that our sample comprising 40 percent of the firms and 54 percent of the output in the industry is representative of the entire pulp and paper industry, except to the extent that there is some relationship between the effectiveness of inspections and either firm size or the decision by states or EPA regional offices to enter discharge data in the PCS data base. If these two factors are for some reason correlated with the effectiveness of inspections, then our results need to be interpreted as estimates of the response to EPA inspections of firms whose discharge levels are regularly reported to the EPA's national data base.

The firms represented in our sample are all located within Standard Industrial Classification (SIC) 26. We have further divided this industry characterization into five four-digit SIC codes (2611, 2621, 2631, 2648, and 2661).[18] For the period from the first quarter of 1982 through the first quarter of 1985 there were 276 inspections of the sources in the sample, of which 43 percent were compliance sampling and 57 percent were compliance evaluation.

In this analysis we use calendar quarters as the unit of analysis. Only rarely was there more than one inspection for a given source in the same quarter. Despite the requirement that sources report DMRs every month to the state enforcement agency or the EPA, for the reasons explained above, some DMR measurements are missing for the sources in our sample. In constructing the quarterly BOD measurements for our statistical analysis, we interpolated to fill in missing values and used averages of the BOD discharge levels within a quarter as the quarterly average BOD discharge levels.

Although the EPA analysts to whom we talked were confident that most of the discharge measurements in the DMRs were reported accurately, permittees do have several opportunities to cheat. They may choose not to report discharge measurements during months with unusually high discharge levels. This behavior would lead to some smoothing of the pattern of reported discharges, eliminating the top end of the

---

[18] The sample of seventy-seven sources matches the full set of pulp and paper sources fairly closely in terms of the distribution of sources across regions, the mix of products across the four-digit SIC codes, and the frequency of plant inspections in each quarter. The only differences of note were that regions 6 and 10 are somewhat overrepresented, while region 2 is underrepresented, and the sample firms were inspected about 25 percent more often.

distribution. More active attempts to mislead EPA include altering the contents of the sample being tested, falsely calibrating the test instruments, and recording false measurements in the DMRs.

Despite these possibilities for sending the EPA misleading or false DMR discharge statistics, there are several incentives to report honest information in the DMRs. The EPA follows the policy of attempting to inspect all major sources at least once a year. Compliance sampling inspections would detect whether most of the reported measurements were inconsistent with the measurements from the inspections, but they could not detect whether outliers were removed from the reports. Compliance evaluation inspections would detect the absence of the required abatement equipment, but would be less useful in evaluating whether the abatement systems were being operated correctly. Of course, the penalties for noncompliance and fraud in reporting also create incentives for truthful reporting of discharge measurements. The possibility of leaks to EPA by disgruntled employees makes this last incentive more compelling to firms considering manipulating their DMR data.

Taking into account the possibility that the DMR measurements may measure true compliance status with some error, it is still instructive to ascertain how well firms comply with the effluent regulations. Recently, the Environmental Protection Agency[19] issued a study of compliance by all the major pulp and paper mills (SIC 2611, 2621, 2631) in the eight southeastern states comprising EPA region 4 over the period from the second quarter of fiscal year 1982 through the first quarter of fiscal year 1984. Eighty-two percent of the measurements fell within the permitted bounds. This compares with 75 percent of the measurements from the pulp and paper firms in our sample being in compliance. The EPA further defines significant noncompliance for BOD as violations of the monthly average permit limits for any two months in a six-month period that exceed the limit by 40 percent, or violations of the monthly average limits for any four months in a six-month period. Using this definition, 94 percent of the measurements indicated discharge levels not in significant noncompliance. The study also showed that four out of the fifty-six mills created most of the instances of significant noncompliance.[20]

[19] U.S. Environmental Protection Agency, Study of Pulp and Paper Industry in Region IV (1986).

[20] In light of the low fines assessed and the relative infrequency of inspections, some readers may question the reasons for the high compliance rates. While our study addresses only the incremental effect of inspections and associated enforcement actions on compliance, we can speculate on the explanation for the high base rate of compliance.

In a well-functioning regulatory system, one would not expect to see frequent use of strong sanctions, such as fines, for firms complying with regulations in order to avoid the sanctions. It is only necessary that firms *believe* they will be sanctioned if they fail to

TABLE 2

MEANS AND STANDARD DEVIATIONS OF VARIABLES DESCRIBING SEVENTY-SEVEN PLANTS
IN SAMPLE (1982:1–1985:1)

Variable	Mean	Standard Deviation
MQAVG (pounds per day)	5,758.288	8,919.173
MVIO (1 = out of compliance)	.252	.434
IQTR1 (1 = inspection one quarter prior to measurement)	.248	.432
IQTR2	.273	.446
IQTR3	.273	.446
IQTR4	.281	.450
IQTR5	.300	.458
IQTR6	.295	.456
REGN1 (1 = source located in region 1)	.095	.293
REGN2	.002	.039
REGN3	.064	.244
REGN4	.154	.361
REGN5	.039	.193
REGN6	.435	.496
REGN7	.000	.000
REGN8	.000	.000
REGN9	.000	.000
REGN10	.213	.410
SIC11 (1 = pulp mill)	.241	.428
SIC21 (1 = paper mill excluding building)	.432	.496
SIC31 (1 = paperboard mill)	.253	.435
SIC48 (1 = 7 stationary products)	.014	.117
SIC61 (1 = building paper or paperboard mill)	.048	.214
TONS (daily output rate)	794.156	587.083

Sample Characteristics

Table 2 summarizes the means and standard deviations for the sample of the variables used in our analysis. The sample is a pooled time series and cross section of seventy-seven plants followed on a quarterly basis from 1982:1 to 1985:1. The first two variables represent the pollution outcome measures that will be of primary interest as dependent variables

---

comply. Despite their infrequent use, there are a variety of punishments that the EPA can impose, short of judicial fines. For firms that do not comply, the agency can raise the frequency and intensity of inspections, write permits using stricter interpretations of the regulations (for example, using average rather than maximum production rates to calculate allowed discharge levels), deny operating permits, subject the firm to bad publicity, and engage in protracted haggling, and possibly prolonged litigation, which imposes high costs in terms of legal fees, management time, and general uncertainty about being allowed to operate.

in different equations. The variable MQAVG is a continuous measure of the extent of pollution. It measures the number of pounds of BOD discharged per day, where this amount is averaged over the quarter. Although the amount of pollution is a variable of substantial economic interest, it is not the sole variable of concern. Different firms may have different permitted pollution levels so that, for example, a large plant may be in compliance with a high BOD level whereas a small plant may be in violation of its permit even though its discharge is less. Analyzing the effect of inspections on total discharges is, however, one of the most important ways of assessing the benefits of the EPA's regulatory enforcement.

The second pollution variable, MVIO, is a discrete zero or one variable that takes on a value of one if the pollution source is in noncompliance with its BOD discharge permit in any of its monthly measurements that quarter. This variable best captures whether the firm's performance is in compliance with its water pollution permit, but it does not reflect the extent of noncompliance. Unfortunately, it is not possible to construct a reliable measure of the amount of pollution in excess of the permitted amount since data pertaining to the level specified in the permit are not available from the PCS data base. Instead, we are restricted to MQAVG and MVIO rather than a hybrid of a continuous pollution measure and discrete compliance measure.

The next set of variables is a series of zero or one dummy variables pertaining to whether the firm was inspected in a particular quarter. The variable IQTRJ is of the general form in which it takes on a value of one if the pollution source received an inspection J quarters previous to the pollution measurement in the current quarter, where J takes on a value from one to six. It is quite striking that the rate of inspection is quite high, on the order of 25–30 percent per quarter.

This relatively high inspection rate distinguishes the EPA enforcement effort from that of OSHA. Not only does the EPA receive regular discharge monitoring reports from firms, but it also undertakes water pollution inspections at a rate of about one inspection annually per major pollution source. In contrast, OSHA has no automatic data feedback mechanism, and it has a much more sporadic inspections effort. In OSHA's early years, some analysts equated OSHA's inspection frequency to other rare events such as the annual chance of seeing Halley's comet. At present, the OSHA inspection rate is much lower than this amount—on the order of 1/200 for each firm in any year.[21] The intensity of

---

[21] See W. Kip Viscusi, Reforming OSHA Regulation of Workplace Risks, in Regulatory Reform: What Actually Happened 259 (L. Weiss and M. Klass eds. 1986).

EPA inspections consequently dwarfs that of OSHA inspections so that there is no reason to believe that the lack of efficacy of OSHA's minimal enforcement operation has any adverse implications for the EPA's chances of success.

The variables of the form REGNJ are zero or one dummy variables for the EPA region J in which the plant is located. These variables will be utilized to ascertain whether there are any important regional differences in pollution patterns. It should be noted that there are no pulp and paper mills located in three of the EPA regions (7, 8, and 9) and there are no PCS data on mills in region 2.

The next set of six variables are of the form SICJK, which represents a dummy variable for the plant's four-digit SIC industry code 26JK, where JK takes on the values 11, 21, 31, 48, and 61. Although all firms are in the pulp and paper industry, it was desirable to also include refined industry-group dummy variables that reflect the firm's specific operations and technology. For example, pulp mills (SIC 2641) have different operations than converted paper plants (SIC 2649).

The final variable listed is TONS. It measures the number of tons of pulp and paper produced daily at the plant. Unlike the other variables in the data set, this variable was not included in the PCS data base. We matched each firm to a capacity measure using data provided in a published industry directory.[22]

IV. The Effect of Inspections on Pollution

The major purpose of this article is to measure empirically the effects of inspections, along with their associated enforcement actions, on the behavior of firms in the pulp and paper industry. We will concentrate on an econometric approach that relates the conduct of an inspection in a given quarter to two measures of the firm's BOD abatement effort: (1) its absolute rate of effluent discharge (MQAVG), and (2) whether its discharge rate falls below its permitted level (MVIO). We also examine the effect of plant inspections on reducing the incidence of nonreporting of DMR data. To the extent that firms purposely refrain from reporting discharge levels during periods of noncompliance, the first two measures of the effect of inspections would be biased toward less effect than what actually occurred. This third measure allows us to determine whether inspections improved the completeness of the EPA's discharge monitoring system that presumably lead to more discovery of noncompliance and, through

---

[22] See Lockwood's Directory of the Paper & Allied Trades (1983 ed.).

subsequent enforcement efforts, further reductions in pollutant discharge levels.

Empirical Framework for Measuring Abatement Effects

The underlying economic framework is straightforward, as pollution levels are governed by a capital investment process relating to the pollution control technology, as well as by the efficiency levels at which the abatement equipment is operated. The role of EPA inspections is to raise the expected cost of noncompliance, boosting the incentives for pollution reduction and compliance with the permit. Since the underlying theoretical basis is straightforward, we will proceed directly to the estimating equations.[23]

The equations to be estimated will be of the same general form whether the pollution variable is MQAVG or MVIO. To illustrate this general form, let $POLLUTION_{it}$ be the value of the pollution variable MQAVG or MVIO for pollution source i in period t. Some additional notation is needed before we can write down the equation to be estimated. The variable $IQTRJ_{it}$ is the zero or one inspection variable for whether pollution source i was inspected in period $t - J$, $TONS_i$ is source i's capacity measure, SIC_i is a vector of four-digit SIC-code dummy variables for pollution source i, $REGN_i$ is a vector of dummy variable for the EPA regions for source i, and $QUARTER_t$ is a vector of dummy variables for the quarters. The resulting estimating equation is of the form

$$POLLUTION_{it} = \alpha + \beta_1 POLLUTION_{it-4} + \sum_{k=1}^{n} \gamma_k\, IQTR_{t-k}$$

$$+ \beta_2 TONS_i + \beta_3 SIC_i + \beta_4 REGN_i$$

$$+ \beta_5 QUARTER_t + v_{it},$$

where v_{it} is a random error term. In the case of the continuous pollution measure, MQAVG, ordinary least squares is the appropriate estimator, whereas for the discrete compliance variable, MVIO, a logistic estimation procedure is employed. With some modifications, this equation is in the same general spirit as similar equations estimated for safety regulations.[24]

The first variable included is the lagged dependent variable, with the noteworthy distinction that the lag is four quarters rather than one. The

---

[23] The model implicit here is articulated more fully for the analogous job safety case in W. Kip Viscusi, The Impact of Occupational Safety and Health Regulation, 10 Bell J. Econ. 117 (1979). More generally, see Richard Posner, Economic Analysis of Law (3d ed. 1986).

[24] The equation bears closest similarity to those in Viscusi, *supra* note 8.

variable POLLUTION$_{it-4}$ is a proxy for the firm's stock of capital related
to pollution control and for the general character of its abatement technology. Firms with high levels of pollution in the past are likely to continue to
have high levels in the future because the nature of their control technology makes it costly to achieve pollution reductions. A four-quarter lag is
utilized rather than a single-quarter lag to capture the seasonality that
often plays an important role in a firm's operations. The products produced, stream flow conditions, and the pollution permit amount may vary
by season.

The lagged dependent variable serves an additional role with respect to
regression-to-the-mean effects. It is possible that firms with an abnor-
mally high pollution level in period t due to stochastic factors will be
inspected in period $t + 1$ and improve their performance compared with
period t, wholly apart from any true inspection effect. Because the lagged
values captures pollution levels, or compliance status, four quarters earlier, however, they are less susceptible to leading to inspection variable
results that simply capture regression-to-the-mean effects.

The next set of variables is a distributed lag on past EPA inspections.
Evidence for OSHA suggests that there is generally a lag before firms can
make the required capital investments to alter their performance level.[25]

Even if compliance only entails changes in operating procedures following an inspection, an effect may not be apparent until the next quarter.
Consider a situation in which the firm files its DMR data for the first
month of the quarter in the middle of the second month of the quarter.
Even if the EPA undertakes an inspection immediately, which is not
usually the case, the sampling will not be completed until the middle of
the final month of the quarter. Thus, under this best-case scenario only
half a month, or one-sixth, of the pollution discharges for the quarter will
be affected by the inspection. Because of time lags before the EPA receives the DMR data, the time needed before the EPA can schedule an
inspector to make a plant visit, the rather lengthy inspection process, and
the time needed before the EPA makes its report to the firm and the firm
can take action on it, no contemporaneous effect is expected.

Before requiring that any inspection effect enter with a lag, we empirically tested for whether the inspection variable led to a contemporaneous
negative effect on pollution. Rather than a negative effect, we observed a
strong and statistically significant positive influence, consistent with the
reverse causality hypothesis. We explored the causality issue in greater
detail. Based on a Hausman[26] specification test, we were able to reject the

[25] *Id.*

[26] Jerry Hausman, Specification Tests in Econometrics, 46 Econometrica 1251 (1978).

hypothesis that the IQTR$_{it}$ variable is exogenous. Attempts to replace IQTR$_{it}$ ($t = 0$) by an instrumental variable estimator also led to positive coefficients, suggesting that the primary relationship between the two variables is through high current levels of pollution leading to EPA inspections, rather than inspections causing immediate reductions in pollution discharge levels. These results allow us to use only lagged inspection variables without losing any of the effects of the inspections on compliance or creating a bias in our estimated coefficients.

The next variable, TONS$_i$, pertains to the capacity of the firm. Other things being equal, firms with larger capacity should produce more pollution, MQAVG, but need not necessarily be more likely to be in or out of compliance with EPA standards. There may be economies of scale with respect to pollution control that would tend to make large firms less likely to be out of compliance. Similarly, the TONS variable may pick up factors related to the vintage of the technology to the extent that larger plants are newer and have less polluting technologies. If these large plants are considerably more efficient in controlling pollution, the absolute levels of pollution may be lower than smaller and more outmoded facilities.

Technological factors of this type will also be captured in the SIC-code dummy variables, implying that differences in technologies and standards across parts of the pulp and paper industry will be taken into account. The regional dummy variables REGNJ also capture firm characteristics to some extent since plants in some regions tend to be older than those in other regions. These regional variables also reflect regional differences in standard setting and the nature of enforcement. These differences may be considerable due to the prominent role that the states have in the enforcement process.

The final set of variables is a series of twelve quarterly dummy variables for all but one of the quarters represented. This formulation was chosen over a simple time-trend variable because of its greater flexibility. Not only do the QUARTER$_t$ variables capture any possible uniform time trend, but they also capture other quarter-specific effects such as any seasonal and cyclical fluctuations in production levels and water flows. Although some quarterly dummy variables were statistically significant, these coefficients are not reported since there was no apparent pattern evident in the results. In addition, we regressed MQAVG against both a continuous TIME variable and its square but found no significant relationships.

Regression and Maximum Likelihood Results

Table 3 reports the ordinary least squares (OLS) results for the continuous pollution measure, MQAVG, and Table 4 reports the maximum likeli-

TABLE 3

Regression Equations for MQAVG (Quarterly Average BOD Discharge Levels in Pounds per Day)

INDEPENDENT VARIABLES	COEFFICIENTS			
	(1)*	(2)	(3)	(4)
INTERCEPT	−434.029	−460.454	−494.034	−213.905
	(1,683.935)	(1,650.046)	(1,592.309)	(1,557.062)
MQAVG4	.983	.983	.983	.982
	(.021)	(.021)	(.020)	(.020)
IQTR1	−1,174.689	−1,059.423	−1,064.031	−1,148.911
	(517.225)	(511.525)	(497.787)	(487.430)
IQTR2	575.256	381.999	398.665	. . .
	(495.099)	(481.687)	(469.908)	. . .
IQTR3	−198.047	−155.912	. . .	. . .
	(467.133)	(463.305)	. . .	. . .
IQTR4	77.479	59.709	. . .	. . .
	(468.403)	(450.159)	. . .	. . .
IQTR5	374.924	. . .	. . .	. . .
	(468.248)	. . .	. . .	. . .
IQTR6	−584.136	. . .	. . .	. . .
	(440.411)	. . .	. . .	. . .
TONS	.322	.320	.320	.329
	(.438)	(.439)	(.437)	(.437)
SIC11	414.177	382.955	410.222	310.442
	(1,408.440)	(1,408.522)	(1,394.943)	(1,389.408)
SIC21	262.356	219.433	252.081	112.177
	(1,418.355)	(1,414.941)	(1,393.285)	(1,382.926)
SIC31	−205.950	−278.427	−253.484	−365.172
	(1,426.645)	(1,424.948)	(1,410.265)	(1,403.533)
SIC48	31.976	−41.814	17.162	−241.752
	(2,806.433)	(2,789.052)	(2,719.955)	(2,701.675)
REGN1	248.482	225.870	213.567	322.009
	(909.025)	(895.892)	(862.661)	(852.791)
REGN3	−499.882	−500.628	−533.588	−360.471
	(1,864.535)	(1,823.241)	(1,764.428)	(1,751.873)
REGN4	230.897	219.572	204.147	310.894
	(890.368)	(846.406)	(807.680)	(797.493)
REGN5	59.067	107.966	115.888	147.361
	(1,298.116)	(1,299.413)	(1,295.674)	(1,294.613)
REGN6	276.987	269.784	265.636	307.939
	(625.214)	(611.374)	(597.714)	(595.387)
Adjusted R^2	.903	.903	.904	.904
N	373	373	373	373

Note.—Each equation also includes twelve quarterly dummy variables. Standard errors are in parentheses.

* Equation (1) uses a second-order polynomial distributed lag formulation for IQTR1–IQTR6.

TABLE 4

Maximum Likelihood Equations for MVIO (Noncompliance with BOD Standards)

INDEPENDENT VARIABLE	COEFFICIENTS			
	(1)	(2)	(3)	(4)
INTERCEPT	−7.872	−7.648	−7.991	−8.012
	(23.008)	(22.884)	(22.884)	(23.113)
MVIOT4	2.650	2.637	2.640	2.641
	(.362)	(.359)	(.356)	(.356)
IQTR1	−1.12	−1.019	−.920	−.914
	(.442)	(.429)	(.418)	(.413)
IQTR2	−.063	−.134	−.037	...
	(.421)	(.411)	(.396)	...
IQTR3	−.606	−.644	...	...
	(.398)	(.396)	...	...
IQTR4	−.030	−.141	...	...
	(.387)	(.369)	...	...
IQTR5	.448	...	...	...
	(.389)	...	...	...
IQTR6	.071	...	...	...
	(.360)	...	...	...
TONS	-5.07×10^{-4}	4.971×10^{-4}	-5.127×10^{-4}	-5.124×10^{-4}
	(4×10^{-4})	(3.956×10^{-4})	(3.913×10^{-4})	(3.91×10^{-4})
SIC11	6.321	6.263	6.396	6.405
	(22.998)	(22.875)	(23.108)	(23.106)
SIC21	5.800	5.754	5.958	5.968
	(22.999)	(22.876)	(23.109)	(23.107)
SIC31	5.352	5.306	5.423	5.431
	(23.00)	(22.877)	(23.110)	(23.109)
SIC48	2.506	2.404	3.064	3.084
	(23.077)	(22.951)	(23.178)	(23.175)
REGN1	1.709	1.791	1.540	1.531
	(.746)	(.736)	(.690)	(.683)
REGN3	2.188	2.474	2.033	2.015
	(1.481)	(1.412)	(1.336)	(1.319)
REGN4	1.098	1.316	1.101	1.094
	(.685)	(.655)	(.621)	(.615)
REGN5	1.835	1.868	1.951	1.950
	(.888)	(.889)	(.877)	(.876)
REGN6	−.531	−.404	−.530	−.535
	(.524)	(.505)	(.482)	(.480)
$-2(\log L)$	281.30	282.64	285.35	285.39
N	374	374	374	374

Note.—Each equation also includes twelve quarterly variables. Asymptotic standard errors are in parentheses.

hood estimates for the noncompliance variable, MVIO. Because of the
close similarity of the findings, we discuss each of the variables in turn for
both of the tables.

The four-quarter lagged pollution variable has the expected strong positive effect on the current pollution status, which suggests that past pollution levels predict current discharge levels accurately because of the
slowness of the capital expenditures process needed to transform their
status. Since the MVIO variable has been altered by the logistic transformation, the results for the continuous pollution measure, MQAVG, can
be interpreted more readily. It is quite striking that the weight placed on
the four-quarter lagged pollution value is in excess of 0.98 in each of the
four equations. Thus, there is almost complete replication of the pollution
experience across time. All else being equal (in particular, controlling for
inspections), past pollution performance is close to a perfect predictor of
current pollution levels.

The next set of variables pertains to the set of lagged inspection variables. Consider the continuous discharge measurements in Table 3. In
equation (1) there is a second-order polynomial distributed lag over inspection variables for the preceding six quarters, equation (2) is a free-
form lag over four quarters, equation (3) is a free-form lag over two
quarters, and equation (4) includes only a single lagged value. The pattern
is strikingly similar in all four equations. There is a consistently significant
and substantial influence of IQTR on reducing discharge levels that occurs with a one-quarter lag. Lagged values of more than a quarter are not
consequential. The discrete compliance status equations in Table 4 con-
vey the same influence of inspections; that is, they cause significant reductions in the rate of noncompliance in the subsequent quarter.[27]

The magnitude of the inspection effect is substantial. Consider equation
(4) in Table 3. Each inspection reduces the value of MQAVG by 1,149
pounds per day, which represents about a 20 percent reduction in the
mean value of BOD discharges.[28] Since the coefficients of subsequent
IQTR variables are never significantly positive, there is no evidence of a

[27] Our results suggest that inspections tend to induce reduced discharge levels and enhanced compliance through immediate attention to better plant operation and maintenance,
rather than longer-term capital investments. This finding is consistent with the observation
in Wasserman, *supra* note 10, that the EPA's main enforcement problems in the water
pollution area involve failure to operate and maintain treatment systems already in place
rather than investment in new treatment systems.

[28] A paper by Jonathan S. Feinstein (Detection-controlled Inference (working paper,
M.I.T., Dep't of Econ., 1986) provides an econometric argument for why the coefficients of
the inspection variables would be biased downward if detection of noncompliance were
masked by the nonsubmittal of DMR data. Thus, our results about the effect of inspections
provide a lower bound on their true magnitude.

significant postinspection rebound in pollution discharge levels. These results imply that a permanent improvement in discharge levels takes place as a consequence of the inspection and all associated enforcement actions. Further, the 1,149 pounds per day reduction in BOD in period t is reflected in an approximately equal reduction four quarters hence because the coefficient of MQAVG4 is 0.982. Thus, inspections substantially reduce BOD discharges after about one quarter, and they have a permanent effect on reducing the firm's future pollution levels.[29]

The compliance status results from Table 4 also indicate a large effect of the inspections, and their associated enforcement actions, on noncompliance rates. The coefficients of IQTR1 in equations (1)–(4) average -1.0, implying that had the source not been inspected its odds of being in noncompliance would have been about double. Since most plants in the sample were inspected about once a year and the average rate of noncompliance is 25 percent, the coefficients from the table suggest that without an inspection this noncompliance rate would have been 48 percent.

Finally, the TONS measure has the expected sign in each case, as firms with larger capacity have higher total levels of pollution and lower chances of being out of compliance. Neither effect is statistically significant, however. Similarly, the SIC and regional dummy variables fail to yield any statistically significant effects.

One might expect that the magnitude of the inspection effect would vary with the firm's present noncompliance status. To test this hypothesis, Table 5 presents the key coefficients for equations (3) and (4) from Table 3 in which the inspection variables have been interacted with the compliance status at the time of the inspection. In particular, IPVIO1 equals one in a quarter when a source was inspected one quarter earlier *and* had a permit violation when it was inspected; otherwise it equals zero. IPNOVIO1 equals one in a quarter when a source was inspected one quarter earlier and did not have a permit violation during the quarter of the inspection, and zero otherwise. Similar definitions apply to IPVIO2 and IPNOVIO2, except that these variables have a two-quarter lag.

The results in Table 5 indicate effects of inspections with a one-quarter lag, but no significant effects with a two-quarter lag. Both the IPVIO1 variable (1 percent confidence level, one-tailed test) and the IPNOVIO1 variable (5 percent confidence level, one-tailed test) are statistically sig-

---

[29] When the inspection variables were redefined to separate the effects of compliance sampling inspections from the effects of compliance evaluation inspections (without sampling), we found no significant differences between the effects of the two types of inspections. While care must be taken in interpreting this result because the sample size is relatively low, it suggests that sampling inspections may not be worth their added costs.

TABLE 5

COMPARISON OF EFFECTIVENESS IN REDUCING DISCHARGE
LEVELS (MQAVG) OF INSPECTIONS ON SOURCES OUT OF
COMPLIANCE VERSUS SOURCES IN COMPLIANCE

VARIABLE*	COEFFICIENTS	
	Equation (1)	Equation (2)
IPVIO1	−1,436.733	−1,606.712
	(861.402)	(850.242)
IPNOVIO1	−766.114	−922.998
	(572.448)	(560.809)
IPVIO2	676.362	...
	(810.234)	...
IPNOVIO2	712.996	...
	(567.250)	...

NOTE.—Standard errors are in parentheses.
* For ease of exposition, the coefficients of all the other variables in
the equation are not recorded. These variables are identical to those in
Table 3.

nificant, so that the inspections reduce pollution levels irrespective of the
compliance status.

The point estimates are consistent with one's expectations concerning
the relative magnitude of the effects, as the reductions achieved for firms
out of compliance are almost double those that are produced for firms not
in violation of their permit. For example, from equation (2) we have the
result that a source out of compliance was associated with a 684 (= 1,607
− 923) pound per day greater decrease in BOD discharge levels than a
source in compliance. It should be noted, however, that the standard
errors of the coefficients imply that the 95 percent confidence intervals for
the IPVIO1 coefficient and the IPNOVIO1 coefficient overlap, so that this
result should be treated with appropriate caution.

Effects on the Incidence of DMR Nonreporting

While our econometric results in the beginning of this section clearly
point to the conclusion that plant inspections cause firms to both reduce
their pollutant discharge levels and come more closely into compliance
with their discharge permits, inspections do serve other purposes as well.
One of these is to induce firms to report more regularly their discharge
levels to the EPA or the designated state enforcement agency. We now
examine whether inspections tended to reduce the incidence of DMR

TABLE 6

MEAN DIFFERENCE BETWEEN THE NUMBER OF DMR REPORTS BEFORE AN INSPECTION AND THE NUMBER OF DMR REPORTS AFTER AN INSPECTION

Number of Months of Possible DMR Data prior to and after Inspection	Mean Difference Averaged across All Inspections in Period
1. Four months:	
a. May 1977–November 1984	−.386 (.060)
b. May 1982–November 1984	−.425 (.108)
2. Six months:	
a. July 1977–September 1984	−.714 (.090)
b. July 1982–September 1984	−.868 (.173)
3. Twelve months:	
a. January 1978–March 1984	−2.107 (.196)
b. January 1983–March 1984	−1.693 (.477)

NOTE.—Standard errors of the mean are in parentheses.

nonreporting as measured by the fraction of months without DMR entries in the PCS data base.[30]

Table 6 suggests that there is such a reporting effect for the firms in our sample. The first line in the table measures the difference between the number of months that DMR data was submitted in the four months prior to an inspection and the number of months with DMR data in the four months immediately following the inspection, averaged across all inspections in one of two periods, May 1977–November 1984 and May 1982–November 1984. The second line reports the analogous differences for a six-month period before and after the inspections, while the third line provides results for a twelve-month period of DMR data. All six mean differences are negative and more than two standard errors away from zero, indicating high levels of statistical significance. Thus, the completeness of DMR reporting is clearly higher after inspections.

We must add one note of caution in interpreting these statistics because the mean differences are not adjusted for the trend of increased reporting of DMR data. Still, this trend could not explain much of this difference.

[30] As the discussion in Section III explained, missing DMR data can result either from the failure of firms to report the data to EPA regions or the states or from the failure of the regions or states to enter the reports in the PCS data base. While the first type of failure is probably more closely related to noncompliance than the second type, both reasons for missing PCS data on the DMRs make the PCS system less useful for monitoring enforcement.

To be conservative, consider the first line of the table reporting four months of DMR data, where the trend ought to be least important. For both the long and short periods, the mean difference averages about -0.10 reports per month, which implies that inspections cause one additional month of DMR data to be reported out of every ten months. If the underlying trend of increased reporting accounted for, say, half of this difference (that is, -0.05), then less than twenty months would have to pass before *no more* nonreporting of DMR data would occur. Since the period from May 1977 to November 1984 (line a) contains eighty-four months, the underlying trend must be negligible relative to the rates of increased reporting of DMR data implied by the mean differences in Table 5.

Thus, inspections did tend to cause increased reporting of DMR data into the PCS data base by the firms in our sample, which in turn allows the EPA to more accurately monitor, and therefore enforce, its water pollution standards.

V. EXPLORATORY BENEFIT-COST ANALYSES

One might conclude that EPA inspections are successful because all three of our measures of firms' responses to inspections show significant effects. From a social welfare perspective, however, this question requires valuing the benefits of the effluent reductions induced by an inspection and comparing these benefits to the full costs of each inspection. In what follows, we provide a preliminary exploration of the components of such a benefit-cost analysis. Unfortunately, the existing estimates of the benefits per ton of BOD eliminated per year are only approximate, and we could find no estimates of the compliance costs due to an inspection. As a result, this exercise is highly imprecise. Nevertheless, it does provide some perspective on the welfare consequences of the EPA inspection program for industrial water pollution.

Vaughan and Russell[31] have estimated the national benefits from the improvements in freshwater quality due to the BPT standards at $683 million (in 1980 dollars). While this estimate includes both the out-of-pocket expenses and the opportunity costs of the time of fishermen, it does not include the aesthetic benefits of fishing on cleaner waters, or other benefits such as those from swimming and boating. Development Planning and Resource Associates[32] estimated that the BPT standards

---

[31] William J. Vaughan & Clifford S. Russell, Freshwater Recreational Fishing: The National Benefits of Water Pollution Control 161 (1982).

[32] Development Planning and Resource Associates, Inc., National Benefits of Achieving the 1977, 1983, and 1985 Water Quality Goals (1976).

would reduce BOD discharges by 3,390,233 tons per year, which together with the previous estimate implies an average value of benefits per ton of BOD removed due to the BPT standards of $201.46.

Using equation (4) in Table 3, each inspection will tend to cause a reduction in BOD discharges of 1,148 pounds per day, or 209.51 tons per year. Given the previous benefits estimate of $201.46 per ton, this implies that an average inspection produces $42,208 of benefits every year.[33]

Given the 0.982 coefficient of the MQAVG variable lagged four quarters in equation (4) in Table 3, the effectiveness of an inspection in maintaining lower effluent discharge levels decays at a negligible rate. Accepting the linear form of the equation and rounding this coefficient to 1.0, the equation implies that any BOD reductions from an inspection remain in force for years after the inspection. Thus, we can approximate the annualized benefits per inspection at $42,208.

Given the mix of inspections in our sample of 43 percent compliance sampling inspections (requiring approximately thirty days) and 57 percent compliance evaluation inspections (requiring approximately three days), an average inspection required 14.6 days. Assuming the full cost of inspectors to be $50,000 per year over 220 working days yields a cost of $227 per day, or $3,315 per inspection. Figuring this inspection cost at a 10 percent discount rate gives an annual cost of $332.[34] Calculating the net inspection cost from the benefits gives an adjusted annualized benefit of $41,876 per inspection.

Consider now whether the average annual compliance costs incurred due to inspections are likely to exceed $41,876 per inspection. Since 75 percent of the firms sampled were already in compliance, we would expect them to spend little or nothing after an inspection. Thus, each non-complying firm must spend at least four times $41,876, or $167,504, per year in order that the costs associated with an inspection exceed their benefits.

Whether compliance costs exceed this threshold probably hinges on whether the firm must make a capital investment to attain compliance or whether a change in operating procedures will suffice. Although detailed

---

[33] This calculation assumes that the average benefits of each pound of BOD removed due to an inspection equal the nationwide average benefits of the BPT standards. This simplifying assumption ignores the fact that the effluent reductions at some plants induced by inspections will yield benefits much greater than the average, whereas inspections at other plants, even if they result in lower emissions, will improve water quality much less than for an average inspection. Without more disaggregated information about benefits, we were forced to make this simplifying assumption.

[34] The use of a 10 percent discount rate is required by the Office of Management and Budget, but other more realistic rates would not significantly affect our conclusions.

cost data are not available for all portions of the pulp and paper industry, some suggestive statistics are available for the costs of an activated sludge treatment system used to comply with the BPT standards in the wastepaper-molded products subcategory of the industry.[35]

For a concrete example, focus on the intermediate-size plant (45 kg/day). Compliance for these firms entails an annual operation and maintenance outlay of $113,000, annual energy cost of $19,000, and an average annual capital cost of $339,000, leading to a total annual cost of $471,000.[36] If compliance following an inspection involves only the operation and maintenance costs, the expenditure of $132,000 is somewhat below the value of benefits less inspection costs. However, if a capital investment is required, the costs exceed the pollution reduction benefits net of enforcement costs by a factor of almost three.

For small plants, with a total annual compliance cost (including amortized capital costs) of $288,000, the compliance costs outweigh benefits once capital costs are included. For large wastepaper-molded products plants with average annual compliance costs of $879,000, even the operation and maintenance costs of $176,000 exceed the pollution reduction benefits.[37]

To the extent that the rough estimates in this particular case reflect the costs and the benefits for other industry subcategories, the following conclusion holds. If inspections lead firms to make substantial capital investments, then the costs of compliance exceed the benefits. Once having made these investments, firms may be more likely to undertake the appropriate operating procedures to maintain their compliance status as a result of an inspection. This promotion of continued vigilance on the part of firms that have already made the required capital investment is more likely to pass a benefit-cost test.

VI. CONCLUSION

Compared with other health, safety, and environmental regulations, EPA water pollution regulations for the pulp and paper industry represent an unusual success story. The EPA sets standards for which compliance

---

[35] While wastepaper-molded product is only one of many subcategories in the industry, the activated sludge treatment system represents a standard technology for biological treatment of pulp and paper mill wastes.

[36] All the cost estimates are found in U.S. Environmental Protection Agency, Development Document for Effluent Guidelines and Standards for the Pulp, Paper, and Paperboard and the Builders' Paper and Board Mills (1982).

[37] See U.S. EPA, *supra* note 36.

is feasible and then enforces these standards relatively vigorously, with inspections averaging one per year for our sample. This mix is the opposite of OSHA's, which has stringent standards coupled with weak enforcement. The coupling of regulations for which compliance is feasible with stringent enforcement is likely to create strong incentives for compliance, and the available evidence bears this out. Inspections and their associated enforcement actions have a strong effect on both pollution levels and rates of compliance with the permit levels. In addition, inspections are associated with less nonreporting of pollutant discharge levels. Judged with respect to its legislative mandate to improve water quality, this effort is clearly a success.

In view of the evident effectiveness of the water pollution enforcement effort, one might well ask whether U.S. water quality should not have improved overall as a result of such efforts. This may not be a meaningful test, however, since the real issue is whether water quality would have been worse in the absence of EPA enforcement, not whether the overall level of water quality has improved. With economic growth, the baseline rate of total discharge of BOD should be increasing. Coupled with a fixed assimilative capacity of any body of water, this growth implies that there should be deteriorating water quality in the absence of EPA actions.

Available information on national surface water quality trends shows modest improvements in the late 1970s and early 1980s. The Council on Environmental Quality reported in their fifteenth annual report that "significant progress has been achieved in cleaning up the nation's waters."[38] This conclusion is based on an assessment by the Association of State and Interstate Water Pollution Control Administrators (ASIWPCA) of the degree to which beneficial uses were supported in surface waters assessed in each state. The ASIWPCA reported that in 1982 64 percent of assessed surface waters supported their designated uses, compared with 36 percent of 1972.[39] In their 1986 Report to Congress, the EPA reported that 74 percent of assessed river miles fully supported their designated uses.[40]

These results are consistent with the conclusions of the ASIWPCA and the EPA in their report, *The States' Evaluation of Progress, 1972–1982*. Based on an evaluation of 42 percent of the nation's streams, they re-

[38] Council on Environmental Quality, Environmental Quality: 15th Annual Report 85 (1984).

[39] Council on Environmental Quality, Environmental Quality: 17th Annual Report C-42 (1986).

[40] U.S. Environmental Protection Agency, National Water Quality Inventory: 1986 Report to Congress 2 (1986).

ported that, between 1972 and 1982, 47,000 miles of assessed streams improved in quality while 11,000 miles declined.[41]

Available data are insufficient to assess changes in ambient BOD levels of an annual or regional basis; however, the Council on Environmental Quality reported that BOD discharges from pulp and paper mills decreased from 706.8 thousand tons in 1974 to 207.6 thousand tons in 1984.[42]

One might raise the more general issue not treated by the EPA's enabling legislation: whether the benefits accruing from this pollution reduction are commensurate with their costs. This calculation needs substantially better data to refine it, but some preliminary observations are in order. If one includes only the operation and maintenance cost associated with pollution control, then the benefits of inspections may exceed their costs. If capital costs are included as well, the results are probably reversed. One major difficulty associated with this calculation is that we cannot distinguish which incremental pollution control expenditures are associated with the effect of the inspections. Notwithstanding these caveats, it appears that the EPA water pollution regulations represent a dramatic departure from the apparent impotence of most other forms of health, safety, and environmental regulation. The remaining challenge is to set standards at a level that will ensure that the regulations are in society's best interests.

[41] Council on Environmental Quality, *supra* note 38, at 82–83.
[42] Council on Environmental Quality, *supra* note 39, at C-38.

[31]
Monitoring and Enforcement

Clifford S. Russell

How often does one hear it said, "In the United States, if a problem appears, all we ever seem to do is pass a new law"? In some cases this complaint may imply that the speaker has an alternative approach in mind. But in many situations what seems to be meant is that we are wont to pass laws and then lose interest in the results. What is missing in such situations is a commitment of resources to checking up on whether those covered by the law and regulations are doing (or not doing) what is required of (or forbidden to) them—this is *monitoring*; and to taking actions that force violators to mend their ways and that provide visible examples to encourage others in the regulated population to maintain desired behavior to avoid a similar fate—this is *enforcement*.

This chapter deals with monitoring and enforcement of the nation's environmental regulations. It argues that this area of public policy shows signs of the sort of half-hearted commitment just described. Efforts to monitor regulated behavior appear to have been inadequate to the task—a very difficult task in many instances—and typical enforcement practices appear to have been insufficiently rigorous. Together these inadequacies seem to have encouraged widespread violations of environmental regulations. Unfortunately, however, the very lack of monitoring means that the evidence on compliance or lack thereof is spotty at best and nonexistent at worst.

243

We begin by drawing distinctions among the different monitoring problems posed by the range of environmental regulation. Some of the very real difficulties in monitoring are then discussed, and to balance the critical comments, notice is taken of recent efforts at improvement. The chapter's final section offers several specific suggestions for policies—and even legislative actions—that should help make further progress easier. It will be useful to begin with some distinctions and elaborations on the simple definitions of monitoring and enforcement set out above.

DISTINCTIONS AND DEFINITIONS

The complexity and the technological bases of U.S. environmental regulation imply that several different kinds of monitoring and enforcement activities have claim to our attention. The air and water pollution legislation of the 1970s stressed the issuance of technology-based discharge or emissions standards. (As discussed by Freeman in chapter 4, this focus represented a reaction against the ineffectiveness of previous efforts to assert control on the basis of ambient quality standards or problems.) The practical result for monitoring and enforcement was concentration on determining that the required technology (pollution control equipment) was in place and capable of performing in such a way as to meet the terms written into discharge permits. This is known as *initial compliance* monitoring. Much of the discussion of compliance rates, successes, and failures over the past decade and a half has referred to initial compliance. Enforcement actions involving failures of initial compliance have been roughly of two sorts: penalties for failing to make any effort to purchase and install the technology; and denial of permission to operate if the equipment, once installed, fails to produce the desired results.

Clearly, however, the purpose of technology-based regulations ought to be to affect the actual discharges from the sources, and to do so on a continuing basis. Checking up on this involves monitoring for *continuing compliance*. Enforcement actions, in principle, follow from findings of failure to live up to permit terms, and may take many forms, from a slap on the wrist to a full-blown criminal prosecution, depending on how serious the violation is thought to be and how cooperative or aggressive the violator appears. Current environmental laws and regulations contemplate that a common form of enforcement action will entail the collection of an administrative or civil penalty designed to capture any cost saving reaped by the source from the violation, plus the extraction of something in the nature of a fine.[1]

The ultimate goal of the laws and regulations is to improve (or in some cases maintain) the day-to-day air and water quality experienced by citizens generally. This quality is the result of the interaction of the discharges of pollutants by sources with the natural systems that receive them—the atmosphere, water courses, land and underground aquifers. These systems dilute, transport, and even transform the pollutants they receive. The complexities of such changes in form, location, and concentration are so great that ambient conditions could not with any confidence be deduced from discharges alone (even if we knew discharges quite exactly—which we do not). Rather, ambient quality must be checked directly if we want to know what we are getting for our pollution control spending. This is *ambient quality* monitoring. The three types of monitoring and their relations to each other are shown schematically in figure 7-1.

This chapter concentrates on monitoring and enforcement of continuing compliance of individual sources. In passing, it will be necessary to cite evidence on initial compliance because this has been dominant in the Environmental Protection Agency's discussions of its activities. But very little will be said about ambient quality monitoring. Our focus should not be taken to imply either that ambient quality monitoring is easy or that it is being done as well as it might be. On the contrary, many observers and analysts argue that far too little effort is made in both air and water quality monitoring.[2] With respect to air quality, the number of monitoring stations and their placement implies a dearth of information on rural pollution levels; in the case of water pollution, roughly the opposite is true—there is a dearth of consistent records for the lower, often urbanized, stretches of rivers and their estuaries. Further, audits of ambient air monitoring stations have suggested that many of them serve their purpose poorly.[3]

Nor is ambient quality monitoring unimportant. One of its functions is to check actual conditions against the ambient standards that express society's judgments about what should be achieved (see chapters 3 and 4). Therefore, ambient monitoring is not simply information-gathering but ultimately is linked to violations and enforcement. The violations, however, are not in general attached to individual sources. Rather, a region or a water course may be found to be failing to meet its ambient quality targets. In the case of air quality, this makes the region a nonattainment area, and as such subject to a variety of special constraints and requirements.

Potentially at least as important, the ambient quality record is the sine qua non for judging the effectiveness of environmental regulation. While inferring effectiveness from such a record is far from trivial (see chapters 3 and 4), nothing can be done in the absence of such a record. Further,

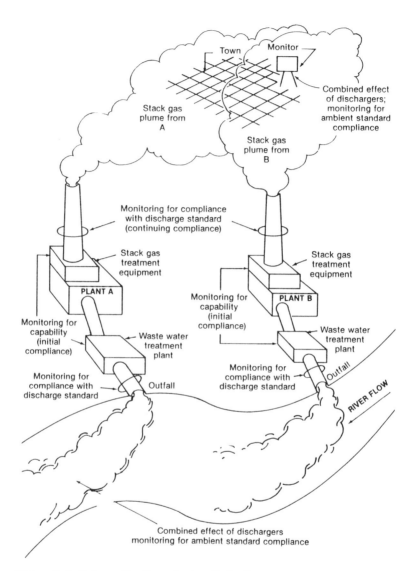

Figure 7–1. Schematic of monitoring distinctions

estimation of the benefits of environmental regulation—an activity important to better policy formulation in the future—almost invariably requires a record of actual environmental quality levels. This has to be matched with base-period data on health status or recreational activities, material damage levels, or agricultural yields in order to estimate statistically the relevant response functions.

The decision to concentrate here on continuing compliance monitoring reflects two judgments: first, that of the kinds of monitoring this has been relatively least discussed in policy analysis; and second, that success in encouraging continuing compliance by sources would produce ambient quality at or near the desired ambient standards.[4]

Within the narrower field of monitoring for continuing compliance, there are two other broad distinctions that are worth setting out. The first is that the quantities and characteristics of the pollutants being measured can differ from source to source (and even between two identical sources located in different jurisdictions). The differences include:

1. Types of pollutants measured. Nitrogen, for example, appears in different chemical states in both waste water and waste (stack) gases. In addition, in discharge permits the compounds being constrained may differ, and so may imply differences in monitoring difficulty.

2. Forms in which the materials enter the environment—as gases in a stack; compounds dissolved or suspended in waste water; hazardous materials in drums or in such spent consumer goods as old appliances; or mercury in batteries or PCBs in old electrical transformers. Again, different monitoring problems are implied. A smokestack, for example, is at a fixed point and visible to the public eye. The dumping of drums or transformers can go on anywhere at any time.

3. How the standard is defined. Alternatives include pollutant concentration (for example, milligrams per liter of a compound dissolved in waste water), mass per unit time, and mass per unit input to or output from the facility in question. Concentration alone is easiest to measure and least useful in protecting the environment. Mass per unit input or output involves the most elaborate measurements but may not provide the desired protection either, since permitted pollution may be allowed to increase with output.

The second broad distinction involves the time dimension. It may best be understood by example. On the one hand, a standard may specify that on *no* day are discharges to exceed 50 tons of SO_2; on the other, daily discharges *on average* over a month might be limited to 50 tons. The second specification does not constrain any particular day's discharge, for amounts of more than 50 tons discharged on some days may be balanced by amounts under that level on others (it does constrain any month's total discharges to be fifty times the number of days in the month, however). This distinction is important, among other reasons, because of its implications for the frequency of monitoring activity as related to inferences about source compliance rates. While this is a very complicated subject, an intuitive appreciation may be gained from the observation that if the responsible agency measures a source's discharge for one day per month, it will have direct evidence only about compliance or violation with respect to a single day's upper limit and only for that day. Inferences about other days and about compliance with monthly limits (daily averages over a month) involve strong assumptions about constancy of source behavior (the representativeness of the sample) *and* exogenous influences such as temperature, volume of production, and input quality.

A SKETCH OF THE CURRENT SYSTEM

For the U.S. environmental monitoring and enforcement system, one might generously describe the present period as one of transition. From the passage of the new clean air and water legislation in the early 1970s until the last few years, the EPA—and perforce the related state agencies—have been interested almost entirely in maintaining progress on the installation of technology required for meeting discharge permit terms. In the words of one EPA observer,

The traditional approach to compliance monitoring and enforcement consists of periodic inspections by government employees, warning letters or notices for some or most violations, followed by additional warnings, or escalation to administrative orders or civil litigation and case referral. Some programs are assisted by information on source compliance achieved through source self-monitoring requirements, but under the traditional approaches, such self-monitoring is imposed more to ensure higher levels of compliance through source self-awareness than it is to manage an enforcement program.

The traditional approach worked fairly well when compliance focused on initial installation of pollution control equipment at a limited number of large facilities. Detection was comparatively straightforward. A periodic inspection could do the job. Heavily publicized selective enforcement actions complemented what were then new pressures and norms of behavior for environmental protection for which there was a large degree of public consensus. Fixed hit lists of known violators proved to be effective in gaining compliance. The traditional approach has run into several problems . . . *In the older established programs such as air and water, the problems have shifted from initial compliance to continuing compliance. . . .* [emphasis in the original][5]

The concentration on initial compliance has had a pervasive influence on the system. Everything from technology design to the laws limiting the agencies' rights to enter a plant to monitor, and from the design of noncompliance penalties to the keeping of program statistics, has reflected it. In particular, monitoring equipment as approved by the EPA has been unwieldy and expensive to set up and operate. For its efficient use, skilled personnel have been required, as well as notice to the source and cooperation from the source. The goal, judging by the standards applied to newer technologies, has been to keep errors of measurement low.

Moreover, noncompliance penalties have been designed primarily to recapture the costs that polluters avoid by failing to meet requirements. Where capital goods purchases are in question, as they are in initial compliance, the largest part of these avoided costs are easily identified. Where only operating costs are involved, as they would be in cases of continuing compliance failures, measurement difficulties may be severe. It may be difficult, for example, even to identify the causes of inadequate performance by wastewater treatment plants, let alone to estimate the cost savings from allowing those causes to occur.

Further, those figures on the percentage of sources in compliance that have been regularly reported by the EPA—and that have shown very high levels of compliance—have historically been based on initial, not continuing compliance. Indeed, as will become clear below, it has been very difficult for the EPA or anyone else to know what levels of continuing compliance are in fact being maintained.[6]

This is not to say that no efforts have been made to ascertain and encourage continuing compliance. The major characteristics of the efforts that have been made are outlined below.

Heavy reliance on self-monitoring by sources. Almost all states require almost all large sources of air and water pollution to monitor their own discharges and report on the results. According to the results of a 1982 survey by Resources for the Future (RFF) of state officials responsible for compliance monitoring, only one of the responding states used no self-monitoring for existing sources.[7] In the case of air pollution, 4 of the 22 reporting agencies required all air pollution sources to self-monitor. Averaged over the responding states, 28 percent of air pollution *sources* were required to self-monitor. As regards water pollution, a self-monitoring requirement was almost universal. Of the 33 state agencies responding to the survey, two-thirds required all water pollution sources to do some self-monitoring. And the overall average percentage of *sources* required to self-monitor was 84.

250   CLIFFORD S. RUSSELL

Continuous self-monitoring of air pollution emissions is required of all
facilities subject to New Source Performance Standards under the Clean
Air Act. And even in the hazardous materials field, some self-monitoring
is in force in the shape of a requirement under the Resource Conservation
and Recovery Act (RCRA) that all generators of hazardous materials keep
track of their wastes after the wastes leave the plant (by using the
complicated shipping manifest system) and report if the wastes fail to
reach an approved final disposal facility. Self-monitoring of groundwater
around land disposal sites is also required under RCRA.[8] (See chapters 4
and 5 for further discussions of self-monitoring.)

Infrequent auditing of the self-reporting sources.[9]   Results from the
RFF survey show that while large sources are audited more frequently
than small ones, on average neither is visited very often (see table 7–1),
especially given the fact that the majority of the standards in question
involved hourly and daily emission limits.[10] Thus outside measurement of
the discharges of large air pollution sources occurs on average only once
every eight and one-half months. For water pollution the frequency is
greater, but still notably low—once every five months.

Table 7–1. Frequency of Audit Visits to Self-Monitored Sources

| | Type of source | | | |
| | Air pollution | | Water pollution | |
Type of visit	Large sources	Small sources	Large sources	Small sources
Without measurement				
Responses (number)	26	15	41	35
Mean (times/yr)	1.70	0.88	3.10	1.44
Standard deviation	1.19	0.64	3.21	1.71
Including measurement				
Responses (number)	18	10	34	26
Mean (times/yr)	1.43	0.89	2.41	1.39
Standard deviation	1.26	0.81	2.91	1.72

Source: Clifford S. Russell, "Pollution Monitoring Survey: Summary Report,"
(Washington, D.C., Resources for the Future, 1983) p. 16. See note 7 to this
chapter.

Audit visits lack the characteristics of a rigorous enforcement effort designed to catch ongoing violations. Most important, those visits that do occur are often announced in advance (see table 7–2).[11] To some extent this is a response to the uncertain state of the law regarding entry for monitoring purposes, in combination with the technological limitations of the approved monitoring methods. Whether the Environmental Protection Agency has the right to make unannounced, warrantless visits for monitoring purposes is an unsettled question.[12] Announcement helps to avoid quarrels and possible litigation about the right of entry.

A related legal question was recently decided by the U.S. Supreme Court, although by a 5 to 4 majority, and with important distinctions drawn that may well encourage future litigation. In this opinion it was found that the remote monitoring instrumentation used by the EPA did not give rise to an unreasonable search. This was a long-running case involving aerial photography of a Dow Chemical plant to check for sources of air pollution, and the District and Appeals courts had disagreed. The Supreme Court, in May 1986, agreed with the Sixth Circuit Court of

Table 7–2. Auditing Self-Monitoring Sources: Conduct and Content of Visits

	All sources	Air pollution sources	Water pollution sources
Conduct of audits			
Responses	$N = 68$	$N = 26$	$N = 42$
Always announced	0.16	0.19	0.14
Never announced	0.19	0.19	0.19
Sometimes announced	0.65	0.62	0.67
Sum of frequencies	1.00	1.00	1.00
Content of audits[a]			
Responses	$N = 68$	$N = 25$	$N = 43$
Inspect records	0.97	0.96	0.98
Inspect equipment	0.93	1.00	0.88
Measure discharges[b]	0.90	0.80	0.95
Other[c]	0.12	0.04	0.16

Source: Russell, "Pollution Monitoring Survey: Summary Report."

[a] The sum of the frequencies for items describing audit contents need not sum to one.

[b] The difference between air and water proportions is significant at 5 percent or better.

[c] Other activities undertaken as part of audits consisted almost entirely of one or another version of a laboratory inspection and a check on analytical methods. Frequencies do not sum to one because in principle every audit could include every content item.

252   CLIFFORD S. RUSSELL

Appeals and upheld the EPA's right to use aerial surveillance for enforcement. This should strengthen the general case for remote monitoring, with implications for the entry question itself. However, the majority opinion drew a distinction between use of a standard commercial camera and "highly sophisticated . . . equipment."[13]

A certain amount of ad hoc invention in the definition of violations. Even though emissions standards are clearly stated under outstanding permits, and even though those standards are based on official documents that take into account discharge variations resulting from changing circumstances beyond the control of the source, another layer of allowance has been tacked on to separate "significant" from "insignificant" violations. These allowances have both a quantitative and a time dimension. Discharges must exceed the actual standard by some amount for some period of time before a significant violation will be said to have taken place.[14]

Infrequent and reluctant use of self-monitoring records as the basis for notices of violation, even though the records show significant violations to be occurring. Thus the General Accounting Office (GAO) observed that, on the basis of a study of water pollution self-monitoring records in six states, formal enforcement action was in some cases not taken for years after noncompliance began.[15]

Indeed, some major environmental groups, feeling that enforcement had become too lax, began a campaign in 1983 to use self-monitoring records—which are public information—as the bases for citizens' lawsuits against the violating polluters.[16] These suits have been successful in at least two ways. First, the courts have ruled against the polluters on most of the technical legal challenges the latter have raised; thus, as part of the overall enforcement armory of society, the device has been found valid under the existing laws.[17] Second, and no doubt more important, the suits have helped embarrass the EPA into rethinking its own enforcement stance, as will be discussed below.

When violations are found and enforcement actions taken, the penalties assessed appear to be so small as to be insignificant in a corporation's (or city's) income statement. Many states claim to pursue a so-called voluntary compliance policy, by which they mean that no penalties are ordinarily levied for violations initially. Rather, if penalties are used, it is to punish sources that refuse to correct violations or otherwise prove notably uncooperative. Even when penalties are assessed, however, they seem on average to be very small. Some evidence of this is presented in table 7–3, which shows both the average penalty per assessment and the average penalty per notice of violation issued. The latter figure can be thought of as the expected penalty per discovered violation. This expected value of the penalty for a violation is closer to the incentive that can be presumed to drive the source's compliance decision than is the average penalty per assessment.[18] The EPA's official civil penalty policy, as mentioned above, is to assess a violator a two-part penalty.[19] The first part is an attempt to remove the profit from violations by recapturing the costs saved by the source. The second part, or gravity component, is intended as punishment pure and simple. This policy has not been in effect long enough to have produced data on its impact.

Before examining the evidence on continuing compliance, such as it is, from the system just described, it is worth adding that at least one other major industrial nation appears to have a similar system. Papers and books about the United Kingdom's activities in pollution control monitoring and enforcement paint a roughly similar picture.[20] Thus in the United Kingdom self-monitoring is important; agency personnel visit infrequently and measure discharges independently even less frequently; and much stress is laid on a version of voluntary compliance, with penalties reserved for recalcitrant violators. The reasons suggested for the evolution of the U.K. system in this direction are about what would be expected: low agency budgets; legal difficulties of bringing a serious prosecution for a "mere" violation of a discharge standard; and the unquantifiable, personal considerations that push toward amiability rather than confrontation when inspectors and plant management know they must live together over the long run.

EVIDENCE ON CONTINUING COMPLIANCE

Since one of the characteristics of the existing system is that very little monitoring is done by the Environmental Protection Agency or its state counterparts, it is inevitable that very few data on actual continuing compliance exist. Some notion of the current situation can be gained from special studies, most frequently those requested by congressional committees and undertaken by the U.S. General Accounting Office.

Air Pollution

In the late 1970s the EPA and the White House Council on Environmental Quality (CEQ) sponsored jointly or singly nine studies examining continuing compliance with air pollution emissions limits by industrial sources.[21]

254   CLIFFORD S. RUSSELL

Table 7–3. State Enforcement Activity (annual averages, 1978–1983)

State	NOVs issued (1)	Civil actions brought (2)	Number of penalties assessed (3)	Penalties ($) Total penalties assessed (4)	Average size of penalties assessed (5) = (4)/(3)	Average penalty per NOV (6) = (4)/(1)
Colorado	124	3.6	0.5	60	120	0.5
Connecticut	800	2.3	21.5	7,800	363[a]	9.8
Indiana	59[b]	NA	21.0	85,000	4,050[c]	1,440
Kentucky	194	5.2	5.2	13,100	2,520[d]	68
Massachusetts	NA	0	0	0	0	0
Minnesota	41	NA	10	109,000	10,900	2,660
Nebraska	59[e]	NA	0.2	400	200	0.07
Nevada	31.5	0.3	2.3	105	45	3.3
New Jersey	1,167	NA	350	500,000	1,430	428
Oregon	197	NA	30.7	21,600	705	110
Pennsylvania	NA	NA	176[f]	260,000	1,480	NA
Rhode Island	5	7.2	0	0	0	0
South Carolina	68[g]	5.5	2.2	53,400	24,250[h]	785
South Dakota	17	1.2	0.3	300	1,000	20
Tennessee	193[i]	8.4	0	0	0	0
Virginia	161	7.8	3	600	200	3.8
Wisconsin	80.5[j]	13.5	7.7	61,200	7,951	760

Note: NOV = Notice of Violation; NA = not available.

Source: Clifford S. Russell, Winston Harrington, and William J. Vaughan, *Enforcing Pollution Control Laws* (Washington, D.C., Resources for the Future, 1986) table 2–7.

[a] Refers to amounts assessed; over the same period actual collections were 62 percent of assessments.

[b] Includes both NOVs and Compliance Orders.

[c] Excludes one penalty of $415,000, which was cancelled when company bought equivalent amount of air pollution equipment.

[d] Excludes performance bonds (two required for total of $45,000).

[e] Excludes Lincoln and Omaha.

[f] Only data from the last quarter of 1983 were readily available. This figure is an extrapolation to an annual average.

[g] No NOVs were issued before April 18, 1983. From July 1, 1983, to March 31, 1984, 51 NOVs were issued.

[h] Excludes one fine of $1,700 per month until compliance was restored and one fine of $250,000 dropped in consideration of a donation of a like sum to a technical college.

[i] Does not include NOVs from continuous monitoring data, which were numerous.

[j] 1980–1983 only.

Two tasks were undertaken. First, state and local agencies were studied and their procedures for pursuing continuing compliance monitoring and enforcement were analyzed. Second, data were gathered on raw pollutant loads, types of control equipment, emissions, and causes of excess emission "incidents." The data from the second task give some (albeit very modest) insight into compliance rates among a small sample of 119 sources from the 23,000 large sources estimated to exist in the United States. Within these limitations, the following data are relevant: percentage of sources in violation—65; percentage of time the sources were in violation—11; excess emissions as a percentage of standards—10. [22]

Before considering what can be learned from the summary document for these studies, it is worth noting why the effort is less valuable than it otherwise might be. To begin with, discussion in the summary document does not make clear what standards form the basis for the "allowable" and "excess" emissions calculations. It is implied that the basis is permitted daily emission limits, or perhaps permitted hourly limits. But the calculation of a "credit" for discharge levels below the total annual permitted amount suggests that it is long-term average performance that is really of interest. While the data are said to be for emissions of various pollutants, no breakdown of incidents or standards by pollutant is offered.

In addition, in the summary document the sources of data are only vaguely described as inspections and records. It does not seem that independent source measurement was involved, suggesting that the results at least do not overstate the size or frequency of violations. Moreover, the causes of violation allowed for in the studies do not include deliberate evasion. Equipment misdesign or malfunction, excusable process accidents (such as a burst pipe or burned-out transformer, beyond operator control) are the only possibilities allowed for. The result is a catalog of "blameless" events. [23]

An even less satisfactory source of data on compliance with air pollution emission standards is provided by a 1979 report from the General Accounting Office. [24] For this report, the GAO staff reviewed documents in the EPA regional offices and in state and local air pollution control agencies. No independent measurements of emissions were undertaken. The principal problem with the report, however, is that it is very unclear on the distinction between continuing and initial compliance. To the extent one can infer what is being discussed from the imprecise language, it appears to deal with initial compliance. A major message of the report does have some relevance for the discussion of continuing compliance, however, because it confirms the characterization of the system given above. The GAO reports that actual emissions monitoring, and even on-site inspection without such monitoring, are very seldom used to determine compliance status. (Only 3 percent of the sources reported to be in full compliance were so characterized on the basis of source tests; only 22 percent had been inspected without a source test.) Reinspection of sources discovered in violation was also found to be spotty. And 50 percent of the sources reported to be on clean-up schedules were found to be in violation of those schedules. A more recent report by the GAO found that in 1984, while some inspection was performed on 95 percent of stationary sources, 40 percent of the inspections performed were inadequate on grounds of either content or infrequency of repetition.[25]

Water Pollution

The General Accounting Office has also prepared two reports on continuing compliance with water pollution control permit terms.[26] These are somewhat more useful than either of the air pollution reports. Though both reports on water pollution control share the understandable weakness of being based entirely on reviews of self-monitoring records, they do at least focus clearly on month-to-month behavior and do supply clear definitions of compliance status. Indeed, since both reports share the same definition of significant noncompliance, it is even possible to make a simple comparison across the space of time separating the reports.[27] Both reports are based on data from major sources, defined as those discharging more than one million gallons of treated water per day.

The total number of incidents of violation found by GAO investigators is shown in the first column of table 7–4. The second, third, and fourth columns show how the total breaks down by length of period of violation: short (one to three months), medium (four to six months), and long (more than six months). It appears that short periods of violation were just as common in 1980/82 as in 1978/79. But long periods of violation, greater than six months, were much less common in the later period. This is an encouraging indicator. On the other hand, when account is taken both of the length and severity of violations, as the GAO did when devising a measure of significant noncompliance, the picture is darker (see table 7–5). At best it would seem that significant noncompliance remained constant between the two periods among large municipal sources. The data do not permit an intertemporal comparison for industrial sources, but the 1980/82 data do indicate that such sources are reporting less significant noncompliance than are municipal treatment plants.[28] This is especially interesting because most of the attention given to pollution problems and violations in the media focuses on industrial sources.

Table 7–4. Extent of Violations Found in Self-Reported Data

Report	Total incidents of violation	Months in violation of at least one limit[a]		
		1–3	4–6	more than 6
1978/79				
Municipal				
N = 242	211	53	39	119
(%)	(87)	(22)	(16)	(49)
1980/82				
Municipal				
N = 274	139	44	29	66
(%)	(51)	(16)	(11)	(24)
Industrial				
N = 257	135	63	24	48
(%)	(53)	(25)	(9)	(19)

Sources: General Accounting Office, *Costly Waste Water Treatment Plants Fail to Perform as Expected,* CED-81-9 (Washington, D.C., 1980); General Accounting Office, *Waste Water Discharges Are Not Complying with EPA Pollution Control Permits,* RCED-84-53 (Washington, D.C., 1983).

[a] For 1980/82, where separate figures were given for concentration and quantity violations, this table gives results for quantities.

The GAO reports on water pollution control include some information on the causes of observed violations, as determined by discussions with knowledgeable state and EPA officials. In the 1978/79 study, five major causes were identified: equipment deficiencies, intake volume overloads, operation and maintenance deficiencies, design deficiencies, and intake pollutant overloads or toxicity. The 1980/82 study found the first three of these were the most prevalent causes of violations in that period. Once again, it is interesting that no explicit account was taken of the possibility of deliberate violation.

It is also worth noting that in the 1980/82 study the General Accounting Office criticized state and EPA enforcement practices for allowing violations to continue uncorrected for long periods of time. It appears that any kind of formal enforcement action may be delayed for months or perhaps years, even after a significant violation is revealed by self-monitoring reports. This may be because taking the enforcement action far enough to assess a civil penalty adds substantially to the required time,

Table 7–5. Significant Violations of Water Pollution Discharge Permits

Report	Plants
1978/79	
Municipal	
N = 242	66
(%)	(27)
1980/82	
Municipal	
N = 274	88
(%)	(32)
Industrial	
N = 257	42
(%)	(16)

Note: Significant violations are defined by size of violation and number of periods the violation continued.

Sources: General Accounting Office, *Costly Waster Water Treatment Plants;* General Accounting Office, *Waste Water Discharges Are Not Complying.*

since the Justice Department has to become involved. Indeed, the GAO concludes that "EPA's current enforcement philosophy for water pollution control still centers around voluntary compliance and the nonconfrontational approach established in 1982."[29]

Air Pollution by Automobiles

Congress has written specific discharge standards for emissions of automobile pollutants (hydrocarbons, carbon monoxide, and nitrogen dioxide) into the Clean Air Act (see chapter 3). The monitoring and enforcement effort for these standards originally was aimed only at the end of the assembly line, where brand new cars were to be tested—an effort analogous to determining initial compliance for stationary sources. As matters have subsequently developed, continuing compliance is now to be monitored through vehicle emissions tests, but only in 30 of 50 states and even there usually only at fixed intervals (an analog to announced monitoring visits).[30]

Two General Accounting Office reports—one issued in 1979 and the other in 1985—on the so-called inspection and maintenance (or I/M) program for cars provide other analogs to experience with stationary source control programs. Under the initial program of prototype testing and certification, very little effort was put into continuing compliance monitoring. The result, as determined by the EPA in the period 1972 to 1976, was that of 2,000 automobiles in use and sampled at that time, about 80 percent failed to meet the exhaust standards for their model year.[31] The readiness with which the failure to meet standards was ascribed by the EPA to conscious cheating by car owners stands in interesting contrast with the stationary source analyses reported above. In particular, it was estimated by the EPA that almost half the vehicle failures were due either to tampering with the pollution control equipment or to improper driving or fuel use. "Maladjustment," the single largest cause of failure, corresponds to operation and maintenance deficiencies in the stationary case. Apparently, in the government's view, individuals may cheat on pollution control for their vehicles but corporations or municipalities will not do so at the plants they operate.

In its 1979 report the GAO recommends expanded and improved monitoring for continuing compliance—at the least, annual vehicle inspections and necessary reinspection to check that needed work has been done. By the time the 1985 GAO report on this subject was issued, such programs were required in 44 metropolitan areas in 30 states and were actually in place in 25 or more.[32] The 1985 report criticizes the in-place programs for ineffective monitoring and enforcement, though the criticisms are not so harsh as those given the water program, and the EPA's efforts to address the problems are recognized. The major criticism in the 1985 GAO report is that enforcement has often been inadequate to spur owners to have their cars inspected. It was entirely too good a gamble from an owner's point of view to ignore the inspection requirement. In addition, some fraud in the issuing of "pass" stickers was observed in undercover visits to inspection stations.

A second criticism might or might not be valid. The problem identified by the GAO was that "too many" cars were passing inspection in some states. The benchmark for this conclusion was the EPA's judgment about how many cars would have to fail and be repaired in order for the program to have the desired effects on emissions and ambient air quality. These projections, however, could only have been made on the basis of the violation rates observed in the pre–I/M small sample checks done in the early 1970s. Whether those rates of violation would continue into an era having an active I/M program is at least doubtful. It may be, therefore, that the program has had the desired effect on continuing compliance, so that the pass rates would be "correct."

In a subsequent (1986) report on the vehicle inspection and maintenance program, the General Accounting Office commented on the EPA's responses to the 1985 report and brought some of the numbers up to date.[33] For example, by the time of the 1986 report, 42 of 44 areas required as of 1982 to produce I/M programs had either started the programs or had produced plans to start such programs by 1987. Until more is known about attainment of ground-level ozone standards, it will not be known how many other areas may have to put I/M programs in place under current regulations. But the EPA has hinted at the need for a nationwide program because of expected deterioration in the performance of the emissions control systems on 1981 and later cars. Some of the same enforcement problems noted in the 1985 report continued to exist in 1986, by the EPA's own admission. The reason appears to be the EPA's unwillingness to push hard for programs that are so unpopular with states and individuals.

Hazardous Waste Disposal

If the problems of checking up on the performance of stationary sources of conventional pollutants have proved difficult for the responsible agencies, it is reasonable to expect even greater difficulties in monitoring and enforcing the laws and regulations designed to produce correct disposal of hazardous materials. Hazardous materials are often produced in small quantities in separable streams within production processes. Thus hazardous materials are often transportable, so that the point of final disposal in the environment is not fixed as it is for waste water and stack gases. The aim of existing legislation is to make sure that the point of final disposal is a carefully designed and regulated facility. The disposal options that will be acceptable constitute a rapidly shrinking universe. Under the RCRA amendments of 1984, all land disposal of untreated wastes will be banned by the beginning of the 1990s, while deep-well injection will only be acceptable if the operator can demonstrate that the wastes will not migrate from the injection zone for 10,000 years or that they will no longer be hazardous by the time they do migrate. The intention is clearly to make it difficult or impossible to contemplate any waste management alternatives other than source reduction, recycling, or high-temperature incineration.[34]

The monitoring system chosen bears some resemblance to that adopted for stationary sources. In particular, self-monitoring is imposed for waste transportation in the form of a system of manifests; and for groundwater contamination by requirements for numbers of test wells, well placement, and water test frequencies. Any generator of a waste who is incapable of disposing of that waste in an acceptable manner on site must fill out a manifest giving details of the amount and characteristics of the waste. This paper accompanies the waste when it is shipped off site, and a copy is supposed to be returned to the generator signed by someone at the authorized disposal facility where the waste ends up. The self-monitoring feature is most clearly seen in the requirement that the generator report to the responsible agency any anomalies, such as nonreceipt of a manifest for a particular shipment.

On its face, such a system would seem open to abuse if sources or shippers have deliberate evasion in mind. For example, a generator of hazardous wastes could dispose of the wastes illegally, thus avoiding the manifest system altogether. Further, auditing the self-monitoring operation would be a considerable job, given the large number of individual sources involved. (It has been estimated that 175,000 small sources of waste were added to the regulated population by the 1984 amendments to RCRA; as a group, however, they add less than one-half of one percent to the quantity of hazardous wastes generated each year.) The requirements generally are too new to have produced a long record of monitoring and enforcement success or failure. And of course the intrinsic logical difficulty of deciding how well a self-monitoring system is working applies here, too. But once again a General Accounting Office study provides some hints.[35]

One finding of this study (issued in 1985) has major implications for the long-term future. It is that of 36 cases of illegal disposal discovered and pursued as enforcement matters in the four states examined by the GAO study team, none were initiated on the basis of the manifest system or related inspections. Thirty-four were the product of information supplied by employees of the firms involved or by concerned citizens. Two were accidentally discovered during other investigations. Thus it should be no surprise that the GAO report is largely a catalog of speculations about the possible extent of illegal disposal and about the expected weaknesses of the monitoring system.

As far as enforcement actions go, in the 36 cases the GAO examined, 28 cases had been completed; in each of those 28, the state had obtained a favorable court opinion. In 10 cases prison sentences were imposed, though in 2 cases that part of the sentence was suspended. In 17 cases, fines were levied, supplemented in 7 cases by damage assessments. In 2 other cases, damages alone were assessed.[36] The fines averaged 30 percent of the maximum allowable penalties in the cases in which they were imposed. State officials told the GAO study team that the penalty record would deter sources from future violations, but one must be skeptical, given the weaknesses in monitoring and hence what must be a low probability of detection.

As for the enforcement of requirements on the handling of hazardous waste at its place of generation, or where it might be subsequently stored or disposed of, the evidence once again, though sparse, suggests problems. A 1987 GAO report characterizes the inspections done on these facilities pursuant to the 1984 RCRA amendments as neither thorough nor complete.[37] And the number of inspections (11,785) performed in fiscal year 1986 suggests that fewer than 10 percent of the facilities that should have been inspected (most annually) were in fact reached at all. Evidence for the charge that inspections were of less than desirable quality came from two tests. In one, 26 actual field inspections were evaluated by experts from the EPA office in charge; only one of these was judged thorough and complete. Overall, for the other inspections, more than 50 percent of the class I violations (those representing an actual or potential release of hazardous waste to the environment) were missed by the inspectors from state agencies, contractors, or the EPA regional offices. The second test consisted of a comparison of sequences of inspection reports for 42 waste handlers. The aim was to see whether earlier inspections missed items—such as the absence of a waste analysis plan—that later inspections turned up and that could not have arisen in the interval between inspections. In ten of the cases reviewed, violations were missed—a total of 95 class I violations. Generally, the report's conclusion is that the very newness of the program, and thus of the inspectors, is a large part of the problem. The major recommendation is for better training of inspectors.

One should add that in at least one respect these inspections do not look so bad. Of the 122 class I violations missed in 25 inspections, only 4 involved the major physical facilities such as tanks, waste piles, and surface impoundments most closely related to actual waste containment or disposal. Twenty others involved the "use and management of containers," while the rest were more in the nature of administrative failures, such as a lack of general written standards or plans.[38]

PRESENT EFFORTS TO IMPROVE MONITORING AND ENFORCEMENT

One should not conclude from this record that the Environmental Protection Agency is oblivious to or cynical about the problems of monitoring and enforcing in its areas of responsibility. To be sure, there was a widely shared perception of slackening zeal during the early years of the Reagan administration.[39] This perception was attacked directly by William Ruckelshaus as part of his campaign to change the agency's, and not incidentally the administration's, environmental image in the period before the 1984 election. For example, in September 1983 a task force on monitoring was set up by the EPA's deputy administrator.[40] While the charge for this group focused on ambient monitoring, enforcement needs were not ignored.

More to the point, Ruckelshaus made explicit his intention to take a tough line against violators. In speeches to the EPA enforcement staff in the summer of 1983, he referred to them as "pussycats" and indicated he wanted them instead to be seen as the "gorilla in the closet"—the bogeyman that state officials could use as a threat in their dealings with recalcitrant polluters.[41] That this kind of personalized pressure had some effect seems clear from the increase in enforcement actions recorded in fiscal year 1984 and later. In table 7–6, the EPA's civil referrals to the

Table 7–6. EPA Civil Referrals to the Department of Justice, Fiscal Years 1972 Through 1988

Fiscal year	Air	Water	Hazardous waste	Toxics, pesticides	Total
1972	0	1	0	0	1
1973	4	0	0	0	4
1974	3	0	0	0	3
1975	5	20	0	0	25
1976	15	67	0	0	82
1977	50	93	0	0	143
1978	123	137	2	0	262
1979	149	81	9	3	242
1980	100	56	53	1	210
1981	66	37	14	1	118
1982	36	45	29	2	112
1983	69	56	33	7	165
1984	82	95	60	14	251
1985	116	93	48	19	276
1986	115	119	84	24	342
1987	122	92	77	13	304
1988	86	123	143	20	372

Source: Based on *Mealey's Litigation Reports: Superfund* vol. 1, no. 18 (December 28, 1988) p. c-1.

264   CLIFFORD S. RUSSELL

Department of Justice over the period fiscal year 1972 through fiscal year
1988 are summarized by problem category. The dip in activity during the
early Reagan years is obvious, as is the higher level of activity in the years
1984–1988. These higher levels were only slightly higher than those
achieved during the notably environmentalist Carter administration,
however. A similar message is conveyed by table 7-7, which shows
administrative actions by the EPA under the six major environmental
laws. The record of administrative actions runs from 1972 through 1988.

A second line of effort within the EPA has been to improve enforcement
by defining a uniform and improved civil penalty policy. This effort
culminated in a document distributed within the agency in February
1984, which set out a policy designed to be flexible and to promote the
goals of deterrence, fair and equitable treatment of the regulated community, and swift resolution of environmental problems.[42] The general
structure of the penalties contemplated under this policy has already been
briefly described: the benefit component is calculated to remove the cost
advantage enjoyed by any noncomplying source; and the gravity component is intended as a penalty tied to the seriousness of the damage
resulting from noncompliance. It should be noted that while calculating
the savings to polluters from noncompliance is difficult, the problem of
attaching dollar penalties to the gravity of a violation is even harder than
assessing the monetized damages of the incident. This is because a
successful violation can be presumed to have some incentive effects on
other polluters, making them slightly more likely to violate the rules
themselves. The EPA document goes on at some length about how to
determine the *relative* gravity of two or more incidents, but offers no
guidance whatever on a benchmark; that is, the relative scale is not
anchored to any dollar number. It will be interesting to see what is made
of this guideline in practice.

A third line of effort in enforcement—undoubtedly the most dramatic if
not the most important—is a new stress on the criminal prosecution of
polluters. Since available penalties for conviction include time in prison
(which might be characterized as the ultimate gravity component) and
since the effort has been accompanied by a strong public relations
exercise, it appears that the EPA wants to plant fear in the hearts of
executives considering whether or not to comply with environmental
regulations.[43] The spectre of the penalty is apparently seen as important
enough to justify the added difficulty of prosecution: obtaining an
indictment from a grand jury, establishing intent, and proving guilt
beyond a reasonable doubt.

While the public stress on criminal prosecutions is a recent phenomenon,
the power to bring such actions is as old as the major environmental laws.

Table 7-7. EPA Administrative Actions Initiated (by Act), Fiscal Years 1972 Through 1988

	Clean Air Act (1970)	Clean Water & Safe Drinking Water acts (1972/1974)	Resource Conservation & Recovery Act (1976)	Superfund (CERCLA) (1980)	FIFRA[a] (1947)	Toxic Substances Control Act (1976)	Totals
1972	0	0	0	0	860	0	860
1973	0	0	0	0	1,274	0	1,274
1974	0	0	0	0	1,387	0	1,387
1975	0	738	0	0	1,641	0	2,352
1976	210	915	0	0	2,488	0	3,613
1977	297	1,128	0	0	1,219	0	2,644
1978	129	730	0	0	762	1	1,622
1979	404	506	0	0	253	22	1,185
1980	86	569	0	0	176	70	901
1981	112	562	159	0	154	120	1,107
1982	21	329	237	0	176	101	864
1983	41	781	436	0	296	294	1,848
1984	141	1,644	554	137	272	376	3,124
1985	122	1,031	327	160	236	733	2,609
1986	143	990	235	139	338	781	2,626
1987	191	1,214	243	135	360	1,051	3,194
1988	224	1,345	309	224	376	607	3,085

Source: Based on *Mealey's Litigation Reports: Superfund* vol. 1, no. 18 (December 28, 1988) p. C-5.
[a] FIFRA = Federal Insecticide, Fungicide, and Rodenticide Act.

266   CLIFFORD S. RUSSELL

Actual use of the power built slowly during the early 1980s.[44] An article in
the *Wall Street Journal* in January 1985 brought the EPA's tiny criminal
investigation unit to national attention, and seems to have been the
opening shot in the public relations effort.[45] At the national level, the
campaign has included attempting to beef up the image of the unit within
the Justice Department that does the actual prosecutions on evidence
obtained by EPA criminal investigation.[46] But reports of actions by state
and local environmental agencies on their use of criminal enforcement
powers are also notably frequent in the press.[47] As of the fall of 1988, 460
individuals or corporations had been indicted for criminal violations of
environmental laws, resulting in more than 300 convictions and the
imposition of over 200 years of jail time, of which more than a quarter
were to be served in federal prisons.[48] Table 7–8 summarizes the EPA's
criminal enforcement efforts over the period fiscal year 1982 through
fiscal year 1988, including referrals to the Department of Justice,
prosecutions, numbers of defendants charged and convicted, the conviction rate, and information on the sentences handed out and served.

Even if the Environmental Protection Agency and related state agencies
are given high marks for improving their images, restructuring and
codifying enforcement penalties, and increasing deterrence by using
criminal actions and the threat of jail terms, it is still fair to ask whether
other strategies are also available. It is especially important to ask this
question in the light of the unsettling evidence on continuing compliance;
that is, even if discovered violations are uniformly and sharply dealt with,
actual deterrence may be negligible since it depends as well on the
probability of detection. The evidence on which to base a judgment about
that probability suggests it must be quite small. Therefore, not only
enforcement but also the prior business of monitoring deserve attention.
Ideas for improvements in both areas are taken up in the next section.

SUGGESTIONS FOR FURTHER IMPROVEMENTS

There are two major paths to an improved system of monitoring and
enforcement, both of which deserve serious attention from the EPA, the
states, and even the U.S. Congress. One path concentrates on improving
agency monitoring (or auditing) by making it less likely that a source will
be able to tinker with discharges before measurement by the agency can
take place. The second attempts to forge a tighter link between monitoring
and enforcement and by so doing allow more efficient use of limited
resources.

Table 7-8. EPA Criminal Enforcement Activities, Fiscal Years 1982 Through 1988

	1982	1983	1984	1985	1986	1987	1988
Referrals to the Department of Justice	20	26	31	40	41	41	59
Cases prosecuted	7	12	14	15	26	27	24
Defendants charged	14	34	36	40	98	66	97
Defendants convicted	11	28	26	40	66	58	50
Conviction rate (%)	78.0	82.4	72.2	100.0	67.3	87.9	51.5
Months sentenced			6	78	279	456	278
per convicted defendant			0.23	2.0	4.2	7.9	5.6
Months served			6	44	203	100	185
per convicted defendant			0.23	1.1	3.1	1.7	3.7
Months probation		534	552	882	828	1,410	1,284
per convicted defendant		19.1	21.2	22.0	12.5	24.3	25.7

Source: Based on *Mealey's Litigation Reports: Superfund* vol. 1, no. 18 (December 28, 1988) p. c-3.

268   CLIFFORD S. RUSSELL

The first path has, itself, two major components: technology and law. Current monitoring technology was developed to determine initial compliance status. Initial compliance involves a well-defined situation in which *capability* is being tested. The source is required to run its production equipment at particular levels and to demonstrate that its pollution control equipment can produce acceptable discharges—discharges consistent with its permit. Advance notice does not give the source any advantage because the test does not involve actual operations. Indeed, notice may be required as a matter of fairness. And there is certainly no need for speed in setting up the measurement equipment. In this setting, complex and awkward measurement technology could be tolerated in the interests of keeping measurement errors low.

The same is not true when continuing compliance is to be monitored. In that context, ease with which equipment can be transported, brevity of set-up time, and economy of operation are more important than small standard errors. For if surprise in inspection can be achieved, multiple samples can be taken to reduce the overall error of the average measured discharge. The ultimate weapon in this regard is remote reading technology, with which no entry onto private premises need be made. Thus the first suggestion for improvement is:

- Concentrate monitoring research money on the development of simple and inexpensive instruments, and especially remote reading methods (such as those based on laser technology) that are capable of measuring stack emissions from outside the factory or utility gates.

The legal component for more effective auditing arises out of the need of state and federal pollution control agencies to have the right to enter private premises unannounced and warrantless for the purpose of measuring pollution discharges. This right is now in place for inspectors of nuclear power plants and mine safety. It seems reasonable that society's commitment to preventing conventional air or water pollution or hazardous waste contamination ought to be as firm as its commitment to preventing accidents at nuclear plants or mines. However, to achieve such an explicit commitment may require legislative action, for the state of the law as determined by the courts is currently unclear. (This same kind of legislative action is needed in the case of remote monitoring equipment as well, even though the Dow case mentioned earlier was decided in the EPA's favor.) Since the constitutionality of surprise inspections in analogous situations does not appear to be in question, it should be possible to achieve the necessary effect by amending existing laws to remove any doubts about the intent of Congress. Thus the second recommendation is:

- Amend as necessary all major environmental statutes to make it clear that Congress contemplates and approves of unannounced and warrantless visits by state or federal agency personnel for the purpose of making independent measurements of discharges. This approval should be explicitly extended to the operation of remote-sensing monitoring instrumentation. There should be no anticipation of privacy in the generation and discharge of pollutants.

The second major path toward improvement is to tighten the linkage between monitoring and enforcement. This could be accomplished straightforwardly by making the future frequency of monitoring depend on the past record of compliance for all significant sources. It can be demonstrated that with such a system in place and appropriate penalties for noncompliance, it is possible to achieve high rates of compliance even with tight monitoring and enforcement budgets.[49]

The necessary system is really quite simple, consisting only of an administrative tripartite grouping of pollution sources according to compliance status. The first group would consist of those sources that have been visited (or audited) and found to be in compliance. The probability of another visit to these sources might be made low enough that they would even feel they could relax their pollution control efforts. The second group would consist of those sources that have been found in violation at their last monitoring visit (if they were in the first group when visited) or that have just been "pardoned" or released from the third group. Sources in group two would not need to be audited any more frequently than sources in group one, but sources in group two that fail audits (that is, are found in violation) would be reclassified into group three, and would have to anticipate staying in that group for a long time. For the third group, audit visits would be conducted frequently—perhaps during every inspection period, but certainly often enough that it would not be worthwhile for any source in the group to attempt to get away with a violation. Thus being in group three would be equivalent to being in compliance. In such a system, properly designed, the vast majority of sources could be expected to end up in group two most of the time; and while in group two they would have an incentive to comply because of the threat of a long sentence to group three (rather than because of the immediate audit frequency for group three).[50]

The final recommendation is therefore:

- Under each major environmental regulatory program, establish a formal system linking expected frequency of monitoring visits (or audits of self-monitoring sources) to past records of discovered violation. The major requirement of such a system would be that

270   CLIFFORD S. RUSSELL

sources found in violation twice in a row should be "sentenced" to frequent audits for a long period.

Together, adoption of these three suggestions would improve the efficiency of use of limited monitoring and enforcement budgets by increasing the probability that an actual violation would be discovered and by improving the way information about past compliance status is used to determine the level of current and future monitoring efforts. These recommendations are based on an assumption that despite the depressing evidence of widespread lack of compliance with strict environmental regulations, better compliance really is the goal of policy. The alternative possibility—that the legislation and accompanying regulations are meant to give the *appearance* of strictness while the reality is reflected by a lack of commitment to monitoring and enforcement—can only be ruled out by a future record that is an improvement on that of the past. The ongoing transition from absorption with initial compliance of conventional point sources of pollution to continuing compliance of hundreds of thousands of generators of a wide variety of wastes will require some fundamental redirection of thinking and effort on the part of the Environmental Protection Agency and state agencies. The three recommendations made here are meant to set the stage for that redirection and give it the greatest chance of success, given that substantial additional money is unlikely to be found in the midst of wars over budget deficits.

NOTES

1. Environmental Protection Agency, "A Framework for Statute-Specific Approaches to Penalty Assessments: Implementing EPA's Policy on Civil Penalties," EPA General Enforcement Policy no. GM-22, February 16, 1984.

2. See, for example, Robert W. Crandall and Paul R. Portney, "Environmental Policy," in Paul R. Portney, ed., *Natural Resources and the Environment: The Reagan Approach* (Washington, D.C., Urban Institute, 1984), especially pp. 49–52.

3. The General Accounting Office (GAO) has published two reports critical of ambient air quality monitoring; see GAO, "Air Quality: Do We Really Know What It Is?" GAO no. CED-79-84, May 31, 1979; and GAO, "Problems in Air Quality Monitoring System Affect Data Reliability," GAO no. CED-82-101, September 22, 1982.

4. The second judgment might turn out to be wide of the mark because in the actual regulatory system the connections made between source discharges and ambient quality results have generally been based on extremely simple models of complex natural systems and events.

5. Cheryl Wasserman, "Improving the Efficiency and Effectiveness of Compliance Monitoring and Enforcement of Environmental Policies: United States: A National Review,"

prepared for the Organization for Economic Co-operation and Development, Environmental Directorate, October 1984.

6. This point is made by Wasserman, albeit in a rather circumspect way. See Wasserman, "Improving the Efficiency and Effectiveness of Compliance Monitoring," pp. VI–5 through VI–7.

7. Clifford S. Russell, "Pollution Monitoring Survey: Summary Report" (Washington, D.C., Resources for the Future, 1983). This mail survey was sent to all 50 states, in some cases to more than one agency within a state. Responses were received from 27 states for air pollution questions and from 36 states for water pollution questions. For 21 states, responses covered both air and water pollution.

8. "New Groundwater Compliance Order Guidance Stresses Phased-in Remedies," *Inside EPA*, September 20, 1985, pp. 4,5.

9. The term audit has commonly been reserved by the EPA to refer to more general compliance checks. One of the enthusiasms of the early 1980s was that of self-audits, in which sources would check themselves over and assert their compliance in return for a reduction in actual agency monitoring.

10. Russell, "Pollution Monitoring Survey," p. 10.

11. This characteristic of standard pollution control monitoring stands in marked contrast to the system run by the Nuclear Regulatory Commission (NRC) to check on the performance of nuclear power plants. In the NRC system, each plant has a resident inspector (during both construction and operation) with freedom to go anywhere in the plant at any time, subject only to the same safety rules that apply to plant personnel. This inspector is encouraged to make rounds at random times and in varying order to add some element of surprise to the knowledge of routine monitoring. The resident inspector is supplemented by outside inspection teams that are encouraged to make their visits unannounced (Nuclear Regulatory Commission, Office of Inspection and Enforcement, *Inspection and Enforcement Manual*, chapter 0300 [January 1983].)

12. Whether or not the EPA and state pollution control agencies must have warrants for monitoring visits has not yet been considered by the U.S. Supreme Court, but the circuit and district courts that have reached the issue have made an analog between the EPA's scheme and that of the Occupational Safety and Health Administration and have thus required warrants. See *Public Service Company of Indiana, Inc. v. U.S. E.P.A.*, 509 F. Supp. 720 (S.D. Ind. 1981); *U.S. v. Stauffer Chemical Company*, 511 F. Supp. 744 (M.D. Tenn. 1981); *Bunker Hill Company v. U.S. E.P.A.*, 658 F. 2d 1280 (9th Cir. 1981). See also Robert W. Martin, Jr., "EPA and Administrative Inspections," *Florida State University Law Review* vol. 7 (Winter 1979) pp. 123–137. The courts have differed as to the type of warrants required for EPA inspections and the level of probable cause required to support them. Several have held that an ex parte warrant is all that is required; see *Bunker Hill Company v. U.S. E.P.A.*, 658 F. 2d at 1285; *U.S. v. Stauffer Chemical Company*, 511 F. Supp. 749. The Tenth Circuit reached the opposite conclusion in *Stauffer Chemical v. EPA*, 647 F. 2d 1075 (10th Cir. 1981). (An ex parte warrant may be granted at a hearing at which the party to be subject to search is not represented.)

13. "Supreme Court Roundup," *New York Times*, June 11, 1985; "Supreme Court Backs the EPA in Dow Dispute," *Wall Street Journal*, May 20, 1986.

14. See, for example, "EPA Will Tolerate Violations from 20% to 40% Over Water Permit Levels," *Inside EPA*, May 11, 1984, p. 7. This article reports on the contents of a draft definition of significant noncompliance then making the rounds within the agency.

15. General Accounting Office, *Waste Water Dischargers Are Not Complying with EPA Pollution Control Permits*, RCED-84-53 (Washington, D.C., 1983).

16. "Ruckelshaus Worried Citizen Suits Will Reveal Poor Enforcement Record," *Inside EPA*, May 11, 1984, p. 1.

17. "District Court Grants Judgment in Citizens Suit on NPDES Violation," *Inside EPA*, November 30, 1984, p. 5; "Court Diverges from Other Decisions in Siding with Citizens in CWA Suits," *Inside EPA*, August 30, 1985, p. 10.

18. Closer but far from identical. The probability of being found in violation, given that the decision to violate is made, is also a necessary deflater in the process of arriving at the expected penalty facing the source when it is deciding whether or not to violate intentionally. This probability cannot be estimated from available data.

19. Environmental Protection Agency, "Policy on Civil Penalties," EPA General Enforcement Policy no. GM-21, February 1984.

20. Genevra Richardson, "Policing the Enforcement Process," and Michael Hill, "The Role of the British Alkali and Clean Air Inspectorate in Air Pollution Control," both in *Policy Studies Journal* vol. 11, no. 1 (September 1982), pp. 153–164 and 165–174, respectively; Genevra Richardson, Anthony Ogus, and Paul Burrows, *Policing Pollution* (Oxford, Clarendon Press, 1982); Keith Hawkins, *Environment and Enforcement* (Oxford, Clarendon Press, 1984).

21. These studies have been summarized by the EPA in two volumes: James S. Vickery, Lori Cohen, and James Cummings, *Profile of Nine State and Local Air Pollution Agencies* (Washington, D.C., Environmental Protection Agency, 1981); Robert G. McInnes and Peter H. Anderson, *Characterization of Air Pollution Control Equipment Operation and Maintenance Problems* (Washington, D.C., Environmental Protection Agency, 1981).

22. Based on McInnes and Anderson, *Characterization of Air Pollution Control Equipment*, appendix A. The 10 percent figure has been calculated as the total excess emissions during violations, divided by the total allowable emissions, both in tons per year.

23. McInnes and Anderson, *Characterization of Air Pollution Control Equipment*. It should be noted that this summary contains some misleading numbers, apparently based on averaging procedures poorly chosen for the problem, though the text's explanation is far from clear on what was actually done with the raw data. The numbers reported above as relevant data were calculated directly from the data appendix provided.

24. General Accounting Office, *Improvements Needed in Controlling Major Air Pollution*, CED-78-165 (Washington, D.C., 1979).

25. General Accounting Office, *Air Pollution: Environmental Protection Agency's Inspections of Stationary Sources*, GAO/RCED-86-1BR (Washington, D.C., 1985).

26. General Accounting Office, *Costly Waste Water Treatment Plants Fail to Perform as Expected* (Washington, D.C., 1980); General Accounting Office, *Waste Water Dischargers Are Not Complying*.

27. The 1980 report is based on twelve months of data from 1978 and 1979, while the 1983 report uses data from October 1980 to March 1982, a period of eighteen months.

28. "Reporting" is a key word here. It is possible, for example, that because of different enforcement attitudes toward industrial and municipal sources, the former have incentives to shade reports toward compliance. This explanation of the contrast in the data competes with the explanation that compliance is actually better among industrial sources. In the absence of independent auditing, there is no way of choosing between the alternatives.

29. General Accounting Office, *Waste Water Dischargers Are Not Complying*, p. 25.

30. The requirement that continuing compliance monitoring (an inspection and maintenance program) be instituted applies to states that requested extensions beyond the 1982 deadlines for compliance with carbon monoxide and ozone NAAQSs. Extensions were

granted only when states could demonstrate that "all reasonably available control measures," if fully implemented, would not allow for attainment. In such situations implementation of, or at least a schedule for, an inspection and maintenance program was required before an extension could be approved when the area involved was urban and had a population greater than 200,000.

31. General Accounting Office, *Better Enforcement of Car Emission Standards—A Way to Improve Air Quality*, CED-78-180 (Washington, D.C., 1979).

32. General Accounting Office, *Vehicle Emissions Inspection and Maintenance Program Is Behind Schedule*, RCED-85-22 (Washington, D.C., 1985). Four states were on compliance schedules in this regard, but the latest agreed-to date for bringing a system in was early 1986.

33. General Accounting Office, *Vehicle Emissions: EPA Response to Questions on Its Inspections and Maintenance Program*, RCED-86-129 BR (Washington, D.C., 1986).

34. William L. Rosbe and Robert L. Gulley, "The Hazardous and Solid Waste Amendments of 1984: A Dramatic Overhaul of the Way America Manages Its Hazardous Wastes," *Environmental Law Reporter* vol. 14, no. 12 (1984) pp. 10458–10467.

35. General Accounting Office, *Illegal Disposal of Hazardous Wastes: Difficult to Detect or Deter*, RCED-85-2 (Washington, D.C., 1985). The manifest requirement became effective in November 1980.

36. Community service requirements were the only penalty imposed in 4 cases; and one conviction resulted only in an order to comply in the future. Ibid., pp. 39, 40.

37. General Accounting Office, *Hazardous Waste: Facility Inspections Are Not Thorough and Complete*, RCED-88-20 (Washington, D.C., 1987).

38. Another GAO report found that the EPA was doing a reasonably good job of checking up on land disposal facilities that had certified their compliance with requirements for groundwater monitoring and financial responsibility (such certification was required to allow continued operation of the facilities after November 8, 1985). Similarly, the EPA was found to have made an effort to check up on the facilities that did not certify compliance and thus were required to cease operating (General Accounting Office, *Hazardous Waste: Enforcement of Certification Requirement for Land Disposal Activities*, RCED-87-60 BR [Washington, D.C., 1987]).

39. "Alm Forms Task Force to Recommend Ways to Improve Monitoring," *Inside EPA*, September 30, 1983, p. 14.

40. Rochelle Stanfield, "Ruckelshaus Casts EPA as 'Gorilla' in States' Enforcement Closet," *National Journal*, May 26, 1984, pp. 1034–1038.

41. Ibid.

42. Environmental Protection Agency, "Policy on Civil Penalties"; and Environmental Protection Agency, "A Framework for Statute-Specific Approaches to Penalty Assessments."

43. Tamar Levin, "Business and the Law," *New York Times*, March 5, 1985. Levin quotes James Oppliger, a deputy district attorney general in California, as saying, "A person facing jail is the best deterrent against wrongdoing."

44. Daniel Riesel, "Criminal Prosecution and Defense of Environmental Wrongs," *Environmental Law Reporter* vol. 15, no. 3 (1985) pp. 10065–10081. Riesel provides some historical background to this practice, but is mainly concerned with a scholarly review of the applicable law.

45. Barry Meier, "Against Heavy Odds, EPA Tries to Convict Polluters and Dumpers," *Wall Street Journal*, January 7, 1985.

46. "Justice Environmental Crimes Unit Wins Record Number of Indictments," *Inside EPA*, September 6, 1985, pp. 5,6.

274   CLIFFORD S. RUSSELL

47. See, for example, Roy J. Harris, Jr., "Mobile Plant Raided in Investigation of Pollution Charge," *Wall Street Journal*, July 29, 1985, describing a criminal investigation raid by Los Angeles County officials; Victoria Churchville, "Executive on Trial in Harbor Dumping Case," *Washington Post*, September 5, 1985, which relishes the contrast between the defendant executive's appearance and his role as an accused criminal; and Tom Vesey, "D.C. Man Jailed for Polluting," *Washington Post*, September 10, 1985, which notes that the defendant, owner of a small dump site, actually is serving time in jail.

48. See Judson W. Starr, "Environmental Enforcement Takes an Ominous New Turn: Managers and Officers Go Directly to Jail," *Environmental News* vol. 2, no. 1 (Fall 1988) pp. 1–3 (publication distributed by Venable, Baetjer, and Howard, Baltimore).

49. See Clifford S. Russell, "Game Theory Models for Structuring Monitoring and Enforcement Systems," *Natural Resource Modeling*, forthcoming. The same end could be achieved by using differential penalties, dependent on past compliance record, for what is necessary is that repeat offenders face very high *expected* penalties. But there is some practical reason to prefer a system that operates through the probability of a monitoring visit, because very high penalties may never be assessed on individual violators.

50. In a simple world without measurement error it would be possible to describe the system more neatly, and the sentence to group three would become indefinite or infinite and no rational source would ever end up in that group. Since measurement error is inevitable, even if all sources try to avoid group three, some will fail and will be sentenced to a term there. If all such terms were indefinite, all sources would eventually be included in group three and the system would break the budget constraint by requiring too many audits per period. Thus it is necessary that some fraction of the sources in group three be pardoned each period. The fraction must be small enough that the sources expect to spend a long time in compliance. How long is long depends on other characteristics of the problem, including the size of fines for violations relative to the costs of compliance.

[32]

JOURNAL OF ENVIRONMENTAL ECONOMICS AND MANAGEMENT **27**, 127–146 (1994)

Guilty until Proven Innocent — Regulation with Costly and Limited Enforcement[1]

JOSEPH E. SWIERZBINSKI

Department of Economics and School of Natural Resources and the Environment, University of Michigan, Ann Arbor, Michigan 48109

Received April 16, 1993; revised July 2, 1993

We consider optimal regulations for a polluting firm when regulators cannot observe emission control costs and can only observe emissions via costly monitoring. Fines (or subsidies) for enforcing compliance are also limited. The optimal regulations resemble a deposit–refund system. The firm reports its emissions and pays an initial tax based on this report. If the firm is monitored, it receives a rebate when actual and reported emissions coincide. The enforcement constraints and the firm's rights determine whether the incentives to reduce emissions are optimally provided by varying the rebate for compliance, the monitoring probability, or the initial tax. © 1994 Academic Press, Inc.

1. INTRODUCTION

Emission or effluent fees are an important part of the system for regulating pollution in several European countries. In textbook discussions, the most important role for such fees, or Pigouvian taxes, is to provide polluters with the correct incentives by internalizing external costs.[2] However, a number of economists have questioned the extent to which several European fee systems actually influence the incentives of polluters.[3] These economists observe that emission fees have typically been set at low levels and that a polluter's payment often depends on an estimated emission level rather than on the polluter's actual emission of pollutants. If a polluter's payment never depends on the amount he pollutes, then the payment obviously provides no incentive to reduce the level of pollution.

European fee systems are actually more complicated. In some systems, some polluters pay a fee based on their actual emissions while others pay a fee that is unrelated to the amount they pollute. It is also noteworthy that, in at least one case, a polluter may request to be monitored and have his estimated fee adjusted if his actual emission of pollutants differs from his estimated emission.[4] This raises the interesting possibility that an incentive to reduce pollution may be provided not by an "up-front" fee, as in the textbook analysis, but rather by variations in a "rebate" which polluters may receive when their actual emissions are monitored.

---

[1] I am grateful to Ken Binmore, Charles Kolstad, Steve Salant, an anonymous referee, and participants in the 1991 Canadian Resource and Environmental Economics Study Group, the 1992 Summer Institute in Game Theory at Stonybrook, and a 1992 conference on economic theory at the University of Michigan for helpful comments and suggestions. Of course, any errors are my own.

[2] See, for example, Baumol and Oates [6].

[3] Several recent surveys of environmental regulation take this view. See p. 692 of Cropper and Oates [18] and pp. 104–105 of Hahn [24].

[4] See Bower *et al.* [13, pp. 128–129] and OECD [35, p. 25].

How should the incentives to reduce pollution best be provided? We consider the optimal regulation of polluting firms in the case where (1) it is costly to monitor the actual emissions of polluters, (2) the regulator is limited in the extent to which he can reward compliance or punish noncompliance, and (3) the regulator, although he knows the distribution of abatement costs, cannot observe the costs of any individual firm. These assumptions appear to be consistent with the characteristics of at least some systems for enforcing pollution controls laws.[5]

Given these limitations on the regulator's information and enforcement power, the following framework provides a reasonable description of the possibilities for regulating pollution via emission fees.[6] (1) A polluter reports what his level of emission will be and pays an initial fee or tax that depends only on his report. (2) The regulator then monitors the pollution level of the firm with a prespecified probability that typically depends on the polluter's reported emission level. For simplicity, we assume that the regulator's monitoring technology is perfectly accurate although costly. (3) If the polluter's emissions are monitored, then the regulator is also precommitted to punish the polluter with a fine or to reward him with a rebate depending on whether the polluter's actual emissions are consistent with his initial report.

We consider two cases. In the bulk of the paper, we assume that the regulator can commit to monitoring a firm with any probability. We show that in this case a random monitoring policy is generally optimal. That is, a firm is typically monitored with a probability that is strictly between 0 and 1. A firm that reports a lower level of emission is more likely to be monitored. As well as levying fines for noncompliance, it is optimal to offer a rebate to a firm that is monitored and found to be producing its reported level of emission. If the maximum possible fine for noncompliance is small compared to the maximum rebate, then almost all of the incentive to reduce emissions is provided by variations in the probability that a firm is monitored. At the other extreme, if the maximum rebate that the regulator is permitted to offer is small compared to the maximum fine, then almost all of the incentive to reduce emissions comes from variations in the initial tax.

We also consider the case where the regulator's ability to commit to a monitoring policy is constrained by the firm's right to be monitored. This is an important issue, since a firm that expects a rebate for compliance will wish to be monitored. One might expect that rebates would not be useful in such circumstances since they encourage firms to demand that the regulator engage in costly monitoring. However, this intuition is only correct if the maximum fine is large enough to provide effective deterrence. If the maximum fine is small, then it will be optimal to provide rebates for at least some reported levels of emission. In the extreme but not implausible case where the maximum fine is very small, all types of firm receive rebates, a firm is always monitored, and, in one scenario, all firms pay the same initial tax and all of the incentive to reduce emissions is provided by variations in the size of the rebate.

[5]Russell [38] observes that the regimes for enforcing pollution control laws in the United States and the United Kingdom are characterized by: (i) a heavy reliance on self-reporting, (ii) infrequent monitoring of polluters, and (iii) low penalties for violations.
[6]Given the structural assumptions of our model, a regulator can achieve at least as large a payoff by using the regulatory framework in our paper as he can by using any of a large class of alternative regulations. This point is discussed further in Section 2 of the paper.

The formal model developed in the paper adopts the approach of Baron and Myerson [5], who consider a regulator that wishes to control the output of a single firm without being fully informed about the firm's costs.[7] In the context of pollution control, a firm's "output" can be interpreted as its reduction in emissions from some base level. However, unlike the present paper, Baron and Myerson's model (and most of the literature on regulation under uncertainty) assumes that the regulator can costlessly observe the firm's output and impose an arbitrary, nonlinear tax as a function of the firm's actual output subject only to participation and incentive compatibility constraints.[8] Note that since there is only one regulatory instrument, the nonlinear tax, in the Baron–Myerson model, the question of which combination of instruments is best suited to provide the incentives for pollution reduction cannot be addressed.

In terms of methodology, the literature on income taxation and auditing, including papers by Border and Sobel [12], Chander and Wilde [16], Melumad and Mookherjee [32], Mookherjee and Png [33], and Reinganum and Wilde [37] is closest to the current paper. In most of this literature, the taxpayer has no choice to make, other than reporting an income level. Hence, the problem addressed in our paper of motivating an agent to both signal information about costs (via the reported emission level) and choose an action appropriately does not arise.[9]

Another literature related to our paper extends Baron and Myerson's [5] model to allow for audits of the firm's production costs. Papers in this literature include Baron [2], Baron and Besanko [4], Chapter 4 of Besanko and Sappington [9], and Demski *et al.* [20]. These papers follow Baron and Myerson in assuming that the regulator can costlessly observe the firm's output. Hence, the opportunity to audit the firm's production costs provides the regulator in these models with the chance to obtain information in addition to that available to the Baron–Myerson regulator. In contrast, the regulator in our model never knows more than the Baron–Myerson regulator and only obtains information that the Baron–Myerson regulator gets for free by engaging in costly monitoring. As a result, the constraints

---

[7]Baron [3], Besanko and Sappington [9], and Caillaud *et al.* [14] survey the large literature on regulated firms that has developed using the Baron–Myerson approach. Dasgupta *et al.* [19] and Spulber [42] develop models of environmental regulation that are similar to Baron and Myerson's model.

[8]Caillaud *et al.* [15] consider a model where a principal can impose a tax based only on a noisy observation of an agent's output. However, the principal in this model cannot pay for better information about the agent's output, and, consequently, many of the issues considered here, such as the optimal monitoring strategy and the optimal tax/rebate once an agent is monitored, do not arise. Ortuno-Ortin [36] develops a model of pollution control where a firm has private information about its pollution control costs and the regulator must engage in a costly inspection to discover the firm's level of emissions. The focus of Ortuno-Ortin's paper is quite different from that of the current paper; for example, rebates for compliance are not allowed in Ortuno-Ortin's model.

[9]Mookherjee and Png [33] consider a model where initially identical taxpayers have income levels that depend on an unobservable choice of effort. Unlike the firm in our model, Mookherjee and Png's agents do not have private information when they choose their effort levels. Moreover, by making structural assumptions which are common in the regulation literature but not in the literature on income taxation, we are able to obtain sharper results concerning the optimal regulations than those obtained by Mookherjee and Png. Other papers that combine auditing and principal–agent models include Baiman and Demski [1], Dye [22], and Evans [23]. The first two papers develop models where the principal costlessly observes output and can, in addition, pay to measure the agent's effort. Evans [23] develops a model where the principal observes nothing and receives a fixed payment unless the agent calls for an audit of both output and effort. As with Mookherjee and Png, the agents in these models do not have private information when they choose their effort levels.

on the optimal regulations are quite different in the two types of models, and even the qualitative properties of the optimal regulations differ. For example, in the limiting case where auditing or monitoring is too costly to be feasible, the Baron–Besanko regulator can still control the firm's output using Baron and Myerson's optimal tax policy. On the other hand, the only options available to the regulator in our model when monitoring is not possible are to shut the firm down with a prohibitive tax or to allow the firm to operate, in which case all types of firm will produce the lowest output or, equivalently, emit the highest level of pollutants.[10] As another example of the differences between the two types of model, in the Baron–Besanko framework it is optimal to audit firms with high costs more frequently, while in our model it is optimal to monitor firms with low costs more frequently.

When an up-front payment is combined with a rebate for compliance, the resulting regulation resembles a deposit–refund system or performance bond. A number of authors have considered the use of deposit–refund systems to control environmental problems.[11] Costanza and Perrings [17] and Cropper and Oates [18] observe that deposit–refund systems may be useful under conditions of uncertainty because they shift the "burden of proof" from the regulator to the polluter. What has been lacking until now is a demonstration that a regulatory system with deposit–refund features is optimal in the context of regulation under uncertainty and an analysis of how such a system interacts with other aspects of regulatory policy, such as the monitoring decision.[12]

The paper proceeds as follows. Section 2 describes the structural assumptions of the model and sets up the optimization that determines the optimal regulations for the case where the regulator can commit to monitoring with any probability. Section 3 discusses these optimal regulations. Section 4 discusses the optimal regulations for the case where the regulator's ability to commit to a monitoring policy is limited by the firm's right to be monitored, and Section 5 contains some concluding remarks.

2. THE MODEL

This section introduces assumptions and notation and describes the model analyzed in the rest of the paper.

Consider a risk-neutral regulator who wishes to control the output of a single firm which has private information about its production costs. In addition to uncertainty about the firm's costs, the regulator must monitor the firm in order to

---

[10] In the opposite limiting case where the cost of monitoring or auditing approaches zero, the audit is perfectly accurate, and the maximum penalties for misreporting approach infinity, the Baron–Besanko regulator achieves the expected net benefits of a perfectly informed regulator who observes both cost and output, and the regulator in our model achieves only the expected net benefits of a Baron–Myerson regulator.

[11] See Bohm [10], Bohm and Russell [11], Costanza and Perrings [17], and Solow [40] for discussions of deposit–refund systems in environmental contexts.

[12] A number of papers including Beavis and Dobbs [7], Beavis and Walker [8], Downing and Watson [21], Harford [25, 26], Harrington [27], Linder and McBride [28], Malik [29, 30], Martin [31], and Seegerson [39] consider issues related to the enforcement of environmental regulations. An important difference between these papers and the current paper is that the current paper allows the firm to have private information about its emission control costs while those of the above papers which consider the design of optimal regulations assume that the regulator and the firm are equally well informed about the costs of controlling the firm's emissions.

observe and, if necessary, penalize deviations from the desired output level. Monitoring is assumed to be accurate but costly.

Let $c(\theta, q)$ denote the total cost to the firm of producing the output level q. The regulator's uncertainty about the firm's costs is summarized by the single, real-valued random variable θ, whose realized value is known to the firm when it chooses q but never becomes known to the regulator. For simplicity, we suppose that θ takes on an arbitrarily large but finite number of values indexed by $k = 1 \dots K$ with $\theta_{k_1} > \theta_{k_2}$ if $k_1 > k_2$. A firm for which $\theta = \theta_k$ will be referred to as a "type k" firm.[13]

In the context of pollution control, let q denote the *reduction* in the firm's emissions from a base level e_0 which would be emitted in the absence of regulation.[14] In this case $c(\theta, q)$ represents the minimum cost to the firm of reducing its emissions by the amount q. This cost may include the cost of operating emission control devices, but it can also include the loss in profit incurred when the firm reduces its emissions by optimally adjusting its mix of inputs and outputs.

It is standard[15] to assume that, for all θ_k and all $q > 0$, $c_q > 0$, $c_{qq} \geq 0$, $c_\theta > 0$, $c_{q\theta} > 0$, $c_{qq\theta} \geq 0$, and $c_{q\theta\theta} \geq 0$, where the subscripts denote partial derivatives. (When $q = 0$, these partial derivatives are also assumed to be nonnegative.) Thus, the firm's marginal cost curve is constant or upward sloping for all values of θ, and both total and marginal costs are increasing functions of θ. For simplicity, we also assume that $c(\theta_k, 0) = 0$ for all k.

Let π_k be the (strictly positive) probability that the regulator assigns to the event that $\theta = \theta_k$. As with similar models, the model in this paper can also be interpreted as one where several firms are regulated and π_k denotes the fraction of the firms for which $\theta = \theta_k$. Under this interpretation, the regulator knows π_k, but the type of a specific firm is either unobservable by the regulator or not verifiable so that different types of firm cannot be regulated separately.

The regulatory regime under which the firm operates consists of three elements. First, the regulator sets a possibly nonlinear payment schedule, $T(\hat{q})$, where $\hat{q}$ is the output level which the firm *reports* that it will produce. Throughout the paper, positive values indicate payments from the firm (e.g., taxes) and negative values indicate payments to the firm. The payment $T(\hat{q})$ is an initial "up-front" tax (or subsidy) that does not depend on the firm's actual output level.

For each reported output level, the regulator also commits to monitoring the firm with probability $\lambda(\hat{q})$. If $\lambda(\hat{q}) < 1$, this implies that the regulator can commit to monitoring the firm with a probability no greater than $\lambda(\hat{q})$. This assumption is relaxed in Section 4. Throughout the paper it is assumed that the regulator can commit to monitoring the firm with a probability no less than $\lambda(\hat{q})$.[16]

---

[13]Using arguments very similar to those in this paper, the optimal regulations for a single firm with a continuum of possible types can be derived. The resulting expressions for the optimal regulations closely resemble the corresponding expressions for a discrete number of types in Sections 3 and 4.

[14]When q is interpreted as a reduction in emissions from a base level e_0, it is natural to impose the constraint $q \leq e_0$. In the rest of the paper, it is assumed, for simplicity, that this constraint is not binding.

[15]See Besanko and Sappington [9].

[16]Since monitoring in this model serves only to deter "cheating" by firms, a regulator would prefer not to carry out his commitment to monitor once cheating has been deterred. A common response is that a regulator will wish to honor present commitments in order to preserve a "reputation" for the future. However, a formal analysis of reputation effects would require a dynamic model that is beyond the scope of the current paper.

132   JOSEPH E. SWIERZBINSKI

The firm's actual output level, q, is not observed by the regulator unless the firm is monitored, in which case the regulator is assumed to measure q with perfect accuracy. When the firm is monitored, it incurs an additional payment, $R(\hat{q}, q)$, that may depend on both the actual and the reported output levels of the firm.

For the purpose of analysis, it is convenient to suppress the reported output levels, which serve only to specify the regulatory regime under which the firm chooses to operate. An equivalent (and standard) reinterpretation of the previously described regulatory regime is that, by reporting an output level, the firm selects an item from a menu of regulatory schemes proposed by the regulator. An item j in the menu is a triplet consisting of: (1) an up-front payment, T_j, (2) a probability of being monitored, λ_j, and (3) a function, $R_j(q)$, that specifies the "ex post" payment incurred by the firm if it is monitored and the actual output is discovered to be q. If there are K types of firm, then, as observed in footnote 18, it will never be optimal for the regulator to offer a menu with more than K items.

In most of the paper, a triplet T_j, λ_j, and $R_j(q)$ constitutes a valid regulatory scheme if it satisfies the restrictions that $0 \le \lambda_j \le 1$ and

$$-\bar{R} \le R_j(q) \le \bar{T}, \tag{1}$$

where $\bar{R}$ and $\bar{T}$ are exogenous, nonnegative parameters. (Section 4 considers a further restriction on the set of valid regulatory schemes.) $\bar{T}$ is the largest tax or "fine" that a regulator can impose on a firm that is monitored and found to be producing q. Similarly, $\bar{R}$ is the largest subsidy or rebate that the regulator can provide to a firm that produces q. In the absence of such constraints on the ex post payments, the regulator could prevent a firm from deviating from its reported output level at an arbitrarily small cost by using an arbitrarily large fine to punish noncompliance.

The process of regulation is modeled as a three-stage game. In the first stage, the regulator chooses a menu of regulations. In the second stage, the firm simultaneously chooses both an item from the regulatory menu and the actual output level which it will produce. Negative output levels are not permitted. In the third stage, the firm is monitored with a probability that is determined by the firm's choice in the second stage, and, if the firm is monitored, the firm incurs a payment that also depends on its choice in the second stage. The optimal menu of regulations and the optimal output levels for each type of firm are determined by a perfect Bayesian equilibrium of this game.

A type k firm is assumed to be risk neutral and to choose the regulatory scheme k and output level q_k in the second stage in order to minimize its expected costs:

$$T_k + \lambda_k R_k(q_k) + c(\theta_k, q_k).$$

The firm is also allowed to choose "quit" in the second stage, in which case the firm incurs a reservation cost, $\bar{c}$. For simplicity, it is assumed that—because of unmodeled social costs associated with quitting—the regulator will never wish to propose a menu that will induce any type of firm to choose "quit" in stage 2.

Let $B(q)$ denote the dollar value of the social benefits obtained when the output q is produced. For all $q > 0$, suppose that $B(q) > 0$, $B'(q) > 0$, and $B''(q) \ge 0$, where a prime denotes the derivative. Suppose also that $B(0) \ge 0$, $B'(0) \ge 0$, and $B''(0) \ge 0$.

The optimal menu for the regulator to propose in the first stage of the regulation game and the equilibrium choices of each type of firm in the second stage can be determined by solving the following equivalent problem where the regulator chooses both a regulatory scheme *and* an output level for each type of firm in order to maximize expected net benefits subject to the constraints imposed by the optimizing behavior of the firm in the second stage. Suppose that the regulator chooses K regulatory schemes, $\{T_k, \lambda_k, R_k(q)\}$, and K output levels, $q_k \geq 0$, in order to

$$\max \sum_{k=1}^{K} \pi_k \left[B(q_k) - c(\theta_k, q_k) + \gamma(T_k + \lambda_k R_k(q_k)) - m\lambda_k \right] \quad (2a)$$

subject to the following constraints which must be satisfied for all $1 \leq k, j \leq K$ and all $q \geq 0$.

$$-\bar{R} \leq R_k(q) \leq \bar{T} \quad (2b)$$

$$0 \leq \lambda_k \leq 1 \quad (2c)$$

$$T_k + \lambda_k R_k(q_k) + c(\theta_k, q_k) \leq \bar{c} \quad (2d)$$

$$T_k + \lambda_k R_k(q_k) + c(\theta_k, q_k) \leq T_j + \lambda_j R_j(q) + c(\theta_k, q), \quad (2e)$$

where k indexes the regulatory scheme and output level that is chosen in equilibrium by a type k firm,[17] and γ and m are positive parameters whose interpretation is discussed in more detail shortly. Cases where the optimal menu contains fewer than K items or where several types of firm find it optimal to choose the same output are handled formally by allowing the output levels and some or all of the elements of the regulatory schemes to be the same for different values of k.[18]

In terms of the original description of the regulatory regime, choosing regulatory scheme k corresponds to reporting output $\hat{q} = q_k$. For all k, $T(q_k) = T_k$ and $R(q_k, q) = R_k(q)$.

---

[17] It is standard to assume that a type k firm will make the choice specified for it by the optimization in Eqs. (2a) through (2e) even if it is indifferent between that choice and other available choices. With this assumption, the regulator can implement a desired Bayesian equilibrium by proposing in stage 1 the menu of K regulatory schemes that results from the optimization in Eqs. (2a) through (2e).

[18] There is no loss of generality in restricting the regulator to menus with only K items and assuming that each type of firm chooses a single regulatory scheme and corresponding output level with certainty.

Consider an arbitrary proposed equilibrium. For each k, select from among the choices of a type k firm a regulatory scheme and corresponding output level that maximizes the term in square brackets in Eq. (2a). Now assume that each type of firm chooses with probability one the regulatory scheme and output selected for it and delete from the original menu any items which are no longer chosen by any type of firm. Since deleting a regulatory scheme and output level does not affect the feasibility of the remaining regulation-output combinations with respect to the inequalities in Eqs. (2b) through (2c), the set of K remaining regulatory schemes and corresponding output levels still represents a possible set of equilibrium choices that would be made by each type of firm in stage 2. Moreover, these choices produce a level of expected net benefits at least as large as the original menu. The optimization in Eqs. (2a) through (2e) maximizes over all such sets; hence, it selects an equilibrium where the regulator's menu contains at most K items, each type of firm chooses a single menu item and output level with certainty, and the level of expected net benefits is at least as great as in the originally proposed equilibrium.

The parameter m denotes the cost of monitoring the firm's output. For simplicity, this cost is assumed to be a positive number that is independent of the firm's type or chosen output level. Hence, the term $m\lambda_k$ in Eq. (2a) represents the expected cost of monitoring conditional on regulatory scheme k being chosen by the firm.

The term $T_k + \lambda_k R_k(q_k)$ in Eq. (2a) is the expected payment made by the firm under regulatory scheme k. γ is a positive parameter that denotes the social value of a dollar transferred from the firm to the regulator. A common interpretation is that γ represents the social cost of raising a dollar of revenue via taxes.[19] γ can also be interpreted as the marginal social value produced by the regulator's expenditure of a dollar obtained from the firm.[20] The term $\gamma(T_k + \gamma_k R_k(q_k))$ represents the conditional expected social benefit of the transfers made under regulatory scheme k.

As previously noted, inequalities (2b) and (2c) are part of the definition of a valid regulatory scheme. Inequalities (2d) and (2e) result from the optimizing behavior of the firm in the second stage of the game. The inequalities in Eq. (2d) are "participation" constraints which ensure that a type k firm will prefer to make its equilibrium choices in the second stage rather than to "quit." The inequalities in Eq. (2e) are "incentive compatibility" constraints which guarantee that each type of firm prefers its equilibrium choice of a regulatory scheme and output level to any other available choice of regulation and output.

An argument similar to those typically associated with the "revelation principle" shows that, given the structural assumptions of our model, the regulator can achieve at least as large a value of the objective in Eq. (2a) by using the regulatory framework which we consider as he can by using any of a large class of alternative regulations. For example, consider a Bayesian equilibrium of any regulation game where (i) the equilibrium strategy for the regulator is a pure strategy, (ii) the equilibrium strategy for each type of firm is either pure or mixed, and (iii) the combination of a pure strategy for the regulator and a pure strategy for a type k firm specifies: (1) an output level, q_k, for the type k firm, (2) a probability, λ_k, that the type k firm is monitored, (3) an initial payment, T_k, that the type k firm makes before it is monitored, and (4) a payment schedule, $R_k(q)$, that specifies an additional payment which the type k firm makes (or receives) if it is monitored and found to be producing q. Depending on the regulation game, these quantities may be determined by one or several stages of play. Moreover, the game may impose constraints on the allowable combinations of payments, outputs, etc. in addition to the ones considered in this paper.[21]

Other than exogenous parameters and functions, the four items in the previous paragraph are the only factors that enter into the payoff functions of either the regulator or the firm. Moreover, as long as the regulator is unable to observe the firm's costs and the firm is free to choose any output, all the combinations

---

[19]Caillaud *et al.* [14] discuss various possibilities for the regulator's objective function, including the one adopted in this paper.

[20]As discussed, for example, in Hahn [24] and OECD [35], some European pollution control agencies may use the funds obtained from regulation to subsidize additional efforts to clean up pollution.

[21]An analysis of the possibility that a regulator might wish to offer lotteries over various combinations of initial taxes, output levels, and so on would be somewhat more complicated. See Myerson [34] for a framework that could be adapted to the model in this paper and used to study this question.

$(q_k, \lambda_k, T_k, R_k(q))$ and $(q_j, \lambda_j, T_j, R_j(q))$ which are realized with positive probability in an equilibrium of the game must also satisfy the constraints in Eqs. (2b)–(2e) for all k, j, and q for the same reasons that these constraints must be satisfied in our model. But the regulatory framework that we consider gives the regulator the greatest freedom to choose the payoff-relevant quantities $q_k, \lambda_k, T_k, R_k(q)$, given the constraints (2b)–(2e). Hence, the expected payoff that the regulator can achieve with our framework must be at least as great as the payoff attainable in the equilibrium of any other game where these payoff-relevant quantities also satisfy Eqs. (2b)–(2e).

3. OPTIMAL REGULATION

This section discusses the optimal regulations specified by the solution to the optimization in Eqs. (2a) through (2e). We particularly wish to note the emergence of rebates as an optimal regulatory response and to compare the optimal regulations in the current environment with the optimal regulations in the benchmark case where the regulator can costlessly observe output and impose an arbitrary, nonlinear tax based on the firm's actual output.

Before considering the optimal regulations, it is helpful to introduce the following notation. Let $\tau_k = T_k + \lambda_k R_k(q_k)$. The quantity τ_k is the total expected payment incurred in equilibrium by a type k firm.

As a point of comparison, consider a model where the structural assumptions, such as the specification of net benefits and the firm's costs, are as described in Section 2, but where the regulator can costlessly impose an arbitrary nonlinear tax that is a function of the firm's actual output. Let τ_k also denote the tax payment incurred by a type k firm in this case. The optimal tax, $\tau(q)$, for the regulator to impose is determined in this model by choosing q_k and τ_k for $k = 1 \ldots K$ to maximize the objective in Eq. (3a) (with $m = 0$) subject to the constraints in Eqs. (3d) and (3e) and the constraint that $q_k \geq 0$. For all k, $\tau(q_k) = \tau_k$. For brevity, we will refer to this benchmark case as the "standard model." Although the structural assumptions in our paper differ somewhat from those in Baron and Myerson's [5] paper, the standard model is closely related to Baron and Myerson's model.

It is convenient to solve the optimization in Eqs. (2a) through (2e) in two stages. In the first stage, the desired output for each type of firm, q_k, is held fixed. For each fixed set of outputs, the optimal regulations to implement these output choices are calculated by solving the constrained optimization in Eqs. (2a) through (2e) for T_k, λ_k, and $R_k(q)$. In the second stage, the optimal output levels are determined.

The next two pages sketch the derivation of the optimal regulations for implementing any fixed, feasible set of output levels. Following this sketch, expressions for these optimal regulations are reported in Eqs. (4a) and (4b) and Eqs. (5a) through (5e).[22]

The derivation of the optimal regulations is simplified by rewriting the optimization in Eqs. (2a) through (2e) in the following equivalent but simpler form.

---

[22]An appendix containing a more detailed derivation of the optimal regulations was omitted at the request of the editor. A working paper which contains this appendix is available upon request from the author.

First note that it must be optimal to let $R_k(q) = \overline{T}$ for all k and all $q \neq q_k$. For when $q \neq q_k$, $R_k(q)$ appears only in Eq. (2b) and on the right-hand side of Eq. (2e). Hence, setting $R_k(q) = \overline{T}$ merely weakens the incentive compatibility constraints on the other variables. In the rest of the paper, suppose that $R_k(q)$ has been set equal to $\overline{T}$ for $q \neq q_k$.

Let $R_k = R_k(q_k)$ denote the equilibrium ex post payment incurred by a type k firm. Once the q_k are set, a regulatory scheme can be specified by the triplet $\{\tau_k, \lambda_k, R_k\}$. Consider a regulator who chooses K such triplets and the corresponding output levels $q_k \geq 0$ in order to

$$\max \sum_{k=1}^{K} \pi_k \left[B(q_k) - c(\theta_k, q_k) + \gamma\tau_k - m\lambda_k \right] \tag{3a}$$

subject to the following constraints which must be satisfied for all $1 \leq k, j \leq K$.

$$-\overline{R} \leq R_k \leq \overline{T} \tag{3b}$$

$$0 \leq \lambda_k \leq 1 \tag{3c}$$

$$\tau_k + c(\theta_k, q_k) \leq \bar{c} \tag{3d}$$

$$\tau_k + c(\theta_k, q_k) \leq \tau_j + c(\theta_k, q_j) \tag{3e}$$

$$\tau_K + c(\theta_K, q_K) \leq \tau_j - \lambda_j R_j + \lambda_j \overline{T}, \tag{3f}$$

where, as before, k indexes the regulatory scheme and output level that is chosen in equilibrium by a type k firm.

Equations (3a) through (3d) correspond to Eqs. (2a) through (2d). Equations (3e) and (3f) are what remain of the incentive compatibility constraints in Eq. (2e) once redundant constraints have been eliminated. The constraints in Eq. (3e) will be referred to as "masquerade" constraints since they guarantee that a type k firm will not wish to masquerade as a type j firm by choosing regulatory scheme j and producing output q_j. The constraints in Eq. (3f) guarantee that a type K firm will never wish to "cheat" by choosing regulatory scheme j and then producing no output.[23]

For a fixed set of outputs, it turns out that the procedure for deriving the optimal τ_k in the current model is the same as the procedure for deriving the optimal nonlinear tax in the standard model.[24] In each case, the participation constraint for the type K firm (Eq. (3d)) and the masquerade constraints with $j = k + 1$ must bind. By solving the resulting system of equations recursively starting with τ_K, one obtains the expressions for the optimal τ_k given in Eqs. (4a) and (5c). Hence, for the same set of output levels, the expected payment incurred

---

[23] If a type k firm is going to deviate from the regulatory scheme and output level intended for it by the regulator, then it does so optimally in one of two ways. First, it can masquerade as a type j firm. Such a masquerade cannot be detected by the regulator since he cannot observe θ. Alternately, the firm may choose to risk the maximum fine, $\overline{T}$, by deviating in a detectable way. In such a case, it is optimal for the firm to choose some regulatory scheme j and then produce nothing. But the cost saving from a given reduction in output is greatest for the type K firm. This turns out to imply that if masquerades are deterred for all types of firm and cheating by producing nothing is deterred for the type K firm, then all other forms of "cheating" by firms will also be deterred.

[24] Spence's [41] derivation of optimal nonlinear prices is an example of a similar calculation.

by a type k firm in our model and the tax paid by such a firm in the standard model are the same. (However, the optimal output for a type k firm will not generally be the same in these two models, so that the actual values of τ_k will also differ in the two cases.)

With the output levels held fixed and optimal values for the τ_k determined as described in the previous paragraph, the optimal values for R_k and λ_k are given by the solution to K uncoupled problems. For each k, λ_k is minimized subject to the constraints in Eqs. (3b), (3c), and (3f). It is clearly optimal to set $R_k = -\overline{R}$ so that the constraint on λ_k in Eq. (3f) is as loose as possible. The optimal value for λ_k is then determined by reducing λ_k until the inequality in Eq. (3f) becomes an equality.

As in the usual analysis of the standard model, it is convenient to write the optimal regulations listed in Eqs. (5a) through (5e) in terms of the following sum of cost terms. For a given set of outputs q_k, let S_k denote the sum specified by

$$S_k = c(\theta_K, q_K) + \sum_{j=k}^{K-1} \left[c(\theta_j, q_j) - c(\theta_j, q_{j+1}) \right], \tag{4a}$$

where the usual convention that a summation equals 0 when the lower limit exceeds the upper limit is used. By collecting the terms in S_k involving q_j, this sum can be rewritten in the form

$$S_k = c(\theta_k, q_k) + \sum_{j=k+1}^{K} \left[c(\theta_j, q_j) - c(\theta_{j-1}, q_j) \right]. \tag{4b}$$

It is well known from the analysis of the standard model that the incentive compatibility constraints in Eq. (3e) cannot be satisfied unless $q_k \geq q_{k+1}$ for all k. In addition, the requirement that $\lambda_k \leq 1$ imposes a further constraint on the feasible output levels; the right-hand side of Eq. (5d) must be less than or equal to one. Any set of output levels which satisfies these two sets of constraints is feasible and can be implemented by the regulations listed in Eqs. (5a) through (5e).

Optimal Regulations for Implementing the Feasible Output Levels q_k

$$R_k(q) = \overline{T}, \quad \text{for } q \neq q_k \tag{5a}$$

$$R_k(q_k) = -\overline{R} \tag{5b}$$

$$\tau_k = \overline{c} - S_k \tag{5c}$$

$$\lambda_k = \frac{S_k}{\overline{R} + \overline{T}} \tag{5d}$$

$$T_k = \overline{c} - \frac{\overline{T}}{\overline{R} + \overline{T}} S_k \tag{5e}$$

Although the parameters m and γ do not appear directly in the above expressions for the optimal regulations, they do affect these regulations indirectly. The quantity S_k is a function of the output levels which the regulator wishes to implement, and, as we shall see shortly, γ and m play an important role in determining the optimal output levels.

Equation (4a) implies that S_k is a weakly decreasing function of k whenever $q_k \geq q_{k+1}$ for all k. Unless $q_k = q_{k+1}$, $S_k > S_{k+1}$. Hence, λ_k is also a (weakly) decreasing function of k. Both T_k, the up-front tax paid by a type k firm, and τ_k, the expected payment incurred by a type k firm, are (weakly) increasing functions of k. In terms of output, a firm with lower costs produces at least as much as a higher cost firm. If the firm produces more (as will often be the case), then it is more likely to be monitored. The expected payment of such a firm is lower than that of a higher cost firm for two reasons. The firm pays a lower up-front tax and is more likely to receive the rebate which occurs when the firm is monitored and found to be producing its reported output.

If a firm that reports q_k "cheats" by producing q instead, then it both looses its rebate and pays a fine, thus incurring the expected cost $\lambda_k[R_k(q) - R_k(q_k)]$. For a given probability of monitoring, deterrence is maximized by making the difference between the fine for noncompliance and a rebate for compliance as large as possible. Hence, Eq. (5a) specifies that any observed deviation from the reported output should optimally be punished with the greatest possible fine. Since fines are not imposed in equilibrium, this threat involves no actual cost.

Equation (5b) specifies that the reward (or rebate) for compliance should also be made as large as possible. Since rebates will actually be paid in equilibrium and transfers to the firm are costly, this policy might appear to involve excessive social costs. However, the up-front tax, T_k, can be increased to cover the expected cost of the rebate.

Since the cost parameter θ is not observable by the regulator, a low-cost firm can always produce less output than the regulator desires by masquerading as a higher cost firm. This consideration determines both the expected payment in our model and the optimal tax in the standard model. In each case, the differences in the payments incurred by lower and higher cost firms provide the smallest incentive needed to induce each type of firm to produce the output which the regulator desires. In particular, the total expected payment incurred in equilibrium by a type k firm is lower than the equilibrium payment of a type $k + 1$ firm by the amount $c(\theta_k, q_k) - c(\theta_k, q_{k+1})$, which is just sufficient to induce the type k firm to produce q_k rather than masquerading as a type $k + 1$ firm.

As noted previously, for the same output levels the formula for τ_k in Eq. (5c) also describes the optimal nonlinear tax in the standard model. The relation between the up-front tax in our model and the tax in the standard model can therefore be studied by combining Eqs. (4a), (5c), and (5e) to produce the following equation which holds for all $k < K$:

$$T_{k+1} - T_k = \frac{\overline{T}}{\overline{R} + \overline{T}}(\tau_{k+1} - \tau_k) = \frac{\overline{T}}{\overline{R} + \overline{T}}[c(\theta_k, q_k) - c(\theta_k, q_{k+1})]. \quad (6)$$

Moreover, $T_k \geq \tau_k$ for all k, and, as long as $\overline{R} > 0$, the inequality is strict for k such that $q_k > 0$. Hence, the up-front tax in our model is "flatter"; that is, it declines more slowly as a function of (reported) output than either the total expected payment in our model or, for the same output levels, the corresponding tax in the standard model.

GUILTY UNTIL PROVEN INNOCENT 139

When $\bar{T}$ is small compared to $\bar{R}$, the optimal up-front tax is almost flat. Hence, in the realistic case where the size of fines for noncompliance is severely limited, the up-front tax will be almost the same for all types of firm, and the incentive for lower cost types to produce more output will be provided almost entirely by variations in the expected rebate that different types of firm receive. Since the optimal rebate is the same for all levels of reported output, the expected rebate varies only because the probability that a firm is monitored varies with the firm's reported output.

At the other extreme, if $\bar{R} = 0$, so that rebates are impossible, then $T_k = \tau_k$. In this case, all of the incentive to produce more output is provided by variations in the up-front tax just as in the standard model.

The size of the up-front tax and the probability that a firm is monitored are related, for if the up-front tax decreases with the reported output level, then the firm has an incentive to cheat by reporting a high output (and so reducing its tax bill) and then producing nothing. In order to deter such cheating, it is optimal to monitor with greater probability a firm that reports a higher output. But this also implies that a firm which reports a higher output receives a higher expected rebate for compliance. The requirements of deterrence, which cause the monitoring probability to increase with reported output, also permit the up-front tax to be flatter than would be the case if this tax were providing all of the incentive for a lower cost firm to produce more output.

By substituting the formulas for the optimal regulations in Eqs. (5a) through (5e) into the objective in Eq. (2a) (or Eq. (3a)) and simplifying the terms involving S_k, one obtains an equation which describes the expected net benefits generated by an optimally regulated firm as a function of the output levels produced by each type of firm,[25]

$$\gamma \bar{c} + \sum_{k=1}^{K} \left\{ \pi_k [B(q_k) - (1 + \tilde{\gamma})c(\theta_k, q_k)] - \tilde{\gamma} \Pi_k [c(\theta_k, q_k) - c(\theta_{k-1}, q_k)] \right\},$$

(7)

where $\tilde{\gamma} = \gamma + m/(\bar{R} + \bar{T})$ and $\Pi_k = \sum_{j=1}^{k-1} \pi_j$. (Note that $\Pi_1 = 0$ and let $c(\theta_0, q) = 0$ for all q.)

Let $q_{K+1} = 0$. With this notation, the optimal output levels for each type of firm are obtained by choosing q_k for $k = 1 \ldots K$ to maximize the objective function in Eq. (7) subject to the "monotonicity" constraints that $q_k \geq q_{k+1}$ for all k and

---

[25] The following equations are useful in simplifying the expression that results from substituting Eqs. (5a) through (5e) into Eq (2a):

$$\sum_{k=1}^{K} \pi_k S_k = \sum_{k=1}^{K} \pi_k c(\theta_k, q_k) + \sum_{k=1}^{K} \pi_k \sum_{j=k+1}^{K} \left[c(\theta_j, q_j) - c(\theta_{j-1}, q_j) \right]$$

$$= \sum_{k=1}^{K} \pi_k c(\theta_k, q_k) + \sum_{k=2}^{K} \Pi_k [c(\theta_k, q_k) - c(\theta_{k-1}, q_k)].$$

The first equality involves a straight substitution from Eq. (4b). The second equality is obtained from the first by collecting all the terms in the double sum that involve q_k.

subject also to the "monitoring" constraint that

$$\lambda_1 = \frac{1}{\bar{R} + \bar{T}}\left[c(\theta_1, q_1) + \sum_{k=2}^{K} [c(\theta_k, q_k) - c(\theta_{k-1}, q_k)]\right] \leq 1. \qquad (8)$$

Since the monotonicity constraints ensure that S_k is a weakly decreasing function of k, the inequality in Eq. (8) is sufficient to guarantee that $\lambda_k \leq 1$ for all k.

The assumption that $c_{qq\theta} \geq 0$ implies that the difference $c(\theta_k, q_k) - c(\theta_{k-1}, q_k)$ is a convex function of q_k for all $k \geq 2$. Hence, the optimization that determines the optimal output levels is a standard problem where a concave objective function is maximized subject to a convex constraint set. We assume in the rest of the paper that a solution to this optimization exists and is characterized by the usual Kuhn–Tucker conditions.

For $1 < k \leq K$, the first-order conditions for the solution to the above problem are given by

$$B'(q_k) = (1 + \tilde{\gamma})c_q(\theta_k, q_k) + \left(\tilde{\gamma}\frac{\Pi_k}{\pi_k} + \frac{\mu}{\pi_k}\right)[c_q(\theta_k, q_k) - c_q(\theta_{k-1}, q_k)]$$

$$+ \frac{\delta_{k-1} - \delta_k}{\pi_k}, \qquad (9a)$$

where $\mu(\bar{R} + \bar{T})$ is the multiplier for the monitoring constraint in Eq. (8) and δ_k is the multiplier for the monotonicity constraint: $q_{k+1} - q_k \leq 0$. For $k = 1$, the first-order condition is given by

$$B'(q_1) = \left(1 + \tilde{\gamma} + \frac{\mu}{\pi_1}\right)c_q(\theta_1, q_1) - \frac{\delta_1}{\pi_1}. \qquad (9b)$$

By setting $m = 0$ in Eq. (3a), substituting for τ_k using Eq. (5c), and simplifying, one can observe that, with $\tilde{\gamma}$ replaced by γ, the optimal output levels in the standard model are also determined by maximizing the objective in Eq. (7) subject to the monotonicity constraints. Hence, if the monitoring constraint in Eq. (8) is not binding, then the optimal outputs both in the current model and in the standard model are determined by the same problem except for the value of a single parameter.

In the rest of this section, we focus mainly on the case where $\bar{R} + \bar{T}$ is sufficiently large that the monitoring constraint does not bind. Note, however, that the first-order conditions imply that, when the monitoring constraint does bind, the result is a reduction in the optimal outputs that resembles the output distortions discussed below. Moreover, if the budget of the agency charged with monitoring and enforcement is sufficiently small, then the right-hand side of the inequality in Eq. (8) should plausibly be reduced from 1 to some lower level $\bar{\lambda}$, which makes it more likely that the monitoring constraint will bind. Thus constraints on enforcement budgets may produce large output distortions even in cases where these distortions would otherwise be thought to be small.

When the monitoring constraint is not binding, it is well known that the optimal outputs can often be obtained by solving an unconstrained problem that ignores the monotonicity constraints. For example, if the differences $\theta_k - \theta_{k-1}$ are chosen to be constant and the ratio Π_k/π_k is a nondecreasing function of k (a discrete analog of a standard assumption in the literature on regulation), then the assumption that $c_{q\theta\theta} \geq 0$ is sufficient to ensure that nonnegative output levels which satisfy the first-order conditions for maximizing the unconstrained objective in Eq. (7) also satisfy the monotonicity constraints. Suppose now that the monotonicity constraints are satisfied by the solution to the unconstrained maximization.

A second useful benchmark is the case of a fully informed regulator who observes θ and the firm's output and can set an arbitrary nonlinear tax based on this information. For each k, such a regulator chooses τ_k and q_k to maximize the objective in Eq. (3a) (with $m = 0$) subject only to the participation constraints in Eq. (3d). Assuming an interior solution, the optimal outputs in this case are given by

$$B'(q_k) = (1 + \gamma)c_q(\theta_k, q_k). \tag{10}$$

Since the term in square brackets in Eq. (9a) is positive, the first-order conditions indicate that, for $k > 1$, the optimal outputs in our model will tend to be smaller than the outputs chosen by a fully informed regulator. It is well known that the regulator's inability to observe costs produces an output distortion in the standard model that reduces the output of all but the lowest cost firm. The size of this distortion depends on γ or, in our model, $\tilde{\gamma}$. In addition, since $\tilde{\gamma} > \gamma$, there is a second distortion in our model that reduces even the output of the lowest cost firm.

If the firm is monitored with probability λ, then $m\lambda$ is the expected cost of monitoring and $\lambda(\bar{R} + \bar{T})$ is the expected penalty for cheating. Hence, the ratio $m/(\bar{R} + \bar{T})$ represents the cost to the regulator of imposing one dollar's worth of expected penalty. The social cost of one dollar of transfers, γ, is augmented in our model by the cost of deterrence. Costly monitoring therefore provides an additional rationale for the output distortions predicted by the standard model.

As $\bar{R} + \bar{T}$ becomes large, the monitoring constraint ceases to bind and $\tilde{\gamma}$ approaches γ. If, in addition, γ becomes small, then the optimal outputs produced in both the current and the standard models approach the output levels that would be chosen by a fully informed regulator. However, even when the optimal levels of output are almost the same in these three cases, we have seen that the form of the optimal regulations will be quite different.

4. THE RIGHT TO BE MONITORED

This section considers the case where the regulator's ability to commit to a monitoring policy is limited by the firm's right to be monitored. In particular, a firm that expects a rebate for compliance can request to be monitored.

When the firm has a right to be monitored, firm types will, in general, divide into two categories. Firm types with sufficiently high costs are monitored with a probability less than one and pay an up-front tax that behaves like the tax in the standard model. In particular, all of the incentive for one type of firm in the category to produce more output than another type in the category is provided by variations in the up-front tax. There are no rebates for these types of firm.

The second category consists of firm types with lower costs. A firm in this category receives a rebate and always exercises its right to be monitored. One optimal tax/rebate policy specifies that all types of firm in the second category pay the same up-front tax. Within this category, all of the incentive to produce more output is then provided by variations in the size of the rebate.

Since monitoring is costly, the regulator will prefer to avoid rebates so that the firm does not request to be monitored. However, when the maximum fine, $\bar{T}$, is small, this fine may need to be supplemented by the threat to withhold a rebate in order to deter the firm from cheating. If $\bar{T}$ is sufficiently small, then all types of firm will be offered rebates and, consequently, fall into the second category mentioned above. On the other hand, if $\bar{T}$ is sufficiently large, then rebates will not be needed for deterrence, and all types of firm will be in the first category.

As in the previous sections, regulation is modeled as a three-stage game. In the first stage of the game, the regulator offers a menu of regulatory schemes. The specification of regulatory scheme j is the same as that in Section 2 except that λ_j is replaced by λ_j^{nr}, which denotes the probability that a firm choosing regulatory scheme j is monitored, but only when the firm does not submit a request to be monitored. (nr stands for "no request".)

In the second stage of the regulation game, the firm simultaneously chooses: (1) an item from the regulatory menu, (2) the actual output level that it will produce, and (3) whether or not to submit a request to be monitored. In the third stage, the firm is monitored with probability λ_j^{nr} if it has chosen regulatory scheme j and not submitted a request to be monitored. If the firm has submitted a request to be monitored, then it is monitored with probability one. Except for the firm's new option to submit a request to be monitored, all the assumptions of the model are the same as those in Section 2.

Whether to submit a request to be monitored is a simple choice for the firm. If a firm chooses regulatory scheme j and output level q_i, then it should request to be monitored if and only if $R_j(q_i) < 0$. (Recall that negative quantities denote payments to the firm.)

Let λ_k denote the equilibrium probability that a type k firm is monitored and R_k denote the equilibrium ex post payment incurred by a type k firm. That is, $R_k = R_k(q_k)$, where q_k is the equilibrium output produced by a type k firm. Given the firm's optimal request policy, the presence of a right to be monitored can be modeled as an additional constraint which the firm's policy imposes on R_k and λ_k. The following inequality must be satisfied for all $1 \leq k \leq K$:

$$R_k(1 - \lambda_k) \geq 0. \tag{11}$$

It is convenient to describe the equilibrium of the above regulation game in terms of an optimization like that in Eqs. (2a) through (2e). In general, the constraints in such an optimization involve both λ_k and λ_k^{nr}. Fortunately, however, we can restrict attention to regulatory schemes for which $\lambda_k = \lambda_k^{nr}$.[26] Hence, there

---

[26] If $R_k \geq 0$, then a type k firm will never submit a request to be monitored, and $\lambda_k = \lambda_k^{nr}$. If $R_k < 0$, then a type k firm will always submit a request to be monitored unless it intends to deviate from its reported output. In this case, it is also optimal for the regulator to set $\lambda_k^{nr} = 1$ since this probability serves only to deter cheating and does not affect the equilibrium level of net benefits.

is no ambiguity in using λ_k to denote both the equilibrium probability that a type k firm is monitored and λ_k^{nr}.

With this notation, the optimal regulations and output levels are described by an optimization where the regulator chooses K combinations, $\{T_k, \lambda_k, R_k(q)\}$, and K nonnegative output levels, q_k, in order to maximize the objective in Eq. (2a) subject to the constraints in Eqs. (2b) through (2e) and the new constraint in Eq. (11). As before, k indexes the regulatory scheme and output level chosen in equilibrium by a type k firm.

In the rest of this section, the optimal regulations for implementing any given, feasible set of output levels are discussed. Many of these regulations are unaffected by the presence of the constraint in Eq. (11). In particular, it is still optimal to let $R_k(q) = \bar{T}$ when $q \neq q_k$, and the optimal values for τ_k are still described by Eqs. (4a), (4b), and (5c) in the current circumstances. Only the expressions for the optimal values of λ_k and R_k need to be changed.

The derivation of the optimal values for λ_k and R_k closely parallels the derivation sketched in Section 3. With the output levels held fixed and the optimal values for τ_k specified by Eqs. (5c), the optimal values for λ_k and R_k are obtained by solving K uncoupled problems. In each problem, λ_k is minimized subject to the inequality constraint on R_k in Eq. (1), the inequality constraint $0 \leq \lambda_k \leq 1$, the inequality in Eq. (11), and the inequality in Eq. (3f), which can be written in the form

$$S_k \leq \lambda_k (\bar{T} - R_k), \tag{12}$$

where, as before, S_k denotes the sum of cost terms in Eqs. (4a) and (4b). Recall that the S_k are functions of the output levels which are to be implemented. The only change from the derivation in Section 3 is the presence of the additional constraint in Eq. (11).

When $\bar{T} \geq S_k$, it is straightforward to solve for the optimal values of λ_k and R_k. It is clearly optimal to make the difference on the right-hand side of Eq. (12) as large as possible by setting $R_k = 0$. The optimal value for λ_k is then obtained by reducing λ_k until the inequality in Eq. (12) becomes an equality. The optimal monitoring probability obtained in this way, $\lambda_k = S_k/\bar{T}$, satisfies the constraint that $\lambda_k \leq 1$. Note that if $\bar{T} \geq S_k$ for all k, then the resulting regulations are the same as those which occur when a firm does not have a right to be monitored but $\bar{R} = 0$ so that rebates are impossible.

An interesting (and plausible) case occurs when $\bar{T}$ is "small" but $\bar{R} + \bar{T}$ is "large" so that $\bar{R} + \bar{T} \geq S_k > \bar{T}$. Since S_k is a weakly decreasing function of k, this situation is more likely to occur for a lower cost firm.[27]

Since $S_k > \bar{T}$, the constraint in Eq. (12) can only be satisfied if $\lambda_k = 1$ so that R_k can be negative. With $\lambda_k = 1$, the inequality in Eq. (12) and the inequalities in Eq. (1) can be combined in the following equation, which must be satisfied by the optimal rebate for a type k firm:

$$-\bar{R} \leq R_k \leq \bar{T} - S_k. \tag{13a}$$

---

[27] If $S_k > \bar{R} + \bar{T}$, then there are no feasible values of λ_k and R_k that satisfy Eq. (12). Hence, a feasible set of output values must satisfy the inequality $S_1 \leq \bar{R} + \bar{T}$, which is just the monitoring constraint from Eq. (8) of Section 3.

144   JOSEPH E. SWIERZBINSKI

The assumptions about $\overline{T}$ and $\overline{R} + \overline{T}$ imply that $-\overline{R} \le \overline{T} - S_k < 0$, so that the optimal policy involves a rebate, and feasible rebates exist.

Combining the optimal formula for τ_k in Eq. (5c) with the accounting identity $T_k = \tau_k - R_k$ produces the following equation for the optimal up-front tax:

$$T_k = \bar{c} - S_k - R_k. \tag{13b}$$

Any combination $\{T_k, R_k\}$ that satisfies Eqs. (13a) and (13b) is an optimal tax/rebate policy for regulating the type k firm.

One reasonable assumption is that the optimal rebate is kept as small as possible so that the regulator avoids collecting money with the up-front tax merely to return it later. In this case, the right-hand inequality in Eq. (13a) becomes an equality and the optimal up-front tax is a constant: $T_k = \bar{c} - \overline{T}$. Every type of firm for which $S_k > \overline{T}$ pays the same up-front tax. Within this category of types, all of the incentive that a lower cost type of firm requires to produce more output is provided by variations in the rebate offered to different types of firm.

As in Section 3, the optimal output levels are obtained by substituting the optimal regulations for each fixed set of outputs into Eq. (2a) and maximizing the resulting objective function subject to the monotonicity constraints that $q_k \ge q_{k+1}$ for all k (letting $q_{K+1} = 0$) and the monitoring constraint in Eq. (8). (See also footnote 27.)

A detailed analysis of the optimal outputs is not necessary to see that if $\overline{T}$ is sufficiently small, for example, $\overline{T} = 0$, then any firm type that produces positive output will be in a category where $S_k > \overline{T}$. On the other hand, if $\overline{T}$ is sufficiently large, then all firms will be in a category where $S_k \le \overline{T}$.

5. CONCLUSION

This paper develops a model of pollution control which recognizes that regulators are likely to be poorly informed about a specific firm's cost of pollution control, that monitoring a firm's emission of pollutants is costly, and penalties for noncompliance are often limited. We demonstrate that in such a regulatory environment, there are many circumstances where the optimal regulations resemble deposit–refund systems. The paper therefore reinforces and extends the message of Bohm [10], Costanza and Perrings [17], Solow [40], and others that such systems deserve consideration as a tool for dealing with environmental problems.

A number of factors not included in this paper may further enhance the desirability of regulations with deposit–refund features. For example, a firm may be able to monitor its own emissions (in a verifiable way) more cheaply than an outside regulator can. A deposit–refund system would encourage firms to engage in such activity. Moreover, as was noted in passing in Section 3, the amount of monitoring that an enforcement agency can do may be constrained by the agency's budget. A deposit–refund system provides a way to work around this budget constraint by encouraging firms to pay for their own monitoring activities.

On the other hand, we have also shown that even broad, qualitative features of the optimal regulations may depend on the details of how enforcement is limited or what the rights of polluters are. This suggests a note of caution. Although regulations with deposit–refund features have an intuitive appeal, the proper design of an effective system need not be trivial.

REFERENCES

1. S. Baiman and J. Demski, Economically optimal performance evaluation and control systems, *J. Accounting Res. Suppl.* **18**, 184–219 (1980).
2. D. Baron, Regulatory strategies under asymmetric information, *in* "Bayesian Models in Economic Theory" (M. Boyer and R. Kihlstrom, Eds.), North-Holland, Amsterdam (1984).
3. D. Baron, Design of regulatory mechanisms and institutions, *in* "Handbook of Industrial Organization" (R. Schmalensee and R. Willig, Eds.), Vol. 2, North-Holland, Amsterdam (1989).
4. D. Baron and D. Besanko, Regulation, asymmetric information, and auditing, *Rand J. Econom.* **15**, 447–470 (1984).
5. D. Baron and R. Myerson, Regulating a monopolist with unknown costs, *Econometrica* **50**, 911–930 (1982).
6. W. Baumol and W. Oates, "The Theory of Environmental Policy," 2nd ed., Cambridge Univ. Press, London/New York (1988).
7. B. Beavis and I. Dobbs, Firm behaviour under regulatory control of stochastic environmental wastes by probabilistic constraints, *J. Environ. Econom. Management* **14**, 112–127 (1987).
8. B. Beavis and M. Walker, Random wastes, imperfect monitoring, and environmental quality standards, *J. Public Econom.* **21**, 377–387 (1983).
9. D. Besanko and D. Sappington, "Designing Regulatory Policy with Limited Information," Harwood Academic, Chur, Switzerland (1987).
10. P. Bohm, "Deposit–Refund Systems: Theory and Applications to Environmental, Conservation, and Consumer Policy," Resources for the Future, Washington, DC (1981).
11. P. Bohm and C. Russell, Comparative analysis of alternative policy instruments, *in* "Handbook of Natural Resource and Energy Economics" (A. Kneese and J. Sweeney, Eds.), Vol. 1, North-Holland, Amsterdam (1985).
12. K. Border and J. Sobel, Samurai accountant: A theory of auditing and plunder, *Rev. Econom. Stud.* **54**, 525–540 (1987).
13. B. Bower, R. Barre, J. Kuhner, and C. Russell, "Incentives in Water Quality Management: France and the Ruhr Area," Resources for the Future, Washington, DC (1981).
14. B. Caillaud, R. Guesnerie, P. Rey, and J. Tirole, Government intervention in production and incentives theory: A review of recent contributions, *Rand J. Econom.* **19**, 1–26 (1988).
15. B. Caillaud, R. Guesnerie, and P. Rey, Noisy observation in adverse selection models, *Rev. Econom. Stud.* **59**, 595–615 (1992).
16. P. Chander and L. Wilde, "A General Characterization of Optimal Income Taxation and Enforcement," Social Science Working Paper 791, California Institute of Technology (1992).
17. R. Costanza and C. Perrings, A flexible assurance bonding system for improved environmental management, *Ecological Econom.* **2**, 57–75 (1990).
18. M. Cropper and W. Oates, Environmental economics: A survey, *J. Econom. Lit.* **30**, 675–740 (1992).
19. P. Dasgupta, P. Hammond, and E. Maskin, On imperfect information and optimal pollution control, *Rev. Econom. Stud.* **47**, 857–860 (1980).
20. J. Demski, D. Sappington, and P. Spiller, Managing supplier switching, *Rand J. Econom.* **18**, 77–97 (1987).
21. P. Downing and W. Watson, The economics of enforcing air pollution controls, *J. Environ. Econom. Management* **1**, 219–236 (1974).
22. R. Dye, Optimal monitoring policies in agencies, *Rand J. Econom.* **17**, 339–350 (1986).
23. J. Evans, Optimal contracts with costly conditional auditing, *J. Accounting Res. Suppl.* **18**, 108–139 (1980).
24. R. Hahn, Economic prescriptions for environmental problems: How the patient followed the doctor's orders, *J. Econom. Perspect.* **3**, 95–114 (1989).
25. J. Harford, Firm behavior under imperfectly enforceable pollution standards and taxes, *J. Environ. Econom. Management* **5**, 26–43 (1978).
26. J. Harford, Self-reporting of pollution and the firm's behavior under imperfectly enforceable regulations, *J. Environ. Econom. Management* **14**, 293–303 (1987).

146 JOSEPH E. SWIERZBINSKI

27. W. Harrington, Enforcement leverage when penalties are restricted, *J. Public Econom.* **37**, 29–53 (1988).
28. S. Linder and M. McBride, Enforcement costs and regulatory reform: The agency and firm response, *J. Environ. Econom. Management* **11**, 327–346 (1984).
29. A. Malik, Markets for pollution control when firms are noncompliant, *J. Environ. Econom. Management* **18**, 97–106 (1990).
30. A. Malik, Self-reporting and the design of policies for regulating stochastic pollution, *J. Environ. Econom. Management* **24**, 241–257 (1993).
31. L. Martin, The optimal magnitude and enforcement of evadable Pigovian charges, *Public Finance* **39**, 347–358 (1984).
32. N. Melumad and D. Mookherjee, Delegation as commitment: The case of income tax audits, *Rand J. Econom.* **20**, 139–163 (1989).
33. D. Mookherjee and I. Png, Optimal auditing, insurance, and redistribution, *Quart. J. Econom.* **103**, 399–415 (1989).
34. R. Myerson, Optimal coordination mechanisms in generalized principal–agent problems, *J. Math. Econom.* **10**, 67–81 (1982).
35. Organization for Economic Cooperation and Development (OECD), "Pollution Charges in Practice," Paris (1980).
36. I. Ortuno-Ortin, Inspections and externalities, unpublished manuscript, Department of Economics, Universidad de Alicante, Alicante, Spain (1992).
37. J. Reinganum and L. Wilde, Income tax compliance in a principal–agent framework, *J. Public Econom.* **26**, 1–18 (1985).
38. C. Russell, Monitoring and enforcement, *in* "Public Policies for Environmental Protection" (P. Portney, Ed.), Resources for the Future, Washington, DC (1990).
39. K. Seegerson, Uncertainty and incentives for nonpoint pollution control, *J. Environ. Econom. Management* **15**, 87–98 (1988).
40. R. Solow, The economist's approach to pollution and its control, *Science* **173**, 498–503 (1971).
41. A. M. Spence, Multi-product quantity-dependent prices and profitability constraints, *Rev. Econom. Stud.* **47**, 821–841 (1980).
42. D. Spulber, Optimal environmental regulation under asymmetric information, *J. Public Econom.* **35**, 163–181 (1988).

Part V
Cost-Effectiveness: Empirical Studies

[33]

JOURNAL OF ENVIRONMENTAL ECONOMICS AND MANAGEMENT 9, 101–121 (1982)

The Empirical Properties of Two Classes of Designs for Transferable Discharge Permit Markets[1]

Department of Economics, University of Wyoming, Laramie, Wyoming 82071

AND

Department of Economics, Colby College, Waterville, Maine 04901

Received January 7, 1981; revised May 1, 1981

Previous work by Atkinson and Lewis (*J. Environ. Econ. Manag.* **1**, 237–250 (1974)) and Anderson *et al.* ("An Analysis of Alternative Policies for Attaining and Maintaining a Short-Term NO$_2$ Standard," MATHTECH, Inc., Princeton, N.J., 1979) has indicated the tremendous cost advantages to be achieved by moving from a policy based on emission standards to one based on marketable emission permits. As Tietenberg (*Land Econ.* **56**, 391–416 (1980)) points out, however, neither of the major permit designs treated in the literature are optimal from all points of view. This has triggered a search for alternative permit designs, which, while they may not minimize compliance costs, have sufficient other virtues as to make them attractive on other grounds. The purpose of this paper is to examine, within the context of an empirical mathematical programming model, the air quality, emission, and cost consequences of two classes of the permit designs which can be implemented in the absence of information on control costs. This case study involves particulate control in St. Louis.

I. INTRODUCTION

Background

The U.S. Environmental Protection Agency (EPA) has embarked on a new regulatory approach to controlling air pollutant emissions through controlled trading of the right to pollute. The principal objective of these reforms is to reduce the rapidly increasing costs of pollution control. The two main existing manifestations of this approach are the "bubble" and "offset" policies.[2] The former allows an emitter to relax the control on one or more sources of a particular pollutant providing it secures an equivalent reduction of the same pollutant from some other nearby source. The latter allows new sources of pollution to enter a region where the standards are already exceeded provided that the new source procures sufficient reductions in emissions from existing sources (over and above their previous legally mandated reductions) to guarantee that the air quality will be improved as a result of the transaction. The EPA is also exploring the possibilities of introducing similar

---

[1] The authors wish to acknowledge the helpful comments received from Cliff Russell and two anonymous referees.

[2] See 44 FR 71780 (11 December 1979) and 40 CFR 51 Appendix S, originally presented in 44 FR 3274 (16 January 1979).

0095-0696/82/020101–21$02.00/0
Copyright © 1982 by Academic Press, Inc.
All rights of reproduction in any form reserved

regulatory innovations in areas which are now cleaner than the standards, but which face the prospect of limited growth in the future.[3]

All of these regulatory initiatives involve restrictions on the trades which can take place. There is sympathy within EPA for eliminating some of the restrictions on these trades, providing the less restrictive trades involve no threat to air quality. The empirical question which is the focus of this paper is how the permits (which define the property right to be traded) and the permit markets (which govern the trading possibilities) can be defined so as to minimize the compliance costs of achieving air quality standards while imposing tolerable regulatory burdens on the state control authorities. By tolerable regulatory burdens we mean simply those which depend on information which the regulatory authority can reasonably be expected to have. Specifically, we examine systems which can be implemented without any information on control costs.

The theoretical characteristics of two marketable permit (MP) systems—the ambient permit and the emissions permit systems—are well known.[4] If the objective is to produce the desired air quality at N predetermined receptor locations at minimum cost, an ambient permit system involving spatially differentiated permits is appropriate. This system takes into account each source's emission diffusion characteristics, which map emissions to each specific receptor's air quality. If, on the other hand, the objective is to reduce the total amount of emissions in an area at minimum cost, the administratively simpler emissions permit system, relying on spatially undifferentiated permits, is appropriate.

Unfortunately, as the set of policy concerns is expanded beyond a singular focus on the cost of compliance neither MP system is dominant. Both the ambient permit and emissions permit systems have shortcomings associated with them. The chief problem with the ambient permit system is its administrative complexity. In order to ensure the cost-effective allocation of the control responsibility among emitters for the achievement of the ambient standards monitored at N receptor locations, the control authority would have to establish N separate permit markets for each pollutant.[5] Each emitter whose emissions had an appreciable effect on the air pollution level at more than one receptor would have to purchase multiple permits to legitimize a single source of emissions. Though this may be manageable it does create a certain interest in seeing whether administratively simpler permit designs could approximate the ambient permit system.

The ambient permit system also has one other potentially undesirable characteristic—while it controls ambient air quality at the receptor locations it may do so while allowing increases in total regional emissions. While total emissions are clearly of secondary importance in the Clean Air Act, strategies which allow large aggregate emission increases may not be allowed under the current Act.[6] It therefore becomes important to establish the degree to which the cost superiority of the ambient permit

---

[3]See the advanced notice of proposed rulemaking on the Set II pollutants for the prevention of significant deterioration in 45 FR 30088 (7 May 1980).

[4]See Tietenberg (1980).

[5]It may be that at a particular point in time fewer than N are needed because air quality at some receptors will be below the legal threshold. Nonetheless to guarantee that the ambient standards are met at all receptors for all points in time all N permit markets need to be activated.

[6]See, for example, the "reasonable further progress" requirement which is defined in terms of aggregate emission reductions rather than improvements in air quality. 42 USC 7501(1) and 42 USC 7503.

system is due to the fact that it allows more of the pollutant to be emitted. Empirical estimates of the importance of this issue are presented in Section IV.

The emissions permit system is generally not a particularly close substitute for an ambient permit system due to the fact that aggregate emissions (the variable being controlled) is not uniquely related to the ambient air quality targets specified in the Clean Air Act.[7] First in order to insure the ambient standards are met, the aggregate level of allowable emissions has to be reduced sufficiently to ensure that the air quality at the most polluted receptor location meets the standard. This results in substantial over-control of the more distant sources, thus increasing the cost of compliance to achieve specified air quality levels. Work by Atkinson and Lewis (1974) on particulate control in St. Louis and Anderson *et al.* (1979) on nitrogen oxide control in Chicago indicates that the increase in compliance cost from using an emission permit system to achieve the ambient standards can be substantial.

The second reason why an emission permit is not a close substitute for an ambient permit is that even if the control authority can find a feasible initial permit allocation there is nothing to guarantee that it will remain feasible over time. A feasible permit allocation is defined as one which results in a geographic allocation of emissions which satisfies the ambient standards. Suppose that a feasible allocation currently exists and that a new source enters the area. The emissions permit system, if enforced, will guarantee that the total level of emissions in the area will be the same before and after the entry of the new source. This source will have purchased a sufficient number of permits from existing sources (necessitating emission reductions by them) to offset its own increase in emissions to the area. If the permits, however, are purchased from emitters in relatively clean parts of the region and used in relatively polluted parts of the region the result will trigger a violation of the ambient standard in a region which was formerly in compliance. The emission permit system is, by itself, powerless to prevent these "hot spots" from emerging.

The fact that neither of these permit systems is optimal from all points of view has triggered a search for alternative permit designs which, while they may not be optimal, may represent a reasonable, pragmatic compromise. The search is for a permit design with low compliance cost as well as low administrative and enforcement cost which provides reasonable assurance that "hot spots" will not arise over time.

An Overview of the Paper

The purpose of this paper is to examine, within the context of an empirical mathematical programming model, the air quality, emission, and cost consequences of a variety of these limited information permit designs. They are limited information in the sense that the control authority can implement them without any information on the costs of control faced by emitters. The focus on limited information designs is dictated by the pragmatic realization that control authorities rarely have this information.

The next section of the paper introduces the types of permit designs considered and formulates the programming models which simulate their effects on control cost,

---

[7]It can be a good substitute when the location of the emission source is of no consequence. This would be the case, for example, with pollutants which rapidly become uniformly mixed in the atmosphere or with global pollutants (e.g., fluorocarbons).

aggregate emissions, and air quality. Section III presents the data employed to operationalize these programming models. Section IV summarizes the main empirical results and Section V draws some conclusions and speculates on their generality for other geographic areas, other pollutants, and other periods of time.

II. MODELS OF MARKETABLE DISCHARGE PERMIT SYSTEMS

The types of marketable permit (MP) designs considered in this paper are modifications of the two basic permit types discussed above—ambient permits and emission permits. The modification of the ambient permit design involves the use of a single ambient permit market tied to the receptor with the worse air quality. This reduces the administrative complexity of the full ambient permit market, but, of course, it does so at some loss of control over air quality measured at the other receptor locations and increased compliance costs for meeting the air quality standards at all receptor locations. The empirical question is how severe these impacts are.

The various modifications of the emissions permit design involve restricting the trading areas of the permits, using trading zones, to reduce the "hot spot" problem. Restrictions on trading areas, however, have three undesirable effects: (1) they raise compliance cost by eliminating some cost reducing trading possibilities, (2) they make the final outcome sensitive to the initial allocation of permits of the control authority (since the initial allocation determines the aggregate amount of emissions allowed in each zone), and (3) they reduce the number of buyers and sellers which increases the potential for noncompetitive behavior. In light of the second effect a complete specification of the various modified emission permit designs must include some rules for initially allocating the permits to the trading zones.

In general, these various permit markets may be simulated by solving the primal and dual constrained cost-minimization programming problems which describe these systems. The validity of this simulation is compromised to the extent that the following assumptions fail to hold in reality:

(1) Plants choose the least-costly combination of control devices and MPs;

(2) Firms are on long-run cost functions as described by the programming problem objective function;

(3) Buyers and sellers of permits are sufficiently numerous within each zone so that permit prices are independent of individual actions; and

(4) Plant production and consumption functions are independent.

In the primal problem control costs are minimized subject to a set of linear constraints. Solution of the primal problem yields the cost-minimizing level of control to be undertaken by each firm. Subtraction of these values from uncontrolled emission levels yields the volume of permits which each plant must purchase. The dual variables associated with the permit constraints represent the permit prices.

The Ambient Permit System

The first benchmark case is provided by a model of the ambient permit (AP) system. It serves to define the minimum compliance cost for meeting the ambient standards as well as to provide an empirical basis for subsequent discussions of some problems with the ambient permit approach.

Assuming that cost functions are highly nonlinear and convex for all sources, we approximate the AP cost-minimization problem with a quadratic program:

minimize
$$z = \sum_j c_j x_j + \sum_j d_j x_j^2 \tag{1}$$

subject to
$$\sum_j a_{ij} x_j \geq b_i, \qquad i = 1,\ldots,m, \tag{2}$$

$$x_j \geq 0, \qquad j = 1,\ldots,n, \tag{3}$$

where:

b_i = the reduction in particulate concentration required to achieve the standard at the ith receptor   ($i = 1,\ldots,m$),

c_j, d_j = coefficients representing the cost of control per day for the jth source ($j = 1,\ldots,h$),

x_j = the number of tons to be removed per day by the jth source,

a_{ij} = the transfer coefficient which relates all emissions from the jth source to air quality at the ith receptor.

This formulation indicates that we assume, as have other studies, the independence of individual source control and the absence of possible synergistic effects in our transfer coefficients.

The objective function to be minimized in Eq. (1) and the constraints in equation set (3) remain the same for all permit strategies. Only the constraint inequality in (2) changes. For the AP system, (2) guarantees that reductions in ambient concentrations will be at least sufficient to meet the standard at receptor i.

Ambient air quality standard q implies a unique b_i^q ($q = 1,\ldots,t$) at each receptor. The volume of permits to be issued at receptor i for standard q consistent with the solution to cost-minimizing problem in Eqs. (1)–(3) is then determined as

$$p_i^q = u_i - b_i^q, \qquad i = 1,\ldots,m, \tag{4}$$

where:

u_i = uncontrolled air quality at receptor i

and

b_i^q = required improvement in air quality at receptor i for standard q.

The volume of permits to be issued for the other MP strategies are also calculated using Eq. (4).

Assuming an initial distribution of permits, free permit trading is allowed within zonal boundaries, in this case the boundaries of the region. Each source must calculate its uncontrolled emissions, map them into air quality degradation at each receptor i using an agreed-upon diffusion model, and then purchase sufficient ambient permits at each receptor to cover this degradation. The permit price at receptor i equals the dual shadow value for constraint i in Eq. (2). Each source purchases permits only if the expenditure on ambient permits at all receptors required due to an increment of emissions is less than its relevant marginal cost of control, i.e., if the dual constraint corresponding to Eq. (2) fails to hold with equality.

The Emission Permit System

The Emission Permit (EP) system quadratic programming problem is written as:

minimize z

subject to $$\sum_j x_j \geq d \qquad\qquad (5)$$

and the constraints in equation set (3), where the scalar, d, is the aggregate amount of regional emissions which must be removed.

To operationalize this model a value for d is needed which is *ex ante* feasible. Following the procedure utilized by many states to justify the adequacy of their SIPs, we use a rollback model. The rollback model is based on a linear relationship between regional emissions and air quality. A given percentage improvement in air quality at the receptor with the worse regional air quality is assumed to require the same percentage reduction in emissions. Thus

$$d^q/\mathrm{RE} = (B_{\max} - B^q)/(B_{\max} - B_{\mathrm{back}}), \qquad q = 1,\dots,t, \qquad (6)$$

where:

   RE = total regional emissions,

   $B_{\max}$ = existing pollution concentration at the receptor having the highest measured or estimated concentration in the region,

   B^q = air quality standard q,

and

   B_{back} = background pollution concentration.

Equation (6) may be easily solved for d^q. Equation (6) will be completely accurate (i.e., result in *ex post* feasibility) only under the extreme condition that individual source emissions are nonsynergistic and that all sources have the same linear transfer coefficient. We use it here to approximate current regulatory behavior and to derive empirically an estimate of the precision of this behavior.

As with the AP strategy, permit trading is allowed throughout the entire region. However, with the EP system, permits are purchased by sources strictly for emission discharge rather than air quality degradation. Emission permits trade at a uniform price equal to the shadow value corresponding to Eq. (5). Thus, the EP problem requires that sources trade permits to minimize total control costs subject to the constraint in Eq. (5) on total emission removal.

For this emission permit system and for the various modifications which follow, *ex post* air quality levels will, in general, differ from *ex ante* levels. Although the control authority theoretically could iteratively adjust the number of permits in circulation to achieve desired air quality after the initial installation of control devices and purchase of permits, such a policy appears unlikely to be followed because of the uncertainty it would induce. Therefore, for the purpose of this paper, in order to simulate what we consider to be the most likely alternatives we do not iterate.

The Highest Ambient Permit System

The highest ambient permit (HAP) system defines a single market in ambient permits. The permits in this market are defined in terms of the most polluted receptor location. The HAP system separable linear programming problem may be written as:

minimize z

subject to $\sum_j a_j^* x_j \geq b^*$ (7)

and the constraints in equation set (3), where:

$b^* = $ the greatest required reduction in pollution concentration among the i receptors,

and

$a_j^* = $ the transfer coefficients relating emissions from source j to ambient concentrations at the receptor with the greatest required reduction in pollution concentration.

Equation (7) represents the permit market constraint. The major difference between the HAP and the AP systems is reflected in this equation. While both systems allow trading of permits within the entire region to minimize total control costs, for the HAP system b^* is a scalar (reflecting the single permit market) while for the AP system it is a vector (reflecting the many permits involved).

Modified Emission Permit Systems

Three types of multiple-zone modified emission permit systems are simulated, all of which may be considered as generalizations of the emission density zoning systems examined recently by the EPA.[8] The multiple-zone systems include the Zonal Discharge Permit–State Implementation Plan (ZDP–SIP) system, the Zonal Discharge Permit–Rollback (ZDP–R) system, and the Uniform Zone Discharge Permit (UZDP) system. For all three systems trading within the zones is permitted while trading across zonal boundaries is not. Thus, greater control over "hot spots"

---

[8]See Kron *et al.* (1978). Emission density zoning systems as proposed by Kron (EPA) are simply a specialization of the general zonal emission permit model, Eq. (8). Typically, emission discharge is maximized subject to air quality restrictions. However, since no allowance is made for least-cost trading of permits within zones, these systems should be far less cost-effective than any of the zonal systems examined in this paper which allow permit trading.

is obtained but at some (to be determined) higher cost of compliance. For all three of these systems *ex ante* and *ex post* air quality will diverge.

The same quadratic programming problem describes the three multizonal MEP systems:

minimize $\qquad\qquad\qquad\qquad\qquad z$

subject to $\qquad\qquad \sum_j e_{jw} x_j \geq d_w, \qquad w = 1,\ldots,r,$ $\qquad\qquad$ (8)

and the constraints in equation set (3), where:

$\qquad d_w =$ the aggregate required emission reduction in zone w,

and

$\qquad e_{jw} = 1$ if source j is in zone w and 0 otherwise.

The three multizonal emission permit systems differ in either: (1) the calculation of zonal emission reduction, d_w, or (2) the mapping of sources to zones, as specified by the e_{jw}. Free trades are permitted within zones, but not among zones. The permit prices will generally vary among zones and are given by the dual variables associated with (8).

The ZDP–SIP and ZDP–R systems both allow the use of nonuniform zones (the exact configuration is discussed in the next section) and differ only in the initial allocation of permits, among these zones. The ZDP–SIP system determines zonal permits as the sum of SIP allowable emission rates within each zone. The ZDP–R system employs the rollback calculation, Eq. (6), to determine allowable emissions within each zone, where B_{max} in each zone is generally the air quality of the highest concentration receptor in that zone. The UZDP system employs grids of generally uniform size and determines zonal permit volumes using the rollback model.[9]

III. DATA: THE ST. LOUIS CASE STUDY

The data employed in this study comprise the 27 largest point sources of particulate emissions in the St. Louis Air Quality Control Region (AQCR), accounting for approximately 80% of total particulate emissions.

---

[9]Transfer coefficients, whose utilization in the AP strategy is a major source of control cost reduction, generally cannot be employed to allocate regional control burdens to zones or to allocate zonal control burdens to individual sources. Infeasible fractions of total regional emission reductions can be easily allocated to zones (i.e., required zonal reductions exceed uncontrolled zonal emission levels). This could easily occur, for example, if the fraction of regional emissions to be removed within each zone were determined by the relative impact of zonal emissions on the highest-concentration receptor in the entire region.

In addition, substantial infeasibilities in intrazonal permit trading can be created by the use of transfer coefficients to assign individual source control burdens. If, for example, the permits within each zone are purchased by each source for degradation of air quality at the highest-concentration receptor within each zone, infeasible solutions (i.e., required air quality improvements are not met even though all sources control all emissions) can easily result since interzonal pollution flows are not considered in assigning control responsibility.

Control Cost Data

Source–receptor transfer coefficients, employed in the AP and HAP constraint equations and utilized in the diffusion model to map source emissions into air quality for the other strategies, are derived using a Gaussian diffusion model developed by Martin and Tikvart (1968). The meteorological input data required for the model are pollution dispersion characteristics which include location, stack height, average mixing height, stack exit conditions, stability wind rose (speed, direction, and stability class), and pollutant decay rates.[10] The output consists of a matrix which gives the contribution of each of n sources to the predicted annual arithmetic average pollutant ground-level concentrations at each of m receptors. Transfer coefficients, with units of $\mu g/m^3/ton/day$, are obtained by dividing the concentration at the ith receptor due to the jth source by the number of tons emitted by the jth source.

Based on Standard Industrial Classification code and source type, all area sources were excluded from the present analysis and all mobile sources and any other sources too small or too numerous to categorize as point sources were treated as part of the background. Thus, this study examines only major point sources, which comprise three basic categories: stationary fuel combustion plants (primarily industrial and steam-electric power-plant boilers), industrial process sources, and solid-waste disposal sources (incineration and open burning).

Prior to developing control cost data, the applicability of control measures to each source was considered. In order to determine the compatibility of control devices with each source, consideration must be given to the temperature and volume of the effluent gas stream, type and efficiency of existing pollution controls, fuel usage requirements, and the maximum process rate. A number of measures were examined: wet scrubbers, mechanical collectors, electrostatic precipitators, mist eliminators, fabric filters, afterburners, and fuel substitution.

The costs of each device (in 1969 dollars) are obtained from the Control Technique Documents prepared by EPA (1969) and are the same for each control strategy. The total annual cost includes annualized capital and installation cost (based on a rate of interest and rated life of the device), as well as annual operating and maintenance costs.[11]

The industrial classifications for 27 plants considered in the St. Louis region are listed in Table I. The emission rate and control cost data may be found in Atkinson and Lewis (1974). The approximate location of each source and the nine receptors for which air quality predictions are made are shown in Fig. 1. The same control information and source–receptor pattern were used for the cost comparisons of all strategies.

The SIP Strategy

In accordance with the Clean Air Act of 1970, each state has submitted to the Federal Government an SIP which describes its basic air pollution control strategy

---

[10] For a more complete discussion see TRW Systems Group (1970).

[11] Discussions with engineers at EPA indicate that while capital and operating costs of particulate control devices have risen substantially, total costs for each control device have risen by the same factor. Thus, all calculations are carried out in 1969 dollars. While EPA is in the process of developing new cost estimates, these are not yet available.

TABLE I

Sources Controlled under All Strategies

Source No.	Standard Industrial Classification
1	2010; Meat packing, boiler
2	2041; Feed and grain mill
3	2041; Feed and grain mill
4	2041; Feed and grain mill
5	2041; Feed and grain mill
6	2046; Wet corn milling, boiler
7	2082; Brewery, boiler
8	2082; Brewery, boiler
9	2600; Paper products, boiler
10	2800; Chemical plant, boiler
11	2816; Inorganic pigments, boiler
12	2819; Inorganic industrial chemical plant
13	2819; Inorganic industrial chemical plant, boiler
14	2911; Petroleum refinery
15	2911; Petroleum refinery
16	2952; Asphalt batching, boiler
17	3241; Cement plant, dry process
18	3241; Cement plant, dry process
19	4911; Powerplant
20	4911; Powerplant
21	4911; Powerplant
22	4911; Powerplant
23	4911; Powerplant
24	4911; Powerplant
25	4911; Powerplant
26	4911; Powerplant
27	4911; Powerplant

for achieving the federally set ambient air quality standards. For purposes of this study, a set of emission regulations suggested in the SIP guidelines and representative of those employed by many states has been selected to form the SIP control strategy. The particulate standards include a heat input standard for fuel combustion sources (0.30 lb. particulate matter/million Btu), a process weight standard for industrial process sources (46.72 lb/hr of particulates/million lb/hr process weight), and a refuse-charged emission standard for solid waste disposal sources (0.20 lb particulate/100 lb of refuse charged).

The total cost of applying the SIP strategy to the St. Louis model region was determined from the cost of control data by reducing particulate emissions to the SIP strategy levels for all 27 sources. Remaining emissions from the controlled sources were then run through the diffusion model to generate *ex post* concentrations. In order to generate a functional relationship between total regional control costs and various air quality levels, a number of SIP strategies were developed by scaling (up and down) and the levels of the suggested SIP emission regulations. Scale factors employed were 0.25, 0.5, 2, 4, and 10. Since the set of SIP control requirements were representative and not specific to St. Louis, *ex ante* levels of air quality can only be roughly associated with scale factors.

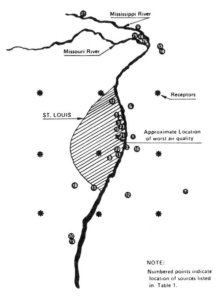

FIG. 1. Map of receptors and sources for St. Louis region.

Separable Cost Functions

A separable linear programming algorithm is utilized to approximate all MEP quadratic programming strategies.[12] Marginal control costs rise rapidly with increasing levels of control, appearing to approach a vertical asymptote at complete pollutant removal. For each of 27 sources, a two-segment piecewise cost function is constructed by tracing out the lower bound of the total cost of the particulate control devices technologically applicable to each source. The objective function for all marketable permit strategies employs the marginal costs of each source's piecewise segments. Since each separable function is convex to the origin and all constraints are linear, all local optima will be global optima.[13]

Zonal Configurations

Figure 2 presents the zonal configurations for the ZDP–SIP and ZDP–R strategies, while Fig. 3 depicts the zonal design employed for the UZDP system. Three zones are employed for the ZDP systems, roughly dividing the St. Louis AQCR into

---

[12] All computations were carried out using an IBM MPSX/370 separable linear programming package 5. For more details concerning the algorithm see IBM (1979).

[13] However, a single control device which removes the optimal tonnage may not exist for all sources. A convex combination of two control devices which border the optimal but nonexistent device must then be determined. Use of these two devices would require technological compatibilities sufficient to achieve splitting of the stack gas stream.

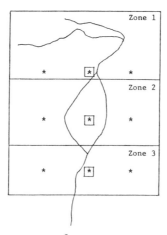

Zones	Sources
1	9, 14, 15, 24, 25, 26, 27
2	1, 2, 3, 4, 5, 6, 8, 10, 13, 16, 17, 21, 22, 23
3	7, 11, 12, 18, 19, 20

FIG. 2. Zonal configurations—ZDP systems.

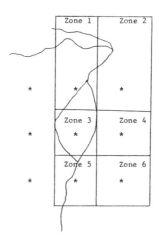

Zones	Sources
1	3, 4, 5, 17, 27
2	6, 9, 14, 15, 24, 25, 26
3	8, 10, 13, 21, 22, 23
4	1, 2, 16
5	11, 18, 19, 20
6	7, 12

FIG. 3. Zonal configurations—UZDP system.

thirds, but including nearly all of the city of St. Louis in Zone 2. Zone sizes are small enough to prevent trading of MEPs with remotely located sources, e.g., Zone 2 sources with the power plants and refineries to the north in Zone 1, while at the same time large enough to achieve substantial cost-saving through least-cost trading of permits. The air quality receptors utilized in the ZDP–R systems, enclosed by a small square in Fig. 2, correspond to the highest concentration receptor in each zone with one exception.[14]

The UZDP system, on the other hand, follows as closely as possible the EPA recommendation of uniform 16 kilometer-square grids—coincidentally the distance separating all St. Louis receptors—with a receptor in the approximate center of each grid. The exceptions are grids 1 and 2 which are each 16 by 32 kilometers. The other four zones conform very closely to the EPA recommendations.

IV. DISCUSSION OF EMPIRICAL RESULTS

There are five major empirical dimensions to be discussed: (1) the cost of compliance with alternative permit market designs, (2) the importance of permit expenditures in the total cost of control, (3) the effects of the various designs on the total amount of emissions removed, (4) the behavior of permit prices in the ambient permit markets, and (5) the magnitude of the divergence between *ex ante* and *ex post* air quality. Each of these five topics is covered in one of the five sections which follow.

The Cost of Compliance

Perhaps the major question to be answered is how permit design affects the cost of compliance. Figure 4 depicts the compliance costs for each of the six permit designs as a function of the level of severity of *ex post* air quality at the worst receptor (with background pollution set to zero for simplicity). In order to focus on the real resource cost of compliance only control expenditures are considered in the figure. Permit expenditures will be discussed in subsequent sections.

A number of important conclusions emerge from an examination of Fig. 4. First, the AP strategy, as expected on theoretical grounds, is the most cost-effective strategy. Furthermore, the relative cost advantage of this system over the SIP strategy is not diminished at very high degrees of control. In contrast the relative cost advantage of the EP system diminishes sharply at high levels of control.

The total cost of the HAP strategy is very close to that of the AP strategy, due to the small number of receptors with positive permit prices in the AP strategy. The multizonal systems and the EP system, as expected, are substantially more expensive than the AP and HAP systems to achieve given levels of air quality. However, the multizonal systems perform somewhat better than the EP systems at stricter levels of control, presumably because the gains due to more precise control of air quality with the multizone systems outweigh the losses due to restricted permit trading.

Among the multizonal systems, the UZDP is the least cost-effective because smaller zone sizes limit cost-minimizing permit trading. Surprisingly, the ZDP–SIP

---

[14] The exception is zone 3 which utilizes the middle receptor while the right-most receptor registers the highest concentration.

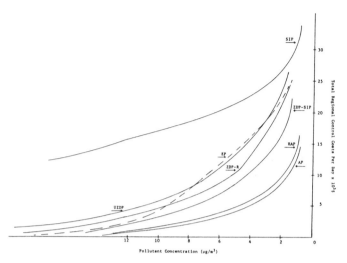

FIG. 4. Total regional control costs as a function of *ex post* air quality ($\mu g/m^3$).

system is somewhat more cost-effective than the ZDP–R system, an unexpected result since the ZDP–R system relies on the rollback model to calculate each zone's control burden.

Finally, as a point of comparison for all MEP strategies, the cost of the SIP strategy is included for various *ex post* air quality levels between 1 and 10 $\mu g/m^3$.

The Magnitude of Permit Expenditures

To the sources which have to buy them, permits represent an extra expenditure which has to be incurred over and above what the sources have to spend installing the control equipment. This is a cost not incurred by sources when the pollution control policy is based upon nontransferable emission standards. Therefore one of the dimensions of interest to the control authorities is how large these permit expenditures are *vis á vis* the control expenditures themselves. The larger they are the more serious the distributional impact of the initial allocation.

To shed light on this issue we have computed both the average cost of a permit per ton emitted as a function of the degree of severity of control and the ratio of permit expenditures to the sum of control cost plus permit expenditures. The first of these calculations is presented as Table II and the second as Table III.

There are several interesting characteristics of these results. The first of these is the striking lack of differences among the permit systems for a given level of control except for the EP system. The EP system generally requires more control than the other systems to achieve a given standard. As demonstrated in Table III this results in very high permit prices. These prices are sufficiently high that even though few permits are purchased under this system total permit expenditures are a higher percentage of total cost.

TABLE II

Average Permit Expenditures ($) per Ton Removed per Day for *Ex Ante* Air Quality Levels

MEP strategy	*Ex ante* air quality ($\mu g/m^3$)					
	1	2	4	6	10	20
AP	225.44	121.64	44.04	30.99	15.13	2.32
EP	15076.07	4941.98	1977.00	1053.99	617.99	132.00
HAP	321.68	82.90	45.02	22.79	15.97	2.26
ZDP–SIP	581.72	341.56	164.77	65.16	36.16	12.05
ZDP–R	1137.74	495.43	268.50	150.38	49.75	7.54
UZDP	562.08	453.14	181.25	145.12	57.83	11.83

TABLE III

Ratio of Permit Expenditures to Total of Permit Expenditures and Control Costs
for *Ex Ante* Air Quality Levels

MEP strategy	*Ex ante* air quality ($\mu g/m^3$)					
	1	2	4	6	10	20
AP	0.34	0.42	0.46	0.53	0.63	0.86
EP	0.75	0.80	0.85	0.87	0.92	0.96
HAP	0.40	0.40	0.50	0.50	0.68	0.76
ZDP–SIP	0.29	0.41	0.50	0.59	0.72	0.73
ZDP–R	0.31	0.36	0.50	0.56	0.60	0.64
UZDP	0.22	0.38	0.43	0.58	0.67	0.80

The second striking result is the trend for permit expenditures to be a less important part of total expenditures as the degree of control becomes more severe. This is true for all six types of permit systems considered. This does not imply, as might appear at first glance, that the demand for permits is elastic. If this were the case then permit expenditures would decrease with increasing severity. We did not find that. What was discovered was that the costs of control rise faster than permit expenditures when the severity of control is increased.

The Quantity of Emissions Removed

In part the differences in the cost of compliance registered by the various permit systems are due to the fact that they require different degrees of emission control. The spatially differentiated systems (AP and HAP) allow more emissions for each microgram per cubic meter of pollution recorded at the receptor locations than the other systems because they reduce those sources having the heaviest impact on the receptor locations most, leaving the other, more distant, sources to emit more. The effect of this is to allow the distant sources an increase in emissions which is greater than the additional reductions imposed on the proximate sources; total emissions increase.

In spite of the Clean Air Act's reliance on ambient air quality standards as the chief policy target it is clear that the courts have interpreted the Act in such a way as to suggest that strategies which achieve the air quality standards while permitting

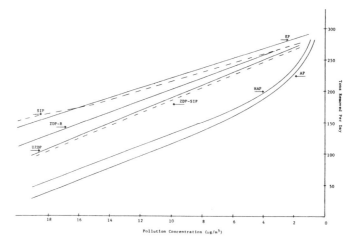

FIG. 5. Tons of pollutant removed per day as a function of *ex post* air quality.

increases in emissions are not to be allowed.[15] In addition it has become increasingly clear that some pollutants are transported long distances (e.g., the acid rain problem) and therefore the total amount of emissions may have policy significance beyond its contribution to the local pollution problem.

Figure 5 presents the tons of pollutant removed per day as a function of *ex post* air quality. It is clear that a substantial portion of the compliance cost savings achieved by the spatially differentiated permit system (AP and HAP) are directly due to the fact that they require less control. At lower levels of control they allow at least twice as many tons of pollutant emitted as the nondifferentiated systems while meeting the same ambient targets. Interestingly this is less true for very high degrees of control when, it may be remembered from Fig. 4, the ADP and HAP systems still maintain a significant cost advantage relative to the SIP strategy.

This finding suggests that EPA and the courts should differentiate between those pollutants which have almost exclusively a local impact (e.g., particulates) and those which travel long distances (e.g., sulfur oxides). For the former the cost savings to be achieved by using differentiated permit systems are substantial while the extra emissions are not harmful (assuming the ambient standards are correct). For the latter the extra emissions may cause serious harm and therefore the control authority may wish to forego these substantial savings as a way of indirectly controlling the long-range transport problem.[16]

[15]See, for example, the ruling in Natural Resources Defense Council, Inc. v. Environmental Protection Agency, 489 F. 2d 390 (1974) reversed, in part, on other grounds in Train v. Natural Resources Defense Council, Inc., 421 U.S. 60 (1975).

[16]An alternative approach would be to define a system which required additional permits for exported pollution. To be cost effective such permits would have to be standardized to take into account the contributions of those exported emissions to concentrations at remote locations, a task requiring models to characterize the interstate or international transport of pollutants. The state of the art in modeling the long distance transport of pollutants is not yet sufficiently advanced to make this a viable alternative yet, though it may well be so in the future.

Permit Prices in the AP Market

The ambient permit market is a complicated one in which each source generally has to purchase several of the receptor-oriented permits. These permit markets are highly interdependent since a reduction in emissions by any particular source will affect a number of receptor locations.

The complexity of this market raises the question as to whether the prices of the permits in each and every market are monotonic with respect to the increasing severity of control. While it is clear that the marginal costs of control for each source will be monotonic with respect to increasing severity of control, it is not clear that a similar monotonicity would characterize the permit prices.

As Table IV demonstrates in this particular market under investigation permit prices are not monotonic. Note, for example, that reversals occur in each of the four markets characterized by positive prices for at least one level of control. In other words for each of these four markets there is at least one point where a higher level of control results in lower permit prices.

Since this result may, at first, seem counterintuitive, it is important to understand why it occurs. The exercise generating the data in Table 4 increases the severity of control at all receptor locations simultaneously. If we had increased the required degree of control at one receptor location while leaving the others unchanged, the expected monotonicity of permit prices would have been achieved for that one receptor location.

As the severity of control is increased the air quality standards will prove more difficult to attain for some receptor locations than for others. In general there will be different troublesome receptor locations at each different degree of control. As responsibility is allocated among the sources to attain the air quality standards at these troublesome locations the resulting emission reductions increase air quality at other locations as well. The most graphic example of this phenomenon can be seen in Table 4. At the highest severity of control considered ($1.0 \ \mu g/m^3$) the degree of control required at receptor location 8 is so high that when the air quality standard at that location is met all of the others are automatically met so their associated permit prices are appropriately zero.

The fact that these prices can exhibit erratic behavior as the supply of permits is changed has not, to our knowledge, been recognized. It is a troubling result. It indicates a kind of instability which could make forward planning by the sources difficult, for example. Though far from conclusive this result seems to suggest that

TABLE IV

Ambient Permit Prices at Various *Ex Ante* Air Quality Levels ($\$/\mu g/m^3/day$)

Receptor	1	2	4	6	10	20
1	0	0	0	0	0	0
2	0	0	0	0	0	0
3	0	0	0	0	0	0
4	0	0	0	0	0	0
5	0	740	129	281	143	29
6	0	2606	653	0	100	0
7	0	0	0	0	0	0
8	7469	181	321	125	15	1.7
9	0	0	0	230	0	0

further research on the behavior of prices in ambient permit markets would be desirable.

The Ex Post/Ex Ante Problem

Because they specify the amount of emissions each source is allowed, command and control systems allow the control authorities to maintain a high degree of control over air quality in their region. When the control responsibility becomes transferable, as it does in marketable permit systems, the ultimate allocation of control responsibility, as determined by the permit market, is not known at the time the system is implemented. Since ambient air quality at the various receptor locations depends on the spatial distribution of emissions as well as the amount emitted, and that spatial pattern cannot be determined in advance, with the exception of the ambient permit system the *ex post* air quality which prevails after the transfers have taken place may be quite different from what the control authority intended *ex ante*.

The ambient permit system allows the control authority to retain complete control over air quality because, by design, the allowable trades do not violate the air quality standards. The number of permits required to achieve the standard can be defined prior to the initiation of the market. This is not true for any other permit system considered in this paper. Therefore for those systems the control authority is required to guess how permits should be issued to insure that the air quality at the worst receptor just meets the standard. In this paper the control authorities are assumed to use the roll back model to determine the number of permits. The question of interest is how far astray this commonly used procedure will lead the attempt to achieve the standard.

The discrepancies between *ex ante* and *ex post* levels of air quality are displayed in Table V. Even strategies which are highly cost-effective in terms of *ex post* air quality may be deemed unacceptable if *ex post* and *ex ante* air quality levels differ substantially. This inability to assure that the air quality standards will be met may well require iterative adjustment of the volume of permits. Since pollution control devices are generally lumpy investments which are often incompatible with larger or smaller devices, adjustment of the volumes of permits would most likely be highly disruptive, costly, and add great uncertainty to future investment decisions.

Only the AP strategy guarantees that *ex ante* air quality levels will be actually achieved, as expected. Among the other systems none uniformly dominates the others. No system is best or worst at achieving all *ex ante* air quality levels. The

TABLE V

Highest Recorded *Ex Post* Concentrations Corresponding to *Ex Ante* Air Quality Levels ($\mu g/m^3$)

MEP strategy	1	2	4	6	10	20
AP	1.00	2.00	4.00	6.00	10.00	20.00
EP	1.61	4.44	6.62	7.14	10.04	18.77
HAP	1.10	2.68	5.22	8.00	11.88	21.81
ZDP–SIP	1.70	2.61	6.80	8.95	12.37	18.45
ZDP–R	1.51	1.93	4.95	7.54	12.21	21.16
UZDP	1.75	2.88	5.25	7.45	13.22	20.02

emissions permit system at low pollutant concentration levels introduces a substantial degree of error. Interestingly, the UZDP system, which relies on small trading zones, does not achieve smaller errors in the attainment of air quality goals as might be expected. This anomaly is apparently due to the inability in such a simple system to consider interzonal pollutant flows in structuring this strategy. It is also interesting to note that the use of the roll back model to determine the number of permits generally results in too little, as opposed to too much, control. Control authorities who use the roll back model would therefore have to build a safety margin into their determination of the number of permits to issue.

V. CONCLUSIONS

This paper has derived a number of empirical conclusions about various permit designs. Since they emanate from a single pollutant, single geographic location model, these results should be considered suggestive, rather than conclusive, until such time as they are replicated for other pollutants and other locations. Nonetheless, to the extent they are subsequently replicated, some specific design characteristics emerge as worthy of further consideration.

The most important of these results, along with their potential policy implications, are:

• Although the ambient permit design, as expected, is the most cost effective, a modified version of that design (the HAP) which has a single permit market defined in terms of the "worst case" receptor location achieves the desired air quality goals at a cost only slightly higher. This is an important result because the HAP system is administratively much simpler than the AP system. The policy implications of this result are not as strong as it might at first seem, however, due to the static nature of the model. To the extent that new sources move into the area and are influenced by permit cost they may tend to cluster at some location far moved from the existing "worse case" receptor. This clustering could then trigger a violation of the air quality standard at this new location. In this case the HAP system would be less desirable; at least two markets would then be necessary.

• All permit designs result in a lower compliance cost at all levels of control than the current SIP approach which focuses on nontransferable emission standards.

• The smaller zone sizes in the UZDP system have a significant positive effect on compliance cost, but they do not, in general, make the hot spot problem less severe. For most *ex ante* air quality levels at least one of the larger zone systems (ZDP–SIP or ZDP–R) has a lower reading at the worst receptor than the UZDP system. These results suggest that any reform designed to restrict geographically the trading areas to zones as small as implied by the UZDP system in order to reduce the "hot spot" problem would be ill-advised.

• The results for the ZDP–SIP system, which allows only within-zone emissions trades but defines the zones to encompass a larger geographic area than the UZDP system, indicate that substantial cost savings can be achieved by moving from the SIP system to this modified version. This is an important result indeed since this is precisely the direction the "bubble" policy is taking, although it remains to be seen how far away sources can be and still be permitted to engage in a "bubble" trade.

- For all permit designs, permit expenditures are more than half of total cost for less severe degrees of control but they become a less significant proportion of total cost as the degree of control is increased. This suggests that the importance of the distributional issues associated with the initial allocation of permits diminishes as the degree of control becomes more severe.
- Permit expenditures as a percentage of total cost are highest for the emission permit system.
- The significant cost advantages of the HAP and AP systems are achieved because they allow substantially more emissions while achieving the predetermined air quality standards. While this may not be a problem for truly local pollutants (e.g., particulates) it will cause problems for pollutants which are transported long distances (e.g., sulfur oxides).
- As the ambient air quality standards are increased simultaneously at all receptor locations, the permit prices in an ambient permit system do not necessarily increase monotonically. This behavior introduces another element of complexity into the ambient permit market when, for example, sources have to meet more stringent secondary standards subsequent to their meeting the primary standards.
- The "hot spot" problem is a serious one. All of the permit systems examined herein except the ambient permit system allow air quality at some receptor locations to exceed the standard. This suggests that all permit systems other than the ambient permit system will have to be designed with some safety margin to insure compliance at all receptor locations.

These results suggest the possibility of a design which is an amalgam of all these designs.[17] Suppose that we define the exchange ratio as the number of tons of pollutant the purchaser is permitted to emit for each ton of the same pollutant that the seller agrees to control further. Further imagine a case by case system in which region wide trades could take place providing the exchange ratio was no larger than the smaller of (a) 1.0 or (b) the ratio of the seller's transfer coefficient to the purchaser's transfer coefficient for the worst receptor (or what would be the worst receptor after the trade). The former constraint insures that emissions do not increase when trades improve local air quality. The latter constraint guarantees that trades do not make air quality worse at the most susceptible receptor location.

Our results indicate that this system would have substantial advantages. Since region wide trades are permitted, the possibilities for cost reduction are enhanced as compared to restricting trades to proximate sources. Yet air quality could be protected without recourse to multiple markets. This system would combine the prevention of emissions increase usually found in an emission permit system with the cost reduction possibilities and lack of complexity of the HAP system with the control over air quality usually associated with an ambient permit system. Based on our results this amalgam would seem a logical next step in EPA's expansion of the controlled trading reforms.

REFERENCES

Anderson, R. J., Jr. *et al.* (1979), "An Analysis of Alternative Policies for Attaining and Maintaining a Short-Term NO$_2$ Standard," a report to the Council on Environment Quality prepared by MATHTECH, Inc., Princeton, N. J.

[17]A similar proposal is advanced in Anderson *et al.* (1979, pp. 7–9 to 7–11).

Atkinson, S. E., and Lewis, D. H. (Nov. 1974), A cost-effective analysis of alternative air quality control strategies, *J. Environ. Econ. Manag.* **1**, 237–250.

Benesh, F. H. *et al.* (March 1977), "Emission Density Zoning," U.S. EPA Report EPA-450/3-77-006.

Brail, R. K. (Feb. 1975), "Land use planning strategies for air quality maintenance," in "Proc. Speciality Conf. on Long-Term Maintenance of Clean Air Standards" (J. J. Roberts, Ed.), pp. 275–289, Lake Michigan States Sec. of Air Pollution Control Assn., Chicago.

Brail, R. K. *et al.* (June 1975), "Emission Density and Allocation Procedures for Maintaining Air Quality," U.S. EPA Report EPA-450/3-75-075.

Cosier, P. C., IV (Fall, 1976), Land use based emission strategies: Their promises and problems, *Planning Comment* **12**(2), 31–47.

CONSAD Research Corporation (July 1971), "The Direct Cost of Implementation Model," Vol. I, Prepared for the U.S. Environmental Protection Agency, Washington, D.C.

Hadley, G. (1964), "Nonlinear and Dynamic Programming," Addison–Wesley, Reading, Mass.

IBM (Dec. 1979), "Mathematical Programming System/370, Program Reference Manual."

Kron, N. F., Cohen, A. S., and Mele, L. M. (Sept. 1978), "Emission Density Zoning Guidebook. A Technical Guide to Maintaining Air Quality Standards Through Land-Use-Based Emission Limits," U.S. EPA Report EPA-450/3-78-048.

Larsen, R. I. (1971), "A Mathematical Model for Relating Air Quality Measurements to Air Quality Standards," U.S. Environmental Protection Agency, Office of Air Programs Publication No. AP-89.

Martin, D. O., and Tikvart, J. A. (1968), A general atmospheric diffusion model for estimating the effects on air quality of one or more sources, *J. Air Pollut. Contr. Assoc.* **18**, 68–148.

Tietenberg, T. H. (1980) Transferable discharge permits and the control of stationary source air pollution: A survey and synthesis, *Land Econ.* **56**, 391–416.

TRW Systems Group (Nov. 1970), "Air Quality Implementation Planning Program," Vol. 1, Operator's Manual, Washington, D.C.

U.S. Department of Health, Education, and Welfare (Jan. 1979), "Control Techniques for Particulate Air Pollutants."

U.S. Environmental Protection Agency, Requirements for Preparation, Adoption, and Submittal of Implementation Plans, Rules and Regulations, *Fed. Reg.* **36**, No. 158.

Zimmer, C. E., and Larsen, R. I. (1965), Calculating air quality and its control, *J. Air Pollut. Contr. Assoc.* **15**, 565–572.

[34]

JOURNAL OF ENVIRONMENTAL ECONOMICS AND MANAGEMENT 10, 346–355 (1983)

Transferable Discharge Permits and Economic Efficiency: The Fox River

WILLIAM O'NEIL, MARTIN DAVID,[1] CHRISTINA MOORE,
AND ERHARD JOERES

Department of Economics, Colby College, Waterville, Maine 04901; Department of Economics, University of Wisconsin at Madison, Wisconsin 53706; Anderson–Nichols, Boston, Massachusetts; and Department of Civil and Environmental Engineering, University of Wisconsin, Madison, Wisconsin 53706.

Received September 31, 1981; revised December 1982

Recent emphasis on reforms of environmental regulation has led to suggestions for strategies which maintain environmental standards but allow the needed flexibility and cost effectiveness. The transferable discharge permit (TDP) is one such strategy for water pollution control recently adopted in Wisconsin. In this article, the potential for substantial cost savings from trading TDPs is demonstrated using data on the Fox River in Wisconsin. A simulation model of water quality (Qual-III) and a linear programming model of abatement costs determine the optimum pattern of discharge. Reaching that optimum from proposed pollution abatement orders is shown to be feasible. Varying conditions of flow and temperature can be accommodated using trade coefficients which can be accurately estimated through interpolation. The calculations demonstrate the value and feasibility of flexible regulations governing water pollution abatement.

I. INTRODUCTION

The Wisconsin Department of Natural Resources approved regulations in March of 1981 which allow dischargers to transfer permits by approved contracts. This makes the Fox River in Wisconsin the first body of water in the United States where cost savings in abatement may be achieved using transferable discharge permits. This paper addresses two of the concerns exposed in the course of discussions about whether or not to adopt this option.

The potential for more cost-effective regulation to modify strict control directives has been discussed in the economics literature starting with Crocker [1] and Dales [2]. The EPA has been actively discussing and developing such policies for air pollution control, but to date there has been no parallel concern in water pollution control. This paper documents the feasibility of using one option for increasing the cost effectiveness of water pollution regulation.

Although the economic efficiency of the market mechanism has been recognized in theory [3], and has been used for the control of air pollution, it has not been used as a water pollution control strategy in the United States prior to its adoption in Wisconsin. This paper provides an evaluation of transferable discharge permits based on a water quality simulation model of the Fox River (Qual-III) to determine water quality effects and a linear programming model to estimate cost effects. Similar studies of the Delaware Estuary were completed by Kneese and Bower [4] and by Ackerman et al. [5]. Although those studies did not explicitly evaluate

[1] To whom correspondence should be addressed: 1180 Observatory Dr., Madison, Wis. 53706.

346

transferable permits, many of the political obstacles to efficient pollution control that were discussed by Ackerman are addressed in the present paper. In particular, the case of the Fox River illustrates how a small number of governmental bodies were able to cooperate in the development of a cost-effective river basin management plan.

II. THE SETTING

The Wisconsin Department of Natural Resources (DNR) has classified the lower Fox River between Lake Winnebago and Green Bay as "water quality limited." This means that the assimilative capacity of the stream is inadequate to maintain water quality standards when industries and municipalities are discharging at the federal maximum uniform treatment requirements—best practicable treatment (BPT) for industries and secondary treatment levels for municipalities. The additional point source abatement needed to meet the water quality standards was handled initially by a central directive requiring proportionate reductions of effluent discharge (biochemical oxygen demanding wastes or BOD) starting from the federal uniform requirements for each discharger. The transferable permit comes into play once this initial allocation of daily pollution discharge has been made.

III. THE WATER QUALITY MODEL

Ten pulp and paper mills and four municipalities discharge effluent into a 22-mile reach of the lower Fox River. The natural effluent decomposition process uses available concentrations of dissolved oxygen (DO) in the stream; and under conditions of low flow and high temperature this process may cause the DO levels to fall below the water quality standard of 6.2 mg/liter.

As an aid in exploring alternative abatement strategies, the state water regulatory agency (DNR) developed a simulation model of the river which allows estimation of the relations between discharger and DO levels at various locations along the stream.

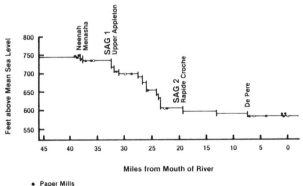

FIG. 1. Location of dams and dischargers on the lower Fox River.

The current version of the model, Qual-III, is a one-dimensional model that accounts for the most important factors influencing the levels, or profile, of DO in a flowing stream. The model simulates levels of DO, via two different rates of BOD decomposition, total phosphorus, organic nitrogen, ammonia, nitrates, nitrites, chlorophyll *a*, and sediment oxygen demand; it can be run in either a steady-state or a dynamic mode [6]. The accuracy with which Qual-III predicts DO levels given input data on flow, temperature, and BOD loadings was tested in several ways [7]. Comparison of Qual-III predictions with actual monitoring data and with the predictions of an ARIMA time-series model convinced both dischargers and regulators that the average prediction error was small enough for political acceptability.

Preliminary simulation analyses confirmed the observation that, because of the location of dischargers, dams, and pools in the lower Fox, effluent discharges cause two local minimums or "sag points" in the levels of dissolved oxygen. Figure 1 depicts the lower Fox showing the location of dischargers, dams, and sag points.

Any allocation of effluent limits which results in the meeting of water quality standards at those sags leads to acceptable water quality throughout the river. Economic analysis of feasible wasteload allocations (WLAs) entails estimation of a 2×14 matrix of linear impact coefficients H from the Qual-III model. H was generated by simulating a series of DO levels associated with increased individual effluent loadings and calculating the changes in the DO at the sag points. Sensitivity analysis confirmed the accuracy of using linear approximation to define the effluent/water quality relationship at different combinations of stream flow and stream temperature [8]. In this analysis linear relations among impact coefficients and flow and temperature were estimated using ordinary least-squares regression techniques. The appropriate model was chosen by examination of the forecast errors.

As Table I shows, historically there have been large variations in the flow and water temperature of the river. These variations cause the impact coefficients to vary

TABLE I

Flow^a and Temperature^b Data for the Lower Fox River (May 1 to October 31, 1934 to 1977)

Month	Mean	Maximum	Minimum	Standard deviation
		Flow data (cfs $\times 10^3$)		
May	5.596	23.6	1.200	3.783
June	4.621	21.3	0.598	3.268
July	3.142	16.2	0.660	2.146
August	2.283	8.1	0.138	1.018
September	2.452	18.0	0.544	2.013
October	2.843	18.2	0.530	2.123
		Temperature data (°F)		
May	58.3	78.0	42.0	6.0
June	69.9	83.0	50.0	4.9
July	75.7	87.0	66.0	3.6
August	74.3	87.0	60.0	3.9
September	65.8	82.0	47.0	5.3
October	54.3	70.0	40.4	5.6

^{a}Measurements taken at the Rapide Croche Dam power station.
^{b}Measurements taken at Appleton and the Bergstrom Paper Company.

TRANSFERABLE DISCHARGE PERMITS   349

significantly from day to day, as illustrated in Fig. 2 for one discharger. An additional complication is that each of the 14 dischargers has different impact coefficients because their wastes decay at different rates. A system which allows one discharger to offset another's pollution has to take these differences into account to be sure that the water quality standards are maintained under any configuration of permissible discharges and any set of flow and temperature conditions in the river.

These observations, together with the knowledge that individual abatement costs differ among dischargers, imply that a fixed central directive is unlikely to be the least-cost approach to maintaining minimum water quality.

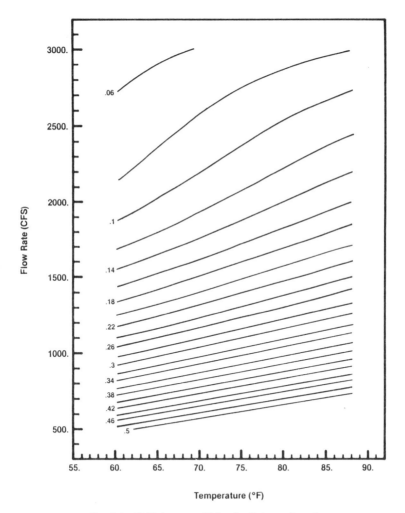

FIG. 2. $h_{31}(F, T)$, impact coefficient for discharger 3, sag 1.

IV. THE ECONOMIC MODEL

Having derived linear impact coefficients h_{ij} for the dischargers, the abatement problem can be characterized mathematically as follows. Let e_i^0 represent federal (BPT) discharge levels, X_i represent additional abatement in pounds of BOD, and q_j^S represent the quantities of DO available for assimilating the effluent at the sag points. The effluent demand for DO at sag j corresponding to federal BPT requirements is

$$q_j^D = \sum_i h_{ij} e_i^0 \tag{1}$$

Abatement, in addition to BPT, is needed if $q_j^D > q_j^S$. Any set of abatement activities X_i which yields

$$\sum_i h_{ij} X_i \geqslant \Delta q_j \equiv \sum_i h_{ij} e_i^0 - q_j^S \qquad \forall j \tag{2}$$

would be sufficient to achive the DO standard.

From an economic perspective, however, it would be preferable to induce an allocation of abatement activities X^* which solves the following "steady-state," constrained cost-minimization problem:

$$\min_{X_i} \sum_i C_i(X_i) \qquad \text{subject to (2) and } X_i \geqslant 0 \tag{3}$$

$C_i(X_i)$ is the abatement cost function for the ith discharger. This form of the problem implies that steady-state "worst case" stream flow and temperature conditions are used to determine H and the level of q_j^S [9].

Table II provides insight into the cost minimization problem. Because of differences in the impact of their wastes, one discharger (No. 3) is more than three

TABLE II
Marginal Cost of Increases in Dissolved Oxygen at Sag $j = 1$

Discharger	Impact[a] coefficient h_{i1}	Marginal abatement cost "end-of-pipe" C_i ($/lb)	Marginal cost of DO increase "at sag" C_i/h_{i1} ($/0.001 mg/l)
1	107	7.20	72
2	189	2.10	11
3	373	1.90[b]	5[b]
4	231	3.10	14
5	184	1.80	11
6	214	7.90	37
7	101	2.60	27

Source. Moore [8, page 32]; O'Neil [9, page 67].
[a] For F, T values 950 cfs, 80°F.
[b] Plant is operating at maximum abatement capacity; the numbers shown in the table are costs for the last unit treated.

times more effective in achieving increased DO than an equal abatement by another (No. 7). In addition, marginal abatement costs differ among dischargers by a factor of four. As a consequence, the cost of increasing DO at the first sag point (the product of the physical impact and the marginal abatement cost) varies sevenfold across the dischargers. This is shown in column 3. (Column 3 is also the "shadow price" of pollution abatement generated from the linear programming problem. It shows that the most effective procedure to reduce abatement cost is to assign more abatement activity to discharger 5 and less to discharger 1, if possible.)

Montgomery [3] has shown that a properly specified market in effluent permits (or DO permits) can yield the least cost allocation as a competitive market equilibrium. The basic conditions of market operation are that all available DO must be allocated initially to permit holders in any feasible WLA scheme, and that subsequent trades of effluent permits must be adjusted according to the ratio of the seller's and buyer's impact coefficients so as to assure no net decrease in dissolved oxygen in the river. (The costs of contracting and supervising the market are assumed zero.) To bound the value of the market option in the case of the Fox River, H from Qual-III was incorporated into (2) and X^* minimizing (3) was determined. This analysis generated estimates of the total and individual abatement costs associated with the initial distribution of permits, by central directive, and with X^*.

Table III, part A, presents these cost results for various water quality standards and stream flow and temperature (F, T) conditions. Columns 3 and 4 list the total annual expenses, including capital costs, of achieving the water quality targets shown in column 1 under the flow–temperature conditions shown in column 2. For example, given a DO target of 6.2 ppm, a stream flow of 950 cfs, and water temperature of 80°F, annual expenditures of $16.8 million would be required with TDP trading allowed. Under the same conditions, the central directive, with no TDP market, would require expenditures of $23.6 million. Thus the TDP market could allow annual savings of about $6.8 million in this case for the Fox River. As can be seen in the table, the potential cost savings vary depending on the DO target and river conditions.

TABLE III

Abatement Cost (Annual Expense)

	Flow/ temperature (cfs/°F)	TDP market ($ million)	Central rule ($ million)
A DO target (ppm)			
2.0	950/80	5.4	11.1
4.0	950/80	10.3	16.1
6.2	950/80	16.8	23.6
6.2	1500/72	9.0	16.5
6.2	2500/64	2.3	6.8
B w/v^a			
0.95	950/80	17.0	23.6
0.90	950/80	17.2	23.6
0.85	950/80	18.0	23.6
0.80	950/80	19.3	23.6

[a]DO target = 6.2 ppm, flow temperature 950 cfs, 80°F.

V. ADMINISTRATION OF A TDP MARKET

The potential value of TDPs has been demonstrated by the optimization. The problem is to implement a mechanism for trading when H varies. The problem can be divided into two parts: estimating H for any given flow and temperature condition and devising a simple mechanism to ensure that the constraint (2) is met.

Knowing that the estimation of H with Qual-III for the full range of possible river conditions (F, T) would be expensive for the dischargers, we approximated $H(F, T)$ by a cubic function

$$h_{ij}(F, T) = a_{ij} + b_{ij}F + c_{ij}T + d_{ij}F^2 + e_{ij}FT + g_{ij}F^3 \qquad (4)$$

The resulting equations fit the data well and showed no systematic forecast error when applied to the range of river conditions for which the approximation was estimated [8].[2] Table IV provides information on the residuals. The last two rows show that extrapolation to other conditions may be precarious.

Devising a mechanism to ensure that the constraint (2) is met is somewhat more complex. Let X_i^p represent the level of abatement entailed by the permit issued to the ith discharger before trading. The difference $(X_i^* - X_i^p)h_{ij}$ reflects the potential supply of DO due to i's activities at the jth sag when the term is positive and the potential demand when the term is negative. A viable sale under fixed F, T of $(X_i^* - X_i^p)$ to buyer k implies that k may decrease abatement by

$$\Delta X_k = (h_{ij}/h_{kj})(X_i^* - X_i^p) \equiv TC_{ik}(X_i^* - X_i^p) \qquad (5)$$

TC_{ik} is the "trading coefficient" relating the two dischargers. It is clear from Eq. (4) that TC_{ik} is a rational polynomial and is easily computed from 12 coefficients. Figure 3 depicts trading coefficient level curves for a pair of dischargers.

TABLE IV

$|h_{ij}(F, T) - h_{ij}(F, T)|$ Residuals from the Polynomial Approximation to Qual-III

Flow/temp[a]	Number of observations within residual ranges (mg/l DO)				
	0.00–0.0020	0.0021–0.0040	0.0041–0.0060	0.0061–0.0100	> 0.0101
	A. Within sample space				
1200/68	8	2	1	2	1
1200/76	7	2	1	2	2
2000/68	2	4	2	1	5
2000/76	0	4	4	2	4
	B. Outside sample space				
3000/68	1	0	1	1	11
3000/76	0	0	1	1	12
Total	18	12	10	9	35

[a] Flow in cfs/temperature in °F.

[2] A multiperiod optimization model was also developed to analyze the case of a time-varying H and q_j^s associated with changes in F, T. See O'Neil [9].

A strategy for trading might then entail one of two alternatives: (A) *Daily computation* of (4) by buyer and seller to assure that DO offsets are equal; (B) *Periodic computation* in which the level of trading is constrained by choosing $TC_{ik} \equiv TCM_{ik}$ for the *likely* river conditions. Augmenting k's discharges will then not exceed the assimilative capacity under those conditions. Choice of an appropriate boundary trading coefficient depends on the damage expected from violation of the DO target. Since no explicit damage function for the Fox River exists, we present only an illustrative analysis of the implications of periodic computation of trading coefficients.

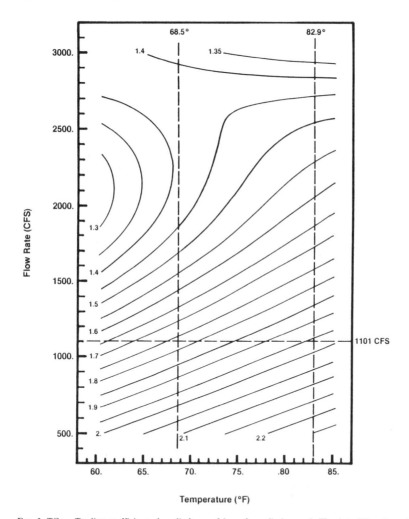

FIG. 3. $TC_{3,5}$, Trading coefficient when discharger 5 buys from discharger 3. The dotted lines bound an area corresponding to flows exceeding the 7-day low flow in 20 years and temperatures expected to occur with a probability of 0.95.

Moore [8] has demonstrated that given the appropriate information, TCM_{ik} can be found and provides a computational algorithm to determine its value on a monthly basis. However, the limit on trading imposed by TCM_{ik} will prevent realization of the full benefits of X^*. Some insight to this problem comes from a recomputation of the programming problem on the assumption that buyers have an impact $vh_{kj}, v > 1$, and sellers have an impact $wh_{ij}, w < 1$. Then $TCM_{ik} \equiv (w/v)TC_{ik} < TC_{ik}$. While w/v can be chosen to transform TCM_{ik} into $\overline{TCM}_{ik}$, it is not the case that those values will define the $\overline{TCM}_{i'k'}$, for another pair of traders i', k'. Hence the solution of the program for a particular v, w is only suggestive of the limitation imposed by periodic computation.

Table III, part B illustrates the cost effects of imposing adjustments of trading coefficients by the factors $w/v = \{0.95, 0.90, 0.85, 0.80\}$. These adjustments clearly decrease the cost savings achievable by trading permits. However, even a safety margin of 20% ($w/v = 0.80$) leaves the cost of the market solution substantially lower with TDPs than the initial centrally determined allocation. These results suggest that the market mechanism may be useful even in restricted trading scenarios.

VI. CONCLUSION

This paper demonstrates that water pollution control strategies allowing transfers between dischargers can be both cost effective and capable of maintaining any desired water quality standard. It shows that transferable permits can be used even in situations where the characteristics of the effluent differ between dischargers and the river conditions vary. The conclusion is that a regulatory system which does not permit and encourage transfers is substantially and needlessly costly. Transferable permits appear to be as appropriate for national water quality as offsets and bubbles have been for air policy. Those interested in regulatory reforms would do well to encourage the development and adoption of these options.

To summarize, there is no obstacle to trading permits *even when* the permit itself varies in proportion to the assimilative capacity of the river. The most straightforward method of trading, a daily trade, requires that buyer and seller make an elementary calculation to determine their permitted discharge. Variants of the daily trade might be the sale of a specified percentage of the seller's permit. Such an arrangement would imply that the seller would need to reduce the level of the initial permit by a fraction while the buyer would apply the relevant trading coefficient to that fraction to determine his permitted discharge. Since both buyer and seller need to review permitted levels of discharge on a daily basis in any case, trading adds flexibility without an excessive increase in administrative complexity. Although use of a permit market does imply increased information requirements for dischargers, conversations with managers have revealed that in the relatively simple market structure of the Fox River transactions costs are not expected to be large in comparison with the potential abatement cost savings. For other cases the design of the market may not be so simple. Clearly the choice of trading rules and particularly the size of the trading area will crucially affect the cost of administering and using the market. Since the potential transactions costs associated with the Fox River market were not estimated explicitly, the Wisconsin DNR wrote the relevant administrative rules so that dischargers have the option to trade permits or not as they see fit. The optional nature of the system was sufficient to neutralize most political arguments opposing the experiment (see Acknowledgments).

ACKNOWLEDGMENTS

We gratefully acknowledge financial support of the Wisconsin Sea Grant Institute, and the cooperation of the Wisconsin Department of Natural Resources in assessing Qual-III.

REFERENCES

1. T. D. Crocker, "The Structuring of Atmospheric Pollution Control Systems", *in* "The Economics of Air Pollution", (Harold Wologin, Ed.), Norton, New York (1966).
2. J. W. Dales, "Pollution, Property and Prices", University Press, Toronto (1968).
3. W. Montgomery, Markets in licenses and efficient pollution control programs, *J. Econ. Theory* 5, 395 (1972).
4. A. V. Kneese and B. T. Bower, "Managing Water Quality: Economics Technology and Institutions", John Hopkins Univ. Press, Baltimore (1968).
5. B. Ackerman *et al.*, "The Uncertain Search for Environmental Quality", Free Press, New York (1974).
6. D. Patterson, "Water Quality Modelling of the Lower Fox River for Wasteload Allocation Department", Wisconsin Department of Natural Resources, Madison (1980).
7. M. Gregory, "Qual-III Model used as a Management Tool", Department of Civil Engineering, University of Wisconsin, Madison (1979).
8. C. Moore, "Implementation of Transferable Discharge Permits When Permit Levels Vary According to Flow and Temperature: A Study of the Fox River, Wisconsin", Department of Civil and Environmental Engineering, University of Wisconsin, Madison (1980).
9. W. B. O'Neil, "Pollution Permits and Markets for Water Quality", unpublished Ph.D. dissertation, University of Wisconsin, Madison (1980).

[35]

JOURNAL OF ENVIRONMENTAL ECONOMICS AND MANAGEMENT 10, 112–124 (1983)

An Empirical Analysis of Economic Strategies for Controlling Air Pollution*

EUGENE P. SESKIN

Bureau of Economic Analysis, U.S. Department of Commerce, Washington, D.C. 20230

ROBERT J. ANDERSON, JR.

MATHTECH, Inc., Princeton, New Jersey, 08540

AND

ROBERT O. REID

Energy and Environmental Analysis, Inc., Arlington, Virginia, 22209

Received April 21, 1981; revised March 1982 and October 1982

In evaluating current environmental protection policy, economists often note that current regulations are more costly than necessary to meet environmental quality standards. While the *a priori* case is strong that current regulatory approaches are resulting in higher-than-necessary costs to attain environmental standards, there is relatively little empirical evidence to support this claim. The purpose of this paper is to supply some of the missing evidence by presenting the results of one study that assesses some of the potential savings associated with implementing economic, rather than command-and-control, regulatory approaches to abate one type of air pollution in one region of the country. Specifically, the paper examines the costs of meeting a prospective short-term standard for nitrogen dioxide under a range of alternative emissions control strategies for stationary sources of nitrogen oxide emissions in the Chicago Air Quality Control Region. The alternative strategies that are considered range from those that might result under current regulatory policy to those that economic policy approaches (such as emissions charges or marketable permits) are designed to implement. The analysis shows that the most efficient program of emissions controls may be more than an order of magnitude less costly than current regulatory strategies, and that economic approaches have additional advantages over more conventional regulatory approaches.

1. INTRODUCTION

In their evaluation of current environmental protection policy, economists have sometimes noted that current regulations are far more costly than strictly necessary to meet environmental quality standards.[2] This conclusion derives, in part, from the

---

*Correspondence should be addressed to: Seskin, BE-52, Bureau of Economic Analysis, U.S. Department of Commerce, Washington, D.C. 20230.

[1] A more detailed and technical presentation of the analysis may be found in [2]. In preparing that study, the authors benefited from the patient and constructive criticism from a number of people including David Tundermann, the Technical Monitor, as well as Allen Basala, Tayler Bingham, Alan Carlin, Alex Christofaro, Toby Clark, John Hoffman, Barbara Ingle, Skip Luken, Willard Smith, Paul Stolpman, and Larry White. The study was subsequently used by EPA together with additional work in [10]. The empirical work reported in the present paper reflects additions made by EPA in the Report to Congress; however, it should not be inferred that EPA necessarily endorses the conclusions expressed in the present paper. Any errors are, of course, the responsibility of the current authors. The opinions and conclusions expressed in this paper are not necessarily those of the organizations with which the authors are affiliated.

[2] See, for example, Kneese and Schultze [5].

112

ECONOMIC STRATEGIES FOR AIR POLLUTION CONTROL   113

observation that regulations often fail to exploit opportunities for polluters to reduce costs while achieving required emissions standards.

The causes adduced for this failure are many. For example, the procedural requirements of current regulatory approaches result frequently in protracted seriatim (source-by-source) negotiations. The standards ensuing from this process are unlikely to reflect a careful balancing of abatement costs with abatement effectiveness. Additional impediments may be imposed politically by legislated environmental laws, especially in cases where little scientific support was available. As a result, sources may be required to adopt specific emissions standards based on proven (or potential) technologies and processes. This, in turn, leaves sources with little incentive to exceed mandated requirements.

While the *a priori* case is strong that current regulatory approaches are resulting in higher-than-necessary costs to attain environmental quality standards, there is relatively little empirical evidence to support this claim. One previous study undertaken by Atkinson and Lewis [3] employed data for the St. Louis Air Quality Control Region (AQCR) to look at alternative strategies to control particulate emissions from 27 major sources of air pollution. Using a somewhat simplified version of the methodology applied in this study, it was found that an air quality control strategy representative of those developed by states in preparing State Implementation Plans (SIPs) was six to ten times as costly as a strategy designed to achieve a prescribed ambient particulate standard at minimum cost.

The purpose of this paper is to supply further empirical evidence of how economic measures could be used to control stationary-source air pollution emissions. Specifically, the paper examines the cost of meeting a prospective "short-term" standard for nitrogen dioxide (NO_2) under a range of alternative emissions control strategies.[3] An "emissions control strategy" refers to any plan specifying emissions limitations to be achieved by the specific sources in a region. The alternative strategies that are considered range from those that might result under current regulatory policy to those that economic policy approaches are designed to implement. The results are strikingly similar to those of Atkinson and Lewis [3]: The most efficient program of emissions controls may be more than an order of magnitude less costly than controls corresponding to current regulatory strategies. Furthermore, economic approaches have several inherent advantages over more conventional regulatory approaches for implementation.

2. METHOD OF ANALYSIS

In proceeding with the analysis, the first question to be answered is: What effect does a control strategy have on ambient concentrations of NO_2? In addressing this question, it is taken as given that an acceptable plan must be effective in meeting and maintaining an ambient short-term, 1-hour standard for NO_2.[4] The second

---

[3]Short-term here refers to an air quality standard based on pollution concentrations averaged over a period of one hour. The EPA has been evaluating alternative short-term, 1-hour NO_2 standards in the range of 250 to 1000 $\mu g/m^3$ because of potential adverse health effects from corresponding exposures.

[4]It should be noted that the results of this analysis will differ according to the standard that is finally set (see below). Furthermore, if the environmental objective was maintenance and attainment of a *long-term* standard based on annual average concentrations, the conclusions would be expected to change because elevated concentrations of NO_2 during short averaging periods occur only sporadically.

question to be addressed is: What effect does the choice of strategy have on the costs of controlling nitrogen oxides? In answering this question, primary emphasis is on the costs incurred by polluters to control their emissions; however, consideration is also given to the costs incurred by polluters to administer their programs and to monitor their emissions as well as costs incurred by the public sector to administer, monitor, and enforce programs.

The Analytical Model

The starting point of the analysis of plans for the control of stationary-source emissions of nitrogen oxides is an examination of alternative emissions control strategies and their effects on the control expenditures necessary to achieve a specified ambient short-term, 1-hour NO_2 standard. Basically, the approach used for this exercise involves the application of a mathematical programming model to the problem.

The specific type of mathematical programming model adopted for the analysis was an integer programming model. This model is basically a variation of what is known in operations research as the Knapsack Problem.[5]

The basic elements of the programming model are as follows:

- Emissions—$E_{i_k} \geq 0$ representing nitrogen oxide emissions of the ith source using control technology level k.
- Cost functions—$C_{i_k}(E_{i_k}) \geq 0$ representing the costs of controlling emissions at the ith source using control technology level k.

Given these elements, the model estimates the total technological costs of emissions controls across all N_s sources:

$$\sum_{i=1}^{N_s} C_{i_k}(E_{i_k})$$

and the corresponding 1-hour ambient concentrations of NO_2 at each of N_r receptors:

$$\sum_{i=1}^{N_s} d_{i_k j} E_{i_k} \quad j = 1, \ldots, N_r,$$

where $d_{i_k j}$ is the contribution of the ith source (with control technology level k in place) of nitrogen oxide emissions to concentrations of NO_2 at the jth receptor.

Using this notation, the least cost optimization problem can be characterized as:

$$\text{minimize} \sum_{i=1}^{N_s} C_{i_k}(E_{i_k})$$

$$\text{subject to} \sum_{i=1}^{N_s} d_{i_k j} E_{i_k} \leq Q \quad j = 1, \ldots, N_r,$$

[5]See, for example, Senju and Toyoda [7], Toyoda [8], and Zanakis [12].

where $Q \geqslant 0$ represents the short-term NO_2 air quality standard that must be met at each of N_r receptors.

For each strategy analyzed, the model simultaneously produces two types of estimates. It computes the total costs of emissions control of nitrogen oxides for all sources on an annual basis. This provides a means for comparing strategies on the basis of these costs. At the same time, it calculates the 1-hour ambient NO_2 concentrations that would result at each receptor. This provides a means for comparing the strategies on the basis of effectiveness in reducing pollution.

The Data

The analysis was performed using data characterizing the Chicago Air Quality Control Region (AQCR) which represents an area approximately 150 by 80 kilometers and includes the following counties: Cook, Lake, Dupage, McHenry, Kane, and Will. Chicago was selected not only because data were available, but also because peak 1-hour NO_2 levels in Chicago were among the highest concentrations observed in any urban area of the United States [9, pp. 22–27].

Because only four continuous monitors of nitrogen oxides existed at the time of the study in the Chicago AQCR, it was necessary to develop an air quality assessment model to evaluate further the extent of the short-term NO_2 problem in the region. For this purpose, a multiple point- and area-source model known as RAM was used.[6] To implement, the model requires data on point sources, area sources, mobile sources, and meteorological conditions.

Point-source information for the Chicago AQCR was obtained and augmented with data from EPA's National Emissions Data System (NEDS) point-source subfile and from the Illinois EPA.[7] For each point source, data included: location, level of nitrogen oxide emissions and associated activity (capacity of unit, hours of operation per year), and characteristics of emissions point (stack height and diameter, exhaust gas temperature and volume). When required data were unavailable, default values typical of normal operating practice were used. Area-source data on small stationary sources emitting less than 10 pounds per hour of nitrogen oxides and mobile-source data primarily on automobiles were apportioned to 634 five-kilometer-square grid cells and entered into the area-source subroutine of RAM. Meteorological data on wind direction, wind speed, stability class, and mixing height were obtained from the National Weather Service and covered stations located in Greenbay, Peoria, Flint, and Dayton.

In addition to the above information, receptor locations were selected to "monitor" the impact of nitrogen oxide emissions across the Chicago AQCR. Their locations were difficult to determine because the pollution concentration at any point is a complex function of meteorological parameters, source locations and

---

[6]RAM is an EPA-approved, Gaussian steady-state model capable of predicting short-term ambient concentrations of relatively stable pollutants from multiple point and area sources. Since NO_2 is primarily a secondary pollutant formed by oxidation of nitric oxide (NO), a dynamic model of NO_2 formation was developed and used in conjunction with RAM to predict nitrogen oxide concentrations at receptors due to point sources and to translate them into NO_2 concentrations.

[7]Specifically, 472 point sources in 1975 were "updated" to 534 sources in 1984. "Point source" here is equivalent to a source classification code (SCC) source. Thus, several "point sources" may be found in a given plant. Since EPA deemed 1984 as the earliest possible date for implementing a short-term NO_x control strategy, EPA "updated" the 1975 source inventory to represent 1984 conditions; see [10, p. A2].

emissions, source overlaps, and so on. Several preliminary analyses revealed the presence of two distinct types of point sources: (1) large plants with tall stacks such as power plants, and (2) plants with a large number of smaller sources with short stacks such as steel mills and refineries. It was found that the diffusion characteristics of emissions from the second category were similar to those of area sources and that together these sources were the dominant cause of high short-term NO_2 concentrations in Chicago. Thus, for the final analyses, 200 high-impact receptors selected by RAM, along with approximately 400 other receptors "blanketing" the region were used.

Finally, information on specific nitrogen oxide control technologies and their associated costs was needed. EPA has identified roughly three hundred separate source categories that emit nitrogen oxides. Nine such categories were sufficient to characterize the important point sources in the Chicago AQCR. These were:

1. Utility Coal-fired Boilers

2. Utility Oil- and Gas-fired Boilers

3. Industrial Coal-fired Boilers

4. Industrial Oil- and Gas-fired Boilers

5. Gas Turbines

6. Large Internal Combustion Engines

7. Industrial Process Units

8. Nitric Acid Plants

9. Municipal Incinerators

For purposes of the analysis that follows, the control technologies applicable to these sources can be categorized in terms of *combinations* of the following techniques: combustion process modification, fuel modification, removal of nitrogen oxides from stack gases (flue gas treatment), and alternative combustion processes. The costs of implementing these controls as well as their effectiveness in terms of reducing emissions of nitrogen oxides were estimated by using data from a number of sources. Information on source size was used to estimate capital costs, while information on source utilization (annual hours of operation) was used to estimate operating and maintenance costs.[8] It is important to note that most of these data were based on engineering analyses of hypothetical plants with only limited "field" experience; hence, the actual cost-effectiveness of these control technologies could vary by a wide margin when applied to individual sources.

3. COMPARING CONTROL STRATEGIES

The relative efficiency (measured in terms of annual emissions control costs) and effectiveness (measured in terms of resulting ambient concentrations) of four basic emissions control strategies were compared. The strategies were:

1. No Control Baseline—which considers only those emissions controls already in place.

_____

[8] The interested reader is referred to chapter 2 of [2] for further details.

ECONOMIC STRATEGIES FOR AIR POLLUTION CONTROL 117

TABLE 1

Analysis of Alternative Emissions Control Strategies for Chicago

Strategy	Number of sources controlled	Number of receptors in violation of $250 \mu g/m^3$ standard	Areawide point-source emission rate reductions (percent)	Annual control costs (millions of dollars)
No Control Baseline	0	36	0	0
SIP (State Implementation Plan)	472	0	21	130
Least Cost	100	0	3	9
Source Category Emissions Controls	472	0	18	66

2. SIP (State Implementation Plan)—which requires similar categories of major polluting sources to meet specific technology-based uniform levels of emissions control.

3. Least Cost—which establishes emissions limits by source based on each source's control costs and impact on ambient air quality.

4. Source Category Emissions Controls—which establishes uniform emissions limits for each source *category* consistent with the air quality objective.

The results of the comparison of these strategies are reported in Table 1.

No Control Baseline

To analyze this case, the model was run under the assumption that no *new* emissions controls were put in place. That is, the 1975 source inventory characteristics updated to 1984 conditions (see Fn. 7) were adopted.[9] As can be seen from Table 1, 36 of the receptor locations used in the final analysis indicated potential violations of a $250 \mu g/m^3$ short-term, 1-hour NO_2 standard. Since no new emissions controls were added it was assumed that no further reduction in nitrogen oxides emissions would take place and no additional costs would be borne under this strategy.

SIP (State Implementation Plan)

This strategy, which simulates the "traditional" approach to pollution control,[10] applies the highest level of emissions control to *all* sources in the three source

[9]Presumably, these characteristics account for any existing pollution controls required to ensure attainment of the current *annual* standard for ambient concentrations of NO_2.

[10]If EPA promulgates a short-term NO_2 standard, many existing State Implementation Plans (SIPs) would have to be revised. Such revisions would involve state regulations prescribing emissions limitations for specific source categories and timetables for compliance. Under the Clean Air Act, states are allowed nine months after a standard is issued to develop their plans, which must demonstrate attainment within three years. Until the area attains the standard, the applicable emissions regulation for an existing source is termed Reasonably Available Control Technology (RACT) and for a new source it is termed Lowest Achievable Emissions Rate (LAER) achieved in practice for that source category, or the most stringent emissions limitation contained in any SIP (whichever is more stringent).

categories that were the major stationary-source polluters in Chicago. The three source categories were: industrial coal-fired boilers, industrial oil- and gas-fired boilers, and industrial process units. Table 1 indicates that for this simulation, controls were placed on 472 sources in these categories, and that these controls were sufficient to attain the ambient standard at all receptor locations. It can also be seen that areawide point-source emissions rates would be reduced by 21% under this strategy. The incremental costs associated with implementing SIP (over and above the No Control Baseline) were estimated to be $130 million per year.

Least Cost

The Least Cost strategy was found by using the programming model described in Section 2 to find a set of emissions controls that simultaneously minimize control costs and meet the short-term 1-hour NO_2 standard at all receptor locations. By definition, under the Least Cost strategy, no receptors would be in violation of the 250 $\mu g/m^3$ standard. Furthermore, as seen in Table 1, only 100 of the point sources would require emissions controls *over and above* those controls associated with the No Control Baseline. These emitters were all from the three most polluting source categories noted above: industrial coal-fired boilers, industrial oil- and gas-fired boilers, and industrial process units.[11] It is also interesting to note that a reduction of only 3% in the areawide emissions rates would be sufficient to attain the short-term standard at all receptor locations. Finally, it can be seen that the annual costs associated with the Least Cost strategy were estimated to be only $9 million more than the No Control Baseline.

Source Category Emissions Controls

This strategy requires the application of a uniform set of emissions controls across all sources in a particular source category. As such, it circumvents the need for the source-by-source emissions limitations necessary to implement the Least Cost option and represents a relatively sophisticated use of current regulatory planning methods. The emissions control level specified for a given source category was that control level—for example, dry selective catalytic reduction (flue gas treatment)—required to ensure that no emitter in the class would violate the short-term, 1-hour ambient standard at any receptor.[12] Again, the three source categories requiring controls were: industrial coal-fired boilers, industrial oil- and gas-fired boilers, and industrial process units. As defined, under this strategy all receptors would be in compliance with the short-term, 1-hour NO_2 standard. The reduction in areawide emissions rates was estimated to be 18% and the total control costs were estimated to be $66 million annually over and above the No Control Baseline (see Table 1).

---

[11] It should be stressed that this does not mean that other sources were not *contributing* to violations of the short-term, 1-hour standard, only that it was not cost-effective to control those sources.

[12] Note, this differs from the SIP strategy in that it does not require that the most stringent control level be applied to each source category in question.

Summary of Findings

The preceding analyses demonstrate that abatement strategies designed to exploit differences in sources' emissions control costs as well as associated meteorological-dispersion characteristics are significantly less costly than those that do not account for such factors. For example, the results for Chicago indicate that a Least Cost strategy is less than one-tenth as costly as a strategy that reflects a more traditional regulatory approach (SIP) and less than one-seventh as costly as a strategy that represents a relatively sophisticated version of current regulatory approaches (Source Category Emissions Controls). In absolute terms, a policy that would lead to the adoption of a scenario approximating the Least Cost strategy to meet a short-term, 1-hour NO_2 standard could save more than $100 million annually in technological control costs in the Chicago Air Quality Control Region alone.

4. IMPLEMENTATION: POLICY INSTRUMENTS AND RELATED ISSUES

It would be premature to conclude that the less costly strategies described would necessarily be superior *in practice* to more traditional regulatory approaches. This follows from the fact that the policy instruments needed to implement the less costly strategies may be unavailable because of legal or political constraints, or may be so costly to administer as to offset the potential savings in emissions control costs. One category of policy instruments—emissions charges—can be examined briefly to shed more light on these issues.[13] In doing so, alternative emissions charge plans will first be described. This will be followed by an examination of informational requirements, their associated costs, and some legal considerations.

Alternative Emissions Charge Plans

The results of a mathematical programming analysis of the type discussed above can be used to formulate emissions charge schemes. In particular, charges were set at the minimum amounts required to ensure that the ambient standard would be achieved at all receptors when the sources acted in their economic self-interest to minimize the sum of annual control costs plus charge payments. It should be noted that the charges were applied to emissions *rates* rather than simply to emissions. This is necessary because a charge scheme based on total emissions would not ensure attainment of a 1-hour standard since sources could reduce their charge liabilities simply by reducing their hours of operation without reducing their *rate* of emissions. The resulting source-by-source charge levels (and the associated annual liabilities for charge payments) were found to vary considerably by source. Total annual liabilities for the charge payments amounted to $4 million or almost 50% of the total control expenditures estimated under the Least Cost option.

---

[13] No explicit consideration is given here to a system of marketable emissions permits for implementing efficient control strategies. However, such policy instruments were analyzed in [2, Chapter 7] and in [10], and were shown to have many favorable attributes and in some ways were thought to be easier to implement and superior to emissions charge systems.

TABLE 2

Effects of Alternative Charge Systems for Chicago

Plan	Number of sources controlled	Areawide point-source emissions rate reductions (percent)	Annual control costs (millions of dollars)	Annual charge payments (millions of dollars)	Annual control costs + charge payments (millions of dollars)
Least cost (source-by-source) charge levels (see text)	100	3	9	4	13
Uniform charge levels ($15,800 per year per pound per hour)	534	84	305	414	719
Source category charge levels[a]	472	18	66	89	155

[a] Industrial coal-fired boilers = $15,800 per year per pound per hour; industrial oil- and gas-fired boilers = $15,300 per year per pound per hour; and industrial process units = $3500 per year per pound per hour.

The marked differences in charge levels and liabilities together with the preceding comparison of control strategies reaffirm that a system based on uniform treatment of sources or even uniform treatment by source classification is not consistent with the implementation of an efficient control program. Nevertheless, it is recognized that there may be practical difficulties in implementing a system in which "seemingly similar" sources are treated differently.[14]

In order to more fully explore the "excess" control costs associated with establishing more uniform charge systems, two alternative schemes were examined. Under the first system, denoted the Uniform Charge plan, a single charge level was levied on all sources. The magnitude of the charge was set at the lowest amount that would ensure compliance with the short-term, 1-hour NO_2 standard at all receptors. That amount was equal to $15,800 per year (per pound of nitrogen oxides per hour).

Under the second system, designated the Source Category Charge plan, three different charge rates were set, one for each of the three source classes that were controlled under the Least Cost strategy.[15] The magnitudes of these charges were set equal to the highest *average* annual control costs (per pound of nitrogen oxides per hour) in each source category under the Least Cost option. The resulting charge levels were $15,800 per year (per pound of nitrogen oxides per hour) for industrial coal-fired boilers, $15,300 for industrial oil- and gas-fired boilers, and $13,500 for industrial process units.

Table 2 presents the results of implementing these alternative schemes together with the basic results from the Least Cost Charge plan (described above). As can be seen, the Uniform Charge plan is associated with estimated nitrogen oxide emissions

[14] For example, it is quite conceivable that the Least Cost strategy would require two plants producing competing brands of the same product to meet substantially different emissions limitations because their locations (and impacts on ambient air quality) differed.

[15] The reader will note similarities between the Source Category Charge plan and the Source Category Emissions Control strategy discussed above.

rate reductions of more than 80% below the levels corresponding to the Least Cost Charge plan, but estimated annual emissions control costs exceed the costs under the Least Cost Charge plan by almost $300 million; annual charge payments are $410 million greater. Under the Source Category Charge plan, emissions rates are reduced by approximately 15% below the levels corresponding to the Least Cost Charge plan but estimated annual emissions control costs are about $57 million greater than those associated with the Least Cost Charge plan and annual charge payments are about $85 million greater. Thus, while the more uniform plans do reduce nitrogen oxide emissions rates substantially below emissions rates under the Least Cost Charge plan, these reductions are exceedingly costly and represent overcontrol in that all figures are based on minimum charge levels necessary to ensure attainment of the short-term, 1-hour NO_2 standard. At the same time, the Least Cost Charge plan appears to impose the smallest *overall* burden on sources (and ultimately society).

Informational Requirements, Associated Costs, and Legal Considerations

As alluded to above, it is sometimes suggested that despite possible Control cost savings associated with economic approaches to pollution control, implementation costs of such approaches would more than offset the potential savings. Therefore, it is useful to explore the feasibility and costs of implementing these approaches.

The Appendix details estimates for the range of administration, monitoring, and enforcement (AME) activities required to implement a regulatory system and a charge system in Chicago. There it is shown that the costs directly attributable to the administration, monitoring, and enforcement of a charge system to control stationary sources of nitrogen oxide emissions are of the same order of magnitude as the costs associated with implementing an effective regulatory policy.[16] The main reason for this is that effective regulation is likely to involve more investigation, negotiation, and litigation than would an equally effective incentive system. At the same time, it is recognized that existing legislation may need to be amended explicitly to allow implementation of such incentive systems. Nevertheless, it does not appear that legal considerations pose serious barriers to the implementation of economic approaches to pollution control such as emissions charge systems. While such approaches and the requisite supporting legislation represent a somewhat new regulatory framework and, as such, have not been fully tested in the courts, a thorough examination of the possible legal bases for implementing these types of economic inventive systems led one study to the conclusion that "[m]any different sources of government power could be invoked to legitimatize the legislature's imposition of charge plans,..." [1, p. 144].

5. CONCLUSIONS

The quantitative analysis of emissions control strategies for Chicago shows that approaches designed to account for differences in sources' incremental costs of

---

[16] While the cost estimates forming the basis for this conclusion are, at best, approximate, it is unlikely that further refinement would change the qualitative results.

controls and incremental contributions to ambient pollution concentrations can achieve a short-term ambient NO_2 standard at significantly lower costs than strategies that do not account for these differences. While acknowledging some of the practical difficulties in implementing such strategies as well as legal and political considerations, these problems do not appear to be insurmountable.

The analysis also suggests that emissions charge plans can provide profit-and-loss incentives to firms sufficient to induce the degree of emissions controls required to attain the short-term NO_2 standard in an economically efficient manner. Furthermore, a charge system provides an effective stimulus to the development and application of new emissions control technology. This, too, is an important practical advantage of the emissions charge approach. It is especially apparent if one recognizes that in the simulation study of Chicago, the emissions control technologies required to meet a short-term, 1-hour NO_2 standard of 250 $\mu g/m^3$ were technologies that are only *projected* to become available (at the earliest) between 1981 and 1985. Clearly, efficient and effective environmental policy must provide incentives for technological development, and economic approaches—in the form of emissions charges or marketable emissions permits—do exactly this. Taken as a whole, then, the magnitude of the potential cost savings associated with simulating such a system in only one region, together with the attributes just noted, appear to provide adequate justification for further experimentation and analysis of economic approaches to control environmental pollution.

APPENDIX: ADMINISTRATION, MONITORING, AND ENFORCEMENT ESTIMATES FOR THE CHICAGO AQCR

To effectively implement the economic strategies described above, the regulatory authority must know sources' emissions rates, the impact of these emissions on ambient air quality, the effects of abatement controls on emissions, and the costs of these controls.[17] One of the most important aspects concerning these informational requirements is the ability of existing monitoring techniques to provide adequate and reliable data. It appears that technically adequate nitrogen oxide emissions monitoring systems are currently available at a cost that is considerably smaller than the total costs of abatement associated with meeting a short-term, 1-hour NO_2 standard of 250 $\mu g/m^3$.[18] This is true even under the Least Cost emissions control strategy in which the average annual control cost per controlled source is approximately $90,000 ($9 million divided by 100; see Table 1). Under the conservative assumption that each controlled source would require a separate monitoring system,[19] annual monitoring costs were estimated to be on the order of $25,000 per source.

It was also noted that under any policy approach to pollution control, sources will bear the burden of some costs of monitoring and reporting their emissions. Thus, one would not expect the costs of these activities to differ very much between a

---

[17]While it has been discussed in the theoretical literature that a regulatory agency could achieve air quality objectives under a "standard-and-charges" program without knowledge of sources' control costs by adjusting charge rates until the desired objectives were achieved (see, for example, [4, p. 144]), in practice, the rigidities imposed by economic, political, and legal constraints make this seem unlikely.

[18]For details see [2], especially Section 2.5.

[19]The assumption is conservative in that many sources are co-located and could therefore share some or all parts of the monitoring system.

ECONOMIC STRATEGIES FOR AIR POLLUTION CONTROL 123

TABLE 3

Administration, Monitoring, and Enforcement Estimates for the Chicago AQCR

Activity	Regulatory system		Charge system	
	Person-years	thousands of dollars[a]	Person-years	thousands of dollars
Initial (one-time) efforts:				
Dispersion modelling	0.24	5	0.24	5
Determining emissions control cost functions	—	—	0.32	6
Equipment expenditures	—	150	—	150
Total	0.24	155	0.56	161
Recurring (annual) efforts:				
Administration	26 to 30	376 to 445	29 to 37	370 to 489
Monitoring	9 to 15	104 to 207	8 to 15	108 to 207
Enforcement	27 to 110	441 to 1950	19 to 44	280 to 631
Total	62 to 155	921 to 2602	56 to 96	758 to 1327

[a] All expenditures in 1978 dollars. Salary information was obtained from [11, p. 40]; estimates do not include overhead and fringe benefits.

regulatory approach and one based on an economic approach such as an emissions charge plan. The main difference would be related to some additional accounting and record-keeping activities necessary under a charge system. Following an examination of alternatives for carrying out these activities, it was concluded that they could be relatively easily and inexpensively integrated into the hardware and software associated with emissions monitoring systems that would be required under conventional regulatory approaches. Specifically, the incremental annual cost was estimated to be on the order of, at most, $2500 per source.[20]

Table 3 details the complete personnel and cost estimates for the range of administration, monitoring, and enforcement (AME) activities required to implement a regulatory system and a charge system in Chicago. Note that there are essentially two stages of effort. Initially, estimates of transfer coefficients (the $d_{i,k,j}$ described on p. 114), background concentrations of pollutants, and (in the case of a charge system) engineering-based emissions control cost functions would be derived. Subsequently, there would be less intense, on-going efforts to improve this information as well as the recurring costs associated with operating the systems.

These estimates were derived by using the format developed in an EPA "manpower planning model" [6]. However, since the model focuses only on state and local agencies, it was necessary to include supplementary information on Federal involvement.[21] Specifically, it was found that in the Chicago AQCR, a major portion of Federal activity involves case development; that is, bargaining and negotiating with sources on acceptable compliance schedules. Since sources can realize substantial economic benefits from delaying compliance, they often challenge agency-desired

[20] Details on system configurations and estimated development and system costs can be found in Section 6.8 of [2].

[21] Much of this information was obtained through conversations and correspondence with Tom Donaldson, USEPA, Control Program Section, Research Triangle Park; Ron Shafer, USEPA, Washington, D.C.; Pat Reape, Enforcement Division, USEPA, Region V; Wayne Jones, Division of Air Pollution Control, Illinois EPA; and Mr. Kason of the Department of Environmental Control, City of Chicago.

124   SESKIN, ANDERSON, JR., AND REID

control techniques or the time frame for compliance. Thus, the agency must have resources available for intensive negotiating sessions as well as for courtroom appearances in formal legal proceedings. It was estimated that such activities could require between 7 and 45 person-years at a cost of between $100,000 and $700,000 annually.[22] The estimated range of total AME costs under a regulatory system were $900,000 to $2.6 million annually, with personnel effort ranging between 62 and 155 person-years.

The specific charge system used in the development of corresponding AME costs estimates was based on the Least Cost Charge plan discussed above. Again, calculations were made using the framework in the EPA manpower planning model. Table 3 indicates that a Least Cost emissions charge program could be implemented in the Chicago AQCR with AME costs running between $750,000 and $1.3 million annually and with personnel effort ranging between 56 and 96 person-years.[23]

REFERENCES

1. F. R. Anderson, A. V. Kneese, P. D. Reed, S. Taylor, and R. B. Stevenson, "Environmental Improvement Through Economic Incentives," Johns Hopkins Univ. Press, Baltimore, 1977.
2. R. J. Anderson, Jr., R. O. Reid, and E. P. Seskin, "An Analysis of Alternative Policies for Attaining and Maintaining a Short-term NO_2 Standard," prepared for the U.S. Environmental Protection Agency, U.S. Council on Environmental Quality, and Council of Economic Advisers, September 17, 1979.
3. S. E. Atkinson and D. H. Lewis, A cost-effective analysis of alternative air control strategies, *J. Environ. Econ. Manag.* **1** (1974), 237–250.
4. W. J. Baumol and W. E. Oates, "The Theory of Environmental Policy," Prentice-Hall, Englewood Cliffs, New Jersey, 1975.
5. A. V. Kneese and C. L. Schultze, "Pollution, Prices, and Public Policy," The Brookings Institution, Washington, D.C., 1975.
6. D. A. Lynn and G. L. Deane, "Manpower Planning Model," Office of Air and Waste Management, Office of Air Quality Planning and Standards, U.S. Environmental Protection Agency, Research Triangle Park, North Carolina, EPA 450/3-75-034, March 1975.
7. S. Senju and Y. Toyoda, An approach to linear programming with 0-1 variables, *Manag. Sci.* **15** (1968), 196–207.
8. Y. Toyoda, A simplified algorithm for obtaining approximate solutions to 0-1 programming problems, *Manag. Sci.* **21**, (1975), 1417–1427.
9. U.S. Council on Environment Quality, "Environmental Quality: The Ninth Annual Report of the Council on Environmental Quality," U.S. Government Printing Office, Washington, D.C., 1978.
10. U.S. Environmental Protection Agency, "An Analysis of Economic Incentives to Control Emissions of Nitrogen Oxides from Stationary Sources," Report to Congress, January 1981.
11. U.S. Environmental Protection Agency, "Methodology Report, Air Pollution Strategy Resource Estimator" 2, 1975.
12. S. H. Zanakis, Heuristic 0-1 linear programming: An experimental comparison of three methods, *Manag. Sci.* **24** (1977), 91–104.

[22] These figures are incorporated in the category of recurring annual enforcement costs of the regulatory system.

[23] It should be noted that these numbers do not reflect Federal-level case development activities, since it is assumed that no compliance schedules are required under a charge system.

Part VI
Uncertainty

[36]

JOURNAL OF ENVIRONMENTAL ECONOMICS AND MANAGEMENT 3, 178–188 (1976)

Uncertainty and the Choice of Pollution Control Instruments

Zvi Adar

Department of Economics, Tel Aviv University, Tel Aviv, Israel

AND

James M. Griffin

Department of Economics, University of Pennsylvania, Philadelphia, Pennsylvania 19174

Received January 12, 1976

This paper compares the relative efficiencies of pollution taxes, pollution standards, and the auctioning of pollution rights when the marginal damage function or marginal control cost are subject to uncertainty. In the first case, we find that all instruments yield the same expected social surplus. In the latter case, the choice of the optimal instrument depends, in general, on the relative elasticities of the marginal damage and marginal expected cost functions, on the way in which uncertainty enters the model, and on the distribution of the error term. Policy conclusions are derived.

I. INTRODUCTION

The choice between alternative pollution control instruments typically centers on comparisons of transaction costs, and administrative and policy costs.[1] While such cost comparisons are no doubt important, we question why the role of uncertainty is generally omitted in welfare calculations. Practitioners in the field of environmental quality recognize that uncertainty concerning the nature of the marginal damage function and the marginal control cost function are major stumbling blocks to selecting Pareto-efficient policies.[2] Perhaps the omission of uncertainty in the instrument choice discussion is due to an implicit assumption that uncertainty, while pervasive, is simply neutral to the instrument choice question. A notable exception is Lerner [6] who recognized that the relative slopes of the marginal damage and control cost functions should affect the choice between price and quantity controls.

As an extension to Lerner's conjecture, this paper delimits the circumstances under which uncertainty does and does not affect the choice between the following three policy instruments: taxes (a flat excise tax per unit of pollution); standards (quantitative controls on the amount of pollution generated by each source or firm[3]); and the Dale's proposal [3] for the auctioning of a fixed quantity of transferable pollution

---

[1] For example, see [3, 5].

[2] For an example concerning SO_2 taxes, see [4], where major uncertainty about the availability of flue-gas desulphurization is a problem.

[3] Note that throughout, we assume standards are set such that the marginal cost of abatement is equalized across firms, i.e., they are Pareto-efficient. Later, we return to this assumption.

178

rights. This paper shows that certain types of uncertainty can affect our choice between the three instruments while other types should not.

In order to delimit the circumstances under which uncertainty matters, it is necessary to distinguish the types and sources of uncertainty. Uncertainty permeates environmental policy-making in terms of the calculations of both marginal damages and marginal control costs. Uncertainty can manifest itself to the individual firm's decision makers and/or to the pollution control agency. At the firm level, uncertainty enters through the firm's marginal control cost function and through uncertainty induced by the control agency. Uncertainty concerning the firm's marginal cost function for pollution abatement is partially technology-induced as the control technology may be unproven or so new that future cost reductions due to potential scale economies, learning-curve phenomena, or technology diffusion rates are uncertain. In addition, uncertainty about future input prices introduces added uncertainty to the firm's marginal control cost function. As firms are well aware, the actions of the control agency can introduce an added source of disturbance. Examples include frequent changes in the pollution tax rate, changes in the pollution standards, or variations in the auction price for tickets.

Uncertainty at the level of the pollution control agency includes many of the same phenomena facing the firm, such as the uncertainty attached to the firm's marginal cost functions which in turn imparts uncertainty to the aggregate marginal control cost function facing the agency (society). Technologic and input price uncertainties prevail at either level of disaggregation. While the agency is presumably free of the uncertainty it confers on the firm through policy changes, it faces two additional sources of uncertainty. First, even if the firm's marginal cost functions are known, the marginal control cost function facing the agency may not be known. Baumol [1] has considered these uncertainties and found them so pervasive that he favors policies which effectively disregard Pareto-efficiency criteria.[4]

A second and perhaps the most perplexing form of additional uncertainty is that the agencies have only vague ideas about the social marginal damage function. The standard errors attached to estimates of health and real estate costs of air pollution are indeed large. Thus a type of measurement uncertainty is connected with the marginal damage function owing to the difficulties of measuring social damage from pollution. Even with correctly measured marginal damage functions, a stochastic component would still enter through ambient air conditions which change continuously depending on climatic conditions. In the case of air pollution, factors varying daily, such as wind velocity and direction, affect the social damage of a given pollutant discharge.

In opposition to Baumol's policy advice, this paper posits that the purpose of policy in the face of uncertainty is to maximize expected welfare. In Section II, we begin by examining uncertainty at the level of the control agency. First, we contrast the expected welfare losses of all three policy instruments when uncertainty enters the marginal damage function. Second, we contrast the three instruments assuming the marginal control cost function is subject to uncertainty. These exercises reveal a fundamental asymmetry in the effects of uncertainty. In Section III, we consider how uncertainty combined with risk aversion by firms might change the previous results. Section IV explores differential uncertainty effects between standards and auctions. In Section V, we recapitulate the major policy implications.

---

[4] Also see [2].

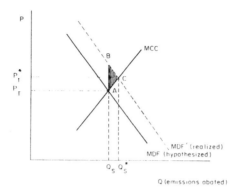

FIG. 1. Uncertain marginal damage function.

II. UNCERTAINTY AT THE AGENCY LEVEL

This section illustrates that there is a basic asymmetry in the effects of uncertainty associated with the marginal damage function *vis-à-vis* the marginal control cost function on the choice between taxes, standards, and auctions. First, consider the case where the marginal control cost function (MCC) is known, and only the marginal damage function (MDF) is subject to uncertainty. In Fig. 1, we measure along the quantity axis, the pollution abated; corresponding to the origin of 0 reductions is the uncontrolled discharge of Q_{max}. Policies based on the hypothesized marginal damage function would result in a tax of P_T, the abatement of Q_S units under a standards policy, or the auction of $Q_{max} - Q_S$ tickets. Due to uncertainty, the realized marginal damage function (MDF) deviated from the hypothesized MDF function in Fig. 1. Equivalent welfare losses will occur under a tax, standards, or an auction. With a tax policy, where the tax is set at P_T, emissions abated will only be Q_S and the welfare loss from a tax is given by the shaded area ABC. Under a standards or auction policy, the emissions abated are again only Q_S, while the optimal reduction is Q_S^*. Likewise, the welfare loss associated with these policies is the area ABC. Thus all three policies yield similar welfare losses. The explanation for this equivalency is that the quantities discharged, irrespective of the instrument, depend solely on the marginal control cost function, which is certain in this case, thereby resulting in identical quantities discharged, irrespective of the policy instrument. Thus the introduction of uncertainty in the damage function has nothing to say about the choice of policy instruments.

While intuition might suggest that similar results hold for the case in which the marginal control cost function is uncertain, this is not the case. Figure 2 illustrates a case in which the optimal tax, P_T, and the equivalent optimal quantity, Q_S, are assigned based upon the hypothesized shape of the marginal control cost function (MCC). However, the actual marginal control cost function turns out to be MCC' as marginal costs are much higher than anticipated.

With a tax of P_T, only Q' emissions were abated, even though at Q' the marginal damage rate ($Q'B$) exceeds the marginal costs ($Q'A$). An optimal tax, P_T^*, would have provided for emission abatement of Q_S^* and would have avoided the welfare loss given by the shaded area (ABC).

With perfect hindsight, the optimal standard would have been Q_S^*. The resulting welfare loss from a standards policy is given by the shaded area CDE. Similarly, since

POLLUTION CONTROL UNDER UNCERTAINTY   181

for an auction only $Q_{max} - Q_S$ tickets would be auctioned, the level of pollution is reduced to Q_S and the welfare loss is equivalent to that of a standards policy (CDE). As Fig. 2 illustrates for the case depicted, the welfare loss from a tax clearly exceeds the welfare loss for either a standards or auction policy. Therefore, we observe a fundamental asymmetry between uncertainty in the MDF, which leads to equivalent welfare losses, and uncertainty in the MCC curve which produces dichotomous results between taxes and quantitative restrictions (either in the form of an auction or standards). The explanation is that in the former case the MCC is known, thus price or quantitative controls are equivalent, while, with a stochastic marginal control cost functions, this uniqueness between prices and quantities no longer holds since quantity under a tax varies with shifts in MCC.

Clearly, the welfare loss, CDE is not equal to the welfare loss ABC. In the case plotted in Fig. 2, the former is smaller, indicating the superiority of quantity restrictions. This, of course, is not a general result; the welfare loss from setting the tax P_T can be approximated by the area of the triangle ABE less the area ACE, and can be expressed by

$$WL_T - ACE = \tfrac{1}{2}\Delta P_T \Delta Q_T, \tag{1}$$

where

$$\Delta P_T = AB \quad \text{and} \quad \Delta Q_T = Q_S - Q'.$$

Substituting the elasticity (e_d) of the MDF curve, (1) can be also written as

$$WL_T - ACE = -(\tfrac{1}{2})(P/Q)(\Delta Q_T)^2(1/e_d). \tag{2}$$

The welfare loss associated with a quantitative restriction can be viewed graphically in Fig. 2 as the triangle ADE less the area ACE. The new welfare loss can be approximated as

$$WL_q - ACE = \tfrac{1}{2}\Delta P_S \Delta Q_S, \quad \text{where} \quad \Delta P_S = DE \quad \text{and} \quad \Delta Q_S = Q_S - Q'. \tag{3}$$

Equation (3) can be restated in terms of price, quantity, and the elasticity of the control cost function (e_c).

$$WL_q - ACE = (\tfrac{1}{2})(P/Q)(\Delta Q_s)^2(1/e_c). \tag{4}$$

Combining Eqs. (2) and (4), and using the fact that $\Delta Q_s = \Delta Q_T$, yields

$$WL_T - WL_q = -(\tfrac{1}{2})(P/Q)(\Delta Q)^2[(1/e_d) + (1/e_c)]. \tag{5}$$

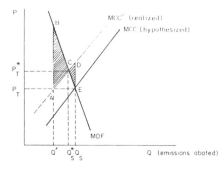

FIG. 2. Uncertain marginal control cost function.

Equation (5) suggests that tax and standards policies will have identical welfare loss properties only when (a) the damage and control cost elasticities are equal in absolute value, or (b) $\Delta Q = 0$, i.e., actual equals the hypothesized marginal control cost function implying zero welfare losses. An important implication of (5) is that standards are preferable when e_d approaches zero and a tax policy is preferable when e_c approaches zero, *ceteris paribus*.

This result suggests a fundamental asymmetry between the uncertainty in the damage function and in the control cost function on the relative effectiveness of tax, standards, and auction policies of abating pollution. When MDF deviates from its hypothesized value, all policies will generate the same deadweight loss. With (vertical) variations in MCC, we should expect welfare losses which depend on the slopes, or elasticities of the MDF and MCC curves. The marginal control cost function forms a behavioral relation for the firm under a tax. If MCC is known, taxes and quantitative instruments yield equivalent reductions in pollution, irrespective of movements in the marginal damage function. In contrast, the actual, as opposed to the hypothesized, marginal damage function, evinces no quantitative response irrespective of the instrument. Thus variations in MDF produce no dichotomy between price (tax) and quantitative (standards or auction) policies.

This asymmetry between uncertainty in the damage and control cost functions carries over to the problem of selecting the optimal policy instrument as well as the level of control under uncertainty.

First, consider the case where MDF is random, and the goal of the agency is to maximize *expected* welfare (i.e., minimizing the expected area ABC in Fig. 2).

Let MDF be given by the relationship

$$\text{MDF}(q, u), \tag{6}$$

where q is the quantity abated and u is a random variable with known density $dF(u)$. By assumption, MCC is a known function depending only on the level of abatement

$$\text{MCC}(q), \tag{7}$$

with quantitative restrictions of either a standards or auction type. An agency attempting to maximize the expected welfare gain will set Q_S such that expected consumers' and producers' surplus is maximized as follows.

$$\mathop{E}_{u} \int_0^{Q_s} [\text{MDF}(q, u) - \text{MCC}(q)]dq. \tag{8}$$

Under a tax policy, the agency will again set a tax, P, so as to maximize the expected consumers' and producers' surplus

$$\mathop{E}_{u} \int_0^{Q_s(P)} [\text{MDF}(q, u) - \text{MCC}(q)]dq. \tag{9}$$

Since $Q_s(P)$ is the single valued relation $Q_s = \text{MCC}^{-1}(P)$, which is known with certainty, maximizing (8) with respect to Q is equivalent to maximizing (9) with respect to P_T. The optimal quantity corresponding to P^* (i.e., Q_S^* or $Q_s(P^*)$] is given by the first-order condition

$$\mathop{E}_{u} \text{MDF}(Q_S^*, u) = \text{MCC}(Q_S^*). \tag{10}$$

POLLUTION CONTROL UNDER UNCERTAINTY   183

The second-order condition invariably holds since we assume a negatively sloped MDF curve and a positively sloped MCC. Substituting optimal Q_s^* into (8) we can find the value of the maximized expected welfare gain. In the linear case with an additive error term shown in Fig. 2, where

$$MDF = a - bq + u, \tag{11}$$

where $E(u) = 0$ and

$$MCC = \alpha + \beta q_s. \tag{12}$$

Note that (10) reduces to

$$Q_s^* = (a - \alpha)/(b + \beta) \quad \text{and} \quad P^* = \alpha + \beta(a - \alpha)/(b + \beta). \tag{13}$$

The expected welfare gain (EWG) depends only on the parameters a, α, b, and β, and is independent of the distribution of u.

$$EWG = \frac{1}{2} \frac{(a - \alpha)^2}{(b + \beta)}. \tag{14}$$

We conclude that when MDF is uncertain and the agency is risk neutral, expected MDF should be used in the selection of optimal control levels. Moreover, taxes, standards, and auctions have the same resulting performance.[5]

Next, consider the case where MDF is known with certainty, but MCC is subject to a stochastic disturbance as given by

$$MCC(q, u), \tag{15}$$

where again u has known density of $dF(u)$. With quantitative controls, we obtain results similar to the previous case. Maximizing with respect to Q_s the expectation

$$\mathop{E}_{u} \int_0^{Q_s} [MDF(q) - MCC(q, u)] dq = \mathop{E}_{u} (\tilde{z}), \tag{16}$$

we conclude that the risk neutral agency should again select Q_s^* to satisfy

$$E[MDF(Q_s^*) - MCC(Q_s^*, u)] = 0, \tag{17}$$

i.e., that Q_s^* where MDF equals expected MCC. This result is independent of $dF(u)$ and of the particular form of uncertainty in MCC.

When a tax policy is used, the quantity of pollution abated with a given tax rate P_T is a random variable

$$\tilde{Q}_s = MCC^{-1}(P_T, u). \tag{18}$$

Writing the expression for expected welfare gain

$$EWG = \mathop{E}_{u} \int_0^{MCC^{-1}(P_T, u)} [MDF(q) - MCC(q, u)] dq, \tag{19}$$

we realize that this source of uncertainty compounds the uncertainty of producer surplus for a given Q_s, and that in general the frequency distribution of u and the form in which it enters MCC will affect both the optimal tax, P_T^*, and the expected welfare gain, EWG. To illustrate this conclusion, consider first the generalization of Fig. 3, which depicts a linear MDF and MCC, and u enters MCC additively. While an additive

----

[5] This is not to say they are equivalent in other respects.

stochastic error term facilitates the exposition, it has considerable economic content since factor price variations in inputs subject to a Leontief production technology would produce additive errors. Specifically, we assume

$$\text{MCC} = \alpha + \beta Q_s + u, \qquad E(u) = 0. \tag{20}$$

Optimal P_T can be derived by solving

$$\frac{d}{dP} \underset{u}{E} \int_0^{\tilde{Q}(P)} [(a - bq) - (\alpha + \beta q + u)]dq = \frac{d}{dP} \underset{u}{E} (\tilde{z}) = 0, \tag{21}$$

where

$$\tilde{Q}(P) = (1/\beta)[P - \alpha - u]. \tag{22}$$

Differentiating, and substituting (22) into (21) we find the optimum condition to be

$$(P^* - \alpha)/\beta = \underset{u}{E} \tilde{Q}(P^*, u) = (a - \alpha)/(b + \beta), \tag{23}$$

which means that P^* should be selected such that MDF equals *expected* MCC at the optimal level of abatement. Note the similarity to our general result for quantity (i.e., standards/auction) setting policy, but note also a difference: substituting into (16) an MCC with an additive error term, the expected welfare gain for the optimal standards or auction policy is

$$E (\tilde{z}) = -\frac{1}{2} \frac{(a - \alpha)^2}{b + \beta}. \tag{24}$$

Substituting (23) into the expectation in (21), we find that with a tax policy, the expected welfare gain for a tax $[E(\tilde{z})]$ is

$$E (\tilde{z}) = (a - \alpha) \underset{u}{E} \tilde{Q} - \left(\frac{b + \beta}{2}\right) E \tilde{Q}^2 - \underset{u}{E} u\tilde{Q},$$

$$= -\frac{1}{2} \frac{(a - \alpha)^2}{(b + \beta)} - \frac{b - \beta}{2\beta^2} \underset{u}{E} u^2, \tag{25}$$

and since $\underset{u}{E} \tilde{Q}(P^*, u)$ for the tax policy equals Q^* for the standards policy,

$$E(\tilde{z}) - E(\tilde{z}) = E u^2[(b - \beta)/2\beta^2]. \tag{26}$$

This is a generalization of the result in (5). It indicates that when the slope of the expected MCC is steeper than MDF, a tax policy is preferred to quantity restrictions, and vice versa. Note that for given MCC and MDF, the difference in performance between the two policies is proportional to the variance of u,[6] but the choice of optimal policy is independent of it.

A multiplicative disturbance term in the marginal control cost function is interesting to contemplate since, like a heteroscedastic error term, its variance increases with marginal costs. This is not unrealistic as technology for high cost, low abatement technology is no doubt subject to greater uncertainty. When MCC contains a multi-

---

[6] If MCC were nonlinear, it would depend on higher moments of $dF(u)$ as well.

plicative error term, i.e.,

$$M\tilde{C}C = (\alpha + \beta q)u, \qquad E(u) = 1,$$
$$\tilde{C}(q) = \alpha uq + \tfrac{1}{2}(\beta u)^2 q^2, \tag{27}$$

our conclusions change. The expected welfare gain of standards or an auction is

$$\mathop{E}_{u}(\tilde{z}) = \mathop{E}_{u} \int_0^Q [(a - bq) - (\alpha + \beta q)u]dq, \tag{28}$$

and maximizing $\mathop{E}_{u}(\tilde{z})$ with respect to Q still yields the familiar first-order condition

$$Q_s^* = (a - \alpha)/(b + \beta). \tag{29}$$

Again, the quantity should be selected where MDF equals *expected* MCC.

However, when a tax policy is used, this result should be modified. Differentiating (28) with respect to P where the upper bound of the integral is

$$\tilde{Q}(P, u) = (P - u\alpha)/u\beta \tag{30}$$

yields the first-order condition

$$\mathop{E}_{u}[(a/u) - \alpha] = \mathop{E}_{u}\{\tilde{Q}(P, u)[(b/u) + \beta]\}, \tag{31}$$

or

$$\mathop{E}_{u}\frac{\partial Q}{\partial P}\,MDF(Q) = \mathop{E}_{u}\frac{\partial Q}{\partial P}\,MCC(\tilde{Q}, u), \tag{32}$$

which is different, of course, from the quantity conditions. Nevertheless, we can still maintain that the economic meaning of the optimal condition is equating *expected* marginal damage to *expected* marginal cost, but in this case, with respect to price rather than quantity.

In a manner similar to the additive case, we can now compare the maximized expected welfare gain under a tax policy and a standards/auction policy. Not surprisingly, the result depends again on parameters of MCC and MDF,

$$\mathop{E}_{u}(\tilde{z}) - \mathop{E}_{u}(\tilde{z}) = \frac{a\beta + \alpha b}{2\beta}\left[\mathop{E}_{u}\tilde{Q} - \frac{a - \alpha}{b + \beta}\right] = \frac{a\beta + \alpha b}{2\beta}[\mathop{E}_{u}\tilde{Q} - Q^*], \tag{33}$$

where $E_u\tilde{Q}$ is expected quantity abated under the optimal tax, and Q^* is the optimal quantity abated under a standards/auction scheme. Since $\tilde{Q} = (P - \alpha u)/\beta u$, the choice of optimal policy depends on $dF(u)$.

We conclude that when MCC is random, the effects of taxes and quantity controls of pollution will differ, and that the choice of the optimal instrument depends on (a) the parameters of MDF and *expected* MCC, (b) the particular way in which the random element u enters the MCC, and (c) the frequency distribution of u.

Admittedly, the difference in the expected welfare gains is more cumbersome to calculate for policy analysis than the simple case where the disturbance is additive and we need only measure the relative elasticities, or slopes and the variance of u. Nevertheless, we believe such calculations are possible and instructive. Policy analysts definitely have some knowledge about the parameters of MDF and MCC [point (a)], since this is presumably the basis for current decision making. By the types of economic rationales offered here for additive and multiplicative disturbance terms, we feel that policy analysts can distinguish the types of uncertainty [point (b)] most relevant to the case at hand. Finally, as for the frequency distribution of u [point (c)], our suggestion is simply to test for sensitivity using several alternative distributions as the policy implications may turn out to be quite robust with respect to distributional changes in u.

III. UNCERTAINTY AT THE FIRM LEVEL AND RISK AVERSION

The results of the previous section hold for cases in which (1) uncertainty exists only at the agency level or (2) uncertainty exists both at the agency and firm level and firms are risk neutral. The first situation is obvious, but the second involves the critical question of whether the expected marginal control cost function evinces a behavioral relationship as well as the social costs of abatement. Throughout this paper, we treat the private and social costs of abatement as equivalent. If the expected MCC does in fact measure private abatement costs, we will face the question of whether it describes a behavioral relation. For the risk neutral firm, it is well known that decisions are based on expected marginal costs. Therefore, under risk neutrality, the expected marginal control cost curve facing the agency is nothing more than the summation of firms' expected marginal cost functions for abatement.

For the risk averse firm, the agency's expected marginal control cost curve describes the relevant expected social costs; however, it does not represent a behavioral relationship. The introduction of risk aversion in firm behavior complicates our results and adds another interesting asymmetry. We introduce risk aversion in the firm's behavior by hypothesizing that firms behave as if they possess a well-behaved utility of profits function, $U(\pi)$ with $U'(\pi) > 0$, $U''(\pi) < 0$, and act to maximize the mathematical expectation of utility.[7] Thus, when the firm knows only the frequency distribution of P_T when making its pollution decision it is assumed to maximize with respect to Q the following expected utility of profits:

$$\underset{\tilde{P}_T}{E}\ U[\tilde{P}_T Q - C(Q)]. \tag{34}$$

When P_T is certain but $C(Q)$ is uncertain, we replace $C(Q)$ by the random cost relationship $C(Q, u)$. The firm maximizes

$$\underset{u}{E}\ U[P_T Q - C(Q, u)]. \tag{35}$$

Analysis of the first- and second-order conditions of these maximization problems usually reveals that the firm will not operate where its expected marginal cost equals the expected price as we established for the risk neutral firm.

When P_T is uncertain, Sandmo has shown that the firm will produce the output, where $\bar{P}_T > C'(Q)$. In a similar manner, when marginal costs are stochastic, the first-order condition

$$EU'(\tilde{\pi})[P_T - C'(Q, u)] = 0, \qquad \tilde{\pi} = P_T Q - C(Q, u), \tag{36}$$

implies that under risk aversion,

$$P_T > \underset{u}{E}\ C'(Q, u). \tag{37}$$

[7] See [7] for a model of the competitive firm facing an uncertain price. This assumption has been widely used in other studies of the economics of the firm under uncertainty.

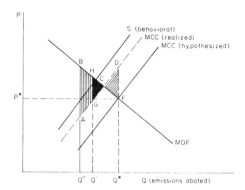

FIG. 3. Uncertain marginal control costs and risk aversion.

As a consequence of the fact that $P > C'(Q, u)$ and $\bar{P} > C'(Q)$, we can no longer use the expected MCC curve to evaluate both the social cost of pollution abatement and to determine the supply response of the polluting firm to a given policy.

For example, in Fig. 3, MDF and the expected MCC curve still have the same normative meaning as in Section II above, but now we represent the firm's behavior by the "supply" curve S. In order to maximize the expected welfare, the agency concludes that price P^* and quantity Q^* are optimal. Following the exposition in Fig. 2, we assume that the actual MCC differs from the expected MCC. Under risk neutrality, the tax P^* would have resulted in the abatement of Q' with welfare loss of CHG; however, risk aversion lead to a reduction of Q'' and a welfare loss of ABC. Note however, that the quantitative restriction (either standards or an auction) resulted in a loss of CDF, the same loss as under risk neutrality. In sum, the asymmetry between taxes and quantitative restrictions for uncertain MCC still holds. The expected welfare loss in Eq. (26) between taxes and quantitative standards would, however, need to be altered for the effects of risk aversion.

IV. SOME CAVEATS AND ADDITIONAL INSIGHTS

Up until now the careful reader might ask why the analysis has been concluded in terms of three policy instruments when the affects of uncertainty have so far fallen into two dichotomous groups—price controls (taxes) and quantity controls (standards and auctions). The answer is that despite the above, standards and auctions are not identical with respect to uncertainty. One of the primary motivations for auctions in place of standards is that they reduce the information requirements of the agency since for standards to be applied efficiently, the agency must know the individual firm's expected marginal cost function for abatement. Obviously, the auction reveals this information in a much cheaper manner to the agency. As a consequence, the auction avoids the uncertainty attendant to the more complex informational requirements of standards.

While auctions have less informational uncertainty attached to them than standards, it is conceivable that under certain conditions some offsetting welfare effects may exist. In the case where the sole source of uncertainty is being induced by the auction via large price variation, the expected auction price will lie above the expected marginal control costs for the risk averse firms. If the difference between the auction price and marginal control costs (a type of risk premium) is merely a pecuniary payment to risk, then expected MCC continues to measure expected social abatement costs and the auction generates no attendant welfare loss. However, if the firm expends real resources to avoid this risk premium,[8] then social costs may be increased in an auction relative to more certain government policies such as taxes or standards. We feel that in most cases the welfare effects induced by the auction's uncertain price are likely to be small relative to the greater informational uncertainty attached to standards. On the other hand, the welfare effects of large variations in the auction price would appear to offer some explanation for why auctions could not replace standards on offshore oil production, nuclear reactor design, and other cases where the price for tickets might be subject to large variations.

V. SUMMARY AND POLICY IMPLICATIONS

This paper reaches the following three major conclusions regarding differential welfare effects between taxes, standards, and auctions.

(1) Uncertainty in the marginal damage function has absolutely no effect on the choice between the three policy instruments.

(2) Uncertainty in the marginal control cost function will yield different expected welfare losses between taxes or quantitative restrictions (standards or auctions) depending on the variance of the stochastic error term and the slopes or elasticities of the MDF and MCC functions.

(3) Even under risk aversion, this asymmetry still holds, however, now the welfare loss also depends on the degree of risk aversion.

In conclusion, in situations of uncertain marginal control costs where the marginal damage function tends to be very price elastic, such as believed to occur for SO_2 emissions, a strong case can be made for taxes. On the other hand, where the marginal damage function is very inelastic, quantitative restrictions of an auction or standards type appear desirable.

In cases where quantity restrictions are appropriate, auctions and standards are not equivalent. Unlike auctions, standards introduce added informational uncertainty, which seems likely to dominate possible welfare losses due to wide variations in the auction price.

REFERENCES

1. W. J. Baumol, On taxation and the control of externalities, *Amer. Econ. Rev.* **62**, 307–322 (1972).
2. W. J. Baumol and W. E. Oates, The use of standards and pricing for protection of the environment, *Swed. J. Econ.* **73**, 42–54 (1971).
3. J. H. Dales, "Pollution, Property and Prices," Univ. of Tronto Press, Toronto (1968).
4. J. M. Griffin, An econometric evaluation of sulphur taxes, *J. Polit. Econ.* **82**, 669–688 (1974).
5. A. V. Kneese and B. T. Bower, "Managing Water Quality: Economics, Technology, Institutions," Johns Hopkins Press, Baltimore (1968).
6. Abba P. Lerner, The 1971 report of the President's Council of Economic Advisers: Priorities and efficiency, *Amer. Econ. Rev.* **61**, 527–530 (1971).
7. Agnar Sandmo, On the theory of the competitive firm under price uncertainty, *Amer. Econ. Rev.* **61**, 65–73 (1971).

[8] For example, to avoid the losses associated with wide variations in ticket prices, the firm might install more flexible abatement equipment providing a flatter short-run cost function over a wider range. Such a technology may not, however, be of least cost at the abatement level corresponding to the standard.

[37]

Journal of Public Economics 5 (1976) 193–208. © North-Holland Publishing Company

EFFLUENT CHARGES AND LICENSES UNDER UNCERTAINTY*

Marc J. ROBERTS and Michael SPENCE**

Harvard University, Cambridge, MA 02138, U.S.A.

Received September 1974, revised version received November 1975

This paper is concerned with pollution control when the regulators are uncertain about firms' cleanup costs. Under these circumstances, the regulatory authority can reduce expected total social costs (consisting of damages from pollution and cleanup costs) below the levels achievable with either effluent fees or licenses. The reduction is achieved by the use of licenses supplemented by an effluent subsidy and a finite penalty, when effluents are below or above the levels permitted by licenses. The mixed system retains the property of efficiently distributing cleanup among firms.

1. Introduction

The purpose of this paper is to explore, in the context of a simple model, what kind of policy might be used to control pollution, when the regulatory authority is uncertain what the actual costs of pollution control will be. In posing the problem as we do, we are rejecting the idea that the government can iteratively 'feel out' the 'optimum' by successively announcing and revising its policies in light of the responses of waste sources. Much of the investment that will be made in any pollution control program will take several years to plan and complete and will be largely irreversible once in place. Thus the response to all subsequent policies will be heavily dependent on previous history. Indeed the cycle time may be so great as to prevent convergence, since the 'correct' solution will be constantly changing. Given these circumstances, we have chosen to explore the once-and-for-all problem, where the government seeks to achieve a comparative static maximum in expected utility terms.

The principal point of the paper is that a mixed system, involving effluent charges and restrictions on the total quantity of emissions via marketable licenses, is preferable to either effluent fees or the licenses used separately.[1]

---

*This work was supported by National Science Foundation Grant GS–39004 and by the Ford Foundation, Office of Resources and Environment. The authors are grateful to Robert Dorfman, Charles Untiet and the referees for helpful comments.

**Department of Economics, Stanford University, Stanford, CA 94305, U.S.A.

[1]Some of the previous treatments of effluent fees and marketable licenses include Kneese and Bower (1968), Jacoby, Schaumberg, and Gramlech (1972) and Montgomery (1972).

This follows because a mixed system permits the implicit penalty function imposed upon the private sector to more closely approximate the expected damage function for pollution at each level of total waste output.

In setting up this model, we are fully conscious of the differences between the formal structures we will use and real situations, and we will call attention to some of them as we proceed. The point of this exercise is not to 'prove' one or another approach 'better.' Rather, by exploring and manipulating some simplified conceptualizations, we hope to develop some insights and formulations which will prove to be useful in formulating policy.

The problem is posed as one of choosing a control scheme so as to minimize expected total social costs, these being the sum of (1) expected damages from pollution and (2) cleanup costs. In order to actually implement any policy, the regulatory authority must quantify its uncertainty about cleanup costs in the form of subjective probabilities. Given these probabilities, the calculation of the optimal parameters for the sort of mixed scheme we will develop is sufficiently straightforward that we believe it could be made even with limited analytical resources.

Effluent charges and marketable licenses have the virtue of inducing the private sector to minimize the costs of cleanup. But in the presence of uncertainty, they differ in the manner in which the ex post achieved results differ from the socially optimal outcome. Effluent charges bring about too little cleanup when cleanup costs turn out to be higher than expected, and they induce excessive cleanup when the costs of cleanup turn out to be low. Licenses have the opposite failing. Since the level of cleanup is predetermined, it will be too high when cleanup costs are high and too low when costs are low.

Given that effluent charges and license outcomes deviate from the optimum in opposite ways, which kind of imperfection is preferable? It turns out, plausibly enough, that the answer depends upon the curvature of the damage function. When the expected damage function is linear, an effluent charge equal to the slope of the damage function always leads to optimal results, regardless of what costs turn out to be, while licenses do not. On the other hand, if *marginal* damages increase sharply with effluents, licenses are relatively more attractive and yield lower expected total costs than the fee system.

Licenses and effluent charges can be used together further to reduce expected total costs. Each can protect against the failings of the other. Licenses can be used to guard against extremely high levels of pollution while, simultaneously, effluent charges can provide a residual incentive to clean up more than the licenses required, should costs be low.

In what follows, the model is described and the mixed effluent fee license scheme set forth and analyzed. In an appendix, we argue that one can come arbitrarily close to minimum expected total costs with the use of multiple licenses supplemented by a carefully constructed schedule of effluent fees.

2. Notation

To simplify the exposition we assume all waste dischargers have the same impact on ambient conditions at the one point we monitor. We will not consider multiple monitoring points, or substances, though the analysis could be generalized in that direction.[2] Thus we can use a single variable, x, to indicate both the total pollution discharged and the resulting quality of the environment. Damages from pollution are measured in dollars. Expected total damages are denoted by $D(x)$. There are, of course, significant uncertainties associated with damages. And if risk aversion were assumed, the monetary equivalents of the damages associated with various policies would rise. The analysis to follow, which focuses upon costs and cleanup, could be amended to account for risk aversion. For expositional clarity, we will deal only with expected damages.

The current level of output of the pollutant of firm i is $\bar{x}_i$. The costs of cleanup for firm i are uncertain from the point of view of the regulators. This uncertainty is summarized by a random variable, ϕ. The costs of cleanup for firm i are stated as a function of its output of pollution, x_i, and the random variable ϕ, and are denoted by $c^i(x_i, \phi)$. These costs represent reductions in total profits. Adjustment in cleanup may be accompanied by changes in the levels of outputs and inputs of the firm. Our assumption here is that this reduction in profits accurately reflects the social cost of cleanup, which can be shown to be correct if markets are competitive.[3] By definition, when there is no cleanup, $x_i = \bar{x}_i$ and $c^i(\bar{x}_i, \phi) = 0$.

Total cleanup costs, $c(x, \phi)$, are simply the sum of the individual firm costs. Again we can simply use ϕ to parameterize our uncertainty. However, in what follows, whenever we write c, we do so only to refer to circumstances where the cleanup is distributed among firms in a cost minimizing manner, so that by definition,

$$c(x, \phi) = \sum_i c^i(x_i, \phi),$$

---

[2] Montgomery (1972) considers the problem of multiple points of concern.

[3] The argument is as follows. Let $P(q)$ be the inverse demand for the firm, and $d(q, x)$ its costs. The effluent charge is e. The surplus generated by the market is

$$T = \int_{p}^{q} P(s)\, ds - d(q, x) - ex.$$

Differentiating with respect to x, we have

$$\frac{dT}{dx} = (P - d_q)\frac{dq}{dx} - (d_x + e).$$

A profit maximizing firm will set $d_x + e = 0$. At that point $dT/dx = 0$ only if either $P = d_q$ (price equals marginal cost – the industry is competitive) or $dq/dx = 0$. The latter occurs when $d_{xq} = 0$. Therefore, when a competitive industry maximizes profits or costs or profit losses, the social optimum is achieved. But if the firm has market power $p > d_q$, there will be too much or too little cleanup depending on the sign of dq/dx.

196   *M.J. Roberts and M. Spence, Effluent charges under uncertainty*

where $x = \sum_i x_i$, and for all i and j,

$$c_x^i(x_i, \phi) = c_x^j(x_j, \phi).$$

The following assumptions are carried throughout: $D''(x) > 0$, so that $D(x)$ is convex, and $c_x < 0$, $c_{xx} > 0$; marginal cleanup costs increase at an increasing rate. The random variable ϕ represents 'states of the world.' It simply captures all the relevant uncertainty about cleanup costs. It can be thought of as an exhaustive labeling of the possible cleanup cost functions for all polluters. The reader may find it easier to think in terms of a large, but finite, exhaustive list. However, to facilitate the following analysis, we will assume that $c_\phi > 0$ and that $c_{x\phi} < 0$. This means that as ϕ shifts, both absolute and marginal costs shift in the same directions for all values of x. In particular, members of the family of aggregate costs do not cross.

The regulatory authority's decision problem is to choose a pollution control scheme to minimize expected total costs. Their subjective distribution for ϕ is represented by $f(\phi)$. Expected total costs are

$$T = \int [D(x) + c(x, \phi)] f(\phi) \, d\phi = E[D(x) + c(x, \phi)].$$

In general, x will be a function of ϕ. The function will vary with the scheme being used for controlling pollution. It is assumed that firms know or can find out their cleanup cost functions. The uncertainty therefore attaches to the regulatory authority.

3. Controlling via mixed effluent charges and licenses

The control mechanism we want to put forward has three components. First there is a finite set of transferable licenses that are issued by the regulatory authority, and are bought and sold in a market. The quantity of licenses is l. The number of licenses held by firm i is denoted by l_i. Second, there is a unit effluent subsidy, denoted by s. It is paid to any firm whose license holdings, l_i, exceed its emissions, x_i. Thus if $l_i > x_i$, the firm receives $s(l_i - x_i)$. Finally, if a firm's emissions exceed its holdings of licenses, so that $x_i > l_i$, then it is assessed a per unit penalty of p, or a total penalty of $p(x_i - l_i)$. The three components then are licenses, l, an efficient subsidy, s, and an effluent penalty, p.

We want to demonstrate that this approach has several properties. First, it allocates cleanup among polluting firms efficienctly.[4] Second, it is preferable to either a pure effluent fee or a pure license scheme. Expected total costs

---

[4]If there is just one polluter, one could set a nonlinear effluent charge equal to marginal damages, $D'(x)$. This would lead to the optimum. But when there is more than one polluter, a nonlinear effluent charge is inconsistent with either decentralization or cost minimization, and possibly both.

(cleanup and damages from pollution) are lower. Third, the system operates as if there were just one polluting firm confronted with a piecewise linear penalty function with one kink in it. This is demonstrated below.

The economic rational for this scheme is the following. One wants to limit effluents; this is done by issuing marketable licenses. But if cleanup costs have been significantly overestimated, one wants a residual incentive to cleanup. This is provided by the subsidy, s. On the other hand, if cleanup costs turn out to be very high, one wants an escape valve from the restriction imposed by the licenses. This escape valve is provided by having a finite penalty, p, for exceeding levels of effluents permitted by licenses. It is assumed that $p \geq s$.

Formally, the functioning of the system is represented as follows. Let q be the market price of the licenses. It is determined as part of the equilibrium in the market for licenses. The total costs for firm i consist of (1) cleanup costs, (2) license costs, and (3) penalties or subsidies when applicable. These costs are

$$c^i(x_i, \phi) + q l_i - s(l_i - x_i) \quad \text{if} \quad x_i \leq l_i, \tag{1}$$

and

$$c^i(x_i, \phi) + q l_i + p(x_i - l_i) \quad \text{if} \quad x_i \geq l_i. \tag{2}$$

The firm minimizes these by selecting x_i and l_i appropriately. In addition, in an equilibrium,

$$\sum_{i=1}^{N} l_i = l.$$

We turn now to the properties of the equilibrium. Suppose first that $q < s$. Then from (1) every firm could reduce costs indefinitely by buying licenses. This is clearly inconsistent with equilibrium in the license market. Thus q cannot be less than s. Now suppose that $q > p$. Then from (2), every firm would set $l_i = 0$, and this is inconsistent with equilibrium in the license market. Therefore, q cannot exceed p. The subsidy s and the penalty p place bounds on the equilibrium value of q: $s \leq q \leq p$.

The next step is to show that $c_x^i(x_i, \phi)$ is always equal to $-q$. Suppose first that $s = q$. Then the firm will set $l_i \geq x_i$ (in fact it is indifferent about the level); and then set $c_x^i(x_i, \phi) = -s = -q$. Next suppose $s < q < p$. Then from (1) and (2) firm i will set $x_i = l_i$. Thus its costs are

$$c^i(x_i, \phi) + q x_i.$$

These are minimized when $c_x^i(x_i, \phi) + q = 0$. Finally if $q = p$, the firm will set $l_i \leq x_i$, and then minimize with respect to x_i by setting $c_x^i(x_i, \phi) + q = 0$. Thus in all possible cases, $c_x^i(x_i, \phi) + q = 0$. This fact has the immediate implication that $c_x^i(x_i, \phi) = c_x^j(x_j, \phi)$ for all i and j, so that cleanup is efficiently distributed among polluters.[5] In addition q is bounded by the effluent subsidy s and the penalty p.

Since marginal cleanup costs are minimized, the condition

$$c_x(x, \phi) + q = 0 \tag{3}$$

is always satisfied.

The remaining question is what determines the levels of q and x? If $s < q < p$, then $x_i = l_i$ for all i, and hence $x = l$. Condition (3) will be satisfied if

$$s < -c_x(l, \phi) < p. \tag{4}$$

Inequality (4) will hold for some intermediate range of costs of cleanup. If cleanup costs are very high, then q will be driven up to the level of the penalty p. At that point, effluents will exceed licenses: $x > l$. The equilibrium condition is

$$c_x(x, \phi) + p = 0. \tag{5}$$

Finally if costs are low, so that $c_x(l, \phi) + s < 0$, then $x < l$ and $q = s$. The level of effluents actually achieved will be given by

$$c_x(x, \phi) + s = 0. \tag{6}$$

In summary: (1) if $c_x(l, \phi) + s > 0$, then $c_x(x, \phi) + s = 0$ and $q = s$; (2) if $s < -c_x(l, \phi) < p$, then $x = l$ and $q = -c_x(l, \phi)$; and (3) if $c_x(l, \phi) + p < 0$, then $c_x(x, \phi) + p = 0$ and $q = p$.

The interesting feature of the mixed effluent–license is that it produces levels of the effluents, conditional on costs, that reproduce exactly the effluents that would occur if (1) the polluting firms were merged (and made cleanup decisions centrally) and (2) they faced a piecewise linear penalty function of the form,

$$P(x) = sx + p \text{ Max } (x - l, 0).$$

If the firms collectively were to minimize the sum of penalties and cleanup costs, $P(x) + c(x, \phi)$, they would act as follows: if $s < -c_x(l, \phi) < p$, they would set $x = l$; if $-c_x(l, \phi) < s$, they would set $c_x(x, \phi) + s = 0$; and if $-c_x(l, \phi) > p$, they would set $c_x(x, \phi) + p = 0$. But this is exactly what the decentralized system does.

The pure efficient fee and pure license systems are special cases of the mixed system. The pure effluent fee is obtained by setting $s = p$, at which point the

[5]Note that $c_x(x, \phi) = c'_x(x_i, \phi)$, for all i, when x is distributed among polluters in a cost minimizing manner.

level of l becomes irrelevant. The implicit penalty function is then linear. If $s = 0$ and $p = +\infty$, then we have a pure license system. It is not therefore surprising that the more flexible mixed system can achieve lower expected total costs.

The mixed system implicitly approximates the expected damage function by a piecewise linear penalty (see fig. 1). The same point can be seen in the context

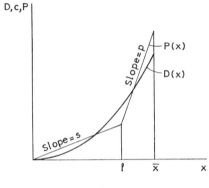

Fig. 1

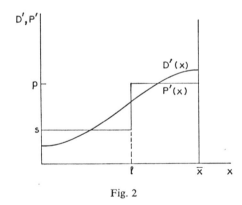

Fig. 2

of the marginal damages (see fig. 2). The mixed effluent–license system approximates the marginal damage function with a step function.

It is worth noting that the implicit penalty function $P(x)$, does *not* correspond exactly to the payments by firms for licenses, plus or minus penalties and subsidies. The actual payments depend upon the parameter ϕ that determines costs, and not just upon x, the final level of effluents. But if we plot ex post payments as a function of effluents, the result is as in fig. 3.

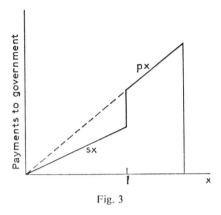

Fig. 3

4. The regulatory authority's optimizing problem

The decision variables for the regulators are s, p, and l. The objective is to minimize expected total costs, consisting of damages from pollution and cleanup costs. For given levels of s, p and l, there will be two critical levels of the cost determining parameter ϕ. The first, ϕ_1, is the level of cost such that

$$c_x(l, \phi_1) = s = 0. \tag{7}$$

Here the marginal cleanup costs are just equal to the effluent subsidy when $x = l$. The second value, $\phi_2 > \phi_1$, is defined by

$$c_x(l, \phi_2) + p = 0. \tag{8}$$

Here costs are almost high enough to cause the system to have effluents exceed licenses.

Let $[0, b]$ be the support of the distribution $f(\phi)$. We define $x_1(\phi, s)$ and $x_2(\phi, p)$ by

$$c_x(x_1(\phi, s), \phi) + s = 0,$$

and

$$c_x(x_2(\phi, p), \phi) + p = 0.$$

Expected total costs are

$$T(s, p, l) = \int_0^{\phi_1} [D(x_1(\phi, s)) + c(x_1(\phi, s), \phi)] f(\phi) \, d\phi$$

$$+ \int_{\phi_1}^{\phi_2} [D(l) + c(l, \phi)] f(\phi) \, d\phi$$

$$+ \int_{\phi_2}^{b} [D(x_2(\phi, p)) + c(x_2(\phi, p), \phi)] f(\phi) \, d\phi.$$

These expected total costs are minimized when the partial derivatives, T_s, T_p and T_l, are zero, or when the following conditional expectations hold:

$$E\left(\frac{D'(x_1)-s}{c_{xx}(x_1, \phi)}\middle| \phi \leq \phi_1\right) = 0, \tag{9}$$

$$E(D'(l)+c_x(l, \phi)|\phi_1 \leq \phi \leq \phi_2) = 0, \tag{10}$$

$$E\left(\frac{D'(x_2)-p}{c_{xx}(x_2, \phi)}\middle| \phi \geq \phi_2\right) = 0. \tag{11}$$

With perfect information about costs, the authority would set

$$D'(x)+c_x(x, \phi) = 0, \tag{12}$$

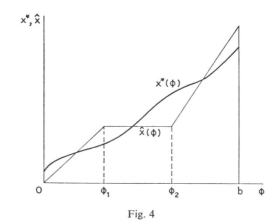

Fig. 4

for all ϕ. Let the optimal schedule of effluents, defined by (12), be $x^*(\phi)$. Let $\hat{x}(\phi)$ be the effluent levels achieved with the optimal mixed system described above. The relationship between $x^*(\phi)$ and $\hat{x}(\phi)$ is depicted in fig. 4. The schedule $\hat{x}(\phi)$ crosses $x^*(\phi)$ three times, once in each interval.

The optimizing conditions, (9) through (11), are simply conditions for optimal pure effluent fees or licenses on each of the three intervals. For example, eq. (9) is the condition for s to be the optimal pure effluent fee assuming costs vary only on the interval $[0, \phi_1]$. A pure effluent fee schedule crosses the optimal schedule once. Hence, the mixed schedule crosses $x^*(\phi)$ once in each of three intervals. Notice that pure effluent fees induce excessive cleanup when costs are low and too little cleanup when costs are high. This occurs because at low levels of pollution the effluent fee exceeds marginal damages, and conversely. The pure license scheme has the opposite property. It is insensitive to variations in clean-up costs.

The superiority of the mixed scheme is simply a result of its ability to better approximate the optimal relationship between pollution levels and damages. The exception occurs when the damage function is linear. In that case, $\phi_1 = 0$, $\phi_2 = b$ and $p = s$. The pure effluent fee system is optimal.

5. Expected gains from using a mixed system

It is not possible in a short paper to comment extensively on the quantitative benefits of the mixed scheme. However, one can isolate the circumstances under which it is likely to yield significant gains. There are two conditions which make the mixed schemed attractive. First, the marginal damages must vary consider-ably with total effluents. Otherwise the pure effluent fee performs quite well. Second, there must be significant uncertainty about the cleanup costs. Otherwise, the pure license scheme performs well. It is perhaps worth noting that when

Table 1

Control scheme	Expected total costs	Percentage above the optimum
Optimum (also mixed system)	12.416	0
Pure effluent fee	20.6	66
Pure licenses	18.25	46

marginal cleanup costs do not vary greatly with quantity, an effluent fee system performs poorly even with small amounts of uncertainty. The reason is that actual levels of cleanup may vary wildly with small shifts in the cost function.

The following numerical example illustrates the potential benefits of the mixed system. It assumes there is a threshold level of pollution, $\bar{l}$, below which marginal damages are one, and above which they are six. Costs are assumed to have the form $(\phi/2)(\bar{x}-x)^2$, where ϕ takes on the values 0.12 and 2.0 with probabilities of one half. A mixed system yields the optimum for this kind of damage function. Table 1 summarizes the results for the various control schemes.

6. Conclusions

When the regulatory authority is uncertain about pollution control costs, the usefulness of monetary incentives to decentralize pollution control decisions is limited by our inability to pick the correct price. That price should be equal to marginal damages and thus depends upon the level of pollution. But it is not known exactly what pollution will be as a function of price because control costs are known only imperfectly. With a nonlinear damage function, and uncertain irreversible costs, we would like to find some way of confronting each firm with incentives to cleanup that in fact depend upon marginal damages, and hence on total waste output. The combination of the license scheme with subsidies and penalties permits one simultaneously to ensure that all firms face the same marginal costs, but to have that cost vary (within limits) depending on what the aggregate costs of cleanup actually turn out to be. The level of pollution also varies with the aggregate cleanup costs.

The authority has three parameters to manipulate: the subsidy, the penalty and the stock of licenses. The authority knows that pollution will equal the stock of licenses provided the market price turns out to be between the subsidy and the penalty. The subsidy provides a residual incentive for firms to clean up even more when costs are low. The finite penalty provides an escape valve in case costs are very high. The aggregate damage function is approximated by a piecewise linear penalty function. But once the equilibrium in license prices is established, each firm effectively faces a linear penalty function whose slope is the price of the license. As a result, marginal cleanup costs are equalized and total cleanup costs are minimized.

How useful is this formulation in the real world? First, we do not believe that limiting our attention to regions of increasing marginal damages is a major practical limitation. There are real cases in which marginal damages may decline – adding more waste to a river which is already an open sewer may have few environmental costs. But in general, even damage functions which exhibit such regions also often appear to be characterized by other regions in which marginal costs are increasing. For example, as the organic material in a river increases, and dissolved oxygen levels decline, we appear to move successively through several thresholds as we lose additional species and human uses. And, intuition suggests that output controls are more likely to be favorable in a region of increasing, rather than decreasing, marginal damages.

In practice, the scheme amounts to setting an ambient target (similar to the ambient standards widely used today) and working back to the magnitude of the discharges allowed by that constraint. Then the regulatory authority has to develop some notion about marginal damages in the regions above and below that point in order to set the subsidy and the penalty fee. Even if the regulatory authority does not quantify its uncertainty and compute an optimal schedule, the rough and ready approach should lead to a reasonable set of policies. Afterall, in a second-best world with imperfectly maximizing waste sources, the formal optimality of a policy scheme is not necessarily proof of what its actual impact will be.

Like any decentralized approach to pollution control, our scheme has certain serious limitations. It will not provide for efforts to act directly on the environment as opposed to on a waste source. Nor does it ensure that all economies of scale in treatment will be exhausted unless waste sources agree to appropriate joint ventures among themselves. We have also not discussed what should be done in the face of natural variations in climate which make it uncertain what damages will in fact result from any waste discharge.[6] All this suggests that a good deal of detailed work would be required to develop a viable set of policies and institutions for any specific circumstances. For example, could we vary policy seasonally or with actual natural conditions?

In theory, we would want a separate system of licenses to control each polluting substance we are concerned with at each geographic point of interest. Since administrative costs will rise with the complexity of the entire scheme, at some point we will need to make a (perhaps crude) compromise between the costs and benefits of additional elaboration and fine-tuning of the system. Note too that we have to construct our markets such that each has enough participants to ensure relatively competitive functioning. Nevertheless, even viewed as a practical measure designed to move us into a better, if not the best, position, we believe the mixed scheme we have proposed has significant merit. Perhaps the next important step is to consider how to set the penalty function in the presence of risk-aversion relative to damages.

Appendix: A generalized decentralization proposition

In the body of the paper, it was argued that expected total costs could be reduced by the use of both licenses and effluent fees, while maintaining the property of efficiently distributing cleanup among polluters. It was pointed out that the system operated as if the firms made a centralized decision against a penalty function with two facets and one kink. We want to argue now that if one is prepared to introduce more than one kind of license, the penalty function can be made to approximate any convex damage function arbitrarily closely. More precisely, by the use of multiple licenses, the system can be made to efficiently distribute costs and implicitly respond to a penalty function with as many kinks as there are types of licenses.

Let l^j be the number of licenses of type j. Assume that $l^0 \leq l^1 \leq l^2 \leq \ldots \leq l^n$ and that $l^0 = 0$. Let $s_0, s_1, \ldots, s_{n+1}$, be an increasing sequence of numbers with $s_0 = 0$. Define a penalty function $P(x)$, in the following way:

$$P(x) = \sum_{j=0}^{n} (s_{j+1} - s_j) \, \text{Max} \, (x - l^j, 0). \tag{A.1}$$

The function $P(x)$ is depicted in fig. 5. It is piecewise linear with kinks at $l^1, \ldots, l^n$. The slopes of the facets are $s_1, \ldots, s_{n+1}$, respectively.

The question then, is whether a system of licenses and effluent subsidies and penalties can induce the firms to act collectively as if the penalty function $P(x)$

---

[6]For a discussion of some of those issues, see Roberts (1975).

M.J. Roberts and M. Spence, Effluent charges under uncertainty   205

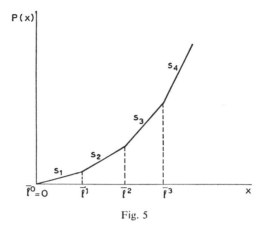

Fig. 5

had been imposed. Let q_j ($j = 1, \ldots, n$) be the market price of the jth type of license. Let x_i be the ith firm's effluents, let l_i^j be the holdings of the jth type of license by the ith firm, and let $c_i(x_i)$ be the cleanup cost function for the ith firm. Having identified the ith firm's variables, we shall suppress the subscript i in what follows. It should be remembered that the following formulae apply to single firms.

The required technique is to confront each firm with the following total cost function:

$$c(x) + \sum_{j=1}^{n} (q_j - s_j) l_j + \sum_{j=0}^{n} (s_{j+1} - s_j) \operatorname{Max} (x - l^j, 0). \tag{A.2}$$

The last term looks very much like the earlier penalty function. The first term represents cleanup costs. The second term is special. The cost function can be interpreted as follows. The firm pays for cleanup. It also pays for the licenses it purchases, but it receives a rebate of s_j per license of type j that it holds. Then, having selected the licenses, the firm pays a penalty given by the piecewise linear function,

$$\hat{P}(x) = \sum_{j=0}^{n} (s_{j+1} - s_i) \operatorname{Max} (x - l^j, 0).$$

The locations of the kinks in this function are determined by the firm, through its license purchases. It is the second term in (A.2) that is crucial, for as we shall see, it has the effect of placing bounds on the license prices, q_j.

It remains to show that firms, in maximizing (A.2), efficiently distribute costs and act as if they were one firm facing the penalty function (A.1).

The first step to show that

$$s_j \leqq q_j \leqq s_{j+1} \tag{A.3}$$

for all j. Suppose first that $q_j < s_j$. Then expand l^j so that $l^j > x$. It follows that the term in (A.2) involving l^j is

$$(q_j - s_j)l_j.$$

By allowing l^j to increase without limit, costs are reduced indefinitely. But that is inconsistent with equilibrium. Now suppose $q_j > s_{j+1}$. Reduce l^j so that $l^j < x$ and the term involving l^j becomes

$$(q_j - s_j)l^j - (s_{j+1} - s_j)l^j = (q_j - s_{j+1})l^j.$$

Hence costs are minimized when $l^j = 0$. If all firms do this, there cannot be an equilibrium in the market for j-type licenses. Therefore

$$s_j \leq q_j \leq s_{j+1},$$

for all $j = 1, \ldots, n$.

The next step is to show that if $q_j < s_{j+1}$, then $q_{j+1} = s_{j+1}$. Suppose that $q_j < s_{j+1}$. We show that $l^j \leq x$. Suppose to the contrary that $l^j > x$. Then the part of costs involving l^j is, from (A.2),

$$(q_j - s_{j+1})l^j.$$

Hence l^j should be contracted. Therefore if $q_j < s_{j+1}$, then $l^j \geq x$, and $l^{j+1} > l^j \geq x$, so that x is less than l^{j+1}. But if $l^{j+1} > x$, then q_{j+1} must equal s_{j+1}. For if $q_{j+1} > s_{j+1}$, then the part of costs involving l^{j+1} is

$$(q_{j+1} - s_{j+1})l^{j+1},$$

and l^{j+1} would be reduced. Hence if $l^{j+1} > x$, then $q_{j+1} = s_{j+1}$. This proves the assertion that if $q_j < s_{j+1}$, then $q_{j+1} = s_{j+1}$.

These arguments tell us a considerable amount about the equilibrium. Only one license price q^j can be in the interior of $[s_j, s_{j+1}]$. The remainder are on the boundaries – upper or lower depending upon whether the corresponding license has a lower or higher index than j, respectively.

We now take a typical interval $[s_j, s_{j+1}]$ and assume q_j is the interior of $[s_j, s_{j+1}]$. From the preceding argument, we know that $l^j = x$, that $q_k = s_{k+1}$ for $k < j$, and that $q_k = s_k$ for $k > j$. Thus the costs for the firm are, from (A.2),

$$c(x) + \sum_{k=0}^{j-1}(s_{k+1} - s_k)l^k + \sum_{k=j+1}^{n}(s_{k+1} - s_k)l^k$$

$$+ \sum_{k=0}^{j-1}(s_{k+1} - s_k)(x - l^k) + (q_j - s_j)x + (s_{j+1} - s_j)(0)$$

$$= c(x) + q_j x.$$

Similarly, if $q_j = s_j$, or $q_j = s_{j+1}$, and if $q_k = s_{k+1}$ for $k < j$ and $q_k = s_k$ for $k > j$, then (A.2) implies that the firm's costs (with licenses optimized out) are

$$c(x) + q_j x.$$

In an equilibrium, the costs for every firm will be

$$c(x) + q_j x,$$

for some j and some equilibrium value q_j. Thus when firms minimize, with respect to x, they set

$$c'(x) + q_j = 0. \tag{A.4}$$

In particular, marginal cleanup costs, $c'(x)$, are the same for every firm. Therefore cleanup costs are efficiently distributed in an equilibrium. Let $C(x)$ be the aggregate cost function, where x is now the sum of the effluents from all firms. In an equilibrium, (A.4) implies that

$$C'(x) + q_j = 0.$$

Moreover, if $s_j < q_j < s_{j+1}$, then $x = l^j$, where l^j is the fixed total number of j-type licenses. If $q_j = s_j$, then $l^{j-1} < x \leq l^j$, and if $q_j = s_{j+1}$, then $l^j < x \leq l^{j+1}$. The equilibrium level of x is therefore determined by the level of costs. If

$$s_j < -C'(l^j) < s_{j+1}, \tag{A.5}$$

then $x = l^j$ in an equilibrium. If $-C'(l^j) < s_j$ and $-C'(l^{j-1}) > s_j$, then $C'(x) = s_j$ in the equilibrium.[7]

The system therefore simply acts so as to minimize

$$C(x) + \sum_{j=1}^{n} (s_{j+1} - s_j) \, \text{Max} \, (x - l^j, 0).$$

This is what we set out to show.

The implication of the preceding argument is that any convex damage function can be approximated to any desired degree of accuracy through the introduction of markets for different kinds of licenses. The private sector can be confronted with a nonlinear damage function without sacrificing efficiency in the distribution of cleanup.

---

[7]Note that (A.5) can only hold for one type of license because as j increases, and $-C'(l^j)$ falls, s_{j+1} rises.

As a practical matter, in the pollution context, the cost of the additional license markets may not be justified by the reduction in expected total cost. But it is perhaps a matter of some intellectual interest, both here and in other decentralization problems, that a carefully designed set of markets for options to buy or sell commodities at various prices can solve the problem of reconciling the competing demands of efficiency and decentralization. This subject is probably worthy of further investigation.

References

Jacoby, H., G. Schaumberg and F. Gramlech, 1972, Marketable pollution rights (M.I.T., Cambridge, MA) unpublished.

Kneese, A.V. and B. Bower, 1968, Managing water quality (Resources for the Future, Johns Hopkins Press, Baltimore).

Montgomery, W.C., 1972, Markets in licenses and efficient pollution control programs, Journal of Economic Theory 5, no. 3, 395–418.

Roberts, M.J., 1975, Environmental protection: The complexities of real policy choice, in: Fox and Swainson, eds., Water quality management: The design of institutional arrangements (University of British Columbia Press, Vancouver).

[38]

Prices *vs.* Quantities [1,2]

MARTIN L. WEITZMAN

Massachusetts Institute of Technology

I. INTRODUCTION

The setting for the problem under consideration is a large economic organization or system which in some cases is best thought of as the entire economy. Within this large economic organization resources are allocated by some combination of commands and prices (the exact mixture is inessential) or even by some other unspecified mechanism. The following question arises. For one particular isolated economic variable that needs to be regulated,[3] what is the best way to implement control for the benefit of the organization as a whole? Is it better to directly administer the activity under scrutiny or to fix transfer prices and rely on self-interested profit or utility maximization to achieve the same ends in decentralized fashion? This issue is taken as the prototype problem of central control which is studied in the present paper. There are a great many specific examples which fit nicely into such a framework. One of current interest is the question of whether it would be better to control certain forms of pollution by setting emission standards or by charging the appropriate pollution taxes.

When quantities are employed as planning instruments, the basic operating rules from the centre take the form of quotas, targets, or commands to produce a certain level of output. With prices as instruments, the rules specify either explicitly or implicitly that profits are to be maximized at the given parametric prices. Now a basic theme of resource allocation theory emphasizes the close connection between these two modes of control. No matter how one type of planning instrument is fixed, there is always a corresponding way to set the other which achieves the same result when implemented.[4] From a strictly theoretical point of view there is really nothing to recommend one mode of control over the other. This notwithstanding, I think it is a fair generalization to say that the average economist in the Western marginalist tradition has at least a vague preference toward indirect control by prices, just as the typical non-economist leans toward the direct regulation of quantities.

That a person not versed in economics should think primarily in terms of direct controls is probably due to the fact that he does not comprehend the full subtlety and strength of the invisible hand argument. The economist's attitude is somewhat more puzzling. Understanding that prices can be used as a powerful and flexible instrument for rationally allocating resources and that in fact a market economy automatically regulates itself in this manner is very different from being under the impression that such indirect controls are generally preferable for the kind of problem considered in this paper. Certainly a careful reading of economic theory yields little to support such a universal proposition.

1 *First version received August* 1973; *final version accepted January* 1974 (*Eds.*).

2 Many people have made helpful comments about a previous version of this paper. I would like especially to thank P. A. Diamond and H. E. Scarf for their valuable suggestions. The National Science Foundation helped support my research.

3 Outside the scope of this paper is the issue of *why* it is felt that the given economic activity must be regulated. There may be a variety of reasons, ranging all the way from political considerations to one form or another of market failure.

4 Given the usual convexity assumptions. Without convexity it may not be possible to find a price which will support certain output levels. In this connection it should be mentioned that non-convexities (especially increasing returns) are sometimes responsible for regulation in the first place.

Many economists point with favour to the fact that if prices are the planning instrument then profit maximization automatically guarantees total output will be efficiently produced, as if this result were of any more than secondary interest unless the prices (and hence total output) are optimal to begin with.[1] Sometimes it is maintained that prices are desirable planning instruments because the stimulus to obtain a profit maximizing output is built right in if producers are rewarded in proportion to profits. There is of course just as much motivation, e.g. to minimize costs at specified output levels so long as at least some fraction of production expenditures is borne by producers. With both modes of control there is clearly an incentive for self-interested producers to systematically distort information about hypothetical output and cost possibilities in the pre-implementation planning phase. Conversely, there is no real way to disguise the true facts in the implementation stage so long as actual outputs (in the case of price instruments) and true operating costs (in the case of quantity instruments) can be accurately monitored. For the one case the centre must ascertain *ceteris paribus* output changes as prices are varied, for the other price changes as outputs are altered.

A reason often cited for the theoretical superiority of prices as planning instruments is that their use allegedly economizes on information. The main thing to note here is that generally speaking it is neither easier nor harder to name the right prices than the right quantities because in principle exactly the *same* information is needed to correctly specify either. It is true that in a situation with many independent producers of an identical commodity, only a single uniform price has to be named by the centre, whereas in a command mode separate quantities must be specified for each producer. If such an observation has meaningful implications, it can only be within the artificial milieu of an iterative *tâtonnement* type of " planning game " which is played over and over again approaching an optimal solution in the limit as the number of steps becomes large. Even in this context the fact that there are less " message units " involved in each communication from the centre is a pretty thin reed on which to hang claims for the informational superiority of the price system. It seems to me that a careful examination of the mechanics of successive approximation planning shows that there is no principal informational difference between iteratively finding an optimum by having the centre name prices while the firms respond with quantities, or by having the centre assign quantities while the firm reveals costs or marginal costs.[2]

If there were really some basic intrinsic advantage to a system which employed prices as planning instruments, we would expect to observe many organizations operating with this mode of control, especially among multi-divisional business firms in a competitive

[1] An extreme example may help make this point clear. Suppose that fulfilment of an important emergency rescue operation demands a certain number of airplane flights. It would be inefficient to just order airline companies or branches of the military to supply a certain number of the needed aircraft because marginal (opportunity) costs would almost certainly vary all over the place. Nevertheless, such an approach would undoubtedly be preferable to the efficient procedure of naming a price for plane services. Under profit maximization, overall output would be uncertain, with too few planes spelling disaster and too many being superfluous.

[2] The " message unit " case for the informational superiority of the price system is analogous to the blanket statement that it is better to use dual algorithms for solving a programming problem whenever the number of primal variables exceeds the number of dual multipliers. Certainly for the superior large step decomposition type algorithms which on every iteration go right after what are presently believed to be the best instrument values on the basis of all currently available information, such a general statement has no basis. With myopic gradient methods it is true that on each round the centre infinitesimally and effortlessly adjusts exactly the number of instruments it controls, be they prices or quantities. But who can say *how many* infinitesimally small adjustments will be needed? Gradient algorithms are known to be a bad description of iterative planning procedures, among other reasons because they have inadmissably poor convergence properties. If the step size is chosen too small, convergence takes forever. If it is chosen too large, there is no convergence. As soon as a finite step size is selected on a given iteration to reflect a desire for quick convergence, the " message unit " case for prices evaporates. Calculating the *correct* price change puts the centre right back into the large step decomposition framework where on each round the problem of finding the best iterative prices is formally identical to the problem of finding the best iterative quantities. For discussion of these and various other aspects of iterative planning, see the articles of Heal [4], Malinvaud [5], Marglin [7], Weitzman [9].

environment. Yet the allocation of resources within private companies (not to mention governmental or non-profit organizations) is almost never controlled by setting administered transfer prices on commodities and letting self-interested profit maximization do the rest.[1] The price system as an allocator of internal resources does not itself pass the market test.[2]

Of course, all this is not to deny that in any *particular* setting there may be important *practical* reasons for favouring either prices or quantities as planning instruments. These reasons might involve ideological, political, legal, social, historical, administrative, motivational, informational, monitoring, enforcing, or other considerations.[3] But there is little of what might be called a system-free character.

In studying such a controversial subject, the only fair way to begin must be with the tenet that there is no *basic* or *universal* rationale for having a general predisposition toward one control mode or the other. If this principle is accepted, it becomes an issue of some interest to abstract away all " other " considerations in order to develop strictly " economic " criteria by which the comparative performance of price and quantity planning instruments might be objectively evaluated. Even on an abstract level, it would be useful to know how to identify a situation where employing one mode is relatively advantageous, other things being equal.

II. THE MODEL

We start with a highly simplified prototype planning problem. Amount q of a certain commodity can be produced at cost $C(q)$, yielding benefits $B(q)$.[4] The word " commodity " is used in an abstract sense and really could pertain to just about any kind of good from pure water to military aircraft. Solely for the sake of preserving a unified notation, we follow the standard convention that goods are desirable. This means that rather than talking about air pollution, for example, we instead deal with its negative—clean air.

Later we treat more complicated cases, but for the time being it is assumed that in effect there is just one producer of the commodity and no ambiguity in the notion of a cost curve. Benefits are measured in terms of money equivalents so that the benefit function can be viewed as the reflection of an indifference curve showing the trade-off between amounts of uncommitted extra funds and output levels of the given commodity. It is assumed that $B''(q) < 0$, $C''(q) > 0$, $B'(0) > C'(0)$, and $B'(q) < C'(q)$ for q sufficiently large.

---

[1] Strictly speaking, this conclusion is not really justified because there may be important externalities or increasing returns within an organization (they may even constitute its *raison d'être*). Nevertheless, the almost universal absence of internal transfer pricing within private firms strikes me as a rather startling contradiction with the often alleged superiority of indirect controls.

[2] About a decade ago, Ford and GM performed a few administrative trials of a limited sort with some decentralization schemes based on internal transfer prices. The experiments were subsequently discontinued in favour of a return to more traditional planning methods. See Whinston [10].

[3] As one example, if it happens to be the case that it is difficult or expensive to monitor output on a continuous scale but relatively cheap to perform a pass-fail litmus type test on whether a given output level has been attained or not, the price mode may be greatly disadvantaged from the start. The pollution by open-pit mining operations of nearby waterways presents a case in point. It would be difficult or impossible to record how much pollutant is seeping into the ground, whereas it is a comparatively straight-forward task to enforce the adoption of one or another level of anti-pollution technology. Another realistic consideration arises when we ask who determines the standards under each mode. For example, if an agency of the executive branch is empowered to regulate prices but the legislature is in charge of setting quantities, that by itself may be important in determining which mode is better for controlling pollution. The price mode would have greater flexibility, but might carry with it more danger of caving in to special interest groups. As yet another realistic consideration, equity arguments are sometimes put forward in favour of price (the supposed " justice " of a uniform price to all) or quantity (equal sharing of a deficit commodity) control modes.

[4] It might be thought that an equivalent approach would be to work with demand and supply curves, identifying the consumers' (producers') surplus area under the demand (supply) curve as benefits (costs) or, equivalently, the demand (supply) curve as the marginal benefit (cost) function. The trouble with this approach is that it tends to give the misleading impression that the market left to itself could solve the problem, obscuring the fact that some key element of the standard competitive supply and demand story is felt to be missing in the first place.

The planning problem is to find that value q^* of q which maximizes

$$B(q) - C(q).$$

The solution must satisfy

$$B'(q^*) = C'(q^*).$$

With

$$p^* \equiv B'(q^*) = C'(q^*),$$

it makes no difference whether the planners announce the optimal price p^* and have the producers maximize profits

$$p^*q - C(q)$$

or whether the centre merely orders the production of q^* at least cost. In an environment of complete knowledge and perfect certainty there is a formal identity between the use of prices and quantities as planning instruments.

If there is any advantage to employing price or quantity control modes, therefore, it must be due to inadequate information or uncertainty. Of course it is natural enough for planners to be unsure about the precise specification of cost and benefit functions since even those most likely to know can hardly possess an exact account.

Suppose, then, that the centre perceives the cost function only as an estimate or approximation. The stochastic relation linking q to C is taken to be of the form

$$C(q, \theta),$$

where θ is a disturbance term or random variable, unobserved and unknown at the present time. While the determination of θ could involve elements of genuine randomness,[1] it is probably more appropriate to think primarily in terms of an information gap.

Even the engineers most closely associated with production would be unable to say beforehand precisely what is the cheapest way of generating various hypothetical output levels. How much murkier still must be the centre's *ex ante* conception of costs, especially in a fast moving world where knowledge of particular circumstances of time and place may be required. True, the degree of fuzziness could be reduced by research and experimentation but it could never be truly eliminated because new sources of uncertainty are arising all the time.[2]

Were a particular output level really ordered in all seriousness, a cost-minimizing firm could eventually grope its way toward the cheapest way of producing it by actually testing out the relevant technological alternatives. Or, if an output price were in fact named, a profit maximizing production level could ultimately be found by trial and error. But this is far from having the cost function as a whole knowable *a priori*.

While the planners may be somewhat better acquainted with the benefit function, it too is presumably discernable only tolerably well, say as

$$B(q, \eta)$$

with η a random variable. The connection between q and B is stochastic either because benefits may be imperfectly known at the present time or because authentic randomness may play a role. Since the unknown factors connecting q with B are likely to be quite different from those linking q to C, it is assumed that the random variables θ and η are independently distributed.

As a possible specific example of the present formulation, consider the problem of air pollution. The variable q could be the cleanliness of air being emitted by a certain type of source. Costs as a function of q might not be known beyond doubt because the technology, quantified by θ, is uncertain. At a given level of q the benefits may be unsure since they depend among other things on the weather, measured by η.

[1] Like day-to-day fluctuations.
[2] For an amplification of some of these points, see Hayek [3].

Now an *ideal* instrument of central control would be a contingency message whose instructions depend on which state of the world is revealed by θ and η. The ideal *ex ante* quantity signal $q^*(\theta, \eta)$ and price signal $p^*(\theta, \eta)$ are in the form of an entire schedule, functions of θ and η satisfying

$$B_1(q^*(\theta, \eta), \eta) = C_1(q^*(\theta, \eta), \theta) = p^*(\theta, \eta).$$

By employing either ideal signal, the *ex ante* uncertainty has in effect been eliminated *ex post* and we are right back to the case where there is no theoretical difference between price and quantity control modes.

It should be readily apparent that it is infeasible for the centre to transmit an entire schedule of ideal prices or quantities. A contingency message is a complicated, specialized contract which is expensive to draw up and hard to understand. The random variables are difficult to quantify. A problem of differentiated information or even of moral hazard may be involved since the exact value of θ will frequently be known only by the producer.[1] Even for the simplest case of just *one* firm, information from different sources must be processed, combined, and evaluated. By the time an ideal schedule was completed, another would be needed because meanwhile changes would have occurred.

In this paper the realistic issue of central control under uncertainty is considered to be the " second best " problem of finding for each producer the single price or quantity message which optimally regulates his actions. This is also the best way to focus sharply and directly on the essential difference between prices and quantities as planning instruments.

The issue of prices *vs.* quantities has to be a " second best " problem by its very nature simply because there is no good *a priori* reason for limiting attention to just these two particular signals. Even if stochastic contingency messages were eliminated on *ad hoc* grounds as being too complicated, there would still be no legitimate justification for not considering, say, an entire expected benefits schedule, or a " kinked " benefit function in the form of a two-tiered price system, or something else. The reason we specialize to price and quantity signals is that these are two *simple* messages, easily comprehended, traditionally employed, and frequently contrasted.[2]

The optimal quantity instrument under uncertainty is that target output $\hat{q}$ which maximizes expected benefits minus expected costs, so that

$$E[B(\hat{q}, \eta) - C(\hat{q}, \theta)] = \max_q E[B(q, \eta) - C(q, \theta)],$$

where $E[.]$ is the expected value operator. The solution $\hat{q}$ must satisfy the first order condition

$$E[B_1(\hat{q}, \eta)] = E[C_1(\hat{q}, \theta].   \qquad ...(1)$$

When a price instrument p is announced, production will eventually be adjusted to the output level

$$q = h(p, \theta)$$

which maximizes profits given p and θ. Such a condition is expressed as

$$ph(p, \theta) - C(h(p, \theta), \theta) = \max_q pq - C(q, \theta),$$

implying

$$C_1(h(p, \theta), \theta) = p.   \qquad ...(2)$$

---

[1] So that it may be inappropriate, for example, to tell him to produce less if costs are high unless a very sophisticated incentive scheme goes along with such a message. For an elaboration of some of these points see Arrow [1], pp. 321-322.

[2] There are real costs associated with using more complicated signals. At least implicitly, we are assuming that the magnitude of such costs is sufficiently large to make it uneconomical to consider messages other than prices or quantities. It would be nice to incorporate these costs explicitly into the model, but this is hard to do in any meaningful way.

If the planners are rational, they will choose that price instrument $\tilde{p}$ which maximizes the expected difference between benefits and costs given the reaction function $h(p, \theta)$:

$$E[B(h(\tilde{p}, \theta), \eta) - C(h(\tilde{p}, \theta), \theta)] = \max_{p} E[B(h(p, \theta), \eta) - C(h(p, \theta), \theta)].$$

The solution $\tilde{p}$ must obey the first order equation

$$E[B_1(h(\tilde{p}, \theta), \eta) . h_1(\tilde{p}, \theta)] = E[C_1(h(\tilde{p}, \theta), \theta) . h_1(\tilde{p}, \theta)],$$

which can be rewritten as

$$\tilde{p} = \frac{E[B_1(h(\tilde{p}, \theta), \eta) . h_1(\tilde{p}, \theta)]}{E[h_1(\tilde{p}, \theta)]}. \qquad \ldots(3)$$

Corresponding to the optimal *ex ante* price $\tilde{p}$ is the *ex post* profit maximizing output $\tilde{q}$ expressed as a function of θ,

$$\tilde{q}(\theta) \equiv h(\tilde{p}, \theta). \qquad \ldots(4)$$

In the presence of uncertainty, price and quantity instruments transmit central control in quite different ways. It is important to note that by choosing a specific mode for implementing an intended policy, the planners are at least temporarily locking themselves into certain consequences. The values of η and θ are at first unknown and only gradually, if at all, become recognized through their effects. After the quantity $\hat{q}$ is prescribed, producers will continue to generate that assigned level of output for some time even though in all likelihood

$$B_1(\hat{q}, \eta) \neq C_1(\hat{q}, \theta).$$

In the price mode on the other hand, $\tilde{q}(\theta)$ will be produced where except with negligible probability

$$B_1(\tilde{q}(\theta), \eta) \neq C_1(\tilde{q}(\theta), \theta).$$

Thus neither instrument yields an optimum *ex post*. The relevant question is which one comes closer under what circumstances.[1]

In an infinitely flexible control environment where the planners can continually adjust instruments to reflect current understanding of a fluid situation and producers instantaneously respond, the above considerations are irrelevant and the choice of control mode should be made to depend on other factors. Similar comments apply to a timeless *tâtonnement* milieu where iterations are costless, recontracting takes place after each round, and in effect nothing real is presumed to happen until all the uncertainty has been eliminated and an equilibrium is approached. In any less hypothetical world the consequences of an order given in a particular control mode have to be lived with for at least the time until revisions are made, and real losses will be incurred by selecting the wrong communication medium.

Note that the question usually asked whether it is better to control prices or quantities for *finding* a plan is conceptually distinct from the issue treated in this paper of which mode is superior for *implementing* a plan. The latter way of posing the problem strikes me as more relevant for most actual planning contexts—either because there is no significant informational difference between the two modes in the first place, or because a step in the *tâtonnement* planning game cannot meaningfully occur unless it is really implemented, or because no matter how many iterations have been carried out over time there are always spontaneously arising changes which damp out the significance of knowing past history. In the framework adopted here, the planners are at the decision node where as much information as is feasible to gather has already been obtained by one means or another and an operational plan must be decided on the basis of the available current knowledge.

---

[1] We remark in passing that the issue of whether it is better to stabilize uncertain demand and supply functions by pegging prices or quantities can also be put in the form of the problem analysed in this paper if benefits are associated with the consumers' surplus area under the demand curve and costs with the producers' surplus area under the supply curve.

III. PRICES *vs.* QUANTITIES

It is natural to define the *comparative advantage of prices over quantities* as

$$\Delta \equiv E[(B(\tilde{q}(\theta), \eta) - C(\tilde{q}(\theta), \theta)) - (B(\hat{q}, \eta) - C(\hat{q}, \theta))]. \qquad \text{...(5)}$$

The loss function implicit in the definition of Δ is the expected difference in gains obtained under the two modes of control. Naturally there is no real distinction between working with Δ or with $-\Delta$ (the comparative advantage of quantities over prices).

The coefficient Δ is intended to be a measure of *comparative* or *relative* advantage only. It goes without saying that making a decision to use price or quantity control modes in a specific instance is more complicated than just consulting Δ. There are also going to be a host of practical considerations formally outside the scope of the present model. Although such external factors render Δ of limited value when isolated by itself, they do not necessarily diminish its conceptual significance. On the contrary, having an objective criterion of the *ceteris paribus* advantage of a control mode is very important because conceptually it can serve as a benchmark against which the cost of " non-economic " ingredients might be measured in reaching a final judgment about whether it would be better to employ prices or quantities as planning instruments in a given situation.

As it stands, the formulation of cost and benefit functions is so general that it hinders us from cleanly dissecting equation (5). To see clearly what Δ depends on we have to put more structure on the problem. It is possible to be somewhat less restrictive than we are going to be, but only at the great expense of clarity.

In what follows, the amount of uncertainty in marginal cost is taken as sufficiently small to justify a second order approximation of cost and benefit functions within the range of $\tilde{q}(\theta)$ as it varies around $\hat{q}$.[1] Let the symbol " $\cong$ " denote an " accurate local approximation " in the sense of deriving from the assumption that cost and benefit functions are of the following quadratic form within an appropriate neighbourhood of $q = \hat{q}$:

$$C(q, \theta) \cong a(\theta) + (C' + \alpha(\theta))(q - \hat{q}) + \frac{C''}{2}(q - \hat{q})^2 \qquad \text{...(6)}$$

$$B(q, \eta) \cong b(\eta) + (B' + \beta(\eta))(q - \hat{q}) + \frac{B''}{2}(q - \hat{q})^2. \qquad \text{...(7)}$$

In the above equations $a(\theta)$, $\alpha(\theta)$, $b(\eta)$, $\beta(\eta)$ are stochastic functions and C', C'', B', B'' are fixed coefficients.

Without loss of generality, $\alpha(\theta)$ and $\beta(\eta)$ are standardized in (6), (7) so that their expected values are zero:

$$E[\alpha(\theta)] = E[\beta(\eta)] = 0. \qquad \text{...(8)}$$

Since θ and η are independently distributed,

$$E[\alpha(\theta) . \beta(\eta)] = 0. \qquad \text{...(9)}$$

Note that the stochastic functions

$$a(\theta) \cong C(\hat{q}, \theta)$$

$$b(\eta) \cong B(\hat{q}, \eta)$$

translate different values of θ and η into pure vertical shifts of the cost and benefit curves. Differentiating (6) and (7) with respect to q,

$$\left. \begin{array}{l} C_1(q, \theta) \cong (C' + \alpha(\theta)) + C'' . (q - \hat{q}) \\ B_1(q, \eta) \cong (B' + \beta(\eta)) + B'' . (q - \hat{q}). \end{array} \right\} \qquad \begin{array}{l} \text{...(10)} \\ \text{...(11)} \end{array}$$

---

[1] Such an approximation can be rigorously defended along the lines developed by Samuelson [8].

Employing the above equations and (8), the following interpretations are available for the fixed coefficients of (6), (7):

$$C' \triangleq E[C_1(\hat{q}, \theta)]$$

$$B' \triangleq E[B_1(\hat{q}, \eta)]$$

$$C'' \triangleq C_{11}(q, \theta)$$

$$B'' \triangleq B_{11}(q, \eta).$$

From (1),

$$B' = C'. \qquad \qquad ...(12)$$

It is apparent from (8) and (10) that stochastic changes in $\alpha(\theta)$ represent pure unbiased shifts of the marginal cost function. The variance of $\alpha(\theta)$ is precisely the mean square error in marginal cost

$$\sigma^2 \equiv E[(C_1(q, \theta) - E[C_1(q, \theta)])^2] \triangleq E[\alpha(\theta)^2]. \qquad \qquad ...(13)$$

Analogous comments hold for the marginal benefit function (11) where we have

$$E[(B_1(q, \eta) - E[B_1(q, \eta)])^2] = E[\beta(\eta)^2].$$

From (10) and (2),

$$h(p, \theta) \triangleq \hat{q} + \frac{p - C' - \alpha(\theta)}{C''} \qquad \qquad ...(14)$$

implying

$$h_1(p, \theta) \triangleq \frac{1}{C''}. \qquad \qquad ...(15)$$

Substituting from (15) into (3) and cancelling out C'' yields

$$\tilde{p} \triangleq E[B_1(h(\tilde{p}, \theta), \eta)]. \qquad \qquad ...(16)$$

Replacing q in (11) by the expression for $h(\tilde{p}, \theta)$ from (14) and plugging into (16), the following equation is obtained after using (8)

$$\tilde{p} \triangleq B' + \frac{B''}{C''}(\tilde{p} - C'). \qquad \qquad ...(17)$$

From (12) and the condition $B'' < 0 < C''$, (17) implies

$$\tilde{p} \triangleq C'. \qquad \qquad ...(18)$$

Combining (4), (14), and (18),

$$\tilde{q}(\theta) \triangleq \hat{q} - \frac{\alpha(\theta)}{C''}. \qquad \qquad ...(19)$$

Now alternately substitute $q = \hat{q}$ and $q = \tilde{q}(\theta)$ from (19) into (6) and (7). Then plugging the resulting values of (6), (7) into (5), using (8), (9), and collecting terms,

$$\Delta \triangleq \frac{\sigma^2 B''}{2C''^2} + \frac{\sigma^2}{2C''}. \qquad \qquad ...(20)$$

Expression (20) is the fundamental result of this paper.[1] The next section is devoted to examining it in detail.

---

[1] In the supply and demand context B'' is the slope of the (linear) demand curve, C'' is the slope of the (linear) supply curve, and σ^2 is the variance of vertical shifts in the supply curve.

IV. ANALYSING THE COEFFICIENT OF COMPARATIVE ADVANTAGE

Note that the uncertainty in benefits does not appear in (20).[1] To a second-order approximation it affects price and quantity modes *equally* adversely. On the other hand, Δ depends linearly on the mean square error in marginal cost. The *ceteris paribus* effect of increasing σ^2 is to magnify the expected loss from employing the planning instrument with comparative disadvantage. Conversely, as σ^2 shrinks to zero we move closer to the perfect certainty case where in theory the two control modes perform equally satisfactorily.

Clearly Δ depends critically on the curvature of cost and benefit functions around the optimal output level. The first thing to note is that the sign of Δ simply equals the sign of $C'' + B''$. When the sum of the " other " considerations nets out to a zero bias toward either control mode, quantities are the preferred planning instrument if and only if benefits have more curvature than costs.

Normally we would want to know the magnitude of Δ and what it depends on, as well as the sign. To strengthen our intuitive feeling for the meaning of formula (20), we turn first to some extreme cases where there is a strong comparative advantage to one control mode over the other. In this connection it is important to bear in mind that when we talk about " large " or " small " values of B'', C'', or σ^2, we are only speaking in a relative sense. The absolute measure of any variable appearing in (20) does not really mean much alone since it is arbitrarily pegged by selecting the units in which output is reckoned.

The coefficient Δ is negative and large as either the benefit function is more sharply curved or the cost function is closer to being linear. Using a price control mode in such situations could have detrimental consequences. When marginal costs are nearly flat, the smallest miscalculation or change results in either much more or much less than the desired quantity. On the other hand, if benefits are almost kinked at the optimum level of output, there is a high degree of risk aversion and the centre cannot afford being even slightly off the mark. In both cases the quantity mode scores a lot of points because a high premium is put on the rigid output controllability which only it can provide under uncertainty.

From (20), the price mode looks relatively more attractive when the benefit function is closer to being linear. In such a situation it would be foolish to name quantities. Since the marginal social benefit is approximately constant in some range, a superior policy is to name it as a price and let the producers find the optimal output level themselves, after eliminating the uncertainty from costs.

At a point where the cost function is highly curved, Δ becomes nearly zero. If marginal costs are very steeply rising around the optimum, as with fixed capacity, there is not much difference between controlling by price or quantity instruments because the resulting output will be almost the same with either mode. In such a situation, as with the case $\sigma^2 = 0$, " non-economic " factors should play the decisive role in determining which system of control to impose.

It is difficult to refrain from noticing that although there are plenty of instances where

---

[1] This is because the *expected* benefit function (see equation (7)) does not depend on the variance of marginal benefits so long as costs and benefits are independently distributed. If they are *not*, so that

$$\sigma_{bc}^2 \equiv E[\{C_1(q, \theta) - E[C_1(q, \theta)]\} \cdot \{B_1(q, \eta) - E[B_1(q, \eta)]\}] = E[\alpha(\theta) \cdot \beta(\eta)] \neq 0,$$

(20) must be replaced by: $\Delta \cong \dfrac{\sigma^2 B''}{2C''^2} + \dfrac{1}{2C''}(\sigma^2 - 2\sigma_{bc}^2)$. The sole effect of having costs and benefits correlated with each other is embodied in the term σ_{bc}^2. When marginal costs are positively correlated with marginal benefits, the *ceteris paribus* comparative advantage of the quantity mode is increased. If prices are used as a control mode, the producer will tend to cut back output for high marginal costs. But with σ_{bc}^2 positive, this is the very same time that marginal benefits tend to be high, so that a cutback may not really be in order. In such situations the quantity mode has better properties as a stabilizer, other things being equal. The story is the other way around when σ_{bc}^2 is negative. In that case high marginal costs are associated with low marginal benefits, so that the price mode (which decreases output for high marginal costs) tends to be a better mode of control other things being equal.

the price mode has a good solid comparative advantage (because $-B''$ is small), in some sense it looks as if prices can be a *disastrous* choice of instrument far more often than quantities can. Using (20), $\Delta \to -\infty$ if either $B'' \to -\infty$ or $C'' \to 0$ (or both). The only way $\Delta \to +\infty$ is under the thin set of circumstances where simultaneously $C'' \to 0$, $B'' \to 0$, and $C'' > -B''$. In a world where C'' and B'' are themselves imperfectly known it seems hard to avoid the impression that there will be many circumstances where the more conservative quantity mode will be preferred by planners because it is better for avoiding very bad planning mistakes.[1]

Having seen how C'' and B'' play an essential role in determining Δ, it may be useful to check out a few of the principal situations where we might expect to encounter cost and benefit functions of one curvature or another. We start with costs.

Contemporary economic theory has tended to blur the distinction between the traditional marginalist way of treating production theory with smoothly differentiable production functions and the activity analysis approach with its limited number of alternative production processes. For many theoretical purposes convexity of the underlying technology is really the fundamental property.

However, there are very different implications for the efficacy of price and quantity control modes between a situation described by classically smooth Marshallian cost curves and one characterized by piecewise linear cost functions with a limited number of kinks. In the latter case, the quantity mode tends to have a relative advantage since $\Delta = -\infty$ on the flats and $\Delta = 0$ at the elbows. Of course it is impossible to use a price to control an output at all unless some hidden fixed factors take the flatness out of the average cost curve. Even then, Δ will be positive only if there are enough alternative techniques available to make the cost function have more (finite difference) curvature than the benefit function in the neighbourhood of an optimal policy.

What determines the benefit function for a commodity is contingent in the first place on whether the commodity is a final or intermediate good. The benefit of a final good is essentially the utility which arises out of consuming the good. It could be highly curved at the optimum output level if tastes happen to be kinked at certain critical points. The amount of pollution which makes a river just unfit for swimming could be a point where the marginal benefits of an extra unit of output change very rapidly. Another might be the level of defence which just neutralizes an opponent's offence or the level of offence which just overcomes a given defence. There are many examples which arise in emergencies or natural calamities. Our intuitive feeling, which is confirmed by the formal analysis, is that it doesn't pay to " fool around " with prices in such situations.

For intermediate goods, the shape of the benefit function will depend among other things on the degree of substitutability in use of this commodity with other resources available in the production organization and upon the possibilities for importing this

---

[1] This idea could be formalized as follows. Consider two generalizations of formulae (6) and (7):

$$C(q, \theta) \triangleq a(\theta) + (C' + \alpha(\theta))(q - \hat{q}) + \frac{C''}{2f(\theta)}(q - \hat{q})^2$$

$$B(q, \eta) \triangleq b(\eta) + (B' + \beta(\eta))(q - \hat{q}) + \frac{B''g(\eta)}{2}(q - \hat{q})^2.$$

The only difference with (6), (7) is that now $1/C_{11}(q, \theta)$ and $B_{11}(q, \eta)$ are allowed to be uncertain. The change in the profit maximizing output response per unit price change is now stochastic, $h_1(p, \theta) = f(\theta)/C''$. Without loss of generality we set

$$E[f(\theta)] = E[g(\eta)] = 1.$$

Note that increasing the variance of $f(g)$ is a mean preserving spread of C_{11} (B_{11}). Suppose for simplicity that f and α are independent of each other. Then we can derive the appropriate generalization of (20) as

$$\Delta \triangleq \frac{B''\sigma^2(1+\delta^2)}{2C''^2} + \frac{\sigma^2}{2C''},$$

where $\delta^2 \equiv E[\{f(\theta) - E[f(\theta)]\}^2]$ is the variance of $f(\theta)$. The above formula can be interpreted as saying that other things being equal, greater uncertainty in $1/C_{11}(q, \theta)$ increases the comparative advantage of the quantity mode.

commodity from outside the organization. These things in turn are very much dependent on the planning time horizon. In the long run the benefit function probably becomes flatter because more possibilities for substitution are available, including perhaps importing. Take for example the most extreme degree of complete " openness " where any amount of the commodity can be instantaneously and effortlessly bought (and sold) outside the production organization at a fixed price. The relevant benefit function is of course just a straight line whose slope is the outside price.

There is, it seems to me, a rather fundamental reason to believe that quantities are better signals for situations demanding a high degree of coordination. A classical example would be the short run production planning of intermediate industrial materials. Within a large production organization, be it the General Motors Corporation or the Soviet industrial sector as a whole, the need for balancing the output of any intermediate commodity whose production is relatively specialized to this organization and which cannot be effortlessly and instantaneously imported from or exported to a perfectly competitive outside world puts a kink in the benefit function. If it turns out that production of ball bearings of a certain specialized kind (plus reserves) falls short of anticipated internal consumption, far more than the value of the unproduced bearings can be lost. Factors of production and materials that were destined to be combined with the ball bearings and with commodities containing them in higher stages of production must stand idle and are prevented from adding value all along the line. If on the other hand more bearings are produced than were contemplated being consumed, the excess cannot be used immediately and will only go into storage to lose implicit interest over time. Such short run rigidity is essentially due to the limited substitutability, fixed coefficients nature of a technology based on machinery.[1] Other things being equal, the asymmetry between the effects of overproducing and underproducing are more pronounced the further removed from final use is the commodity and the more difficult it is to substitute alternative slack resources or to quickly replenish supplies by emergency imports. The resulting strong curvature in benefits around the planned consumption levels of intermediate materials tends to create a very high comparative advantage for quantity instruments. If this is combined with a cost function that is nearly linear in the relevant range, the advantage of the quantity mode is doubly compounded.[2]

V. MANY PRODUCTION UNITS

Consider the same model previously developed except that now instead of being a single good, $q = (q_1, \ldots, q_n)$ is an n-vector of commodites. The various components of q might represent physically distinct commodities or they could denote amounts of the same commodity produced by different production units. Benefits are $B(q, \eta)$ and the cost of producing the ith good is $c^i(q_i, \theta_i)$. As before, for each i the two random variables η and θ_i are distributed independently of each other.

Suppose the issue of control is phrased as choosing either the quantities $\{\hat{q}_i\}$ which maximize

$$E\left[B(q, \eta) - \sum_1^n c^i(q_i, \theta_i) \right],$$

---

[1] The existence of buffer stocks changes the point at which the kink occurs, but does not remove it. For a more detailed treatment of this entire topic, see Manove [6].

[2] Note that in the context of an autarchic planned economy, such pessimistic conclusions about the feasibility of using Lange-Lerner price signals to control short run output do not carry over to, say, agriculture. The argument just given for a kinked benefit function would not at all pertain to a food crop, which goes more or less directly into final demand. In addition, the cost function for producing a given agricultural commodity ought to be much closer to the classical smooth variety than to the linear programming type with just a few kinks.

or the prices $\{\tilde{p}_i\}$ which maximize

$$E[B(h(p, 0), \eta) - \Sigma c^i(h_i(p_i, \theta_i), \theta_i)],$$

where $\{h_i(p_i, \theta_i)\}$ are defined analogously to (2).

Naturally the coefficient of comparative advantage is now defined as

$$\Delta_n \equiv E\left[\left\{B\left(\tilde{q}(0), \eta\right) - \sum_1^n c^i(\tilde{q}_i(\theta_i), \theta_i)\right\} - \left\{B(\hat{q}, \eta) - \sum_1^n c^i(\hat{q}_i, \theta_i)\right\}\right].$$

Assuming locally quadratic costs and benefits, it is a straightforward generalization of what was done in Section III to derive the analogue of expression (20),

$$\Delta_n \triangleq \sum_{i=1}^{n} \sum_{j=1}^{n} \frac{B_{ij}\sigma_{ij}^2}{2c_{11}^i c_{11}^j} + \sum_{i=1}^{n} \frac{\sigma_{ii}^2}{2c_{11}^i}, \qquad \qquad ...(21)$$

where

$$\sigma_{ij}^2 \triangleq E[\{c_1^i(q_i, \theta_i) - E[c_1^i(q_i, \theta_i)]\}\{c_1^j(q_j, \theta_j) - E[c_1^j(q_j, \theta_j)]\}]. \qquad ...(22)$$

To correct for the pure effect of n on Δ_n, it is more suitable to work with the transformed cost functions

$$C^i(x_i, \theta_i) \equiv nc^i(x_i/n, \theta_i). \qquad \qquad ...(23)$$

The meaning of C^i is most readily interpreted for the situation where n different units are producing the same commodity or a close substitute with similar cost functions. Then C^i is what total costs would be as a function of total output if each production unit were an identical replica of the ith unit. When " other things being equal " n is changed, it is more appropriate to think of C^i being held constant rather than c^i.

With C^i defined by (23), we have

$$C_1^i = c_1^i \qquad \qquad ...(24)$$

$$C_{11}^i = \frac{c_{11}^i}{n}. \qquad \qquad ...(25)$$

Relation (24) means that in the quadratic case the coefficients of the marginal cost variance-covariance matrix for the $\{C_1^i\}$ are the same as those given by (22) for the $\{c_1^i\}$. Substituting (25) into (21),

$$\Delta_n \triangleq \frac{1}{n^2} \sum_{i=1}^{n} \sum_{j=1}^{n} \frac{B_{ij}\sigma_{ij}^2}{2C_{11}^i C_{11}^j} + \frac{1}{n} \sum_{i=1}^{n} \frac{\sigma_{ii}^2}{2C_{11}^i}. \qquad ...(26)$$

The above formula shows that in effect the original expression for Δ holds *on the average* for Δ_n when there is more than one producer. Naturally the generalization (26) is more complicated, but the interpretation of it is basically similar to the diagnosis of (20) which was just given in the previous section.

There is, however, a fundamental distinction between having one and many producers which is concealed in formula (26). With some degree of independence among the distributions of individual marginal costs, less weight will be put on the first summation term of (26). Other things being equal, in situations with more rather than fewer independent units producing outputs which substitute for each other in yielding benefits, there is a correspondingly greater relative advantage to the price mode of control. Although this point has general validity, it can be most transparently seen in the special regularized case of one good being produced by many micro-units with symmetrical cost functions. In such a case

$$B_{ij} = B'' \qquad \qquad ...(27.i)$$

$$C_{11} = C'' \qquad \qquad ...(27.ii)$$

$$\sigma_{ii}^2 = \sigma^2 \qquad \text{...(27.iii)}$$

$$\sigma_{ij}^2 = \rho\sigma^2, \quad i \neq j, \quad -1 \leq \rho \leq 1. \qquad \text{...(27.iv)}$$

The coefficient ρ is a measure of the correlation between marginal costs of separate production units. If all units are pretty much alike and are using a similar technology, ρ is likely to be close to unity. If the cost functions of different units are more or less independent of each other, ρ should be nearly zero. While in theory the correlation coefficient can vary between plus and minus unity, for most situations of practical interest the marginal costs of two different production units will have a non-negative cross correlation.

Using (27), (26) can be rewritten as

$$\Delta_n \underset{=}{\circ} \rho\left(\frac{B''\sigma^2}{2C''^2} + \frac{\sigma^2}{2C''}\right) + (1-\rho)\left(\frac{1}{n}\frac{B''\sigma^2}{2C''^2} + \frac{\sigma^2}{2C''}\right). \qquad \text{...(28)}$$

If the marginal costs of each identical micro-unit are perfectly correlated with each other so that $\rho = 1$, it is as if there is but a single producer and we are exactly back to the original formula (20). With $n > 1$, as ρ decreases, Δ_n goes up. A *ceteris paribus* move from dependent toward independent costs increases the comparative advantage of prices, an effect which is more pronounced as the number of production units is larger. If there are three distinctly different types of sulphur dioxide emitters with independent technologies instead of one large pollution source yielding the same aggregate effect, a relatively stronger case exists for using prices to regulate output.

When it is desired to control different units producing an identical commodity by setting prices, only a single price need be named as an instrument. The price mode therefore possesses the *ceteris paribus* advantage that output is being produced efficiently *ex post.* With prices as instruments

$$c_1^i(\tilde{q}_i, \theta_i) = c_1^j(\tilde{q}_j, \theta_j) = \hat{p},$$

whereas with quantities

$$c_1^i(\hat{q}_i, \theta_i) \neq c_1^j(\hat{q}_j, \theta_j)$$

except on a set of negligible probability.

Using prices thus enables the centre to automatically screen out the high cost producers, encouraging them to produce less and the low cost units more. This predominance in efficiency makes the comparative advantage of the price mode go up as the number of independent production units becomes larger, other things being equal. The precise statement of such a proposition would depend on exactly what was held equal as n was increased—the variance of *individual* costs or the overall variance of *total* costs. For simplicity consider the case of completely independent marginal costs, $\rho = 0$. Then (28) becomes

$$\Delta_n \underset{=}{\circ} \frac{1}{n}\frac{B''\sigma^2(n)}{2C''^2} + \frac{\sigma^2(n)}{2C''}, \qquad \text{...(29)}$$

where $\sigma^2(n)$ is implicitly some (given) function of n. If the " other thing " being equal is the constant variance of marginal costs for each individual producing unit, then $\sigma^2(n) \equiv \sigma^2$. If the variance of total costs is held constant as n varies, $\sigma^2(n) \equiv n\sigma^2$. Either way Δ_n in (29) increases monotonically with n and eventually becomes positive.

It is important to note that such *ceteris paribus* efficiency advantages of the price mode as we have been considering for large n are by no means enough to guarantee that Δ_n will be positive in a particular situation for any *given* n. True, what aggregate output is forthcoming under the price mode will be produced at least total cost. But it might be the wrong overall output level to start with. If the $\{-B_{ij}\}$ are sufficiently large or the $\{C_{11}^i\}$ sufficiently small, it may be advantageous to enjoy greater control over total output by setting individual quotas, even after taking account (as our formula for Δ_n does) of the losses incurred by the *ex post* productive inefficiency of such a procedure.[1]

Returning to the general case with which this section began, we note that the basic difference between benefits and costs becomes somewhat more transparent in the n commodity vector formulation. Only the centre knows benefits. Even if it could be done it would not help to transmit $B(.)$ to individual production units because benefits are typically a non-separable function of *all* the units' outputs, whereas a particular unit has control only over its *own* output. In any well formulated mode of decentralized control, the objective function to be maximized by a given unit must depend in some well-defined way on *its* decisions alone. For the purposes of our formulation B need not be a benefit and the $\{c^i\}$ need not be costs in the usual sense, although in many contexts this is the most natural interpretation. The crucial distinction is that B is in principle knowable only by the centre, whereas c^i is best known by firm i.[2]

When uncertainties in individual costs are unrelated so that the random variables θ_i and θ_j are independently distributed, the decision to use a price or quantity instrument to control q_i alone is decentralizable. Suppose it has already been resolved by one means or another whether to use price or quantity instruments to control q_j for each $j \neq i$. To a quadratic approximation, the comparative advantage of prices over quantities for commodity i is

$$\Delta^i \triangleq \frac{\sigma_{ii}^2 B_{ii}}{2c_{11}^{i^2}} + \frac{\sigma_{ii}^2}{2c_{11}^i}, \qquad \qquad ...(30)$$

which is exactly the formula (20) for this particular case.

In some situations, " mixed " price-quantity modes may give the best results. As a specific example, suppose that q_1 is the catch of a certain fish from a large lake and q_2 from a small but prolific pond. Let q_1 be produced with relatively flat average costs but q_2 have a cost function which is curved at the optimum somewhat more than the benefit function. The optimal policy according to (30) will be to name a quota for q_1 and a price for q_2.

REFERENCES

[1]　Arrow, K. J. " Research in Management Controls: A Critical Synthesis ", pp. 317-327, in Bonini, Jaedicke and Wagner (eds.), *Management Controls* (New York: McGraw-Hill, 1964).

[2]　Dales, J. H. *Pollution, Property and Prices* (Toronto: University of Toronto Press, 1968).

[1] An even better procedure from a theoretical point of view in the case where an identical output is produced by many firms would be to fix *total* output by command and subdivide it by a price mechanism. This kind of solution is proposed by Dales [2] who would set up a market in " pollution rights ", the fixed supply of which is regulated by the government. In effect, such an approach aggregates the individual cost functions, and we are right back to a single cost function. Note that a basic question would still remain: is it better to fix the total amount by a quantity or price control mode?

[2] An interesting application of the ideas of this section is provided by the problem of choosing a control mode for best distributing a deficit commodity in fixed supply (say gasoline). In this case what we have been calling an individual cost function, $c^i(q_i, \theta_i)$, would really be the negative of a user's benefit function (as measured by the area under his demand curve). Our B function (of total demand) would just reflect the opportunity loss of having a surplus or shortage of the implied amount when only a fixed supply is available. All the considerations of this section would apply in determining the coefficient of comparative advantage. Accurately characterizing the B function seems especially difficult in the present context. If the commodity can be bought from or sold to the outside world, the B function would just embody the terms of this opportunity (in particular, it would be flat if any amount of the commodity could be bought or sold at some fixed price). Under autonomy, the shape of the B function would depend on what is done in a surplus or deficit situation. With a surplus (from naming too high a price), it would depend on the value to future allocation possibilities of the excess supply, relative to what welfare was lost at the present

[3] Hayek, F. A. " The Use of Knowledge in Society ", *American Economic Review*, **35**, 4 (September 1945), pp. 519-530.

[4] Heal, G. " Planning Without Prices ", *Review of Economic Studies*, **36** (1969), pp. 347-362.

[5] Malinvaud, E. " Decentralized Procedures for Planning ", ch. 7 in Malinvaud (ed.), *Activity Analysis in the Theory of Growth and Planning* (New York: Macmillan, 1967).

[6] Manove, M. " Non-Price Rationing of Intermediate Goods in Centrally Planned Economies ", *Econometrica*, **41** (September 1973), pp. 829-852.

[7] Marglin, S. " Information in Price and Command Systems of Planning ", in Margolis (ed.), *Conference on the Analysis of the Public Sector* (Biarritz, 1966), pp. 54-76.

[8] Samuelson, P. A. " The Fundamental Approximation Theorem of Portfolio Analysis in Terms of Means, Variances and Higher Moments ", *Review of Economic Studies*, **37** (October 1970), pp. 537-542.

[9] Weitzman, M. " Iterative Multilevel Planning with Production Targets ", *Econometrica*, **38**, 1 (January 1970), pp. 50-65.

[10] Whinston, A. " Price Guides in Decentralized Institutions ", Ph.D. thesis, Carnegie Institute of Technology, 1962.

---

time from having demand less than supply. With a deficit (from naming too low a price), the loss of welfare hinges on how shortages are actually distributed among consumers. If shortages result in some people doing completely without the product, the overall welfare losses may be very great and $|B''|$ could be large. If there is some inherent reason to believe that shortages will automatically be evenly distributed, then $|B''|$ may not be so big. In addition to redistribution losses, there will always be waiting time losses in a shortage. Finally, note that if the amount of the fixed supply is known, a superior policy to naming prices or quantities is to distribute ration tickets (instead of quantities), allowing them to be resold at a competitively determined market price.

Name Index